Plumbing

Level One

Trainee Guide
Third Edition

PEARSON

Upper Saddle River, New Jersey
Columbus, Ohio

NCCER
President: Don Whyte
Director of Product Development: Daniele Dixon
Plumbing Project Manager: Daniele Dixon
Production Manager: Debie Ness
Quality Assurance Coordinator: Jessica Martin
Desktop Publishers: Jessica Martin and Laura Parker
Editors: Tara Cohen, Brendan Coote, Bethany Harvey, and Natalie Smith

Writing and development services provided by EEI Communications, Alexandria, Virginia.

Pearson Education, Inc.
Product Manager: Lori Cowen
Production Editor: Stephen C. Robb
Design Coordinator: Karrie M. Converse-Jones
Text Designer: Kristina D. Holmes
Cover Designer: Kristina D. Holmes
Copy Editor: Sheryl Rose
Scanning Coordinator: Karen L. Bretz
Scanning Technician: Janet Portisch
Production Manager: Pat Tonneman

This book was set in Palatino and Helvetica by Carlisle Communications, Ltd. It was printed and bound by Courier Kendallville, Inc. The cover was printed by Phoenix Color Corp.

This information is general in nature and intended for training purposes only. Actual performance of activities described in this manual requires compliance with all applicable operating, service, maintenance, and safety procedures under the direction of qualified personnel. References in this manual to patented or proprietary devices do not constitute a recommendation of their use.

Copyright © 2005, 2000, 1992 by NCCER, Alachua, FL 32615 and published by Pearson Education, Inc., Upper Saddle River, NJ 07458. All rights reserved. Printed in the United States of America. This publication is protected by copyright, and permission should be obtained from NCCER prior to any prohibited reproduction, storage in a retrieval system, or transmission in any form or by any means, electronic, mechanical, photocopying, recording, or likewise. For information regarding permission(s), write to: NCCER Product Development, 13614 Progress Boulevard, Alachua, FL 32615.

Pearson Prentice Hall™ is a trademark of Pearson Education, Inc.
Pearson® is a registered trademark of Pearson plc
Prentice Hall® is a registered trademark of Pearson Education, Inc.

Pearson Education Ltd.
Pearson Education Singapore Pte. Ltd.
Pearson Education Canada, Ltd.
Pearson Education—Japan
Pearson Education Australia Pty. Limited
Pearson Education North Asia Ltd.
Pearson Educación de Mexico, S.A. de C.V.
Pearson Education Malaysia Pte. Ltd.

10 9
ISBN 0-13-109178-6

Preface

TO THE TRAINEE

Most people are familiar with plumbers who come to their home to unclog a drain or install an appliance. In addition to these activities, however, plumbers install, maintain, and repair many different types of pipe systems. For example, some systems move water to a municipal water treatment plant and then to residential, commercial, and public buildings. Other systems dispose of waste, provide gas to stoves and furnaces, or supply air conditioning. Pipe systems in power plants carry the steam that powers huge turbines. Pipes also are used in manufacturing plants, such as wineries, to move material through production processes.

Plumbers and their associated trades constitute one of the largest construction occupations, holding about 550,000 jobs. Theirs is also among the highest paid construction occupations. Even better, job opportunities are expected to be excellent as demand for skilled craftspeople is expected to outpace the supply of trained plumbers.[1]

We wish you success as you embark on your first year of training in the plumbing craft and hope that you'll continue your training beyond this textbook. As most of the half-million craftspeople employed in this trade can tell you, there are many opportunities awaiting those with the skills and the desire to move forward in the construction industry.

NEW WITH *PLUMBING LEVEL ONE*

NCCER and Prentice Hall are pleased to present the third edition of *Plumbing Level One*. This edition features two new modules: *Plumbing Safety* and *Corrugated Stainless Steel Tubing*. To see what kinds of work you might be doing in the plumbing trade, check out the opening pages of each of the thirteen modules contained in this textbook.

Repairing big chillers, installing domestic and chilled water piping, placing fixtures and faucets, fitting and joining carbon-steel piping and even corrugated stainless steel, reading blueprints, and working with a variety of hand tools—there's *much* more to plumbing than unclogging a drain!

We invite you to visit the NCCER website at **www.nccer.org** for the latest releases, training information, newsletter, and much more. You can also reference the Pearson product catalog online at **www.crafttraining.com.** Your feedback is welcome. You may email your comments to **curriculum@nccer.org** or send general comments and inquiries to **info@nccer.org.**

NCCER STANDARDIZED CURRICULA

NCCER is a not-for-profit 501(c)(3) education foundation established in 1995 by the world's largest and most progressive construction companies and national construction associations. It was founded to address the severe workforce shortage facing the industry and to develop a standardized training process and curricula. Today, NCCER is supported by hundreds of leading construction and maintenance companies, manufacturers, and national associations. The NCCER Standardized Curricula was developed by NCCER in partnership with Prentice Hall, the world's largest educational publisher.

Some features of NCCER's Standardized Curricula are as follows:

- An industry-proven record of success
- Curricula developed by the industry for the industry
- National standardization, providing portability of learned job skills and educational credits
- Compliance with Apprenticeship, Training, Employer, and Labor Services (ATELS) requirements for related classroom training (CFR 29:29)
- Well-illustrated, up-to-date, and practical information

NCCER also maintains a National Registry that provides transcripts, certificates, and wallet cards to individuals who have successfully completed modules of NCCER's Standardized Curricula. *Training programs must be delivered by an NCCER Accredited Training Sponsor in order to receive these credentials.*

[1]U.S. Department of Labor, Bureau of Labor Statistics. *Occupational Outlook Handbook, 2004–05 Edition.* www.bls.gov/oco/

Special Features of This Book

In an effort to provide a comprehensive user-friendly training resource, we have incorporated many different features for you to use. Whether you are a visual or hands-on learner, this book will provide you with the proper tools to get started in the plumbing industry.

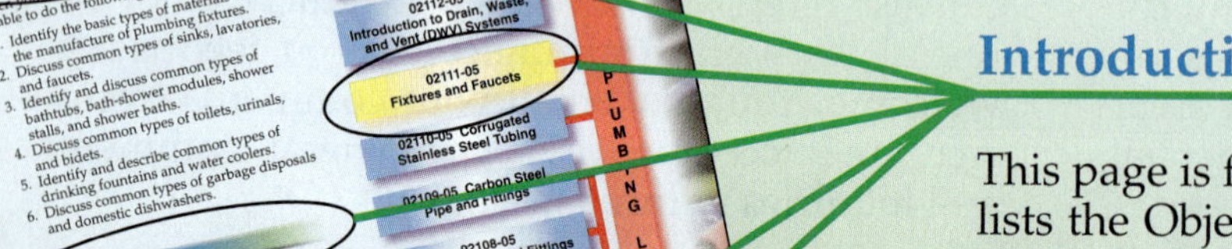

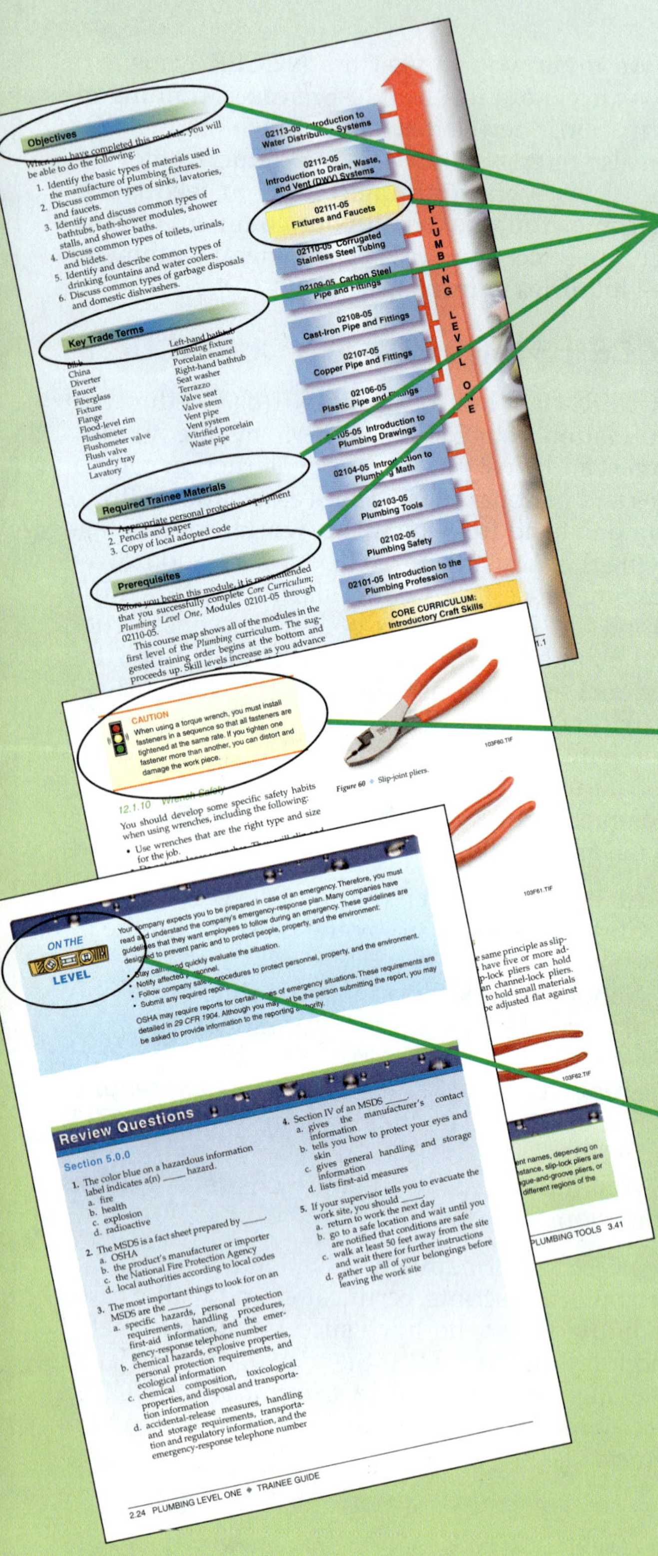

Introduction Page

This page is found at the beginning of each module and lists the Objectives, Key Trade Terms, Required Trainee Materials, Prerequisites, and Course Map for that module. The Objectives list the skills and knowledge you will need in order to complete the module successfully. The list of Key Trade Terms identifies important terms you will need to know by the end of the module. Required Trainee Materials list the materials and supplies needed for the module. The Prerequisites for the module are listed and illustrated in the Course Map. The Course Map also gives a visual overview of the entire course and a suggested learning sequence for you to follow.

Notes, Cautions, and Warnings

Safety features are set off from the main text in highlighted boxes and organized into three categories based on the potential danger of the issue being addressed. *Notes* simply provide additional information on the topic area. *Cautions* alert you of a danger that does not present potential injury but may cause damage to equipment. *Warnings* stress a potentially dangerous situation that may cause injury to you or a co-worker.

On the Level

The *On the Level* features offer technical hints and tips from the plumbing industry. These often include nice-to-know information that you will find helpful. *On the Level* also presents real-life scenarios similar to those you might encounter on the job site.

Did You Know?

The *Did You Know?* features introduce historical tidbits or modern information about the plumbing industry. Interesting and sometimes surprising facts about plumbing are also presented.

Color Illustrations and Photographs

Full-color illustrations and photographs are used throughout each module to provide vivid detail. These figures highlight important concepts from the text and provide clarity for complex instructions. Each figure is denoted in the text in *italic type* for easy reference.

Step-by-Step Instructions

Step-by-step instructions are used throughout to guide you through technical procedures and tasks from start to finish. These steps show you not only how to perform a task but also how to do it safely and efficiently.

Review Questions

Review Questions are provided to reinforce the knowledge you have gained. This makes them a useful tool for measuring what you have learned.

Key Trade Terms

Each module presents a list of *Key Trade Terms* that are discussed within the text, defined in the *Glossary* at the end of the module, and reinforced with a *Key Trade Terms Quiz*. These terms are denoted in the text with **blue bold type** upon their first occurrence. To make searches for key information easier, a comprehensive Glossary of Key Trade Terms from all modules is found at the back of this book.

Profile in Success

Profile in Success shares the experiences of and advice from successful professionals in the plumbing field, from contractors and company presidents to inspectors and construction engineers to corporate training managers and association executives. All of these professionals got their start in the plumbing trade.

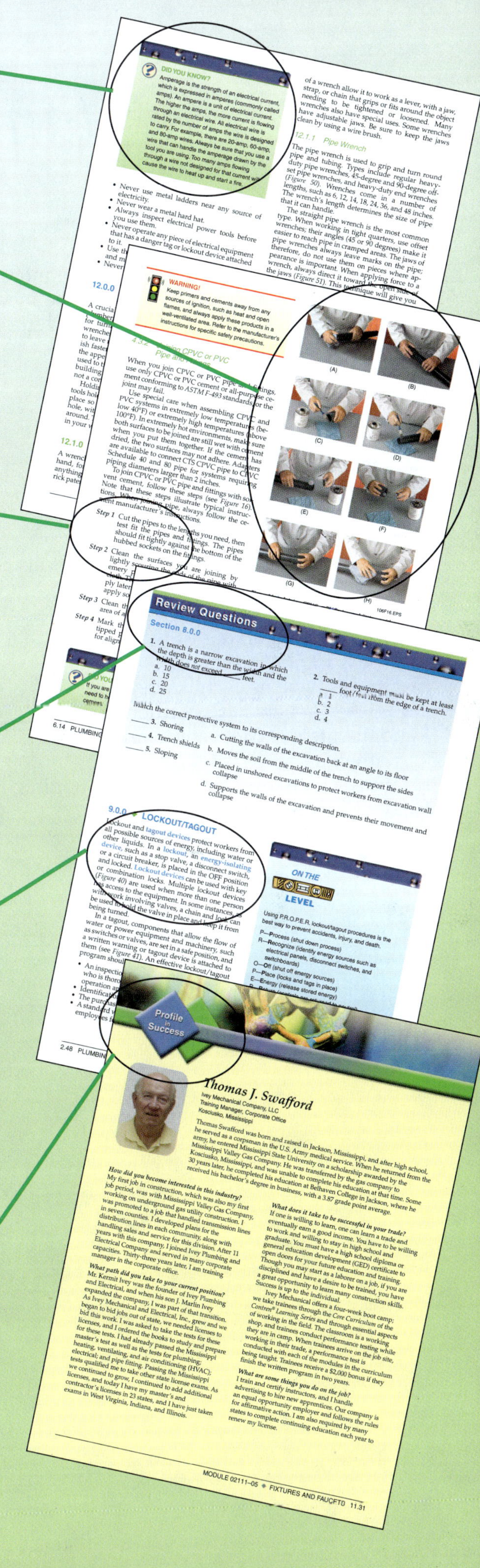

NCCER Standardized Curricula

NCCER's training programs comprise more than 40 construction, maintenance, and pipeline areas and include skills assessments, safety training, and management education.

Boilermaking
Carpentry
Carpentry, Residential
Cabinetmaking
Concrete Finishing
Construction Craft Laborer
Construction Technology
Core Curriculum: Introductory Craft Skills
Currículum Básico
Electrical
Electrical, Residential
Electrical Topics, Advanced
Electronic Systems Technician
Exploring Careers in Construction
Fundamentals of Mechanical and Electrical Mathematics
Heating, Ventilating, and Air Conditioning
Heavy Equipment Operations
Highway/Heavy Construction
Instrumentation
Insulating
Ironworking
Maintenance, Industrial
Masonry
Millwright
Mobile Crane Operations
Painting
Painting, Industrial
Pipefitting
Pipelayer
Plumbing
Reinforcing Ironwork
Rigging
Scaffolding
Sheet Metal
Site Layout
Sprinkler Fitting
Welding

Pipeline

Control Center Operations, Liquid
Corrosion Control
Electrical and Instrumentation
Field Operations, Liquid
Field Operations, Gas
Maintenance
Mechanical

Safety

Field Safety
Safety Orientation
Safety Technology

Management

Introductory Skills for the Crew Leader
Project Management
Project Supervision

Acknowledgments

This curriculum was revised as a result of the farsightedness and leadership of the following sponsors:

The Arkord Company
Associated Builders and Contractors of Southern California
EBL Engineers, LLC
G.J. Hopkins, Inc.
Ivey Mechanical Company, LLC
JF Ingram
John J. Muller Plumbing & Heating, Inc.
LeDuc & Dexter, Inc.
M. Davis and Sons
Norfolk Technical Vocational Center
Universal Plumbing & Heating Company
Wat-Kem Mechanical, Inc.

This curriculum would not exist were it not for the dedication and unselfish energy of those volunteers who served on the Authoring Team. A sincere thanks is extended to the following:

Jonathan Byrd
Charlie Chalk
Steve Guy
Richard Kerzetski
Bob Kordulak
Tom LeDuc
Radford Mitchell, Sr.
Bob Muller
Bill Overstreet
Charles Owenby
Charles Robbins
Thomas J. Swafford, CPC
Ray G. Thornton

NCCER PARTNERING ASSOCIATIONS

American Fire Sprinkler Association
American Petroleum Institute
American Society for Training & Development
American Welding Society
Associated Builders & Contractors, Inc.
Association for Career and Technical Education
Associated General Contractors of America
Carolinas AGC, Inc.
Citizens Democracy Corps
Construction Industry Institute
Construction Users Roundtable
Design-Build Institute of America
Electronic Systems Industry Consortium
Merit Contractors Association of Canada
Metal Building Manufacturers Association
National Association of Minority Contractors
National Association of State Supervisors for Trade and Industrial Education
National Association of Women in Construction
National Insulation Association
National Ready Mixed Concrete Association
National Systems Contractors Association
National Utility Contractors Association
National Technical Honor Society
North American Crane Bureau
North American Technician Excellence
Painting & Decorating Contractors of America
Portland Cement Association
SkillsUSA
Steel Erectors Association of America
Texas Gulf Coast Chapter ABC
U.S. Army Corps of Engineers
University of Florida
Women Construction Owners & Executives, USA
Youth Training and Development Consortium

Contents

Introduction to the
Plumbing Profession
02101-05

02101-05

Introduction to the Plumbing Profession

Topics to be presented in this module include:

Overview

Plumbers protect the health, safety, and comfort of the nation. Plumbers who are well trained are able to install safe and efficient plumbing systems, which reduce the spread of disease and improve sanitation. Successful plumbers develop the skills and knowledge they need to carry out a wide range of tasks. They must also develop professional work habits; convey a positive, cooperative attitude; and practice good safety habits every day.

Plumbers also must have critical thinking skills and a firm grasp of mathematics. Plumbing apprenticeship programs ensure that trainees are able to perform job-related tasks safely. The National Center for Construction Education and Research (NCCER) developed a program to help standardize construction training. This means that students completing their training through an accredited sponsor can receive credentials that are portable and recognized nationwide.

Plumbing apprentices are able to choose from multiple career paths. Skills, interests, and the ability to obtain the right training and remain apprised of industry developments will dictate how far a plumber can progress in the trade. Whatever path is chosen, plumbers should never take chances with their own safety or the safety of others, always demonstrate professionalism, and adhere to relevant codes.

Focus Statement

The goal of the plumber is to protect the health, safety, and comfort of the nation job by job.

Code Note

Codes vary among jurisdictions. Because of the variations in code, consult the applicable code whenever regulations are in question. Referring to an incorrect set of codes can cause as much trouble as failing to reference codes altogether. Obtain, review, and familiarize yourself with your local adopted code.

Objectives

When you have completed this module, you will be able to do the following:

1. Describe the history of the plumbing profession.
2. Identify the responsibilities of a person working in the construction industry.
3. State the personal characteristics of a professional.
4. Identify the stages of progress within the plumbing profession and its positive impact on society.

Key Trade Terms

Aboveground rough-in
Appurtenances
Aqueduct
Backflow
Backflow preventer
Chlorine
Code
Cross-connection
Disinfection
Drain, waste, and vent (DWV)
Ethics
Filtration
Finish
Fixture
Journey plumber
Model code
On-the-job training (OJT)
Plumbarius
Plumber
Plumbing
Plumbum
Polyvinyl chloride (PVC)
Potable
Softening
Stack-out
Thermoplastic
Thermoset
Top-out
Trim finish
Trim-out
Underground rough-in

Required Trainee Materials

1. Appropriate personal protective equipment
2. Paper and pencil
3. Copy of local adopted code

Prerequisites

Before you begin this module, it is recommended that you successfully complete *Core Curriculum.*

This course map shows all of the modules in the first level of the *Plumbing* curriculum. The suggested training order begins at the bottom and proceeds up. Skill levels increase as you advance on the course map. The local Training Program Sponsor may adjust the training order.

101CMAP.EPS

1.0.0 ◆ INTRODUCTION

The National Center for Construction Education and Research (NCCER) designed and developed this training program to meet the needs of the construction industry. It is the only nationally accredited, competency-based construction training program in the United States. A competency-based program requires the trainee to demonstrate the ability to perform specific job-related tasks safely to receive credit.

The primary goal of NCCER is to standardize construction craft training throughout the country so that employers and employees will benefit from it, no matter where they are located. As a trainee in an NCCER program, you will get many benefits to help you prepare for your career in the **plumbing** profession. You will become part of the National Registry. You will receive a certificate for each level of training you complete. If you apply for a job with any participating contractor in the country, a transcript of your training will be available for that contractor to help verify your qualifications. If your training is incomplete when you make a job transfer, you can pick up where you left off, because every participating training center is using the same training program. Many technical schools and colleges also use this program.

A good apprenticeship program effectively combines competency-based, hands-on training with classroom instruction. This combination is the most effective way for a beginning **plumber** to learn and advance through the plumbing profession. Successful apprentices will develop the skills and knowledge they need to carry out a wide range of tasks. They must also develop good work habits; convey a positive, cooperative attitude; and practice good safety habits every day.

If you possess these professional and personal skills, you can become a top-notch, highly productive **journey plumber**. Journey plumbers are plumbers who have successfully completed an apprenticeship training program. The term comes from the 15th-century word *journeyman*. Journeymen were apprentices who left their masters after learning their craft and worked alone or for someone else.

Successful journey plumbers who practice their craft with pride and skill are valued employees and co-workers. Once you have mastered the professional and personal skills necessary, you will be a good candidate to advance to an exciting and challenging career as a master plumber, superintendent plumber, plumbing supervisor, instructor or inspector, project manager, or contractor and business owner.

2.0.0 ◆ A BRIEF HISTORY OF PLUMBING

Plumbing has profoundly influenced the development of modern society by improving public health and safety. Plumbing systems allow people to have safe, healthy, fresh water for drinking, washing, cooking, and other uses. Plumbing also reduces the spread of disease by safely draining away wastewater that contains harmful organisms. Improved sanitation contributes to longer life expectancies for men, women, and children. Plumbing systems also allow people to fight fires, water their lawns, fill their pools, and do many daily activities.

But this was not always so. Plumbing as we know it today is the result of thousands of years of improvements, inventions, and innovations. *Figure 1* illustrates how plumbing systems have evolved since ancient times.

2.1.0 Advancements through the Ages

Archaeologists have determined that rudimentary plumbing systems were in use as early as 2900 B.C.E. Earthenware pipes, masonry sewers, water closets, and drainage systems have been found in Mesopotamia (modern-day Iraq) to prove this.

In 312 B.C.E., the Romans began bringing water into Rome through **aqueducts**. Most aqueducts were open, stone-lined trenches that used gravity to move water downhill. The more famous arched aqueducts were not nearly as commonly used as the simpler trench systems. By 100 C.E., the aqueduct system was so advanced that Rome built and maintained public bathhouses and fountains throughout the city. The aqueducts were also used to drain wastes and discharge them into the river downstream from the city.

The Romans also gave us the word that we use

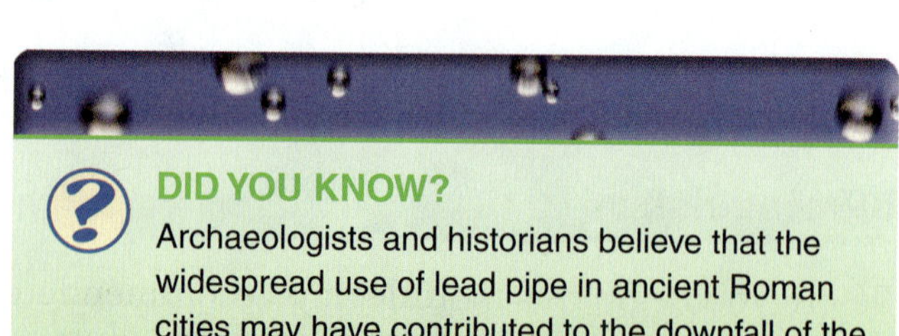

DID YOU KNOW?

Archaeologists and historians believe that the widespread use of lead pipe in ancient Roman cities may have contributed to the downfall of the Roman Empire. Lead is harmful to the brain, the nervous system, and vital organs such as the liver. Water carried through lead pipes may have gradually poisoned many people in the Roman Empire.

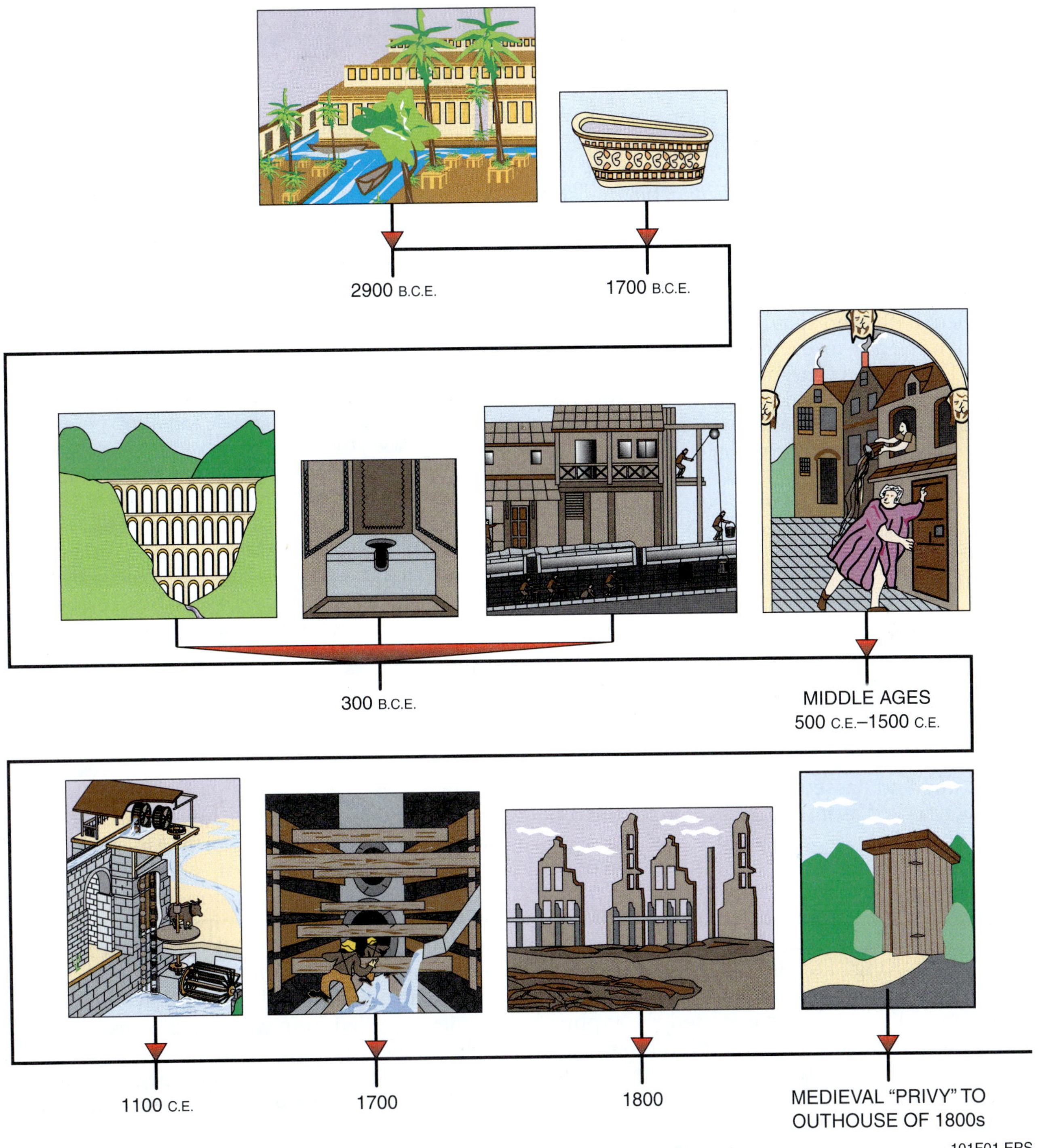

Figure 1 ◆ The evolution of plumbing.

today to describe people who install and maintain water supply and waste systems. Many Roman cities used lead pipe in their plumbing systems. The Latin word for lead is ***plumbum***, and a person who worked with lead was called a ***plumbarius***. Over the centuries, this ancient word has come down to us as a word you know very well: plumber.

During the fifth century C.E., the Goths invaded Rome and the Roman Empire came to an end. Roman cities and their infrastructure fell into disrepair. Much of the science, math, and technology of the Roman and Greek golden age were gradually forgotten. The period after the fall of Rome (500 C.E. to 1400 C.E.) is often referred to as the Dark Ages because of the social and cultural decline that resulted from this loss.

During that time, people neglected sanitation and hygiene. They emptied their sewage into the streets and did not store or prepare their food in a healthful way. As a result, more people died during those years from disease than from wars. Some historians estimate that one-third of the population of Europe died from diseases spread by unsanitary living conditions. Bubonic plague and smallpox killed thousands across the continent.

By the mid-1300s, however, sanitation began to return to the cities of Europe. London, England, laid its first water supply pipe during that time. The ancient baths and spas built by the Romans in England were put to use again during the 16th century. During that time, plumbing reemerged as a profession. In 1625, the first apprenticeship laws for plumbing went into effect in England. Other countries began to develop water supply systems. A new era of sanitary awareness slowly emerged.

One of the first known water closets was built in England in 1596 by a godson of Queen Elizabeth I, Sir Jon Harrington. However, it was another 200 years before the use of this invention became widespread. In 1775, Alexander Cumming invented the S trap, the precursor of the toilet traps used today. Three years later, Joseph Bramah patented an improved version of Cumming's device that incorporated two hinged valves. The original is still in use today in the House of Lords in London's Parliament building. In 1848, England passed the *National Public Safety Act*, which served as model legislation for the rest of the world. The Act required every house to have sanitary facilities, whether a flush toilet or an outhouse.

The plumbing profession received a royal boost in 1871. That year, Albert, the Prince of Wales (later King Edward VII), almost died from typhus. Later investigation of the prince's illness found that he had contracted the typhus from contact with contaminated plumbing lines. The problem was corrected before they could cause more illness. A relieved and healthy Prince of Wales was later quoted as saying, "If I could not be a prince, I'd rather be a plumber."

Inspiration and innovation played a large role in the development of modern plumbing, but so did disasters. In 1666, after the Great Fire of London nearly destroyed the city, its citizens organized the first firefighting brigades and expanded the city's water supply system. In the United States, deadly cholera epidemics in Philadelphia in 1793 and New York City in 1832 spurred construction of improved water supply systems. Earthquakes in California led to the modernization of municipal water systems there as well.

DID YOU KNOW?

Invention of Flush Toilets

London plumber Thomas Crapper has been credited with the invention of the self-contained flush toilet. He was a plumber who was responsible for many innovations in his profession. During the late 19th century, he was granted patents for everything from improvements to drains to manhole covers to pipe joints. Around the same time, the first "modern" toilet appeared: a U-bend siphoning system to flush the pan where waste was deposited. Records show that the patent for this device was actually issued to Albert Giblin. It is likely that Crapper bought the patent rights from Giblin and then marketed the device himself. Crapper also served as the sanitary engineer for many members of the English royal family. Crapper's association with the device may have originated during World War I, when soldiers used his name as a synonym for the toilets they saw as they passed through England on the way to the front lines in Europe.

The United States, in fact, has been responsible for many of the most notable advances in plumbing technology as well as in health and safety standards. Early settlers of North America brought with them the idea that baths and spas had curative properties. In 1652, Boston developed the first waterworks, using wood pipes for both firefighting and private use. In 1804, Philadelphia began using cast-iron piping for its water mains. By the mid-1800s, American homes often had outdoor privies supplied from wells or cisterns.

It was not until the middle of the 19th century, however, that the United States began to develop practical water and sewage systems. In 1857, an engineer named John Adams designed a sewer system for Brooklyn, New York. When it proved to be successful, he published his design, and other cities adopted it.

Wastewater treatment, however, still proved to be a problem for many cities. The first advancement in water purification was a slow sand filter installed in Richmond, Virginia, in 1832. Scientists later discovered that treating water supplies with **chlorine** (a heavy, greenish-yellow gas) would kill deadly bacteria. Following that discovery, cities in the United States and England began treating wastewater with chlorine.

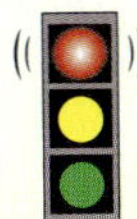

WARNING!

Chlorine is highly toxic and corrosive. It also poses a serious fire risk. Chlorine will react violently with many chemicals, including water. Always wear appropriate personal protective equipment (PPE) when working with liquid or gaseous chlorine. Always consult the material safety data sheet (MSDS) before moving, storing, or working with chlorine.

Despite these advancements, by the mid-1800s plumbing systems were neither complex enough nor large enough to keep up with the rapid growth of cities and populations spurred by the Industrial Revolution. By 1855, for example, Chicago still relied on individual wells and the Chicago River to support a population of 75,000. Densely populated factory cities in the United States threatened to deplete local water supplies and overwhelm rivers and streams with wastewater. The plumbing profession had to change with the times to meet these new demands.

2.2.0 A New Professional Organization

By the late 19th century, plumbing technology and practices were recognizably modern. Manufacturers and wholesale dealers of new plumbing devices sold them over the counter as separate components. They ignored the fact that these components would have to be combined and installed into a properly designed plumbing system in order to work. Dealers claimed no responsibility for the proper installation of plumbing systems. The results were predictable: People suffered from shoddy and unsafe plumbing systems that sometimes made sanitary conditions worse instead of better. And plumbers were usually blamed—unfairly—when things went wrong. Plumbers had the knowledge and ability to install safe and efficient plumbing systems, but the manufacturers and dealers completely dominated the trade.

In 1883, a group of master plumbers united to take action and address the situation. They convened a meeting in New York City and there formed the National Association of Master Plumbers (NAMP), the forerunner of today's Plumbing-Heating-Cooling Contractors—National Association (PHCC). The association believed that proper public sanitation could be ensured only if a single person—the contractor—was responsible for acquiring plumbing materials and using them to build effective, safe, and complete sanitary systems.

Accordingly, at the association's second annual convention in Baltimore in 1884, the attendees adopted a set of professional guidelines that became famous as the "Baltimore Resolutions." The resolutions succeeded at last in putting plumbers in charge of the plumbing profession. They helped ensure that plumbing installations in the United States would be safe and sanitary, and that plumbers would be held to rigorous professional standards.

2.3.0 Plumbing in the 20th Century

Sanitation, along with medical science, continues to be largely responsible for the maintenance of public health. In the United States, great progress in the development of plumbing methods and technologies has been made since 1910. The reliability of traditional piping materials such as copper and cast iron has been dramatically improved. New plastic compounds called **thermoplastics** and **thermosets** have been developed. The physical properties of these plastics make them ideal for use in sanitary systems. **Polyvinyl chloride (PVC)**, developed in the 1930s, was the first plastic used in plumbing. PVC is widely used in cold water systems. It can last indefinitely under most conditions. You will learn about different types of pipe materials elsewhere in *Plumbing Level One*. Manufacturers have also improved plumbing **fixtures**, which are devices that receive water from a water supply line. Common fixtures include sinks, faucets, shower stalls, and toilets.

In 1911, the Kohler Company developed the first one-piece recess bath. Before this, baths had been built in two separate sections: the tub itself and the surrounding apron. Kohler's one-piece tub was much more sanitary and attractive. In 1926, Kohler introduced the electric sink, a combination of a conventional sink and electric dishwasher.

Other inventors established companies whose names are familiar to today's plumbers. Al Moen, who held more than 75 patents, invented the single-handle mixing faucet. In 1906, William Sloan developed a flush valve in which water flows under pressure from the supply pipe directly to the fixture, so that it is always ready to be used. Halsey W. Taylor established the company that bears his name in 1912 to manufacture and sell hygienic drinking fountains.

The Baltimore Resolutions

Whereas, The manufacturing and wholesale firms in plumbing materials persist in selling to consumers, to our injury and detriment, placing us toward our customers in the light of extortionists, causing endless trouble; and

Whereas, The system of protecting us from this wrong, which draws in its wake other wrongs, is ineffective; it is absolutely necessary to perfect such a system, by united action, which will remove these evils from which we have suffered for years; therefore be it

Resolved, That we withdraw our patronage from any firm manufacturing or dealing in plumbers' material selling to others than master plumbers.

Resolved, That the manufacturers of gas fixtures selling to consumers shall not receive the patronage of any master plumber.

Resolved, That the master plumbers shall demand of the manufacturers and wholesale dealers in plumbing materials to sell goods to none but master plumbers.

Resolved, That this association shall keep a record of all journeymen and plumbers who place in buildings plumbing material bought by consumers of manufacturers or dealers.

Resolved, That any manufacturing or wholesale dealers, dealing in wrought iron pipe, who sell to consumers shall not receive our patronage.

Resolved, That a committee be appointed by this association in every state and county, for the purpose of reporting to the proper officer at its head in the state any violation of these resolutions.

Resolved, That these measures are just and necessary to our welfare, and a rigid enforcement is demanded.

Resolved, That this convention endorse the above, and urge upon the National Association to perfect and adopt a uniform system of protection for the trade over their entire jurisdiction.

Source: *A Heritage Unique* (PHCC, 1999)

3.0.0 ◆ THE PLUMBING PROFESSION

Plumbing is practically everywhere: in kitchens and bathrooms; at golf courses; in parks; and at municipal sanitary, sewage, and water systems. Because our man-made environment relies on plumbing, plumbers play a very important role in society. The systems that plumbers install do many things, including the following:

- Make drinking water safe and pure
- Protect homes and businesses from fires
- Help keep basements dry
- Help keep lawns and gardens green

The plumbing profession is made up of talented men and women who draw on a set of skills that are well suited to the field. Plumbers are good at working with their hands. They can use precise measuring and testing tools with skill. They get a lot of satisfaction from solving complex problems. Plumbers take pride in their work, and they get along with all kinds of people.

A plumber must have critical thinking skills to be able to read and interpret construction drawings, review manufacturer's specifications, understand and apply legal standards, rough-in entire plumbing systems, and troubleshoot problems. Plumbers must have a firm grasp of basic mathematics. They need to be fairly fit and flexible, because their work can be physically demanding. Plumbers often work in cramped spaces and do a lot of work standing up.

Plumbers need these skills because they perform a wide range of tasks. Plumbers design, install, repair, and maintain **potable** (drinkable) water supply lines and piping layouts; drainage systems; **drain, waste, and vent (DWV)** systems; and gas systems. There are many professions related to plumbing that require the same base knowledge that a plumber has, including the following:

- Sprinkler fitters, who install fire-sprinkler and other fire-protection systems in buildings
- Pipefitters, who install piping for steam, water heating and cooling, lubrication, and other uses
- Steamfitters, who install pipe systems to move liquids or gases under high pressure
- Irrigation system installers, who design, install, and maintain irrigation systems for small areas such as residential gardens and large areas such as golf courses

Plumbers also specialize in commercial and residential design, construction, remodeling, renovation, and maintenance. Plumbers can own and operate their own contracting business. They can also branch out into related fields by becoming plumbing inspectors or instructors, plumbing estimators, or safety managers.

Many plumbers specialize in servicing specific types of systems, such as air conditioning, heating, and water distribution systems. They become experts in troubleshooting and repairing malfunctioning equipment. If you specialize in the service and repair of specific systems, you will always be in demand.

Whatever their specialty, plumbers work with the latest technology, tools, and equipment. Being a plumber means keeping up-to-date on the technological developments in plumbing, on better methods to do plumbing work, on the qualities and advantages of new materials, and on changes to the **codes** and standards that govern plumbing. A code is a legal document adopted by a jurisdiction that establishes the minimum acceptable standards, rules, and regulations for all materials, practices, and installations used in buildings and building systems.

Most states and localities require plumbers to be licensed as a journey plumber, master plumber, and/or plumbing contractor. Refer to your local code for the guidelines that apply where you work.

3.1.0 What Is Plumbing?

Plumbing has different meanings to different people. The *National Standard Plumbing Code* (NSPC) defines plumbing as follows:

> *The practice, materials, and fixtures within or adjacent to any building structure or conveyance, used in the installation, maintenance, extension, alteration, and removal of all piping, plumbing fixtures, plumbing appliances, and plumbing* **appurtenances** *connected with the following:*
>
> *a. sanitary drainage systems and related vent systems;*
> *b. storm water drainage facilities and venting systems;*
> *c. public or private potable water supply systems;*
> *d. the initial connection to a potable water supply upstream of any required* **backflow** *prevention devices and the final connection that discharges indirectly into a public or private disposal system;*
> *e. medical gas and medical vacuum systems;*
> *f. indirect waste piping, including refrigeration and air conditioning drainage; and*
> *g. liquid waste or sewage, and water supply, of any premises to their connection with the approved water supply system or to an acceptable disposal facility.*

The NSPC also specifically excludes the following types of piping systems from its definition of plumbing:

- All piping, equipment, or material used exclusively for environmental control
- Piping used for the incorporation of liquids or gases into any product or process for use in the manufacturing or storage of any product, including product development
- Piping used for the installation, alteration, repair, or removal of automatic sprinkler systems installed for fire protection only
- The related appurtenances or standpipes connected to automatic sprinkler systems or overhead or underground fire lines beginning at a point where water is used exclusively for fire protection
- Piping used for lawn sprinkler systems downstream from backflow prevention devices

3.2.0 Protecting Public Health, Safety, and Comfort

The goal of plumbers is to protect the health, safety, and comfort of the nation. Plumbers accomplish this goal in two ways:

- By installing safe, reliable plumbing systems
- By properly maintaining and repairing existing systems

Federal agencies such as the Environmental Protection Agency (EPA) enforce rules and regulations related to environmental issues. Laws such as the *Clean Water Act* and the *Clean Drinking Water Act* have been passed to develop programs to help ensure the enforcement of these rules and regulations. State and local health agencies also establish their own complementary standards and regulations that apply to their jurisdictions.

Your responsibility as a plumber is to follow all of the rules and safeguards related to your work. You should also report any accidents or events that may be hazardous to your health or to the environment.

One of the biggest health and safety issues facing plumbers today is contamination caused by an improper **cross-connection** between plumbing systems. A cross-connection is an arrangement between a potable water system and a nonpotable water system in which an accidental pressure differential between the two systems causes backflow of contaminated water into the freshwater system. Backflow can force wastes through a cross-connection. Several things can cause backflow:

- Cuts or breaks in the water main
- Failure of a pump
- Injection of air into the system
- Accidental connection to a high-pressure source

When wastewater or other liquids are siphoned into the fresh water supply, they can cause contamination, sickness, and even death. *Escherichia coli*, also known as *E. coli*, and *Legionella* (the bacteria that causes Legionnaire's disease) can be spread through backflow from contaminated systems. Safety devices called **backflow preventers** keep wastewater from entering the water supply system. You will learn more about backflow and backflow prevention later in the *Plumbing* curriculum.

To prevent the spread of harmful organisms, chemicals, and materials through water supply systems, water is made safe through **disinfection**, **filtration**, and **softening**. To disinfect water means to destroy harmful organisms in the water. Filtration is the process of cleansing water to remove particles and chemicals. Softening removes magnesium and sodium salts that cause scale on the inside of pipes and fittings. You will learn more about these methods of water treatment later in the *Plumbing* curriculum.

The safety and health of the plumber is just as important as that of the customer. One of the dangers facing plumbers is injury caused by scalding. For many years, plumbing codes have usually required water to be available at temperatures of up to 140°F. The American Society of Plumbing Engineers (ASPE) has found that this temperature is necessary to prevent the growth of *Legionella* bacteria. At this temperature, however, water can scald exposed skin almost immediately. In response to this danger, many codes include specific requirements to reduce the threat of scalding.

ON THE

LEVEL

Low-Flow Fixtures

As you have learned, one of the most important goals of the plumbing profession protects the health, safety, and comfort of the nation. Another important goal is to act as a steward of the environment. Water conservation is one of the most important environmental issues in the 21st century, and water-conserving fixtures such as toilets and shower heads help preserve this vital natural resource.

One of the most promising (and most controversial) types of water-conserving fixtures is the low-flow toilet. Toilets consume nearly 40 percent of the water in a typical household. In 1992, the federal government required that low-flow toilets be installed in all new and renovated plumbing systems. These toilets use 1.6 gallons per flush, about half as much as older toilets. Advocates of low-flow toilets claim that the water savings from low-flush toilets can save millions of gallons a year and extend the operating life of water treatment facilities.

However, these fixtures must be properly sized and installed to work. When low-flow toilets are connected to DWV systems that are designed for higher flow rates, the systems may clog and overflow, or they may require multiple flushes to empty the bowls, defeating the purpose of the fixture. The installation of low-flow fixtures may even require replacement of the sewer lines to accommodate the lower fluid flows: an expensive proposition. Contractors report that they receive more callbacks regarding low-flow toilets than for any other reason.

Always refer to your local code when sizing and installing plumbing fixtures. Read the manufacturer's instructions and follow the installation instructions carefully. Getting the installation right the first time will help save the contractor time and money, and will ensure satisfied customers.

CAUTION

Incorrectly sized pipe in a plumbing system can cause the system to work incorrectly and could even damage the pipe and fixtures. Ensure that all pipe used in a plumbing system is sized according to the applicable local code. Codes include sizing charts that should be used for reference.

3.3.0 The Three Phases of a Plumbing Project

Plumbing can be divided into three broad phases:

- **Underground rough-in**
- **Aboveground rough-in** (also called **top-out** or **stack-out**)
- **Finish** (also called **trim-out** or **trim finish**)

During the underground rough-in phase, the plumber locates all supply and waste connections from the building systems to public utilities, and establishes where these systems will enter or leave the building. During the aboveground rough-in phase, the plumber cuts holes in walls, ceilings, or floors to attach or hang pipes for connection to fixtures. Then the plumber installs the pipe for the building's various supply and waste systems. To join pipe runs, a plumber might use welding tools, soldering equipment, or special chemicals for plastic pipes. The plumber may operate power threading machines, propane torches, and other power tools during this phase.

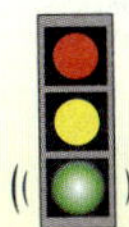

NOTE

Many plumbing terms are used interchangeably in some areas. Be sure you are using the terms the way they are typically used where you work.

Finally, in the finish phase, the plumber installs fixtures such as sinks, showers, and toilets and appliances such as dishwashers, water purification systems, and water heaters. The plumber may be called on to install the automatic controls that regulate pressurized pipe systems as well.

During each of the three stages, you must ensure that the system is tight and safe. Do not let your customers discover leaks behind the walls after you have completed your work. Callbacks are expensive and can hurt your reputation as a professional.

Regardless of how well a plumbing system is installed, it must be maintained and periodically repaired. You could even think of service and maintenance as a fourth phase of a plumbing project. Plumbers perform a variety of tasks as part of service and maintenance, including the following:

- Checking lubrication levels in pumps, test gauges, and meters
- Repairing faulty fixtures and components
- Verifying operating systems
- Regulating flow and usage rates

3.4.0 Plumbing Codes and Licenses

Plumbing codes have grown increasingly comprehensive to accommodate new developments in technology and materials. Municipalities throughout the United States adopt ordinances based on suggested national plumbing standards, called **model codes**. They often modify or interpret these model codes to reflect local conditions. For instance, a city that is prone to earthquakes, such as Los Angeles, would incorporate standards to protect installations from earthquake damage. Likewise, a city that is subject to flooding, such as Houston, would add special requirements to guard against contamination from floodwater.

The three primary model codes associated with plumbing are the *International Plumbing Code* (IPC), the *Uniform Plumbing Code* (UPC), and the *National Standard Plumbing Code* (NSPC). You will learn how codes are developed and implemented later in the *Plumbing* curriculum.

Although model codes act as guidelines, municipal ordinances have the force of law. As an apprentice plumber, you will need to know the applicable plumbing code that applies to the area where you work. Your local plumbing code covers the installation, alteration, repair, replacement, additions to, and use of plumbing systems. It establishes standards for many things, including the following:

- Materials commonly used in plumbing systems
- Fixtures, joints, and connections
- Hangers and supports
- Water supply and distribution
- Vents
- Storm water and sanitary drainage systems
- Potable water supply systems

All states require plumbers to be licensed. Many states require completion of an apprenticeship program and/or **on-the-job training (OJT)** before testing for a license. When you apply for a license (see *Figure 2*), you will be tested on your knowledge of both your profession and the local plumbing ordinances.

101F02.EPS

Figure 2 ◆ Examples of plumbing licenses.

Review Questions

Sections 1.0.0–3.0.0

1. Which of the following is *not* a benefit of participating in an NCCER training program?
   a. Training transcripts are available for your employer.
   b. You are listed in the National Registry.
   c. You are certified as a master plumber.
   d. You receive certificates for each completed training level.

2. The modern word *plumber* is derived from the Latin word for _____.
   a. the water channels that supplied Rome
   b. a person who works with lead
   c. doctors who treated lead poisoning
   d. a person who repaired bath houses

3. The NSPC definition of plumbing includes piping systems used for the installation, alteration, repair, or removal of automatic sprinkler systems for fire protection.
   a. True
   b. False

4. Above all other codes and ordinances, the _____ code carries the force of law.
   a. local
   b. *Uniform Plumbing*
   c. model
   d. *National Standard Plumbing*

Match the description of the work to be done to the correct phase of the plumbing project.

5. Plumbers cut holes in walls, ceilings, or floors to attach or hang pipes for connection to fixtures. Then they install the pipe for the building's various supply and waste systems.
6. Plumbers install fixtures such as sinks, showers, and toilets and appliances such as dishwashers, water purification systems, and water heaters. Plumbers may also be called on to install the automatic controls that regulate pressurized pipe systems.
7. Plumbers locate all supply and waste connections from the building systems to public utilities and establish where these systems will enter or leave the building.

a. Finish
b. Underground rough-in
c. Estimating
d. Aboveground rough-in

4.0.0 ◆ KEYS TO PROFESSIONAL SUCCESS

To be a successful plumber, you must be able to produce finished products of high quality in a minimum amount of time. You have probably heard the phrase "on time and on (or under) budget." Take a moment to really think about what this phrase means. It means that the job is completed by the deadline, and that it does not cost more than what was originally expected. It pays to keep this saying in mind—literally.

Successful plumbers must be flexible enough to adapt their methods of working to the particular needs of each job. As you gain experience as a plumber, you will develop the flexibility necessary to meet changing job requirements. You will also improve your knowledge of the different tools and materials needed to accomplish certain jobs, and you will learn more about the technologies and methods that are available. Over time, you will improve your ability to do your job on time, on budget, and to the satisfaction of your customers. Satisfied customers mean repeat business.

Keep current with the latest methods, materials, and equipment in the plumbing profession. This knowledge will help you and the company you work for stay competitive. You should never stop learning. When you complete your apprenticeship, you will simply be moving into a bigger classroom: the world.

Never take chances with your own safety or the safety of others. Always demonstrate a high level of professionalism. Remember, you are part of a proud heritage, and you represent a profession that is vital to modern society. As a plumber, you will want your customers to think well not only of your work as an individual, but of the profession in general.

Construction professionals adhere to a code of professional **ethics**. Ethics are principles and values that guide conduct. As a plumber, you are responsible for understanding and following the code of ethics for your chosen profession. *Appendix A* lists the ethical principles that are followed by members of the construction trades.

Successful plumbers exhibit all of the following qualities:

- Positive attitude
- Honesty
- Loyalty
- Willingness to learn
- Responsibility
- Cooperation
- Attentiveness to rules and regulations
- Promptness and reliability

You were introduced to many of these concepts in *Core Curriculum*. In this module, you will learn how they apply specifically to the plumbing trade. In the classroom and on the job, you will see that the successful plumbers you meet will demonstrate all of these qualities. Learn from them, and become a better plumber.

4.1.0 Positive Attitude

A positive attitude entails being energetic, motivated, attentive, and alert. It means understanding the importance of safety and responding flexibly to change. A positive attitude will help you fulfill your obligations and take pride in your job.

An employee with a positive attitude contributes to others' productivity by setting an example. You can probably recall a situation where you had to work with someone who was complaining and slacking off. Do you remember how difficult and unpleasant it was to work in a situation like that? Try to keep that in mind when you are on the job, and ask yourself: Am I demonstrating a positive attitude?

A positive work environment not only leads to greater productivity, but also makes the job more interesting and exciting. Employers pay attention to your attitude. Supervisors see your attitude in your approach to the job, in your reactions to instructions, and in the way you handle problems, especially unforeseen ones.

Tips for a Positive Attitude

Here is a short checklist of things that you should keep in mind to help you develop and maintain a positive attitude:

- Remember that your attitude follows you wherever you go.
- Helpful suggestions and compliments are much more effective than negative ones.
- Look for the positive characteristics about your teammates and supervisors.
- Focus on the good things about your job and the place where you work.

4.2.0 Honesty

Honesty and personal integrity are important traits of the successful professional. Honesty means more than just telling the truth and never cheating or stealing. It also means doing a fair day's work for a fair day's pay. It means holding up your side of a bargain. It means backing up promises with action.

You should take pride in performing a job well and you should never cut corners. Never lie to clients, supervisors, or co-workers. Respect the property of others. Ensure that the tools and materials belonging to your employer (or sponsor), customers, and other trades are not damaged or stolen from the shop or the job site. Treat their property with the same respect that you give to your own.

Your reputation will always precede you. A reputation for honesty and integrity will make employers want to hire you because they know that you can be trusted. On the other hand, if you lie, cheat, or steal, that reputation will precede you wherever you apply for work and your career will suffer accordingly.

If you want to be gainfully employed, earn a good salary, and be well regarded by teammates and supervisors, then you should start by being honest. It's that simple.

4.3.0 Loyalty

As an employee, you expect your employer or sponsor to look out for your interests, keep you steadily employed, and promote you as job openings occur. Your employers and sponsors, in turn, have a right to expect something from you: loyalty. Remember, however, that loyalty cannot be demanded from someone; it can only be earned.

Loyalty means different things to different people, but in a professional setting loyalty usually means the following:

- Keeping the interests of your employer or sponsor in mind
- Speaking well of your employer or sponsor to others
- Keeping any minor troubles strictly within the plant or office
- Respecting the confidentiality of all matters that pertain to your employer's or sponsor's business

A loyal employee fulfills commitments and meets obligations. Never go to work for someone else without first ensuring that you have met all of your commitments to your current employer or sponsor. A prospective employer will be impressed with your commitment and be willing to wait for you. If not, you might want to reconsider whether you should work for that company.

4.4.0 Willingness to Learn

A good plumber never stops learning. New machines, tools, materials, and methods are constantly being introduced. Even experienced plumbers need to be willing to learn about these advancements. If you keep your methods and tools up-to-date, your company will be able to use your skills to compete successfully with other companies. The better a company competes, the more profit it makes. The more profit the company makes, the more likely it is to stay in business, provide jobs, and raise salaries, all of which benefit you as an employee.

In addition, each workplace has its own way of doing things. Your employer or sponsor will expect you to learn how things are done where you work. You must be able to adapt to change and be willing to learn new methods and procedures as quickly as possible.

4.5.0 Responsibility

As a trainee, your primary responsibility is to do what is asked of you. As a professional plumber, however, your responsibility will extend beyond assigned tasks. You should be able to see what needs to be done and do it without having to be asked. Your supervisor expects you to do the work well without being asked twice. Assuming responsibility makes a favorable impression on an employer/sponsor.

Always take responsibility for your actions. Do not make excuses or blame others when you make an error. Doing so will cause people to distrust you and lose respect for you. If that happens, the quality of your work, and that of everyone who works with you, will suffer.

4.6.0 Cooperation

As a plumber, you will probably work with other people as part of a team. Every day, you interact with members of your work crew and with your supervisor. The ability to cooperate with your teammates is an important skill in the construction industry. But simply "working and playing well with others" is not enough. The success of any team depends on all of its members doing their parts. Everyone must contribute to the team to ensure that it achieves its goals. Such cooperation will help get projects done on time and on budget.

4.7.0 Attention to Rules and Regulations

Employers want workers who respect the rules and regulations that apply to their company, their work, and their profession. Rules and regulations exist to keep people safe and to keep a project on schedule. Companies establish rules pertaining to work times, administrative issues, and professional conduct. Local, state, and federal authorities develop regulations that apply to a wide range of issues, including safety and health, materials, tools and equipment, and legal issues.

People can work well together only if they all understand the work that needs to be done, when it needs to be done, and who will perform the work. Rules and regulations are necessary in any work situation. The rules might vary from employer to employer, so it is important to make sure you understand your employer's or sponsor's regulations each time you start a new job.

4.8.0 Promptness and Reliability

Imagine that you are a member of a three-person team that is laying pipe. According to the schedule, you must lay a certain amount of pipe in two weeks. If a member of your team were absent the whole time, you and the other team member would have to do that person's work in addition to your own, which means that each of you would have to do 50 percent more work than anticipated over the two-week period. However, if your teammate notified your supervisor, then arrangements could be made to provide another worker or to extend the completion date of the pipe-laying task.

Professionals are always prompt and reliable when they report to work. A prompt employee shows up to work on time; a reliable worker shows up every day. Employees who are chronically late or absent have poor work habits. They are being unprofessional and are showing that they lack commitment.

The clock governs your work life. Everyone is required to be at work and ready to start at a definite time. Failure to get to work on time results in confusion, lost time, and the resentment of your teammates. It could even get you fired. The habit of being late can stand in the way of a promotion; the habit of being early can help you get promoted. The good news, however, is that bad habits can be changed. If you are chronically late for work, then change your morning routine. Leave earlier, for example, or get up earlier to take care of your personal responsibilities. You could also go to bed earlier.

Sometimes you need to take time off from work because you or a family member are sick, or because you have suffered a death in your family; or you might get stuck in a traffic jam or have car trouble. When something like this happens, let your employer or sponsor know as soon as possible. That way, your supervisor can make arrangements to replace you while you are out.

Don't come late to work because you need to take care of personal business or just because you are dissatisfied with your job. Take care of personal business outside of working hours. If you are dissatisfied with your job, talk to your supervisor or employer about it. Don't just fail to show up. You will not get your point across that way, and you may lose your job.

Your employer or sponsor has the right to expect you to be on the job unless you have a good reason. You have an obligation to your supervisor and your teammates to be at your job because they are relying on you. Indeed, you should be proud of this; it means that they trust you and respect you. When you fulfill that obligation, you will be trusted by your co-workers and valued by your supervisor.

5.0.0 ◆ CAREER OPPORTUNITIES IN PLUMBING

The construction industry employs more people and contributes more to the nation's economy than any other industry. Our society will always need new homes, roads, airports, hospitals, schools, factories, and office buildings. This means there will always be a source of well-paying jobs and career opportunities for construction trade professionals, including plumbers. Both large and small contracting firms employ plumbers. Plumbers who master the principles of small-business management and develop and hone their entrepreneurial skills can be self-employed.

Plumbers can choose from a number of career paths depending on their skills and interests (see *Figure 3*). Where you will go will depend on the pride and professionalism you show in your work. How well you succeed will depend on your willingness to develop your skills and obtain the right training. With time and experience, plumbers progress from apprentices to one or more different specialties. Following are some of the many positions and careers in the plumbing profession.

An apprentice is someone who is learning the plumbing trade. As you have learned, when a plumber successfully completes an apprenticeship, he or she becomes a journey plumber. Journey plumbers can remain at that level or advance in the profession. Journey plumbers might supervise less experienced plumbers or be called on to estimate the materials and labor for a specific job. With larger companies and on larger jobs, journey plumbers often become specialists.

A master plumber is someone who consistently demonstrates the highest level of skill at a profession. Master plumbers are proven experts in the plumbing profession. They can act as mentors and teachers of less experienced members of the profession. Master plumbers often start their own businesses.

Foremen are leaders who direct a single team of apprentices and/or journey plumbers. Large construction projects with more than one team require supervisors who oversee the work of all teams. They are responsible for assigning, directing, and inspecting the work of each team.

The safety manager is another key, on-site person. Safety managers are responsible for the safety and health of the workers on a project. They are responsible for developing a safety plan and procedures, training workers in safety procedures, and ensuring that the project is in compliance with applicable health and safety regulations.

Project managers, often called project administrators, control the scope and direction of a plumbing business. Managers ensure that the right workers are being assigned to each task. Managers are also responsible for handling project accounting, which involves managing payroll, taxes, and employee benefits; tracking project finances; and providing administrative support for

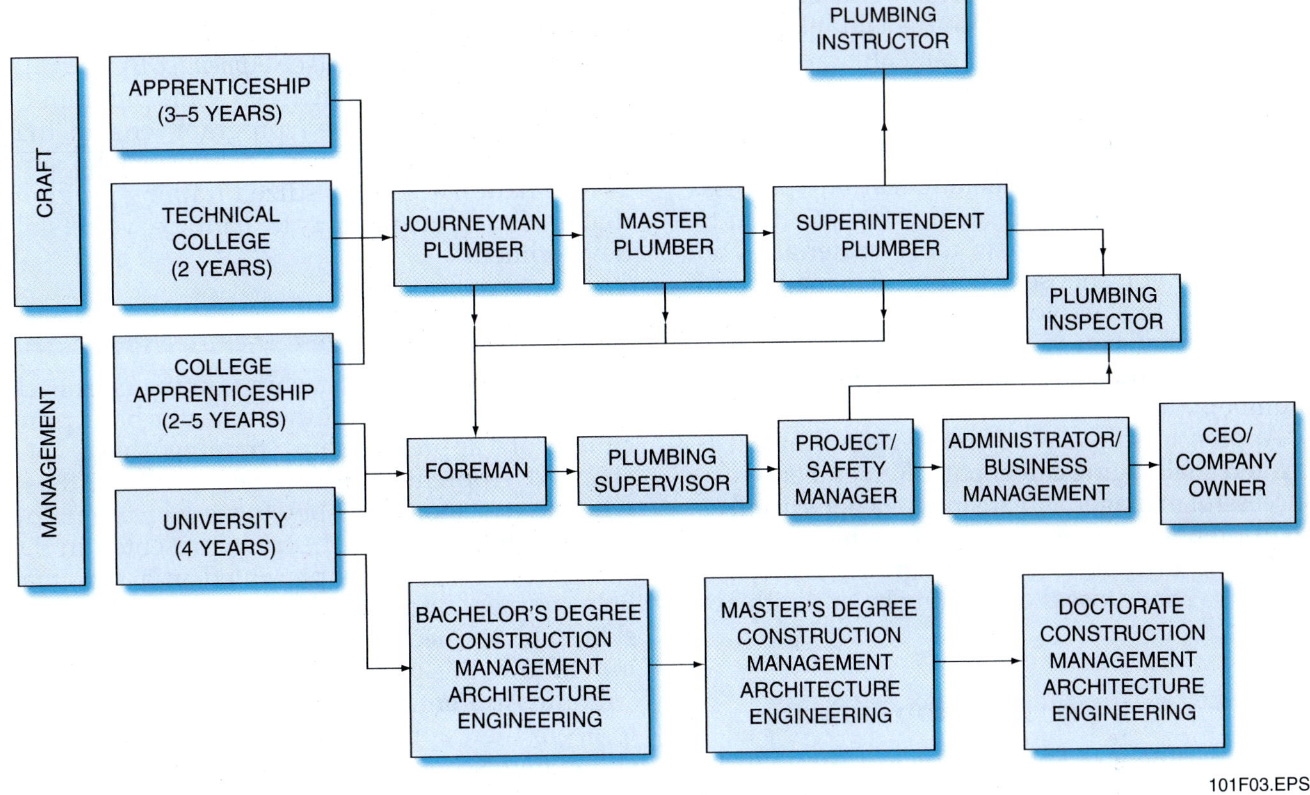

Figure 3 ◆ Opportunities in the construction industry.

the contractor. Large contracting firms may have several project managers.

Plumbing estimators work for contractors and building supply companies. They assess the materials and labor required to complete a project. Contractors submit bids for jobs based on these estimates. Plumbing estimators must have a thorough understanding of construction methods as well as knowledge of the availability and prices of materials and supplies.

Plumbing estimators are experienced plumbers with superior mathematical abilities and the patience to prepare detailed, accurate estimates. Today's estimators also need solid computer skills. Estimators have a great deal of responsibility, because errors in estimates can cost the contractor significant amounts of money. Depending on the size and type of the business, the estimating job may also be done by the contractor or the project manager.

Mechanical engineers research, develop, design, manufacture, and test a wide range of mechanical devices, including the following:

- Machine tools
- Internal combustion engines
- Electric generators
- Steam and gas turbines
- Refrigeration and air conditioning equipment
- Material handling systems
- Elevators and escalators
- Industrial production equipment
- Robots used in manufacturing

Construction contractors and owners establish and run contracting businesses. Typically, they hire apprentices, journey plumbers, and master plumbers. Depending on the size of their business, contractors may work with the crew or may manage the business full-time. Small contractors may have only one or two employees who manage the business, prepare estimates, obtain supplies, and perform the actual contracted work. Subcontractors and contractors/owners may specialize in installing particular systems such as sprinkler systems, high-pressure pipe systems, or municipal water-treatment facilities.

Plumbing inspectors may work for a contractor or for a municipal jurisdiction such as a local or state government. Inspectors ensure that all plumbing installations follow applicable codes and quality requirements. Plumbing inspectors review and approve work at various stages of construction, depending on the installation.

Plumbing instructors train apprentices to become the next generation of journey and master plumbers, managers, inspectors, and engineers. Many dedicated and experienced plumbers find that becoming an instructor is a natural evolution in their career. Master plumbers often become instructors as their careers progress.

No matter what career path you choose, you will be embarking on a lifelong learning process. Effective plumbers along any career path need to keep up-to-date with new tools, materials, and methods. If you choose to become a manager or if you intend to start your own construction business someday, you will need to learn the appropriate skills and continue to hone your skills as a plumber.

Apprenticeship is your first step down an exciting and challenging career path. Remember, every successful manager and business owner started as an apprentice. Successful plumbers have at least one thing in common: a willingness to keep on learning. That process of learning begins with apprentice training.

5.1.0 Formal Construction Training

The Department of Labor's Office of Apprenticeship Training, Employer and Labor Services (ATELS) sets the minimum standards for training programs across the country. ATELS programs rely on mandatory classroom instruction and OJT. ATELS requires 144 hours of classroom instruction per year and 2,000 hours of OJT per year. In a typical ATELS program, trainees spend 576 hours in classroom instruction and 8,000 hours in OJT before receiving journey plumber certificates.

To address the training needs of the professional communities, NCCER developed a four-year plumbing training program. NCCER uses the minimum ATELS standards as a foundation for comprehensive curricula that provide trainees with in-depth classroom and OJT experience.

The NCCER *Plumbing* curriculum provides trainees with industry-driven training and education. It adopts a purely competency-based teaching philosophy. This means that trainees must demonstrate to the instructor that they possess the understanding and the skills necessary to complete the tasks that are covered in each module before they can advance to the next stage of the curriculum.

When the instructor is satisfied that a trainee has successfully demonstrated the required knowledge and skills for a particular module, that information is sent to NCCER and kept in the National Registry. The National Registry can then confirm training and skills for workers as they move from state to state, company to company, or even within a company (see *Appendix B*).

Whether you enroll in an NCCER program or another ATELS-approved program, ensure that you work for an employer or sponsor who supports a nationally standardized training program that includes credentials to confirm your skill development.

5.1.1 Youth Apprenticeship Program

The concept of apprenticeship training goes back thousands of years. Over that time, the basic principles of apprenticeship training have not changed. Apprentices learn their craft from those who have mastered it. They focus on learning by doing. Although some theory is presented in the classroom, it is always presented in a way that helps trainees understand the purpose behind the skill to be learned. The NCCER training curriculum follows this proven approach.

Youth Apprenticeship Programs allow students to begin their apprentice training while they are still in high school. Students entering a plumbing program in the 11th grade may complete as much as one year of the NCCER four-year program by their high school graduation. In addition, the program, in cooperation with local craft employers, allows students to work in the trade and earn money while they are still in school.

This training program is similar to the one used by NCCER-accredited training sponsors across the country. With your official transcripts, you can enter the second year of the program wherever it is offered. You may also apply these credits to a two- or four-year college that offers degree or certificate programs in the construction profession.

You may become an employee at age 16, but the United States Department of Labor (DOL) limits your work to occupations that are not specifically defined as hazardous. Hazardous jobs include such things as explosives manufacturing, mining, meatpacking, and power-driven sheet metal work. You may work on a project site as an apprentice or journey plumber when you are 16, but only if you are enrolled and in good standing in a plumbing apprentice program. Under certain conditions, 17-year-olds are allowed to operate motor vehicles.

You must be at least 18 before you can legally work in a hazardous occupation. When you have graduated from your apprentice program, you can enter the industry at a higher level and with more pay than people who are just starting the apprenticeship program.

5.1.2 Apprenticeship Standards

All apprenticeship standards call for certain work-related training, or OJT. OJT is broken down into specific tasks in which the apprentice receives hands-on training. Each task requires a specified number of hours. The total number of hours in a plumbing apprenticeship is traditionally 8,000, which amounts to four years of training. In a competency-based program, this time may be shortened by testing out of specific tasks through a series of performance exams.

In a traditional program, you may complete the required OJT at the rate of 2,000 hours per year. If you are laid off or get sick, it may take longer.

Sponsors need to receive, log in, and report an apprentice's time records, and send them to the federal government and/or the state apprenticeship program for review. Therefore, apprentices should keep accurate and up-to-date time records; after the first 144 classroom hours and 1,000 hours of related work, and then for each subsequent 1,000 hours, the standards entitle an apprentice to a pay increase.

Apprentices may not always receive your classroom instruction and OJT at the same time. Apprentices with special job experience or other coursework may obtain credit toward their classroom requirements. This will reduce the time they are required to spend in the classroom while meeting the 8,000-hour OJT requirement. These special cases will depend on the type of program and its regulations and standards.

For those entering an apprenticeship program, a high school or technical school education is desirable, with courses in plumbing, shop, mechanical drawing, and general mathematics. Manual dexterity, good physical condition, and a good sense of balance are important. It is essential to be able to solve arithmetic problems quickly and accurately and to work closely with others.

If you want to become an apprentice, you must submit to the Apprenticeship Committee certain information, which may include the following (the actual requirements may vary by state):

- The results of your aptitude test (General Aptitude Test Battery [GATB] or GATB Form Test, normally administered by the local Employment Security Commission)
- Letters of reference and recommendation from your past employers or sponsors
- Proof of your age (e.g., driver's license)
- For veterans of the Armed Forces, a copy of your Form DD-214
- A record of any technical training you received that is related to the construction industry and/or a record of any training you received before applying for the apprenticeship
- Your high school diploma or General Educational Development (GED) credential

Once you are in an apprenticeship program, you must do the following:

- Wear proper safety equipment on the job.
- Buy and maintain tools of the trade as needed and as required by the contractor.
- Submit a monthly OJT report.
- Report any change in your employment status or change in sponsor.
- Attend classroom instruction and follow all regulations for the classroom, correspondence study, or distance learning.

Child Labor Laws

Federal law establishes the minimum standards for workers under the age of 18. Some municipal jurisdictions may enforce stricter regulations. Employers are required to abide by the laws that apply to them.

The Child Labor Provisions of the *Fair Labor Standards Act* forbid employers from using illegal child labor, and also forbid companies from doing business with any other business that does. DOL investigates alleged abuses of the law. In such cases, employers have to provide proof of age for their employees.

In addition to the Child Labor Provisions, employers in the construction trades are required to follow DOL's Child Labor Bulletin No. 101, *Child Labor Requirements in Nonagricultural Occupations Under the Fair Labor Standards Act.* Bulletin No. 101 does the following:

- Explains the coverage of the Child Labor Provisions
- Identifies minimum age standards
- Lists the exemptions from the Child Labor Provisions
- Sets out employment standards for 14- and 15-year-old workers
- Defines the work that can be performed in hazardous occupations
- Provides penalties for violations of the Child Labor Provisions
- Recommends the use of age certificates for employees

5.2.0 What You Should Expect from Your Employer or Sponsor

Once the Apprenticeship Committee selects an applicant to be an apprentice, the employer or sponsor hires the apprentice. The apprentice's job has the same potential for career advancement as a non-apprentice employee. In return, the employer or sponsor requires the apprentice to make satisfactory progress in OJT and related classroom instruction throughout the duration of the apprenticeship program.

The employer or sponsor agrees not to employ the apprentice in a way that may violate the apprenticeship standards. The employer or sponsor also pays a share of the cost of operating the apprenticeship program.

5.3.0 What You Should Expect from a Training Program

Employers and sponsors who take the time and initiative to provide quality training are willing to invest in their workforce and improve the abilities of their workers. Nevertheless, you should take the time to find a program that will train you well.

Select an employer who has an approved apprenticeship program and who is willing to act as your sponsor. The employer's training program should be comprehensive, standardized, and based on developing and demonstrating competencies, not just on the amount of time you spend in a classroom. Before you enroll in a training program, find out how many employers in the area use the same program. The training program should have a well-defined compensation ladder so that your pay will increase in recognition of your new skills and experience.

The program should be nationally recognized and should provide you with transcripts and completion credentials. As an employee in the construction profession, you may end up working for several different contractors over the course of your career. Contractors in other cities and states must be able to recognize the experience and qualifications that you obtained through your training program.

Finally, the training program's curriculum should be complete and up-to-date. The curriculum should feature the latest developments in plumbing techniques, materials, tools, and equipment. The program should also take advantage of modern interactive training technology, such as CD-ROMs and the Internet. As a plumber, you know how important it is to stay current; the same applies to training programs.

5.4.0 What You Should Expect from the Apprenticeship Committee

The Apprenticeship Committee is the local administrative body to which apprentices are assigned. It is responsible for the appropriate training of apprentices. Every apprenticeship program, whether state or federal, is covered by the following standards that have been approved by appropriate government agencies:

- The Committee is responsible for enforcing standards and for making sure that training is conducted properly. This ensures that graduates are fully qualified in those areas of training designated by the standards.
- The Committee screens and selects individuals for apprenticeship and refers them to participating training programs.
- The Committee places apprentices under written agreement for participation in the program.
- The Committee establishes minimum standards for related classroom instruction and OJT and monitors the apprentice to see that these standards are followed during the training period.
- The Committee hears all complaints of violations of apprenticeship agreements, whether by employer/sponsor or apprentice, and takes action within the guidelines of the standards.
- The Committee notifies the registration agencies of all enrollments, completions, and terminations of apprentices.

Review Questions

Sections 4.0.0–5.0.0

1. Project managers are responsible for _____.
   a. assessing the materials and labor needed on projects
   b. ensuring that projects are in compliance with safety regulations
   c. controlling the scope and direction of a plumbing business
   d. instructing all new apprentices on a job site

2. You are offered a job by another contractor before you have fulfilled your current obligations. The professional response is to _____.
   a. ask the contractor to wait until you have met your obligations
   b. notify your current employer of your intention to leave immediately
   c. ask your current employer to let you divide your time between both projects
   d. determine if your current employer is willing to negotiate better terms

3. As a trainee, your primary responsibility is to see something that needs to be done and to do it without having to be asked.
   a. True
   b. False

4. Your supervisor points out that you forgot to complete a step in the assembly of a drain in a DWV system. One of the apprentices that you are supervising was the person installing the drain. The responsible action is to _____.
   a. explain that one of the apprentices was responsible
   b. ask the apprentice to explain to your supervisor
   c. explain to your supervisor why the apprentice did the work that way
   d. take responsibility for work done under your supervision

5. People who develop, design, manufacture, and test mechanical devices are called _____.
   a. contractors or owners
   b. inspectors
   c. mechanical engineers
   d. master estimators

6. Which of the following is *not* a legitimate reason for taking time off work?
   a. You are tired because you were out late.
   b. Your spouse's mother died the previous day.
   c. You are too sick to come to work.
   d. Your child is ill and must be taken to the hospital.

7. _____ sets the minimum standards for training programs in the United States.
   a. NCCER
   b. NEMA
   c. OSHA
   d. ATELS

8. According to the Department of Labor, 17-year-olds are allowed to operate motor vehicles under certain conditions.
   a. True
   b. False

9. Apprentices must submit OJT reports on a _____ basis.
   a. daily
   b. weekly
   c. monthly
   d. quarterly

10. When deciding on a training program, you should consider _____.
   a. how many hours are dedicated to OJT and classroom study
   b. whether it focuses on developing and demonstrating competencies
   c. how many apprentices successfully completed the program during the previous year
   d. the number of students in the classroom

Summary

History has taught us time and time again that plumbing is vital to public health, safety, and comfort. Today, the plumbing profession is more important than ever. As the world's population grows, the built environment is expanding to accommodate it. The world needs more skilled and experienced plumbers to ensure that the public stays safe, healthy, and comfortable.

The men and women who make up the plumbing profession have a range of skills and a broad base of knowledge. They are responsible for installing and maintaining a wide variety of supply and drainage systems. Plumbers carefully follow local codes and regulations to ensure that systems are safe.

Because the plumbing profession is always evolving, successful plumbers never stop learning about new techniques, materials, tools, and standards. In addition to staying current with their technical skills and knowledge, successful plumbers also follow high ethical standards. They maintain a positive attitude, are honest and loyal, and are responsible for their actions. They cooperate with their teammates and supervisors, pay attention to rules and regulations, and are always prompt and reliable employees.

As an apprentice embarking on your plumbing career, you have the right to expect a quality education that will serve you well. Your employer or sponsor and your Apprenticeship Committee will do their best to ensure that you are well prepared for the responsibilities you will take on. Ultimately, though, the decision is yours as to whether you will try hard to become the best professional plumber that you can be.

Notes

Trade Terms Quiz

Fill in the blank with the correct trade term that you learned from your study of this module.

1. During the _______________ phase of a plumbing project, plumbers locate all supply and waste connections from the building systems to public utilities.
2. _______________ is the process of cleansing water to remove chemicals and particles.
3. You can use _______________ thermoplastic pipe as tubing for cold water systems.
4. Someone who installs or repairs plumbing systems and fixtures is called a(n) _______________.
5. Water supply systems provide fresh water; _______________ systems remove wastewater.
6. Always wear appropriate personal protective equipment when disinfecting water with _______________, a heavy, greenish-yellow gas.
7. You are _______________ water when you remove magnesium and sodium salts from it.
8. ATELS requires apprentices to have 2,000 hours of _______________ per year.
9. Water that is safe to drink is referred to as _______________.
10. Construction ordinances written by national construction organizations and based on suggested national plumbing standards are called _______________.
11. During the _______________ phase of a plumbing project, plumbers install fixtures, appliances, water purification systems, water heaters, and controls.
12. When a potable water system and a nonpotable water system are arranged so that an accidental pressure differential between the two systems would cause contaminated water to flow into the freshwater system, it is called a(n) _______________.
13. Plumbers destroy harmful organisms in potable water through _______________.
14. During the Roman Empire, _______________ were used to carry water from rivers and lakes to cities.
15. In ancient Rome, a person who worked with lead was called a(n) _______________.
16. As a construction professional, you must ensure that your conduct is in accordance with your profession's principles and values, which are called _______________.
17. One of the most important safety concerns for plumbers is _______________, which is the flow of contaminated water into the freshwater system through a cross-connection.
18. Plastic that becomes substantially infusible and insoluble when treated by heat or chemicals is called _______________.
19. During the _______________ phase of a plumbing project, plumbers cut holes in walls, ceilings, and floors to attached or hang pipe runs.
20. Plastic that is soft and pliable when heated and hard and rigid when cooled is called _______________.
21. _______________ are legally binding requirements published by state and local governments to establish minimum standards for various types of construction.
22. Sinks, shower stalls, and toilets are common examples of _______________.
23. The Latin word for lead is _______________.
24. Plumbers install _______________ to keep nonpotable water from contaminating a potable supply system.

25. If you are installing piping for a system that is designed exclusively for environmental control, you are not engaged in _______________.

26. To become a(n) _______________, you must first complete an apprenticeship training program.

27. Plumbing _______________ are types of apparatus used in such trades as installing drainage systems.

Trade Terms

Aboveground rough-in
Appurtenances
Aqueduct
Backflow
Backflow preventer
Chlorine
Code
Cross-connection
Disinfection
Drain, waste, and vent (DWV)
Ethics
Filtration
Finish
Fixture
Journey plumber
Model code
On-the-job training (OJT)
Plumbarius
Plumber
Plumbing
Plumbum
Polyvinyl chloride (PVC)
Potable
Softening
Stack-out
Thermoplastic
Thermoset
Top-out
Trim finish
Trim-out
Underground rough-in

Profile in Success

Charlie Chalk

Construction Engineer/Inspector, Mechanical Section
EBL Engineers
Baltimore, Maryland

Charlie Chalk was born in Baltimore, Maryland, and attended City College High School. At 16, he went to work as a plumber's helper and then as a plumbing apprentice for George Hardesty Plumbing, a third-generation business still in operation today. Although he hadn't completed high school, Charlie believed in finishing what he started and completed his General Educational Development certificate while working on his apprenticeship. Today, he is a construction engineer for EBL Engineers and an instructor for PHCC.

How did you become interested in this industry?
My entire family was in the business—my dad was a plumbing foreman and my grandfather a master plumber. I began working in the business as a summer job and kept going. I've now been working in the trade for 43 years, making it the longest summer job ever!

What path did you take to your current position?
After completing my four-year apprenticeship, I got my journeyman's license and my master gas fitter's license. I took a written and practical exam for this license. We went through a house that was set up on the property to determine problems. I received my master plumber's license, which I hold in Maryland and Delaware, and my Washington Suburban Sanitary Commission certification.

I have a certified backflow technical license and a state-certified building inspector's license. Additionally, I am a certified plumbing and mechanical inspector and a certified plan reviewer. I've taken seminars throughout my career, as well as code courses and numerous supply house fixture seminars. I've also taken Hartford Insurance courses for boilers and chillers. I attended Catonsville Community College for instructor training.

I was a Baltimore city inspector for 20 years. Two years ago, EBL Engineers offered me the position of construction manager.

What does it take to be successful in your trade?
The first thing you want to do is get a codebook. If you read one chapter a day, by the time you get to the fourth year of your apprenticeship, you'll be familiar with everything in the book. It will be that much easier to learn the code, because you'll be accustomed to it. You'll review it in the first and second years, and, by the fourth year, you'll have a better understanding of what the instructors are talking about.

Remember that there is always something new to learn. I still go to classes to keep myself abreast of new things and equipment. If people learning the trade don't do that, they are cutting themselves short. You get out of it what you put into it. Stick to it and get both on-the-job and classroom training. If you don't get both, you'll only be partway when you get a license. Some people think that they are done when they finish the four-year apprenticeship. I tell apprentices, "Now you know the basics, and here is where you really start learning your trade."

What are some things you do on the job?
I conduct inspections for work that the office has completed, design plumbing systems, and review materials that come in for jobs. We write specifications for anything that pertains to design. I primarily deal with plans. We have 50 employees in the office. I review plans, assess needs, and determine potential problems. I conduct inspections during construction as well as when the job is complete to ensure that the work is up to code. I check under slab and rough-ins, and conduct close-in ceilings inspections and then final inspections. I oversee 33 jobs at one time, 10 percent of which are high-dollar residential and the rest commercial, including schools, military projects, and hospitals.

I've been a PHCC teacher for 17 years. I still teach three nights a week and sometimes on Saturdays. I teach the plumbing code and the gas fitter course. I also fill in for instructors for the apprenticeship program. As the lead instructor, I teach others about instructing. I've taught inspectors all over the state of Maryland. I conduct a gas fitting review and PHCC seminars for the state. I am a national member of the NSPC Code Committee—we adopt and write code for the national standard, which is revised annually.

What is the most interesting aspect of your profession? What makes your trade stand out from others?
You can talk to anybody in any trade and they will say their trade is the best. Plumbers deal with health and safety issues. Our business is to protect health and safety, but this is taken for granted. Water is just supposed to be there. People have no idea what it takes to get clean potable water to the customer, as well as to remove water without infecting anyone.

What do you like most—and least—about your job?
I like everything. I get involved from the beginning of the design and I get to see the end result. Seeing projects start to finish is very rewarding. When someone says, "Hey, that looks good," it really puts a spark in us.

I don't really know what I like the least. I don't like inclement weather!

What would you say to someone entering the trade today?
You get out of it what you put into it. I've worked with gentlemen who just went along and didn't put anything into it. Later, they regretted what they should have done, but didn't. One man worked his way up and got his license, but he later dropped it. When he was presented with the chance to own his own business, he lost the opportunity. So, keep your licenses and stay current.

You may think, "What do I need that for?" Once you are ready, you should get the license—do not wait. Ordinances change. You might be qualified and competent now, but what you know may go out of date and you'll need new training.

What can an apprentice expect to earn in his/her first years on the job? What can he/she expect to earn after 10-plus years in the industry?
After about six years, you become a journeyman and then you get a master's license. After 10 years, you can be making pretty good money: between $25 and $30 per hour. It is a lucrative business. You have opportunities to advance. Again, you get out of it what you put into it.

Trade Terms Introduced in This Module

Aboveground rough-in: The second phase of a plumbing project. During this phase, holes are cut in walls, ceilings, and floors. Then, supply and waste pipes are attached or hung so they can be connected to fixtures.

Appurtenances: Accessories or apparatus.

Aqueduct: A man-made channel used to carry water.

Backflow: The flow of contaminated water into the freshwater system resulting from a cross-connection between potable and nonpotable water systems.

Backflow preventer: A device that prevents non-potable water from entering a potable supply system.

Chlorine: A heavy, greenish-yellow gas used as a disinfectant in water treatment. Chlorine should be handled only when wearing appropriate personal protective equipment.

Code: A requirement published by state and local governments to establish minimum standards for various types of construction. A code carries the force of law.

Cross-connection: An arrangement between a potable water system and a nonpotable water system in which an accidental pressure differential between the two systems causes backflow of contaminated water into the freshwater system.

Disinfection: The process of destroying harmful organisms in potable water.

Drain, waste, and vent (DWV): A piping system that combines sanitary drainage with venting.

Ethics: A set of principles and values that guide an individual's conduct.

Filtration: The process of cleansing water to remove particles and chemicals.

Finish: The third phase of a plumbing project. During the finish phase, plumbers install fixtures, appliances, water purification systems, water heaters, and controls.

Fixture: A device that receives water from a water supply line. Common fixtures include sinks, shower stalls, and toilets.

Journey plumber: A plumber who has successfully completed an apprenticeship training program.

Model code: A construction ordinance that is written by a national construction organization according to suggested national plumbing standards. Model codes do not have the force of law.

On-the-job training (OJT): Field experience used in conjunction with classroom lessons in an apprenticeship program. ATELS requires 144 hours of classroom instruction per year and 2,000 hours of OJT per year.

Plumbarius: The Roman term for someone who works with lead. The root of the modern word *plumber.*

Plumber: One who installs or repairs plumbing systems and fixtures.

Plumbing: According to the *National Standard Plumbing Code,* plumbing is "the practice, materials, and fixtures within or adjacent to any building structure or conveyance, used in the installation, maintenance, extension, alteration, and removal of all piping, plumbing fixtures, plumbing appliances, and plumbing appurtenances connected with the following: (a) sanitary drainage systems and related vent systems; (b) storm water drainage facilities and venting systems; (c) public or private potable water supply systems; (d) the initial connection to a potable water supply upstream of any required backflow prevention devices and the final connection that discharges indirectly into a public or private disposal system; (e) medical gas and medical vacuum systems; (f) indirect waste piping, including refrigeration and air conditioning drainage; and (g) liquid waste or sewage, and water supply, of any premises to their connection with the approved water supply system or to an acceptable disposal facility."

Plumbum: Latin word for lead.

Polyvinyl chloride (PVC): A thermoplastic material frequently used in tubing for cold water systems and the first type of plastic approved for use in plumbing.

Potable: Water that is safe for cooking and drinking.

Softening: The process of removing magnesium and sodium salts that cause scale on the inside of pipes and fittings.

Stack-out: Another term for aboveground rough-in.

Thermoplastic: A plastic material used in plumbing and sanitary systems that is soft and pliable when heated and hard and rigid when cooled.

Thermoset: A plastic material used in plumbing and sanitary systems that becomes substantially infusible and insoluble when treated by heat or chemicals.

Top-out: Another term for aboveground rough-in.

Trim finish: Another term for finish.

Trim-out: Another term for finish.

Underground rough-in: The phase of a plumbing project during which the plumber locates all supply and waste connections from the building systems to public utilities, and establishes where these systems will enter or leave the building.

Appendix A

Ethical Principles for Members of the Construction Trades

Honesty: Be honest and truthful in all dealings. Conduct business according to the highest professional standards. Faithfully fulfill all contracts and commitments. Do not deliberately mislead or deceive others.

Integrity: Demonstrate personal integrity and the courage of your convictions by doing what is right even where there is pressure to do otherwise. Do not sacrifice your principles because it seems easier.

Loyalty: Be worthy of trust. Demonstrate fidelity and loyalty to companies, employers and sponsors, co-workers, and trade institutions and organizations.

Fairness: Be fair and just in all dealings. Do not take undue advantage of another's mistakes or difficulties. Fair people are open-minded and committed to justice, equal treatment of individuals, and tolerance for and acceptance of diversity.

Respect for others: Be courteous and treat all people with equal respect and dignity.

Obedience: Abide by laws, rules, and regulations relating to all personal and business activities.

Commitment to excellence: Pursue excellence in performing your duties, be well informed and prepared, and constantly try to increase your proficiency by gaining new skills and knowledge.

Leadership: By your own conduct, seek to be a positive role model for others.

Samples of NCCER Apprentice Training Recognition

August 2, 2004

Joseph Smith
c/o Training Sponsor, Inc.

Dear Joseph,

On behalf of the National Center for Construction Education and Research, I congratulate you for successfully completing the NCCER's standardized craft training program.

As the NCCER's most recent graduate, you are a valuable member of today's skilled construction and maintenance workforce. The skills that you have acquired through the NCCER craft training programs will enable you to perform quality work on construction and maintenance projects, promote the image of these industries and enhance your long-term career opportunities.

We encourage you to continue your education as you advance in your construction career. Please do not hesitate to contact us for information regarding our Management Education and Safety Programs or if we can be of any assistance to you.

Enclosed please find your certificate, transcript and wallet card. Once again, congratulations on your accomplishments and best wishes for a successful career in the construction and maintenance industries.

Sincerely,

Donald E. Whyte

Donald E. Whyte
President, NCCER

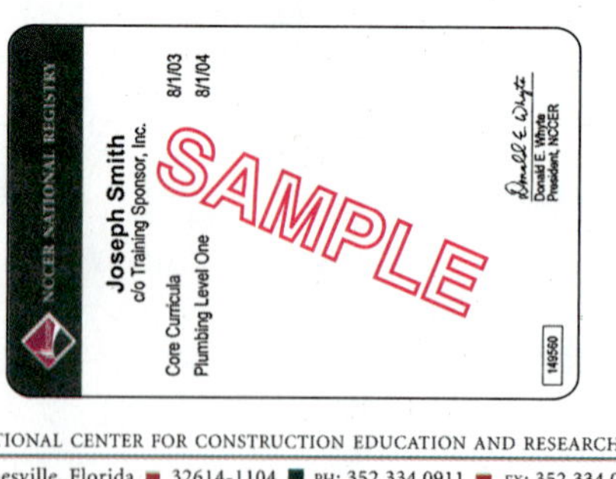

NATIONAL CENTER FOR CONSTRUCTION EDUCATION AND RESEARCH
P.O. Box 141104 ■ Gainesville, Florida ■ 32614-1104 ■ PH: 352.334.0911 ■ FX: 352.334.0932 ■ www.nccer.org

101A01.EPS

Figure A–1 ◆ NCCER National Registry wallet card and letter.

101A02.EPS

Figure A–2 ◆ NCCER National Registry certificate of completion.

NATIONAL CENTER FOR CONSTRUCTION EDUCATION AND RESEARCH
Affiliated with the University of Florida
Post Office Box 141104 Gainesville, Florida 32614-1104
352.334.0911 Fax 352.334.0932 www.nccer.org

8/2/04
Page: 1

Official Transcript

Joseph Smith
c/o Training Sponsor, Inc.

R#: 000105221

Current Employer:

Course / Description		Instructor	Training Location	Date Compl.
00101	Basic Safety		NCCER	8/1/03
00102	Basic Math		NCCER	8/1/03
00103	Introduction to Hand Tools		NCCER	8/1/03
00104	Introduction to Power Tools		NCCER	8/1/03
00105	Introduction to Blueprints		NCCER	8/1/03
00106	Basic Rigging		NCCER	8/1/03
02101	The Plumbing Trade		NCCER	8/1/04
02102	Basic Plumbing Tools		NCCER	8/1/04
02105	Math for Plumbers		NCCER	8/1/04
02106	Introduction to Plumbing Blueprint Reading		NCCER	8/1/04
02107	Reading Residential Plumbing Drawings		NCCER	8/1/04
02201	Copper and Plastic Piping Practices (from HVAC curricula)		NCCER	8/1/04
02202	Soldering and Brazing		NCCER	8/1/04
02203	Cutting and Threading Carbon Steel Pipe		NCCER	8/1/04
02205	Joining Cast-Iron Pipe and Fittings		NCCER	8/1/04

SAMPLE

Donald E. Whyte
President

101A03.EPS

Figure A–3 ◆ NCCER National Registry transcript.

Additional Resources and References

ADDITIONAL RESOURCES

This module is intended to present thorough resources for task training. The following reference works are suggested for further study. These are optional materials for continued education rather than for task training.

The Plumber's Handbook, Revised Edition, 1998. Howard C. Massey. Carlsbad, CA: Craftsman Book.

Plumbing a House, 1994. Peter A. Hemp. Newtown, CT: Taunton Press.

The 2003 National Standard Plumbing Code, 2003. Falls Church, VA: PHCC.

REFERENCES

A Heritage Unique: A History of the National Association of Plumbing-Heating-Cooling Contractors (PHCC), 1999. Falls Church, VA: PHCC Educational Foundation.

Code Check Website, www.codecheck.com.

Dictionary of Architecture and Construction, Third Edition, 2000. Cyril M. Harris, ed. New York: McGraw-Hill.

International Association of Plumbing and Mechanical Officials Website, www.iapmo.org.

International Code Council Website, www.icbo.org, www.iccsafe.org.

PHCC First-Year Instructor's Manual. Falls Church, VA: PHCC.

Plumbing and Mechanical Engineer Magazine. "Greatest Plumbing & Heating Inventions" (July 12, 2000), www.pmengineer.com/pme/cda/articleinformation/features/bnp_features_item/0,,6735,00+en-uss_01dbc.html.

Plumbing and Mechanical Engineer Magazine. "Knollenberg Resurrects Bill to Amend Federal Low Flow Toilet Law" (April 11, 2001), www.pmengineer.com/pme/cda/articleinformation/features/bnp_features_item/0,,24371,00+en-uss_01dbc.html.

Plumbing Trainee Guide, 2003. Falls Church, VA: PHCC.

Supply House Times Magazine. "Consumers Satisfied with Low-Flow Toilets" (May 2000), www.theplumber.com/toiletssatisfiedlowflowmay2000.html.

Thompson Education Direct Website, www.workforcedevelopment.com.

The 2003 National Standard Plumbing Code, 2003. Falls Church, VA: PHCC.

United States Congress. "Child Labor Provisions, Fair Labor Standards Act." 29 USC 8 § 212, www4.law.cornell.edu/uscode/29/212.html.

United States Department of Labor. Employment Standards Administration. Wage and Hour Division. Child Labor Bulletin 101, "Child Labor Requirements in Nonagricultural Occupations Under the Fair Labor Standards Act," September 1991, as amended. Washington, DC: USDOL.

"An Update on Low-Flush Toilets," Mike McClintock, www.theplumber.com/lowflowupdate99.html (reprinted from *The Washington Post*).

NCCER CURRICULA — USER UPDATE

NCCER makes every effort to keep its textbooks up-to-date and free of technical errors. We appreciate your help in this process. If you find an error, a typographical mistake, or an inaccuracy in NCCER's curricula, please fill out this form (or a photocopy), or complete the online form at **www.nccer.org/olf**. Be sure to include the exact module ID number, page number, a detailed description, and your recommended correction. Your input will be brought to the attention of the Authoring Team. Thank you for your assistance.

Instructors – If you have an idea for improving this textbook, or have found that additional materials were necessary to teach this module effectively, please let us know so that we may present your suggestions to the Authoring Team.

NCCER Product Development and Revision

13614 Progress Blvd., Alachua, FL 32615

Email: curriculum@nccer.org

Online: www.nccer.org/olf

❑ Trainee Guide ❑ AIG ❑ Exam ❑ PowerPoints Other ______________________________

Craft / Level: Copyright Date:

Module ID Number / Title:

Section Number(s):

Description:

Recommended Correction:

Your Name:

Address:

Email: Phone:

Plumbing Safety
02102-05

02102-05
Plumbing Safety

Topics to be presented in this module include:

Overview

Unsafe acts and unsafe conditions lead to plumbing-related accidents, and while they may result in death or personal injury, accidents also cost companies time and money. The majority of accidents, however, can and should be prevented. Working to ensure this, the Occupational Safety and Health Administration adopts and enforces safety regulations to protect workers and ensure a safe workplace.

There are many important safety measures that help plumbers remain safe on the job site. Personal protective equipment is designed to protect workers from injury. Personal fall protection equipment prevents falls from elevations greater than 6 feet. In situations where there is a danger of suffocation or breathing hazards, plumbers wear respirators. Material safety data sheets identify hazardous materials, exposure limits, and precautions for safe handling. Recognizing hazards and creating safe work environments are important tasks in the plumbing industry. Hand and power tools are used on job sites every day and, when not used and stored properly, can cause serious injuries. Excavations and confined-space operations require additional precautions and safeguards. Lockout and tagout devices secure equipment, describe problems, and protect workers from possible sources of energy. Despite safety precautions, every job site should have an emergency-response plan for times when accidents do occur. If an accident does occur, it is important to notify a supervisor.

Focus Statement

The goal of the plumber is to protect the health, safety, and comfort of the nation job by job.

Code Note

Codes vary among jurisdictions. Because of the variations in code, consult the applicable code whenever regulations are in question. Referencing an incorrect set of codes can cause as much trouble as failing to reference codes altogether. Obtain, review, and familiarize yourself with your local adopted code.

Objectives

When you have completed this module, you will be able to do the following:

1. Describe the common unsafe acts and unsafe conditions that cause accidents.
2. Describe how to handle unsafe acts and unsafe conditions.
3. Explain how the cost of accidents and illnesses affects everyone on site.
4. Demonstrate the use and care of appropriate personal protective equipment.
5. Identify job-site hazardous work specific to plumbers.
6. Demonstrate the proper use of ladders.
7. Demonstrate how to maintain power tools safely.
8. Explain how to work safely in and around a trench.
9. Describe and demonstrate the lockout/tagout process.

Key Trade Terms

Apparatus
Asbestos
Atmospheric hazard
Benching
Bladed tools
Combustible
Competent person
Confined space
Decibels (dB)
Electrically powered tools
Energy-isolating device
Energy source
Gassy operations
Guards
Guy wires
Hazard Communication (HazCom) Standard
Hypothermia
Impact tools
Liquid-fuel tools
Lockout
Lockout device
Lockout/tagout procedure
Material safety data sheet
NFPA warning diamond
Nonpermit-required confined space
Occupational Safety and Health Administration
Oxygen-deficient atmosphere
Oxygen-enriched atmosphere
Permit-required confined space
Power tools
Protective system
Shield
Shoring
Subsidence
Tagout device

Required Trainee Materials

1. Appropriate personal protective equipment
2. Pencils and paper
3. Copy of local adopted code
4. Copies of *29 CFR 1904, 1910, 1926,* and *1929*

Prerequisites

Before you begin this module, it is recommended that you successfully complete *Core Curriculum; Plumbing Level One,* Module 02101-05.

This course map shows all of the modules in the first level of the *Plumbing* curriculum. The suggested training order begins at the bottom and proceeds up. Skill levels increase as you advance on the course map. The local Training Program Sponsor may adjust the training order.

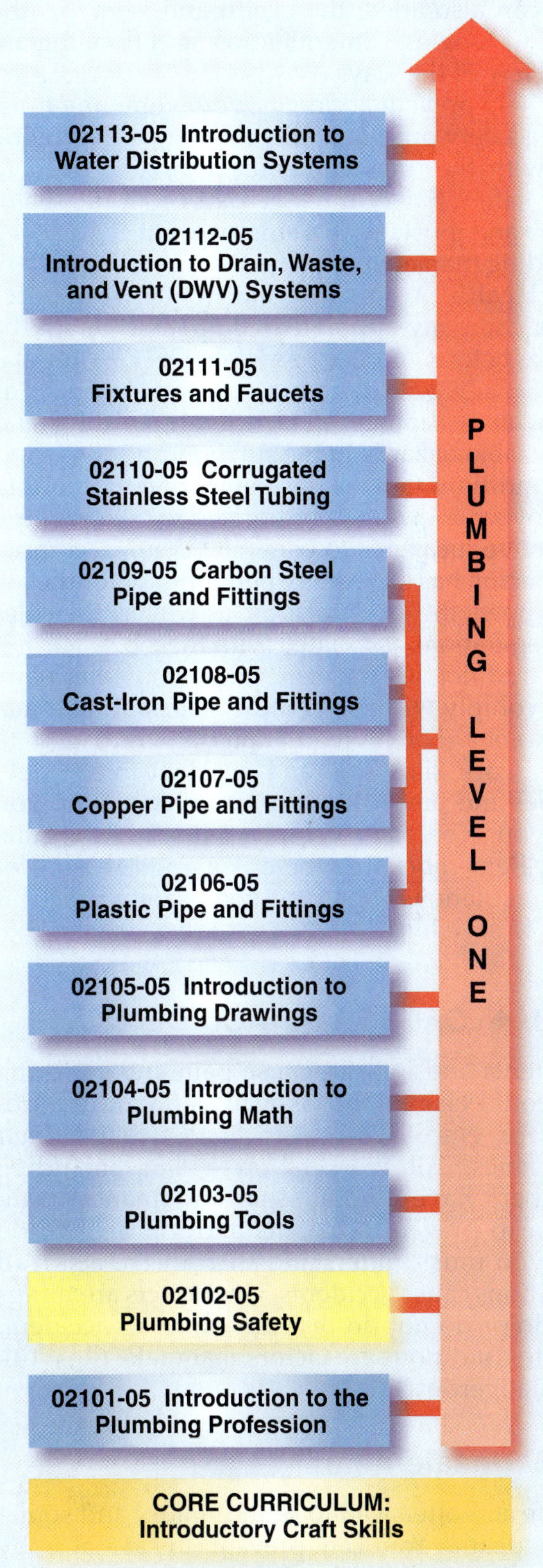

102CMAP.EPS

1.0.0 ◆ INTRODUCTION

Most plumbing-related accidents and injuries are caused by worker carelessness, poor safety planning, lack of training, or failure of the employer or employee to follow safety regulations. Accidents do not only affect plumbers and their employers; they can also affect the health and safety of the public. Diseases, contamination, and flooding are just a few of the ways.

To help prevent accidents, your company must have a safety program. This program will provide you with the rules and safeguards you need to work safely. Safety must be part of all phases of the job and must involve employees at every level, including management. The **Occupational Safety and Health Administration (OSHA)** requires that a company-appointed **competent person** be on site before you start any job. OSHA's regulation *CFR 1926.32* defines a competent person as one who is capable of identifying existing and predictable hazards in the surroundings or working conditions that are unsanitary or dangerous to employees and who is authorized to take prompt, corrective measures to eliminate them. A competent person has experience and training for the job and knows the job's hazards, as well as the rules and regulations associated with the job.

In addition to its own rules, your company must comply with many local, state, and national regulations. Safety regulations are intended to make work sites safe and accident-free. Safety policies and procedures are available to help you and your company comply with these regulations. Remember that there is a good reason for each regulation. Following good safety practices helps to save lives.

2.0.0 ◆ CAUSES OF ACCIDENTS

Accidents and injuries cause pain and suffering that could be avoided, as well as financial hardship for employees, their families, and their companies. All workers on a site, including plumbers, have a moral, legal, and financial obligation to prevent accidents and injuries. To do this, you must understand that unsafe acts and conditions cause accidents. Unsafe acts are things you do or do not do that can cause an accident. Unsafe conditions are factors that make the work area dangerous.

2.1.0 Unsafe Acts

Unsafe acts often lead to serious injury and sometimes death. You can prevent unsafe acts by changing your behavior. It is your responsibility to recognize unsafe acts and stop them immediately. This can mean telling your co-workers to stop what they are doing. If your co-workers do not stop acting in an unsafe manner, stop what you are doing and move as far away from them as possible. In some cases, you may need to inform your supervisor of the problem. Here are some examples of the most common unsafe acts:

- Operating equipment at improper speeds
- Operating equipment without authority
- Using defective equipment
- Disabling a safety device
- Servicing equipment while it is in motion or energized
- Using equipment improperly
- Failing to use personal protective equipment (PPE)
- Failing to warn co-workers of a dangerous or potentially dangerous situation
- Working in an improper position
- Working while impaired by alcohol or illegal drugs
- Operating tools or equipment when taking certain types of prescription drugs
- Lifting loads improperly
- Loading or placing equipment or supplies improperly
- Horseplay (see *Figure 1*)

An unsafe act can also be defined as work that is not done correctly. Workers who fail to use proper PPE, follow safety procedures, or warn co-workers of potentially hazardous conditions can cause or worsen accidents. Keep yourself safe, and look out for the safety of others. Always follow safety rules, and use the right equipment for the job.

2.2.0 Unsafe Conditions

Unsafe working conditions cause many accidents and injuries. Environmental factors such as noise, extreme heat or cold, poor lighting, and poor air circulation can create unsafe conditions by impairing your reactions or limiting your movements. Poor housekeeping is also a hazard. Clutter in walkways, spills, and improper waste disposal can make simply walking on a construction site dangerous. To work safely, you must be able to hear, see, breathe, and keep your balance. Before you start a job, look for and correct any unsafe working conditions. Pick up and properly store loose pipes and fittings, and keep your work area clean and free of debris that could cause accidents. Keep glue pans closed to reduce the amount of toxic (poisonous) fumes in the air. Check the ground for standing water that could

Figure 1 ◆ Horseplay is dangerous.

be an electrical hazard. The toxic, biological, and electrical hazards affecting plumbers are covered later in this module.

3.0.0 ◆ COSTS AND IMPACTS OF ACCIDENTS

Accidents are very costly. When they happen, everyone involved loses, including the company and its employees. You may not believe that all accidents will cost you money. Think about it this way: If there is an accident on a job site, the company will have to pay higher insurance rates, costs for medical care and/or repairs, downtime, and the cost of investigations or fines. If the company is paying for these things, it has less money to spend on raises and performance bonuses and to hire replacement workers. What that means to you is more work without an increase in pay. You are a part of the company (see *Figure 2*), and job-site accidents will affect you, even if you are not directly involved.

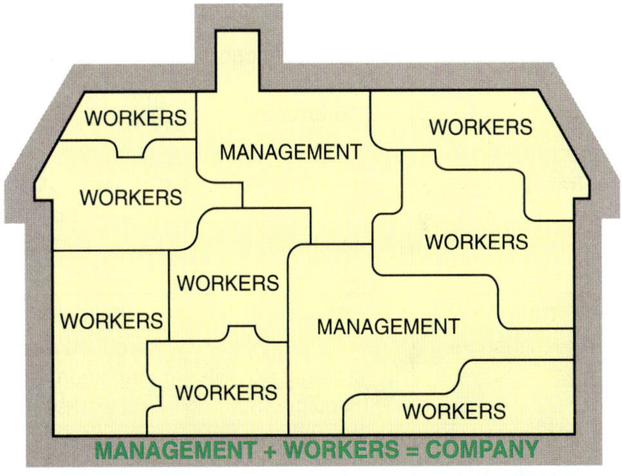

Figure 2 ◆ Companies are a combination of workers and management.

Accident costs can be classified as direct (insured) and indirect (uninsured). The costs associated with accidents can be compared to an iceberg (see *Figure 3*). The tip of the iceberg represents the direct costs, such as medical bills, compensation, and insurance premiums. The larger, indirect costs are unseen. They include property and equipment damage, production delays, and lost time. Injuries that occur off the job may equally delay production and involve other costs for the employers. The indirect costs of accidents are usually two to seven times greater than the direct costs. *Table 1* lists the costs of accidents that are not covered by insurance.

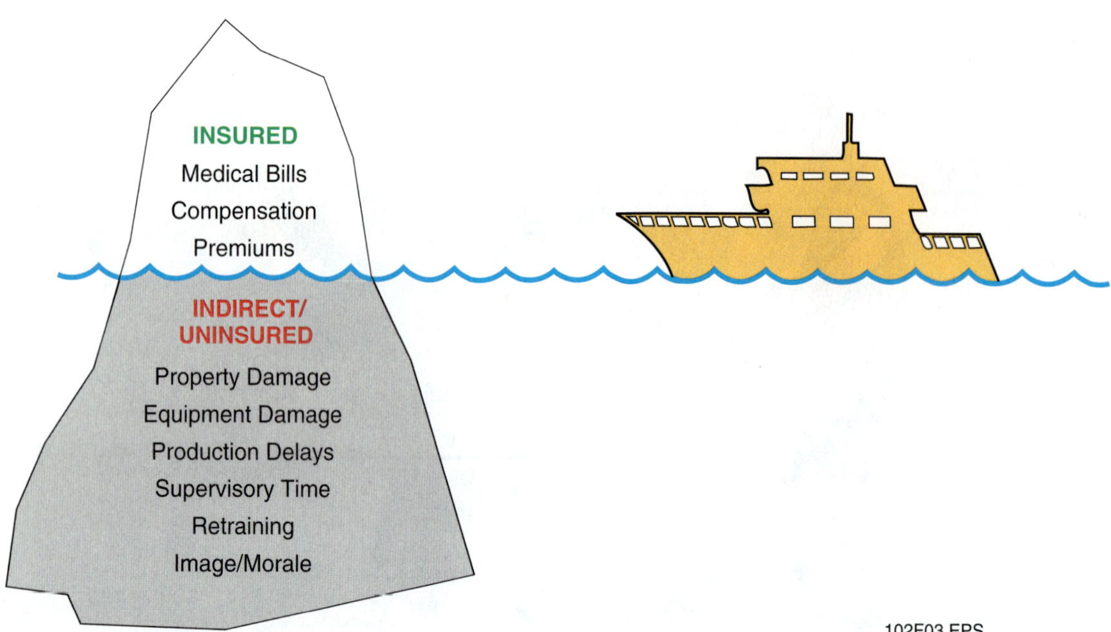

Figure 3 ◆ Hidden costs of accidents.

Table 1 Costs/Impact Not Covered by Typical Employer Insurance Policies

Injuries	Wage Losses	Production Costs	Off-the-Job Accidents	Intangibles	Associated Costs
First aid expenses	Idle time of workers whose work is interrupted	Product spoiled by accident	Cost of medical services	Lowered employee morale	Difference between actual losses and amount recovered
Transportation costs	Time spent cleaning up accident area	Loss of skill and experience	Time spent on injured worker's welfare	Increased labor conflict	Rental of equipment to replace damaged equipment
Costs of investigations	Time spent repairing damaged equipment	Lowered production of replacement worker	Loss of skill and experience	Unfavorable public relations	Surplus worker for replacement of injured worker
Cost of processing reports	Time lost by workers receiving first aid	Idle machine time	Training replacement worker		Wages or other benefits paid to disabled worker
			Decreased production of replacement worker		Overhead costs while production is stopped
			Benefits paid to injured worker or dependents		Loss of bonus or payment of forfeiture for delays

102T01.EPS

DID YOU KNOW?

The average cost of an eye injury is $1,463. That figure includes both the direct and indirect costs, but it does not take into account the long-term effects on the health of the worker, which is priceless.

Source: Occupational Safety and Health Administration

OSHA, part of the U.S. Department of Labor, is a government agency that protects millions of workers each year. Its mission is to ensure a safe and healthy workplace. Since the agency was created in 1971, workplace deaths have been cut in half, and injury and illness rates have declined 40 percent. At the same time, the number of people working has increased. In fact, since 1971, the number of people working in the United States has doubled, from 56 million workers at 3.5 million work sites to 111 million workers at 7 million sites.

OSHA adopts and enforces safety regulations known as standards. Some, like the ***Hazard Communication (HazCom) Standard,*** apply to all industries. Others apply only to a specific industry. In addition to general industry standards, OSHA has a series of standards for plumbing construction. Twenty-six states have their own OSHA programs. The state regulations must be at least as strict as the national standards. Find out which regulations apply to your job site. The federal and state OSHA programs keep workplaces safe through accident investigation, inspections, and outreach.

Review Questions

Sections 1.0.0–3.0.0

1. According to OSHA, a competent person is _____.
   a. a worker who has been trained to use tools properly
   b. a worker who can identify hazards
   c. a worker who can identify hazards and is authorized to take prompt corrective measures
   d. any foreman or supervisor on the job site

2. You see a co-worker fooling around on the work site. Your first course of action is to _____.
   a. tell your supervisor
   b. tell your co-worker to stop
   c. move as far away from your co-worker as possible
   d. complain to other co-workers about this person's actions

3. One simple way to help ensure a safe work site is to _____.
   a. check the job site periodically and remove any clutter
   b. purchase your own first-aid kit
   c. keep foul weather gear stored in your truck
   d. avoid working in cold weather

4. OSHA stands for _____.
   a. *Occupational Standards for Health Act*
   b. Occupational Services and Health Administration
   c. Office of Safety and Health Administration
   d. Occupational Safety and Health Administration

5. OSHA's mission is to protect _____.
   a. the economy
   b. employers
   c. workers
   d. the whole workplace environment

4.0.0 ◆ PERSONAL PROTECTIVE EQUIPMENT

PPE is designed to protect you from injury. Many plumbers are injured on the job because they are not using PPE.

You won't see all the potentially dangerous conditions on a job site just by looking around. Before doing any job, stop and consider what type of accidents could happen. Using common sense and PPE will greatly reduce your chances of getting hurt. The best PPE is of no use unless you do the following four things:

- Inspect it regularly, and replace any PPE that is damaged or worn.
- Care for it properly.
- Use it properly when it is needed.
- Avoid altering or modifying it in any way.

As a plumber, you will most commonly use the following types of PPE:

- Hard hats
- Eye and face protection
- Gloves
- Safety shoes
- Hearing protection
- Fall protection
- Respiratory protection
- Proper clothing

NOTE

The safety harness and lanyard are parts of a system that is known as the personal fall protection system. Workers must know how to properly inspect, don, and maintain their system.

NOTE

Accidents and incidents are defined differently by OSHA. When reporting injuries, always use the word accident, not incident.

4.1.0 Hard Hats

Figure 4 shows a typical hard hat. The outer shell of the hat protects your head during a fall or from a flying object. The webbing inside the hat maintains a space between the shell and your head. When wearing a hard hat, adjust the headband, not the hard hat, so that the webbing fits your head and there is at least 1 inch of space between your head and the shell.

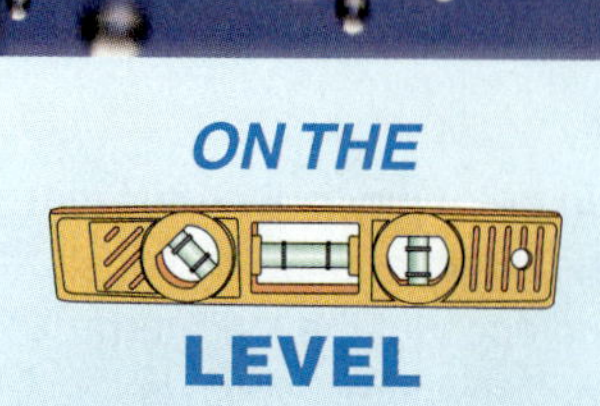

ON THE LEVEL

You won't get the protection that a hard hat is designed to provide if you do not wear it correctly. In one reported accident, a worker who was wearing his hard hat backwards was pulling down a piece of 6-inch pipe. The pipe fell, knocking his hat back off his head. Because the brim of his hat was turned backward, it did not properly deflect the pipe. Instead, the pipe came down square on his forehead. The worker suffered brain damage as a result and could not work again.

Pull back long hair (longer than 4 inches), which can easily get caught in machinery. Hair can also get trapped in chains, belts, suctions, and blowers. You should not be required to cut your hair, but you may need to cover it with a bandana or hair net under your hard hat.

You must wear your hat correctly. Notice that hard hats have a short, sturdy brim at the front. This brim is designed to deflect falling objects. Never wear the hat backward or pushed back from your face. Doing so eliminates the protection.

4.2.0 Safety Glasses, Goggles, and Face Shields

Wear eye protection (*Figure 5*) whenever there is even the slightest chance of an eye injury. Areas where there are potential eye hazards from falling or flying objects are usually identified, but you should always be on the lookout for other possible hazards, such as sewage and pressurized water.

There are different types of eye protection. Safety goggles give your eyes the best protection from all directions. Regular safety glasses will protect you from objects flying at you from the front, such as large chips, particles, sand, or dust. You can add side shields for further protection. Do not wear contact lenses because you will burn your eyes. You may, however, substitute prescription safety glasses if they provide the same protection as regular safety glasses. If welding is part of your job, you must use safety goggles with tinted lenses or a welding hood. Tinted lenses protect your eyes from the bright welding arc or flame.

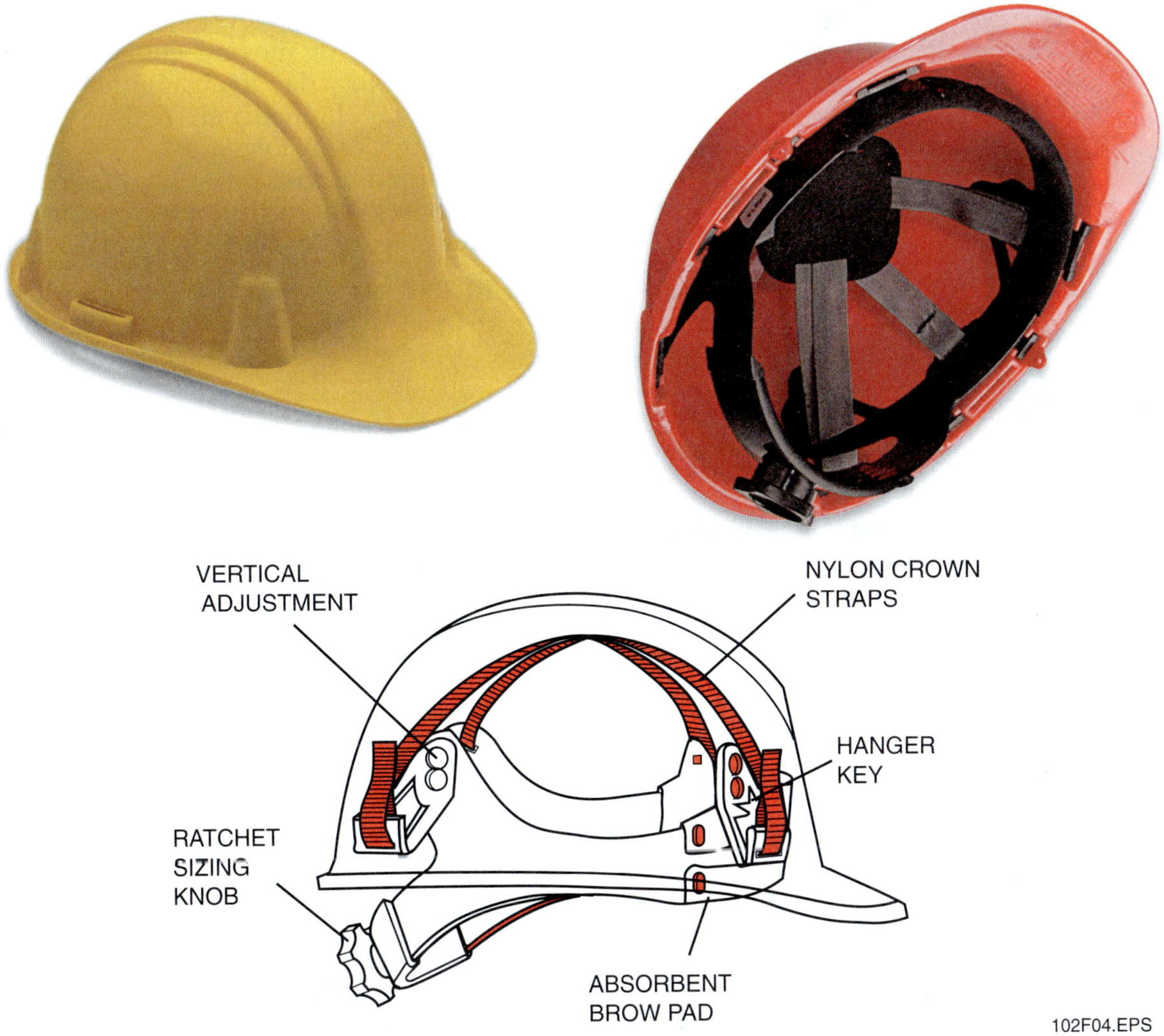

VERTICAL ADJUSTMENT
NYLON CROWN STRAPS
HANGER KEY
RATCHET SIZING KNOB
ABSORBENT BROW PAD

102F04.EPS

Figure 4 ◆ Typical hard hat.

102F05.EPS

Figure 5 ◆ Typical safety goggles and glasses.

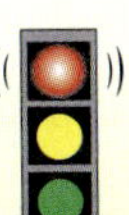

WARNING!

Never wear contact lenses while welding. The ultraviolet rays may dry out the moisture beneath the lens, causing it to stick to your eye.

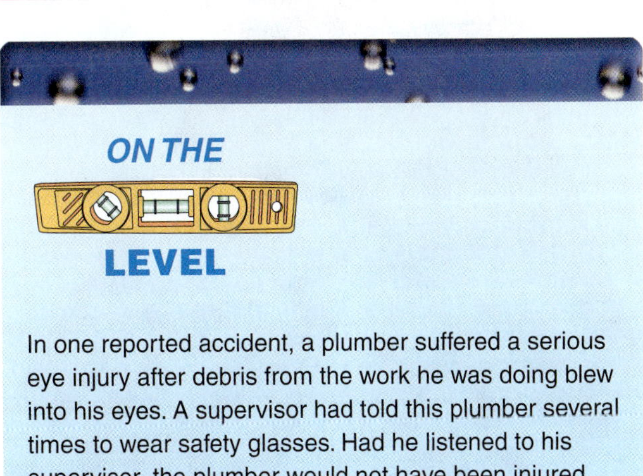

In one reported accident, a plumber suffered a serious eye injury after debris from the work he was doing blew into his eyes. A supervisor had told this plumber several times to wear safety glasses. Had he listened to his supervisor, the plumber would not have been injured, nor would he have had to take time off from work to recover.

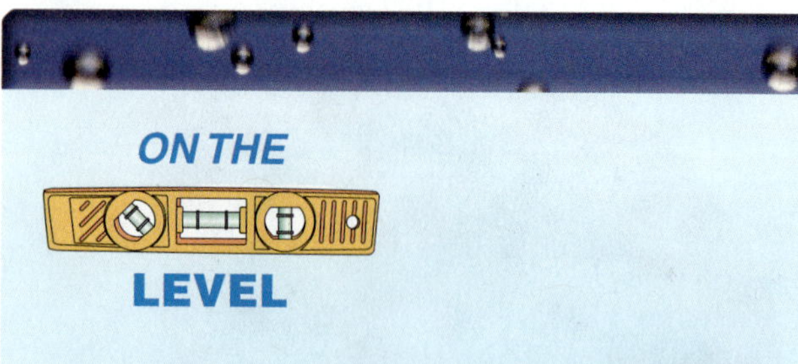

If you must wear corrective glasses, OSHA regulations state that your eyes must be protected by one of the following ways:

- Spectacles whose protective lenses provide optical correction
- Goggles that can be worn over corrective spectacles without disturbing the adjustment of the spectacles
- Goggles that incorporate corrective lenses mounted behind the protective lenses

Your optometrist can provide the required eyewear.

Source: Occupational Safety and Health Administration *CFR 1926.102*

4.3.0 Gloves

On many jobs, you must wear heavy-duty gloves to protect your hands from cuts, chemical burns, or exposure to raw sewage (see *Figure 6*). Work gloves are usually made of cloth, canvas, or leather, but they may also be made of metal mesh or a material called Kevlar®, which protects against metal cuts. Never wear cloth gloves around rotating or moving equipment. Inspect your gloves every day to ensure that they are in good condition. Immediately replace gloves that are worn, torn, or no longer fit.

102F06.EPS

Figure 6 ◆ Work gloves.

In addition to a visual test, an effective way to check rubber gloves is to conduct an air test. The following steps are outlined in *29 CFR 1910.137:*

Step 1 Hold the glove by its cuff, flip it several times to make a seal, and roll the glove toward its fingers. An air pocket will form inside the glove.

Step 2 Hold the rolled portion of the glove tightly. Inspect the inflated exterior of the glove for cracks or degradation of the insulating material surface. Forcing air into the glove will expose damage to the insulation that cannot be seen during visual inspections.

Step 3 Inspect the glove for holes in the insulating material. Hold the glove close to your ear. If you hear air escaping from the glove or if the glove will not hold pressure, the glove is damaged. Damaged gloves must be removed from service.

CAUTION

Protective equipment that is good one day may be damaged the next. Always do a daily visual inspection of PPE to prevent injuries from defective or worn equipment.

WARNING!

Always thoroughly wash and disinfect your hands if you have been exposed to sewage, even if you were wearing gloves. Doing so can prevent disease or illness. In general, it is best to use an alcohol gel instead of an antibacterial cleaner. Recent research indicates that overuse of antibacterial cleaners may make bacteria more resistant.

4.4.0 Safety Shoes

To protect your feet from falling objects and punctures, wear steel-toed, steel-soled safety shoes (*Figure 7*). The steel toe protects your toes from falling objects. The steel sole keeps nails and other sharp objects from puncturing your foot. You are required to wear this type of safety shoe when using tamping equipment or jackhammers. If

102F07.EPS

Figure 7 ◆ Safety shoes.

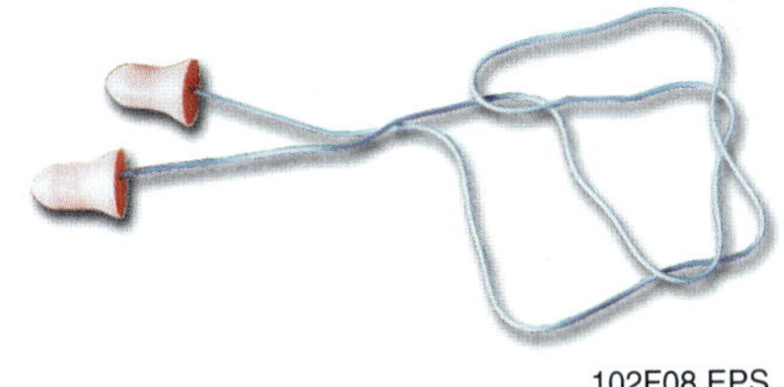

102F08.EPS

Figure 8 ◆ Earplugs.

electrical hazards are present, wear metal-free shoes or boots. Welding jobs call for boots or other sturdy shoes without laces or eyelets to prevent hot metals or sparks from becoming trapped in the shoes. Don't wear oil-soaked shoes if you are welding. Doing so increases the risk of fire.

It's also important to consider how your shoes will be affected by water, especially the soles. Many injuries have occurred as a result of slips on wet surfaces. When working on wet or icy surfaces, wear rubber boots or shoes with skid-resistant soles. Always replace boots or shoes when the sole tread becomes worn or the shoes have holes, even if the holes are on top.

Never wear open-toed shoes, open-heeled shoes, beach sandals, flip flops, sneakers, or other canvas shoes. Such footwear increases the possibility of falls and of injury to the feet. Canvas shoes can catch fire easily and will not protect against chemical spills.

4.5.0 Hearing Protection

Damage to most parts of the body causes pain. Ear damage, however, does not always cause pain. Exposure to loud noise over a long period can cause hearing loss, even if the noise is not loud enough to cause pain. Most companies follow OSHA's rules regarding when workers must use hearing protection. Specially designed earplugs that fit into your ears and filter out noise are one type of hearing protection (see *Figure 8*). You must clean earplugs regularly with soap and water to prevent ear infections.

Another type of hearing protection is earmuffs, which are large, padded covers for the entire ear (*Figure 9*). You must adjust the headband on earmuffs for a snug fit. If the noise level is very high, you may need to wear both earplugs and earmuffs.

You can prevent noise-induced hearing loss by limiting exposure and using appropriate PPE. *Table 2* shows the maximum safe exposure to sound levels rated 90 **decibels (dB)** and higher.

102F09.EPS

Figure 9 ◆ Earmuffs.

Table 2 Maximum Noise Levels

Sound Level (decibels)	Maximum Hours of Continuous Exposure per Day	Examples
90	8	Power lawn mower
92	6	Belt sander
95	4	Tractor
97	3	Hand drill
100	2	Chop saw
102	1.5	Impact wrench
105	1	Spray painter
110	0.5	Power shovel
115	0.25 or less	Hammer drill

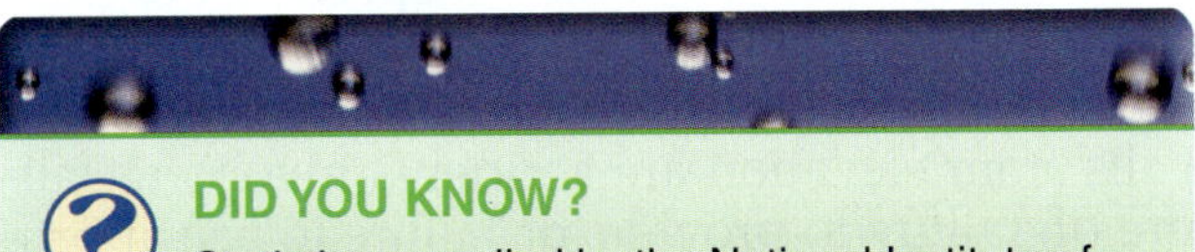

DID YOU KNOW?

Statistics compiled by the National Institute of Occupational Safety and Health (NIOSH) indicate that 48 percent of plumbers report some hearing loss due to the use of noisy electric machinery in tight places.

Always keep a first-aid kit handy. Minor injuries can be treated easily when the necessary items are available. A complete first-aid kit should include the following items:

- Self-adhesive bandages
- Sterile dressing
- Gauze bandages
- First-aid tape
- Moleskin
- First-aid scissors
- Tweezers
- Surgical gloves
- Iodine and/or alcohol swabs
- Ibuprofen or other anti-inflammatory aids
- Antibiotic ointment
- Aspirin or other pain reliever

Some kits also come with tourniquets, splints, safety whistles, safety streamers, safety pins, and garbage sacks. Always keep the first-aid kit stocked. That way, when someone needs to use the kit, the necessary items will be available. It may also be helpful to keep an accident report and a pencil in your first-aid kit.

Never attempt to treat serious injuries yourself. Always call 911 in an emergency. Check your company's policy for administering medications. Some companies may prohibit the distribution of medications.

4.6.0 Fall Protection

OSHA requires workers to wear personal fall protection equipment in certain workplace situations. OSHA developed a Focused Inspection Program to target the four high-hazard areas. One of the four leading hazard groups is falls from elevation. Falls from elevation are accidents involving failure of, failure to provide, or failure to use appropriate fall protection. Wearing personal fall protection equipment can help to prevent these accidents. This equipment usually includes a safety harness and a lanyard.

4.6.1 Equipment

The safety harness (*Figure 10*) is an extra-heavy-duty harness that fits over your body. The harness includes leg, shoulder, chest, and pelvic straps. The lanyard (*Figure 11*) is a length of strong rope with reinforced ends with D-rings snapped onto them. One D-ring attaches to the safety harness. The other is attached to an anchor point above the work area. A qualified person will tell you what a strong anchor point is. The lanyard should be long enough to allow you to perform your tasks but short enough to keep you from falling more than 6 feet.

WARNING!

Never use a safety harness and lanyard for anything except their intended purpose. Always follow the manufacturer's instructions for hooking up a safety harness or lanyard. More than 70 percent of reported job-site accidents are caused by improper use of the lanyard and harness.

4.6.2 Safety

Fall protection includes covers or guardrail systems to prevent you from falling off any surface that is 6 feet or higher above a lower surface or into a hole or excavation that is 6 feet deep or more. Covers for holes also protect you from objects that may fall from above the work area. Use of fall protection equipment is required when working in these areas:

- More than 6 feet above the ground or the next lower surface
- Near openings in floors (for example, a stairwell, elevator shaft, chimney, or plumbing stack opening)
- Near holes or excavations deeper than 6 feet
- Near protruding rebar

Always tie off your fall protection equipment even if you have to lean over by only a foot.

Keep your safety harnesses and lanyards in good working order. Before using this equipment, inspect it for damage, such as wear on the harness straps and buckles or D-rings that are bent or deeply scratched. Check the harness for any cuts or rough spots. Do not use fall protection equipment that shows these signs of wear or damage. If you find any damage, turn in the harness for testing or replacement.

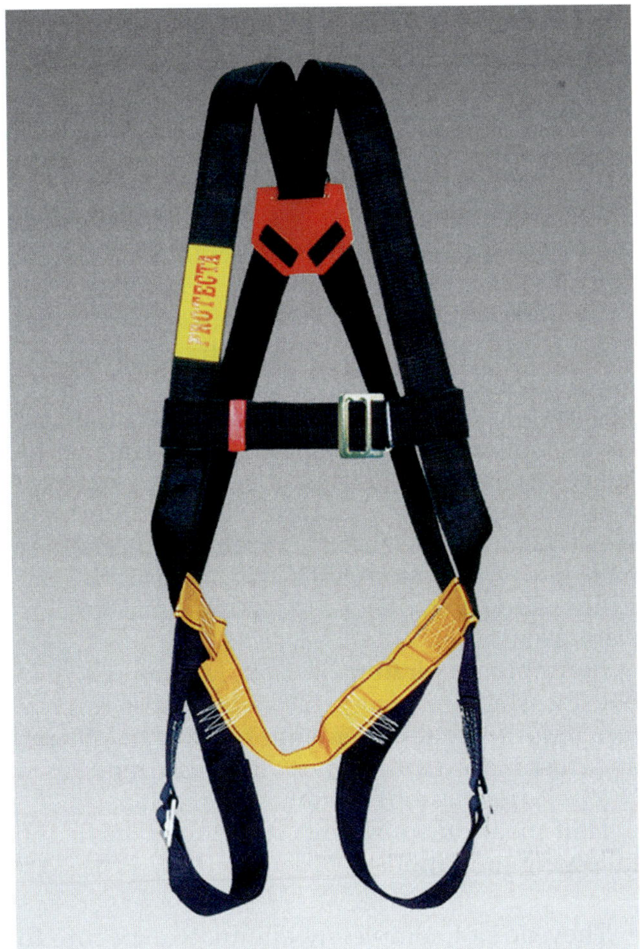

Figure 10 ◆ Typical safety harness.

CAUTION

Always follow the manufacturer's instructions when wearing a harness. Know and follow your company's safety procedures when working on roofs, ladders, and scaffolds.

Figure 11 ◆ Lanyards.

Fall Protection

Effective February 1995, fall protection must be provided for working at an elevation above 6 feet or where a drop into a hole or trench is greater than 6 feet. Supervisors must ensure that all walking and working surfaces have the strength and structural integrity to support the workers. Work conducted on an otherwise unprotected side or edge that is 6 feet or more above a lower level must be protected by the use of guardrail systems, safety-net systems, or personal fall-arrest systems.

Effective January 1998, an acceptable personal fall-arrest system is a body harness with a lanyard attached to a D-ring in the center of the back. Body belts are not acceptable as part of a personal fall-arrest system because they unevenly distribute pressure on the wearer's midsection.

All companies are required by law to have a written Fall Protection Plan that addresses site-specific fall hazards and the steps taken to prevent each hazard. This information will normally be included in your training or regularly scheduled safety meetings. If you are unsure of your company's plan, or have questions about it, ask your supervisor before performing any aerial work.

4.7.0 Respiratory Protection

Wherever there is danger of suffocation or other breathing hazards, you must use a respirator. You must also wear a respirator when working with or near **asbestos** or where hazardous molds are growing. Mandatory special training is required for the use of respirators.

Federal law specifies which type of respirator to use for various breathing hazards. Respirators are grouped into three main types, based on how they protect the wearer from contaminants:

- Air-purifying respirators
- Supplied-air respirators
- Self-contained breathing **apparatus** (SCBA)

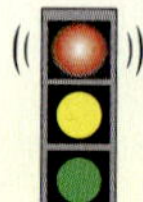

WARNING!

If you encounter asbestos or mold, do not handle them or attempt to remove them. Tell your supervisor about the presence of these materials immediately. Asbestos and mold are linked to long-term illnesses. Only trained personnel who have the proper equipment for handling and disposing of these materials can deal with such situations.

4.7.1 Air-Purifying Respirators

Air-purifying respirators provide the lowest level of protection. They are made for use only in atmospheres that have enough oxygen to sustain life (at least 19.5 percent). Air-purifying respirators are chemical-specific. They use special filters and cartridges to remove specific gases, vapors, and particles from the air. The respirator cartridges contain charcoal, which absorbs certain toxic or deadly vapors and gases. When the wearer detects any taste or smell, that indicates the charcoal's absorption capacity has been reached. This means the cartridge can no longer remove the contaminants. The respirator filters remove particles such as dust, mists, and metal fumes by trapping them within the filter material. Filters should be changed when breathing becomes difficult.

Depending on the contaminants, cartridges can be used alone or with a filter/prefilter and filter cover. Air-purifying respirators should be used for protection against only the types of contaminants listed on the filters and cartridges. Refer also to the NIOSH approval label affixed to each respirator carton and replacement filter/cartridge carton. Respirator manufacturers typically classify air-purifying respirators into four groups:

- No maintenance
- Low maintenance
- Reusable
- Powered air-purifying respirators (PAPRs)

No-maintenance and low-maintenance respirators are typically used for residential or light commercial work that does not call for constant, heavy respirator use. No-maintenance respirators are typically half-mask respirators with permanently attached cartridges or filters. The entire respirator is discarded when the cartridges or filters are spent. Low-maintenance respirators are generally half-mask respirators that use replaceable cartridges and filters. They are not designed for constant use.

Reusable respirators (*Figure 12*) are made in half-mask and full-facepiece styles. These respirators require the replacement of cartridges, filters, and respirator parts. Their use also requires a complete respirator maintenance program.

Half-mask air-purifying respirators have the following limitations:

- They do not completely eliminate exposure to contaminants, although they will reduce the exposure to an acceptable level.
- They do not supply oxygen and must not be used in areas where the oxygen level is below 19.5 percent.
- They must not be used in areas where chemicals are hard to detect, such as those with no taste or odor.

PAPRs (*Figure 13*) are made in half-mask, full-facepiece, and hood styles. They use battery-operated blowers to pull air through the cartridges and filters attached to the respirator. The blower motors can be either mask- or belt-mounted. Depending on the cartridges used, PAPRs can filter particles, dust, fumes, and mists along with certain gases and vapors. Other than units with the blower mounted in the mask, PAPRs have a belt-mounted, powered air-purifier unit connected to the mask by a breathing tube. Many models have an audible and visible alarm that is activated as soon as the airflow falls below the required minimum level. This feature alerts the user to a loaded filter or low battery charge.

102F12.EPS

Figure 12 ◆ Reusable air-purifying respirators.

102F13.EPS

Figure 13 ◆ Powered air-purifying respirator.

ON THE LEVEL

Asbestos is a hazardous, fibrous substance that causes lung diseases, including cancer. It was once used regularly in construction. Pipe insulation, shingles, wallboard, floor covering, and blown-in insulation are just a few products that may contain asbestos.

The federal government stopped production of most asbestos products in the early 1970s. However, installation of these products continued through the late 1970s and early 1980s. Today, asbestos fibers can be released during renovations of older buildings. Breathing asbestos dust can have chronic effects, which are delayed or long-term reactions. A chronic effect can be deadly, such as cancer. Smoking increases the risk of serious illness from asbestos.

Source: Agency for Toxic Substances and Disease Registry

4.7.2 Supplied-Air Respirators

Supplied-air respirators (*Figure 14*) provide air for extended periods through a high-pressure hose connected to an external source of air, such as a compressor, compressed air cylinder, or pump. They provide protection in atmospheres where air-purifying respirators are not adequate. Supplied-air respirators are typically used in toxic atmospheres. Some can be used in atmospheres that are immediately dangerous to life and health (IDLH), as long as they are equipped with an air cylinder for emergency escape. An atmosphere is considered IDLH if it poses an immediate hazard to life or produces immediate, irreversible, and debilitating effects on health. There are two types of supplied-air respirators: continuous flow and pressure demand.

A continuous-flow supplied-air respirator provides air to the user in a constant stream. One or two hoses deliver the air from the air source to the facepiece. Unless the compressor or pump is specially designed to filter the air, or a portable air-filtering system is used, the unit must be located where there is breathable air. The flow of air to the user may be adjusted either at the air source (fixed flow) or on the unit's regulator (adjustable flow). Continuous-flow respirators are made with tight-fitting half-masks or full facepieces. They are also made with hoods.

Figure 14 ◆ Supplied-air respirator.

The pressure-demand supplied-air respirator is similar to the continuous-flow type except that it supplies air to the user's facepiece via a pressure-demand valve as the user inhales. It typically has a two-position exhalation valve that allows the user to switch between the pressure-demand and negative-pressure modes to make it easier to enter, move within, and exit from a work area.

4.7.3 Self-Contained Breathing Apparatus

SCBAs provide the highest level of respiratory protection. They can be used in **oxygen-deficient atmospheres** (below 19.5 percent oxygen), in poorly ventilated or **confined spaces**, and in IDLH atmospheres. These respirators supply air for about 30 to 60 minutes from a compressed-air cylinder worn on the user's back. An emergency escape breathing apparatus (EEBA) is a smaller version of an SCBA cylinder (see *Figure 15*). EEBAs are used to escape from hazardous environments and generally provide a 5- to 10-minute air supply.

OSHA regulations (*29 CFR 1910.134*) require employees who use respirators in the workplace to be trained on an annual basis. Employees must complete the training before they are required to use a respirator. According to OSHA, "The training must be comprehensive, understandable, and recur annually, and more often if necessary." Additional training may be required if earlier training becomes obsolete or if the employee has not mastered the use of a respirator and demonstrates difficulty using a respirator.

An OSHA Letter of Interpretation clarified the annual training requirement for respiratory protection. "Annual means that training and fit testing must be conducted every year, before or on the anniversary date of the employee's previous training and fit test; for example, if the employee is trained or fit tested on February 1, 1999, then the employee must be trained or fit tested before or on February 1, 2000." If the anniversary date is missed, the reason for the delay and the expected date of completion should be noted in the employee's file.

Trained employees must be able to demonstrate knowledge in the following areas:

- Why a respirator is necessary, and the consequences of improper fit, use, or maintenance
- The limitations and capabilities of respirators
- How to use the respirator effectively in emergency situations, including when the respirator malfunctions
- How to inspect, put on, remove, use, and check the seals of the respirator
- The procedures for maintaining and storing respirators
- How to recognize medical signs and symptoms that may limit or prevent the effective use of respirators
- The information in OSHA standard *29 CFR 1910.134*

Sources: Department of Energy website, http://tis.eh.doe.gov/rl/pres/docs/D9903008.htm, reviewed March 8, 2004, and Sherman Safety website, www.shermansafety.com/oshatraining.html, reviewed March 8, 2004

NOTE

You must be certified to wear an SCBA. Training for certification covers maintaining and checking the equipment, as well as correctly putting it on and removing it.

102F15.EPS

Figure 15 ◆ Emergency escape breathing apparatus.

4.8.0 Respiratory Program

A respirator must be selected based on the contaminant present and its concentration level. It must be fitted and used properly, in accordance with the manufacturer's instructions. It must be worn during all times of exposure.

Employers must have a respiratory protection program with the following components:

- Standard operating procedures for selection and use
- Employee training
- Regular cleaning and disinfecting
- Sanitary storage
- Regular inspection
- Annual fit testing
- Pulmonary function testing

During some types of plumbing work, you will need to wear respiratory protection. Always check the cartridge on your respirator to ensure that it is the correct type for the air conditions and contaminants on your job site. In certain concentrations, vapors or fumes can be eliminated by the use of air-purifying devices as long as the oxygen levels are acceptable. Smoke billowing from a fire and the fumes generated when welding are examples.

When selecting a respirator to wear while working with specific materials, you must first determine the hazardous ingredients in the material and their exposure levels. Always read the **material safety data sheet (MSDS)**, which is often located in a binder in the project manager's or supervisor's office. MSDSs identify the hazardous ingredients and should list the type of respirator and cartridge recommended for use with the material. You will learn more about MSDSs later in this module.

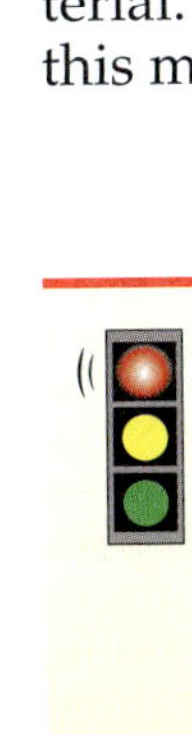

WARNING!

If breathing becomes difficult, if you become dizzy or nauseated, if you smell or taste the chemical, or if you have other noticeable effects of exposure, leave the area immediately. Get to an area with fresh air, and seek any necessary assistance. You must have at least one other worker with you when working in a confined space. You will learn more about working in confined spaces later in this module.

4.8.1 Positive and Negative Fit Checks

Respirators are useless unless properly fit tested to each wearer. To obtain the best protection from your respirator, you must perform positive and negative fit checks each time you wear it. Repeat these fit checks until you have obtained a good face seal.

Use the following steps to perform a positive fit check:

Step 1 Adjust the facepiece for the best fit, then adjust the head and neck straps to ensure good fit and comfort.

Step 2 Block the exhalation valve with your hand or other material.

Step 3 Breathe out into the mask.

Step 4 Check for air leakage around the edges of the facepiece.

Step 5 If the facepiece puffs out slightly for a few seconds, you have a good face seal.

Perform a negative fit check using the following steps:

Step 1 Block the inhalation valve with your hand or other material.

Step 2 Attempt to inhale.

Step 3 Check for air leakage around the edges of the facepiece.

Step 4 If the facepiece caves in slightly for a few seconds, you have a good face seal.

Ensure that your respirator is clean and in good condition and that all of its parts are in place. Otherwise, it will not protect you. Respirators must be cleaned every day. Failure to do so will limit their effectiveness, and they will offer little or no protection. For example, suppose you wore a respirator for two weeks and did not clean it. The bacteria that accumulated as you breathed into the respirator, plus the airborne contaminants that managed to enter the facepiece, would make the inside of your respirator very unsanitary. Continued use could do you more harm than good. Remember, only a clean and complete respirator will give you the necessary protection. Use the following general guidelines for taking care of your respirator:

- Inspect your respirator before and after each use.
- Do not wear a respirator if the facepiece is distorted or if it is worn and cracked. You will not be able to get a proper face seal.
- Do not wear a respirator if any part is missing. Replace worn straps or missing parts before use.
- Do not expose respirators to excessive heat or cold, chemicals, or sunlight.
- Clean and wash your respirator each day.
- Sanitize your respirator each week.

4.9.0 Proper Clothing and Grooming

Except for PPE, OSHA has not set a standard for proper workplace clothing. There is also no OSHA standard for personal grooming. However, proper clothing and good grooming can be as important to your safety as PPE. Treat your work clothing the same way you treat your PPE. Inspect it often for signs of wear that could cause problems, and keep it clean.

Avoid wearing clothing that is too tight or too loose. Tight clothing can restrict your ability to move quickly; loose clothing can catch in machinery or equipment. Make sure that your clothing does not have flaps, strings, or ragged edges that could be tangled in power tools. Don't wear pants that drag on the floor or shirts or jackets with sleeves that are too long. Leg chaps and knee pads can also provide increased protection to your legs. When welding, you must not wear pants with cuffs into which bits of molten metal or sparks may fall.

Safety shoes or high-top sneakers can protect your ankles from cuts and abrasions. Steel toes will protect your feet from injuries caused by falling objects and power tools. Shoes with good traction can help to prevent slips and falls.

Do not wear jewelry, which can catch in machinery. If your hair or beard is long, tie it back or braid it to keep it out of your way while working. Long hair falling into your eyes can affect your ability to see clearly. Like jewelry, long hair can get caught in machinery. Tie back, wrap, or cover long hair. Dressing and grooming in a way that is appropriate for your job will not only help to keep you safe but will also reflect well on you as an employee.

Beards and Respirators

If you wear a beard, you may not be able to seal a respirator facepiece properly. There are, however, loose-fitting respirators that are available for routine or emergency use. According to OSHA, these respirators can be used by bearded wearers because facial hair does not interfere with the facepiece seal of these units. Your employer may determine whether the respirator is acceptable for your use.

ON THE

LEVEL

Clean Your Respirator

Clean your respirator daily. OSHA's Respirator Cleaning Procedures (Mandatory) are located in Appendix B of *CFR 1910.134.* Cleaning your respirator is an easy task that helps to ensure its safe use. Briefly, the procedures include the following steps:

1. Remove filters, cartridges, or canisters. Disassemble facepieces and discard or repair any defective parts.
2. Wash components in warm water with a mild detergent. Use a stiff bristle (not wire) brush to remove dirt.
3. Rinse components thoroughly in clean, warm water. Drain.
4. If your cleaner does not contain a disinfecting agent, immerse components for two minutes in one of the following: hypochlorite solution, aqueous solution of iodine, or other commercially available cleansers of equivalent disinfectant quality.
5. Rinse components thoroughly in clean, warm water.
6. Air dry components or hand dry them with a clean, lint-free cloth.
7. Reassemble facepiece, replacing filters, cartridges, and canisters where necessary.
8. Test the respirator to ensure that all components work properly.

Source: Occupational Safety and Health Administration website, "Regulations (Standards – *29 CFR*), *Respirator Cleaning Procedures (Mandatory) – 1910.134 Appendix B-2,*" www.osha.gov/pls/oshaweb/owadisp.show_document?p_table=STANDARDS&p_id=9782, reviewed March 5, 2004

Review Questions

Section 4.0.0

1. When you wear a hard hat, there should be _____ inch(es) of space between your head and the shell of the hard hat.
   a. $\frac{1}{4}$
   b. $\frac{1}{2}$
   c. 1
   d. 2

2. When welding, you must wear _____.
   a. safety glasses
   b. safety glasses with side shields
   c. tinted goggles or a hood
   d. safety goggles with a wrap-around strap

3. An air-purifying respirator cartridge contains _____, which absorbs certain toxic vapors from the air.
   a. a treated fiber
   b. a baffle chamber
   c. filters filled with baking soda
   d. charcoal

4. The highest level of respiratory protection is provided by _____.
   a. air-purifying respirators
   b. supplied-air respirators
   c. self-contained breathing apparatus
   d. exhaust hoods

5. To obtain the best protection from a respirator, you must perform a positive and negative fit check _____.
   a. the first time you wear it
   b. every time you wear it
   c. after you have worn it for at least 20 minutes
   d. once or twice a week

5.0.0 ◆ HAZARD COMMUNICATION

Exposure to hazardous chemicals and materials such as sealants, asbestos, cleansers, and compressed gas can cause both environmental and health problems. Health problems can include skin or eye irritation, breathing difficulty, allergic reactions, and cancer. Environmental hazards can include fire, corrosion, and reactivity (explosions).

The types of hazardous materials on a site can range from chemicals to radiation. Sewage is also considered hazardous. Radiation is probably the least thought-about hazard because it cannot be seen or tasted. Radiation is present during radiographic testing of welds in piping, vessels, medical equipment, or pumps.

Chemicals present a significant danger because they exist in many different forms. Chemicals are not only liquids. They can also be solids, gases, fumes, and mists. Many common products contain several chemicals. For example, some paints contain cadmium and lead. Many chemicals pose health hazards, such as disease or burns. Others pose physical hazards, including fire or explosion. Some pose both health and physical hazards. A strong HazCom program that includes MSDSs will help you identify the hazards and understand how to protect yourself. You have the right to know the hazards of all the chemicals you will be exposed to on the job.

5.1.0 Right to Know

OSHA directs employers to tell workers about hazardous materials on the job site through HazCom. You may have heard it called the worker right-to-know program. Everyone on site must be educated about the hazardous materials they might use on the job. This is done through a written HazCom program and training. In addition, all materials must have proper labels and MSDSs. Labels and MSDSs provide information about health hazards, safety precautions, and emergency responses. You need to understand this information in order to protect yourself. The final responsibility for your safety rests with you. Plumbers have the following responsibilities when it comes to HazCom:

- Learn to recognize hazardous materials labels.
- Know where MSDSs are kept on your job site.
- Report any hazards you spot on the job site to your supervisor.
- Know the physical and health hazards of the materials you use.
- Know how to protect yourself from hazards.
- Know what to do in an emergency.
- Understand your employer's HazCom program.

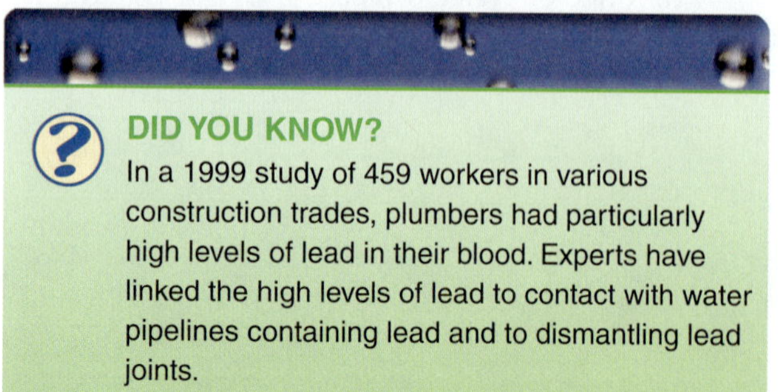

DID YOU KNOW?

In a 1999 study of 459 workers in various construction trades, plumbers had particularly high levels of lead in their blood. Experts have linked the high levels of lead to contact with water pipelines containing lead and to dismantling lead joints.

5.2.0 Labels

On a construction site, all materials in containers must have a label. Labels describe what is in a container. They also warn you of chemical hazards.

The *HazCom Standard* states that hazardous material containers must be labeled, tagged, or marked. The label must include the name of the material, the appropriate hazard warnings, and the name and address of the manufacturer.

OSHA does not require specific labels. Label information can be any type of message using words, pictures, or symbols. However, labels must describe the hazards present. Labels must also be readable and easy to see.

Common HazCom labels come from the National Fire Protection Association (NFPA), the National Paint and Coatings Association (NPCA), and the U.S. Department of Transportation (DOT). The NFPA's hazardous materials classification system is often referred to as the **NFPA warning diamond** (see *Figure 16*). The four-color diamond can be a container label. It is also used on doors to note the hazards in a room or building to aid firefighters and emergency responders. Each section and color represents a hazard: health, flammability, stability, and specific hazards. Numbers from zero to four indicate increasing hazards.

The NPCA's Hazardous Materials Information System (HMIS®) (*Figure 17*), like the NFPA warning diamond, uses a color coding system to identify the hazards associated with a particular product and assigns numbers to each section to reflect the degree of hazard. HMIS® is meant for alerting employees to potential hazards.

The DOT requires labels for hazardous materials shipments (*Figure 17*). Either of these labels can be part of your company's HazCom labeling program. You may see these labels at your job site.

Hazardous materials at the site must be properly labeled. If the material is transferred from a labeled container, the new container must be labeled with all of the information from the original

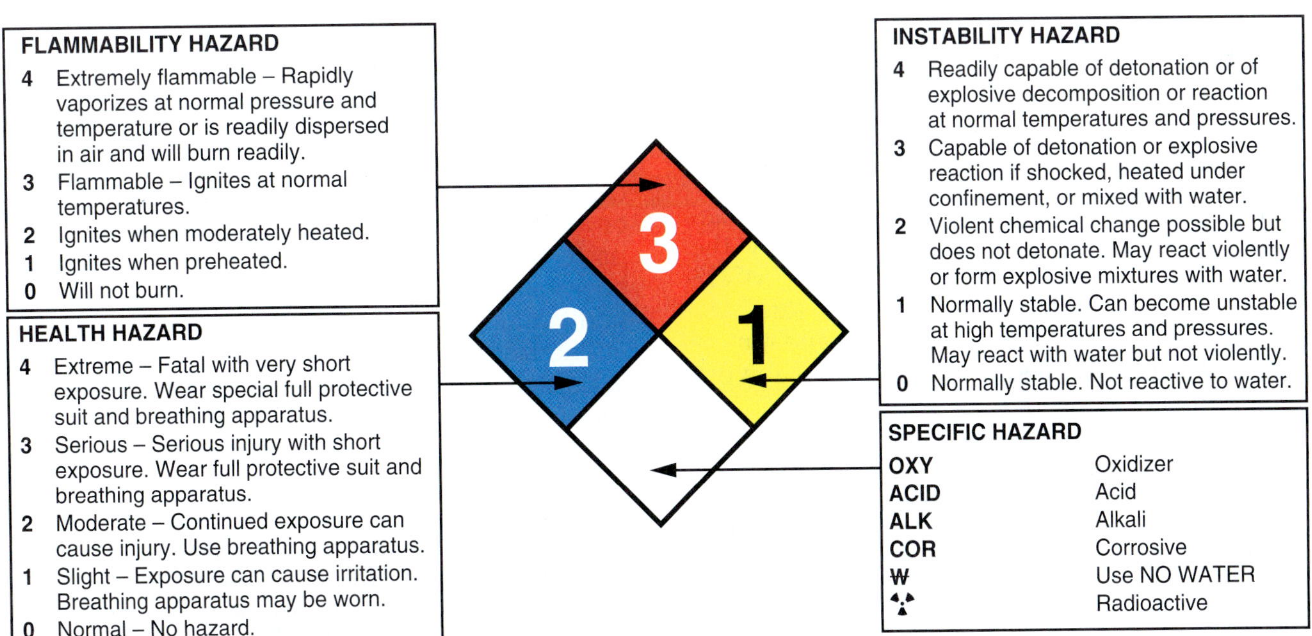

Figure 16 ◆ NFPA warning diamond.

Figure 17 ◆ HazCom labels.

label. Make sure that any materials you work with are labeled. Be sure that you understand your company's labels.

5.3.0 Material Safety Data Sheets

Each product on a construction site must have an MSDS. MSDSs are fact sheets prepared by the product's manufacturer or importer. An MSDS describes the substance and its hazards, safe handling requirements, first-aid needs, and emergency spill procedures. OSHA does not have a mandatory form, but it does require inclusion of specific information. The Chemical Manufacturers Association has developed a standard form that meets national and international standards. Most chemical manufacturers use this form. The sections of the form include the following:

- Chemical product and company information
- Composition/information on ingredients
- Hazard identification
- First-aid measures
- Firefighting information
- Accidental release measures
- Handling and storage
- Exposure controls/personal protection
- Physical and chemical properties
- Stability and reactivity
- Toxicological properties
- Ecological properties
- Disposal considerations
- Transportation information
- Regulatory information
- Other information related to the chemical

An MSDS can be difficult to read. The scientific information is fairly technical. *Figure 18* is an example of an MSDS for a common construction adhesive. Look for the following important information on an MSDS:

- Specific hazards
- Personal protection requirements
- Handling procedures
- First-aid information
- The 24-hour emergency-response telephone number

Using *Figure 18*, try to find the information you would need to use the PVC adhesive described on the MSDS. First locate the hazards. Section III of the MSDS shows that the adhesive is moderately toxic and is harmful if swallowed or the vapors are inhaled. It is also an eye and skin irritant. Next, find out how to minimize these hazards. Section VIII tells you to use an appropriate respirator, chemical-resistant gloves, and chemical goggles or a full face shield. It also recommends ways to protect against spills and splashes. To find out how to handle and store this adhesive, check Section VII.

The HMIS® information and NFPA ratings are in Section XVI. The NFPA label (refer to *Figure 16*) could be used for this product. Now you have the facts you need to protect your health and that of your co-workers.

Section IV lists the first-aid measures for inhalation, eye contact, skin contact, and ingestion. Section V explains fire hazards and firefighting measures. You will find the emergency telephone number in Section I. Now you have the information you need in case of an emergency. All MSDSs must be kept on site. Ask your supervisor to tell you where the MSDSs are and to point out the sections that relate to your job. Your health and safety and that of your co-workers depend on it.

5.4.0 Responding to Emergencies

Every job site should have an emergency-response plan. Planning should be coordinated well and communicated to everyone involved. If you are told by your supervisor to evacuate a work site, go to a safe location, and wait until you are notified that conditions are safe. Being prepared can reduce the severity of an emergency. Planning is especially important if the accident happens in a remote area that does not have a telephone. Make sure you know your employer's emergency plan. Find out whom to call in an emergency and what you need to do to protect yourself. Know where your emergency hospital is and know where you and your co-workers should go in an emergency. Response plans will vary from company to company. At a minimum, all plans should include the name of the nearest emergency hospital, safety meeting points, and an emergency point of contact. If your company does not have a plan, see to it that one is made. You will learn more about emergency response later in this module.

PVC CLEAR ADHESIVE

Material Safety Data Sheet

SECTION I - COMPANY IDENTIFICATION

PRODUCT: PVC CLEAR ADHESIVE

CAT. NO.: PVCA-QP, PVCA-HP, PVCA-P

DISTRIBUTED BY:
Virginia KMP Corporation
4100 Platinum Way
Dallas, Texas 75237

TELEPHONE NUMBERS:
Office: 1-(214) 330-7731
Emergency Only: 1-(800) 424-9300

SECTION II - HAZARDOUS INGREDIENTS

OSHA Hazardous Components (29 CFR 1910.1200)		EXPOSURE LIMITS: 8 HR. TWA OSHA PEL		ACGIH TLV
Tetrahydrofuran (CAS# 109-99-9)		200 ppm		200 ppm
	STEL	250 ppm	STEL	250 ppm
Methyl ethyl ketone (CAS# 78-93-3)		200 ppm		200 ppm
	STEL	300 ppm	STEL	300 ppm
Dimethyl formamide (CAS# 68-12-2)		10 ppm (skin)		10 ppm (skin)
Cyclohexanone (CAS# 108-94-1)		25 ppm (skin)		25 ppm (skin)

SECTION III - HAZARDS IDENTIFICATIONS

EMERGENCY OVERVIEW: DANGER! Moderate toxic. Harmful if swallowed or vapors inhaled. Skin and eye irritant.

POTENTIAL HEALTH EFFECTS:

INHALATION: Harmful, vapors may irritate upper respiratory tract. May have narcotic effects.

EYE CONTACT: Liquid or vapors may irritate eyes immediately on contact.

SKIN CONTACT: Liquid or vapors may irritate skin, aggravate pre-existing skin disorders. Irritation develops slowly after contact. Ingredients may be absorbed through skin.

INGESTION: Moderately toxic. May affect peripheral and central nervous systems.

CHRONIC Effects: Kidney and liver damage possible, contains dimethylformamide. Dimethylformamide is an OSHA listed possible human carcinogen.

NOTE:

CARCINOGENICITY: LISTED IN NTP? No IARC? No OSHA Regulated? Yes

SECTION IV - FIRST AID MEASURES

INHALATION: Remove victim to fresh air and, if needed, immediately begin artificial respiration. Give oxygen if breathing is labored. Get emergency medical help. Contact a physician immediately.

EYE CONTACT: Flush eyes with water for 15 minutes. Get medical attention if symptoms develop and persist.

SKIN CONTACT: Flush with water or soap and water for 15 minutes or until all traces have been removed. Seek medical attention if symptoms develop and persist.

INGESTION: Do not induce vomiting. Rinse mouth out with water. Get immediate medical attention.

SECTION V - FIRE FIGHTING MEASURES

FLASHPOINT (TEST METHOD): 20 F (estimate)

FLAMMABLE LIMITS: **LOWER:** ND **UPPER:** ND

AUTOIGNITION TEMPERATURE: N

GENERAL HAZARD: Vapors may travel to ignition source and flashback.

FIRE FIGHTING INSTRUCTIONS: Approach fire from upwind side. Avoid breathing smoke, fumes, mist, or vapors on the downwind side. Firefighters wear protective clothing and self contained breathing apparatus.

EXTINGUISHING MEDIA: Dry powder, carbon dioxide (CO_2), alcohol foam for large fires.

HAZARDOUS COMBUSTION PRODUCTS: Acrid smoke, CO, CO_2, toxic fumes of NO_x

102F18A.EPS

Figure 18 ◆ MSDS for PVC clear adhesive. (1 of 3)

PVC CLEAR ADHESIVE

SECTION VI - ACCIDENTAL RELEASE MEASURES

LAND SPILL: Emergency response coordinator must have mandated training. Eliminate all ignition sources. SMALL SPILLS: Pick up with absorbent materials and place in non-leaking containers; seal tightly for proper disposal or reuse. LARGE SPILLS: Evacuate the hazard area of unprotected personnel. Wear appropriate respirator and protective clothing. Shut off source of leak if safe to do so. Dike and contain. Remove with vacuum trucks or pump to storage/salvage vessels.

WATER SPILL: Notify proper authorities.

Clean up leaks/spills immediately to prevent soil or water contamination.

SECTION VII - HANDLING AND STORAGE

HANDLING: Avoid contact with skin, eyes, and clothing. After handling this product, wash hands before eating, drinking, or smoking. If contact occurs, remove contaminated clothing. If needed, take first aid action shown in section IV. Launder contaminated clothing before reuse.

STORAGE: Store away from oxidizers. Store in a cool, ventilated area, away from ignition sources.

SECTION VIII - EXPOSURE CONTROLS/PERSONAL PROTECTION

ENGINEERING CONTROLS: Local exhaust ventilation recommended.

PERSONAL PROTECTION: Use NIOSH approved respirator, chemical resistant gloves, chemical goggles or full face shield. Wear boots, aprons, drench showers, eye wash as needed for protection against spills and/or splashes.

SECTION IX - PHYSICAL AND CHEMICAL PROPERTIES

VAPOR PRESSURE @ 0 C:51 mmHg
SPECIFIC GRAVITY (H_2O=1): 0.9112
SOLUBILITY IN WATER: Negligible
pH: NA
BOILING POINT: 149 - 321 F
APPEARANCE & ODOR: Clear slightly viscous liquid, ketone odor.

VAPOR DENSITY (Air=1): >1
EVAPORATION RATE (BuAc=1): 5.5
VOC (LB/GAL): NA
FREEZING POINT: ND

SECTION X - STABILITY AND REACTIVITY

STABILITY: Stable.
CONDITIONS TO AVOID: High temperatures, ignition sources.
MATERIALS TO AVOID: Oxidizers.
HAZARDOUS DECOMPOSITION PRODUCTS: From combustion: acrid smoke, toxic fumes of NO_x, carbon dioxide, carbon monoxide.
HAZARDOUS POLYMERIZATION: Will not occur.

SECTION XI - TOXICOLOGICAL INFORMATION

Tetrahydrofuran	LDLo:	3000 mg/kg	(oral - rat)	
	TCLo:	25000 ppm	(inh - human)	CNS
Methyl ethyl ketone	TCLo:	100 ppm	(inh - human)	TER
	LD50:	2737 mg/kg	(oral - rat)	
Dimethyl formamide		100%/24H	(skin - human)	mild
	LD50:	2800 mg/kg	(oral - rat)	
	LD50:	4720 mg/kg	(skin - rabbit)	
Cyclohexanone	TCLo:	75 ppm	(inh - human)	EYE, PUL
	LD50:	948 mg/kg	(skin- rabbit)	
	LD50	1535 mg/kg	(oral - rat)	

SECTION XII - ECOLOGICAL INFORMATION

Dangerous to aquatic life in high concentrations. Methyl ethyl ketone 5640 mg/L / 48 hr / blue gill / TLm / freshwater

SECTION XIII - DISPOSAL CONSIDERATIONS

Dispose as hazardous waste. Classification and documentation is required before disposal. Follow all local, state and federal regulations.

Prepared by: Virginia KMP Corporation
This information is furnished without warranty, expressed or implied, except that it is accurate to the best knowledge of Virginia KMP. The data on this sheet related only to specific material designated herein. Virginia KPM assumes no legal responsibility for use or reliance upon these data.

102F18B.EPS

Figure 18 ◆ MSDS for PVC clear adhesive. (2 of 3)

PVC CLEAR ADHESIVE

SECTION XIV - TRANSPORTATION INFORMATION

PROPER SHIPPING NAME: Consumer Commodity ORM-D
HAZARD CLASS:
IDENTIFICATION NUMBER:
DOT Emergency Guide #: 128
Reportable Quantity (RQ): Not applicable
International: Adhesives, containing a flammable liquid, 3, UN1133, PG II , LTD QTY

SECTION XV - REGULATORY INFORMATION

TSCA (Toxic Substance Control Act): Components of this product are listed on the TSCA Inventory.
CERCLA (Comprehensive Environmental Response, Compensation and Liability Act): Reportable quantity is 5,000 gallons (methyl ethyl ketone). Contact local authorities for other reporting requirements. **Clean Air Act: Section 112(b)** Hazardous Air Pollutant reportable quantity released to air is 1 lb. dimethyl formamide.
SARA TITLE III (Superfund Amendments and Reauthorization Act):

Section 313, Toxic Materials:	Methyl ethyl ketone	CAS# 78-93-3	<15%

CALIFORNIA PROPOSITION 65: Not listed.

SECTION XVI - OTHER INFORMATION

State Right-to-Know Programs: MA, PA, NJ

NFPA Ratings

Health:	1
Flammability:	3
Reactivity:	0

HMIS Protective Equipment: X See your supervisor
Revised 10/2001

Prepared by: Virginia KMP Corporation
This information is furnished without warranty, expressed or implied, except that it is accurate to the best knowledge of Virginia KMP. The data on this sheet related only to specific material designated herein. Virginia KPM assumes no legal responsibility for use or reliance upon these data.

102F18C.EPS

Figure 18 ◆ MSDS for PVC clear adhesive. (3 of 3)

Your company expects you to be prepared in case of an emergency. Therefore, you must read and understand the company's emergency-response plan. Many companies have guidelines that they want employees to follow during an emergency. These guidelines are designed to prevent panic and to protect people, property, and the environment:

- Stay calm, and quickly evaluate the situation.
- Notify affected personnel.
- Follow company safety procedures to protect personnel, property, and the environment.
- Submit any required reports.

OSHA may require reports for certain types of emergency situations. These requirements are detailed in *29 CFR 1904*. Although you may not be the person submitting the report, you may be asked to provide information to the reporting authority.

Review Questions

Section 5.0.0

1. The color blue on a hazardous information label indicates a(n) _____ hazard.
   a. fire
   b. health
   c. explosion
   d. radioactive

2. The MSDS is a fact sheet prepared by _____.
   a. OSHA
   b. the product's manufacturer or importer
   c. the National Fire Protection Agency
   d. local authorities according to local codes

3. The most important things to look for on an MSDS are the _____.
   a. specific hazards, personal protection requirements, handling procedures, first-aid information, and the emergency-response telephone number
   b. chemical hazards, explosive properties, personal protection requirements, and ecological information
   c. chemical composition, toxicological properties, and disposal and transportation information
   d. accidental-release measures, handling and storage requirements, transportation and regulatory information, and the emergency-response telephone number

4. Section IV of an MSDS _____.
   a. gives the manufacturer's contact information
   b. tells you how to protect your eyes and skin
   c. gives general handling and storage information
   d. lists first-aid measures

5. If your supervisor tells you to evacuate the work site, you should _____.
   a. return to work the next day
   b. go to a safe location and wait until you are notified that conditions are safe
   c. walk at least 50 feet away from the site and wait there for further instructions
   d. gather up all of your belongings before leaving the work site

6.0.0 ◆ WORK ZONES

Plumbing work is often done at construction sites in or near public areas. Creating a clear work zone is an important part of working safely. Barricades, fencing, caution tape, signs, and cones mark a construction work zone. To ensure everyone's safety, you must keep the public and their vehicles away from your work area.

Signs, tags, and color codes in the workplace protect workers from hazardous conditions and help them respond to emergencies (see *Figure 19*). For signs, tags, and color codes to be effective, all workers must understand what they mean and know what action they are required to take. This reduces confusion and ensures their effectiveness.

Signals such as alarms, bells, buzzers, whistles, and horns also communicate hazards to workers. For example, backup alarms are used on forklifts, construction equipment, and trucks. Fire alarms are used to clear work areas. Conveyer belt lines have buzzers, bells, or whistles to let workers know they are about to be started.

Barricades are another way to warn of danger. They are used on construction sites to keep out unauthorized personnel and control traffic (see *Figure 20*). Barricades are also used to control pedestrian traffic outside, or in rooms or hallways that have been recently washed or waxed.

It's important to recognize all of the signs and signals on your job site and make sure they are placed properly and working correctly. Doing this can save a life.

6.1.0 Signs

All work sites have specific markings and signs to identify hazards. Signs can also provide emergency information (see *Figure 21*). These are common types of signs you will see on a site:

- Danger signs
- Caution signs
- Informational signs
- Safety signs
- Safety tags

6.1.1 Danger Signs

Danger signs are usually red, black, and white. They tell workers that an immediate hazard exists, such as high voltage or flammable materials. Danger signs also have specific precautions that must be observed to avoid an accident (*Figure 22*). Examples of danger signs include NO SMOKING and KEEP OUT.

6.1.2 Caution Signs

Caution signs are yellow with black letters. Caution signs warn workers about potential hazards or unsafe practices (*Figure 23*). When you see a caution sign, take action to protect yourself. Common examples include DO NOT OPERATE, KEEP AISLES CLEAR, and ELECTRIC FENCE.

Yellow is the basic color used for caution. It identifies places where physical hazards may be caused by striking against objects; stumbling,

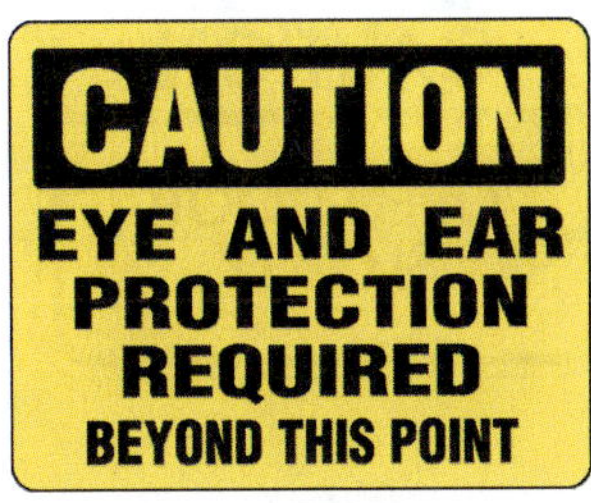

Figure 19 ◆ Work zone signs and tags.

102F20.EPS

Figure 20 ◆ Typical uses of barricades.

102F21.EPS

Figure 21 ◆ Work zone signs.

102F22.EPS

Figure 22 ◆ Typical danger sign.

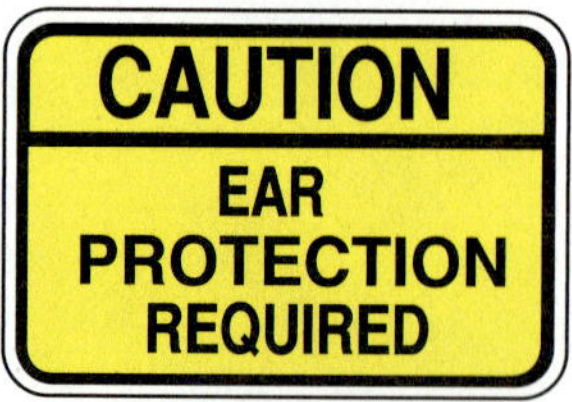

102F23.EPS

Figure 23 ◆ Typical caution sign.

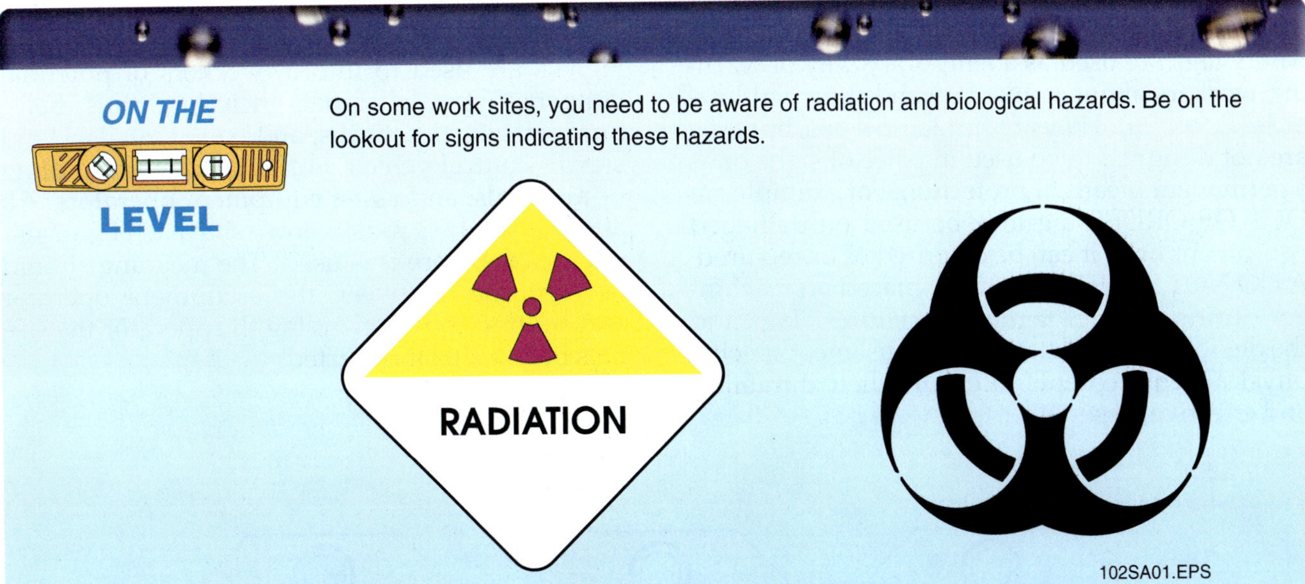

falling, tripping; or being caught between obstacles. Solid yellow, yellow and black stripes, or yellow and black checkers caution workers against these hazards.

Caution signs for piping systems that contain dangerous materials are also yellow. Yellow warns workers against starting machinery under repair. Painted barriers and flags should be at the starting point or power source. They should be displayed so that workers will notice them easily on such things as electrical controls, ladders, scaffolds, vaults, valves, dryers, boilers, elevators, and tanks.

6.1.3 Informational Signs

Informational signs provide general information that is not related to safety (*Figure 24*). The standard color is blue. The background, the entire sign, or just a panel may be blue. Common examples include NO ADMITTANCE, NO TRESPASSING, and EMPLOYEES ONLY.

Informational signs can also be black and white. These signs are used as traffic and housekeeping markers. They identify information such as the following:

- Dead ends of aisles or passageways
- Location of trash cans
- Location and width of aisles
- Rooms or passageways
- Stairways (risers, direction, borders)
- Drinking fountains and food-dispensing machines

Figure 24 ◆ Common informational sign.

6.1.4 Safety Signs

Safety instruction signs are used for general instructions and suggestions related to safety measures (*Figure 25*). The background and lettering on these signs are often white and green, but they can vary depending on the message and the location of the sign. Any letters used against the white background are black. Common examples include REPORT ALL UNSAFE CONDITIONS TO YOUR SUPERVISOR; WALK, DON'T RUN; and HELP KEEP THIS PLANT SAFE AND CLEAN.

Figure 25 ◆ Common safety sign.

6.1.5 Safety Tags

Safety tags are used as a temporary way of warning workers about immediate and potential hazards (*Figure 26*). They are similar to signs, but they are not designed to be used in place of signs or as a permanent means of protection. For example, an OUT OF ORDER tag may be used on damaged equipment until it can be disposed of or repaired. A DO NOT START tag may be placed on machinery during **lockout/tagout procedures**. Tags and the devices used to attach them must meet specific physical requirements to ensure their durability and effectiveness.

6.2.0 Signals

Signals are used to inform workers of potential dangers. Types of signals include alarms, bells, buzzers, whistles, horns, and hand signals. Hand signals control vehicle traffic, guide the handling of materials, and assist equipment operators. All affected workers must know what each hand signal means before it is used. The meaning should be confirmed between the equipment operator and spotter or person giving the operator the signals before a task is started.

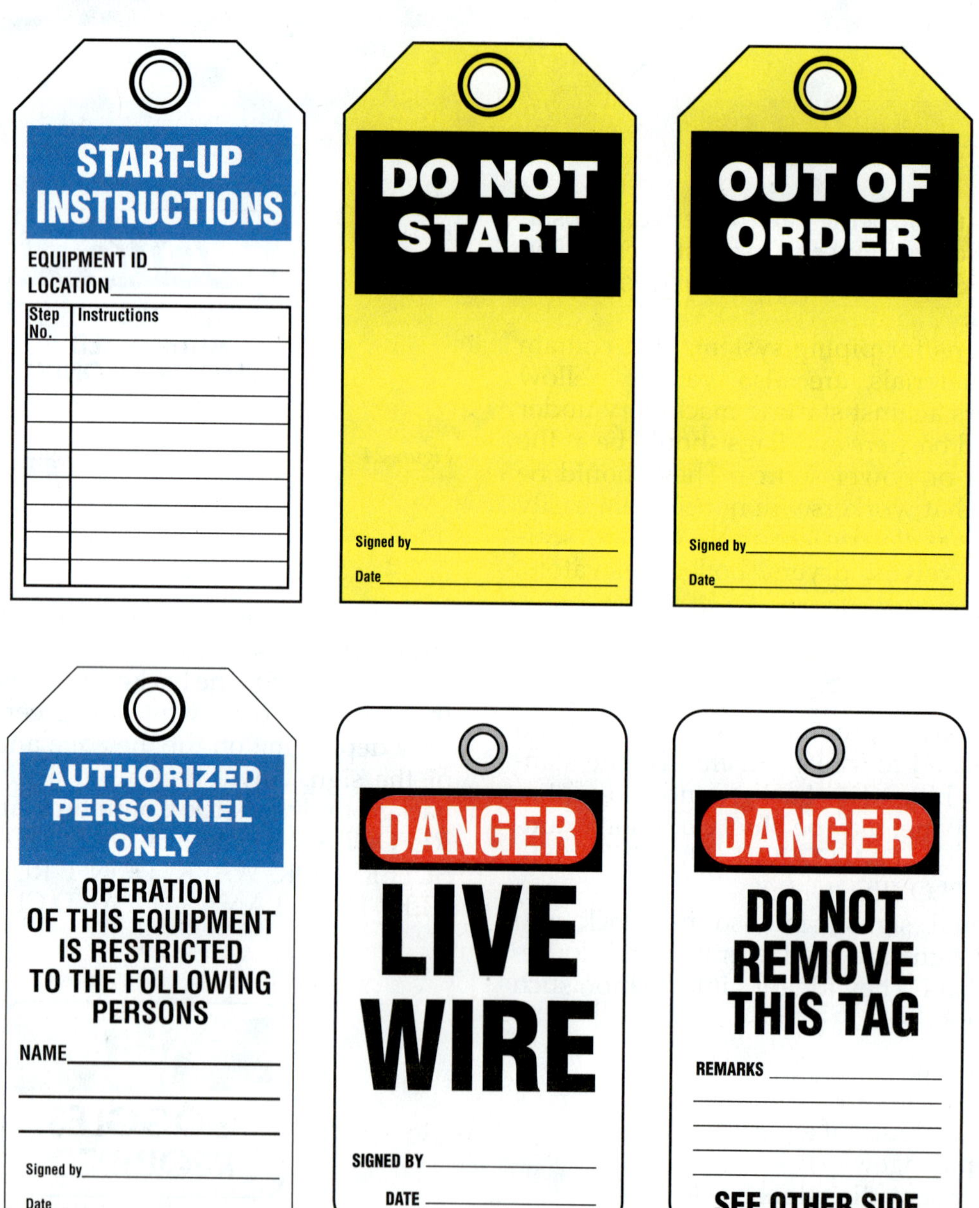

Figure 26 ◆ Examples of safety tags.

6.3.0 Barricades and Barriers

Any opening in a wall, floor, or the ground is a safety hazard. There are two types of protection for these openings: guarded or covered. Cover any hole whenever possible. When it is not practical to cover a hole, use barricades. If the bottom edge of a wall opening is less than 3 feet above the floor and would allow someone to fall 4 feet or more, place **guards** around the opening. For instance, you would place guards around a window opening several stories up to prevent a worker from sitting on the opening and falling out the window. The two most commonly used guard methods are railings and warning barricades. Typical warning barricades are made of plastic tape or rope strung from wire or between posts. The tape or rope is color coded red, yellow, or yellow and purple.

Red means danger. No one may enter an area with a red warning barricade. A red barricade is used when there is danger from falling objects or when a load is suspended over an area.

Yellow means caution. You may enter an area with a yellow barricade, but you must know what the hazard is and be careful. Yellow barricades are used around wet areas or areas containing loose dust. Yellow with black lettering warns of physical hazards, such as bumping into something, stumbling, or falling.

Yellow and purple indicates a radiation warning. No one may pass a yellow and purple barricade. These barricades are often used where piping welds are being X-rayed.

Protective barricades provide both a visual warning and protection from injury. They can be posts, rails, chain, or cable. People cannot get past protective barricades. Blinking lights are placed on barricades so they can be seen at night.

6.4.0 Walking and Working Surfaces

Slips, trips, and falls on walking and working surfaces cause 15 percent of all accidental deaths in the construction industry. Some accidents occur because of environmental conditions, such as snow, ice, or wet surfaces. Others happen because of poor housekeeping and careless behavior, such as leaving tools, materials, and equipment out and unattended. Such accidents can be avoided if workers are aware of their surroundings and follow the rules on the site. It is your responsibility to keep all walking and working surfaces clean and dry.

Walking and working surfaces vary depending on the job. Some workers are exposed to openings in the floors and walls, while others experience icy or wet walking and working surfaces. Ice is dangerous because it's often hard to see. Some common examples of the areas where workers may encounter hazards include the following:

- Floors
- Walls
- Platforms
- Ramps and runways
- Stairs
- Ladders and scaffolding

Always use caution when walking or working on or near these surfaces, especially if there are hazards in the area.

6.4.1 Floors

Floors can be a hazard in a number of ways. Ice, grease, oil, or wet processes can make them slippery. Tools, equipment, materials, or litter can clutter them. Unguarded openings in the floor or ground can cause fatal falls. The following are some of the types of openings to be aware of on a work site:

- Stairs
- Hatches
- Chutes
- Trapdoors
- Manholes

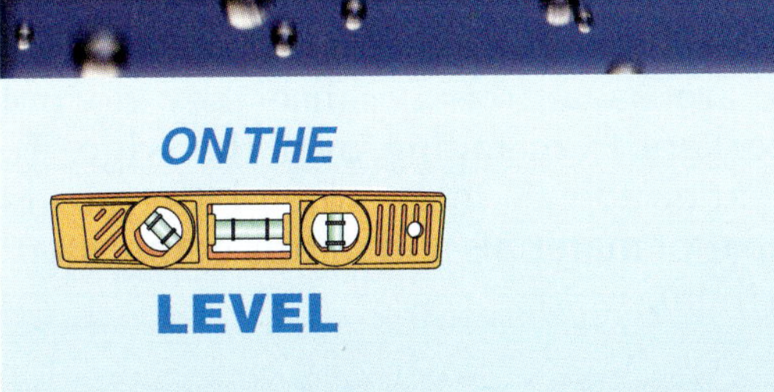

A 21-year-old laborer had been throwing old roofing materials off a roof with six unguarded skylights. During a work break, he sat down on one of the skylights, which began to break under his weight. He tried to raise himself from the skylight with his arms, but the plastic dome collapsed. He was killed when he fell 27 feet to the concrete floor below. The victim had been warned by his supervisor and co-workers not to sit on the skylights.

The Bottom Line: Ignoring safety warnings can be fatal.

Source: National Institute of Occupational Safety and Health

You can avoid slips, trips, and falls on floors by making sure the surface is free of ice, snow, moisture, and clutter. If you cannot remove the ice, wear shoes with skid-resistant cleats. If wet processes are used in the work area, make sure there is proper drainage, grating, and mats. Keep the floor clear of tools, equipment, materials, or litter that you could trip over or would cause you to slip. Avoid openings in floors unless they are properly guarded or covered.

6.4.2 Walls

Openings in walls, such as windows, doors, and chutes, are generally at eye level and seem easy to avoid. However, wall openings are dangerous when they are not protected by guardrails or fences. Any rain entering the opening may cause a slipping hazard. For example, if you lose your balance and fall near an unguarded wall opening, you could slip through the opening and fall to the ground or a lower work area. Tools or materials that fall through the opening can seriously injure those below. If you are working near a wall opening, make sure it is barricaded and that the work area is dry and free of clutter.

6.4.3 Platforms

Platforms are work areas elevated above the floor or ground. They may be above dangerous equipment, such as galvanizing tanks or degreasing units. Many platforms do not have guardrails. These platforms, called open-sided platforms, are hazardous because they do not protect workers from falling over the edge. To help prevent accidents, make sure platforms are dry and clear of materials and debris before stepping onto them.

6.4.4 Ramps and Runways

The hazards of ramps and runways are similar to those of floors. Workers can trip on tools and equipment or slip on wet or icy surfaces. Ramps and runways can be more dangerous, however, because they are sloped. When slips, trips, and falls happen at a downhill angle, the worker slides or rolls down the ramp or runway more quickly and hits the ground harder. The resulting injuries are often more serious. Imagine if a worker were carrying a tool with a sharp blade or edge during such an accident. It is likely that the worker would be injured from the blade or edge of the tool, as well as the fall. You can avoid slips, trips, and falls on ramps and runways by following these guidelines:

- Check the surface before using it. If the ramp or runway is icy or wet, don't use it until it is dry and free of ice. If the ice cannot be removed, wear shoes with skid-resistant cleats.
- Make sure the ramp is clear of tools, equipment, materials, or debris.
- Make sure any tools or equipment you are carrying are turned off and secured. This will help prevent injuries if you fall.

6.4.5 Stairs

Workers use stairs to travel between levels, in and out of pits, and on and off platforms. Depending on the location of the stairs, they can be wet, icy, or slippery. They can also be damaged or cluttered with tools and equipment. All of these conditions can cause workers to slip, trip, or fall. Don't use stairs if you notice any of these conditions or if the stairs do not have a guardrail.

6.4.6 Ladders

The greatest hazard of using a ladder is falling. Workers can fall from the ladder, or the ladder can slip out from under them. This usually happens when the ground or ladder is wet (see *Figure 27*) or icy, or the ladder is not properly secured. Serious injuries or death can result. These safe practices can help prevent slips and falls from ladders:

- Use appropriate fall protection.
- Wear safe, strong work boots that are in good condition.
- Watch where you step. Be sure your footing is secure.
- Maintain clean, smooth walking and working surfaces. Fill holes, ruts, and cracks.
- Clean up slippery material.

102F27.EPS

Figure 27 ◆ Water on a step ladder.

- Pick up litter.
- If you must climb to reach something, use a sound ladder that has been safely set up and properly secured at the top and bottom (see *Figure 28*).
- When climbing a ladder, always face the ladder and use both hands.
- Always maintain three-point contact with the ladder: one hand and two feet or two hands and one foot.
- Don't overreach from a ladder. Climb down and move the ladder to the desired position.
- Don't walk a ladder.

WARNING!

Metal ladders conduct electricity. Never use metal ladders around electrical equipment. Although wooden ladders will not conduct electricity, you must not use them around electrical equipment if the wood is wet or damp. Even a small amount of water will act as a conductor.

6.4.7 Scaffolding

Scaffolding supports workers and materials on elevated platforms. It is generally used on multistory buildings or structures. Weather conditions, poor housekeeping, and carelessness can make working on scaffolding very dangerous. Always refer to *29 CFR* for detailed instructions before setting up scaffolding. Always pay attention and follow these guidelines to avoid injury and death:

- Use appropriate fall protection.
- Keep scaffolding planks clear of extra tools and materials. Clean up any slippery substances that get spilled on scaffolding.
- Anchor freestanding scaffolding with **guy wires** to prevent tipping or sliding. Guy wires are ropes, chains, cables, or rods attached to something as a brace or guide.
- Read the posted safety rules and regulations for scaffolding use.
- When removing objects and tools from the work area, lower them down with a rope. Never throw or drop them from the scaffolding.
- Make sure all personnel are off the scaffolding before moving it.
- Keep tools and materials back from the edge of the scaffolding.
- Use only approved scaffolding.
- Inspect the scaffolding regularly.
- Install all braces and accessories according to the manufacturer's recommendations.
- Ensure that the scaffolding is plumb and level.
- Install all safety rails according to regulations.

Figure 28 ◆ Properly secured ladder.

- Make sure the wheels are locked before climbing or disassembling rolling scaffolding.
- Do not work on scaffolding during storms or high winds or when the scaffolding is covered with ice or snow.
- Do not work on scaffolding that is wobbly or bouncy or can be pulled down easily; it is not safe. Scaffolding must have a sound structure and include toeboards and handrails. Wood floor planks must be free of knots. Broken members must be repaired immediately.

6.4.8 Roofs

Walking and working on roofs presents two different kinds of risks: falling through an opening, such as a skylight or vent, and falling off the roof. Weather hazards, poor housekeeping, or carelessness can cause both to happen. The following guidelines will help ensure your safety when working on a roof:

- Wear appropriate fall protection, even on shallow-pitch roofs.
- Wear boots or shoes with rubber or crepe soles that are in good condition.
- Rain, frost, and snow are all dangerous because they make a roof slippery. If possible, wait until the roof is dry. Otherwise, wear special roof shoes with skid-resistant cleats in addition to wearing fall protection.
- Brush or sweep the roof periodically to remove any accumulated dirt or debris.
- On pitched roofs, install the necessary roof brackets and toeboards (see *Figure 29*) as soon as possible. They can be removed and repositioned as shingle-type roofing is installed.
- Remove any unused tools, cords, and other loose items from the roof. They can be a serious hazard.
- Be alert to any other potential hazards, such as live power lines.
- Use common sense. Taking chances can lead to injury or death.

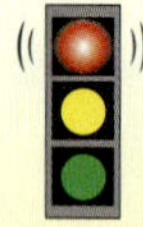

WARNING!

When working 6 feet or more above ground, you must tie off and wear appropriate fall protection. Failure to do so can result in falls that lead to severe injury or even death.

6.5.0 Motorized Vehicles

Your ability to get and keep a job in the plumbing industry depends partly on your ability to maintain an acceptable driving record. According to the Plumbing-Heating-Cooling Contractors–National Association, motor vehicle accidents are

102F29.EPS

Figure 29 ◆ Roof bracket and toeboard.

a top issue for plumbing contractors. Auto liability losses account for 43 percent of all contractor insurance losses and 30 percent of the dollars paid to settle these losses. One of your responsibilities is to operate motorized vehicles safely on construction sites and on the road.

Motorized vehicles include trucks, vans, forklifts, backhoes, and cranes. Motorized equipment includes portable equipment, such as generators, compressors, and pumps, and larger equipment, such as earth-moving equipment and man lifts. A piece of motorized equipment is considered portable if it can be transported from job site to job site or to different areas of the same job site. Whether you are operating one of these vehicles on the construction site, working nearby, or driving a vehicle on the roadways, you must adhere to the safety precautions outlined in this section.

6.5.1 Personal Safety

Many motor vehicle accidents can be prevented. It is your responsibility to be aware of the safety measures you should take to protect yourself and your co-workers. Following are some personal safety guidelines:

- Do not work if you are taking a prescription medicine that could impair your motor or thinking skills.
- Never drink alcohol prior to work or during work hours. Never drink alcohol before or when driving.
- Never, under any circumstances, use illegal drugs.
- Buckle up. Always wear your safety belt and ensure that any passengers wear theirs.
- Ensure that all passengers are seated in a firmly secured seat. Do not allow anyone to ride in a truck bed.

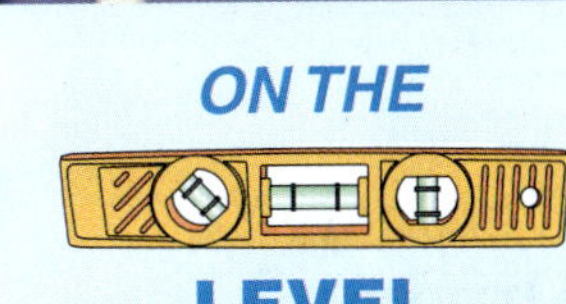

Reckless Driving

Reckless driving is a common cause of accidents. It is also one of the most preventable. The following are some examples of reckless driving:

- Driving too fast for conditions
- Not looking in the direction of travel
- Stopping, starting, or turning suddenly
  To avoid these and other hazards, always observe the following rules:
- Adjust speed for driving conditions. Slow down on wet, bumpy, or slippery surfaces.
- Drive with the load as low as possible and tilted back.
- Look in the direction of travel; this includes backing up.
- Make smooth, deliberate starts, stops, and turns. Sudden movements can cause load shifts or vehicle instability.
- Watch for people and other vehicles.

6.5.2 Vehicle Safety

Just as you routinely check your PPE, you should also routinely check any vehicle you are responsible for operating. Use the following guidelines to ensure vehicle safety:

- Ensure that safety devices such as the horn, backup alarm, mirrors, and brakes are in good working order. (Brakes include trailer brake connections, parking system or hand brakes, and emergency brakes.)
- Keep windshields, side windows, mirrors, and lights clean and functional. Keep windshield washer wells supplied with cleaning fluid.
- Always shut off the engine before fueling.
- Never run motorized vehicles in an enclosed, unventilated area.
- Ensure that the area behind your vehicle is clear before backing up. Safe backing tips are presented later in this section.
- Carry safety equipment, such as flares and fire extinguishers.
- Tie down or secure truck bed loads, and secure rear gates.
- Never remain in a truck that is being loaded by excavating equipment. Stand clear of the area until the loading is completed and the load is secured.
- Ensure that levers used to control hoisting or dumping are latched when not in use to prevent accidental starting or tripping of the mechanism.
- Stand clear of levers used to control hoisting or dumping while these devices are in operation.
- Always turn off the engine and set the brakes before leaving the vehicle.
- Immediately report any unsafe conditions on vehicles to your supervisor.

6.5.3 Company-Owned Vehicles

When you drive a company-owned vehicle, you must use it only for your assigned tasks. Most companies have a written policy governing the use of company-owned vehicles. Employees who do not follow the policy may be, and should be, disciplined or fired. Although specific rules will vary from one company to another, in general, they cover the following areas:

- If you operate a licensed vehicle owned or controlled by the company, you must maintain a current driver's license as required by federal or state regulations.
- Do not transport non-employee passengers. Do not allow non-employees or unqualified employees to use company vehicles, unless an authorized official of the company has given permission.
- Inspect your company vehicles at the beginning of each workday and keep them clean.
- Obey all traffic laws. All fines are your responsibility. Report any traffic citations to your supervisor in writing. Repeated violations may result in suspension or dismissal.
- Wear your seat belt at all times, and be sure that others in the vehicle wear theirs.
- When you leave a vehicle unattended, remove the keys, set the brakes, roll up the windows, and lock the doors.
- Consumption of alcohol or nonprescription drugs is grounds for immediate dismissal, whether ingested before work or while on the job. If you are taking prescribed medication that may affect your ability to perform your duties safely, you must notify your supervisors when you report to work.
- Tell your safety director or supervisor immediately about any incidents involving damage to company property, property of others, personal injury, or injury to others. Failure to report accidents involving a company vehicle is grounds for termination.
- Do not use radar-detection equipment in any company vehicle.
- Be courteous to other motorists. You and the vehicle are a rolling billboard for your company.

- Be trained in and use good defensive driving techniques while operating company vehicles. Employees should follow all insurance requirements.
- Remember that you are also responsible for all tools and equipment assigned to your company vehicle.
- Be sure your vehicle is equipped with an appropriate fire extinguisher and a first-aid kit.

6.5.4 Safe Backing

You must take extra precautions when backing up construction vehicles. Whenever possible, avoid backing to avoid the increased risks that come with reduced visibility behind any vehicle. Find a parking spot that will allow you to leave the site without backing up. However, many sites have tight or constricted turning areas, and you will not always be able to avoid backing up. In those cases, follow these backing safety guidelines:

- Avoid blocking the rearward inside view with equipment and stock.
- Walk completely around the vehicle first, to check for hazards you may encounter when backing up. Don't forget to look up for overhangs and electrical wires.
- Before backing up, roll down the window and turn off the radio. Check all mirrors and look over both shoulders. Sound the horn twice or activate the backup alarm to warn others. Back up slowly.
- Whenever possible, have a co-worker guide you. The use of a radio can also be helpful when communicating with the guide. The guide should stand at the rear driver's side of the vehicle (if there is room) and use full-motion arm signals, not hand signals. If you lose sight of your guide, stop immediately. Do not start again until your guide is visible.
- When parking in an area from which you will have to back up, place orange traffic cones behind the vehicle to ensure that this area remains clear.

WARNING!

Operating a motorized vehicle in an indoor location without proper ventilation can sicken or even kill you. All motorized vehicles emit carbon monoxide as part of their exhaust. You cannot see, smell, or taste carbon monoxide, and you will not know when it is in the air. Never leave your vehicle running when it is not being used. Do not leave keys in an unattended vehicle. Only use approved vehicles indoors, and do so under the supervision of a competent person.

ON THE

LEVEL

Defensive Driving

Your best protection against a motor vehicle accident is to become a defensive driver. Defensive drivers stay focused on the road, on road conditions, and on the other drivers around them. Here are some tips (in addition to the safety guidelines included in this section) to help you stay focused and be a good defensive driver:

- Avoid long cell phone calls while driving. Even with a hands-free unit, you can become distracted. Pull over to complete your calls.
- Get in the habit of checking your blind spots frequently.
- Avoid eating, drinking, talking on the cell phone, or smoking while driving. Hot burning ashes or spills of hot or cold food or liquids can all distract you momentarily, which is enough time for an accident to happen.
- Clean off the dashboard and seats. Papers, clipboards, empty paper cups, small tools, and other debris on your dashboard and seats will slide around while you are driving and can distract you. If you have to make a sudden stop, those items will become airborne and could cause injuries.
- Know where you are going before you start your vehicle. If you must consult a map, pull over. Do not try to navigate and drive at the same time.
- Become a weather-wise driver. When the weather includes ice, snow, fog, or wet pavements, change your driving pattern to accommodate the conditions. Slow down and be especially watchful of the cars around you. Be prepared to take defensive actions to protect yourself.

Remember, most accidents happen while the vehicle is being backed up. Always check your rearview and sideview mirrors, use backup alarms, and be sure that your path is clear.

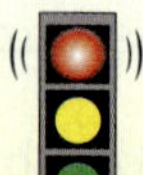

WARNING!

Construction sites can be extremely noisy. Noise on the site may prevent you from hearing a backup alarm. Use your eyes as well as your ears when operating or working around motorized vehicles to prevent being struck by a motorized vehicle, lever, crane, or load.

Review Questions

Section 6.0.0

Match the type of sign or tag with the corresponding description.

_____ 1. Danger

_____ 2. Caution

_____ 3. Informational

_____ 4. Safety

a. This sign or tag is used to tell workers about general information not related to safety.

b. This sign or tag is used for general instructions and suggestions related to safety measures.

c. This sign or tag is used to tell workers that an immediate hazard exists and that specific precautions must be observed to avoid an accident.

d. This sign or tag is used to tell workers about potential hazards or unsafe practices.

e. This sign or tag is used to describe pertinent regulatory requirements.

5. _____ are the two types of protection for openings in walls, floors, or the ground.
   a. Guards and covers
   b. Poles and screens
   c. Ropes and planks
   d. Blocks and berms

6. Slips, trips, and falls cause _____ percent of all accidental deaths in the construction industry.
   a. 10
   b. 15
   c. 20
   d. 25

7. When job site surfaces are wet or icy, you should _____.
   a. delay the job until conditions improve
   b. wear shoes with cleats
   c. spread salt or sand
   d. move to another work area

8. Walking and working surfaces that are located aboveground and sometimes over large equipment are called _____.
   a. ceilings
   b. floors
   c. platforms
   d. ramps

9. To remove tools from scaffolding, it is best to _____.
   a. lower them to the ground using a rope
   b. lower the height of the scaffolding to hip level and then remove the tools
   c. use the buddy system, and pass each tool from one person to another
   d. toss them gently into a strong net set up below the scaffold

10. When working on a pitched roof, it is necessary to _____.
   a. build a flat holding pen to store necessary tools and equipment
   b. erect a wooden wind break
   c. install roof brackets and toeboards
   d. mount temporary lightning rods on all gable ends

7.0.0 ◆ HAND AND POWER TOOLS

Tools are used every day on a job site, and it is easy to forget that they can pose serious safety risks. The types of injuries that occur vary. Some are the result of workers tripping on tools, electrical shock, sharp edges, or fire. Others are due to carelessness or lack of training. The following list demonstrates the types of injuries that occur as a result of hand and power tool accidents:

- Burns
- Cuts
- Sprains
- Electrical shock
- Eye injuries
- Hearing loss
- Broken bones

7.1.0 Hand Tools

Hand tools are nonpowered tools that require the worker's strength and force to operate. Hand tools become dangerous when they are misused or improperly maintained. **Bladed tools** and **impact tools** present a great risk of injury because of the force needed to operate them. The best way to reduce hand-tool accidents is to inspect and maintain tools regularly and always wear appropriate PPE.

7.1.1 Bladed Tools

Bladed tools (*Figure 30*), such as saws and utility knives, present the risks of cutting, slicing, snipping, and stabbing. These injuries can be as minor as a small cut or as severe as an amputation or stab wound. To minimize these risks adhere to the following:

- Use bladed tools with the blades and points aimed away from yourself and others.
- Direct bladed tools away from aisles.
- Store bladed tools properly; use the sheath or protective covering if there is one.
- Keep blades sharp and inspect them regularly. Dull blades are hard to use and control and can be far more dangerous than sharp blades.

7.1.2 Impact Tools

Impact tools (*Figure 31*), such as chisels and punches, are meant to be struck by a hammer. The most common types of injuries caused by impact tools are hammer strikes and eye injuries from flying tool fragments. Getting struck by a hammer is

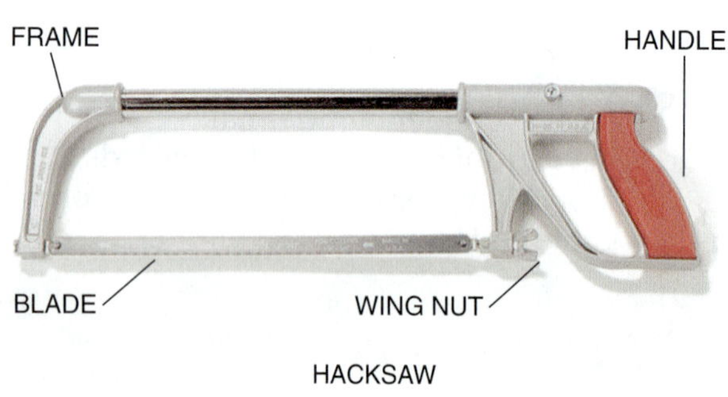

HACKSAW

TUBE CUTTER

UTILITY KNIFE

ALL PURPOSE (PVC) SAW

102F30.EPS

Figure 30 ◆ Bladed tools.

Figure 31 ◆ Impact tools.

usually the result of carelessness or not paying attention. If you are using a hammer, keep your eyes and your mind on what you're doing.

You can prevent injuries caused by tool fragments flying off damaged tools by inspecting the tool before using it. Look for mushroomed heads, cracks, chips, or other signs of damage or weakness. Wear proper eye protection while using impact tools.

7.1.3 Dust and Suspended Particles

Some hand tools create a lot of airborne dust and particles. Sandpaper, planes, files, and saws, for example, may create a lot of sawdust or drywall powder. Short-term exposure to these particles may cause minor irritation of the eyes, nose, or sinuses. In some people, exposure may trigger a more serious reaction, such as an asthma attack. Long-term exposure over the course of a career can affect your lungs, possibly causing lung disorders. To help prevent these conditions, wear filtering masks and eye protection when you are using a tool that creates dust.

7.2.0 Power Tools

Power tools are powered by electricity, pressurized air, or fuel (see *Figure 32*). They can be hazardous when they are improperly used or poorly maintained. Most of the risks associated with hand tools are also associated with power tools. Adding a power source to a tool increases the risk factors. For example, a radial saw arm is far more dangerous than a hand-powered saw.

Power tools are powered by different sources. Power sources for those used in plumbing include electricity and fuel, such as propane gas.

You should know the safety rules and operating procedures and be properly trained for each tool you use. The user's manual supplied by the tool's manufacturer provides specific operating procedures and safety rules. Before operating any power tool for the first time, always read the manual to familiarize yourself with the tool. If the manual is missing, contact the manufacturer for a replacement.

It's important to understand the general safety rules that apply when using all power tools, regardless of type. Use the following safeguards at all times to prevent accidents and injury:

- Never carry a tool by the cord.
- Keep cords away from heat, oil, and sharp edges.
- Always wear appropriate PPE.
- Wear clothes that won't get caught in tools.
- Do not distract others or let anyone distract you.
- Do not engage in horseplay.
- Do not run or throw objects.
- Never leave a power tool running unattended.
- Secure work with clamps or a vise, leaving both hands free to operate the tool.
- Be sure that an electrical power tool is properly grounded and connected to a ground fault circuit interrupter (GFCI) before using it (see *Figure 33*).
- Do not use dull or broken tools or accessories.
- Never use tools with frayed cords.
- Never use a power tool with guards or safety devices removed or disabled.
- Never operate a power tool if your hands or feet are wet.
- Keep the work area clean at all times.
- Keep a firm grip on the power tool at all times.
- Use electric extension cords of sufficient size to service the power tool you are using.
- Report unsafe conditions to your instructor or supervisor.
- Remove damaged tools from use and tag out with DO NOT USE tags.

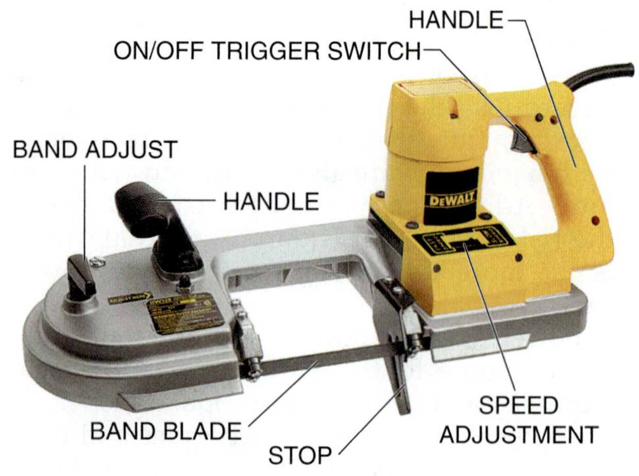

PORTABLE BAND SAW

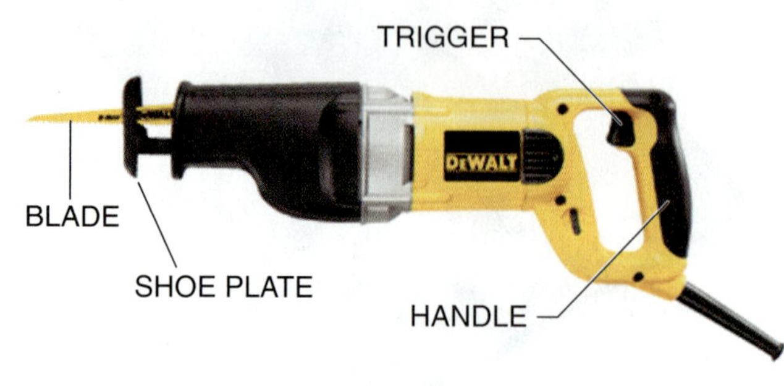

RECIPROCATING SAW

HAMMER DRILL

ROTARY HAMMER DRILL

102F32.EPS

Figure 32 ◆ Power tools.

102F33.EPS

Figure 33 ◆ Ground fault circuit interrupter.

7.3.0 Electrically Powered Tools

The most serious danger in using **electrically powered tools** is electrocution. Electricity can cause burns, shocks, explosions, electrocution, and fires. Electrical shocks can be minor and uncomfortable or they can be severe, causing burns or death. Even a small amount of current can cause the heart to stop pumping in rhythm. If not corrected, this condition will result in death. Electrical shock can also cause a loss of balance, muscle control, or consciousness, which could then cause the victim to fall or drop a tool. A fall from a ladder or scaffolding can be quite serious. To prevent electrical shock, tools must provide at least one of the following types of protection:

- *Double insulated*—Double insulation is more convenient than three-wire cords. The user and tools are protected in two ways: by normal insulation on the wires inside and by a housing that cannot conduct electricity to the user in the event of a malfunction.
- *Powered by a low-voltage isolation transformer*—If your electrically powered tools do not have either a ground plug or double insulation, check with your supervisor to make sure that you are protected by a low-voltage isolation transformer.
- *Grounded with a three-wire cord*—Three-wire cords have two current-carrying conductors and one grounding conductor. You are probably familiar with the three-prong plug common on electrically powered tools. When there is a three-prong cord, that means the tool is powered by a grounded three-wire cord, and it should only be plugged into a three-prong, grounded receptacle. Never remove the third prong (grounding conductor) from a plug. If you are using a three-prong extension cord, make sure that it is properly grounded at its source. The tool being used must be protected by a GFCI or a generator.
- *Twist locks*—Twist locks allow the power source to match an identical voltage rating, using a color-coded scheme. The terminals hold conductors and the brass mounting system provides a low-resistance ground path. Twist locks have a high impact resistance.

You can avoid accidents and injury when using electrically powered tools. Follow these guidelines to protect yourself and your co-workers:

- Use the right tools for the job, and use them the right way.
- Wear all appropriate PPE, such as gloves and safety footwear.
- Store power tools in a dry place when not in use.
- Never use electrically powered tools in damp or wet places.
- Work only in well-lit work areas.
- Consult local codes before working with electrical equipment.

Because malfunctioning electrically powered tools can cause sparks, these tools can cause fires and explosions. Make sure you are aware of any fire hazards in your work area. Avoid using electrically powered tools around flammable materials, fumes, and gases.

Finally, electrical cords and extension cords pose a tripping hazard. Extension cords should be brightly colored to make them more visible. Cords and cables should be run somewhere other than walkways, or at least along a wall, rather than in the middle of a walkway or across a walkway. Avoid running cables and cords across elevated work areas and scaffolding. Occasionally, it may be necessary to run a cord or cable across a walkway. If so, either tape the cord down and put a carpet over it, or place it in a cord runner designed to minimize the tripping hazard. When electrical and extension cords are no longer in use, always hang the cords according to OSHA standards for keeping floors clear. To avoid creating tripping hazards, workspaces, walkways, and similar paths must be clear of cords. Never use worn or frayed cables (see *Figure 34*). Turn in any frayed cords that you encounter to your supervisor.

CAUTION

Do not run a cord through doorways or through holes in ceilings, walls, and floors, which might pinch the cord. Check to see whether there are sharp corners along the cord's path. Any of these situations will lead to cord damage. Extension cords are a tripping hazard. They should never be left unattended and should always be put away when not in use.

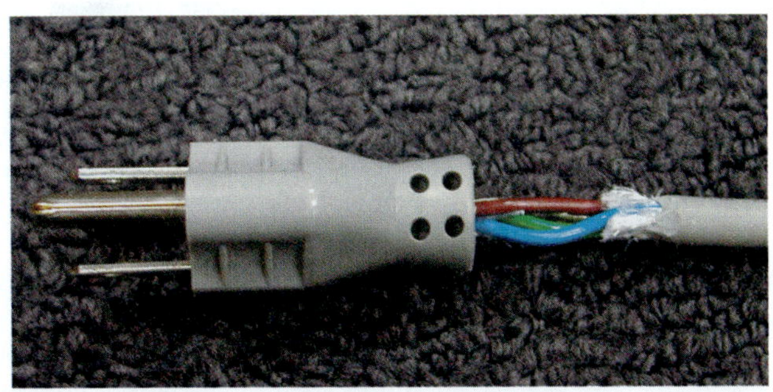

102F34.EPS

Figure 34 ◆ Never use damaged cords.

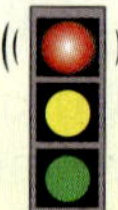

WARNING!

Frayed electrical cords can cause electrocution. Never use a power tool if the cord is frayed or damaged. Never use frayed or damaged extension cords. Follow your company's procedures for removing equipment with damaged or frayed cords from service.

7.4.0 Liquid-Fuel Tools

Some power tools, like torches (see *Figure 35*), are powered by a liquid fuel, such as propane. Whenever a torch is used, a fire extinguisher must be nearby. The most serious hazard with fuel-powered tools comes from fuel vapors that can burn or explode. Burning liquid fuel also gives off exhaust fumes, which can be dangerous. Here are a few tips on using **liquid-fuel tools** safely:

- Always wear the appropriate PPE, including eye protection, gloves, and respirators, if necessary.
- Handle, transport, and store the fuel only in approved flammable-liquid containers.
- Before refilling the tank for a liquid-fuel powered tool, shut down the engine and allow it to cool. This will reduce the risk of a hot tool igniting fuel vapors.
- If you are using a liquid-fueled tool inside a closed area, there must be adequate ventilation and/or respirators in use so you can avoid breathing dangerous exhaust fumes.
- When you are using a liquid-fuel powered tool, make sure fire extinguishers are available nearby.
- When using torches, a fire watch should be maintained.

WARNING!

Propane gas is heavier than air. When there is no ventilation, leaking propane will settle in low places, such as along floors or at the bottom of a trench. You must not strike a match or operate electrical equipment near a propane leak, as this may cause an explosion. Propane will replace oxygen and may cause suffocation in confined spaces.

Figure 35 ◆ Torch kit (with striker) and tank.

ON THE LEVEL

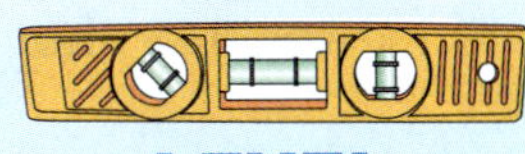

Added Oxygen Causes Death

A welder entered a 24-inch-diameter steel pipe to grind a bad weld on a valve about 30 feet from the entry point. Before he entered, other crew members decided to add oxygen to the pipe near the bad weld to make sure the air was safe. The welder had been grinding off and on for about five minutes when a fire broke out. The fire covered his clothing. He was pulled from the pipe, and the fire was put out. The burns were so serious that the welder died the next day.

The Bottom Line: This accident could have been avoided. It happened because of poor communication among workers and unsafe work practices.

Source: Occupational Safety and Health Administration

Review Questions

Section 7.0.0

1. A bladed hand tool is safest if it is _____.
   a. slightly dull to prevent serious cuts
   b. inspected and maintained regularly
   c. well oiled
   d. kept warm to prevent shattering

2. A _____ blade can be far more dangerous than a sharp blade.
   a. dull
   b. broken
   c. pointed
   d. narrow

3. Chisels, wedges, and punches are designed to be used with a _____.
   a. wedge
   b. spud wrench
   c. hammer
   d. burring tool

4. When inspecting an impact tool, _____ indicates that the tool may be dangerous.
   a. a mushroomed head
   b. a missing guard
   c. a two-prong plug
   d. no insulation

5. Liquid-fuel tools could most likely start a fire when _____.
   a. flammable fuel vapor comes into contact with the hot tool
   b. the power cord shorts out and causes sparks
   c. the operator is using the tool while standing in water
   d. the air hose is not grounded

8.0.0 ◆ TRENCHING

Safety is crucial during any excavation job. Safety precautions must be exercised at all times to prevent injury to yourself or other workers. Always use the most recent edition of the OSHA manual as the governing standards for excavation safety.

Excavations are done for a number of reasons, including laying pipe and locating utility and sewage lines. During an excavation, earth is removed from the ground, creating a trench. A trench is a narrow excavation made below the surface of the ground in which the depth is greater than the width and the width does not exceed 15 feet. The soil that is removed from the ground is called spoil. When soil is removed from the ground, extreme pressures may be generated on the trench walls. If the walls are not properly secured by **shoring**, sloping, or shielding, they will collapse. The collapse of unsupported trench walls can instantly crush and bury workers. This type of collapse happens because not enough material is available to support the walls of an excavation.

DID YOU KNOW?

Federal and state safety regulations have established standards for protecting those who work in excavations. OSHA *29 CFR 1926* defines the trench protective devices that are acceptable. OSHA mandates how the protective devices must be used. In accordance with OSHA regulations, using appropriate manufactured and engineered trench-shielding and -shoring devices or sloping trench walls to angles that eliminate the risk of cave-ins other than in a stable location is required for excavations 5 feet or deeper.

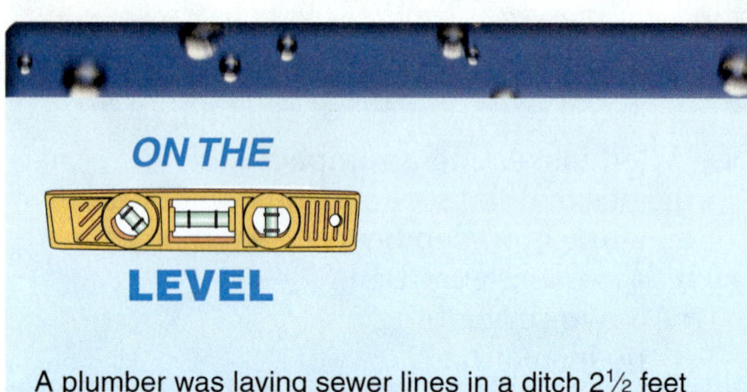

A plumber was laying sewer lines in a ditch 2½ feet wide and 8 feet deep. When he stood up to stretch his legs, the ditch collapsed. Dirt from the inadequately supported ditch buried him past his hat. Luckily, another plumber was nearby to uncover his head quickly. A backhoe was then used to free him.

The plumber was seriously injured. He suffered a skull fracture, broken jaw, damaged hips, and minor brain damage. He recovered from the accident but now has two artificial hips and a neurological disorder.

The Bottom Line: This plumber would not have been injured if he had followed OSHA's requirements for trenching. Be aware of your surroundings and follow OSHA regulations.

8.1.0 Trenching Hazards

Working in and around excavations is one of the most hazardous jobs you will ever do. The design of the excavation may not be your responsibility, but you should be aware of the safety hazards involved in the placement and design of the excavation. You must take safety precautions at all times to prevent injury to yourself and others. Some of the hazards you may encounter during an excavation include the following:

- Surface encumbrances (buildings, vegetation, rocks, or other objects along the surface area of the trench that can hinder operations and block sightlines)
- Underground installations (sewer, telephone, fuel, electric, water lines)
- Flooding from broken water or sewer mains
- Hazardous atmospheres (toxic liquid or gas leaks)
- Cave-ins due to trench failure
- Falls from employees working too close to the trench edge
- Electrical shock from striking electrical cable in the trench or striking overhead lines
- Auto traffic, if the excavation site is near a highway
- Exposure to falling loads
- Collapse of walls or buildings adjacent to the trench
- Loose soil or rock inside the trench
- Exposure to hazardous atmospheres

The type of soil in and around a trench contributes to the collapse of trench walls. Soil type is a major factor to consider in trenching operations. Although you should be aware that soil type plays a role in the safe construction of trenches, only a competent person has enough experience, training, and education to determine whether the soil in and around a trench is safe and stable.

"Get Out of That Trench!"

That's the order an OSHA inspector gave to a plumber in an unshored, unsloped trench. It's a good thing the inspector was paying attention because, 30 seconds after the plumber left the trench, the wall near where he had been standing collapsed. The plumber avoided serious injury or death because he listened to the warning and left the trench.

The Bottom Line: Be aware of your surroundings, and abide by your company's safety procedures.

Source: Occupational Safety and Health Administration

8.2.0 Guidelines for Working in and around a Trench

Working in a trench exposes a worker to many potential hazards. The most obvious hazard is a trench failure. However, there are other situations you should be aware of when working in a trench. You must always use caution when working near the edge of a trench. A trench wall could suddenly give way, sending you to the bottom of the trench.

When you are working in a trench, flooding can be a concern. If a water or sewer main is ruptured while excavating, the trench can quickly fill with water. A little less obvious but just as dangerous are those conditions where a natural water supply has been dammed off. Tons of water may be held back, and if the dam fails, the trench could be quickly flooded.

Electrical shock is also a potential hazard in the trench. While excavating you could accidentally pierce the insulation of an electrical cable in the trench. If overhead electrical lines are near the trench, there is always the danger that the excavator will come into contact with the electrical lines.

When working near highways, auto traffic is a potential hazard. Always check for traffic as you move about the job site. Wear appropriate PPE at all times. Sometimes, trenches can fill with dangerous gases, or oxygen can be displaced from the bottom of the trench. This can happen if the trench is near facilities that store large volumes of chemicals. The liquids or gases could flow or leak into the trench.

Finally, you must be aware of the possibility of injury from objects falling into the trench. Chunks of dirt from the excavator bucket, parts of the bucket, or dirt and stone from the excavation can unexpectedly fall into the trench. Never enter the trench without appropriate PPE.

OSHA requires that a competent person inspect excavations, the adjacent areas, and fall protection and shielding devices daily. The competent person must perform these inspections at the following times:

- Before the start of work
- As needed throughout the shift
- After every rainstorm or other occurrence that increases hazards to workers

When working in or around any excavation or trench, you are responsible for your personal safety. You are also responsible for the safety of others in the trench. The following guidelines must be enforced to ensure everyone's safety:

DID YOU KNOW?

Trenching regulations are now determined by state regulations rather than by OSHA. Consult your local applicable code, and follow the instructions of a competent person.

- Never enter an excavation without the approval of the designated competent person.
- Never enter an excavation until it has been inspected.
- Wear protective clothing and equipment, such as hard hats, safety glasses, work boots, and gloves. Use respirator equipment if necessary.
- Wear traffic warning vests that are marked with or made of reflective or highly visible material if you are exposed to vehicle traffic.
- Get out of the trench immediately if water starts to accumulate in the trench.
- Do not walk under loads being handled by power shovels, derricks, or hoists.
- Stay clear of any vehicle that is being loaded.
- Be alert. Watch and listen for possible dangers.
- Do not work above or below a co-worker on a sloped or benched excavation wall.
- Barricade access to excavations to protect pedestrians and vehicles.
- Check with your supervisor to see if workers entering the excavation need excavation entry permits.
- Make sure someone is on top to watch the walls when you enter a trench.
- Make sure you are never alone in a trench. Two people can cover each other's blind spots.
- Keep tools, equipment, and the excavated dirt at least 2 feet from the edge of the excavation.
- Make sure shoring, trench boxes (see *Figure 36*), **benching,** or sloping are used for excavations and trenches over 5 feet deep.
- Stop work immediately if there is any potential for a cave-in. Do not start work again until the designated competent person completes an inspection and verifies that the area is safe.

Workers need a safe and reliable way of entering and exiting excavations. OSHA refers to this as access (getting into a trench) and egress (getting back out). Several methods are available, including ramps, stairways, and ladders (see *Figure 37*).

When structural ramps are used, a competent person must design them. The ramps must be

102F36.EPS

Figure 36 ◆ Trench box.

102F37.EPS

Figure 37 ◆ Ladder in a trench.

constructed so that each section fits together well to prevent displacement of the soil in the excavation. *Table 3* shows the categories of rock and soil as determined by OSHA. In addition to proper displacement of soil, the surface of ramps must provide enough traction so that workers will not slip. Guardrails on ramps and stairways must be smooth enough to protect workers from punctures, lacerations, and snagging of clothing. Following are additional safety measures for access/ egress:

- Ladders, ramps, or stairways are required in trench excavations that are more than 4 feet deep.
- Ladders, ramps, or stairways used as exits must be located every 25 feet in any trench that is more than 4 feet deep.
- Ladder side rails must extend a minimum of 3 feet above the landing or top of the trench.
- Ladders must have nonconductive side rails if work will be performed near equipment or systems using electricity.
- Two or more ladders must be used where 25 or more workers are working in an excavation in which ladders are the primary means of entry and exit or are used for two-way traffic in and out of the trench.
- All ladders must be inspected before each use for signs of damage or defects.
- Damaged ladders should be labeled DO NOT USE and removed from service until repaired.
- Use ladders only on stable or level surfaces.
- Secure ladders when they are used in any location where they could be displaced by excavation activities or traffic.
- While on a ladder, do not carry any object or load that could cause you to lose your balance.
- Exercise caution whenever using a trench ladder.

8.3.0 Indications of an Unstable Trench

A number of stresses and weaknesses can occur in an open trench or excavation. For example, increases or decreases in moisture content can affect the stability of a trench or excavation. The follow-

Table 3 Determination of Soil Type

Category/Type	Description
Stable rock	Natural solid mineral matter that can be excavated with vertical sides and remain intact while exposed. It is usually identified by a rock name such as granite or sandstone. Determining whether a deposit is of this type may be difficult unless it is known whether cracks exist and whether or not the cracks run into or away from the excavation.
Type A soils	Cohesive soils with an unconfined compressive strength of 1.5 tons per square foot (tsf) (144 kPa) or greater. Examples of Type A cohesive soils are often clay, silty clay, sandy clay, clay loam, and, in some cases, silty clay loam and sandy clay loam. (No soil is Type A if it is fissured; is subject to vibration of any type; has previously been disturbed; is part of a sloped, layered system where the layers dip into the excavation on a slope of four horizontal to one vertical [4H:1V] or greater; or has seeping water.)
Type B soils	Cohesive soils with an unconfined compressive strength greater than 0.5 tsf (48 kPa) but less than 1.5 tsf (144 kPa). Examples of other Type B soils are angular gravel; silt; silt loam; previously disturbed soils unless otherwise classified as Type C; soils that meet the unconfined compressive strength or cementation requirements of Type A soils but are fissured or subject to vibration; dry unstable rock; and layered systems sloping into the trench at a slope less than 4H:1V (only if the material would be classified as a Type B soil).
Type C soils	Cohesive soils with an unconfined compressive strength of 0.5 tsf (48 kPa) or less. Other Type C soils include granular soils such as gravel, sand and loamy sand, submerged soil, soil from which water is freely seeping, and submerged rock that is not stable. Also included in this classification is material in a sloped, layered system where the layers dip into the excavation or have a slope of four horizontal to one vertical (4H:1V) or greater.
Layered geological strata	Where soils are configured in layers (i.e., where a layered geologic structure exists), the soil must be classified on the basis of the soil classification of the weakest soil layer. Each layer may be classified individually if a more stable layer lies below a less stable layer (i.e., where a Type C soil rests on top of stable rock).

Source: Occupational Safety and Health Administration website. "OSHA Technical Manual, Section V, Chapter 2. Excavations: Hazard Recognition in Trenching and Shoring," www.osha.gov/dts/osta/otm/otm_v/otm_v_2.html#4, reviewed February 23, 2004

ing sections discuss some of the more frequent causes of trench failure. These conditions are illustrated in *Figure 38*.

Tension cracks usually occur from one-quarter to half the distance from the top of a trench. Sliding or slipping may occur as a result of tension cracks. In addition to sliding, tension cracks can cause toppling. Toppling occurs when the trench's vertical face shears along the tension crack line and topples into the excavation. An unsupported excavation can create an unbalanced stress in the soil, which in turn causes **subsidence** at the surface and bulging of the vertical face of the trench. If uncorrected, this condition can cause wall failure and trap workers in the trench or greatly stress the **protective system**. Bottom heaving is caused by downward pressure created by the weight of adjoining soil. This pressure causes a bulge in the bottom of the cut. Heaving and squeezing can occur even when shoring and shielding are properly installed.

Another indication of an unstable trench is boiling. Boiling is when water flows upward into the bottom of the cut. A high water table is one cause of boiling. Boiling can happen quickly and can occur even when shoring or trench boxes are used. If boiling starts, leave the trench immediately.

8.4.0 Trench Failure

The most common hazard during an excavation is trench failure or cave-in. Using common sense and following all applicable safety precautions will make the trench a safer place to work.

To understand the seriousness of trench failure, consider what can happen when there is a shift in the earth that surrounds an unsupported trench. Workers could be buried when any of the following events happen:

- One or both edges of the trench cave in
- One or both walls slide in
- One or both walls shear away and collapse

Failure of unsupported trench walls is not the only cause of burial. Tons of dirt can be dumped on workers if the spoil pile or excavated earth slides into the trench. Such slides occur when the pile is placed too close to the edge of the trench or

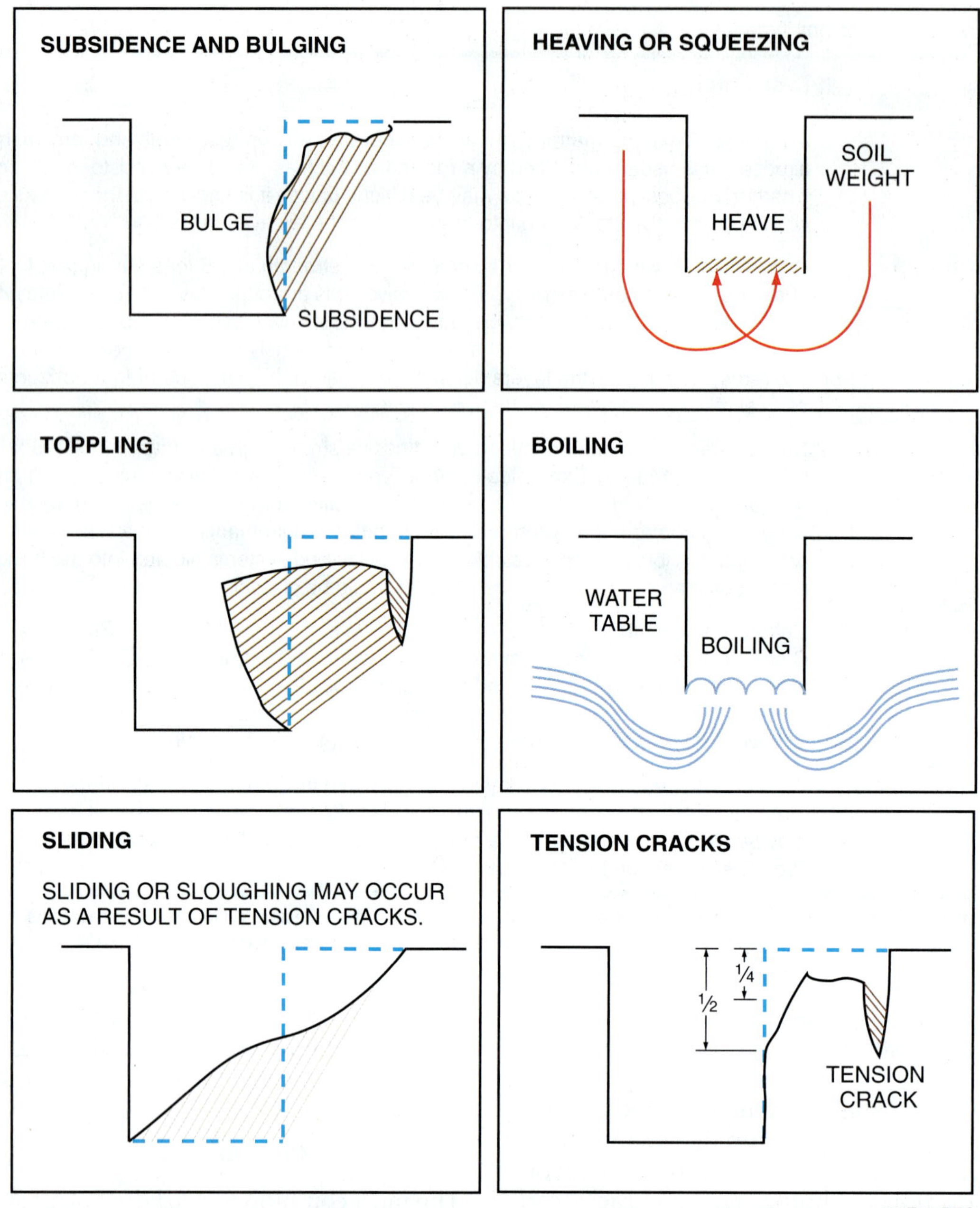

Figure 38 ◆ Indications of an unstable trench.

when the ground beneath the pile gives way. There must be a minimum of 2 feet between the trench wall and the spoil pile. This area must also be kept free of any tools and materials.

The following conditions will likely lead to a trench cave-in. If you notice any of these conditions, immediately inform your supervisor. The conditions are listed in order of seriousness:

- Disturbed soil from previously excavated ground
- Trench intersections where large corners of earth can break away
- A narrow right-of-way, causing heavy equipment to be too close to the edge of the trench
- Vibrations from construction equipment, nearby traffic, or trains
- Increased subsurface water that causes soil to become saturated and therefore unstable
- Drying of exposed trench walls, which causes the natural moisture that binds together soil particles to be lost
- Inclined layers of soil dipping into the trench, causing layers of different types of soil to slide on each other and cause the trench walls to collapse

8.5.0 Making the Trench Safer

There are several ways to make the trench a safer place to work (see *Figure 39*). Trench shoring, shielding, and sloping are used to protect workers and equipment. It is important that you recognize the differences among them.

- *Shoring*—Shoring supports the walls of a trench and prevents their movement and collapse. Shoring does more than provide a safe environment for workers in a trench. Because it restrains the movement of trench walls, shoring also stops the shifting of adjacent soil formations containing buried utilities or on which sidewalks, streets, building foundations, or other structures are built.
- *Trench* **shields**—Trench shields, also called trench boxes, are placed in unshored excavations to protect workers from wall collapse. They provide no support to trench walls or surrounding soil, but for specific depths and soil conditions, they will withstand the side weight of a collapsing trench wall.
- *Sloping*—Sloping an excavation means cutting its walls back at an angle to its floor. OSHA regulations (*CFR 1926*) give companies two options for setting a safe and appropriate angle: Companies may use the tables and charts recommended by OSHA, which are included in *CFR 1926*, or they may use a sloping design that has been developed by a registered, professional engineer.

8.5.1 Fall Protection

Many large excavations require the use of walkways so that workers can cross from one side of the excavation to the other. Where these walkways are 6 feet or more above the ground or lower work levels, OSHA requires the installation of guardrails and, when necessary, safety nets. *CFR 1926.502* outlines the specifications for the type of guardrail systems to be used for various types of projects. In general, guardrails must comply with the following provisions:

- The top rail must be 42 inches (plus or minus 3 inches) above the walking/working level.
- When midrails are used, they must be installed midway between the top rail and the walking/working level.
- When screens and mesh are used, they must extend from the top rail to the walking/working level and along the entire opening between rail supports.
- Balusters between posts must not be more than 19 inches apart.
- The surface of guardrail systems should be smooth to prevent punctures, lacerations, and snagging of clothing.
- Where safety nets are required, the nets must be inspected at least once a week for wear or damage and must meet the requirements set forth in *CFR 1926.502*.

TRENCH BOX

SHORING AND SLOPING

SHORING

102F39.EPS

Figure 39 ◆ Shoring methods.

Review Questions

Section 8.0.0

1. A trench is a narrow excavation in which the depth is greater than the width and the width does *not* exceed _____ feet.
   a. 10
   b. 15
   c. 20
   d. 25

2. Tools and equipment must be kept at least _____ foot/feet from the edge of a trench.
   a. 1
   b. 2
   c. 3
   d. 4

Match the correct protective system to its corresponding description.

_____ **3.** Shoring

_____ **4.** Trench shields

_____ **5.** Sloping

a. Cutting the walls of the excavation back at an angle to its floor

b. Moves the soil from the middle of the trench to support the sides

c. Placed in unshored excavations to protect workers from excavation wall collapse

d. Supports the walls of the excavation and prevents their movement and collapse

9.0.0 ◆ LOCKOUT/TAGOUT

Lockout and **tagout devices** protect workers from all possible sources of energy, including water or other liquids. In a **lockout**, an **energy-isolating device,** such as a stop valve, a disconnect switch, or a circuit breaker, is placed in the OFF position and locked. **Lockout devices** can be used with key or combination locks. Multiple lockout devices (*Figure 40*) are used when more than one person has access to the equipment. In some instances, as with work involving valves, a chain and lock can be used to hold the valve in place and keep it from being turned.

In a tagout, components that allow the flow of water or power equipment and machinery, such as switches or valves, are set in a safe position, and a written warning or tagout device is attached to them (see *Figure 41*). An effective lockout/tagout program should include the following:

- An inspection of equipment by a trained person who is thoroughly familiar with the equipment operation and associated hazards
- Identification and labeling of lockout devices
- The purchase of locks, tags, and blocks
- A standard written operating procedure that all employees follow

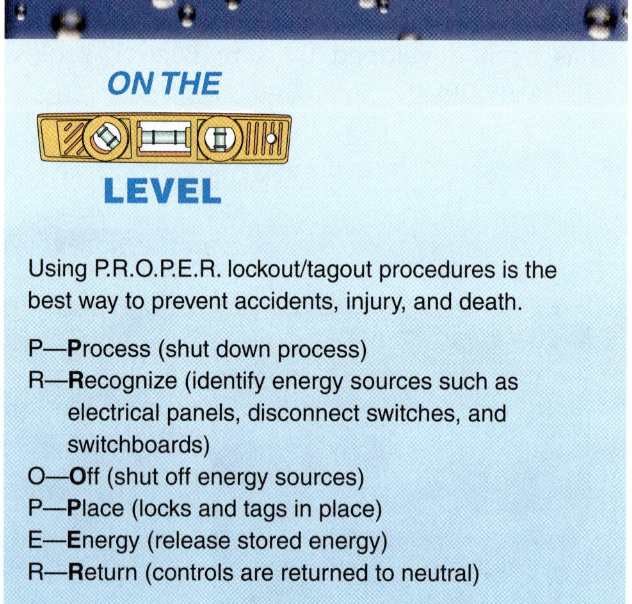

Using P.R.O.P.E.R. lockout/tagout procedures is the best way to prevent accidents, injury, and death.

P—**P**rocess (shut down process)
R—**R**ecognize (identify energy sources such as electrical panels, disconnect switches, and switchboards)
O—**O**ff (shut off energy sources)
P—**P**lace (locks and tags in place)
E—**E**nergy (release stored energy)
R—**R**eturn (controls are returned to neutral)

9.1.0 Lockout/Tagout Procedures

The exact procedures for lockout/tagout may vary at different companies and job sites. Ask your supervisor to explain the lockout/tagout

102F40.EPS

Figure 40 ◆ Multiple lockout device.

procedure on your job site. You must know and follow this procedure. This is for your safety and the safety of your co-workers. If you have questions about lockout/tagout procedures, ask your supervisor. Each foreman should have a kit.

A typical lockout/tagout procedure is made up of these three steps:

- Sequence for lockout/tagout
- Restoring energy
- Emergency removal authorization

9.1.1 Sequence for Lockout/Tagout

A typical lockout/tagout program sets out a sequence of events that must occur during the procedure. A typical lockout/tagout follows these steps:

Step 1 Check the procedures to ensure that no changes have been made since you last used a lockout/tagout.

Step 2 Identify all authorized and affected employees involved with the pending lockout/tagout.

102F41.EPS

Figure 41 ◆ Placing a lockout/tagout device.

Step 3 Notify all authorized and affected personnel that a lockout/tagout is to be used and explain why it is necessary.

Step 4 Shut down the equipment or system using the normal OFF or STOP procedures.

Step 5 Lock out all **energy sources**, and test disconnects to be sure they cannot be moved to the ON position. Next, open the control cutout switch. If there is no cutout switch, block the magnet in the switch OPEN position before working on electrically operated equipment or apparatus such as motors or relays. Remove the control wire.

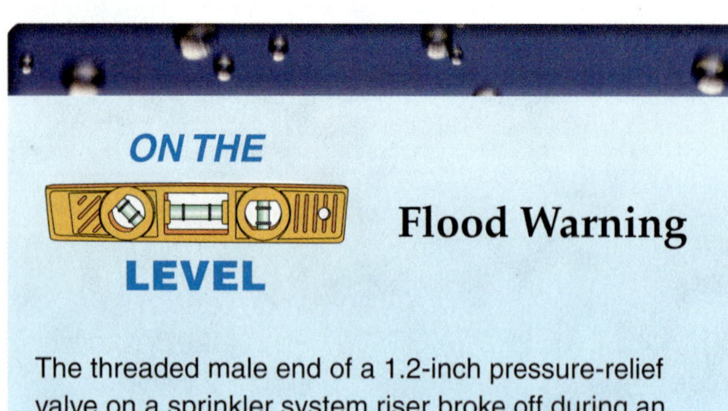

Flood Warning

The threaded male end of a 1.2-inch pressure-relief valve on a sprinkler system riser broke off during an attempt to re-pipe the discharge side of the valve. This caused approximately 550 gallons of water to be discharged into a utility closet and surrounding office areas. There were no injuries as a result of this accident, but there was significant property damage.

The Bottom Line: A lockout/tagout should have been performed on the sprinkler system.

Step 6 Lock and tag the required switches or valves in the OPEN position. Each employee who is authorized to use the equipment has to have a lock.

Step 7 Dissipate any stored energy by attaching the equipment or system to a ground.

Step 8 Verify that the test equipment is functional using a known power source.

Step 9 Confirm that all switches are in the OPEN position, and use test equipment to verify that all components are de-energized.

Step 10 If you have to leave the area temporarily, upon returning retest to ensure that the equipment or system is still de-energized.

9.1.2 Restoration of Energy

Once work is done on the machinery or equipment, energy can be restored. The following steps are typically used to restore energy:

Step 1 Completely reassemble and secure the equipment or system.

Step 2 Confirm that all equipment and tools, including shorting probes, are accounted for and removed from the equipment or system.

Step 3 Replace and reactivate all of the safety controls.

Step 4 Remove the locks and tags from the isolation switches. Each employee must remove his or her own lock and tag. Notify all affected personnel that the lockout/tagout has ended and that the equipment or system will be re-energized.

Step 5 Operate or close the isolation switches to restore energy.

9.1.3 Emergency Removal Authorization

OSHA regulations (*CFR 1019.147*) state that only the employee who applied a lockout/tagout device is authorized to remove it. However, there are times when emergency removal of a lockout/tagout device is required. For example, the employee who applied the device may be absent. Therefore, OSHA allows an exception to this rule, provided the company follows these additional rules:

- Verify that the authorized employee who applied the lockout/tagout device is not at the facility.
- Make a reasonable effort to inform the authorized employee that the lockout/tagout device has been removed.
- Ensure that the authorized employee knows this before resuming work at the facility.

OSHA regulations require employers to have specific procedures and training in place to handle emergency removal. This training and procedures should be part of the company's energy control program. Some companies may have additional requirements. For example, a company may require that a written record of the emergency removal procedure be made and signed by authorized personnel.

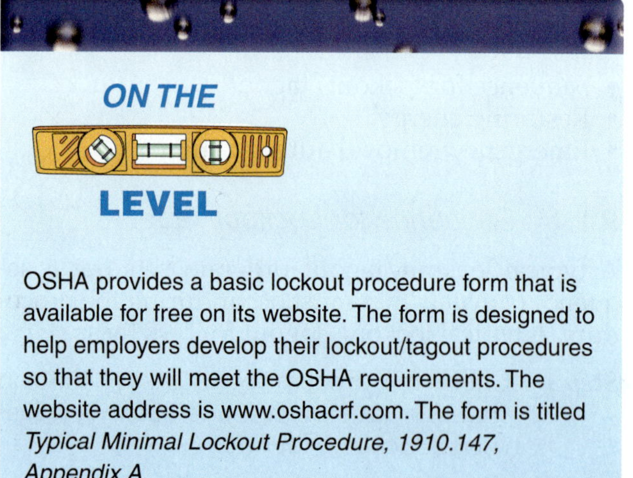

OSHA provides a basic lockout procedure form that is available for free on its website. The form is designed to help employers develop their lockout/tagout procedures so that they will meet the OSHA requirements. The website address is www.oshacrf.com. The form is titled *Typical Minimal Lockout Procedure, 1910.147, Appendix A.*

9.2.0 Safeguards

It is important to know how to protect yourself and your co-workers on the job. Everyone must be aware of the activities taking place on the site and understand how to perform them safely. Some common safeguards to keep the job site safe include the following:

- Never operate any device, valve, switch, or piece of equipment that has a lock or a tag attached to it.
- Use only tags that have been approved for your job site.
- If a device, valve, switch, or piece of equipment is locked out, make sure the proper tag is attached.
- Lock out and tag all electrical systems.
- Lock out and tag pipelines containing acids, explosive fluids, and high-pressure steam.
- Lock and tag motorized vehicles and equipment when they require repair or are being replaced. Disconnect or disable any starting devices.

Review Questions

Section 9.0.0

1. Stop valves, disconnect switches, and circuit breakers are examples of _____ devices.
   a. energy-isolating
   b. energy-removal
   c. lockout/tagout
   d. multiple lockout

2. Lockout and tagout devices are used to _____.
   a. prevent the theft of construction tools and equipment
   b. help keep inventory of construction tools and equipment
   c. protect workers from all possible sources of energy
   d. allow only inspectors to gain access to energy sources

3. When preparing for lockout/tagout, you should _____.
   a. purchase or request any additional lockout/tagout devices that may be required for the procedure
   b. examine your supply of lockout/tagout devices for signs of wear or damage
   c. have a meeting about the upcoming lockout/tagout procedure
   d. check the procedures to ensure that no changes have been made since the last lockout/tagout

4. The second step in the lockout/tagout sequence is to _____.
   a. shut down the equipment
   b. dissipate any stored energy
   c. verify that the test equipment is functional
   d. identify all authorized and affected personnel

5. When locking and tagging motorized vehicles, you should also _____.
   a. drain and purge the gas tank
   b. seal the doors with red tape
   c. remove and store the license plates
   d. disconnect or disable the starting device

10.0.0 ◆ CONFINED SPACES

Spaces on a job site are considered confined when their size and shape restrict the movement of anyone who must enter, work in, and exit the space. Confined spaces often have poor ventilation and are difficult to enter and exit. For example, employees who work in process vessels generally must squeeze in and out through narrow openings and perform their tasks in a cramped or awkward position. In some cases, confinement itself creates a hazard.

Confined spaces are entered for inspection, equipment testing, repair, cleaning, or emergencies. They should be entered only for short periods of time. Some of the confined spaces you may work in include the following:

- Manholes
- Boilers
- Trenches
- Tunnels
- Sewers
- Underground utility vaults
- Pipelines
- Pits
- Air ducts
- Process vessels

In a confined space, hazards such as poor air quality, toxins, explosions, fire, and moving machinery parts tend to be far more deadly than they are in open spaces. Confined spaces may also contain unknown hazards. In one instance, a worker was lowered into a 21-foot-deep manhole on a looped chain seat. Twenty seconds after entering the manhole, he started gasping for air and fell. He landed face down in the water at the bottom of the manhole. An autopsy determined that he died from lack of oxygen.

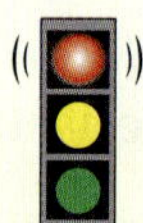

WARNING!

Often more than one worker is killed in manhole accidents. Never enter a manhole to rescue a co-worker. You may also be overcome or killed by toxic gases. To keep safe when working in or around manholes, follow these rules:

- The worker inside the manhole should wear a safety harness so that co-workers can quickly pull him or her out if necessary.
- The worker inside the manhole should whistle, sing, or keep up a steady stream of conversation so that co-workers know that person is all right.
- If the worker inside the manhole falls silent, co-workers should immediately pull that worker out.
- Always test for gases before going into a manhole.

Most confined spaces have restricted entrances and exits. Workers are often injured as they enter or exit through small doors and hatches. It can also be difficult to move around in a confined space, and workers can be struck by moving equipment. Escapes and rescues are much more difficult in confined spaces.

Management typically develops written confined-space entry programs to protect workers.

ON THE LEVEL

The Dangers of Gas in Confined Spaces

At the end of a workday, a worker put away his Presto-Lite acetylene torch without completely shutting the regulator valve. The torch remained sealed in its box, as the gas slowly leaked through the night. The next morning, as the worker lit a cigarette, he also opened the box. When the lid opened, the gas escaped and caught the flame, causing an explosion. Fortunately, the worker was only singed, but he was lucky.

Always check gas valves to ensure that they are completely closed. Failure to do so can result in severe burns or death.

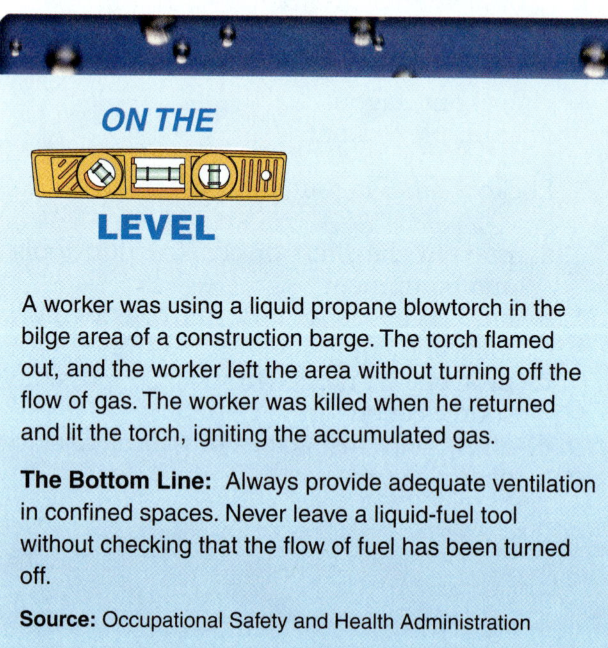

ON THE LEVEL

A worker was using a liquid propane blowtorch in the bilge area of a construction barge. The torch flamed out, and the worker left the area without turning off the flow of gas. The worker was killed when he returned and lit the torch, igniting the accumulated gas.

The Bottom Line: Always provide adequate ventilation in confined spaces. Never leave a liquid-fuel tool without checking that the flow of fuel has been turned off.

Source: Occupational Safety and Health Administration

These programs identify the hazards and specify the equipment or support that is needed to avoid injury. All industrial and some construction sites have written confined-space entry programs. It is your responsibility to know and follow your company's program.

10.1.0 Confined-Spaces Classification

All confined spaces must be inspected before work can begin. Inspection helps to identify possible hazards. After an inspection by a company-authorized person, the confined space is classified based on any hazards that are present. The two classifications are nonpermit required and permit required *(CFR 1910.146).*

10.1.1 Nonpermit-Required Confined Space

A **nonpermit-required confined space** is a workspace free of any mechanical, physical, electrical, or **atmospheric hazards** that can cause death or injury. After a space has been classified as nonpermit required, workers can enter using the appropriate PPE for the type of work to be performed. Always check with your supervisor if you are not sure which PPE is required.

10.1.2 Permit-Required Confined Space

A **permit-required confined space** has actual or possible hazards. These hazards can be atmospheric, physical, electrical, or mechanical. OSHA *CFR 1910.146* defines a permit-required confined space as having one or more of the following characteristics:

- Contains or has the potential to contain a hazardous atmosphere
- Contains a material that has the potential for engulfing (overwhelming) an employee who enters the confined space
- Has an internal configuration such that an employee entering the space could be trapped or asphyxiated by inwardly converging walls or by a floor that slopes downward and tapers to a small cross-section
- Contains any other recognized serious safety or health hazard

The job-site supervisor must issue and sign an entry permit before anyone enters the confined space. No one is allowed to enter a confined space without a valid entry permit. The permit is to be kept at the confined space while work is being done. Always check with your supervisor if you are not sure whether you need a permit to enter a confined space.

10.2.0 Entry Permits

Confined spaces can be extremely dangerous. Entry into the space begins when any part of your body passes the entrance or opening of the space. Before entering a permit-required confined space, you must have an entry permit (*Figure 42*).

An entry permit is a job checklist that verifies that the space has been inspected. It also tells everyone on the site about the hazards of the job. The supervisor must fill out and sign all entry permits before anyone enters the space. The permit must be posted at the entrance to the site and be available for workers to review. Entry permits must include the following information:

- A description of the space and the type of work to be done
- The date the permit is valid and how long it lasts
- Test results of all atmospheric testing, including oxygen, toxin, and flammable material levels
- The name and signature of the person who did the tests
- The name and signature of the entry supervisor
- A list of all workers, including supervisors, who are authorized to enter the site
- The means by which workers and supervisors will communicate with each other
- Special equipment and procedures that are to be used during the job
- Other permits needed for work done in the space, such as welding
- The contact information for the emergency-response rescue team

10.3.0 Hazards

You need to be aware of the hazards that make confined spaces dangerous. The main hazards are atmospheric, **combustible**, and toxic. All of these hazards are made greater by the poor airflow and restricted movement in a confined space.

10.3.1 Atmospheric Hazards

Atmospheric hazards are the most common type of hazard in a confined space. In a hazardous atmosphere, the air can have either too little or too much oxygen, be explosive or flammable, or contain toxic gases. The atmosphere in a confined space must be monitored constantly. Special meters are used to detect atmospheric hazards (see *Figure 43*). These meters must be calibrated and operated according to the manufacturer's instructions. In addition, meters must be able to detect oxygen and combustible gases at the levels specified in OSHA *29 CFR 1910.146.*

A confined space that does not have enough oxygen is called an oxygen-deficient atmosphere. A confined space with too much oxygen is called an **oxygen-enriched atmosphere**. For safe working conditions, the oxygen level in a confined space must range between 19.5 and 23.5 percent by volume, with 21 percent considered the normal level. Oxygen concentrations below 19.5 percent by volume are considered oxygen-deficient; those

Attachment 16 – 2
Confined-Space Entry Permit

Master Card / Safe Work Ticket No. ____________

1. **Work Description:** ____________
   Equip. Name / Number & Location or Area ____________
   Purpose of Entry ____________
   Valid Start Date ____________ Duration Time ________ to ________

2. **Hazardous Materials:**
   What did the equipment last contain? ____________
   Will the work generate a hazardous atmosphere? ☐ Yes ☐ No If yes, specify hazards and controls.
   ____

3. **Rescue Requirements:**
   ☐ External, by attendant ☐ Complex Rescue, by rescue team at point of entry
   ☐ Non-IDLH and/or Simple Rescue, by rescue team on-site ☐ IDLH, by rescue team at point of entry
   Has the rescue team been notified of the entry? ☐ Yes ☐ N/A Time of notification ________
   How will the rescue team be summonsed for an emergency? ☐ Radio Channel: ____ ☐ Other: ________

4. **Gas Test Requirements:**
   LEL/0_2 - Instrument Mfg./No. ______/______ Bump Check Time/Gas Tester - ______/______
   Toxicity - Instrument Mfg./No. ______/______ Bump Check Time/Gas Tester - ______/______
   Frequency of Testing: ☐ Continuous ☐ Other - Specify - ____________
   ∞ Continuous monitoring results must be recorded every three hours

Acceptable Levels	Results								
Oxygen: 19.5% - 23.5%									
Combustible Gas: %LEL - <10%									
Other ________ < PEL* ______									
Other ________ < PEL* ______									
Other ________ < PEL* ______									

∞ Entry in excess of the PEL will require appropriate PPE.

5. **Ventilation / Exhaust Equipment:**
   ☐ None required, natural ventilation adequate ☐ Forced air ventilation ☐ Exhaust ventilation
   Equipment Type: ☐ Air powered horn ☐ Electric blower Volume Required - ________ cfm

6. **Personal Protection:**
   ☐ Gloves (type) ____________ ☐ Respirator (type) ____________
   ☐ Goggles or face shield ☐ Self Contained Breathing Equipment
   ☐ Lifelines Attached to Harness ☐ Other, specify: ____________
   ☐ Chemical Resistant Suit, Specify Type ____________

7. **Fire Protection:** ☐ None required ☐ Portable Fire Extinguisher – type and size: ________
   ☐ Fire Watch ☐ Other, specify: ____________

102F42A.EPS

Figure 42a ◆ Entry permit.

8. Condition of Area and Equipment:

Required Yes	N/A	THESE KEY POINTS MUST BE CHECKED
		a. Equipment locked and tagged out?
		b. Piping is disconnected, capped or plugged and/or blinded.
		c. Equipment emptied, washed, purged & ventilated?
		d. Low voltage or GFCI protected equipment provided?
		e. Explosion proof electrical equipment provided?
		f. Provisions are made to barricade or post signs at entry points when attendant is not on duty.

Other Requirements:

9. Special Instructions: ☐ None ☐ Check with issuer before starting work

10. Approval	Permit			Permit Acceptance		
	Supt. / Area Supv.	Date	Time	Maint. Supv. / Engineer / Contractor Supv.	Date	
Issued by						
Endorsed by						
Endorsed by						

11. Individual Review / Entrant Roster: I have been instructed in the proper Work Permit, Confined Space Entry, Lockout/Tagout Procedures, associated physical and atmospheric hazards and have reviewed the gas testing results.

Entrants	Date	Time In / Out		Time In / Out		Time In / Out		Time In / Out		Time In / Out	

I have been informed of the duties and responsibilities for an attendant, the associated physical and atmospheric hazards, and have reviewed the gas testing results.

Attendants	Date	Time On / Off		Time On / Off		Time On / Off	

12. Job Completion:

☐ Yes ☐ N/A Has the rescue team been notified?
☐ Yes ☐ No Is the work on equipment complete & the confined space ready to return to service?
☐ Yes ☐ No Has the worksite been cleaned and made safe?
Workers answering above questions: ____________________

13. Post Job Review: Were any hazards encountered or created during entry operations?

☐ Yes ☐ No If yes, describe: ____________________
Possible solutions: ____________________

Forward to job file within 7 days of job completion.

102F42B.EPS

Figure 42b ◆ Entry permit.

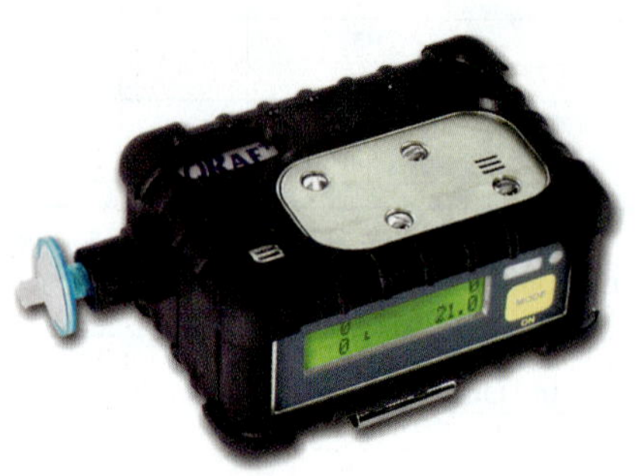

102F43.EPS

Figure 43 ◆ Detection meters.

above 23.5 percent are considered oxygen-enriched. Portable, battery-operated gas meters can be used to measure oxygen levels.

Many of the processes that occur in a confined space use oxygen and may reduce the percentage of oxygen to an unsafe level. These processes include the following:

- Burning
- Rusting of metal
- Breaking down of plants or garbage
- Oxygen mixing with other gases

When the oxygen in a confined space is reduced, it becomes harder to breathe. The symptoms of insufficient oxygen happen in this order:

1. Fast breathing and heartbeat
2. Impaired mental judgment
3. Extreme emotional reaction
4. Unusual fatigue
5. Nausea and vomiting
6. Inability to move your body freely
7. Loss of consciousness
8. Death

Too much oxygen in a confined space is a fire hazard and can cause explosions. Materials like clothing and hair are highly flammable and will burn rapidly in oxygen-enriched atmospheres. Fires can start easily in a confined space with an oxygen-enriched atmosphere.

10.3.2 Combustible Atmospheres

Air in a confined space becomes combustible when chemicals or gases reach a certain concentration. Flammable gases can be trapped in confined spaces. Flammable gases include, but are not limited to, acetylene, butane, propane, and methane. Dust and work by-products from spray painting or welding can also form a combustible atmosphere.

Some flammable gases are lighter than air and concentrate at the top of a confined space. Vapors from fuels are generally heavier than air and will concentrate at the bottom of the space. A spark or flame will cause an explosion in a combustible atmosphere.

10.3.3 Toxic Atmospheres

Toxic gases and vapors come from many sources. They can be deadly when they are inhaled or absorbed through the skin above certain concentration levels. In spaces with no ventilation, high concentrations can gather and quickly become toxic. Even in lower doses, some chemicals can seriously affect your breathing and brain functions.

The effects of toxic gases and vapors vary. Many toxic gases, such as carbon monoxide, cannot be seen or smelled. Some toxic gases have harmful effects that may not show up until years

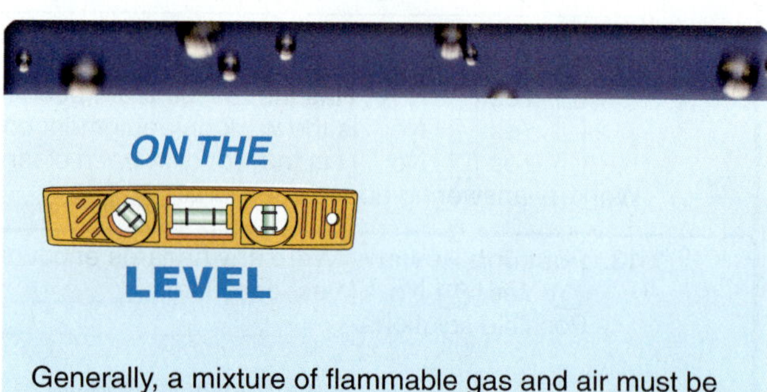

Generally, a mixture of flammable gas and air must be 5 to 15 percent gas and 85 to 95 percent air for the gas to explode. A mixture of gas and air that does not have enough gas to explode is below the lower explosive limit. This mixture is too lean to explode. A mixture with more than the maximum amount of gas is above the upper explosive limit. This mixture is too rich to explode.

after contact. Others, such as nitric oxide, can kill quickly.

Special meters, which are calibrated to detect toxic gases and vapors, will sound an alarm to alert workers of a toxic atmosphere. These meters must be set and operated according to the manufacturer's instructions.

10.4.0 Additional Hazards

In addition to atmospheric hazards, there are several other physical and environmental hazards in confined spaces. Be aware of and take precautions against the following dangers:

- Electric shock
- Purging
- Falling objects
- Engulfment
- Extreme temperatures
- Noise
- Slick or wet surfaces
- Moving parts

10.4.1 Electric Shock

Electric shock can occur when power tools and line cords are used where there are wet floors or surfaces. Always ground tools and equipment, or use a GFCI when working in a confined space.

10.4.2 Purging

Purging happens when toxic, corrosive, or natural gases enter and mix with the air in a confined space. Purging most often occurs in pipelines. Purging gases are used to clean pipelines. These gases can create an oxygen imbalance in the space that will suffocate workers almost immediately. Once purging is complete, appropriate ventilation must be established to render the atmosphere safe. Air monitoring is necessary to verify air quality.

10.4.3 Falling Objects

Materials or equipment can fall into confined spaces and strike workers. This usually happens when another worker enters or exits a confined space with a topside opening, such as a manhole. Vibrations can also cause materials or tools to fall.

10.4.4 Engulfment

Engulfment occurs when a worker is buried alive by a liquid or material that enters a confined space. Small, loose material stored in bins and hoppers, such as grain, sand, or coal, can engulf and suffocate a worker. Heavier materials or liquids that enter a confined space can crush or strangle a worker. Even when a worker is engulfed only up to the waist, death may result from the massive pressure of the materials and constriction of the body.

10.4.5 Extreme Temperatures

Confined spaces that are too hot or too cold can be hazardous. Spaces that are too hot can cause heat stroke or heat exhaustion. Spaces that are too cold can cause **hypothermia**, a condition in which your body temperature is too low. Another temperature-related hazard in a confined space is steam. Steam is extremely hot and can cause severe or fatal burns.

10.4.6 Noise

Noise in a confined space can be very loud. This is because the sound bounces off the walls of a small space. Too much noise, or noise that is too loud, can permanently damage hearing. It can also prevent workers from communicating. If this happens, you could miss an evacuation warning.

10.4.7 Slick or Wet Surfaces

Workers can be seriously hurt by slips and falls on slick or wet surfaces. Wet surfaces also add to the chance of electrocution from electrical circuits, equipment, and power tools.

10.4.8 Moving Parts

Workers can get struck or trapped by moving parts, such as augers or belts. This usually happens when a worker slips or falls. It also can happen if the operator of the moving machinery is wearing loose clothing or jewelry.

10.5.0 Safeguards

It is important to understand how to protect yourself and your co-workers in a confined space. Everyone must be aware of what is happening on the site and understand how to work safely. These are the most common safeguards to follow during confined-space operations:

- Monitoring and testing
- Ventilation
- PPE
- Communications
- Training

10.5.1 Monitoring and Testing

The air in a confined space must be tested before any workers enter it and continuously monitored while workers are in the space. Testing must be done by a properly trained, qualified person, called the confined-space attendant. This person must be company-approved or otherwise designated. The confined-space attendant tests the air for oxygen content, explosive gases or vapors, and toxic gases or vapors.

The atmosphere in a confined space may need to be monitored during the entire job. This is done by attaching monitors to workers entering the confined space or by using outside devices. When the atmosphere is monitored, workers can be assured that the air quality is good and that they will know immediately about changes in the atmosphere that would require them to leave. The confined-space attendant always remains outside the workspace to monitor the workers and their environment. The attendant does not enter the space even to perform rescues but is responsible for calling for evacuation when needed.

Workers who are trained and authorized to monitor the air in a confined space must do the following:

- Read and understand the operating and maintenance manual provided by the monitor's manufacturer.
- Maintain monitors according to the manufacturer's instructions.
- Inspect, test, and calibrate monitors according to the manufacturer's instructions.
- Verify that batteries for battery-operated monitors are working properly.
- Test the atmosphere in a confined space before employees enter the space.
- Continuously monitor the air in the confined space while employees are in the space.
- Carry out the company's safety procedures regarding hazardous atmospheres or emergency situations that may arise.
- Complete any reports required by company policy or OSHA standards (*CFR 1904*).
- Seek retraining or update training as necessary.

10.5.2 Ventilation

If the air in a confined space is hazardous or may become dangerous, the space must be ventilated immediately to remove toxic gases or vapors and replace lost oxygen. Ventilators blow clean air into the space. They must stay on as long as workers are in the space.

Just because air is being blown into or out of the space, it doesn't mean that the space is being ventilated. Toxic gases can hide in confined spaces. Make sure the attendant has carefully tested the entire space before you enter it.

10.5.3 Personal Protective Equipment

Every job requires some type of PPE. Standard PPE includes hard hats, safety goggles and glasses, boots, and gloves. In a confined space, the following items may also be needed:

- Full body harness
- Lifelines
- Air-purifying respirator
- Air-supplying respirator

10.5.4 Communication

All workers on a site must be able to communicate with one another. It is especially important for attendants and workers entering the site to be able to communicate. This allows attendants to warn workers about dangers and order an evacuation when necessary. Communication among workers is another way to monitor the confined space.

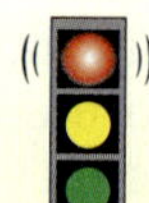

WARNING!

When combustible gases or vapors are present, sparks from communication devices can cause an explosion. Therefore, you must not operate electrical equipment, including telephones or cell phones, that is not classified as intrinsically safe/explosion-proof.

10.5.5 Training

Entering a confined space requires specialized training. No one is allowed to enter a confined space without the proper training and authorization from the site supervisor to do so. Training gives workers the knowledge to complete their jobs safely and efficiently. If you have not been properly trained, do not enter any confined space.

ON THE

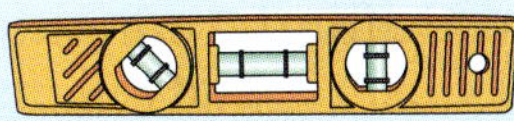

LEVEL

Some tasks require workers to use intrinsically safe/explosion-proof electrical equipment. An intrinsically safe (I-safe) communication device is one that has been certified for use in hazardous environments. This equipment is designed to protect workers from explosions caused by sparks. Such equipment includes certified I-safe phones, mobile computers, and portable data terminals. It must be certified to meet the following requirements of the *National Electrical Code®*:

- Outside air cannot penetrate the casing.
- The casing must be strong enough to contain an explosion inside the equipment without igniting the surrounding air.
- The interior must be filled with nonflammable powder or inert gas to prevent ignition.

Review Questions

Section 10.0.0

1. The two classifications for confined spaces are _____.
 a. nonpermit required and permit required
 b. nonhazardous and hazardous
 c. one worker and two worker
 d. height restricted and width restricted

2. The _____ must issue and sign an entry permit before anyone enters the confined space.
 a. OSHA representative
 b. job-site supervisor
 c. project manager
 d. general contractor

3. Entry into a confined space begins when _____.
 a. the permit is issued
 b. the confined-space attendant arrives
 c. any part of your body passes through the entrance of a confined space
 d. the first worker enters the confined space

4. The most common type of hazard in a confined space is _____.
 a. atmospheric
 b. electrical
 c. temperature
 d. noise

5. For working conditions in a confined space to be considered safe, the oxygen level must range between _____ by volume.
 a. 6.5 and 10.5 percent
 b. 11.5 and 14.5 percent
 c. 19.5 and 23.5 percent
 d. 24.5 and 28.5 percent

11.0.0 ◆ UNDERGROUND SAFETY

Tunnels, shafts, chambers, and passages located underground have their own special safety requirements. Because underground workspaces are completely enclosed except for access and equipment tunnels, the results of an accidental release of gas, fire, and other hazards are magnified accordingly. Plumbers installing lines in underground locations should follow the OSHA requirements outlined in *29 CFR 1926.800* very closely.

Before going underground, ensure that the proper safety precautions are in place both aboveground and in the workspace. Once underground, familiarize yourself with the location of safety equipment, potential hazards, and escape routes. Be prepared, so that in an emergency you can take the correct action immediately to protect yourself and your co-workers from injury and death.

Employers are responsible for ensuring that their employees who work underground receive proper training on underground hazards and safety, including the following elements when appropriate:

- Access
- Air monitoring
- Ventilation
- Illumination
- Communications
- PPE
- Explosives
- Fire prevention and protection
- Emergency procedures

11.1.0 Access

OSHA regulations specify the number and placement of access shafts for personnel and equipment. Hoisting mechanisms are used in shafts to provide access for workers. These hoists must be able to operate safely and effectively in an emergency. They must be able to operate even if the power fails in the rest of the underground workplace. Independent power sources for personnel hoists should meet all safety requirements for the underground workspace.

11.2.0 Air Monitoring and Ventilation

Fresh air must be supplied to underground work areas in sufficient quantities to prevent the build-up of harmful dust, fumes, mist, vapors, or gases. OSHA regulations specify the use of mechanical ventilation in cases where natural ventilation is insufficient to provide fresh air to the workers. Refer to your local applicable code for the minimum safe airflow requirements. Ensure that all machines used in an underground space are properly vented. OSHA regulations require a minimum of 200 cubic feet of fresh air per minute for each employee underground.

The site supervisor is responsible for designating a competent person who will perform air-monitoring tests in an underground work site. The atmosphere in an underground workplace must be the same content and pressure as normal surface air. Air-monitoring tests check for oxygen content and contaminants. If flammable or noxious gases or other contaminants are present at higher-than-permitted levels, OSHA regulations specify the correct response.

Working conditions are classified as **gassy operations** when air monitoring shows one or more of the following:

- The level of methane or explosive gases in the air exceeds minimum acceptable levels.
- A gas ignition has occurred previously in the underground space.
- The underground work area is connected to another area that has been classified as a gassy operation.

Vent all fan and compressor exhausts to ensure that they do not contaminate the underground air. If air monitoring shows that conditions are dangerous to life, evacuate the underground work area, post signs by the entrances, and take all other necessary precautions as specified in OSHA regulations and in your emergency plan.

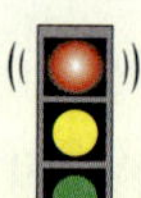

WARNING!

If the levels of methane or other explosive gases reach 5 percent of the lower explosive limit (LEL), increase ventilation to remove the gases. If the levels reach 10 percent of the LEL, immediately suspend all hot work in an underground workspace. If the levels reach 20 percent of the LEL, immediately evacuate all workers who are not required to deal with the situation, and disconnect all non-emergency electrical equipment. Otherwise, a fire or explosion could result, causing injury or death.

11.3.0 Illumination

Refer to your local code and the OSHA regulations for proper lighting requirements. Use only approved lighting equipment when working underground. All employees working underground should have a portable lamp (either handheld or attached to a hardhat) for emergency use, unless there is sufficient natural light or an emergency lighting system is in place.

11.4.0 Communications

Use radios, walkie-talkies, or other approved communications equipment when voice communications are not effective within the workspace or between the workspace and the surface. OSHA regulations require at least two communication methods, one of which must be voice, for use near shafts when hoisting or transferring personnel. Ensure that powered communications systems are able to operate even if one of the units breaks down. Test radios and other communications equipment before going into the underground workplace.

A specially designated person must be on duty on the surface whenever a worker or workers are in the underground workplace. That person is responsible for calling in aid if needed and keeping an accurate count of all employees working underground.

At the end of your shift, notify the incoming shift about any safety occurrences on your shift, such as cave-ins, gas leaks, or equipment failures. Ensure that people on the rest of the job site coordinate their activities with the underground work so as to avoid conflicts.

11.5.0 Personal Protective Equipment

Refer to your local applicable code and site safety regulations to determine the appropriate PPE to be used. Use only NIOSH-approved respirators when working underground. Ensure that all respirators and other safety equipment are properly maintained according to the manufacturer's specifications. Employers are responsible for training all workers in the proper use of PPE. PPE and safety equipment that is not worn, such as respirators, must be located in easily accessed areas near work stations where smoke or gas is a risk.

11.6.0 Explosion and Fire Hazards

Open flames and fires are not permitted underground. Smoking is also prohibited, and all ignition sources, such as matches and lighters, must be collected before workers are allowed to descend underground. Ensure that appropriate warning signs are posted where fire and explosion hazards are present. Never use or store gasoline underground. Diesel fuel can be used, though you must be sure to follow local code and the manufacturer's specifications when pumping diesel fuel to and from an underground space.

When using gas such as acetylene and liquid petroleum gas (LPG), for hot work, refer to your local applicable code and the OSHA regulations for proper storage and handling. When fuels and gases are not in use, ensure that all fuel containers are sealed and gas tank valves are shut tight. On the surface, flammable and explosive materials must be stored at least 100 feet away from underground entrances.

Clearly mark all entrances and ventilation shafts with signs or other approved indicators. When hot work is being performed underground, post fire watches according to OSHA regulations.

11.7.0 Emergency Procedures

Employers must ensure that rescue teams are on site or are available to respond to an emergency within 30 minutes. According to OSHA regulations, when 25 or more people are working underground, two 5-person rescue teams must be available. One team must be on site or no more than 30 minutes away. The other team cannot be more than 2 hours away. With fewer than 25 workers, only one team is required on site or within 30 minutes of the work site.

Emergency team members must be qualified in rescue procedures, in the use of breathing apparatuses, and in firefighting. Employers are responsible for ensuring that rescue teams are familiar with the layout of the work site.

Review Questions

Section 11.0.0

1. Safety requirements for underground work are covered in OSHA regulations outlined in _____.
   a. *29 CFR 800.1926*
   b. *29 CFR 1926.800*
   c. *29 CFR 1910.800*
   d. *29 CF 800.1910*

2. OSHA regulations specify the use of mechanical ventilation when _____.
   a. natural ventilation is insufficient
   b. more than 25 workers are present
   c. rescue teams are not located on site
   d. diesel engines are being used

3. OSHA regulations require a minimum of _____ of fresh air for each employee working underground.
   a. 100 cubic feet per minute
   b. 100 cubic feet per hour
   c. 200 cubic feet per hour
   d. 200 cubic feet per minute

4. Provide at least _____ communication methods near shafts when hoisting or transferring personnel.
   a. two
   b. three
   c. four
   d. five

5. Gasoline should never be used or stored underground.
   a. True
   b. False

12.0.0 ◆ EMERGENCY RESPONSE

You can avoid accidents, injuries, or contamination by working safely and following the rules. When accidents do happen, it is important to stay calm and tell your supervisor immediately. A worker who is trained in first aid should evaluate the injury and decide on the best treatment plan. For example, if the injury is a minor cut, a bandage can be applied. If the cut is deep and requires stitches, the worker should be taken to the hospital. If the cut is severe and an extreme loss of blood or amputation is involved, immediately call 911 or your local emergency-response number.

Many companies, regardless of their size, have an emergency action plan (*Figure 44*). An emergency action plan describes the actions employees should take to ensure their safety in an emergency situation. A company-appointed competent person or emergency action plan coordinator is responsible for creating an emergency action plan. Plans will vary by company. In general, most plans will include some or all of the following items:

- Conditions under which an evacuation would be necessary
- A clear chain of command
- Person authorized to order an evacuation
- Evacuation procedures and emergency escape route assignments (see *Figure 45*)
- Designated meeting area outside of the building
- Procedures for helping visitors to evacuate
- Procedures for helping those with disabilities or difficulty understanding English
- Procedures for workers who remain to operate critical plant operations before evacuating
- Procedures for accounting for workers after the evacuation
- Designated rescue and first-aid personnel
- The location and telephone numbers of local emergency agencies, such as hospitals and fire and police departments. (Generally, calling 911 is the most appropriate response.)

If you are unsure about the action plan at your job site, ask your supervisor to explain it to you. It can mean the difference between life and death.

EMERGENCY ACTION PLAN

In the event of a fire or other emergency situation on the project that warrants emergency response or evacuation, the following procedures will be strictly adhered to:

1. **SOUND THE ALARM** by shouting loudly and repeatedly the description of the emergency situation you have observed, for example:
   - **FIRE, FIRE, FIRE**
   - **EMPLOYEE INJURED, EMPLOYEE INJURED, EMPLOYEE INJURED**
   - **NEED PARAMEDICS, NEED PARAMEDICS, NEED PARAMEDICS**
   - **EVACUATE, EVACUATE, EVACUATE**

   Make sure that at least one other employee has heard and understood the alarm.

2. **CALL 911 AND NOTIFY THE EMERGENCY OPERATOR.** The employee first observing the emergency will travel to the closest telephone or cellular phone, **DIAL 911** and **REPORT THE EMERGENCY.** State the following:

   THERE IS AN EMERGENCY SITUATION AT ______________________________

   IN THE EVENT OF A REQUEST FOR PARAMEDIC RESPONSE, THE EMPLOYEE REPORTING THE EMERGENCY SHALL TRAVEL TO THE PROJECT ENTRY POINT AND STAND BY TO DIRECT EMERGENCY RESPONSE PERSONNEL.

3. In the event of an **INJURED EMPLOYEE ALARM,** personnel trained, capable, and prepared to render first aid may respond to the scene of the accident and render the assistance required to sustain the injured employee, until such time as the paramedics arrive. Assist if and ONLY when your assistance may be rendered without personal risk to you, the first-aid provider, due to hazards present in the accident area.

4. In the event of an uncontrollable fire or other emergency that warrants evacuation, **EVACUATE!** Select the closest, safest route to exit the building and proceed in an orderly and expeditious manner outside. While en route to the exit, assist in the notification of other employees by re-sounding the alarm to evacuate. Yell loudly and repeatedly:

   FIRE, EVACUATE! FIRE, EVACUATE! FIRE, EVACUATE!

5. **ASSEMBLE:** As you exit the building, select the closest and safest route, and proceed in an orderly manner to the DESIGNATED ASSEMBLY AREA, where a head count will be taken to ensure that everyone has safely evacuated. Remain in the assembly area until released to return to work or instructed otherwise.

   DESIGNATED ASSEMBLY AREA FOR THE JOB SITE:

   ____

Note: Each individual subcontractor is responsible for taking head counts for his or her employees and for reporting to emergency response personnel if or when employees are unaccounted for.

102F44.EPS

Figure 44 ◆ Sample Emergency Action Plan.

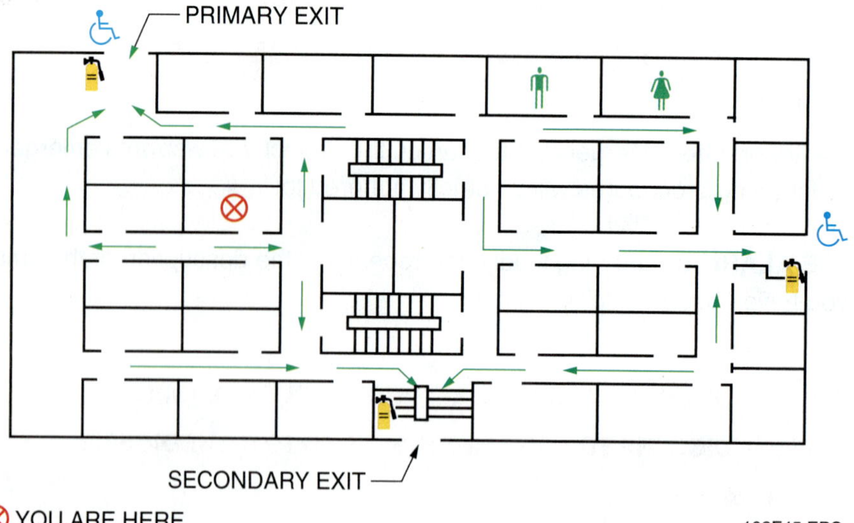

Figure 45 ◆ Typical emergency escape plan.

If you are working in a facility where escape maps are necessary, study the map carefully. This is especially important if you are working in a large building. Memorize the route. It is best, of course, to use an exit nearest your location. However, the closest exit may be blocked by fire or debris. Therefore, always know alternate escape routes so that you have more than one way to evacuate.

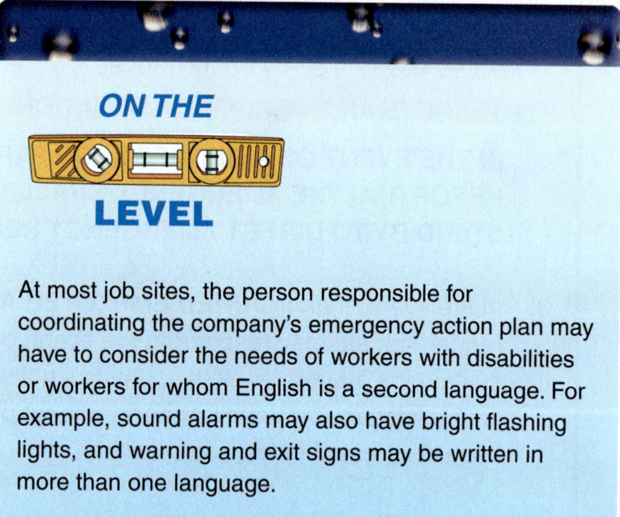

At most job sites, the person responsible for coordinating the company's emergency action plan may have to consider the needs of workers with disabilities or workers for whom English is a second language. For example, sound alarms may also have bright flashing lights, and warning and exit signs may be written in more than one language.

Review Questions

Section 12.0.0

1. When accidents happen at work, it is important to _____.
   a. evacuate the site immediately
   b. find the telephone number for the nearest hospital
   c. stay calm and tell your supervisor immediately
   d. quickly review the company's emergency response plan

2. _____ is responsible for creating an emergency action plan.
   a. A company-appointed competent person or coordinator
   b. The president of the company
   c. The job-site supervisor in meetings with the work team
   d. The human resources director or deputy director

3. If you are unsure about the emergency action plan at your job site, ask your _____ to explain it to you.
   a. OSHA representative
   b. local fire department
   c. co-workers
   d. supervisor

4. When you are evacuating a building using an escape route, you should _____.
   a. run as fast as possible
   b. be prepared to use an alternate route if the main escape route is blocked
   c. first gather up all your personal belongings
   d. quickly recheck your escape route map before evacuating

Summary

Serious accidents and injuries can occur on a work site. Whether you work for a large contractor or you are a subcontractor, you have a responsibility to work safely and follow the rules and regulations of the site. You also have a responsibility to be aware of what's going on around you and to notice the behavior of your co-workers. Personal injury and damage to equipment is less likely to happen when you do so. It is also important that everyone on a job site knows and follows the company's emergency-response plan.

Notes

Trade Terms Quiz

Fill in the blank with the correct trade term that you learned from your study of this module.

1. ______________ is a condition that occurs when body temperature is too low.
2. A(n) ______________ provides information about hazardous chemicals and materials.
3. A(n) ______________ is a structure used in unshored trenches to protect workers from cave-ins.
4. The ______________ is a four-color label placed on containers or doors to alert people to specific safety hazards in a product, room, or building.
5. ______________ operate using a power source.
6. A confined workspace that is free of any mechanical, physical, electrical, or atmospheric hazards is a(n) ______________.
7. When working with or near ______________, or hazardous mold, you must always wear a respirator.
8. ______________ restrains the movement of trench walls.
9. Acetylene, butane, propane, and methane are considered ______________ gases.
10. Torches are considered ______________.
11. An unbalanced stress in the soil causes ______________ at the surface.
12. Electrocution is the most serious hazard of using ______________.
13. ______________ is a method of cave-in protection that differs from shoring and shielding.
14. ______________ are written warnings to protect workers from energy sources.
15. Safety standards are enforced by the ______________.
16. Hazards in a(n) ______________ can include chemical, mechanical, electrical, and physical hazards.
17. Subsidence can cause great stress on a trench's ______________.
18. ______________ tools have sharp edges.
19. When an atmosphere has less than 19.5 percent oxygen, it is considered to be a(n) ______________.
20. Tags are placed on machinery and equipment during a(n) ______________ procedure.
21. A stop valve is considered a(n) ______________.
22. A(n) ______________ has the training, experience, and authority to stop a job if it is not being done safely.
23. A confined space containing too much oxygen is referred to as a(n) ______________.
24. Hammer strikes are a common injury when using ______________.
25. The ______________ requires employers to tell workers about hazardous materials on the job site.
26. ______________ can be used with key or combination locks.
27. Too much or too little oxygen in a confined space creates a(n) ______________.
28. Ventilation is essential if you are working in ______________.
29. ______________ should be locked out and disconnected to make sure they are not powered on.
30. SCBA stands for self-contained breathing ______________.
31. ______________ are also called barricades and barriers.

32. When air monitoring shows that an underground work area contains higher-than-minimum levels of methane or explosive gases, the work area is classified as a(n) _______________.

33. Sound is measured in _______________.

34. Freestanding scaffolding must be anchored with _______________.

35. A stop valve, a disconnect, or a circuit breaker is placed in the OFF position during a(n) _______________.

Trade Terms

Apparatus
Asbestos
Atmospheric hazard
Benching
Bladed tools
Combustible
Competent person
Confined space
Decibels (dB)
Electrically powered tools
Energy-isolating device
Energy source
Gassy operations
Guards
Guy wires
Hazard Communication (HazCom) Standard
Hypothermia
Impact tools
Liquid-fuel tools
Lockout
Lockout device
Lockout/tagout procedure
Material safety data sheet (MSDS)
NFPA warning diamond
Nonpermit-required confined space
Occupational Safety and Health Administration (OSHA)
Oxygen-deficient atmosphere
Oxygen-enriched atmosphere
Permit-required confined space
Power tools
Protective system
Shield
Shoring
Subsidence
Tagout device

Profile in Success

Steve Guy

President
Wat-Kem Mechanical, Inc.

How did you choose a career in the plumbing field?
My grandfather and father were both master plumbers, and I grew up in the trade for as far back as I can remember. I worked during summer vacations from school with my dad. I tried college (chemical engineering) and found that I missed working in the field. After military duty, I went into the apprenticeship program for plumbing and pipefitting.

Tell us about your apprenticeship experience.
I completed a four-year registered plumber and pipefitter apprenticeship program through the UA Local #162. It was a good training program, which prepared me for a career in construction. However, the *Contren*® apprenticeship training programs now rival or surpass the organized trades apprenticeship programs.

What positions have you held in the industry?
Laborer, helper, apprentice, journeyman, foreman, superintendent, project manager, and corporate management. In other words: the full spectrum.

What would you say is the primary factor in achieving success?
Hard work and learning something every day. Training and preparation can make the road to success a shorter journey.

What does your current job involve?
Managing the business operations and managing individual projects.

Do you have any advice for someone just entering the trade?
Start each day with a good attitude and a willingness to learn. Be at work on time and ready to work. Ask questions and learn from your mistakes. Be a part of the solution, not a part of the problem.

Trade Terms Introduced in This Module

Apparatus: One tool, which combines a variety of functions, to perform a particular job.

Asbestos: A fibrous, fire-resistant substance used in pipe insulation, shingles, wallboard, floor coverings, and certain types of insulation. Now banned by government regulation as a health hazard.

Atmospheric hazard: A potential danger in the air or a condition of poor air quality.

Benching: A method of protecting workers from cave-ins by excavating the sides of an excavation to form one or a series of horizontal levels or steps, usually with vertical or near-vertical surfaces between levels.

Bladed tools: Tools that use sharp edges to accomplish their tasks. Bladed tools include saws, knives, scissors, tin snips, and wire cutters.

Combustible: Air or materials that can explode and cause a fire.

Competent person: An individual who is capable of identifying existing and predictable hazards or working conditions that are hazardous, unsanitary, or dangerous to employees, and who has authorization to take prompt, corrective measures to eliminate or control these hazards and conditions.

Confined space: A space that, by design and/or configuration, has limited openings for entry and exit, has unfavorable natural ventilation, may contain or produce hazardous substances, and is not intended for continuous employee occupancy.

Decibels (dB): A measure of sound intensity or loudness. The higher the decibel level, the louder and more potentially damaging the sound is.

Electrically powered tools: Tools that use electrical current to operate.

Energy-isolating device: Any mechanical device that physically prevents the transmission or release of energy. Can include manually operated electrical circuit breakers, disconnect switches, line valves, and blocks.

Energy source: Any source of electrical, mechanical, hydraulic, pneumatic, chemical, thermal, or other energy.

Gassy operations: Working conditions in which one or more of the following conditions exist: higher than minimum levels of methane or explosive gases are present; a gas ignition has previously occurred there; or the area is connected to an underground area designated a gassy operation.

Guards: Devices that protect tool operators from dangerous moving parts, such as blades, gears, and pulleys.

Guy wires: Ropes, chains, cables, or rods attached to something as a brace or guide.

Hazard Communication (HazCom) Standard: A federal OSHA regulation requiring employers to educate and inform workers about chemical hazards on the job site (*29 CFR 1910.1200*).

Hypothermia: A life-threatening condition caused by exposure to very cold temperatures.

Impact tools: Tools that must strike or be struck to accomplish their task. They include hammers, chisels, and taps.

Liquid-fuel tools: Tools that use a liquid fuel, such as gasoline or liquid propane, to operate.

Lockout: The placement of a lockout device on an energy-isolating device, in accordance with an established procedure, ensuring that the energy-isolating device and the equipment being controlled cannot be operated until the lockout device is removed.

Lockout device: Any device that uses positive means such as a lock to hold an energy-isolating device in a safe position, thereby preventing the energizing of machinery or equipment.

Lockout/tagout procedure: A process for identifying hazardous equipment, locking it so that no workers can use it until it is certified for safe use, and placing a tag on the equipment that describes the problem and warns against use.

Material safety data sheet (MSDS): A document that must accompany any hazardous material. The MSDS identifies the material and gives the exposure limits, the physical and chemical characteristics, the kind of hazard it presents, precautions for safe handling and use, and specific control measures.

NFPA warning diamond: A four-color diamond label placed on containers or doors to alert people to specific safety hazards in a product, room, or building.

Nonpermit-required confined space: A confined workspace free of any atmospheric, physical, electrical, and mechanical hazards that can cause injury or death.

Occupational Safety and Health Administration (OSHA): The division of the U.S. Department of Labor mandated to ensure a safe and healthy environment in the workplace.

Oxygen-deficient atmosphere: An atmosphere in which there is not enough oxygen to support life. Usually considered less than 19.5 percent oxygen by volume.

Oxygen-enriched atmosphere: An atmosphere in which there is too much oxygen. Usually considered more than 23.5 percent oxygen by volume.

Permit-required confined space: A confined space that has actual or possible hazards. These hazards can be atmospheric, physical, electrical, or mechanical.

Power tools: Tools that require a power source, such as electricity, hydraulics, or pneumatics, to operate.

Protective system: A method of protecting employees from cave-ins, from material that could fall or roll from an excavation face or into an excavation, or from the collapse of adjacent structures. Protective systems include support systems, sloping and benching systems, and shielding systems.

Shield: A structure that is able to withstand the forces imposed on it by a cave-in and thereby protect employees within the excavation. Shields can be permanent structures or portable and moved along as work progresses. Shields can be either premanufactured or job-built in accordance with *29 CFR 1926.652 (c)(3)* or *(c)(4)*.

Shoring: A structure such as a metal hydraulic, mechanical, or timber system that supports the sides of an excavation and is designed to prevent cave-ins.

Subsidence: A depression in the earth that is caused by unbalanced stresses in the soil surrounding an excavation.

Tagout device: Any prominent warning device, such as a tag and a means of attachment, that can be fastened securely to an energy-isolating device in accordance with an established procedure. The tag indicates that the machine or equipment to which it is attached is not to be operated until the tagout device is removed in accordance with the energy-control procedure.

Additional Resources and References

ADDITIONAL RESOURCES

This module is intended to present thorough resources for task training. The following reference works are suggested for further study. These are optional materials for continued education rather than for task training.

Core Curriculum, 2004. NCCER. Upper Saddle River, NJ: Prentice Hall.

Pipefitter Level 1, 1998. NCCER. Upper Saddle River, NJ: Prentice Hall.

Pipefitter Level 2, 1998. NCCER. Upper Saddle River, NJ: Prentice Hall.

Pipelayer Level 1, 1999. NCCER. Upper Saddle River, NJ: Prentice Hall.

Environmental Protection Agency website: www.epa.gov

Interactive Plumbing Network website: www.plumbnet.com

National Safety Council website: www.nsc.org

OSHA website: www.osha.gov

REFERENCES

Dave Cameron of Federated Insurance, Owatonna, MN, personal communication (background materials on motorized vehicle safety).

Occupational Safety and Health Administration. *Technical Manual, Section V, Chapter 2* (definitions of *competent person* and *confined space*), www.osha.gov/dts/osta/otm/otm_v/otm_v_2.html, reviewed on February 23, 2004.

NCCER CURRICULA — USER UPDATE

NCCER makes every effort to keep its textbooks up-to-date and free of technical errors. We appreciate your help in this process. If you find an error, a typographical mistake, or an inaccuracy in NCCER's curricula, please fill out this form (or a photocopy), or complete the online form at **www.nccer.org/olf**. Be sure to include the exact module ID number, page number, a detailed description, and your recommended correction. Your input will be brought to the attention of the Authoring Team. Thank you for your assistance.

Instructors – If you have an idea for improving this textbook, or have found that additional materials were necessary to teach this module effectively, please let us know so that we may present your suggestions to the Authoring Team.

NCCER Product Development and Revision

13614 Progress Blvd., Alachua, FL 32615

Email: curriculum@nccer.org
Online: www.nccer.org/olf

❑ Trainee Guide ❑ AIG ❑ Exam ❑ PowerPoints Other ____________________

Craft / Level: Copyright Date:

Module ID Number / Title:

Section Number(s):

Description:

Recommended Correction:

Your Name:

Address:

Email: Phone:

Plumbing Tools
02103-05
RIDGID

02103-05
Plumbing Tools

Topics to be presented in this module include:

Overview

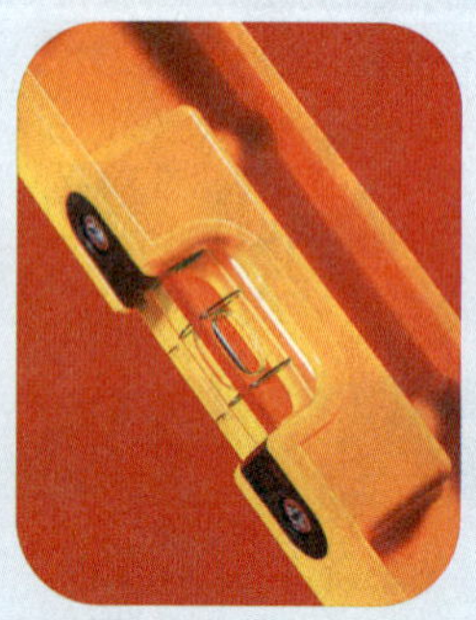

A plumber's tools are a plumber's livelihood. Learning to use them safely and maintain them properly is essential to the trade. As plumbers advance in their trade, it is important to invest in quality tools that pertain to the work they perform and skills they develop. Apprenticeship training teaches trainees how to properly use hand and power tools with expertise and precision.

Precise measurements are fundamental to plumbing, and plumbers must learn to accurately use tools for measuring, laying, and leveling. Structural changes are often required before installing plumbing materials and fixtures. To complete these changes, plumbers use instruments for cutting, drilling, and soldering. Tooth-edged cutting tools, smooth-edged cutting tools, and drilling tools require accuracy and safety to avoid serious injuries. Electrical pipe-threading machines and soldering tools should be used only after proper training and with appropriate supervision.

Plumbers also use the tools that most construction professionals use. Tools for assembly and holding, hammers, and screwdrivers are available in multiple styles for a variety of uses. Each tool is designed for specific tasks and should never be used for anything other than its intended purpose. Expert use of plumbing tools improves the quality of work and enhances plumbers' reputation as professionals.

Focus Statement

The goal of the plumber is to protect the health, safety, and comfort of the nation job by job.

Code Note

Codes vary among jurisdictions. Because of the variations in code, consult the applicable code whenever regulations are in question. Referencing an incorrect set of codes can cause as much trouble as failing to reference codes altogether. Obtain, review, and familiarize yourself with your local adopted code.

Objectives

When you have completed this module, you will be able to do the following:

1. Identify the basic hand and power tools used in the plumbing trade.
2. Demonstrate the proper use of plumbing tools.
3. Demonstrate the ability to know when and how to select the proper tool(s) for tasks.
4. Demonstrate the proper maintenance for caring for hand and power tools.
5. Demonstrate how to prepare a surface for tool use.
6. Describe the safety requirements for using plumbing tools.

Key Trade Terms

Amperage
Bar stock
Burr
Diameter
Electrical ground
Ferrous
Flux
Joist
Keel crayon
Kerf
Level
Miter
Nonferrous
Offset
Plumb
Reaming
Scribe
Soapstone
Solder
Soldering
Spirit level
Straightedge
Temper
Tolerance
Torque
Tube stock
Voltage drop

Required Trainee Materials

1. Appropriate personal protective equipment
2. Pencil and paper
3. Copy of local adopted code

Prerequisites

Before you begin this module, it is recommended that you successfully complete *Core Curriculum; Plumbing Level One,* Modules 02101-05 and 02102-05.

This course map shows all of the modules in the first level of the *Plumbing* curriculum. The suggested training order begins at the bottom and proceeds up. Skill levels increase as you advance on the course map. The local Training Program Sponsor may adjust the training order.

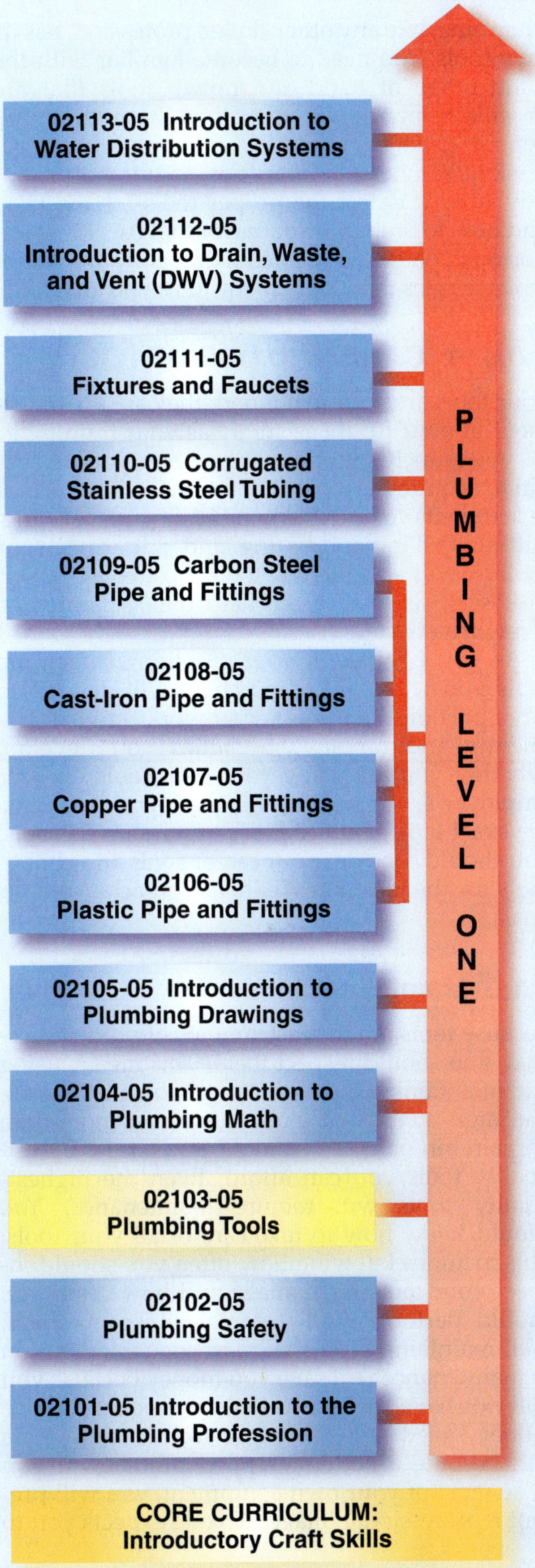

103CMAP.EPS

1.0.0 ◆ INTRODUCTION

Plumbing, like any other skilled profession, has its own tools. You need to become familiar with the wide range of hand and power tools that are specific to plumbing, as well as the many other general-use tools necessary in the plumbing profession. You need to know how to use tools properly, how to select the proper tools for the task, and how to prepare surfaces for tool use. Most important, you must practice all the safety and maintenance rules that apply to the tools you are using.

2.0.0 ◆ CARE AND USE OF TOOLS

Your tools are your livelihood; they are an investment in your future as surely as your training is. As you launch your plumbing career, you will begin to acquire the tools of your profession, such as instruments for measuring and layout, leveling, cutting, drilling and boring, and soldering. Your toolbox will also include tools most construction professionals use: wrenches, pliers, hammers, screwdrivers, and vises.

As you advance in your plumbing career, the tools you will want to purchase will depend in large part on the projects you work on most frequently and on any specialties you develop. Whether you are using your own tools or your employer's, you will want to make sure you have the exact tools you need. When purchasing tools, you should consider your career goals and choose tools for the jobs that will help you achieve your goals.

2.1.0 Maintenance

Because tools are the instruments of your profession, you should respect them. This means using and maintaining them properly. Tools cost money, and high-quality tools obviously cost more than bargain-bin ones. You should purchase the best-quality tools you can afford. Even the highest-quality tools will require maintenance. You should know how to take care of all your tools. This includes knowing how often you should inspect your tools for damage and how often you should perform regular maintenance on them. Well-maintained tools will last longer. Attention to maintenance will save you money because you will not have to replace tools as frequently. The money saved will allow you to buy more tools and become a better-equipped plumber. If you have most of your own equipment, you will present a professional image. No one expects you to be fully equipped just starting out, but you should always be looking to add to your toolbox as your personal finances allow.

You should treat the tools your company lends you with the same care as your own tools. Just as you would never abuse your own drill, you should not abuse your company's drill. Just as you would keep your own cutting tools sharp, you should take good care of the cutting tools your company provides for you. Showing respect for your company's tools and your own demonstrates pride and a good work ethic.

2.2.0 Apprentice Toolbox

There is a saying: "Craftworkers are only as good as their tools." That saying holds as true for the plumbing profession as it does for any other construction-related craft. Tools allow you to perform your job. All the knowledge, training, and skills you acquire will mean little if you cannot do your job because you do not have the right tools.

You are responsible, as a plumbing professional, for selecting and maintaining your tools. You must know how to choose high-quality and durable instruments, and how to take care of them. Preventive maintenance prolongs the life of any tool, and tools in good condition enhance your productivity. Cheap tools break easily and will have to be replaced. You will be better off buying the best tools you can afford now, rather than scrimping on your tool purchases and hoping to get by with inferior products.

The plumbing tasks you take on as an apprentice will require supervision by more experienced workers. This direction will include how to use and care for tools properly and safely. Use this as an opportunity to learn more about your craft. Learning how to use your tools goes hand in hand with learning how to install, repair, and maintain plumbing systems and fixtures. This learning should become habitual. Working with and observing experienced and accomplished plumbers will enhance your own abilities. Someday, perhaps, it will be you teaching the next generation of apprentice plumbers. You might also find yourself working with other construction professionals, such as carpenters, electricians, and technicians. You should familiarize yourself with what they do and which tools they use (often the same ones plumbers use); you will find it useful to know the basics of these associated professions.

As an apprentice plumber, you should have the basic tools shown in *Figure 1* in your toolbox.

Figure 1 ◆ Plumber apprentice toolbox.

3.0.0 ◆ SAFETY

Safe work habits are important to all workers. Plumbing professionals must know how to use tools safely. Many of the tools you use as a plumber can be dangerous if used improperly or if you are unfamiliar with the tool. This is one reason why your apprenticeship training is so important: it teaches you proper use of the tools you will need to do your job.

Safety on the job site protects everyone from injury and possible death. All workers need to follow safety procedures not only for their own sake, but for the sake of their fellow workers. Developing safe work habits is more than just knowing specific dos and don'ts. You should develop an attitude wherein safety is always the foremost consideration. You will see signs posted on job sites that urge workers to "Think Safety." These signs are a constant reminder of the importance of safety to you, your co-workers, and the job. In general, safe operation of tools means careful operation of tools. If you are using a tool safely, you are likely doing a top-quality job as well.

As a plumber, you should follow the same safety precautions in your use of tools as other construction professionals do. That means always using the appropriate personal protective equipment for the tools you use. When using a potentially dangerous tool, you must never skip taking safety preparations in order to save a few minutes or to speed up the work. An injury that results from your failure to take those safety precautions will cost you far more time and money than a few minutes of preparation. As an apprentice plumber, create a work ethic in which safety is of utmost importance.

Listing every possible safety precaution would be impractical. Yet you can start by becoming familiar with and adhering to the basic safety principles outlined in the following sections. You may also want to refer back to *Plumbing Level One, Plumbing Safety.*

3.1.0 Personal Protective Equipment

Using personal protective equipment (PPE) is the first step in ensuring a safe work environment:

- Wear the PPE for the task you are doing, such as a hard hat, non-slip safety shoes, gloves, or safety goggles.
- Keep your PPE clean and in good condition.
- Inspect your PPE often for signs of wear or damage.

3.2.0 Work Habits

Following good work habits is essential to preventing accidents and injuries on the job. Exercise these guidelines as you develop safe and professional work habits:

- Keep your work area neat. Short lengths of pipe and other debris scattered around can cause you or someone else to fall. It also makes the job site look unprofessional.
- Concentrate on your work. Avoid distracting fellow workers. Give your undivided attention to what you are doing. That way, you will not make careless mistakes that could hurt you or your co-workers or that could slow down the job.
- Do not hold or contort your body in such a way that a tool's operation might hurt you. Avoid using hand tools with your wrist bent, if possible. Always have a firm grasp on tools, especially power tools.

3.3.0 Tools

Consider the following safety guidelines when working with hand and power tools.

- Know which tools are available and what each tool has been designed for.
- Use each tool as it is supposed to be used (for instance, do not use wrenches as hammers or screwdrivers as chisels).
- Inspect tools for defects before each use, and repair or replace defective tools.
- Do not apply excessive force or pressure when using tools.
- Do not cut toward yourself when using cutting tools. Always aim the blade away from your body.
- Carry hand tools to and from the work site in a strong, secure toolbox. Never carry a sharp tool or blade in your pockets.
- Do not throw hand tools to co-workers; hand them (handle first, if the tool has a handle) to co-workers.
- Point the tips of sharp tools, such as saws, chisels, and knives, that are lying on benches or tables away from aisles or places where people are walking. Make sure tool handles do not extend over the edge of the workbench or worktable. A job site should not be littered with unattended or unused tools.
- Check electrical tools periodically for damage to power cords. Also, test three-prong electrical plugs with a meter to make sure the safety

ground is properly connected. Three-prong electrical plugs are effective only if properly grounded. A supervisor should oversee any repairs.
- Do not use electrical tools in damp or wet locations, even if the tools are equipped with a safety ground.
- Do not use electrical tools near flammable liquids; one spark is all it takes to ignite these liquids and start a fire.
- Do not take risks when using power tools, such as removing safety guards from equipment.
- Store tools properly after use.

3.4.0 Tool Maintenance

Maintenance and safety go hand in hand. Ensure the safe use of hand and power tools by caring for them and maintaining them properly:

- Check to see that cutting edges are sharp, and keep them sharp by using a flat file followed by a whetstone.
- Keep tools clean and lubricated (as appropriate).
- Prevent tools from rusting by applying light oil to metal parts, especially when the tool is in storage for long periods.
- Replace steel wedges and blades as needed.
- Check to see that tool handles are secure.
- Replace worn jaws on wrenches, pipe tools, and pliers.
- Restore **burred** or mushroomed heads of striking tools.

Never use tools that are in poor condition. This is another vital safety precaution. A well-maintained tool is far less likely to cause an injury than a tool in poor condition.

Safety can be expressed in two words: *common sense*. As you work, use your common sense. Your life might depend on it.

4.0.0 ◆ MEASURING AND LAYOUT TOOLS

A plumber's basic instruments include measuring and layout tools. You need to take precise measurements before installing plumbing piping, systems, and fixtures. Measuring tools allow you to determine the length, width, height, and **diameter** of objects such as pipes. These tools also allow you to determine if a surface is **level** (straight on the horizontal plane) or **plumb** (straight on the vertical plane). Layout tools are used to make accurate lines, circles, and curves.

Accurate measurements are fundamental to plumbing. Inaccurate measurements lead to poorly designed or useless plumbing systems. Getting the right measures at the beginning is essential to productivity in plumbing work. To get the right measures, you need to have the right tools (rules, tape measures, squares) and know how to use them.

4.1.0 Folding Rule

Folding rules are usually made of wood and come in 6-foot and 8-foot lengths (*Figure 2*). They often come equipped with a 6-inch-long brass extension that can be used to take inside measurements between two walls or to measure the depth of an opening. The brass extension allows you to measure in tight places or to measure short distances beyond the end of the folding rule.

You can extend the folding rule above your head to measure heights. This allows one person to take measurements that would otherwise require two people and a ladder. To maintain a folding rule, lubricate the joints lightly with oil or silicone. Be careful when opening or closing the rule because it is easy to break a folding rule at its joints. Try not to drop it, because that might loosen the joints enough to give inaccurate measurements. Small inaccuracies in measuring inches can lead to much larger errors over a span of many feet.

4.2.0 Plumber's Rule

A variation of the wooden folding rule is the plumber's rule (*Figure 3*). It has measurements on one side and a 45-degree scale on the other. You use it to measure **offsets** when you are installing a pipe or vent around an obstacle. Using elbows or bends, an offset changes the direction of a run of pipe (for example, to clear an obstacle) and then returns the pipe to a path that is parallel to its original direction.

103F02.TIF

Figure 2 ◆ Folding rule.

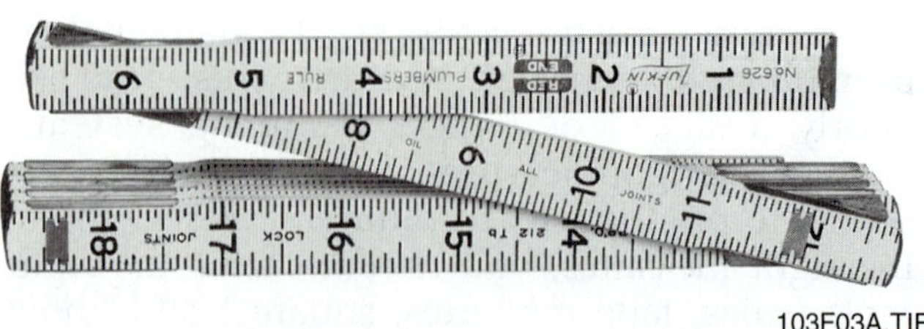

103F03A.TIF

103F03B.EPS

Figure 3 ◆ Plumber's rule with 45-degree scale.

4.3.0 Steel Tape Measure

Steel tape measures (*Figure 4*) are valuable in many professions, including plumbing. Steel tape measures are compact, flexible, and easy to carry. They can clip onto a belt or fit into special leather holders. Retractable steel tape measures are available in lengths up to 30 feet. Plumbers most often use steel tape measures that retract into a casing at the push of a button.

When measuring long pipe runs, a 25-foot, 50-foot, or 100-foot-long steel tape can be used (*Figure 5*). These lengths of tape are wound back into their cases by a hand crank.

Steel tape measures come in various widths. Most plumbers prefer a steel tape that is 1 inch wide, because it is stiffer and easier to work with. You can use it to take longer measurements without having to call on a co-worker to help. A 1-inch

103F04.TIF

Figure 4 ◆ Retractable steel tape measure.

103F05.TIF

Figure 5 ◆ Steel tape measure.

tape is bulkier, but its stiffness lets it extend farther without kinking or bending than, say, a ½-inch tape.

All steel tape measures must be kept clean, dry, and free of kinks. Dirt and sand will wear away the markings on the tape. Water will cause the tape to rust, especially if it gets into the casing that holds the rolled-up tape. Kinks in the tape that are not straightened out will prevent it from rewinding. Wipe away water and dirt before rewinding a steel tape. Apply a light machine oil or penetrating oil periodically to keep the tape lubricated and operating smoothly.

4.4.0 Squares

Straight, accurate lines and angles are necessary for any construction work, including plumbing. A man named Laroy Starrett recognized the importance of straight lines and true angles (usually 90 degrees, with one edge perpendicular to another) and invented the combination square in 1877. Because of its versatility, the square is an essential tool for tradespeople.

Square tools are used to make sure lines are straight and angles are true. Squares are used for marking, testing, and measuring. The type of square to be used depends on the job and the plumber's preference. Square tools are not square-shaped at all; squares that you will use most often are the speed square (shaped like a triangle), the combination square, and the framing square.

4.4.1 Speed Square

The speed square, also known as a rafter angle, magic square, or protractor square, is used for marking and measuring. As shown in *Figure 6*, the speed square is actually shaped like a triangle, but it functions as a square. The speed square is made of cast aluminum. It combines a try **miter**, protractor, and framing square in one tool.

103F06.TIF

Figure 6 ◆ Speed square.

One leg of the triangle has degree gradations marked on its face, allowing for fast and easy layout. The other leg has a raised edge that allows it to fit into the workspace. The legs of a speed square range from 6 to 12 inches in length. The longest side of the triangle, also called the hypotenuse, has degrees marked on it (from 0 to 90) to help in measuring and in marking miter cuts. A speed square with 12-inch-long legs will have a hypotenuse of just less than 17 inches (16.97 inches, to be exact).

Use a speed square to see if a cut or joint is square, to mark cutoff lines, or as a **straightedge,** to see if a surface has been warped or cupped. You can use a speed square as a straightedge because it does not bow, twist, or warp along its length. You can also use it as a cutting guide when using a handheld circular saw. The speed square is small, making it easy to store and carry on the job site. Smaller speed squares, in fact, fit into the pocket of most tool belts.

4.4.2 Combination Square

The combination square (*Figure 7*) is used to establish whether two sides of an object join in a true 90-degree angle. The combination square is adjustable; it has a rigid steel blade 12 inches long with a slot in the middle of it that allows the head to slide along the length of the blade and be fixed to any point along it. The head has both a 45-degree and a 90-degree angle measure. Some combination squares have a small **spirit level** in their handle for leveling and a carbide **scribe** for marking metal.

The combination square is useful in marking and checking 90-degree crosscuts and miter cuts. When the head is set at the end of the rule, the combination square can measure height and can also be adjusted to measure depth. Like the speed square, the combination square can be used as a saw guide. Combination squares are precision tools and are very useful for work that requires extremely accurate measurements. The best combination squares will not lose their accuracy even when used a lot. To maintain combination squares, keep them clean and dry, avoid dropping them, and apply a little machine oil to them periodically.

4.4.3 Framing Square

The framing square (*Figure 8*) is not square-shaped either. This tool is L-shaped, made of flat steel or aluminum, with a 24-inch-long, 2-inch-wide blade that is perpendicular to a 16-inch-long, 1½-inch-wide tongue. The framing square has a variety of uses: marking and laying out patterns, testing for squareness, and measuring. The framing square can also be used as a straightedge for testing the flatness of a surface. Most framing squares are marked in inches and fractions of an inch, down to one-eighth.

4.4.4 Care

Be careful not to drop a square or strike it hard enough to change the angle between the blade and head (of a combination square) or tongue (of a framing square). Keep squares dry to prevent them from rusting. A periodic coat of light machine oil or penetrating oil will make these tools last longer.

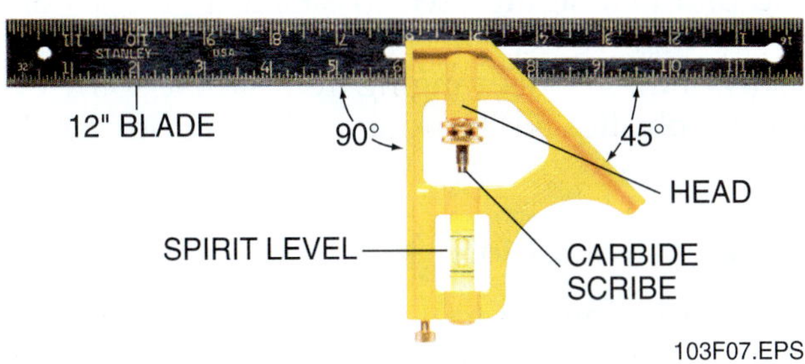

103F07.EPS

Figure 7 ◆ Combination square.

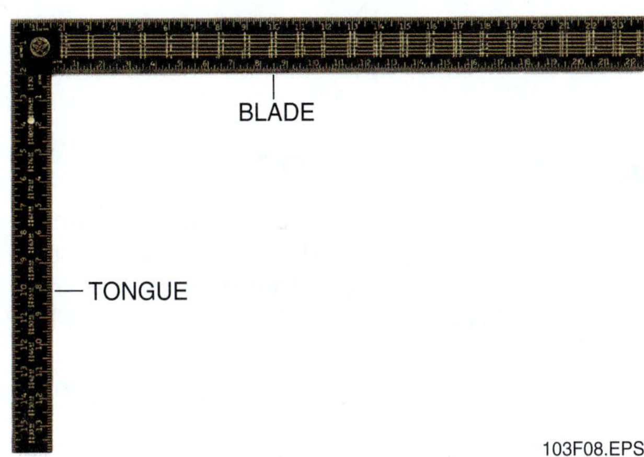

103F08.EPS

Figure 8 ◆ Framing square.

Review Questions

Sections 2.0.0–4.0.0

1. Attention to _____ will save money because tools will not have to be replaced as frequently.
   a. maintenance
   b. brand names
   c. professionalism
   d. work habits

2. Avoid using hand tools with your _____ bent.
   a. waist
   b. wrist
   c. arms
   d. legs

3. Carry hand tools to and from the work site in a _____.
   a. durable, canvas bag
   b. properly gloved hand
   c. heavy, plastic bin
   d. strong, secure toolbox

4. Point the tips of sharp tools that are lying on benches or tables _____ aisles or places where people are walking.
   a. toward
   b. away from
   c. parallel to
   d. facing

5. Prevent tools from rusting by applying _____ to metal parts.
   a. water
   b. grease
   c. light oil
   d. petrol

6. The brass extension on a folding rule _____.
   a. protects the rule from damage when the tool is folded
   b. protects the rule from damage when the tool is extended
   c. must be oiled periodically to prevent rust
   d. allows you to measure in tight places

7. A plumber's rule has _____ markings to measure offsets.
   a. 30-degree
   b. 45-degree
   c. 60-degree
   d. 90-degree

8. Most plumbers prefer working with a 1-inch-wide steel tape because it _____.
   a. is the most economical size for plumbing jobs
   b. can be used to measure long distances without help from a co-worker
   c. is easier to read the numbers on this size tape than on any other size
   d. requires less maintenance than tapes of other sizes

9. A _____ square includes a try miter, protractor, and framing square in one tool.
   a. framing
   b. carpenter's
   c. plumber's
   d. speed

10. The _____ is used to establish whether two sides of an object join in a true 90-degree angle.
    a. combination square
    b. speed square
    c. plumber's rule
    d. folding rule

5.0.0 ◆ LEVELING TOOLS

Squares are useful for measuring small rectangular objects. For larger, fixed structures, however, a square is not large enough to determine true horizontal and vertical lines. This is where leveling tools become useful.

The term *level* describes the straightness of horizontal members. The term *plumb* describes the straightness of vertical members. Level and plumb members intersect at 90-degree angles, and level and plumb tools are used to establish precise horizontal and vertical lines, respectively. There are many of these tools, ranging from simple spirit levels to sophisticated electronic and laser instruments.

5.1.0 Spirit Level

The most common tool used to check both level and plumb is the spirit level (*Figure 9*). Spirit levels are made of wood, aluminum, or a lightweight alloy. Plumbers generally buy aluminum or magnesium spirit levels because they are resistant to moisture.

The essential element of the spirit level is a sealed glass or plastic tube containing a specific amount of alcohol. This tube, or vial, is slightly curved, and the outside of each vial is marked with two parallel lines in its center. This vial is mounted into the body of the level tool. The tube is almost completely filled with the liquid, except for an air bubble, upon which the entire function of the instrument depends. When the frame of the tube is precisely level horizontally or precisely plumb vertically, the air bubble will be centered between the two parallel lines.

Spirit levels generally contain three bubble vials: the vial mounted at the center of the tool is used to establish true horizontal (and verify level), and one vial is mounted transversely at each end for establishing true vertical (and verifying plumb). Vials are replaceable. Spirit levels come in a variety of sizes. The 24-inch-long spirit level is standard, but spirit levels also come in 28-inch, 48-inch, 72-inch, and 78-inch-long sizes.

Spirit levels are accurate only up to their lengths. The longer the level tool is, the greater its accuracy. For accuracy over greater lengths, you must use a longer level. But if you are working in a cramped space, as plumbers often do, a level tool that is too long is useless. These considerations—the length of plumb or level to be measured and the amount of room you will have when taking these measures—should guide your choice of spirit level for a job.

One way to increase the accuracy of a spirit level is with a straightedge. A straightedge is a length of wood or metal that does not bow or twist. Place the straightedge across the span to be checked, and put the spirit level in the center of the straightedge.

5.2.0 Plumber's Level

The engineer's level or plumber's level (*Figure 10*) can measure slope (angle) as well as plumb and level. One of the vials is adjustable to different angles. When you adjust the vial and hold the level at the desired angle, the bubble in the vial is centered. Check one level against another to ensure the reading is correct.

You can check the accuracy of your level by placing the level on your work. Note the position of the bubble. Place the level end to end and check the bubble again. The readings should be identical.

103F09.TIF

Figure 9 ◆ Spirit level.

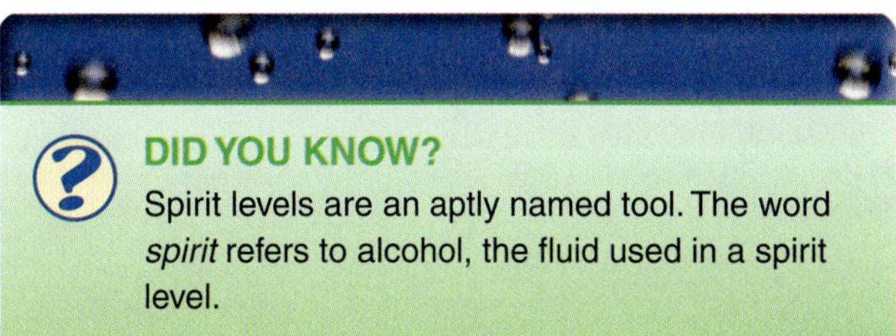

DID YOU KNOW?
Spirit levels are an aptly named tool. The word *spirit* refers to alcohol, the fluid used in a spirit level.

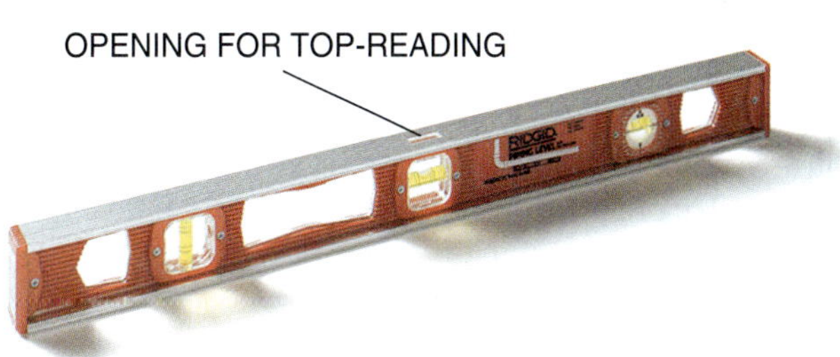

103F10.TIF

Figure 10 ◆ Plumber's level.

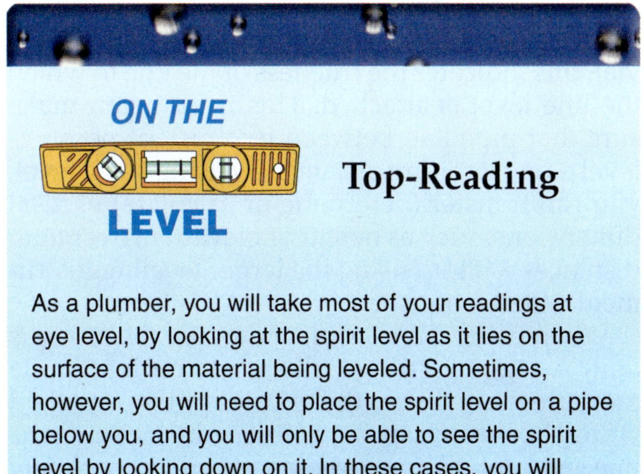

ON THE LEVEL

Top-Reading

As a plumber, you will take most of your readings at eye level, by looking at the spirit level as it lies on the surface of the material being leveled. Sometimes, however, you will need to place the spirit level on a pipe below you, and you will only be able to see the spirit level by looking down on it. In these cases, you will need to use a spirit level that allows the user to top-read. This type of spirit level has an opening on top that allows you to see the vial and the air bubble from above.

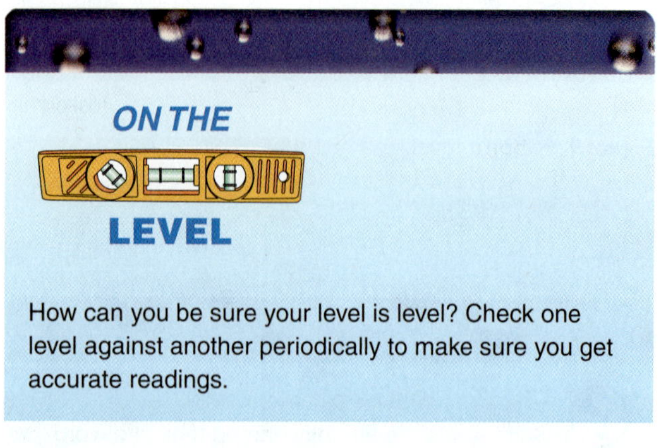

How can you be sure your level is level? Check one level against another periodically to make sure you get accurate readings.

5.3.0 Torpedo Level

The torpedo level (*Figure 11*) is a small spirit level that is useful in confined areas or for leveling short runs of pipe. Most torpedo levels have a vial that indicates 45-degree angles. Plumbers often use a torpedo level with a magnetic base. This is handy because the magnetic base will stick to iron pipe, making maneuvering in a confined space easier. Torpedo levels are usually 9 inches long and can fit comfortably into a pants pocket. They are also tapered at the ends.

5.4.0 Line Level

Plumbers use a line level by hooking it on a tightly stretched, nylon string or line between two points to be leveled with each other (*Figure 12*). Hooks at either end of the line level attach to the line at about the midpoint of the reach. When the air bubble is centered between the markings on the vial, this indicates the trueness of the line to which the line level is attached. This is how you make sure that pipe laid between two distant points is level along its entire distance. By using a line level, you can transfer (determine or calculate) vertical dimensions, such as height or elevation, over long distances without using the larger leveling instrument called a transit.

When using a line level, make sure the string is taut. A sagging string will lead to an inaccurate measurement. Even the most tightly stretched string, however, can sag a little, which means the line level, for all its convenience, is not completely accurate. It should therefore not be used in situations where precise measures are required.

5.5.0 Precision and Distance Leveling Instruments

If distances are too great to be measured accurately by the levels mentioned above, or if more accuracy is required, you will need to use a builder's level (also called a surveyor's level) or a transit level with a leveling rod (*Figure 13*). For these instruments to work, they must be calibrated properly. You will learn more about calibration later in this module.

Both the builder's level and the transit level work on the principle that a line of sight is perfectly straight. A telescope is mounted on a tripod and rotated in a complete circle without changing its horizontal position. You then look, or sight, through the telescope to a stadia (or graduated) rod mounted or held vertically by a co-worker some distance away. The stadia rod is a long, brightly painted rod with graduated measurements (usually in feet and tenths of a foot). The rod is located at a point of interest and sighted through the telescope to determine vertical dimensions.

The main difference between a builder's level and a transit level is that the telescope with the builder's level is fixed, but the telescope with a transit level can move up and down. A builder's level, then, can work only on the horizontal plane; the transit level can work in either the horizontal or the vertical plane. You can use either the builder's level or the transit level to measure differences in elevation between two distant points.

The transit level is a precision instrument, calibrated to indicate not only true horizontal but also the angle of inclination in degrees, minutes, or seconds. The telescope is mounted on a horizontal circular plate; attached to this plate are tools that balance and orient the telescope. A compass permits the user of the transit to approximate the bearing of the scope. Once the transit is set up on a tripod, the legs of the tripod must not be moved, because doing so will affect the accuracy of the measurement. The telescope is then pivoted on a horizontal axis to point in any direction. When a co-worker positions the stadia rod, you view the rod through the telescope and then determine the relative height of the grade (also known as slope) or the object upon which the rod stands.

5.6.0 Laser Level Tools

Another tool you will use is a cold beam laser (*Figure 14*). This leveling device projects a thin, 90-degree beam of low-wattage light. Laser leveling instruments are more expensive than other leveling tools (although they are becoming more affordable), but they offer several advantages. Lasers are more accurate over longer distances. Lasers are very easy to use when establishing vertical or horizontal lines and are used for measuring or laying out openings.

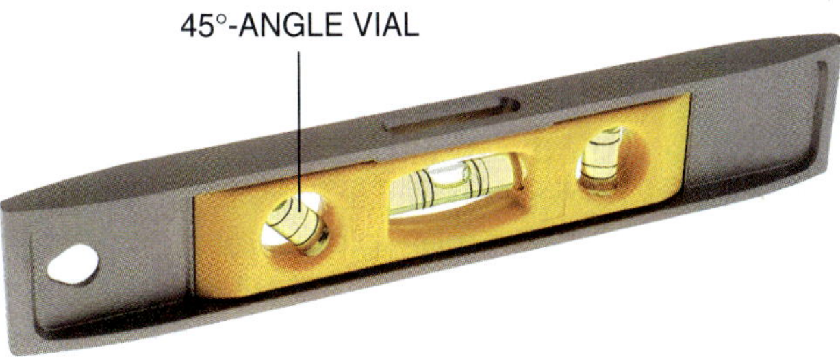

Figure 11 ◆ Torpedo level.

Figure 12 ◆ Line level.

ON THE

LEVEL

Calibrating a Builder's Level

A leveling vial is used on the builder's level. It is attached directly to the telescope. To make sure the telescope measures a grade accurately, you must level it by using the attached leveling vial. The vial is a spirit level tool design, but it has been manufactured to be much more accurate.

Four leveling screws are used to level the vial. You rotate the telescope until it lies over two opposing leveling screws. You then adjust the two screws in opposite directions until the bubble is centered in the leveling vial. After rotating the telescope 90 degrees, you adjust the other two screws. This process should be repeated at least once to make sure that the level is accurately calibrated. Once the level is properly set, you are ready to measure grade.

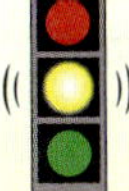

CAUTION

Use cold beam lasers with extreme care. Although low-power lasers can be used, arcs can still be unpredictable. Always wear appropriate personal protective equipment and refer to the manufacturer's specifications.

CAUTION

Don't lean on your level. These precision tools can lose their calibrated accuracy if you do so and will need to be reset, which wastes time on the job.

Figure 13 ◆ Precision levels.

103F14.TIF

Figure 14 ◆ Cold beam laser leveling device.

Some laser level tools can project beams up to 1,500 feet to hit a specific point. They can have a variety of components designed to get the most accurate possible readings:

- Large vertical and horizontal indicator bubbles
- Lens refractors that show line or dot images
- Dual diodes for increased beam visibility
- Wall, floor, or tripod mounts, brackets, or clamps
- Built-in bump sensors that will automatically notify the user, or shut off the device, if the instrument has been bumped hard enough to affect the accuracy of its readings
- Electronic level vials that dampen vibrations from conditions at the job site, to ensure the stability of the laser beam over distance
- Remote control capability
- Optional receivers that increase the laser's operating range still more

Using this tool, you can tell immediately if something has been aligned. If you are installing gravity-flow pipelines, for instance, you can use this tool as an alignment guide and lay the pipe along the beam of light emitted by the laser. Other applications of cold beam lasers include installing and aligning such building elements as walls, partitions, and access floors.

Portable, handheld lasers that feature plumb (both up and down), level, and square capabilities can greatly enhance productivity, which saves time and money. Lasers can be set up more quickly than other leveling devices and, because of their accuracy, require fewer repeat measures. The time saved when using a laser leads to cost savings. The initial expense of a laser is quickly made up in increased productivity.

These devices require relatively little maintenance. Like other levels, lasers should be cleaned off after use and stored in a safe, dry place. Once a year, the unit should be cleaned, checked, and adjusted by a professional. Heat, cold, and moisture—conditions common to plumbing sites—can affect a laser's operating performance. The degree to which these conditions affect the laser's stability and accuracy will depend on the quality of the instrument itself. Higher-quality lasers are generally more resilient to adverse environmental conditions. Some lasers are specifically designed and tested to endure shock, moisture, and temperature changes. The performance of all lasers, however, begins to deteriorate rapidly at temperatures of 110°F or higher.

The light emanating from a cold beam laser will not instantly injure humans, but long-term exposure could injure the eyes. Therefore, never point a laser level at another person. Only qualified and trained plumbers should use this equipment. Cold beam lasers must meet all of the Electronic Product Radiation Control provisions of the federal *Food, Drug, and Cosmetic Act*.

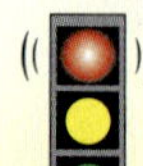

WARNING!

Never look directly into the laser beam or point the beam at a co-worker. Over time, exposure to the beam can injure eyes.

5.7.0 Maintenance

Levels and transits are precision instruments, and they are very expensive. Take care not to drop them or bump them against other materials, because loose or broken vials make these tools useless. You also need to keep levels and transits clean and dry so they will retain their true measure.

In some situations, such as in bright sunlight, it may be difficult to see the laser beam. Enhancement goggles are available that allow the user to see the laser beam more easily. Some laser level instruments are specifically designed for outdoor use and have beams that are easily visible in sunlight.

5.8.0 Plumb Bob

A plumb bob (*Figure 15*) is a precise, yet simple, measuring tool used to align points vertically, from one floor to another floor below. To establish this exact vertical line, the plumb bob uses the law of gravity. A plumb bob generally consists of a specially designed weight attached to one end of a nylon thread. Nylon is used because it is more resistant to dampness. A precision plumb bob has a pointed tip.

When you are setting a vertical run of pipe, you will need to know the center point on that run of pipe. To find this center point, dangle the plumb bob from a point in the top plate of the framed ceiling. The bob is allowed to swing freely on the string. When the bob stops swinging and is hanging straight and free, the point of the bob will be directly below the point where the string has been affixed above. The point on the level plane directly below the tip of the bob, then, will line up exactly vertically with the point above, and the level plane will also be exactly perpendicular to the string holding the bob. You can then pass a segment of pipe through the level planes at precisely the same point from floor to floor.

Although the plumb bob is not an intricate tool, it needs to be handled carefully. Be sure not to drop the plumb bob. If the tip of the plumb bob point becomes bent or rounded, it will provide inaccurate readings. If the string is not properly attached to the center point of the plumb bob, it will give inaccurate readings because the bob will be hanging at a slight angle to the string (*Figure 16*). If the tip becomes damaged or out of alignment, or if the string is not properly attached, replace them. Before using a plumb bob, always check the condition of the tip and the alignment of the string. Laser plumb bobs create an accurate and highly visible plumb reference instantly. Laser plumb bobs have magnetically dampened stands that allow the plumb bob to settle and be used much faster than the traditional plumb bob with string.

103F15.TIF

Figure 15 ◆ Plumb bob.

103F16.EPS

Figure 16 ◆ String attached to a plumb bob.

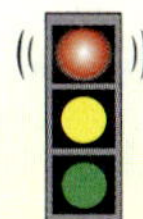

WARNING!

The point on a plumb bob is sharp. It is particularly easy to stab your toe with this instrument. Handle this tool with care to avoid hurting yourself or a co-worker.

5.9.0 Chalkline

The chalkline (*Figure 17*) is used to lay out long, straight lines of chalk on smooth or semi-smooth surfaces. You pull the chalkline taut between two points, then carefully stretch the line away from the surface and let it snap back to produce a straight line of chalk. To maintain a chalkline, add powdered chalk, and replace the string line when it is worn.

Because of the way most chalklines are designed, they can be used instead of a plumb bob. When you hang the chalkline overhead from the metal clip and allow the box to hang free and straight, the pointed end at the base of the chalkline indicates the exact point below on the vertical plane. You then transfer the mark to the floor to match the mark on the overhead framing member.

103F17.EPS

Figure 17 ◆ Chalklines.

Review Questions

Section 5.0.0

1. Level and plumb members intersect at _____ angles.
   a. 30-degree
   b. 45-degree
   c. 90-degree
   d. 180-degree

2. The proper function of a spirit level depends on _____.
   a. a bubble
   b. precise calibration
   c. the ability of the tool to withstand rust
   d. its length

3. The _____ level is a small tool useful in confined areas or on short runs of pipe.
   a. spirit
   b. laser
   c. torpedo
   d. plumber's

4. A _____ has a movable telescope to allow measurements in both the horizontal and vertical planes.
   a. carpenter's level
   b. builder's level
   c. plumb bob
   d. transit level

5. The _____ is used for laying out long, straight lines on smooth or semi-smooth surfaces.
   a. chalkline
   b. plumb bob
   c. transit
   d. framing level

6.0.0 ◆ TOOTH-EDGED CUTTING TOOLS

Sometimes plumbers need to make changes to a commercial or residential structure before installing plumbing materials and fixtures. Plumbers use tooth-edged cutting tools to make these changes. In these cases, plumbers are often doing work similar to that of carpenters. Knowing about both professions' types of tools will be helpful in your work.

Preparation is essential to making accurate and safe cuts. Before starting to cut, determine what kind of saw and what kind of blade is most appropriate for the material you are cutting. Carefully measure and mark the dimensions of the cut (its distance, angle, and depth), and then adjust the saw controls to match those measurements. Whatever saw you choose, make sure it is in good working condition, with sharp teeth and a tight handle.

CAUTION

Walls and floors may conceal wiring, pipes, conduits, or supporting beams. Before you saw into a wall or floor, check to find out what is behind or below the surface.

6.1.0 Hacksaw

The hacksaw is a multipurpose tool used to cut metals. Hacksaw types (*Figure 18*) include PVC saws and Sheetrock® saws. Mini-hacksaws often come with 6-inch-long blades called Tiny Tims, which are commonly used by plumbers (*Figure 19*).

Hacksaw blades come in different styles for different uses. Blades differ in length, flexibility, the set (or bend) of their teeth, and coarseness (number of teeth per inch), depending on the material they are made to cut. The hacksaw blade most often used to cut pipe is a flexible-back blade. Its teeth are set, which means each tooth is bent to alternating sides so the blade cuts a **kerf** (the groove made by a saw blade). The kerf is slightly wider than the blade is thick (*Figure 20*). This prevents the blade from binding, and it reduces the friction that results from cutting. Blades are typically very coarse (*Figure 21*). Keep the blades sharp to make a good cut.

The more teeth per inch of the blade, the finer the cut the blade will make. Generally, a blade with 18, 24, or 32 teeth per inch is used for cutting pipe. A saw with more teeth per inch will generate more debris, which can clog the kerf. When choosing a saw, remember that a larger saw, while it might have a larger blade that is better for cutting tougher material, is also less portable, which is always a consideration for a plumber.

You should not use a hacksaw for square cuts. A crooked cut can cause problems later during threading and other operations that require accurate measurements. Close **tolerance** (the allowable variations in a given measurement or quantity) means there is little room for extra space. If a plumbing fixture must be installed to meet close tolerances, a crooked cut can make it nearly impossible to fit the fixture in the prepared space.

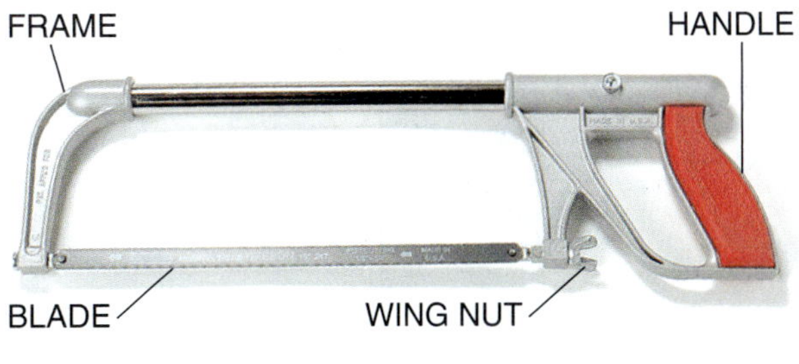

HACKSAW

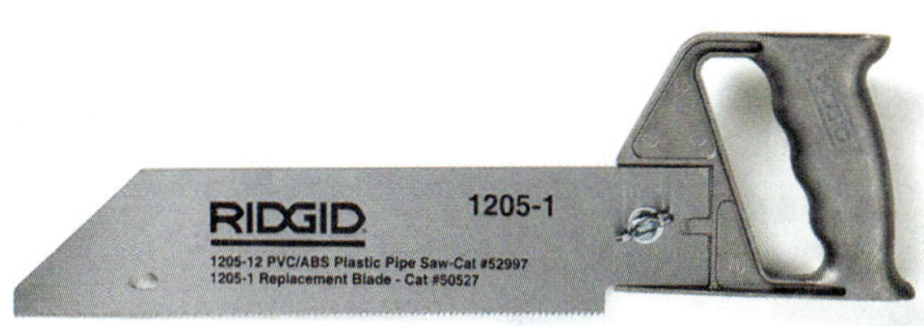

ALL-PURPOSE (PVC) SAW

KEYHOLE SAW

103F18.EPS

Figure 18 ◆ Hacksaws.

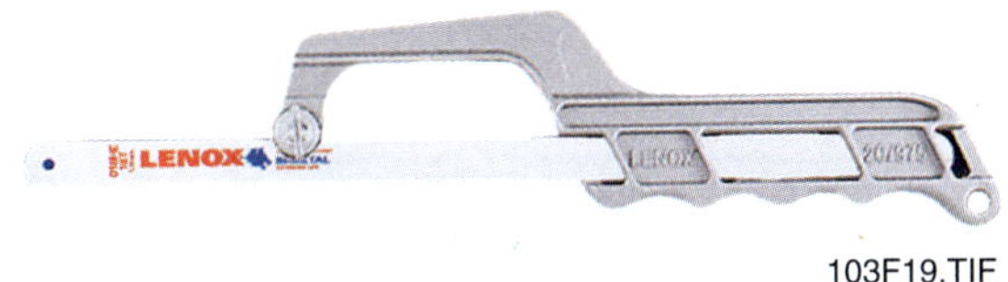

103F19.TIF

Figure 19 ◆ Mini-hacksaw.

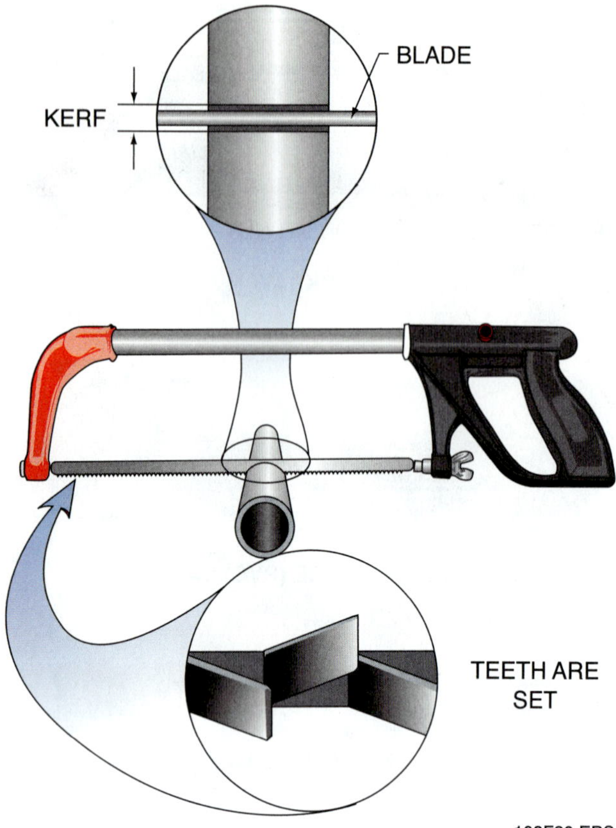

Figure 20 ◆ Hacksaw blade kerf cut.

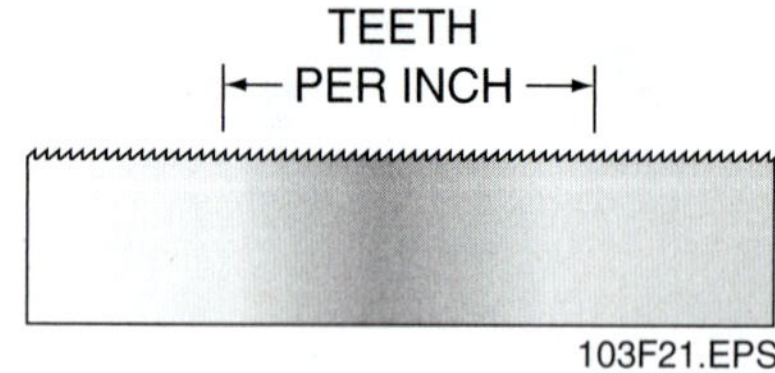

Figure 21 ◆ Blade coarseness.

When using a hacksaw, follow these safety and maintenance instructions:

- Select a saw blade suitable to the material you are cutting.
- Place the blade in the hacksaw frame so that the blade's teeth point toward the end of the frame and away from the handle (and from you). Tighten the blade firmly, and make sure the hacksaw frame is properly aligned.
- Do not try to make straight cuts with crooked hacksaw frames or loose hacksaw blades; this might cause the blade to buckle, twist, or break, which is dangerous.
- Do not apply too much pressure or twisting when sawing; this might cause the hacksaw blade to break, which could result in injury.
- Replace dull hacksaw blades. Worn and dull blades are more likely to cause accidents.

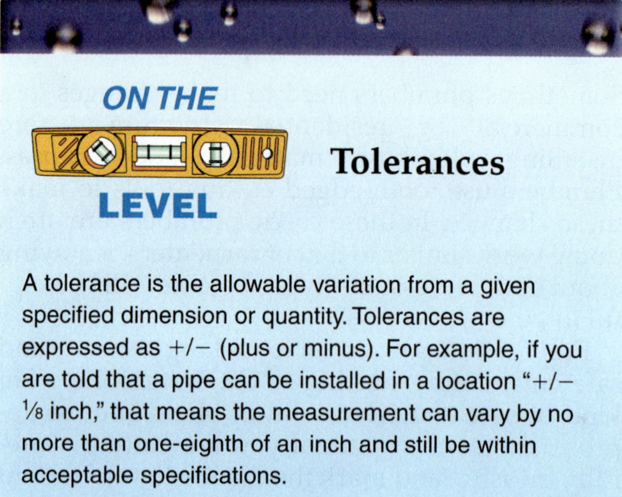

Tolerances

A tolerance is the allowable variation from a given specified dimension or quantity. Tolerances are expressed as +/− (plus or minus). For example, if you are told that a pipe can be installed in a location "+/− ⅛ inch," that means the measurement can vary by no more than one-eighth of an inch and still be within acceptable specifications.

- Make the cuts away from you, and use long, straight, and steady strokes, using almost the entire length of the blade, when cutting.
- Ease pressure on the hacksaw on the backward stroke to avoid dulling the blade's teeth.

6.2.0 Reciprocating Saw

The reciprocating saw (*Figure 22*) is electrically powered and can be used with a variety of blades. The choice of blade depends upon the material to be cut. A shaft pushes the blade forward and back. On the backstroke, it also lifts the blade to clear the cut and keep the saw's teeth from being worn. Single-speed, two-speed, and variable-speed models are available. The strokes per minute made by a reciprocating saw range from 1,700 (the slowest speed on two-speed models) to 2,400 (the highest speed on most variable-speed models).

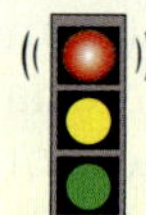

WARNING!

Never reverse a saw while it is running. If the blade is pinched, it can cause a powerful kickback, which can jerk the saw toward the operator. Severe injuries, lacerations, and even death may result. Always use a safety chain and read the manufacturer's instructions.

Source: City of Hampton website. "Office of Emergency Management." www.hampton.va.us/eoc/documents/chainsawsafe.pdf, reviewed March 6, 2004.

The reciprocating saw brings brute force to bear on a task and is designed for use on hard-to-cut materials. It is used in practically every construction-related profession. You will use this powerful saw to cut cast iron, metal, plastic, and wood. It can saw

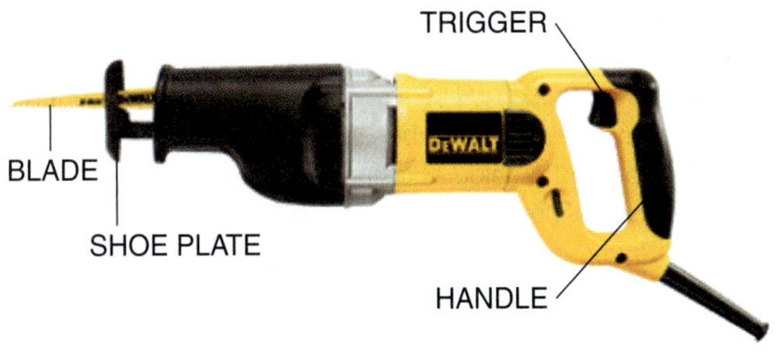

Figure 22 ◆ Reciprocating saw.

through walls or ceilings to create openings for plumbing lines. The reciprocating saw is a necessity in any demolition work.

At the front of every reciprocating saw is a shoe plate. Never use a reciprocating saw without the shoe plate because, for the saw to make a good cut, the shoe plate must rest against the work surface. Offset blade adapters can be used to move the blade to the top of the shoe plate for cutting flush or even with the material. Always be sure to use the types of blades specified by the manufacturer for the kind of work you are doing with the reciprocating saw.

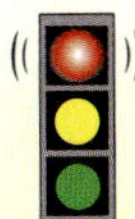

WARNING!

Always unplug any saw and/or remove the battery before you change the blades.

The reciprocating saw, like all electrical tools, must be grounded with an **electrical ground**, which is a conductive connection that provides a path for an electric current to pass from an electrical appliance to the earth. Inspect the body and the cord for damage before you operate the saw. Do not operate any electric tools when you are in contact with water or near flammable gases.

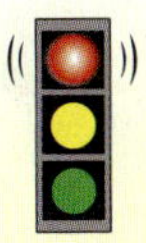

WARNING!

Do not remove the safe-ground pin on the plug of the reciprocating saw.

Some manufacturers make a cordless reciprocating saw. This saw works the same way as a regular reciprocating saw, but it uses battery power instead of an electrical current. The lack of a cord can be an advantage in tight spaces. You must use all the same precautions, however, when operating a cordless tool as you would with an electric tool. When you are using any battery-powered tool, make sure you have a spare battery handy so that you do not have to waste time looking for one if the battery you are using dies or needs recharging.

Choosing the Proper Saw Blade

There are several different types of saw blades. Steel blades are generally the least expensive, but they dull faster. If you do not use your power tools frequently, these blades might be good enough. If you use your power saws a lot, however, carbide-tipped blades are a better choice. These blades are more expensive than steel ones, but they also last longer.

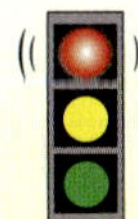

WARNING!

Be sure not to stand in water when using any power tool. Doing so greatly increases the chances of electrocution, which can cause serious injuries and even death.

6.3.0 Portable Band Saw

The portable band saw (*Figure 23*) is used to cut a variety of materials, including pipes both with and without iron. These types of pipes are referred to as **ferrous** and **nonferrous** pipe, respectively. Other types of pipes and materials that are cut

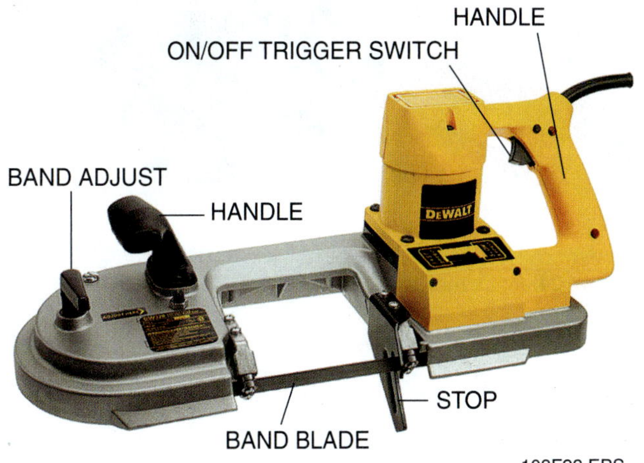

Figure 23 ◆ Portable band saw.

with a portable band saw are **bar stock** and **tube stock**. Plastics and materials with irregular shapes are also cut using a portable band saw. This saw's blade, which is tooth-edged like a hacksaw, revolves continuously around guides located at both ends of the saw. For best results, lubricate the blade to reduce friction and prolong the life of the saw.

6.4.0 Abrasive Saws

You might also use abrasive saws to cut pipe and other materials. These saws use a special wheel that can slice through either metal or masonry. The most common types of abrasive saws are the demolition saw and the chop saw. The main difference between the two is that the demolition saw is not mounted on a base.

6.4.1 Demolition Saw

Demolition saws (*Figure 24*) run on either electricity or gasoline. These saws are effective for cutting through most materials found at a construction site. Although all saws are potentially hazardous, the demolition saw is a particularly dangerous tool that requires the full attention and concentration of the operator.

CAUTION

When sawing, be sure to use the appropriate PPE for your eyes, ears, and hands. If you have long hair, be sure to tie it back or cover it properly.

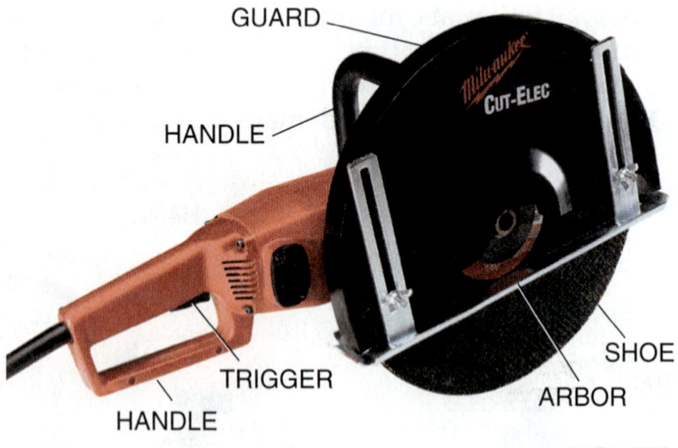

Figure 24 ◆ Demolition saw.

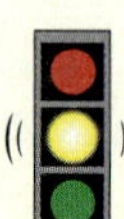

CAUTION

OSHA *29 CFR 1910.106* regulates the amount of flammable or combustible liquids that may be located outside of or inside of a gas storage cabinet or in any one fire area of a building. The standard prohibits more than 60 gallons of Class I or Class II liquids or more than 120 gallons of Class III liquids in one storage cabinet.

Gas storage cabinets must be designed to limit the internal temperature not to exceed 325°F. Metal cabinets should be constructed of 18-gauge sheet metal and should be double walled. For added safety, gas storage cabinets should be self-closing. Cabinets must be labeled in conspicuous lettering with the statement "Flammable—Keep Fire Away."

Source: Occupational Safety and Health Administration website. "Regulations (Standards–*29 CFR*), *Flammable and Combustible Liquids–1910.106*." www.osha.gov/pls/oshaweb/owadisp.show_document?p_id=9752&p_table=STANDARDS, reviewed March 30, 2004.

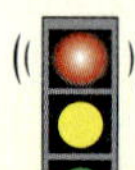

WARNING!

Gas-powered demolition saws must be handled with caution. Using a gas-powered saw where there is little ventilation can cause carbon monoxide poisoning. Carbon monoxide gas is hard to detect, but it is deadly and can kill you! Use fans to circulate air, and have a trained person monitor air quality.

Source: Centers for Disease Control and Prevention website. www.cdc.gov/elcosh/docs/d0100/d000021/d000021.html, reviewed March 5, 2004.

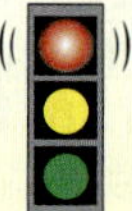

WARNING!

Because a demolition saw is handheld, there is a tendency to use it in positions that can easily compromise your balance and footing. Be very careful not to use the saw in any awkward positions.

6.4.2 Chop Saw

The chop saw is a versatile and accurate tool. It is a lightweight circular saw mounted on a spring-loaded arm that pivots and is supported by a metal base (*Figure 25*). Because of its size and weight, it is often portable. This is a good saw to use to get exact square or angled cuts. It uses a wheel to cut pipe, channel, tubing, conduit, or other light-gauge materials. The most common wheel houses an abrasive metal blade, but blades made for other materials such as PVC or wood can also be used. A vise on the base of the chop saw holds the material securely and pivots to allow miter cuts. Some can be pivoted past 45 degrees in either direction.

Chop saws are sized according to the diameter of the largest abrasive wheel they accept. Two common sizes are 12-inch and 14-inch wheels. Each wheel has a maximum safe speed. Never exceed that speed. Typically, the maximum safe speeds are 5,000 revolutions per minute (rpm) for 12-inch wheels and 4,350 rpm for 14-inch wheels.

6.4.3 Abrasive Saw Safety

Abrasive saws, such as demolition and chop saws, require extreme care during use. To avoid injuring yourself, follow these guidelines:

- Wear PPE to protect your eyes, ears, and hands.
- Use abrasive wheels that are rated for higher rpms than the tool can produce. This way, you will never exceed the maximum safe speed for the wheel you are using. The blades are matched to a safety rating based on their rpms.

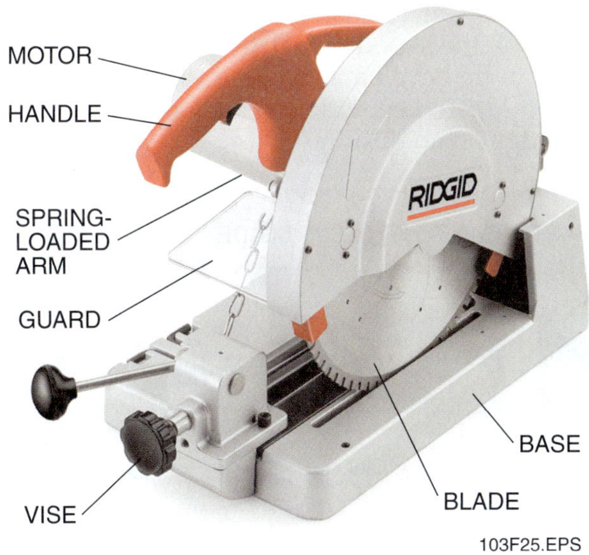

Figure 25 ◆ Chop saw.

- Be sure the wheel is secured on the arbor before you start using a demolition saw (refer to *Figure 24*). Point the wheel of a demolition saw away from yourself and others before you start the saw.
- Use two hands when operating a demolition saw.
- Secure the materials you are cutting in a vise. The jaws of the vise hold the pipe, tube, or other material firmly and prevent it from turning while you cut. (Vises are discussed later in this module.)
- Be sure the guard is in place and the adjustable shoe is secured before using a demolition saw.
- Keep the saw and the blades clean.
- Inspect the saw each time you use it. Never operate a damaged saw. Ask your supervisor if you have a question about the condition of a saw.
- Inspect the blade before you use the saw. If you see damage, throw the wheel away (*Figure 26*).

6.4.4 Ground Fault Circuit Interrupters and Power Tool Safety

Ground fault circuit interrupters (GFCIs) are fast-acting circuit breakers designed to shut off electric power almost instantaneously in the event of a ground fault. Ground faults are the most common form of electrical-shock hazard at a job site. GFCIs, however, will not protect you from all electrical-contact hazards. GFCIs can be tripped and electric current flow interrupted by wet connectors or tools. Even with GFCIs around, you should make it a habit to protect electric tools from excessive moisture.

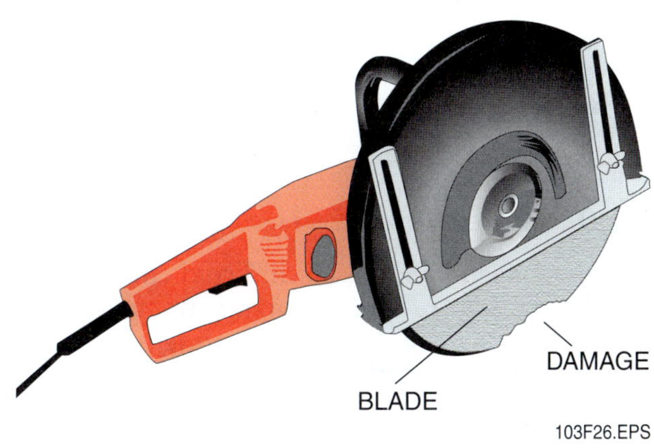

Figure 26 ◆ Damaged saw blade.

Review Questions

Section 6.0.0

1. On a hacksaw, the more teeth per inch on a blade, _____.
   a. the coarser the cut and the more debris
   b. the finer the cut and the more debris
   c. the coarser the cut with less debris
   d. the finer the cut with less debris

2. Do not use a hacksaw when you need a _____ cut.
   a. fast
   b. square
   c. crooked
   d. flexible

3. A _____ is a necessity for any demolition work because of its brute power.
   a. hacksaw
   b. portable band saw
   c. reciprocating saw
   d. chop saw

4. The _____ saw has a continuous blade that revolves around guides located at both ends of the saw.
   a. portable band
   b. reciprocating
   c. chop
   d. demolition

5. A _____ on the base of the chop saw holds the material securely and pivots to allow miter cuts.
   a. vise
   b. clip
   c. channel
   d. lock pin

7.0.0 ◆ SMOOTH-EDGED CUTTING TOOLS

Smooth-edged cutting tools have a smooth cutting edge instead of teeth. Chisels, pipe cutters, and shovels are the smooth-edged tools most commonly used by plumbers.

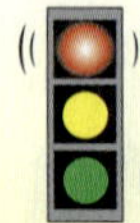

WARNING!

Rusty or dull cutting tools not only perform poorly but can also cause serious accidents if they slip.

7.1.0 Chisels

Plumbers use wood chisels and cold chisels. Both types of chisels are made from heat-treated steel to increase the hardness of their cutting edges. You force the chisel's cutting edge into and through the material you are working on by hitting the chisel's handle with a hammer.

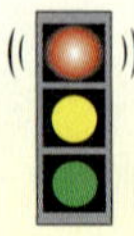

WARNING!

To avoid injury to your hands, face, and eyes, always wear PPE when sharpening chisels. Appropriate eye protection is of the utmost importance.

7.1.1 Wood Chisel

The wood chisel (*Figure 27*) is used to make openings or notches in wooden structural material so that pipes can be installed through these openings. For the wood chisel to cut well, its blade needs to be beveled at a precise 25-degree angle (*Figure 28*). To keep it keen, hone the chisel's cutting edge on an oilstone.

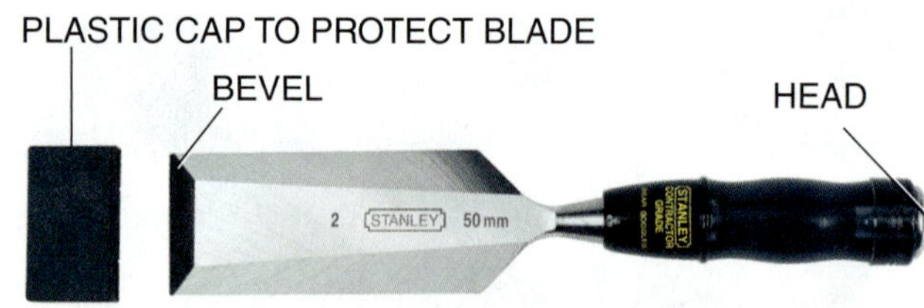

Figure 27 ◆ Wood chisel.

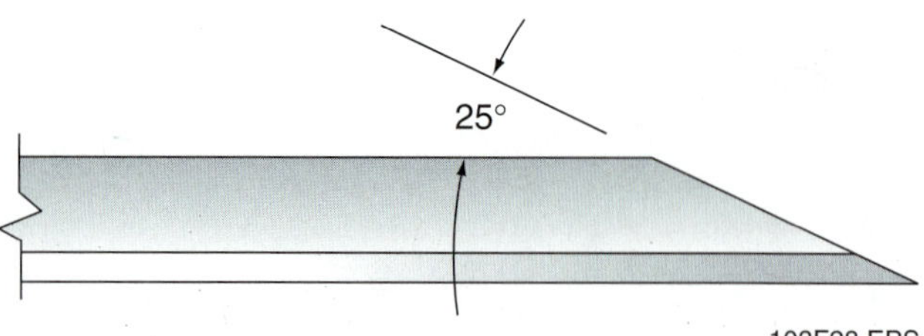

Figure 28 ◆ 25-degree angle cutting edge.

7.1.2 Cold Chisel

The cold chisel (*Figure 29*) is used for cutting, shaping, and removing cold metal softer than the chisel's cutting edge, such as cast iron, wrought iron, bronze, copper, and steel. It is also used for enlarging holes. The blade of a cold chisel is beveled on both sides to a 60-degree angle (*Figure 30*). If the blade of the cold chisel is dull, do not use the tool.

Mushrooming of the chisel head (or handle) is a common problem with cold chisels. Mushrooming of the chisel head results from the hammering it takes. Hammering on a mushroomed head can be dangerous because it can cause metal fragments to fly off the chisel head, seriously injuring the user's eyes or hands. Never use a chisel that has a mushroomed head. When the handle end of a chisel mushrooms, grind off or trim the mushroomed part of the chisel at a slight bevel (*Figure 31*).

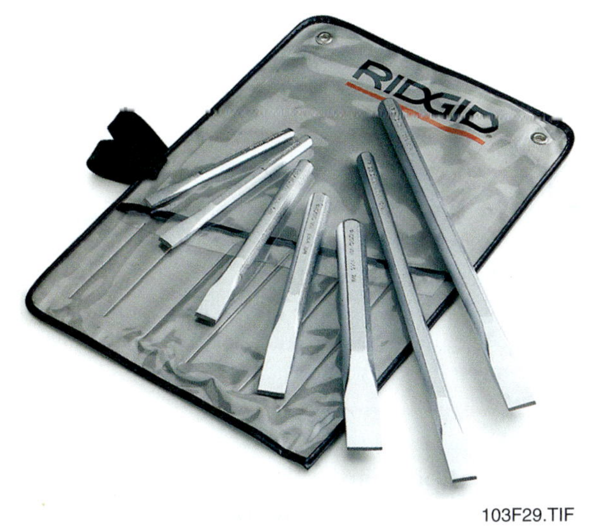

Figure 29 ◆ Cold chisels.

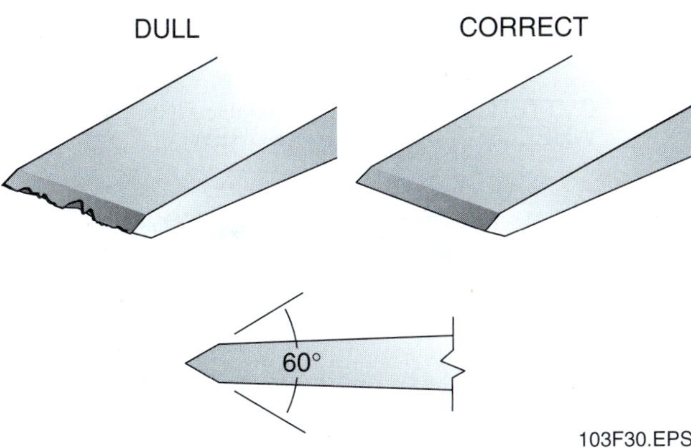

Figure 30 ◆ 60-degree angle cutting edge and dull versus correct chisel blades.

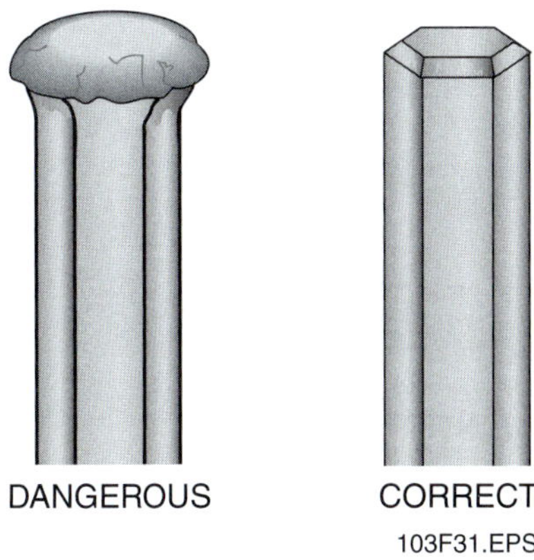

Figure 31 ◆ Repairing a mushroomed chisel head.

Cold chisels can also be used for trimming concrete, but never use these tools to cut or split stone or concrete. When striking a chisel head, point the tool away from your body. Do not use a chisel as a prying or wedging tool, as its brittle steel can break and cause injury. Wear gloves and a face shield or goggles when using chisels.

7.2.0 Manual Pipe Cutters

Pipe-cutting tools come in a variety of sizes, shapes, and designs. Common sizes range from ⅛ inch to 4 inches. The three most commonly used manual pipe cutters are the tube cutter, the pipe cutter, and the soil pipe cutter.

7.2.1 Tube Cutter

The tube cutter (*Figure 32*) is used for cutting copper tubing or thin-wall conduit that is ⅛ inch to 4 inches in diameter. It also is effective for cutting other malleable (pliable) building materials, such as brass and aluminum.

A conventional tube cutter has four movable parts: a cutter wheel, an adjusting screw, and at least two guiding wheels. The blade on the cutter wheel slices into the pipe as you make revolutions around the pipe, gradually tightening the cutter.

When doing this, you should take care to keep the blade in the same groove and to make steady and repetitive revolutions with the cutter, in order to obtain a clean cut without damaging the pipe. The tube cutter also has a reamer feature, which allows you to remove any metal burrs from the inside of the pipe once the cut is completed.

A specialized midget tube cutter is used for cutting copper tubing ⅛ inch to ⅞ inch in diameter

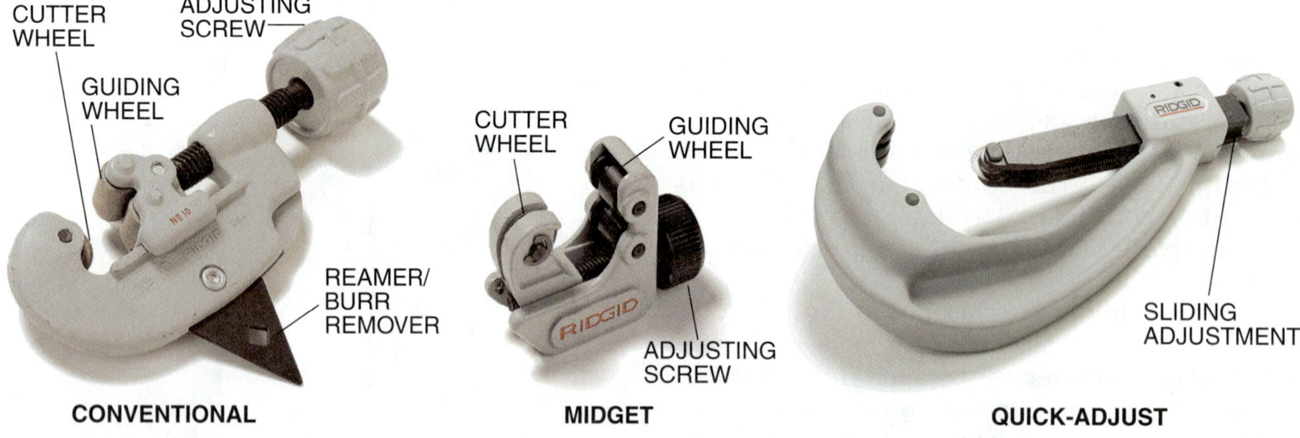

Figure 32 ◆ Tube cutters.

and is handy when working in tight places. Another special tool of this type, a quick-adjust tube cutter, has a sliding adjustment instead of an adjusting screw. You will often use a waxy crayon, also called a **keel crayon**, to mark a cutting line on the surface of the tube. In other cases, you will use **soapstone** (a soft, talc-based rock) to mark the pipe. Keel produces a narrower line than soapstone. You will then set the cutting wheel of the tube cutter on the cutting line and follow that line when making the cut.

7.2.2 Pipe Cutter

The pipe cutter (*Figure 33*) is used to cut steel pipe. This tool is basically the same as the tube cutter, but it is heavier and it often has more than one cutter wheel. It is used to cut large-diameter pipes. Use a four-wheeled pipe cutter if there is not enough space to swing a single-wheel pipe cutter completely around the pipe.

7.2.3 Soil Pipe Cutter

Soil pipe cutters are used to cut cast iron or clay pipe. There are two types, the ratchet cutter and the snap cutter (*Figure 34*). Soil pipe cutters are especially appropriate to use in confined spaces or when cutting hard-to-reach piping.

The soil pipe ratchet cutter has a length of chain that wraps around the pipe. The chain is attached to a handle so that ratcheting the handle tightens the chain, which in turn pushes cutters into the pipe and creates a clean break.

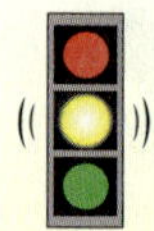

CAUTION

When using a soil pipe ratchet cutter, ensure that the chain is placed around the pipe correctly. The chain should be perpendicular to the pipe, not at an angle to it. Chains placed improperly will cause the pipe to shatter when you attempt to make the cut.

The soil pipe snap cutter, which is more common, cuts with a scissors action. Two long handles provide leverage to the cutting chain wrapped around the pipe. The chain is set with cutting rollers at each link. Tighten the chain, and turn it

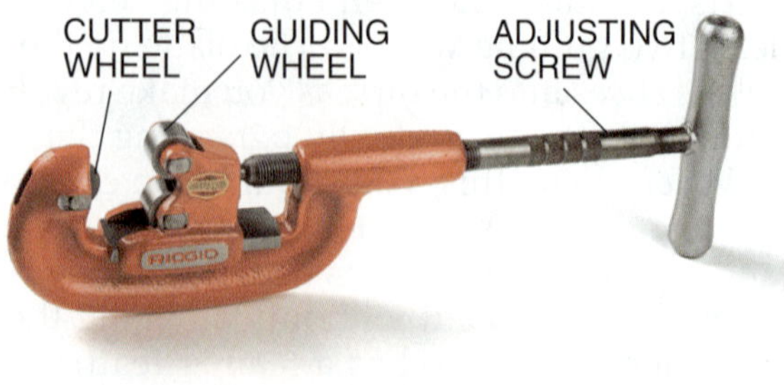

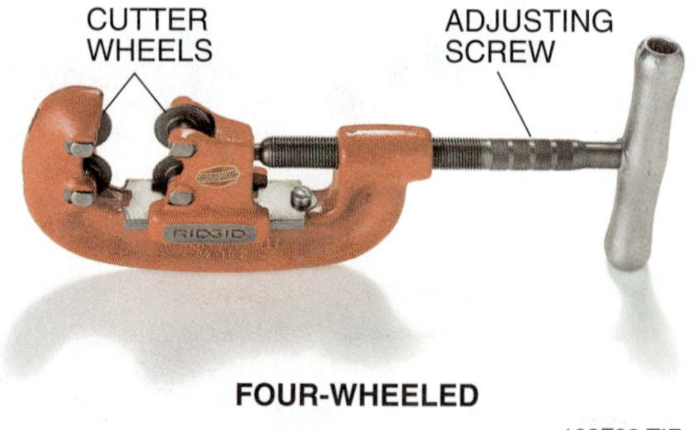

Figure 33 ◆ Pipe cutters.

Figure 34 ◆ Soil pipe cutters.

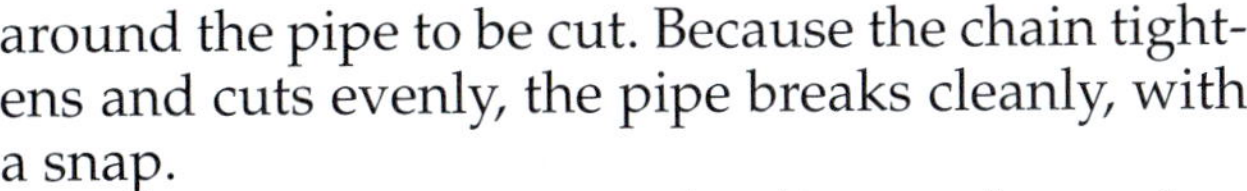

around the pipe to be cut. Because the chain tightens and cuts evenly, the pipe breaks cleanly, with a snap.

You will sometimes mark a line on the surface of the pipe with soapstone to guide the cut. Make sure that the line is perpendicular, otherwise you will shatter the pipe when you make the cut.

7.2.4 Care of Cutters

If the tube or pipe you are cutting looks as if it has been mashed after you cut it, you probably need to replace the cutting wheel or wheels. Replace the cutter wheels that are nicked or otherwise damaged. Apply lubricating oil regularly to all movable parts of the pipe cutter so that it will operate smoothly.

7.3.0 Shovels

Plumbers use a variety of shovels, including handheld trench shovels, to dig trenches for water pipes. Before digging, you must always identify the location of all nearby gas, electrical, water, and waste and sewage lines and systems. Shovels hitting underground utility wires and pipes can cause a number of injuries and expensive damage. When digging around tree roots, use shovels with shock-absorbing handles. The size of the shovel you use should be dictated by the task and by your own size. Shovels with long handles, for instance, are easier for taller plumbers to use than short-handled shovels. When digging in a confined area, such as a soak well, however, you will need to use shovels with extra-short handles, whatever your size may be.

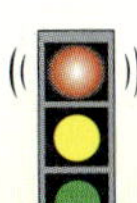

WARNING!

You must alert local utility companies before doing any digging to avoid damaging underground cables and piping. The utility company will mark the area so that you can dig safely. In addition, all states maintain a One-Call notification system. Construction companies can call this number to have pipelines that carry petroleum or natural gas marked before any digging takes place.

Review Questions

Section 7.0.0

1. To cut into material using a chisel, you _____.
   a. strike the flat part of the chisel blade with a hammer
   b. strike the top of the handle with a hammer
   c. first cut a starter channel and then strike the upper side of the chisel blade with a hammer
   d. first clamp the chisel to the material and then strike the handle with a hammer

2. For the _____ chisel to cut well, the blade needs to be beveled at a precise 25-degree angle.
   a. wood
   b. mushroomed
   c. cold
   d. snap

3. _____ is a common result of hammering a cold chisel.
   a. Smashing
   b. Mushrooming
   c. Spreading
   d. Splaying

4. A _____ is used for cutting small-diameter, soft materials such as copper, brass, and aluminum tubing or thin-wall conduit.
   a. pipe cutter
   b. snap cutter
   c. reamer
   d. tube cutter

5. To make a clean break in cast iron pipe, it is best to use a _____.
   a. hammer and cold chisel
   b. snap cutter
   c. chop saw
   d. hacksaw

8.0.0 ◆ DRILLING AND BORING TOOLS

Electric drilling and boring tools are versatile instruments that have changed the plumbing profession, greatly shortening the amount of time it takes to install pipe. Few things have enhanced the plumber's productivity more than the electric drill, and die tools have made piecing pipe together easier. Drilling tools allow you to make holes in a building structure quickly.

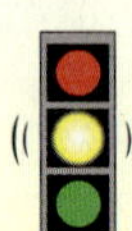

CAUTION

Never assume you know how to use a power tool. Always thoroughly read and understand the operator's instructions before using and starting power tools.

Picking up a drill and pressing the trigger is not difficult, but using a drill with skill and care requires training and experience. To match a drill bit to a task, you will need to know both the capabilities of the drill bits and the properties of the materials into which you are drilling. While drills are less dangerous than some tools, like saws, using or maintaining drills improperly can lead to injury.

Successful drilling depends on preparation. If possible, place the work piece you plan on drilling in a clamp or a vise so that both your hands are free to operate the drill. This will also prevent the work piece from moving around as you drill. Before you drill a hole in metal, you will need to establish and mark a precise center point on the metal's surface. If you are drilling a large hole, you might want to drill a pilot hole with a smaller drill bit.

When you start drilling, place the drill bit, which is the part of the drill that does the cutting, on the center point you have marked on the work piece. First run the drill at a slow speed so that the drill bit does not stray from your center point. Never force a drill. If you apply too much pressure to a drill, you are likely to break the drill bit. If you are drilling all the way through a work piece, slow the drill speed as the drill bit approaches the other side so that when the bit breaks through, it will leave a smooth exit hole. If the drill speed is too high or you are applying pressure to the drill, the exit hole will be rough.

If the drill seems to jam or threatens to jam, withdraw the drill. Doing so will allow the drill bit to clear any debris from the hole that you have made. Then you can resume where you left off. Unlike other cutting tools, drill bits do not allow for an exit path for the debris they create; the side-wall of the hole blocks the debris caused by the drilling. It is usually necessary, especially when drilling deep holes, to back the drill bit out of the hole in order to clear out the debris. This also helps to dissipate the heat that is generated by the constant friction between the cutting edge of the drill bit and the work piece.

Drilling is something of an art form. To do it well, you have to consider several factors: the drill's rotational speed (which is defined by its rpm), the type of drill bit, the diameter and depth of the hole being drilled, and the material properties of the work piece. These factors must be balanced in order to avoid too much heat buildup, which could lead to tool failure. These factors are discussed in the following sections.

8.1.0 Portable Electric Drills

The portable electric drill (*Figure 35*), sometimes called a pistol drill because of its shape, is used to drill holes in structural materials so that pipe can be run and fixtures can be installed. This basic drill has a motor built into the pistol-shaped body. The chuck is a gripping mechanism that houses the drill bit. After the drill bit is inserted into the jaws of the chuck, the bit is locked in place by tightening the chuck. The chuck is tightened either by a key or, in newer models, by hand. Always unplug the drill when tightening and locking a drill bit into place.

If you need to drill between **joists** or in tight corners where a pistol drill will not fit, a right-angle (offset) portable electric drill can come in handy (see *Figure 36*). This drill's offset design allows you to drill holes where space is limited. Drill models with extended handles allow you to get a good grip so you can better guide the drill when boring through material. Drilling with a right-angle portable drill can be dangerous, so these drills should have a clutch that will kick in if the drill bit sticks in a hole. This feature prevents you from losing control of the drill and possibly getting hurt. Milwaukee's Hole Hawg® is an offset drill used in tight places. This brand of drill is legendary for its **torque,** or the twisting or turning force applied in a rotating motion.

Two other kinds of drills are the hammer drill and the rotary hammer drill (*Figure 37*). The hammer drill combines the force of a hammer with the boring capability of a drill. Its hammering action penetrates dense material such as concrete or masonry, and the rotating bit removes the chips. A variety of drill bits are available for different types of jobs. The rotary hammer drill is used for larger jobs; the regular hammer drill is used for drilling holes to set small anchors that secure piping to a structure.

8.1.1 Uses

Pistol drills are perhaps the most widely used and most versatile of all power tools. Their primary function is to drill holes in steel, wood,

Figure 35 ◆ Portable electric drill (pistol drill).

Figure 36 ◆ Offset drills.

HAMMER DRILL

ROTARY HAMMER DRILL

103F37.EPS

Figure 37 ◆ Hammer drills.

concrete, and plastic—in fact, in virtually any material used in building. Pistol drills come with many accessories that allow them to function as grinders, sanders, polishers, sheet-metal cutters, and, with the addition of wire or flapper wheels, metal finishers.

Many types of pistol drills are available, made by a variety of manufacturers. Some are durable, industrial-quality tools meant for heavy use, while others are better suited to occasional use. Naturally, price often reflects quality as well as adaptability to various functions. An industrial-strength drill model designed for continuous use can be a wise investment.

Though there are many different styles of pistol drills, each one has a specified capacity (the maximum size of the chuck cavity) that dictates the size of the drill bit the chuck cavity can hold. Common capacities are ¼-inch, ⅜-inch, and ½-inch.

8.1.2 Chuck

The chuck is the mechanism at the end of the drill motor (see *Figure 38*). Chucks usually contain three jaws that open and close simultaneously to release or grip a drill bit or other accessory. The chuck mechanism is worked with a chuck key or, in newer models, by hand. The chuck key is a gear that meshes with another gear at the base of the drill chuck. The key is attached to the drill with a piece of rubber.

The size of the pistol drill is stated in terms of the diameter of the chuck cavity when the chuck jaws are fully open. For instance, if a pistol drill's chuck cavity measures ¼ inch in diameter when the jaws are fully open, that drill is called a ¼-inch drill. It uses drill bits with shanks that are up to one-quarter of an inch in diameter.

8.1.3 Capacity

Each drill motor has a maximum no-load rating, which is stated in rpm. "No-load" means the pistol drill is not doing any actual drilling and the

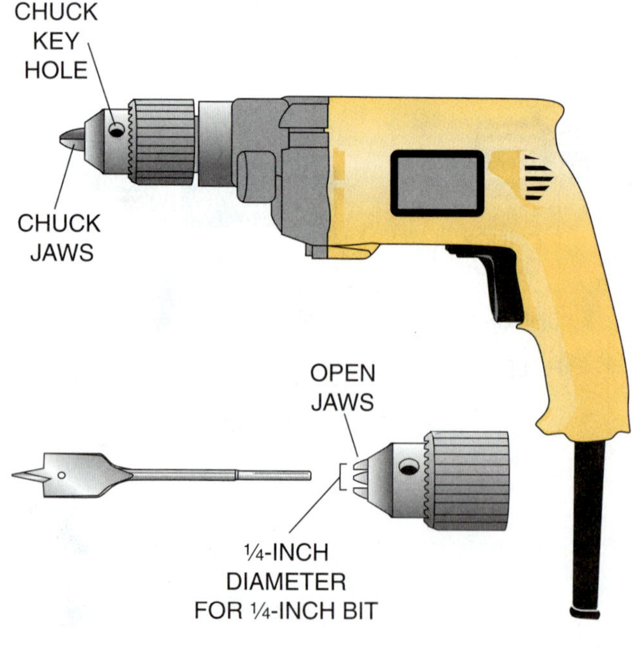

103F38.EPS

Figure 38 ◆ Drill chuck.

motor is running freely at its maximum speed. No-load ratings vary depending on the drill's capacity. On all drills, as drill capacities increase, drill speeds decrease. Simply stated, larger drill bits turn more slowly. For example, a ¼-inch drill may have a no-load rating between 2,000 and 3,500 rpm. A ⅜-inch drill may have a rating between 650 and 1,200 rpm. A ½-inch drill may have a rating between 350 and 850 rpm. These ratings vary slightly by manufacturer. The rpm of the Hole Hawg® drill, for instance, is geared down to allow it to use larger, self-feeding drill bits.

The reason an increase in drill capacity means a decrease in drill speed is that, as drill capacity increases, so does the drill's torque. Torque is a measure of a tool's ability to do work. It requires more torque, for example, to drill a ½-inch hole through an I-beam than it does to drill a ⅛-inch hole through a metal stud. Therefore, you would choose a larger drill for the I-beam than for the metal stud.

A larger-capacity drill motor will turn more slowly than a smaller-capacity motor. A larger drill bit cannot tolerate the high speeds of a smaller-capacity drill motor without overheating and losing its **temper**. Temper is the drill bit's resilience and strength. This is one reason why, when drilling through thick steel, you should periodically withdraw the drill bit to prevent it from overheating.

8.1.4 Rotation

Variable-speed drills can operate in a wide range of rotational speeds, from slightly above zero to the maximum speed of the drill's motor. The pressure applied to the drill's trigger determines the rotational speed: the more pressure, the higher the speed. Manufacturers realize that drilling applications are not standard and that different applications require different speeds. Holes in some materials are best drilled at high speeds, while holes in other materials are best drilled at lower speeds.

A variation on the basic drill motor is the reversing motor. These drill motors enable either clockwise or counterclockwise rotation of the drill bit. This is useful if you need to back the drill bit out of a hole when it has become pinched. You switch the direction of rotation by toggling a lever or switch on the body of the drill. Most reversing drill motors also have variable speed capabilities. Most manufacturers designate these tools as VSR, for variable-speed reversing.

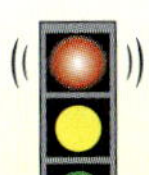

WARNING!

Never reverse drill motors while they are running. You can damage the drill or the work piece or injure yourself.

8.2.0 Cordless Drills

Cordless drills are becoming more popular because of their versatility and convenience over the traditional portable drill. A cordless drill is especially useful when working in areas where a power source is difficult to find or where it would be awkward or hazardous to run an electrical cord.

ON THE LEVEL — Tempering

Certain metals and special glass are tempered (heated and cooled) to enhance their hardness and resilience. Alternately heating and cooling a metal, such as steel, pre-stresses the metal, greatly increasing its strength. Drill bits are often tempered to make them stronger than the materials they drill into. Overheating a tempered metal, however, causes it to lose its enhanced hardness.

Battery-operated tools are appearing more frequently at construction sites. They eliminate the need for extension cords and electrical outlets. Most of these tools come with a replacement battery and a battery charger. When a battery runs low, you replace it with a fully charged battery, and recharge the battery you had been using. When possible, use lithium batteries. These batteries are popular for having more power, constant voltage, and longer life.

It is important to properly handle and dispose of batteries when necessary. Improper disposal is harmful to the environment. Follow the manufacturer's recommendations for proper battery disposal.

In cordless drills, a rechargeable battery pack runs the motor (see *Figure 39*). The removable pack is attached to the drill during use and can be plugged into a battery charger when the drill is not in use. Some battery chargers can recharge the pack in an hour; others require more time. If you use a cordless drill a lot, always have a spare battery pack on hand so you can continue to use your cordless drill while your other battery pack recharges.

All the features of traditional pistol drills are also found on the cordless models. Many cordless drills, for instance, have a VSR capability. Some also have adjustable clutches that shift the force (torque) to allow the drill to double as a power screwdriver. Cordless drills generally cost more than traditional drills, but that expense may be more than offset by their flexibility.

8.3.0 Bits

Drill bits are the cutting tools used in electric drills, traditional and cordless alike. Drill bits are mounted in the drill chuck cavity and then locked into place. There are several different types of drill bits: twist, spade, hole saw, self-feeding, auger, and masonry. Each is shaped differently, and each has special properties or special uses (*Figure 40*).

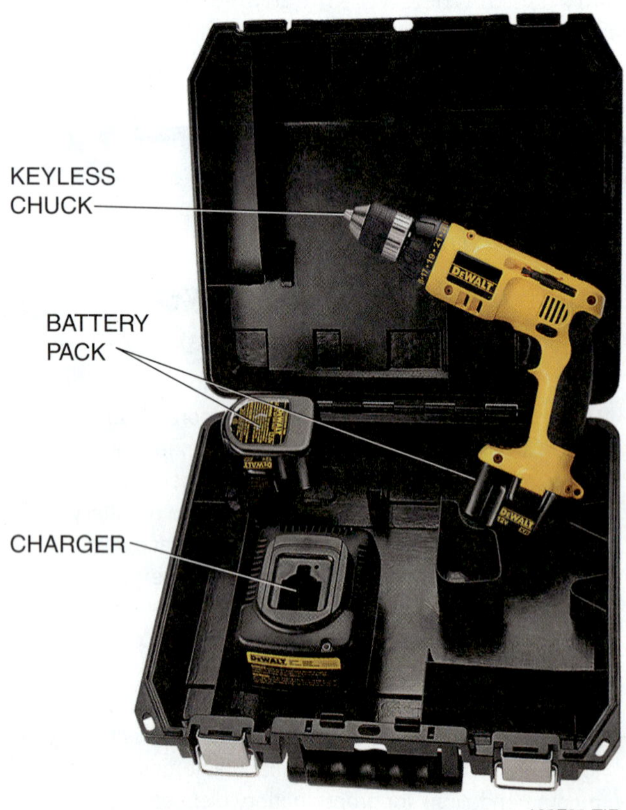

Figure 39 ◆ Cordless drill.

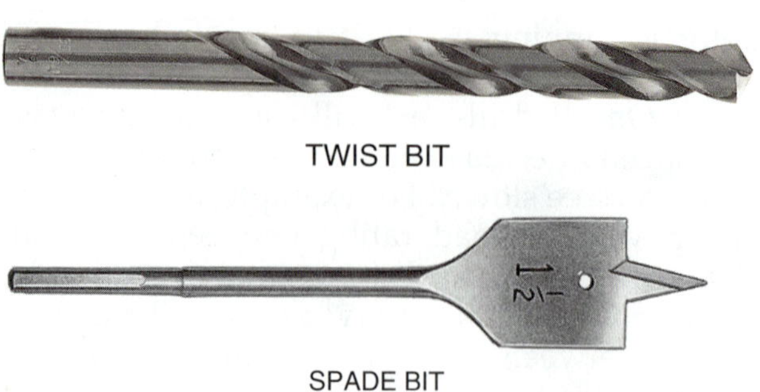

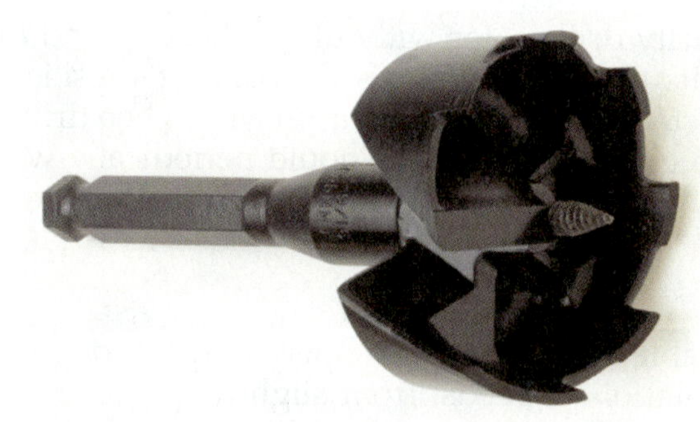

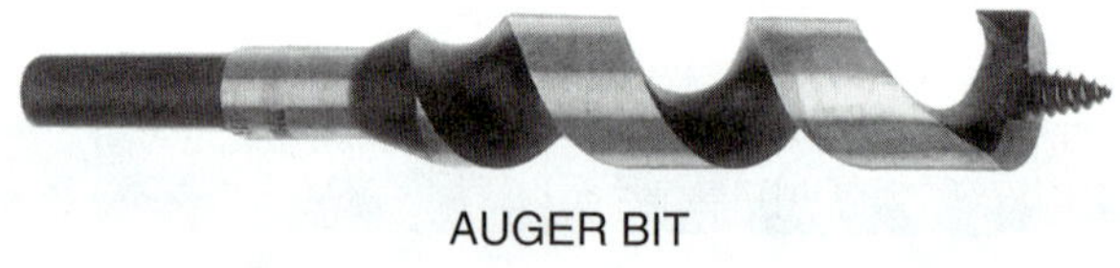

Figure 40 ◆ Types of drill bits.

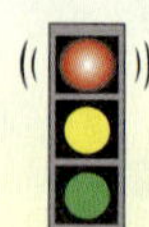

WARNING!

Only trained personnel should sharpen bits. Do not sharpen drill bits without training; you could hurt yourself and damage the bit.

NOTE

Always consult the manufacturer's specifications before using drill bits. Ensure that you are using the correct bit for proper application.

8.3.1 Twist Bit

The twist bit is a versatile drill bit that can be used to drill into almost any material. Twist bits are most commonly used for drilling small holes before installing plumbing fixtures. Twist bits are commonly sized at ½ inch.

Two types of twist bits are available: those made from high-speed steel and those made from carbon steel. Twist bits made from high-speed steel are of higher quality, can better withstand high temperatures without "burning up," and are used to drill metal (they can also drill wood, plastics, and composites). These twist bits have a slightly blunted nose that will not deform under pressure, but they also have a tendency to spin away from the intended hole center. The cheaper, high-carbon steel twist bits are designed specifically for drilling wood. They are not suitable for drilling hard metals such as iron and steel.

8.3.2 Spade Bit

Spade bits are relatively inexpensive. As their name suggests, these bits have a spade-shaped blade with a sharp triangular tip. The spade is flat, and the sharp tip acts as a guide, centering the hole. Most of the drilling is done by the honed cutting edge at the shoulders of the spade. Spade bits cut fast and are long enough to drill through fairly thick work pieces. These bits are designed to cut wood and some plastics. Do not use them to drill metal. Spade bits and the holes they produce range from ¼ inch to 2 inches in diameter.

8.3.3 Hole-Saw Bit

The hole-saw bit has saw-like teeth arranged around a hollow cylinder at the end of the bit. This blade is attached to a shaft, which is inserted into the drill chuck like the shaft of any other drill bit. As the drill turns the bit, the teeth on the blade cut through the material. Use these bits to drill holes between 1 inch and 3½ inches in diameter. Hole-saw bits are appropriate for drilling wood, plastics, and a variety of metals, such as iron, steel, and aluminum. Hole-saw bits are very useful for making holes to install piping, tubing, and conduit.

8.3.4 Self-Feeding Bit

Self-feeding bits, or multispur bits, bore holes in wood quickly. These bits come in diameters ranging from ½ inch to 4⅝ inches. They are good for drilling through floor joists and wall studs and for drilling in hard-to-reach areas because you do not need to move the drill to bore the hole.

CAUTION

Self-feeding drill bits should be used only at right angles to the surface being drilled; otherwise, the bits might slip.

8.3.5 Auger Bit

Auger bits come in diameters ranging from ¾ inch to 1½ inches. They are designed to cut only wood. Do not use them to cut metal; it will damage the bit.

8.3.6 Masonry Bit

A masonry bit has a carbide tip that allows for easy drilling through masonry materials, such as concrete, brick, stone, and plaster, that would quickly dull or fracture most other drill bits. Do not use these specially hardened bits on wood—they are designed only for masonry. Masonry bits should be used at relatively low drill speeds to prevent them from overheating. These bits cut relatively small holes (generally ¾ inch in diameter or less) in masonry through which to run piping.

8.3.7 Care of Bits

Protect drill bits from moisture and keep them sharp. A thin coat of all-purpose oil can help prevent rust. If you notice that a bit is not sharp, replace it or have a qualified person sharpen it.

8.4.0 Pipe Reaming

After you cut a pipe, you will find a burr on the inside of the pipe. A pipe reamer is used to remove the burr (*Figure 41*) in a process called **reaming**. If the burr is left in the pipe, deposits will begin to collect inside the pipe and clog it, slowing the flow of liquid within the pipe (*Figure 42*). Because reamers are tapered, one reamer can deburr many sizes of pipes. Be sure, however, not to overream the inside of a pipe. One type of reamer is a deburr tool (sometimes called a pencil reamer because it is about the size of a pencil), which is used to remove burrs from copper tubing (*Figure 43*).

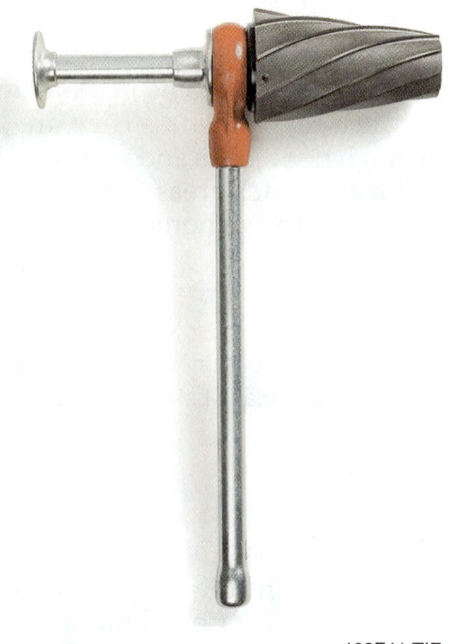

103F41.TIF

Figure 41 ◆ Pipe reamer.

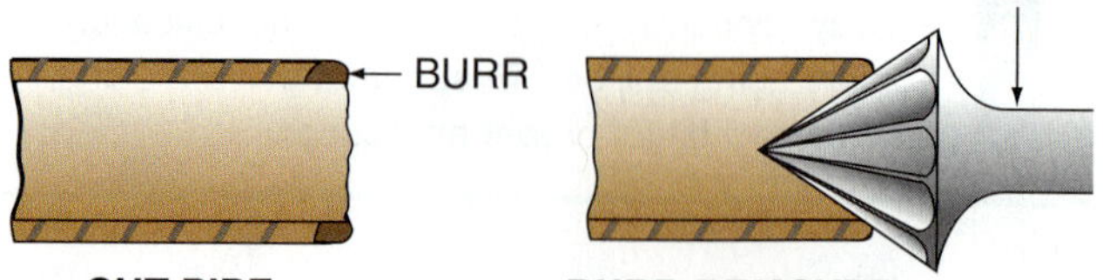

PIPE REAMER
BURR
CUT PIPE
BURR REMOVED

103F42.EPS

Figure 42 ◆ Burr and burr removal by reaming.

103F43.TIF

Figure 43 ◆ Deburr tool.

Review Questions

Sections 8.0.0–8.4.0

1. A _____ is a mechanism at the end of a drill motor that is used to grip or release a drill bit.
   a. vise
   b. chuck
   c. key
   d. setscrew

2. _____ drills are useful because some materials are drilled best at high speeds, others at lower speeds.
   a. Alternating-speed
   b. Electric-speed
   c. Variable-speed
   d. Reversible-speed

3. _____ bits will bore holes in wood quickly and are good for drilling through floor joists and wall studs and for drilling in hard-to-reach areas.
   a. Masonry
   b. Hole-saw
   c. Twist
   d. Self-feeding

4. _____ bits are best for cutting wood and some plastics and should not be used to cut metal.
   a. Self-feeding
   b. Hole-saw
   c. Auger
   d. Spade

5. The _____ bit combines a drill bit with a cylindrical saw blade.
   a. hole saw
   b. twist
   c. variable
   d. pistol

8.5.0 Dies

One method of joining pieces of pipe together is to screw one pipe end into another screwed fitting, such as male and female fittings and couplings. For pipe segments to be joined in this way, the ends of the segments must be prepared by threading. Threading entails passing something through or around a pipe segment to form grooves that allow another segment to be screwed onto or into it. Galvanized pipes and black-iron (cast-iron) pipes need to be threaded on the outside. To make clean cuts of these threads, use die tools (*Figure 44*). Dies must be sharp so that the threads they cut will be correctly tapered.

Applying cutting oil to the die tool will reduce the friction and heat caused by cutting. If there is too much friction or heat, the die tool will push the metal that is being threaded instead of making a clean cut. Heat and friction can also break the teeth off the die.

DIE-TOOL SET

103F44A.TIF

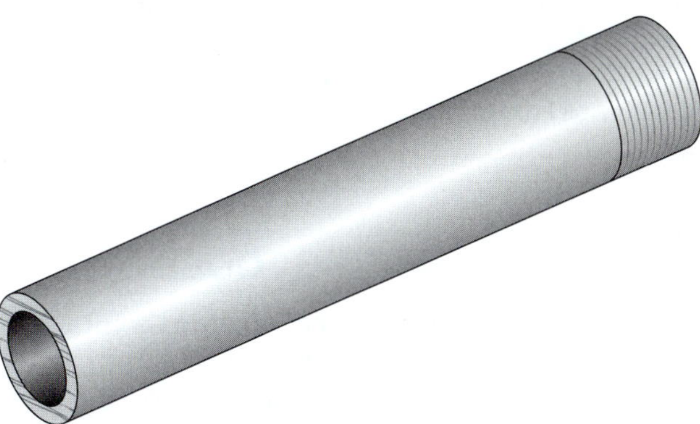

PIPE WITH THREADED END

103F44B.EPS

Figure 44 ◆ Die-tool set and threaded pipe.

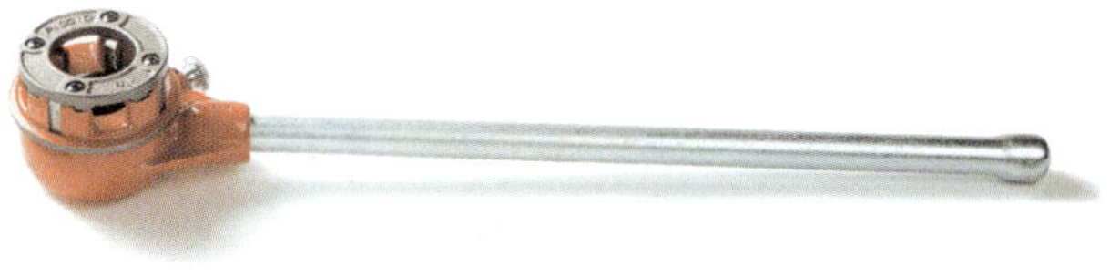

103F45.TIF

Figure 45 ◆ Die and ratchet stock.

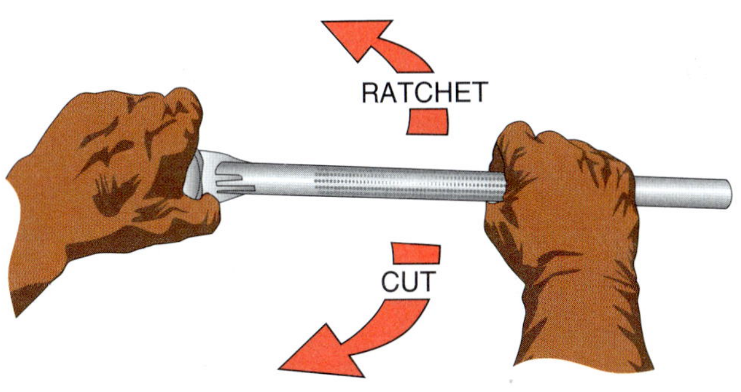

103F46.EPS

Figure 46 ◆ Operating a die and ratchet stock.

Die tools are made to fit all standard pipe sizes and piping materials. Dies cut threads in a variety of sizes, typically in ½-inch, ¾-inch, 1-inch, and 1¼-inch sizes. Be sure the die tools are appropriate for use with the material of the pipe to be threaded. Never use steel-pipe die tools to thread stainless-steel pipe.

The die tool is locked or clamped into a ratchet stock, where the ratcheting mechanism is attached to a handle (*Figure 45*). The ratchet stock helps to turn the die by increasing leverage on it. The ratcheting mechanism makes it easy to operate the stock. To cut the thread, apply pressure in a downward direction and then ratchet the handle upward to prepare for another downward stroke (*Figure 46*).

9.0.0 ◆ ELECTRIC PIPE-THREADING MACHINE

The electric pipe-threading machine rotates, threads, cuts, and reams pipe (*Figure 47*). The pipe is inserted into the chuck of the machine, and the chuck is tightened. A foot pedal controls the machine. Use your foot to operate the pedal. Do not, for example, place a piece of wood or other object on the pedal to depress it. For trouble-free operation, be sure to follow the preventive maintenance schedule provided by the machine's manufacturer.

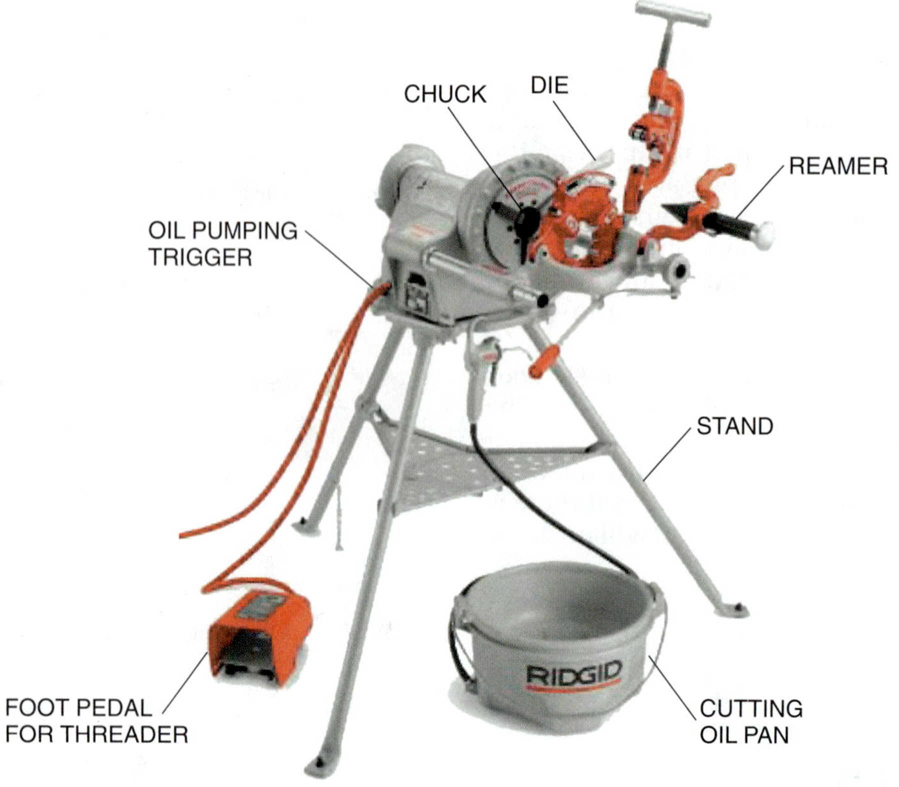

Figure 47 ◆ Electric pipe-threading machine.

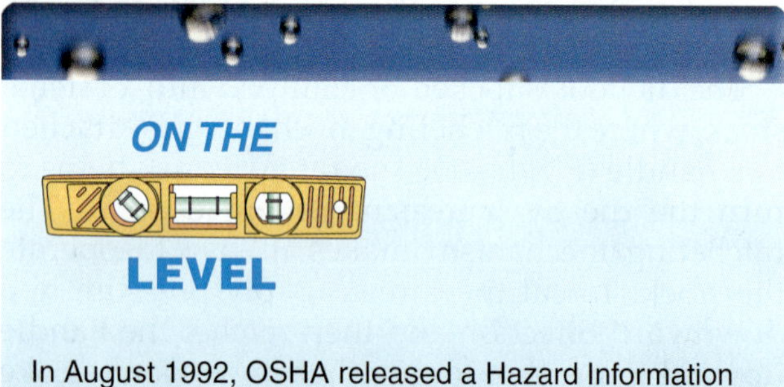

In August 1992, OSHA released a Hazard Information Bulletin (HIB) about an accident involving a Ridgid Model 400A pipe-threading machine. The machine did not have a deadman foot switch and, as a result of the accident, one person died. You can contact Ridgid Tool Company for a list of older models whose foot switches can be retrofitted. Additionally, the company can provide you with information about the proper use of foot switches on their pipe-threading machines.

Source: Occupational Safety and Health Administration website. "OSHA Hazard Information Bulletins: Ridgid Pipe Threading Machines." www.osha.gov/dts/hib/hib_data/hib19920817.html, reviewed February 23, 2003.

WARNING!

When you are working with a power-threading machine, do not wear loose clothing, jewelry, or gloves that could easily be caught in the rotating parts of the machine. Be sure to wear appropriate PPE. Be sure to tie back long hair to prevent it from being entangled in the machine during operation. Use a piece of cloth or cardboard to absorb any oil that spills during the machine's operation, so the oil does not get on the bottom of your work shoes and you do not track the slippery oil around the job site.

10.0.0 ◆ SOLDERING TOOLS

Soldering is the use of heat to form joints. You generally use soldering to join rigid copper pipe and fittings. The filler material, called **solder**, is distributed evenly between the close-fitting surfaces of the pipes when the pipes are heated to the necessary temperature. You use a small, portable gas torch for soldering (*Figure 48*). Propane and acetylene are the two most commonly used gases. The torch heats the metal and the **flux** (also known as soldering paste) to join the pipe. The droplets of solder are molten metal. They will severely burn you if they touch you. Always wear goggles and protective clothing when you are using a welding torch. Weld only under proper supervision and only after proper training.

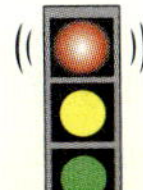

WARNING!

Acetylene has extreme ignition characteristics. Always refer to the manufacturer's specifications for appropriate tank pressure. Avoid contact with copper, silver, and mercury. Acetylene cylinders should remain upright at all times.

Source: Murray State University website. "Department of Chemistry." http://campus.murraystate.edu/academic/faculty/Beth.Brubaker/Gases.PDF, reviewed March 8, 2003.

103F48.EPS

Figure 48 ◆ Propane gas torch.

CAUTION

Acetylene and propane gases are under extreme pressure, can cause explosions, and can release toxic gas into the atmosphere. Follow these general precautions when working with compressed gases.

1. Do not drop cylinders or strike them with force.
2. Leave the cylinder cap on until the cylinder is ready for use.
3. Transport cylinders using an appropriate cart. Do not roll, drag, or slide cylinders.
4. Do not tamper with safety devices on cylinders, regulators, or valves.
5. Do not subject compressed gas cylinders to temperatures exceeding 125°F.
6. Do not allow a flame to contact a compressed gas cylinder.
7. Use compressed gases in well-ventilated areas.
8. Use the smallest cylinder size available that is adequate for your application.
9. Use a trap or check valve to prevent liquid from backing into the regulator or cylinder when discharging gas into a liquid.
10. Refer to the label information and MSDS before handling any compressed gas.
11. Wear appropriate PPE.
12. Use caution when removing a cap from a cylinder that has been stored outdoors. Cylinders stored outside can invite wasps and other insects.
13. Turn off the cylinder valve when the cylinder is not in use.

Source: Murray State University website. "Department of Chemistry." http://campus.murraystate.edu/academic/faculty/Beth.Brubaker/Gases.PDF, reviewed March 8, 2003.

The torch flame produced by a soldering tool has different temperatures, all of them hot, naturally, but some hotter than others. The hottest part of any flame is at the tip of the inner cone, where the pale flame meets the deeper-colored outer flame.

Propane torches are used for soldering copper pipes. Heat between 400°F and 800°F can solder plumbing connections. Temperatures of 3,000°F can braze metals and cut or weld iron and mild steel. Be sure to match the torch's heat output to the task. While propane will burn on its own by drawing oxygen from the air, supplementing the propane gas from a handheld torch with oxygen from a separate tank can generate even more heat.

You can use a carbon-arc torch to solder copper, tinned, and galvanized metal parts. Overlap the parts to be soldered for best results. Begin by cleaning the surfaces to be soldered. Cover the surfaces with soldering flux or acid core solder to prevent oxidation and ensure proper adhesion. Start the arc on the area to be soldered. When the joint reaches a temperature high enough to melt the solder, feed the solder into the joint. Do not use excessive heat. This will cause the solder to boil.

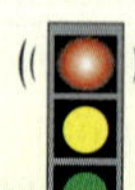

WARNING!

Working with soldering tools can be dangerous. Soldering irons can easily burn you and objects around you. Never touch the tip and heater while the unit is on. Allow plenty of time for the iron to cool before changing the tip. Use a stand for your soldering iron and keep it a safe distance from flammable materials.

CAUTION

If you suffer from any chest problems, be sure to work in a well-ventilated area. Soldering irons contain rosin-based fluxes, which can irritate the respiratory system. Rosin-free solder is also available.

ON THE LEVEL

Brazing

In plumbing, a process called brazing is used to join pipe and fittings for pipelines that carry saltwater and oil and for refrigeration and chemical-handling systems, vacuum and air lines, and low-pressure steam lines. Brazing requires much hotter temperatures than those needed for soldering copper pipe.

To achieve a sufficiently hot flame for brazing, use an oxyacetylene torch unit capable of generating temperatures of 1,400°F (or 760°C) or higher. Brazing requires a torch with the correct tip, regulator valves, and both oxygen and acetylene tanks. These tanks come in two sizes. Plumbers commonly use the smaller B-tank, which has two compartments to hold the pressurized acetylene and oxygen.

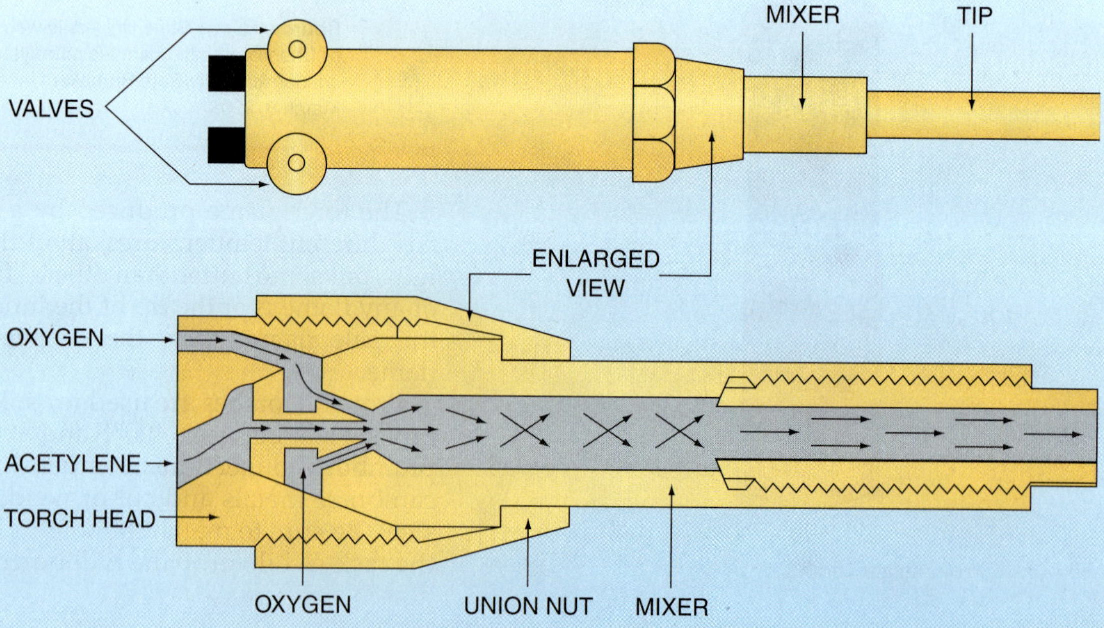

MIXER
TIP
VALVES
ENLARGED VIEW
OXYGEN
ACETYLENE
TORCH HEAD
OXYGEN
UNION NUT
MIXER

103SA01.EPS

You can also perform soldering using a single bare carbon electrode placed in the electrode holder. Begin by sharpening the carbon electrode at one end to resemble a pencil with a dull point. Then insert the carbon electrode in the electrode holder. Adjust the DC welding machine to minimum output. Set it for straight polarity. Attach the work clamp to the parent metal to be soldered. Clean and flux the metal.

To begin soldering, touch the carbon electrode to the metal, but do not draw an arc. Hold the electrode in contact with the metal until the point becomes red hot. When the metal becomes hot enough to melt the solder, drag the carbon electrode slowly over the metal. Feed in more solder as needed. Do not lift the carbon electrode while soldering. When you are finished, quickly remove the carbon from the work. Avoid creating an arc.

A striker or a pack of flints (*Figure 49*) can be used to light torches that do not have self-igniting capability.

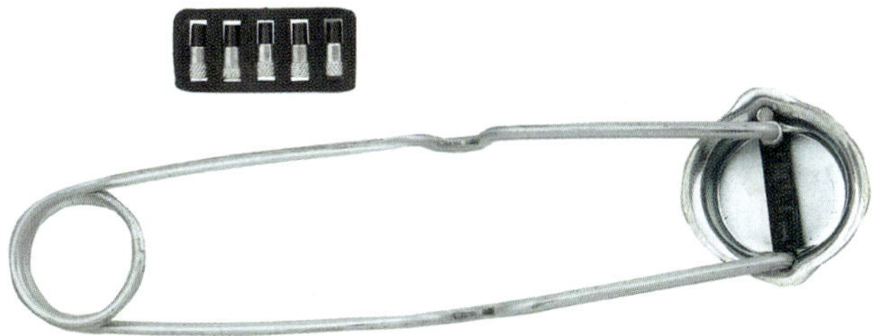

103F49.TIF

Figure 49 ◆ Flint spark lighter with flints.

NOTE

Dispose of propane fuel tanks properly to avoid harming the environment. Many communities have hazardous waste disposal areas set up to handle the disposal of such items.

Review Questions

Sections 8.5.0–10.0.0

1. _____ are used to cut threads into galvanized and black-iron (cast-iron) pipes.
   a. Augers
   b. Bits
   c. Dies
   d. Tube cutters

2. A _____ helps to turn a die tool by increasing the leverage on it.
   a. torque wrench
   b. ratchet stock
   c. threader
   d. chuck key

3. An electric pipe-threading machine rotates, cuts, threads, and _____ pipe.
   a. joins
   b. drills
   c. ratchets
   d. reams

4. Plumbers use a propane or acetylene gas torch when they are soldering to _____.
   a. thread the pipe
   b. oxidize the pipe and the fitting
   c. heat the pipe and the fitting
   d. remove any burrs

5. Temperatures between _____ and _____ can solder plumbing connections.
   a. 100°F, 300°F
   b. 200°F, 600°F
   c. 400°F, 800°F
   d. 600°F, 1,000°F

11.0.0 ◆ EXTENSION CORDS

To function properly, a power tool must receive enough electrical energy. As electricity travels the length of an extension cord, it tends to lose voltage. This loss is called **voltage drop**. When voltage drops, the tool has to draw more **amperage**, causing the motor to run hotter. If such unusual amperage draw continues for a prolonged period, the motor can overheat and be ruined.

Table 1 shows the wire gauge required for extension cords at given lengths for given amperage draws. The lower the gauge number, the heavier the wire. Be sure that the extension cords you use, especially long ones, are heavy enough to minimize voltage drop.

Receptacles on the ends of extension cords are not part of the permanent wiring of the building. Therefore, receptacles must be protected by GFCIs, according to OSHA regulations, whether or not the extension cord is plugged into the permanent wiring. Even when GFCIs are in use, always protect extension cords from excessive moisture.

OSHA and your company have specific policies and procedures to keep the workplace safe from electrical hazards. You can do many things to reduce the chance of an electrical accident. If you ever have any questions about electrical safety on the job site, ask your supervisor. Adhere to these basic, job-site electrical safety guidelines:

- Use three-wire extension cords, and protect them from damage. Never fasten them with staples, hang them from nails, or suspend them from wires. Never use damaged cords.
- Make sure that panels, switches, outlets, and plugs are grounded.
- Never use bare electrical wire.

Electrical Cord Safety

Electrical cords are frequently seen on construction sites, yet they are often overlooked. Use the following safety guidelines to ensure your safety and the safety of other workers.

- Every electrical cord should have an Underwriters Laboratories (UL) label attached to it. Check the UL label for specific wattage. Do not plug more than the specified number of watts into an electrical cord.
- A cord set not marked for outdoor use is to be used indoors only. Check the UL label on the cord for an outdoor marking.
- Do not remove, bend, or modify any metal prongs or pins of an electrical cord.
- Extension cords used with portable tools and equipment must be three-wire type and designated for hard or extra-hard use. Check the UL label for the cord's use designation.
- Avoid overheating an electrical cord. Make sure the cord is uncoiled, and that it does not run under any covering materials, such as tarps, insulation rolls, or lumber.
- Do not run a cord through doorways or through holes in ceilings, walls, and floors, which might pinch the cord. Also, check to see that there are no sharp corners along the cord's path. Any of these situations will lead to cord damage.
- Extension cords are a tripping hazard. They should never be left unattended and should always be put away when not in use.

Table 1 Amperage Ratings

Amps Rating (on nameplate)	**0–2.0**	**2.1–3.4**	**3.5–5.0**	**5.1–7.0**	**7.1–12.0**	**12.1–16.0**	
Ext. Cord Length	**Wire Size**						
25'	18	18	18	18	16	14	
50'	18	18	18	16	14	12	
75'	18	18	16	14	12	10	
100'	18	16	14	12	10	8	
150'	16	14	12	12	8	8	
200'	16	14	12	10	8	6	**Not normally available as flexible extension cords**
300'	14	12	10	8	6	4	
400'	12	10	8	5	4	4	
500'	12	10	8	5	4	2	
600'	10	8	6	4	2	2	
800'	10	8	6	4	2	1	
1,000'	8	6	4	2	1	0	

DID YOU KNOW?

Amperage is the strength of an electrical current, which is expressed in amperes (commonly called amps). An ampere is a unit of electrical current. The higher the amps, the more current is flowing through an electrical wire. All electrical wire is rated by the number of amps the wire is designed to carry. For example, there are 20-amp, 60-amp, and 80-amp wires. Always be sure that you use a wire that can handle the amperage drawn by the tool you are using. Too many amps flowing through a wire not designed for that current will cause the wire to heat up and start a fire.

- Never use metal ladders near any source of electricity.
- Never wear a metal hard hat.
- Always inspect electrical power tools before you use them.
- Never operate any piece of electrical equipment that has a danger tag or lockout device attached to it.
- Use three-wire cords for portable power tools, and make sure they are properly connected.
- Never use worn or frayed cables.

12.0.0 ◆ TOOLS FOR ASSEMBLY AND HOLDING

A crucial tool for assembly is the wrench. As a plumber, you will use a variety of wrenches for turning pipe, fittings, and fasteners. Some wrenches, such as a strap wrench, are designed to leave no marks. They are used for turning finish fasteners, where the slightest mark will ruin the appearance of the finish. Other wrenches are used to tightly grip materials that are hidden in a building's structure, where marks left behind are not a concern.

Holding tools include vises and pliers. These tools hold plumbing materials and work pieces in place so that you can do something, like drill a hole, without the material or work piece moving around. These holding tools help you to be precise in your work.

12.1.0 Wrenches

A wrench is a familiar tool, usually operated by hand, for tightening or loosening bolts, nuts, or anything that needs to be turned. Solymon Merrick patented the first wrench in 1835. The physics of a wrench allow it to work as a lever, with a jaw, strap, or chain that grips or fits around the object needing to be tightened or loosened. Many wrenches also have special uses. Some wrenches have adjustable jaws. Be sure to keep the jaws clean by using a wire brush.

12.1.1 Pipe Wrench

The pipe wrench is used to grip and turn round pipe and tubing. Types include regular heavy-duty pipe wrenches, 45-degree and 90-degree offset pipe wrenches, and heavy-duty end wrenches (*Figure 50*). Wrenches come in a number of lengths, such as 6, 12, 14, 18, 24, 36, and 48 inches. The wrench's length determines the size of pipe that it can handle.

The straight pipe wrench is the most common type. When working in tight quarters, use offset wrenches; their angles (45 or 90 degrees) make it easier to reach pipe in cramped areas. The jaws of pipe wrenches always leave marks on the pipe; therefore, do not use them on pieces where appearance is important. When applying force to a wrench, always direct it toward the open side of the jaws (*Figure 51*). This technique will give you the best grip and leverage.

A variation on the straight pipe wrench is the compound-leverage pipe wrench (*Figure 52*). This design increases the leverage that can be applied on a pipe. This wrench is generally used on pipe joints that are frozen or are locked together.

Another variation of the pipe wrench is the chain wrench (*Figure 53*). This tool has a length of chain permanently attached to the wrench handle

DID YOU KNOW?

The pipe wrench was invented by a steamboat firefighter named Daniel Stillson. Stillson had suggested to Walworth, a heating and piping firm, that it design a wrench that could screw pipes together. The firm's owner told Stillson to take a crack at making a prototype that would twist off the pipe. To the everlasting gratitude of future plumbers, Stillson was successful and his prototype twisted off the pipe. In 1870, Stillson's design was patented, and Walworth began manufacturing the wrench. Even today, pipe wrenches are sometimes called Stillson wrenches.

Source: Inventors website. Mary Bellis. "Pipe Wrench." http://inventors.about.com/library/inventors/blwrench.htm, reviewed January 22, 2004.

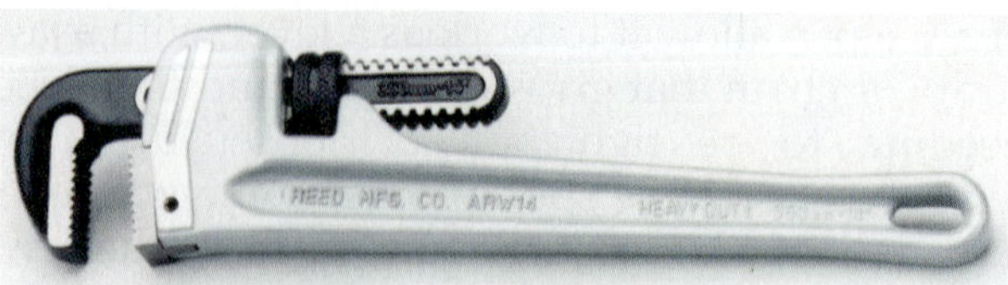
STRAIGHT PIPE WRENCH

HEAVY-DUTY PIPE WRENCH
45-DEGREE OFFSET WRENCH

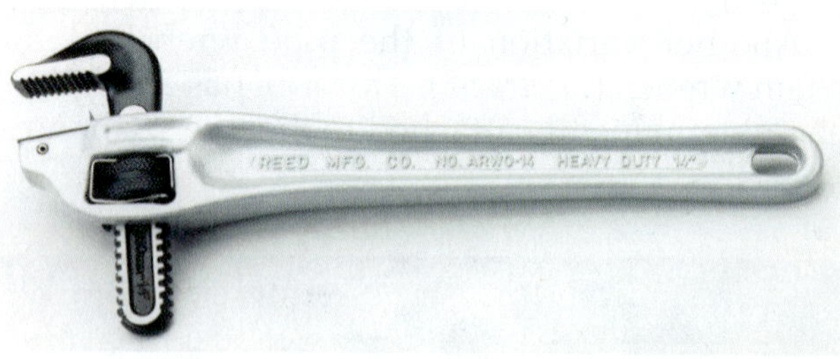
90-DEGREE OFFSET PIPE WRENCH

HEAVY-DUTY END WRENCH

103F50.EPS

Figure 50 ◆ Pipe wrenches.

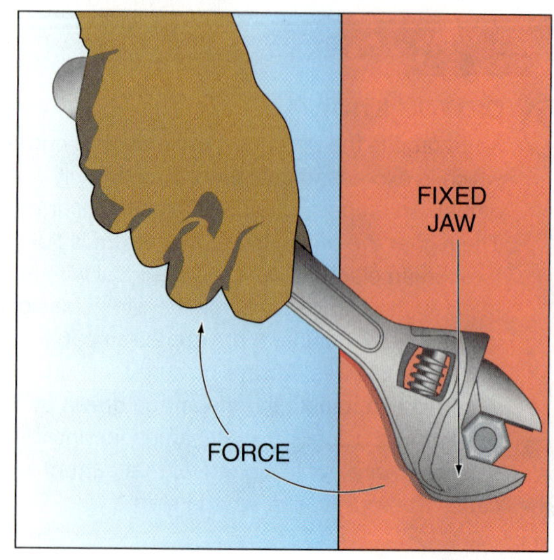

103F51.EPS

Figure 51 ◆ Correct method of applying force to a wrench.

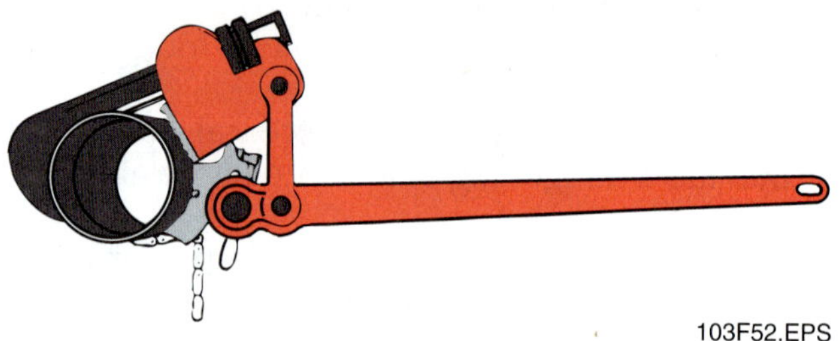

103F52.EPS

Figure 52 ◆ Compound-leverage pipe wrench.

103F53.TIF

Figure 53 ◆ Chain wrench.

at one end. The chain is looped around the pipe to grip and secure it, and the other end of the chain can be secured to various positions on the wrench's handle. Oil the chain frequently to prevent it from becoming stiff or rusty.

12.1.2 Pipe Tongs

Pipe tongs are the wrenches typically used on large pipe (*Figure 54*). Pipe tongs also have chains, which need to be oiled frequently. They provide extra leverage needed for tougher, more heavy-duty jobs.

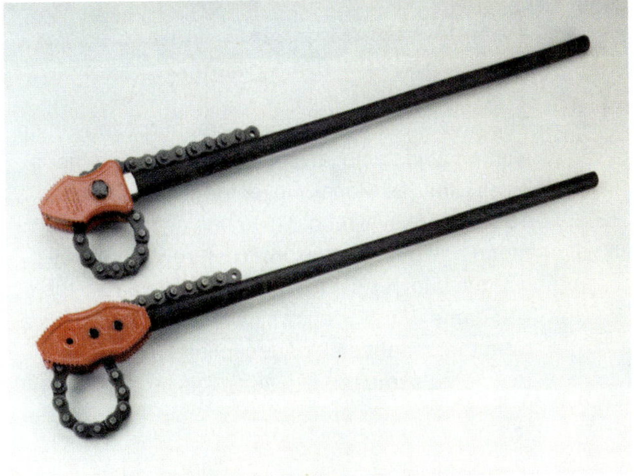

103F54.TIF

Figure 54 ◆ Pipe tongs.

CAUTION

Never use "cheaters" (pipes used to extend wrenches) because they can cause equipment damage. Always use the right-sized wrench to avoid damaging the pipe.

12.1.3 Strap Wrench

The strap wrench (*Figure 55*) is used to hold chrome-plated or other types of finished pipe. The strap wrench does not leave jaw marks or scratches on the pipe because it grips with a strap rather than a jaw or chain. With some strap wrenches, rosin needs to be applied to the strap so that it does not slip. Other strap wrenches use vinyl straps that do not need rosin.

12.1.4 Spud Wrench

The spud wrench is similar to a pipe wrench, except that it has no teeth in its jaws (*Figure 56*).

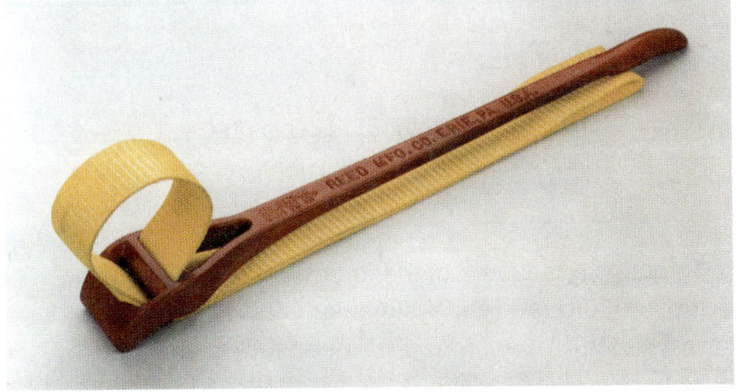

103F55.TIF

Figure 55 ◆ Strap wrench.

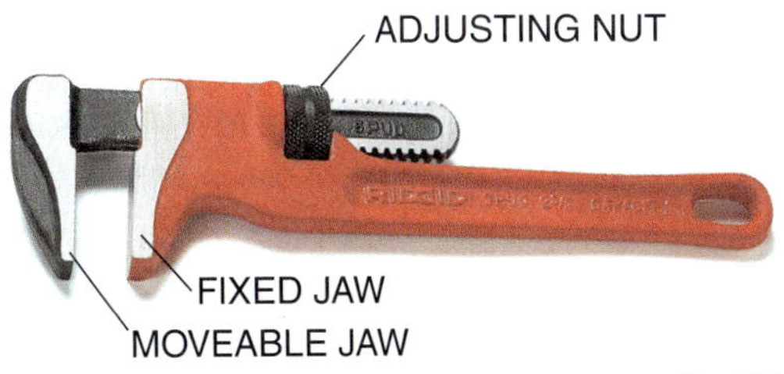

103F56.TIF

Figure 56 ◆ Spud wrench.

Spud wrenches loosen and tighten fittings on drain traps, sink strainers, and toilet connections and are also used on large, oddly shaped nuts. The wrench's narrow jaws allow it to fit into tight places. The spud wrench is difficult to adjust, though, and plumbers often use adjustable wrenches and open-end wrenches instead.

12.1.5 Open-End Wrench

The open-end wrench (*Figure 57*) usually has two different-sized openings, each on one end of the tool. The size of each opening is stamped on the handle. This wrench is usually used for assembling fittings and fasteners that are less than 1 inch across.

12.1.6 Adjustable Wrench

The adjustable wrench, like the Crescent® wrench in *Figure 58*, is similar to the open-end wrench, except that it has an adjustable jaw. A roller at the

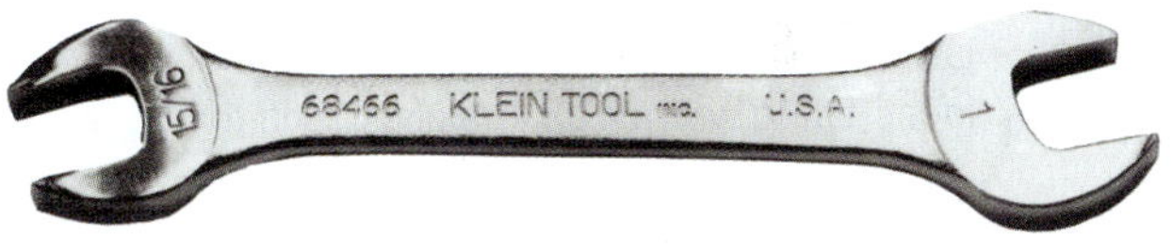

103F57.TIF

Figure 57 ◆ Open-end wrench.

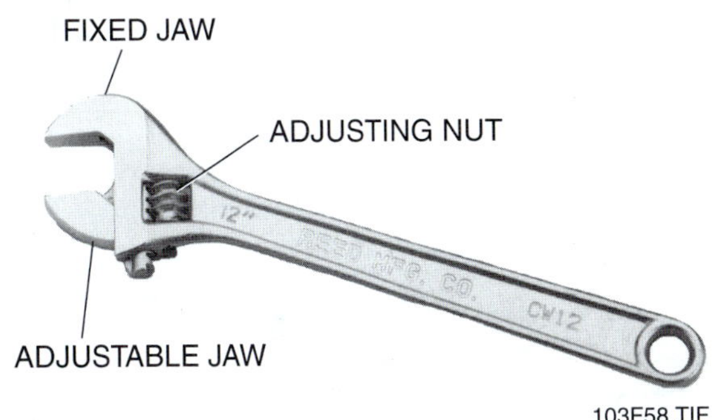

103F58.TIF

Figure 58 ◆ Adjustable Crescent® wrench.

stem of the jaw adjusts the size of the opening. Because it is so easy to adjust, this wrench is found in virtually every toolbox. Various sizes of adjustable wrenches are available, from 6 to 18 inches in length.

Adjustable wrenches have smooth jaws, making these tools ideal for turning nuts, bolts, small pipe fittings, and chrome-plated pipe fittings. Using an adjustable wrench can save you considerable time, because you will not have to constantly switch wrench sizes as you work with different sizes of nuts and bolts.

12.1.7 Basin Wrench

The basin wrench (*Figure 59*) was developed to work in tight, hard-to-reach areas such as recesses where sink and lavatory faucets are secured. This specialty wrench, with its offset jaws, permits you to turn a nut or a fastener that you could not reach with other wrenches.

12.1.8 Monkey Wrench

The monkey wrench is a flat-jawed wrench with one movable jaw that is adjusted by a screw. Because it can fit around various sizes of nuts, this wrench is commonly used when an open-end wrench does not fit properly. The jaws of a modern monkey wrench can be finely adjusted to fit nonstandard bolts.

CAUTION

Do not use a monkey wrench as a hammer. Doing so can damage the wrench, causing its jaws to loosen and making it difficult to adjust. A loose-fitting wrench can easily slip off the nut and cause injuries.

12.1.9 Torque Wrench

The torque wrench has a gauge attached to it, which indicates the torque (the force that is being applied to the work piece). This wrench is known for its accuracy and reliability. There are several types of torque wrenches. The no-hub design does not have a gauge; instead, it is designed to disengage at a preset torque limit. The clicker-type wrench makes adjustments easy because it has a sliding scale with a locking mechanism. Other torque wrenches have ratcheting heads that can be removed. These wrenches allow you to remove the head and re-insert it to torque in the left-hand direction. Because of the low-torque range, this wrench is popular with transmission mechanics.

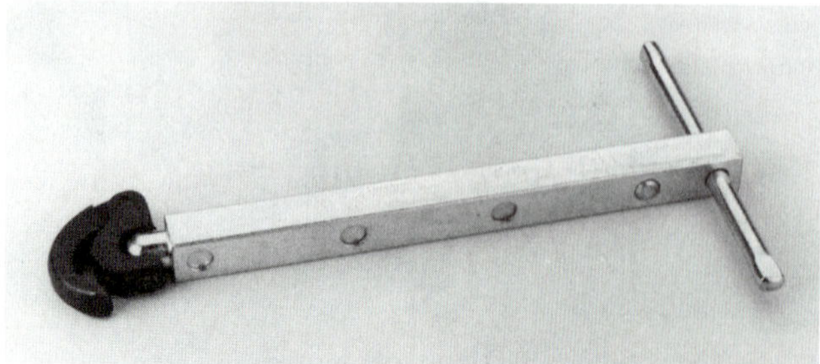

103F59.TIF

Figure 59 ◆ Basin wrench.

DID YOU KNOW?

The monkey wrench is a general-purpose wrench with an adjustable jaw for turning nuts of various sizes. Charles Moncky invented this tool around 1858. "Monkey wrench" also has an informal meaning: to disrupt, as in, "He threw a monkey wrench into our plans." *Monkey* comes from the last name of the wrench's inventor. *Wrench* was a term that meant trick or deception long before the tool of the same name was developed.

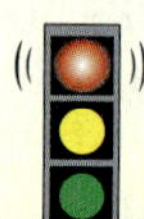

WARNING!

When using a torque wrench, be careful not to force the tool. Battery-operated wrenches can cause just as much damage as electric ones. Using excess torque on either model can break your wrist.

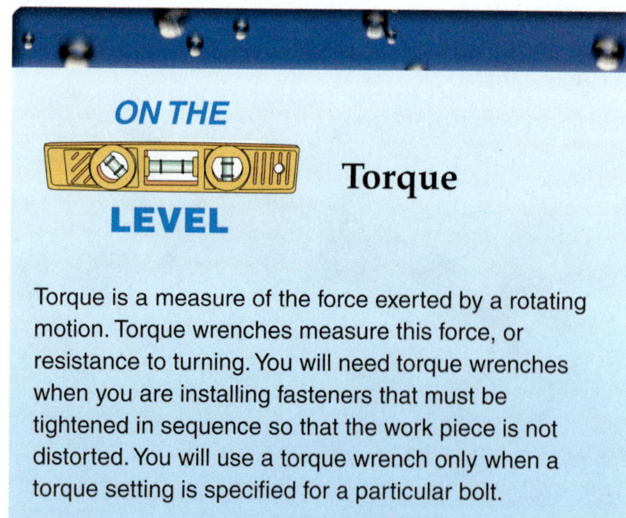

Torque

Torque is a measure of the force exerted by a rotating motion. Torque wrenches measure this force, or resistance to turning. You will need torque wrenches when you are installing fasteners that must be tightened in sequence so that the work piece is not distorted. You will use a torque wrench only when a torque setting is specified for a particular bolt.

CAUTION

When using a torque wrench, you must install fasteners in a sequence so that all fasteners are tightened at the same rate. If you tighten one fastener more than another, you can distort and damage the work piece.

12.1.10 Wrench Safety

You should develop some specific safety habits when using wrenches, including the following:

- Use wrenches that are the right type and size for the job.
- Do not use loose wrenches. They will slip and can cause serious injury. A loose wrench can also ruin the corners of a cap screw or nut.
- Take care when working in a tight space that your grip on the wrench will not injure your hand as you turn the wrench.
- Pull up on the wrench; do not push it. If you push on a wrench, slippage can cause injury to your hand, or even to your face and body. If you cannot avoid pushing a wrench, do so with a stiff arm, holding your face and body back.
- Do not place too much strain on a small wrench.
- Do not use a wrench as a hammer.
- Switch to a larger wrench if you need more leverage.

12.2.0 Pliers

Pliers are a type of adjustable wrench and are used for bending, gripping, and holding. Some pliers can also be used to cut wires. Pliers come in many different head styles, depending on their use. Do not use pliers on exposed pipe or fasteners, because they will leave jaw marks.

12.2.1 Slip-Joint Pliers

Slip-joint pliers are very popular (*Figure 60*). They are used for bending wire and for gripping and holding during assembly operations. The adjustable jaws make it easy to change the distance between them. There are two jaw settings: one for small materials and one for larger materials.

12.2.2 Lineman's Pliers

Lineman's pliers (*Figure 61*), also known as side cutters, have wider jaws than slip-joint pliers. Lineman's pliers are used mainly for cutting heavy or large-gauge wire and for holding. The shape of the jaws reduces the chances that wires will slip, and the hook bend in the handles provides a better grip.

103F60.TIF

Figure 60 ◆ Slip-joint pliers.

103F61.TIF

Figure 61 ◆ Lineman's pliers.

12.2.3 Slip-Lock Pliers

Slip-lock pliers work on the same principle as slip-joint pliers, but their jaws have five or more adjustments (*Figure 62*). Slip-lock pliers can hold much larger materials than channel-lock pliers. Do not use slip-lock pliers to hold small materials because the jaws cannot be adjusted flat against each other.

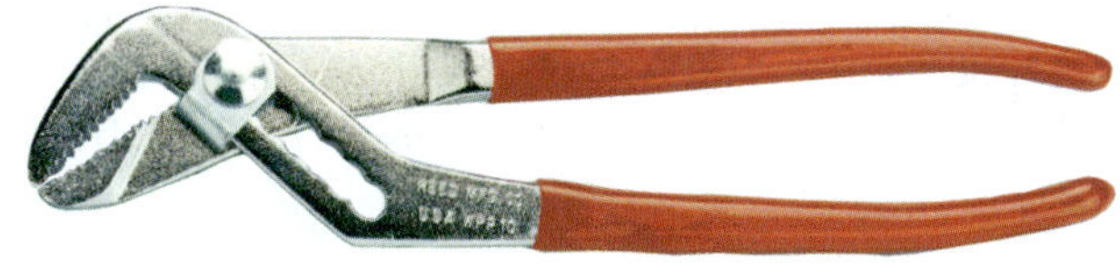

103F62.TIF

Figure 62 ◆ Slip-lock pliers.

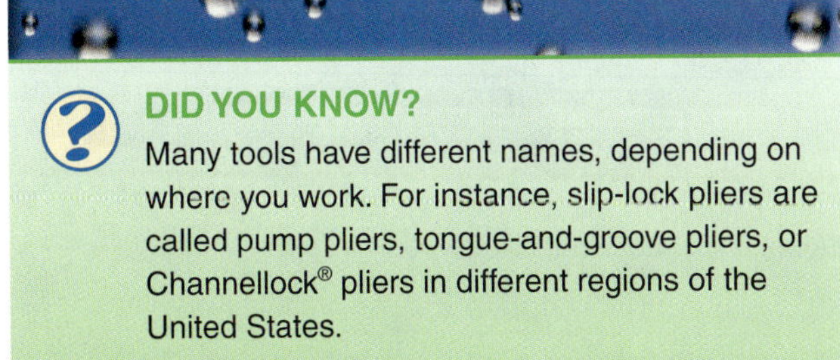

DID YOU KNOW?

Many tools have different names, depending on where you work. For instance, slip-lock pliers are called pump pliers, tongue-and-groove pliers, or Channellock® pliers in different regions of the United States.

12.2.4 Locking Pliers

These handy, adjustable pliers can be used both as pliers and as a handheld vise or clamp (*Figure 63*). They are also known as plier wrenches, lever-wrench pliers, and by their proprietary name, Vise-Grips®. The jaws of these pliers can be adjusted by turning a screw attached to one of the handles, to lock firmly onto a work piece (usually one made of metal) in a vise-like grip. A trigger on the other handle loosens the grip. The standard design for locking pliers has serrated, straight jaws, but these pliers also have other designs (long-nosed, flat-jawed, C-clamp jaws). Because of their fine jaw adjustment and tremendous clamping power, locking pliers are an excellent tool for removing bolts with heads that have been rounded off.

12.2.5 Care of Pliers

Pliers need very little preventive maintenance. A little oil on all movable joints and some oil on the locking pliers' adjustment screw will prevent rust and keep them working smoothly. Do not expose pliers to extreme heat, which can melt their plastic handles. To ensure the longevity of your pliers, use them only for their intended purpose. Do not use pliers to hammer nails or to turn nuts or bolts.

12.2.6 Safety

The following are the specific safety guidelines for pliers:

- Pliers must not be used as a substitute for wrenches, because pliers cannot hold a work piece as securely as wrenches. Using pliers as wrenches will lead to slippage, which could cause injury.
- Insulated pliers must be used for electrical work.

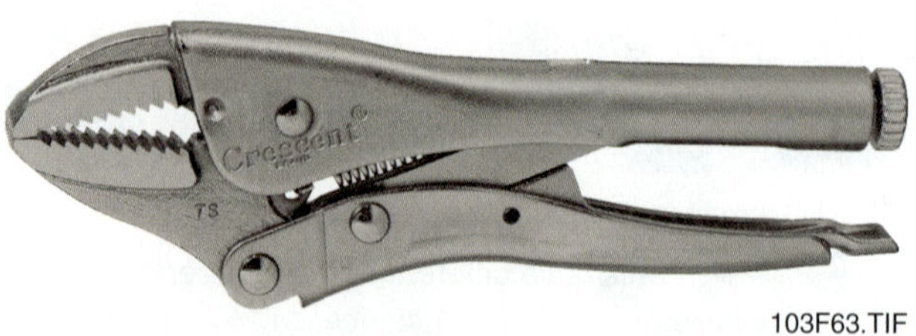

Figure 63 ◆ Locking Vise-Grips®.

13.0.0 ◆ HAMMERS

Plumbers, like all other skilled workers in the construction profession, use hammers. There are many types of hammers, including claw hammers, ball peen hammers, sledgehammers, and mauls.

Hammers come in many sizes. The size of a hammer is determined by the weight of its head. Usually a 14- or 16-ounce hammer is considered medium weight (the weight is stamped on the hammer head). Some claw hammers are as heavy as 32 ounces, and, of course, sledgehammers can weigh several pounds.

Hammers also come with different types of handles. Hammers with wooden (usually hickory wood) handles are the least expensive, and the handles are comfortable, but the hammer head can easily work loose from the handle and may need periodic reshimming. Wooden handles can also break from vigorous nail pulling or if subjected to repeated overstriking. Steel handles are very strong (the head and the handle are one forged tool) but they transfer more of the shock of the hammer blow to your hand than hammers with wooden or fiberglass handles do. Fiberglass-reinforced plastic handles can be a good compromise. Fiberglass handles are almost as comfortable as wooden handles, and they are also durable and shock absorbent.

Always wear eye protection when using a hammer. Check to make sure handles are securely wedged onto the hammer head before use, and do not use hammers that have cracked, splintered, or badly worn handles. Never strike one hammer face with or against another hammer or a hatchet. Never strike nail pullers, steel chisels, or other hardened objects with a nail hammer, because the face of the hammer may chip. Nail hammers are intended for driving or pulling common, unhardened nails only.

13.1.0 Claw Hammer

The most common type of hammer is the carpenter's curved claw hammer (*Figure 64*). The hammer's head has a face for striking nails and, on the opposite side of the hammer head, a two-pronged claw for pulling nails. The head is steel. The handle can be fiberglass, wood, or steel. Fiberglass and steel hammers usually have shock-absorbing rubber or vinyl coatings on the handle to improve the grip. Curved-claw hammers can be 7 or 16 ounces.

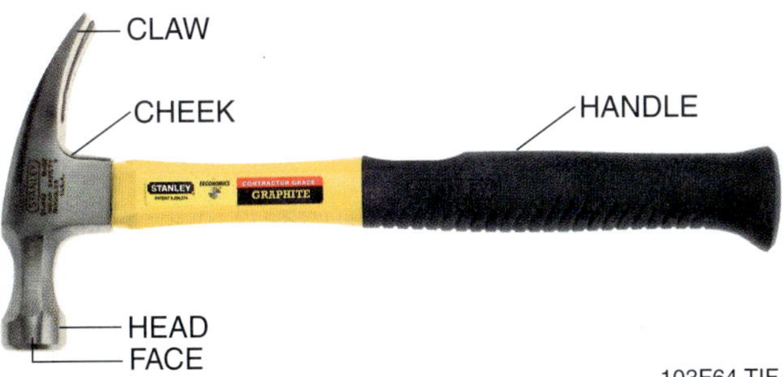

Figure 64 ◆ Claw hammer.

Figure 66 ◆ Stanley® jacketed graphite nail rip-claw hammer.

To pull a nail, slip the claw of the hammer under the nail head and pull until the handle is nearly straight up (vertical) and the nail is partly drawn out of the wood. Then, pull the nail straight up from the wood (*Figure 65*). To gain added leverage when pulling out long nails, firmly secure a block underneath the hammer's head and on the same surface from which you are pulling the nail.

Variations on the claw hammer are straight-claw hammers and rip-claw hammers (*Figure 66*). Straight-claw hammers are better at prying out nails, but curved-claw hammers are more compact and better at pulling nails. Rip-claw hammers are 16 or 20 ounces, and the handles can be made of jacketed graphite, jacketed fiberglass, or tubular steel.

WARNING!

When working in close quarters, it is best to use a ball peen hammer instead of a claw hammer. In a tight space, you could accidentally claw yourself with the curved end of the hammer.

13.2.0 Ball Peen Hammer

The ball peen hammer (*Figure 67*), also known as the plumber's hammer, has a flat face for striking and, on the other side of its head, a rounded face that can align brackets and drive out bolts. This hammer is also classified by weight. Ball peen hammer heads range in weight from 6 ounces to 2½ pounds. Ball peen hammers are designed for

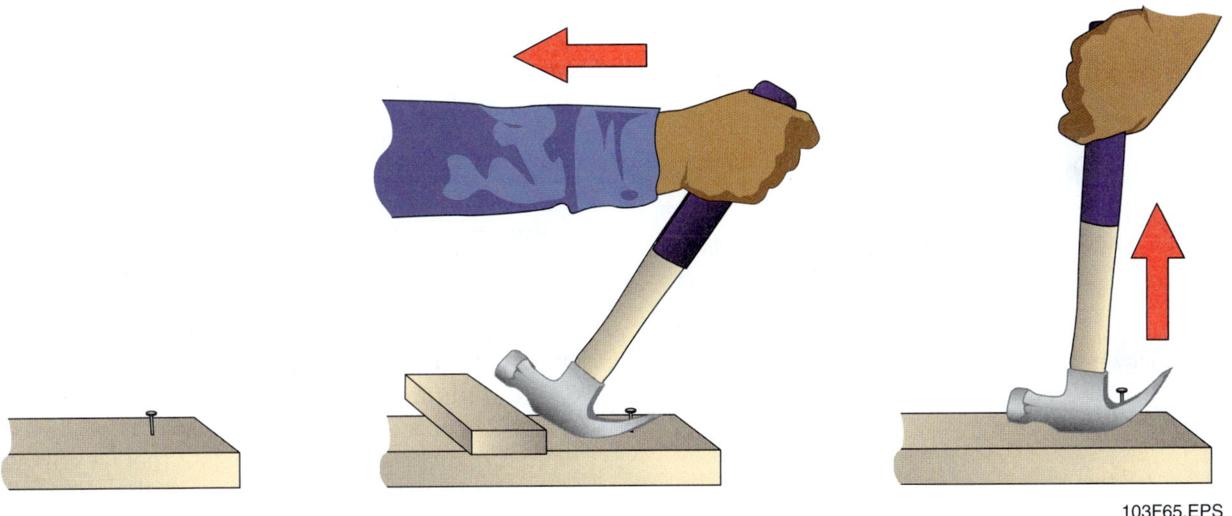

Figure 65 ◆ Using a claw hammer to pull a nail.

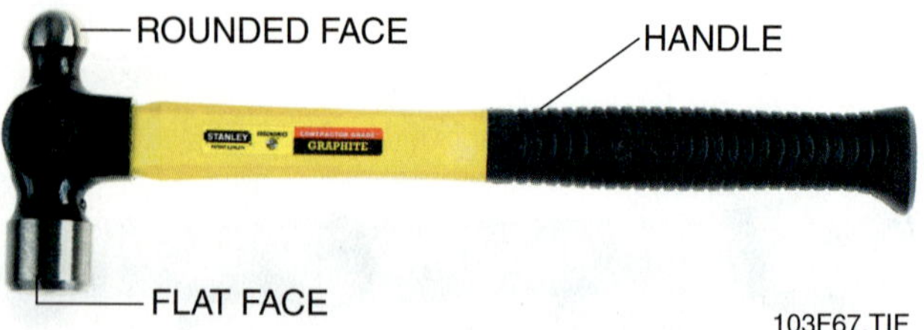

Figure 67 ◆ Ball peen hammer.

striking chisels and for riveting, shaping, and straightening unhardened metal. When using a ball peen hammer to strike a chisel or other tools that are designed to be struck, make sure the hammer's face that is doing the striking has a diameter at least ⅜ inch larger than the face of the tool that is being struck.

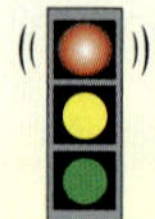

WARNING!

When using a ball peen hammer, strike squarely at the chisel. Glancing blows off the chisel can cause the edge of the hammer's face to chip, which could lead to serious injury, especially to the eyes.

13.3.0 Sledgehammer

A sledgehammer is a particularly heavy hammer (*Figure 68*). The sledgehammer is a heavy-duty tool used for driving posts or other large stakes. It can also be used for breaking up cast iron or concrete. The head of the sledgehammer is made of high-carbon steel and weighs from 2 to 20 pounds. The shape of the sledgehammer's head and the length of its handle depend on the job it is designed to do.

Always wear eye protection when using a sledgehammer, and make sure the handle is securely wedged to the sledgehammer's head. Never attempt to strike an object at or above shoulder level with a sledgehammer. Instead, use a platform to raise yourself above the object you want to strike. If a co-worker is holding a stake, nail, pin, or other object to be driven by the sledgehammer, that person should stand at a right angle to the direction of the sledgehammer, grip the item being driven with a holding device, and wear gloves.

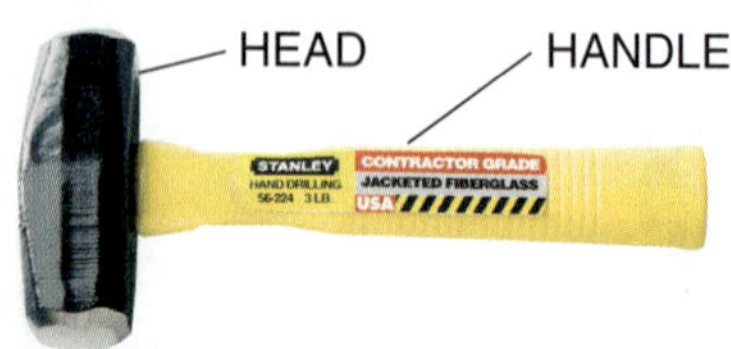

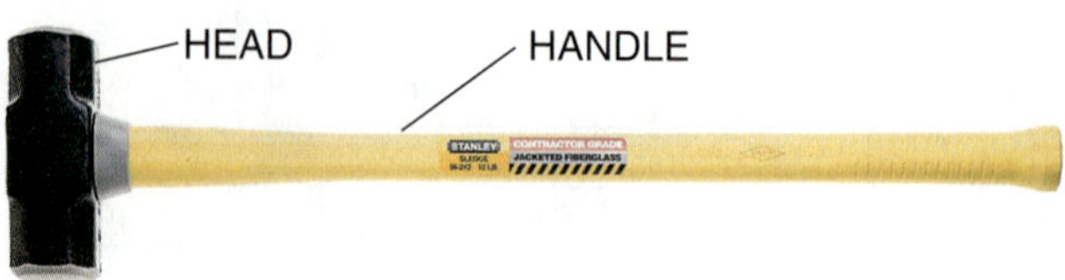

Figure 68 ◆ Sledgehammers.

13.4.0 Maul

Mauls are designed for splitting wood and are also used along with wood-splitting wedges. When using a wedge with a maul, first make a notch with the splitting edge, and then drive the wedge with the maul's striking face.

Never use mauls to strike concrete, metal, or other hard materials. Never drive one maul by striking it with another maul or with another type of hammer or striking tool. Never use a maul with a loose or damaged handle. Replace any maul or wedge that is dented, cracked, chipped, mushroomed, or shows other signs of excessive wear.

14.0.0 ◆ SCREWDRIVERS

Screwdrivers are used to install and remove a wide variety of screws: slotted, Phillips®, clutch-drive, and Allen. Screwdrivers have generally the same design: a steel shank, the screwdriver blade at the tip of the shank, and a handle. Because of the wide variety of screws available, there are many types of screwdriver blades and handles. The most common types are the standard screwdriver and the Phillips® head screwdriver (*Figure 69*). The screwdriver's handle is most often bulb-shaped and made of wood or plastic. The length of its blade and the width of its tip determine a screwdriver's size. Screwdrivers range in length from 2½ inches to as much as 2 feet. A correct fit between the screwdriver tip and the screw head will prevent the marring of the screw head. Use the correct screwdriver size to prevent the tip from slipping out of the screw head and possibly stripping the screw.

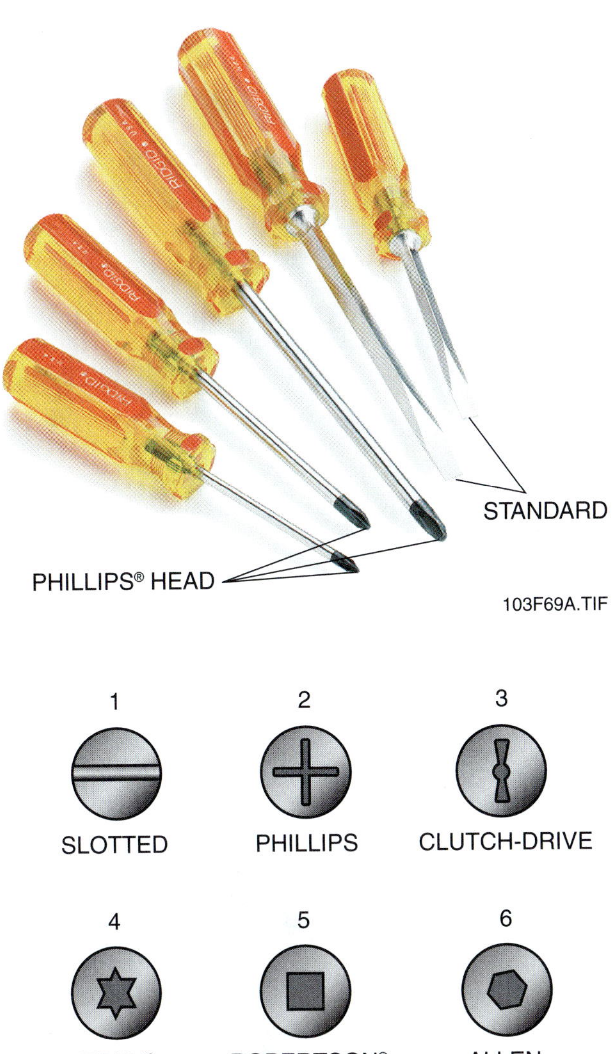

Figure 69 ◆ Common screwdrivers.

WARNING!

Do not use a screwdriver as a substitute for a hammer, wedge, nail puller, lever, chisel, or punch. To protect screws, always replace worn screwdriver tips. When using a screwdriver, secure the work piece in a vise. Do not hold the work piece in your hand unless the screw turns easily. Use only screwdrivers with insulated handles for electrical work. Do not carry exposed screwdrivers in your pockets.

14.1.0 Nut Driver

A nut driver (*Figure 70*) is similar to a screwdriver. Instead of a blade, the nut driver has a socket end for tightening hex-head screws and bolts. It comes in all bolt sizes.

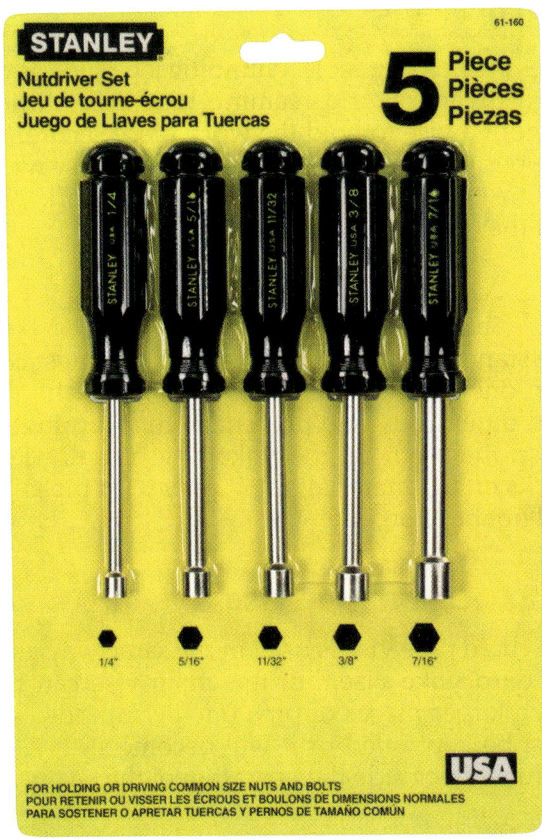

Figure 70 ◆ Nut driver set.

14.2.0 Hollow-Shank Screwdriver

Some screwdrivers come with interchangeable bits (*Figure 71*). You simply remove one bit from the end of the shank and replace it with another bit. The bits are stored in the handle of the screwdriver. The bits typically have a flat blade on one end and a Phillips® head on the other. Specialty bits are also included.

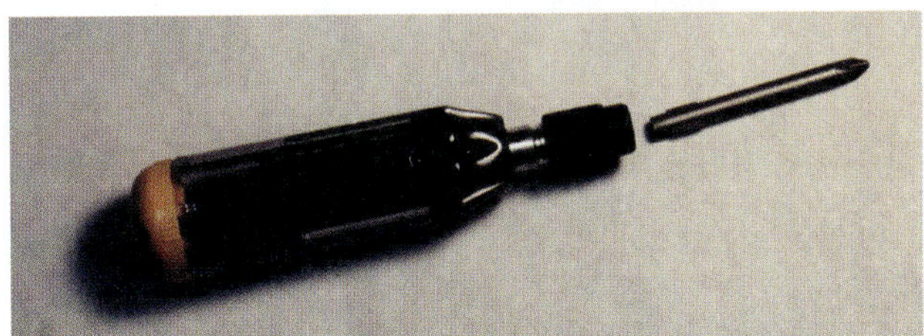

Figure 71 ◆ Hollow-shank screwdriver.

15.0.0 ◆ VISES

Vises are holding tools. Plumbing jobs such as cutting, threading, and reaming would be difficult without a vise to hold the work piece secure. The vise permits you to do work that otherwise would require two people (one to hold the work piece and the other to cut, thread, or ream it).

15.1.0 Standard Yoke Vise

The standard yoke vise is probably the most common one used by plumbers (*Figure 72*). Its jaws hold pipe firmly and prevent it from turning. Because the teeth of the yoke vise usually leave marks on the material, use it only with pipes that will not be exposed.

15.2.0 Chain Pipe Vise

The chain pipe vise is used in the same way as the standard yoke vise, but the chain vise can hold much larger pieces of pipe (*Figure 73*). The chain must be kept oiled or it will become stiff. A stiff chain will degrade the operation of the chain vise.

15.3.0 Bench Vise

The bench vise has two sets of jaws: one to hold flat work and another to hold pipe (*Figure 74*). Rotating the T-handle screw tightens or loosens the vise. You will not see this vise very often on the job site, but you might find it mounted in some plumbing trucks.

103F72.TIF

Figure 72 ◆ Yoke vise.

103F73.TIF

Figure 73 ◆ Chain pipe vise.

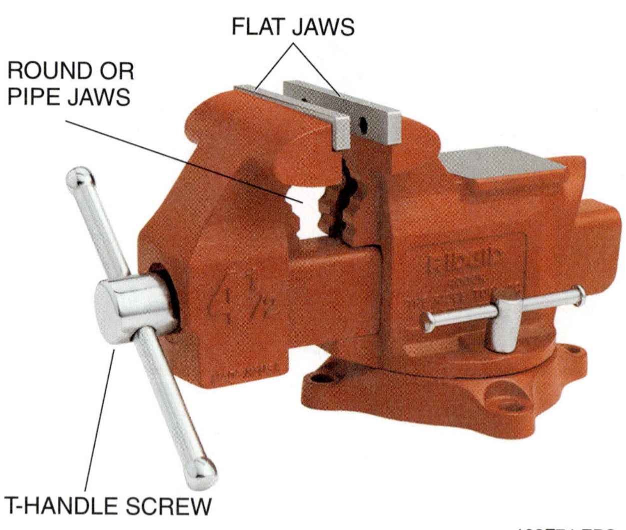

103F74.EPS

Figure 74 ◆ Bench vise.

16.0.0 ◆ CALCULATOR

You will be often called upon to choose the right combination of pipes, pumps, filters, and associated devices, especially if you are participating in the design and construction of sewage or water systems for new construction or renovations to those systems for existing structures. You will need to make quick calculations when doing other jobs as well, so you should carry a calculator. The traditional battery-powered, handheld calculator that helps you do basic math problems is one option. Specialized calculators can help you perform many types of construction calculations.

Review Questions

Sections 11.0.0–16.0.0

1. Before using extension cords, be sure to check for _____.
   a. voltage drop
   b. color match
   c. cable length
   d. worn or frayed cables

2. When working in tight quarters, it is best to use a(n) _____ wrench.
   a. offset
   b. straight pipe
   c. spud
   d. adjustable

3. The _____ wrench is used to hold chrome-plated or other types of finished pipe.
   a. spud
   b. pipe
   c. strap
   d. open-end

4. _____ wrenches are smooth jawed for turning nuts, bolts, small pipe fittings, and chrome-plated pipe fittings.
   a. Basin
   b. Strap
   c. Pipe
   d. Adjustable

5. _____ pliers are used mainly for cutting heavy or large-gauge wire and for holding work.
   a. Locking
   b. Slip-joint
   c. Lineman's
   d. Slip-lock

6. The size of a hammer is determined by the _____ of the head.
   a. length
   b. width
   c. weight
   d. shape

7. The _____ hammer has a flat face for striking and a round face for aligning brackets and driving out bolts.
   a. claw
   b. drywall
   c. ball peen
   d. sledge

8. A _____ holds pipe securely and permits one plumber to do work that otherwise would require two workers.
   a. winch
   b. threader
   c. vise
   d. ratchet

9. The chain pipe vise is used to hold _____ pieces of pipe.
   a. small
   b. large
   c. several
   d. two

10. You must put _____ on vise chains to keep them from becoming stiff and unable to wrap around a pipe.
    a. water
    b. oil
    c. solder
    d. rosin

Summary

Plumbing is a sophisticated and complex profession, full of challenges. Being a professional involves learning how to do all the different tasks plumbers perform. That does not just mean fixing sinks and unclogging toilets. It means designing, installing, and maintaining new systems, renovating existing systems, and repairing broken systems. To do all these jobs, you need a variety of tools, and you need to be adept in their use. Your expertise with your tools will directly affect the quality of your work.

As you begin to learn your profession, you will become familiar with the tools used in plumbing. The hand tools you will use every day include many types of measuring, marking, and cutting tools. Each tool has a specific use, safety requirements, and care requirements. Learn these until they are second nature to you. Maintenance is not just fixing or replacing a tool after it breaks but servicing and maintaining your tools to prevent them from breaking or wearing out prematurely.

Hand tools can be as hazardous as power tools and can cause equally serious injuries. You must not be careless in the use of any tool, whether hand or power. Knowing all of a particular tool's safety precautions, and getting in the habit of taking a few minutes to make sure the tool is in proper working condition before you start using it, can save you from serious injury or worse.

Some tools are designed to help you be more productive by combining several functions into one instrument. A combination square is a good example: You can use it to measure both 45- and 90-degree angles, to check for square and level, and to mark measurements. Take advantage of the productivity enhancements multifunction tools offer by becoming expert in their use. It will pay off in the future.

You will spend a lot of your time measuring, marking, cutting, and installing different types of piping and tubing. You will need to learn how to lay pipe that is straight and even. To do this, you may use simple tools like a plumb bob and a spirit level. You will use special tools designed to cut specific materials, like cast iron, various types of plastic, and copper. Each type of tool is designed for a specific material and a specific task. You must learn which tool to choose for which task and for which material. You will also need to know how to care for these tools. All these factors will directly affect the quality of your work.

Plumbing is closely related to a number of other construction trades. Your work will affect how the carpenters and electricians do their jobs. You will work with heating, ventilating, and air conditioning technicians to install systems in new buildings. To be a successful plumber, you will develop some of the same skills, including the use of tools, as these other craftspeople have. After a few years, you will find that you know a bit about each of these crafts. When other craftspeople find out that you not only know your own profession inside out but also know something about theirs, it will make the job easier, earn you their respect and even gratitude, and enhance your reputation as a professional.

You may find yourself using sophisticated electronic and laser equipment for some employers. As technology advances, new tools will be developed, and you will need to learn to use them in order to keep up-to-date and competitive. The successful professional makes a habit of continuing to learn. New technologies and processes will invariably lead to new tools that you, as a successful plumber, will want to use.

Notes

Trade Terms Quiz

Fill in the blank with the correct trade term that you learned from your study of this module.

1. ______________ are used to mark metal.
2. ______________ is the substance that helps join heated metal pipe.
3. A right-angle portable electric drill is useful when drilling between ______________.
4. Pipes that do not contain iron are called ______________ pipes.
5. ______________ is uncut tubing material on hand for use on a project.
6. A surface that is straight vertically is considered ______________.
7. ______________, or ______________, are used to mark a cutting line on the surface of a tube.
8. A(n) ______________ can be found in the handle of some combination squares for leveling.
9. A(n) ______________ is a combination of elbows or bends that brings one section of the pipe out of line but into a line parallel with the other section.
10. The process of removing burrs in a pipe is pipe ______________.
11. Bar material that is uncut on a work site is ______________.
12. A surface that is straight horizontally is considered ______________.
13. Heads of striking tools that have ______________, or are mushroomed, must be replaced or restored.
14. ______________ is the twisting or turning force applied in a rotating motion.
15. ______________ is the filler material used to connect joints during soldering.
16. Speed squares combine a(n) ______________, protractor, and framing square in one tool.
17. The ______________ and length of a pipe are determined by using measuring tools.
18. A speed square can be used to mark cutoff lines or as a(n) ______________.
19. ______________ describes the strength of the electricity used by electrically-powered tools.
20. The device used to ground tools is a(n) ______________.
21. An example of a(n) ______________ is +/−2.
22. The groove made by a saw blade is called a(n) ______________.
23. Portable band saws are used to cut both ______________ and nonferrous pipes.
24. The loss of voltage is called ______________.
25. When drilling through thick steel, periodically withdraw the drill bit to avoid a loss of ______________.
26. Measuring tools allow you to calculate the width, height, and ______________of pipes.
27. ______________ is the use of heat to form joints.

Trade Terms

Amperage
Bar stock
Burr
Diameter
Electrical ground
Ferrous
Flux
Joist
Keel crayon
Kerf
Level
Miter
Nonferrous
Offset
Plumb
Reaming
Scribe
Soapstone
Solder
Soldering
Spirit level
Straightedge
Temper
Tolerance
Torque
Tube stock
Voltage drop

Profile in Success

Bob Kordulak

Retired President
The Arkord Company
Belmar, New Jersey

Bob Kordulak is the retired president of The Arkord Company, a plumbing and heating contractor. Bob was born in Jersey City, New Jersey, and went to Dickinson High School. During the early 1940s, he went to work with his father, who was a plumbing contractor. After his father passed away, the company separated and Bob opened his own business. He worked in new construction, commercial work, and government contracting for nearly 50 years.

How did you become interested in this industry?
Once I started in the field, I became interested. I went to work as an apprentice for my father, a plumbing contractor. My five brothers, father, and uncle all went into business together. Originally it was just a plumbing business, but eventually we went into plumbing and heating contracting. We worked from Maine to Virginia, and as far west as Ohio. We covered a lot of ground.

What path did you take to your current position?
I served my apprenticeship at my father's company for two years before joining the service. I was in the U.S. Army for two years. I started as a journeyman, got into estimating, and then moved into construction management. While working, I took courses in estimating, business law, accounting, and air conditioning to advance. I attended New York College and the Newark College of Engineering. I picked up the rest on my own. I ran the work at my father's company and then started my own business after disbanding with my brothers.

What does it take to be successful in your trade?
The best thing is to plan. You need good planning skills and you need to be flexible. Always take responsibility for what you do. You have to anticipate what may happen if you do one thing versus another. Have some foresight. When I look at a job, I don't look at the beginning of the job, I look at the end product. I look at the drawings and I don't just see lines, but rather the end result and what problems I may encounter. Have goals and roll with the punches. Remember that you are responsible for whatever you do.

What are some things you did on the job?
I worked on everything from layouts and fabrication to oil lines, chemical lines, and welding. I ran fabrication welding shops, where we fabricated the materials, brought them to the job site, installed them, and tested them.

I started with small commercial buildings and school work. I moved up to working on blending plants for Standard Oil and chemical plants for Union Carbide. I worked on the Homeport in Staten Island, New York. Special hazardous materials training was required; we set up fire watches. You have to be very cautious with everything you do on an oil refinery.

What is the most interesting aspect of your profession? What makes your trade stand out from others?
Human beings use plumbing every day of their lives. You can do without electricity, oil, and so on, but every person alive deals with plumbing. It is a tremendous field. Everyone in the world is involved with plumbing, because everyone has to use a disposal system. You can do without lights and heat, but you have to use a plumbing system every day in one way or another.

What do you like most—and least—about your job?
I most like challenges. Every day there is something new. I like that you face new aspects of what you are doing and working on. For example, there might be a new piece of material that comes out. You have to consider whether you'll use it, how you'll use it, and how it will affect you. There is so much involved in this industry, so there is always something new.

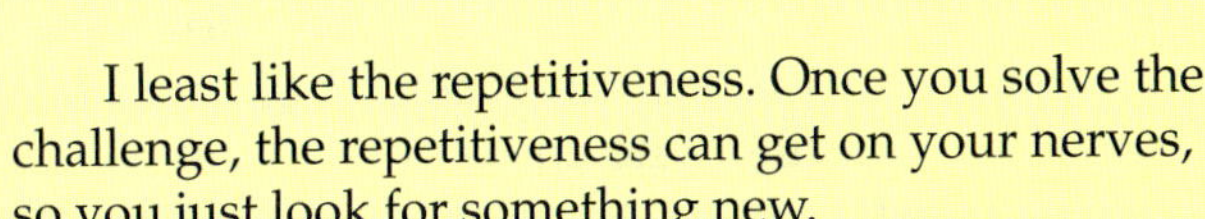

I least like the repetitiveness. Once you solve the challenge, the repetitiveness can get on your nerves, so you just look for something new.

What would you say to someone entering the trade today?
You should be aware of everything that's going on in the industry. Get involved in the industry and meet the people you work with. There is so much going on. Keep up with everything new so that you can satisfy the customer. Be aware and reach out. Get information in trade and organization journals. There is all sorts of information out there that you have to be aware of in this industry.

Also, stick to it. Some kids come out of school where they had problems, but the counselors only stressed college. Consider the trades rather than college. There are tremendous opportunities in this industry.

What can an apprentice expect to earn during the first years on the job in your area? What can that person expect to earn after 10-plus years in the industry?
It depends on the area that you work in, but an apprentice can start at about $10 per hour. Others may start higher or lower. In this area, apprentices are starting at $15 per hour. Journeymen can make between $55 and $60 per hour, again depending on the area of the country. Apprentices receive a 10 percent increase until their fifth or sixth year; then they move up into the journeymen wages. With more experience, you make more money.

Trade Terms Introduced in This Module

Amperage: A measure of electrical current.

Bar stock: Uncut bar material on hand for use on a project.

Burr: Uneven or jagged edge left on metal by certain cutting tools.

Diameter: The distance across the center of a circle.

Electrical ground: A conductive connection that provides a path for electrical current to pass from an electrical component into the earth.

Ferrous: Containing iron.

Flux: A water-soluble substance that facilitates the fusion (joining) of metals and helps prevent surface oxidation (rusting, tarnishing) during welding, brazing, and soldering. Also called soldering paste.

Joist: A piece of lumber used horizontally as a support for a ceiling or a floor.

Keel crayon: A waxy crayon used to mark a cutting line on the surface of a tube. Also called soapstone.

Kerf: The cut or groove made by a saw blade, determined by the way the teeth are set on the blade.

Level: Straight on a horizontal plane.

Miter: A surface forming the beveled end or edge of a piece where a joint is made by cutting two pieces at an angle and then fitting these pieces together.

Nonferrous: Not containing iron and therefore not magnetic.

Offset: A combination of elbows or bends that brings one section of the pipe out of line but into a line parallel with the other section.

Plumb: Straight on a vertical plane.

Reaming: A process that removes burrs from pipes after they have been cut.

Scribe: A sharply pointed and hardened steel tool used for marking a surface to be cut by etching a line or a point into the surface.

Soapstone: Another term for keel crayon.

Solder: An alloy (tin plus antimony, copper, and silver) with a low melting point used to join metals or seal joints.

Soldering: A method of joining metals or sealing joints using solder and heat.

Spirit level: A level in which the adjustment to the horizon is shown by the position of a bubble in liquid contained in a nearly horizontal glass tube or a circular box with a glass cover.

Straightedge: A length of wood or metal that does not bow or twist along its length.

Temper: The strength and resilience of a metal.

Tolerance: Allowable variation in a given measurement or quantity.

Torque: Twisting or turning force applied in a rotating motion. Measurements are given in either inch-pounds or foot-pounds.

Tube stock: Uncut tubing material on hand for use on a project.

Voltage drop: The tendency of electricity traveling through the length of an extension cord to lose voltage.

Additional Resources and References

ADDITIONAL RESOURCES

This module is intended to present thorough resources for task training. The following reference works are suggested for further study. These are optional materials for continued education rather than for task training.

Contractor Books, website, www.contractorbooks.com

Core Curriculum, 2004. NCCER. Upper Saddle River, NJ: Prentice Hall.

Sheet Metal Level Three, 2003. NCCER. Upper Saddle River, NJ: Prentice Hall.

Council Tools, website, www.counciltool.com/tools.html

Hand Tool Safety Workers Compensation Fund, website, www.wcf-utah.com/SafetyResources/WCF1008HandToolSafety.pdf

REFERENCES

National Standard Plumbing Code, 2003. Chapter 1, Definitions (definition of *offset*). Falls Church, VA: Plumbing-Heating-Cooling Contractors—National Association.

NCCER CURRICULA — USER UPDATE

NCCER makes every effort to keep its textbooks up-to-date and free of technical errors. We appreciate your help in this process. If you find an error, a typographical mistake, or an inaccuracy in NCCER's curricula, please fill out this form (or a photocopy), or complete the on-line form at **www.nccer.org/olf**. Be sure to include the exact module ID number, page number, a detailed description, and your recommended correction. Your input will be brought to the attention of the Authoring Team. Thank you for your assistance.

Instructors – If you have an idea for improving this textbook, or have found that additional materials were necessary to teach this module effectively, please let us know so that we may present your suggestions to the Authoring Team.

NCCER Product Development and Revision

13614 Progress Blvd., Alachua, FL 32615

Email: curriculum@nccer.org

Online: www.nccer.org/olf

❑ Trainee Guide ❑ AIG ❑ Exam ❑ PowerPoints Other ______________________________

Craft / Level: Copyright Date:

Module ID Number / Title:

Section Number(s):

Description:

____

____

____

____

Recommended Correction:

____

____

____

____

Your Name:

Address:

____

Email: Phone:

Introduction to Plumbing Math

02104-05

02104-05
Introduction to Plumbing Math

Topics to be presented in this module include:

Overview

Plumbers use math to read plans, calculate pipe length, and lay out fixtures. Consequently, the ability to develop accurate math skills is essential to advance in the plumbing profession. Understanding whole numbers, fractions, and decimals—and how to convert them from one form to another—is essential for efficiency and productivity.

Plumbing fixtures are often available in English and metric sizes. Plumbers may encounter situations when they need to convert metric measurements for weight, length, or volume to the English system. Plumbing calculations often require square numbers or the square roots of numbers. Plumbers must learn to determine these values with and without the use of a calculator.

Plumbers must be able to measure pipe quickly and accurately. Being able to identify the basic parts of a fitting allows a plumber to define the beginning and ending points of the measurement. Plumbers must also be able to determine fitting allowance—the distance from the end of the pipe that goes into the fitting. Plumbers can calculate this distance or refer to manufacturers' tables. Three pipe-measuring techniques—center-to-center, end-to-center, and face-to-face—are used to calculate the length of pipe to cut. Skilled plumbers understand which method to use based on the available information and the type of installation.

Focus Statement

The goal of the plumber is to protect the health, safety, and comfort of the nation job by job.

Code Note

Codes vary among jurisdictions. Because of the variations in code, consult applicable code whenever regulations are in question. Referencing an incorrect set of codes can cause as much trouble as failing to reference codes altogether. Obtain, review, and familiarize yourself with your local adopted code.

Objectives

When you have completed this module, you will be able to do the following:

1. Add, subtract, multiply, and divide whole numbers.
2. Add, subtract, multiply, and divide fractions.
3. Add, subtract, multiply, and divide decimals.
4. Convert decimals to percentages and percentages to decimals.
5. Convert fractions to decimals and decimals to fractions.
6. Explain what the metric system is and how it is important in the plumbing trade.
7. Square various numbers and take square roots of numbers, with and without a calculator.
8. Identify the parts of a fitting and use common pipe-measuring techniques.
9. Use fitting dimension tables to determine fitting allowances and thread makeup.
10. Calculate end-to-end measurements using fitting allowances and thread makeup.

Key Trade Terms

Back
Center
Centerline
Center point
Face
Fitting allowance
Remainder
Schematic drawings
Thread makeup
Throat

Required Trainee Materials

1. Appropriate personal protective equipment
2. Sharpened pencils and paper
3. Copy of local adopted code
4. Calculator

Prerequisites

Before you begin this module, it is recommended that you successfully complete *Core Curriculum; Plumbing Level One,* Modules 02101-05 through 02103-05.

This course map shows all of the modules in the first level of the *Plumbing* curriculum. The suggested training order begins at the bottom and proceeds up. Skill levels increase as you advance on the course map. The local Training Program Sponsor may adjust the training order.

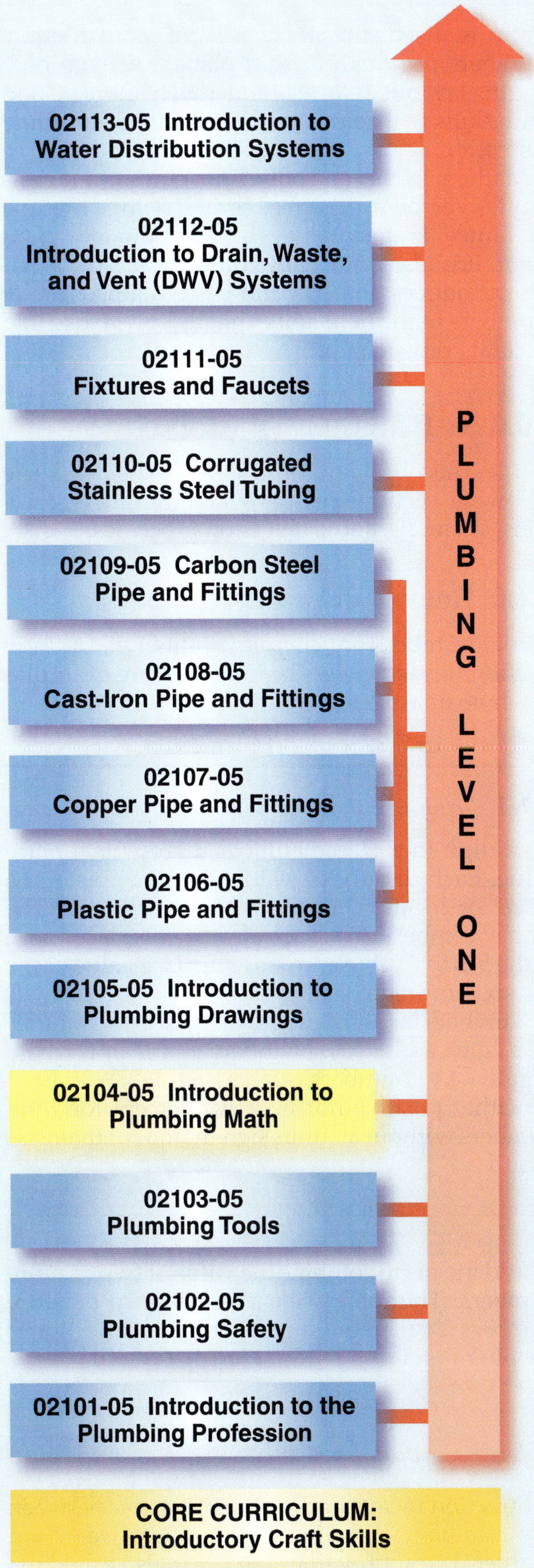

104CMAP.EPS

1.0.0 ◆ INTRODUCTION

Math is used in all phases of construction. Plumbers use math to read plans, calculate pipe length, lay out fixtures, and much more. Good math skills will help you advance in the plumbing profession.

Mathematics is one of the most important tools you'll ever own. It's free, comes with a lifetime guarantee, and actually gets better the more you use it. It is also your daily companion. This module introduces some of the basic math used by plumbers in the field and explains how to use it to calculate pipe length.

2.0.0 ◆ BASIC MATH REVIEW

You learned basic math in *Core Curriculum*. The following sections review whole numbers, fractions, decimals, and conversion processes.

2.1.0 Review of Whole Numbers

Whole numbers are complete units without any fractions or decimals. The following are examples of whole numbers:

15 32 60 144 2,436

2.1.1 Place Value

Each digit in a whole number represents a place value. Each digit has a value that depends on its place, or location, in the whole number. In the whole number 5,679, for example, the place value of the 5 is five thousand, and the place value of the 6 is six hundred.

Numbers larger than zero are positive (+) numbers (such as 1, 2, 3). Numbers less than zero are negative (−) numbers (such as −1, −2, −3). Zero is neither positive nor negative. Except for zero, numbers without a minus sign in front of them are positive.

2.1.2 Addition

To add means to combine the value of two or more numbers. The total when you add two or more numbers is called the sum. The sign for addition is the plus sign (+).

$$56 + 32 = 88$$

2.1.3 Subtraction

Subtraction means finding the difference between two numbers, or taking one number away from another. The subtraction sign (−) is also called the minus sign. The result (answer) of a subtraction problem is called the difference.

$$56 - 32 = 24$$

2.1.4 Multiplication

Multiplication is the quick way to add the same number together many times. The symbol for multiplication is the × sign. For example, you must deliver five boxes each to eight job sites. How many boxes do you deliver in all?

$$5 \times 8 = 40 \text{ boxes}$$

$$5 + 5 + 5 + 5 + 5 + 5 + 5 + 5 = 40 \text{ boxes}$$

2.1.5 Division

Division is the opposite of multiplication. The symbol for division is the ÷ sign. Instead of adding a number several times as in multiplication, when you are dividing you subtract it several times. You can solve a problem faster by using division instead of subtracting the same number over and over. For example, you have 40 boxes to deliver to 8 different job sites, and you need to find out how many boxes go to each site.

$$40 \div 8 = 5$$

$40 - 8 - 8 - 8 - 8 - 8 = 0$ (You had to subtract 8 five times, so the answer is 5.)

2.2.0 Review of Fractions

A fraction divides whole units into parts. Fractions are written as two numbers separated by a slash or by a horizontal line, like this:

$$1/2 \text{ or } \frac{1}{2}$$

Think of a fraction as a division problem. The lower number (denominator) of the fraction tells you the number of parts by which the upper number (numerator) is being divided. The slash or horizontal line means the same thing as the ÷ sign. The fraction ½ means 1 divided by 2, or one divided into two equal parts. Read this fraction as "one-half."

2.2.1 Equivalent Fractions

Equivalent fractions have the same value, or are equal. For example, ½, 2/4, 4/8, and 8/16 are equivalent fractions. If you cut off a piece of wood 8/16-inch long and a trainee cuts off a piece ½-inch long, the two pieces would be the same length.

You need to know how to find equivalent fractions so that you can compare, add, and subtract fractional measurements. For example, to find out how many eighths of an inch are equal to ½ inch, you multiply the numerator and denominator by the same number. Ask yourself what number you would multiply by 2 to get 8. The answer is 4, so you multiply the numerator and denominator by 4.

$$\frac{1}{2} \times \frac{4}{4} = \frac{4}{8}$$

The answer is that 4/8 inch is equivalent to ½ inch.

2.2.2 Lowest Terms

When you are working with fractions, it is best to reduce them to lowest terms. To reduce a fraction, ask yourself, "What is the largest number I can divide evenly into both the numerator and the denominator?" If you're unsure about the largest number that divides evenly into both the numerator and the denominator, try using the number 2 or 3 to start, and then continue until the fraction is in its lowest form. If no number (other than 1) will divide evenly into both numbers, the fraction is already in its lowest terms. Look at the following examples:

2/4 = ½
4/16 = ¼
8/32 = ¼
3/8 = 3/8 (already in lowest terms)
7/16 = 7/16 (already in lowest terms)

2.2.3 Common Denominator

A common denominator means that the bottom number, or denominator, in a group of fractions is the same. For example, ¼, 2/4, and ¾ have common denominators. Before you can compare fractions, they must have the same denominators. Which of these fractions is larger?

$$\frac{3}{4} \text{ or } \frac{2}{3}$$

To compare, you need to find a common denominator for the fractions. The common denominator is a number that both denominators can go into evenly. Here's one way to find a common denominator.

Step 1 Multiply the two denominators (4 × 3 = 12). The result is a common denominator for ¾ and 2/3. You found a common denominator so that you can easily compare the fractions.

Step 2 Convert the two fractions so they will have the same denominator by multiplying the numerator and denominator of each fraction by the denominator of the other fraction. Convert ¾ and 2/3 to fractions having the common denominator of 12.

$$\frac{3 \times 3}{4 \times 3} = \frac{9}{12}$$

$$\frac{2 \times 4}{3 \times 4} = \frac{8}{12}$$

Now it's easy to compare the two fractions to see which is larger: 9/12, or ¾, is larger than 8/12, or 2/3.

In construction, many fractions have denominators of 2, 4, 8, or 16. If one of the denominators is a multiple of the other, the larger number is the common denominator. For example, the common denominator for ¾ and 5/8 is 8. Since the lowest common denominator is 8, you would then multiply ¾ by a number that will convert the denominator to an 8. That number is 2.

$$\frac{3 \times 2}{4 \times 2} = \frac{6}{8}$$

Now you can see that ¾, which is equivalent to 6/8, is more than 5/8.

2.2.4 Improper Fractions

An improper fraction has a numerator larger than the denominator. For example, 11/8 is an improper fraction.

To reduce the improper fraction 11/8 to its lowest terms, convert it to a mixed number. A mixed number is a combination of a whole number and a fraction.

$$11/8 = 8/8 + 3/8 = 1\,3/8$$

To change a whole number into an improper fraction, simply place the number over 1. Remember, fractions express division. When you divide any number by 1, the result is the same number. Here is an example:

$$4 = \frac{4}{1}$$

To change a mixed number (for example, 2 1/8) to an improper fraction, follow these steps:

Step 1 Multiply the whole number by the denominator.

$$2 \times 8 = 16$$

Step 2 Add the result to the numerator. Put this total over the denominator.

$$16 + 1 = 17$$

$$\frac{17}{8}$$

2.2.5 Addition

How many total inches will you have if you add ¾ inch to ⅝ inch? To answer this question, you will have to add the two fractions using the following steps.

Step 1 Find the common denominator of the fractions you wish to add. Since 8 is a multiple of 4, it is the common denominator.

Step 2 Convert the fractions to equivalent fractions with a common denominator.

$$\frac{3 \times 2}{4 \times 2} = \frac{6}{8}$$

Step 3 Add the numerators of the fractions. Place this sum over the common denominator.

$$\frac{6}{8} + \frac{5}{8} = \frac{11}{8}$$

Step 4 Reduce the fraction to its lowest terms. For this example, it is 1⅜.

2.2.6 Subtraction

Subtracting fractions is very much like adding fractions. You must find a common denominator before you subtract. For example, subtract the following fractions:

$$\frac{7}{8} - \frac{1}{4}$$

Step 1 Find the common denominator. Convert the fractions to equivalent fractions with the same denominator. Multiply the numerator and denominator by 2 to get a new fraction with the common denominator of 8.

$$\frac{1 \times 2}{4 \times 2} = \frac{2}{8}$$

Step 2 Rewrite the equation and subtract the numerators. The difference is ⅝.

$$\frac{7}{8} - \frac{2}{8} = \frac{5}{8}$$

2.2.7 Multiplication and Division

Multiplying and dividing fractions is different from adding and subtracting them. You do not have to find a common denominator when you multiply or divide fractions. Using ¾ × ⅚ as an example, follow these steps:

Step 1 Multiply the numerators to get a new numerator. Multiply the denominators to get a new denominator.

$$\frac{3 \times 5}{4 \times 6} = \frac{15}{24}$$

Step 2 Reduce the fraction to its lowest terms.

$$\frac{15}{24} = \frac{5}{8}$$

Dividing fractions is very much like multiplying fractions, with one difference. You must invert, or flip, the fraction you are dividing by. Using ⅜ ÷ ¾ as an example, follow these steps:

Step 1 Invert the fraction you are dividing by.

$$\frac{3}{4} \text{ becomes } \frac{4}{3}$$

Step 2 Change the division sign to a multiplication sign.

$$\frac{3 \div 3}{8 \div 4} = \frac{3 \times 4}{8 \times 3}$$

Step 3 Multiply the fractions.

$$\frac{3 \times 4}{8 \times 3} = \frac{12}{24}$$

Step 4 Reduce the fraction to its lowest terms.

$$\frac{12}{24} = \frac{1}{2}$$

If you are working with a whole number or a mixed number, you must convert it to an improper fraction before you invert it. Using 5¾ ÷ 4 as an example, follow these steps:

Step 1 Convert 5¾ to an improper fraction. Multiply the whole number by the denominator. Add the result to the numerator. Put this total over the denominator.

$$5 \times 4 = 20$$

$$20 + 3 = 23$$

$$5\tfrac{3}{4} = \frac{23}{4}$$

Step 2 Convert 4 to an improper fraction.

$$4 = \frac{4}{1}$$

Step 3 Now you have the following equation. Invert 4/1 and multiply the fractions.

$$\frac{23 \div 4}{4 \div 1} = \frac{23 \times 1}{4 \times 4} = \frac{23}{16}$$

Step 4 Reduce the fraction to its lowest terms.

$^{23}/_{16} = 1^{7}/_{16}$

Practice Exercises

2.2.8 Whole Numbers and Fractions

Solve the following problems. Remember to show all your work.

1. 178
   568
   + 10

2. 923
   −598

3. 9
   ×6

4. 7
   ×45

5. 20 ÷ 5 = _____

6. Find the equivalent of the following measurement: $^{5}/_{8}$ inch = _____/32 inch.

7. Reduce $^{108}/_{288}$ to its lowest terms. _____

8. The lowest form of the fraction $^{5}/_{16}$ is _____.

9. Of $^{15}/_{16}$ and $^{5}/_{8}$, which fraction is larger? _____

10. Reduce $^{57}/_{16}$ to its lowest terms. _____

11. $^{1}/_{2} + ^{4}/_{16}$ = _____

12. $^{1}/_{2} - ^{4}/_{16}$ = _____

13. $^{5}/_{32} \times ^{1}/_{4}$ = _____

14. $^{17}/_{32} \div ^{3}/_{4}$ = _____

15. $^{5}/_{8} \div ^{1}/_{2}$ = _____

2.3.0 Review of Decimals

Decimals represent values less than one whole unit. They are fractions expressed in a different form. You are already familiar with decimals in the form of money.

25¢ = 0.25 or $^{25}/_{100}$
10¢ = 0.10 or $^{10}/_{100}$
50¢ = 0.50 or $^{50}/_{100}$

On the job, you may need to use decimals to read instruments or calculate flow rates. You will also encounter them when reading civil drawings and calculating water pressure levels.

The following chart compares whole number place values with decimal place values:

Whole Numbers	Decimals
1 ones	
10 tens	0.1 tenths
100 hundreds	0.01 hundredths
1000 thousands	0.001 thousandths

To read a decimal, say the number as it is written and then the name of its place value. For example, read 0.05 as "five hundredths." Mixed numbers also appear in decimals. You would read 1.05 as "one and five hundredths," for example.

2.3.1 Addition and Subtraction

There is one major rule to remember when adding and subtracting decimals: Keep your decimal points lined up.

Suppose you want to add 4.76 and 0.834. Line up the problem like this:

4.760
+0.834
5.594

You can add a 0 as a place holder to help keep the numbers lined up.

The same is true for subtraction. Suppose you want to subtract 2.724 from 5.6. Line up the decimal points, and add two zeros to the end of the first number to make it easier to subtract.

5.600
−2.724
2.876

2.3.2 Multiplication

To multiply decimals, set up the problem as you would with whole numbers.

3.6
× 9

Step 1 Multiply.

3.6
× 9
324

Step 2 Once you have the answer, count the number of digits to the right of the decimal point in both numbers being multiplied. In this example, there is only one decimal point and only one number to the right of it.

Step 3 In the answer, count over the same number of digits (from right to left) and place the decimal point there.

```
  3.6   Count one total digit to the right of the
×   9   decimal point in the two numbers.
 32.4   Count one digit from right to left in the
        answer, and place the decimal point there.
```

2.3.3 Division

Three types of division problems involve decimals:

- Those that have a decimal point in the number being divided (the dividend)

  22)44.5

- Those that have a decimal point in the number you are dividing by (the divisor)

  0.22)4450

- Those that have decimal points in both numbers (the dividend and the divisor)

  0.22)44.5

Solve the first problem: How many 22-inch lengths of pipe can you cut from a piece of pipe measuring 44.5 inches?

Step 1 Place a decimal point directly above the decimal point in the dividend.

22)44.5

Step 2 Divide as usual.

```
     2.0
 22)44.5
   −44
    00.5
    − 0
    00.5r
```

How many 22-inch pieces of pipe will you have? The answer: two, with a little left over. This leftover portion is called the **remainder** and is expressed as *r*. In this case, $r = 0.5$.

Now solve the second problem: If couplings cost $0.22 each, how many can you buy with $4,450?

Step 1 Move the decimal point in the divisor to the right until you have a whole number.

0.22)4450 *When you move the decimal point in the divisor two places to the right, 0.22 becomes 22.*

Step 2 Move the decimal point in the dividend the same number of places to the right. You may have to add zeros first. Then divide as usual.

```
     20227.2
 22)4450.00 0
   −44
    0500
   −0044
    00060
   −00044
    000160
   −000154
    0000060
   −0000044
    0000016r
```

After you have added the zeros and moved the decimal in the dividend two places to the right, the number becomes 445,000.

Now solve the third problem: You have $44.50 to buy 200 nuts costing $0.22 each. Can you stay within budget?

Step 1 Move the decimal point in the divisor to the right until you have a whole number.

0.22)44.5 *After you have moved the decimal point in the divisor to the right, 0.22 becomes 22.*

Step 2 Move the decimal point in the dividend the same number of places to the right (so 44.5 becomes 4,450). Then divide as usual.

```
     202
 22)4450
   −44
    050
    −44
     06r
```

To find the remainder in terms of money, you have to undo the changes you made to the decimals in the original problem. In this case, move the decimal point back two places to the left: 0.06.

You will be able to buy 202 nuts with the $44.50, with $0.06 left over, so, yes, you will stay within budget.

2.3.4 Rounding Decimals

Often, calculations with decimals produce very precise answers, such as 25.29411764. But in most practical measurements, you probably need an answer only to the nearest tenth (0.1). For this exercise, you will round 25.29411764 to the nearest tenth:

Step 1 Underline the place to which you are rounding.

25.<u>2</u>9411764

Step 2 Look at the digit one place to its right.

25.<u>2</u>**9**411764

Step 3 If the digit to the right is 5 or more, you will round up by adding 1 to the underlined digit. If the digit is less than 5, leave the underlined digit the same. In this example, the digit to the right is 9, which is more than 5, so you round up by adding 1 to the underlined digit.

25.<u>3</u>**9**411764

Step 4 Drop all other digits to the right.

25.3

Practice Exercises

2.3.5 Decimals

Solve the following problems. Remember to show all your work and round to the nearest tenth if necessary:

1. 7.8
   13.4
   +0.8

2. 129.6
   −54.9

3. 8.9
   ×3.5

4. 785 ÷ 6.7 = _____

5. 23000.50 ÷ 7.5 = _____

2.4.0 Review of Conversion Processes

In some situations, you need to convert the numbers you want to work with so that all your numbers are in the same form. For example, you may have some numbers that appear as decimals, some that appear as percentages, and some that appear as fractions. Decimals, percentages, and fractions are all just different ways of expressing similar things. The decimal 0.25, the percentage 25%, and the fraction ¼ all mean the same thing. To work successfully with the different forms of numbers like these, you will need to know how to convert them from one form into another.

2.4.1 Decimals to Percentages and Percentages to Decimals

What are percentages? Think of a whole number divided into 100 parts. You can express any part of the whole as a percentage. Percentages are an easy way to express parts of a whole. Decimals and fractions also express parts of a whole. Let's look at the relationship among percentages, decimals, and fractions.

To convert a decimal to a percentage, simply multiply the decimal by 100 and add a percent sign. Try converting 7.35 to a percentage.

Step 1 Multiply the decimal by 100.

$$7.35 \times 100 = 735$$

Step 2 Add a percent sign.

735%

Suppose you are preparing a gallon of cleaning solution. The mixture should contain 10 to 15 percent of the cleaning agent. The rest should be water. You have 0.12 gallon of cleaning agent. Will you have enough to prepare a gallon of the solution? To answer the question, you must convert a decimal (0.12) to a percentage.

Step 1 Multiply the decimal by 100. (Move the decimal point two places to the right.)

$$0.12 \times 100 = 12$$

Step 2 Add a percent sign.

12%

You have enough cleaning agent to make the solution. Recall that the mixture should be 10 to 15 percent cleaning agent. You have 12 percent.

You may also need to convert percentages to decimals, as illustrated in the following example. Let's say that another mixture should contain 22 percent of a certain chemical by weight. You're making 1 pound of the mixture. You weigh the ingredients on a digital scale. How much of the chemical should you add? To answer this, you must convert a percentage (22%) to a decimal:

Step 1 Drop the percent sign.

22

Step 2 Divide the number by 100. (Move the decimal point two places to the left.)

$$22 \div 100 = 0.22$$

So you would add 0.22 pounds of the chemical to make a 22 percent mixture.

2.4.2 Fractions to Decimals

To convert a fraction to a decimal, do what the fraction tells you to do; that is, divide the numerator by the denominator. Convert 7⁄9 to its equivalent decimal.

Step 1 Set up the division problem: Divide the numerator by the denominator.

$$9\overline{)7}$$

Step 2 In this example, you need to put the decimal point and a zero after the number 7 because you need a number large enough to divide by 9. Put the decimal point directly above its location in the number within the division symbol.

$$9\overline{)7.0}\quad .?$$

Step 3 Once the decimal point is in its proper place above the line, you can divide as you normally would. The decimal point holds everything in place.

$$\begin{array}{r} .777 \\ 9\overline{)7.00} \\ -6.3 \\ \hline 7.0 \\ -6.3 \\ \hline 0.70 \\ -0.63 \\ \hline 0.7r \end{array}$$

Dividing to three places is satisfactory for plumbing problems.

Suppose you need 1¾ of a pound of material and need to know its decimal equivalent.

Step 1 Convert the proper fraction into an improper fraction. Multiply the whole number by the denominator and add the numerator. Place this number over the denominator.

$$1\tfrac{3}{4} = \tfrac{7}{4}$$

Step 2 Set up the division problem: Divide the numerator by the denominator.

$$4\overline{)7.0}$$

Step 3 Put the decimal point directly above its location within the division symbol.

$$4\overline{)7.0}\quad .?$$

Step 4 Divide.

$$\begin{array}{r} 1.75 \\ 4\overline{)7.00} \\ -4 \\ \hline 30 \\ -28 \\ \hline 20 \\ -20 \\ \hline 0 \end{array}$$

The fraction 7⁄4 converted to a decimal is 1.75, so 1¾ pounds is the same as 1.75 pounds.

2.4.3 Decimals to Fractions

Not only will you need to convert fractions to decimals, you must know how to go the other way and convert decimals into fractions. For example, what fraction of a pound is 0.25?

Step 1 Say the decimal in words.

0.25 is expressed as "twenty-five hundredths."

Step 2 Write the decimal as a fraction.

0.25 written as a fraction is 25⁄100

Step 3 Reduce it to its lowest terms.

$$\frac{25}{100} = \frac{25 \div 25}{100 \div 25} = \frac{1}{4}$$

You can see that 0.25 converted to a fraction is ¼.

Let's say you have 17.35 pounds of nails. What fraction of a pound is that?

Step 1 Say the decimal in words.

17.35 is expressed as "17 and 35 hundredths."

Step 2 Write the decimal as a fraction.

$$17.35 = 17\tfrac{35}{100}$$

Step 3 Reduce the fraction portion to its lowest terms.

$$\frac{35}{100} = \frac{35 \div 5}{100 \div 5} = \frac{7}{20}$$

So 17.35 pounds of nails is the same as 17 7⁄20 pounds of nails.

When calculating the cut length of pipe, plumbers need to be able to convert a decimal to the nearest 1⁄16 of an inch. Here are the steps for converting to sixteenths of an inch, using 0.91 feet as an example.

Step 1 Multiply the decimal part of a foot by 12, which is the number of inches in a foot. Thus, 0.91 equals 10 inches with a remainder of 0.92.

$$0.91 \times 12 = 10.92$$

Step 2 Now convert the remainder to sixteenths of an inch by multiplying it by 16.

$$0.92 \times 16 = 14.72$$

Because this answer is not a whole number, round it. So, 14.72 rounds up to 15. Hence, 0.91 feet = 10 15⁄16 inches.

Convert 13.67 feet into the nearest sixteenth of an inch. You already know you have 13 feet. You need to convert 0.67 feet into inches.

Step 1 Multiply the decimal part of a foot by 12.

$$0.67 \times 12 = 8.04$$

Thus, 0.67 equals 8 inches with a remainder of 0.04.

Step 2 Now multiply the remainder by 16.

$$0.04 \times 16 = 0.64$$

Because this answer is not a whole number, round it. 0.64 rounds up to 1. The answer is 13 feet 8 1/16 inches.

An easier way to convert a decimal into sixteenths (or other fractions) of an inch is to use a conversion table (see *Table 1*). For instance, if you have 0.86 feet, find the number closest to that number in the "Decimals of a Foot" column (0.859) and look in the same row of the corresponding "Inches" column. So 0.86 feet = 10 5/16 inches.

Table 1 Inches Converted to Decimals of a Foot

Inches	Decimals of a Foot	Inches	Decimals of a Foot	Inches	Decimals of a Foot	Inches	Decimals of a Foot
1/16	0.005	3 1/16	0.255	6 1/16	0.505	9 1/16	0.755
1/8	0.010	3 1/8	0.260	6 1/8	0.510	9 1/8	0.760
3/16	0.016	3 3/16	0.266	6 3/16	0.516	9 3/16	0.766
1/4	0.021	3 1/4	0.271	6 1/4	0.521	9 1/4	0.771
5/16	0.026	3 5/16	0.276	6 5/16	0.526	9 5/16	0.776
3/8	0.031	3 3/8	0.281	6 3/8	0.531	9 3/8	0.781
7/16	0.036	3 7/16	0.286	6 7/16	0.536	9 7/16	0.786
1/2	0.042	3 1/2	0.292	6 1/2	0.542	9 1/2	0.792
9/16	0.047	3 9/16	0.297	6 9/16	0.547	9 9/16	0.797
5/8	0.052	3 5/8	0.302	6 5/8	0.552	9 5/8	0.802
11/16	0.057	3 11/16	0.307	6 11/16	0.557	9 11/16	0.807
3/4	0.063	3 3/4	0.313	6 3/4	0.563	9 3/4	0.813
13/16	0.068	3 13/16	0.318	6 13/16	0.568	9 13/16	0.818
7/8	0.073	3 7/8	0.323	6 7/8	0.573	9 7/8	0.823
15/16	0.078	3 15/16	0.328	6 15/16	0.578	9 15/16	0.828
1	0.083	4	0.333	7	0.583	10	0.833
1 1/16	0.089	4 1/16	0.339	7 1/16	0.589	10 1/16	0.839
1 1/8	0.094	4 1/8	0.344	7 1/8	0.594	10 1/8	0.844
1 3/16	0.099	4 3/16	0.349	7 3/16	0.599	10 3/16	0.849
1 1/4	0.104	4 1/4	0.354	7 1/4	0.604	10 1/4	0.854
1 5/16	0.109	4 5/16	0.359	7 5/16	0.609	10 5/16	0.859
1 3/8	0.115	4 3/8	0.365	7 3/8	0.615	10 3/8	0.865
1 7/16	0.120	4 7/16	0.370	7 7/16	0.620	10 7/16	0.870
1 1/2	0.125	4 1/2	0.374	7 1/2	0.625	10 1/2	0.875
1 9/16	0.130	4 9/16	0.380	7 9/16	0.630	10 9/16	0.880
1 5/8	0.135	4 5/8	0.385	7 5/8	0.635	10 5/8	0.885
1 11/16	0.141	4 11/16	0.391	7 11/16	0.641	10 11/16	0.891
1 3/4	0.146	4 3/4	0.396	7 3/4	0.646	10 3/4	0.896
1 13/16	0.151	4 13/16	0.401	7 13/16	0.651	10 13/16	0.901
1 7/8	0.156	4 7/8	0.406	7 7/8	0.656	10 7/8	0.906
1 15/16	0.161	4 15/16	0.411	7 15/16	0.661	10 15/16	0.911
2	0.167	5	0.417	8	0.667	11	0.917
2 1/16	0.172	5 1/16	0.422	8 1/16	0.672	11 1/16	0.922
2 1/8	0.177	5 1/8	0.427	8 1/8	0.677	11 1/8	0.927
2 3/16	0.182	5 3/16	0.432	8 3/16	0.682	11 3/16	0.932
2 1/4	0.188	5 1/4	0.438	8 1/4	0.688	11 1/4	0.938
2 5/16	0.193	5 5/16	0.443	8 5/16	0.693	11 5/16	0.943
2 3/8	0.198	5 3/8	0.448	8 3/8	0.698	11 3/8	0.948
2 7/16	0.203	5 7/16	0.453	8 7/16	0.703	11 7/16	0.953
2 1/2	0.208	5 1/2	0.458	8 1/2	0.708	11 1/2	0.958
2 9/16	0.214	5 9/16	0.464	8 9/16	0.714	11 9/16	0.964
2 5/8	0.219	5 5/8	0.469	8 5/8	0.719	11 5/8	0.969
2 11/16	0.224	5 11/16	0.474	8 11/16	0.724	11 11/16	0.974
2 3/4	0.229	5 3/4	0.479	8 3/4	0.729	11 3/4	0.979
2 13/16	0.234	5 13/16	0.484	8 13/16	0.734	11 13/16	0.984
2 7/8	0.240	5 7/8	0.490	8 7/8	0.740	11 7/8	0.990
2 15/16	0.245	5 15/16	0.495	8 15/16	0.745	11 15/16	0.995
3	0.250	6	0.500	9	0.750	12	1.000

104T01.EPS

2.4.4 Inches to Decimal Equivalents in Feet

Sometimes you will need to convert inches to their decimal equivalents in feet. To do so, first express the inches as a fraction that has 12 as the denominator. You use 12 because there are 12 inches in a foot. Then reduce the fraction and convert it to a decimal.

For example, 3 inches equals what decimal equivalent in feet?

Step 1 First, place the number of inches over 12.

$$3/12$$

Step 2 Reduce the fraction to its lowest terms.

$$3/12 = 1/4$$

Step 3 Convert the fraction 1/4 to a decimal by dividing the 4 into 1.00:

$$\begin{array}{r} 0.25 \\ 4\overline{)1.00} \\ -0.8 \\ \hline 0.20 \\ -0.20 \\ \hline 0 \end{array}$$

Thus, 3 inches converts to 0.25 feet.

Now try a problem that includes decimals: 14.6 inches equals what decimal equivalent in feet?

Step 1 Convert the decimal to a fraction. If this fraction is in proper form, change it to improper form.

$$14.6 = 14\,6/10 = 146/10$$

Step 2 Multiply by 1/12.

$$\frac{146}{10} \times \frac{1}{12} = \frac{146}{120}$$

Step 3 Convert the fraction 146/120 to a decimal by dividing the denominator (120) into the numerator (146) and rounding the answer if necessary. In this case, round the answer to the nearest thousandth.

$$\begin{array}{r} 1.2166 \\ 120\overline{)146.0000} \\ -120 \\ \hline 260 \\ -240 \\ \hline 200 \\ -120 \\ \hline 800 \\ -720 \\ \hline 800 \\ -720 \\ \hline 80 \end{array}$$

Thus, 14.6 inches converts to 1.217 feet.

Practice Exercises

2.4.5 Conversions

Round answers to the nearest thousandth where necessary.

1. Convert 0.85 to its equivalent percentage. _____
2. Convert 4.35 to its equivalent percentage. _____
3. Convert 20 percent to its equivalent decimal. _____
4. Convert 150 percent to its equivalent decimal. _____
5. Convert 3/8 to its equivalent decimal. _____
6. Convert 1 1/2 to its equivalent decimal. _____
7. Convert 6/4 to its equivalent decimal. _____
8. Convert 0.75 to its equivalent fraction. _____
9. Convert 13.4 to its equivalent fraction. _____
10. Convert 133.365 to its equivalent fraction. _____
11. Convert 0.3 feet into the nearest sixteenth of an inch. _____
12. Convert 11.86 feet into the nearest sixteenth of an inch. _____
13. Convert 75.8 feet into the nearest sixteenth of an inch. _____
14. Convert 16 inches to its decimal equivalent in feet. _____
15. Convert 115.6 inches to its decimal equivalent in feet. _____

2.5.0 Metric System

The metric system is a system of measurement that uses a base-ten method of determining weight, length, volume, and temperature. In other words, all measurements are counted in tens.

There are only seven basic units of measurement in the modern metric system, including the meter, liter, and gram. Multiples or fractions of the basic units are expressed as powers of 10. A standard set of prefixes is used to denote these larger or smaller numbers (see *Table 2*). That way, you will always know that a kilometer is 1,000 meters just by looking at the prefix.

2.5.1 Converting within the Metric System

To convert from one metric unit to another, simply move the decimal point the required number of spaces. For instance, to find how many millimeters are in 5 meters, move the decimal point three places to the right (multiply by 1,000). If you want to know how many meters equals 1,500 centimeters, you would move the decimal point two places to the left (divide by 100). See *Figure 1* for an easy way of doing these calculations. For every jump on the number line, move the decimal one place in the same direction. Add zeros if necessary. For example, if you have 100 centiliters and want to convert to hectoliters, you move the decimal four places to the left, adding a zero (or divide by 10,000). So 100 centiliters equals 0.01 hectoliter.

Table 2 Metric Prefixes

deka = 10	deci = 0.1
hecto = 100	centi = 0.01
kilo = 1,000	milli = 0.001
mega = 1,000,000	micro = 0.000001

2.5.2 Converting Measurements

Some countries use the metric system, and others use the English system. The English system uses measurements such as feet and pounds. Plumbing fixtures are often available in both English and metric sizes. You may encounter situations where pipe measurements need to be converted. For example, imported materials may come in metric measurements, and you may need to convert them to the English system. Also, federal agencies use the metric system. To convert between the English and metric systems, you can use a conversion table such as *Table 3*.

CAUTION

Be sure to express measurements using the correct system. Errors caused by using the wrong system of measurement can cost time and money. Never use a metric tool on English-system pipe and fittings, or vice versa. You could damage both the tool and the fitting.

To convert a measurement from one system to the other, multiply the measurement by the number in the far-right column of *Table 3*. For example, you may want to know how many kilograms equal 250 pounds. Looking at the table, you find pounds in the "To convert" column and kilograms in the "Into …" column, and you see that the number must be multiplied by 0.45 (250 × 0.45). The answer is 112.5 kilograms.

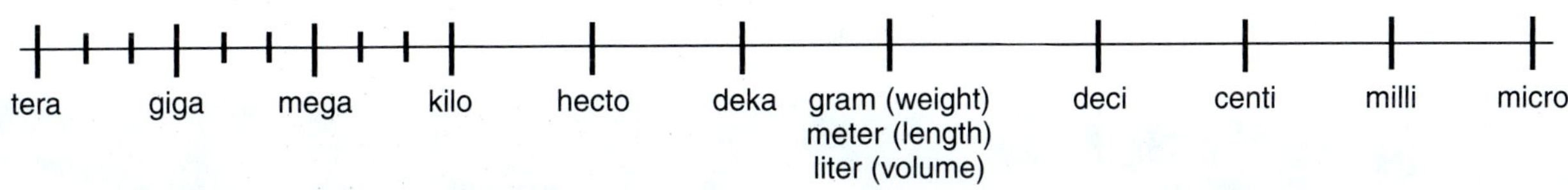

10^0 = 1	BASE UNIT: GRAM (WEIGHT), METER (LENGTH), LITER (VOLUME)				10^0 = 1
MULTIPLIER	**PREFIX**	**MEANING**	**PREFIX**	**MEANING**	**MULTIPLIER**
10^1 = 10	deka-	ten	deci-	tenth	10^{-1} = 0.1
10^2 = 100	hecto-	hundred	centi-	hundredth	10^{-2} = 0.01
10^3 = 1,000	kilo-	thousand	milli-	thousandth	10^{-3} = 0.001
10^6 = 1,000,000	mega-	million	micro-	millionth	10^{-6} = 0.000001
10^9 = 1,000,000,000	giga-	billion	nano-	billionth	10^{-9} = 0.000000001
10^{12} = 1,000,000,000,000	tera-	trillion	pico-	trillionth	10^{-12} = 0.000000000001

104F01.EPS

Figure 1 ◆ The decimal scale.

Table 3 Conversion Table

A. English to Metric

	To convert . . .	Into . . .	Multiply the English unit by . . .
LENGTH	Inches	Millimeters	25.40
	Feet	Centimeters	30.00
	Yards	Meters	0.90
	Miles	Kilometers	1.60
AREA	Square inches	Square centimeters	6.50
	Square feet	Square meters	0.09
	Square yards	Square meters	0.80
	Square miles	Square kilometers	2.60
	Acres	Hectares	0.40
MASS and WEIGHT	Fluid ounces	Grams	28.00
	Pounds	Kilograms	0.45
	Short tons	Megagrams	0.90
LIQUID MEASURE	Ounces	Milliliters	30.00
	Pints	Liters	0.47
	Quarts	Liters	0.95
	Gallons	Liters	3.80

B. Metric to English

	To convert . . .	Into . . .	Multiply the metric unit by . . .
LENGTH	Millimeters	Inches	0.040
	Centimeters	Feet	0.400
	Meters	Yards	1.100
	Kilometers	Miles	0.620
AREA	Square centimeters	Square inches	0.160
	Square meters	Square yards	1.200
	Square kilometers	Square miles	0.400
	Hectares	Acres	2.500
MASS and WEIGHT	Grams	Fluid ounces	0.035
	Kilograms	Pounds	2.200
	Megagrams	Short tons	1.100
LIQUID MEASURE	Milliliters	Ounces	0.034
	Liters	Pints	2.100
	Liters	Quarts	1.060
	Liters	Gallons	0.260

104T03.EPS

Practice Exercises

2.5.3 Metric Conversion

Convert the following measurements:

1. 75 meters = _____ centimeters
2. 0.25 milliliters = _____ dekaliters
3. 675,000 micrometers = _____ hectometers
4. 8.9 kilograms = _____ dekagrams
5. 1,011 millimeters = _____ centimeters
6. 68 inches = _____ millimeters
7. 4.5 fluid ounces = _____ grams
8. 35 centimeters = _____ feet
9. 875 liters = _____ gallons
10. 1,568 millimeters = _____ inches

2.6.0 Squares and Square Roots

Plumbing calculations may require you to square numbers as well as take square roots of numbers.

To square a number, multiply the number by itself. To show that a number is being squared, write a superscript 2 to the right of the number.

$$3 \text{ squared} = 3^2 = 3 \times 3 = 9$$

To take the square root of a number means to find the number that, when multiplied by itself, results in the original number. The radical sign ($\sqrt{\ }$) indicates a root of a number.

$$\text{The square root of } 9 = \sqrt{9} = 3$$

The easiest way to square numbers and take square roots of numbers is to use a calculator (see *Figure 2*). To square a number using a calculator, simply enter the number, press the $\times$ key, and then enter the number again and press =. With more complicated numbers, you can save time by using the special square function. Enter the number and then press the x^2 key on the calculator. The answer will appear in the display. You do not need to press the = sign.

To find the square root of a number, enter the number. Then press the $\sqrt{\ }$ key. The answer will appear in the display. You do not need to press the = sign.

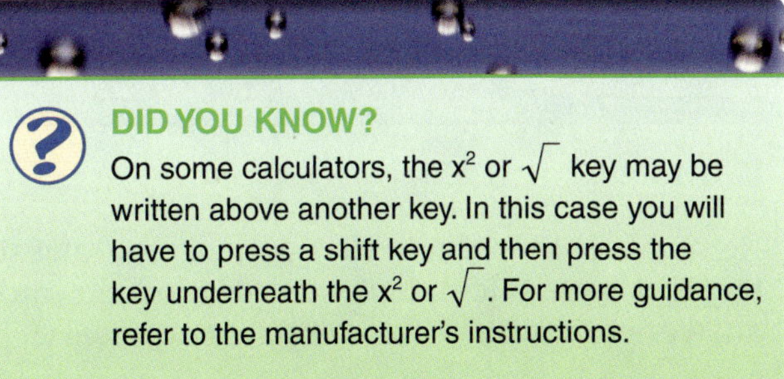

DID YOU KNOW?

On some calculators, the x^2 or $\sqrt{\ }$ key may be written above another key. In this case you will have to press a shift key and then press the key underneath the x^2 or $\sqrt{\ }$. For more guidance, refer to the manufacturer's instructions.

104F02.EPS

Figure 2 ◆ Construction Master calculator.

Practice Exercises

2.6.1 Squares and Square Roots

Use a calculator when necessary to solve the following problems. Round your answers to the nearest tenth.

1. 8 squared is _____
2. $12^2 =$ _____
3. The square root of 156 is _____
4. $\sqrt{3{,}136}$
5. $\sqrt{54}$

Review Questions

Section 2.0.0

Remember to reduce all fractions to lowest terms. Use a calculator when needed. Round answers to the nearest tenth when necessary.

1. Find the equivalent of the following measurement: $\frac{3}{8}$ inch = _____/16 inch.
   a. 6
   b. 8
   c. 10
   d. 12

2. The lowest form of the fraction $\frac{4}{16}$ is _____.
   a. $\frac{1}{8}$
   b. $\frac{1}{4}$
   c. $\frac{1}{2}$
   d. 1

3. What is the common denominator for $\frac{2}{6}$ and $\frac{3}{8}$? _____
   a. 6
   b. 12
   c. 16
   d. 24

4. What is the lowest common denominator for $\frac{4}{12}$ and $\frac{8}{12}$? _____
   a. 3
   b. 4
   c. 6
   d. 8

5. $\frac{5}{8} + \frac{1}{16} =$ _____
   a. $\frac{11}{16}$
   b. $\frac{14}{16}$
   c. $\frac{10}{24}$
   d. $\frac{24}{24}$

6. $\frac{12}{16} - \frac{4}{8} =$ _____
   a. $\frac{1}{16}$
   b. $\frac{1}{8}$
   c. $\frac{1}{4}$
   d. $\frac{4}{16}$

7. $\frac{3}{4} \times \frac{5}{8} =$ _____
   a. $\frac{5}{6}$
   b. $\frac{6}{7}$
   c. $\frac{15}{32}$
   d. $\frac{21}{12}$

8. $\frac{6}{8} \div 3 =$ _____
   a. $\frac{1}{8}$
   b. $\frac{1}{4}$
   c. $\frac{3}{24}$
   d. $\frac{9}{8}$

9. $$\begin{array}{r} 2.8 \\ 4.2 \\ +6.0 \\ \hline \end{array}$$
   a. 12.8
   b. 13.0
   c. 113
   d. 130

10. Yesterday, a supply yard contained 6.7 tons of sand. Since then, 2.3 tons were removed. The supply yard now contains _____ tons of sand.
    a. 0.9
    b. 4.4
    c. 9
    d. 44

11. You already have 2,032.25 pounds of bricks. If you buy 7,525.5 more pounds, you will have _____ total pounds of bricks.
    a. 2,758
    b. 2,775.80
    c. 9,557.75
    d. 95,577.5

12. A length of pipe measures 14.25 inches. You cut off a piece measuring 6.825 inches. The remaining pipe measures _____ inches.
    a. 6
    b. 7.425
    c. 8.425
    d. 21.075

13. If cast-iron pipe weighs 4.75 pounds per foot, 129.4 feet weighs _____ pounds.
    a. 614.65
    b. 1,770.40
    c. 6,146.50
    d. 17,704

14. If you have to cut 18 lengths of pipe, each of which must measure 75.1875 centimeters, you will need a pipe measuring a total of _____ centimeters.
    a. 93.1875
    b. 676.6875
    c. 1,353.375
    d. 2,353.375

15. If you have a length of pipe measuring 1,780 inches, you can cut _____ 12-inch pieces with _____ inches left over.
    a. 148; 3
    b. 148; 4
    c. 151; 4
    d. 151; 8

16. 55.35 ÷ 18 = _____
    a. 0.307
    b. 3.075
    c. 30.75
    d. 307.5

17. 90 ÷ 0.45 = _____
    a. 2.20
    b. 20.02
    c. 22
    d. 200

18. 28.9 ÷ 3.4 = _____
    a. 0.85
    b. 8.5
    c. 85
    d. 850

19. If tubing costs $4.20 per foot and you pay a total of $120.95, you have purchased _____ feet of tubing.
    a. 2.8
    b. 28.6
    c. 28.8
    d. 287.9

20. 0.42 = _____%
    a. 0.42
    b. 4.2
    c. 42
    d. 420

21. Converted to a decimal, $\frac{3}{16}$ equals _____.
    a. 0.1875
    b. 0.316
    c. 1.875
    d. 3.187

22. Converted to a fraction, 0.12 equals _____.
    a. $\frac{12}{10}$
    b. $\frac{6}{5}$
    c. $\frac{0.12}{100}$
    d. $\frac{3}{25}$

23. Convert 2.8 to its equivalent fraction and reduce it to its lowest terms. _____
    a. $2\frac{1}{4}$
    b. $2\frac{4}{5}$
    c. $2\frac{8}{10}$
    d. $2\frac{8}{25}$

24. Convert 0.75 feet into the nearest sixteenth of an inch. _____
    a. 9 inches
    b. $9\frac{1}{16}$ inches
    c. $9\frac{9}{16}$ inches
    d. 72 inches

25. Convert 88.47 feet into the nearest sixteenth of an inch. _____
    a. $5\frac{10}{16}$ inches
    b. 88 feet $5\frac{17}{25}$ inches
    c. 88 feet $5\frac{5}{8}$ inches
    d. 88 feet 10 inches

26. Convert 4 inches to its decimal equivalent in feet. _____
    a. 0.25
    b. 0.33
    c. 0.50
    d. 0.60

27. Convert 6.85 inches to its decimal equivalent in feet. _____
    a. 0.50
    b. 0.57
    c. 0.60
    d. 0.75

28. Convert 736 millimeters to hectometers. _____
    a. 0.00736
    b. 0.0736
    c. 0.736
    d. 73.6

29. Convert 85.3 kilometers to centimeters. _____
    a. 8,530
    b. 85,300
    c. 853,000
    d. 8,530,000

30. Convert 98.8 inches to millimeters. _____
   a. 3.952
   b. 2,509.52
   c. 2,964
   d. 9,880

31. Convert 2 millimeters to inches. _____
   a. 0.08
   b. 8
   c. 50.8
   d. 80

32. Convert 12.5 gallons to liters. _____
   a. 3.25
   b. 11.875
   c. 47.5
   d. 1,250

33. 25^2 = _____
   a. 3.87
   b. 30
   c. 125
   d. 625

34. The square root of 144 is _____.
   a. 12
   b. 14
   c. 288
   d. 20,736

35. $\sqrt{15,764}$ = _____
   a. 0
   b. 125.55
   c. 157
   d. 248,503,696

3.0.0 ◆ MEASURING PIPE

A plumber must be able to measure pipe accurately and quickly. Measuring is basic to the plumbing profession and lies at the heart of many other trade skills as well. Pipe, like other construction materials, such as 2 × 4 lumber, comes in nominal sizes. Nominal sizes are used for the purpose of general identification. The actual size of the piece will be approximately the same as the nominal size, but won't be exactly the same.

In the fabrication and installation of piping systems, plumbers often use the center-to-center (C–C) measurements between two fittings. The lengths of pipe and fittings are measured along **centerlines.** The extensions of the centerline of the pipe and the centerline of the fitting meet inside the fitting to create a **center point.**

3.1.0 Parts of a Fitting

The basic parts of a fitting are shown in *Figure 3.* The terms **face, center,** and **back** are used to describe these parts. These elements are important when you measure pipe length because they define the beginning and ending points of the measurement. You will often use a face-of-fitting to center-of-fitting measurement. The **throat** of the fitting, also shown in *Figure 3,* is not used as often in measuring pipe.

Figure 4 shows the various methods of measuring pipe lengths. You will have a chance to practice some of these methods later in the module.

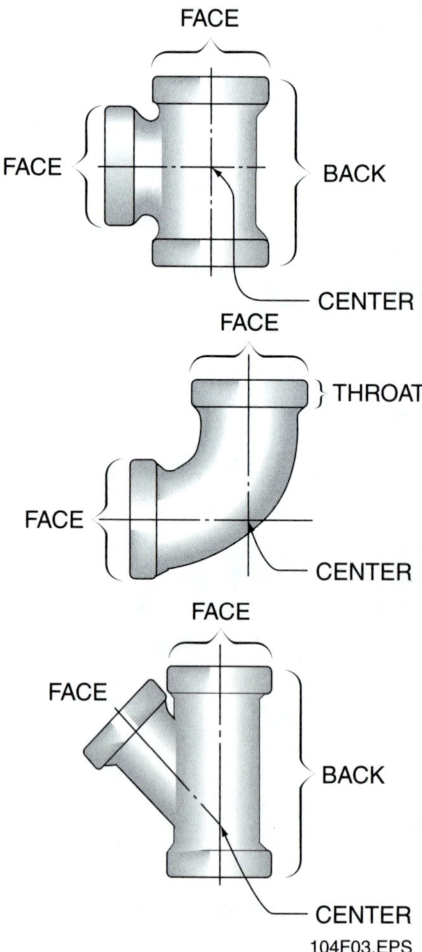

Figure 3 ◆ Basic fitting parts.

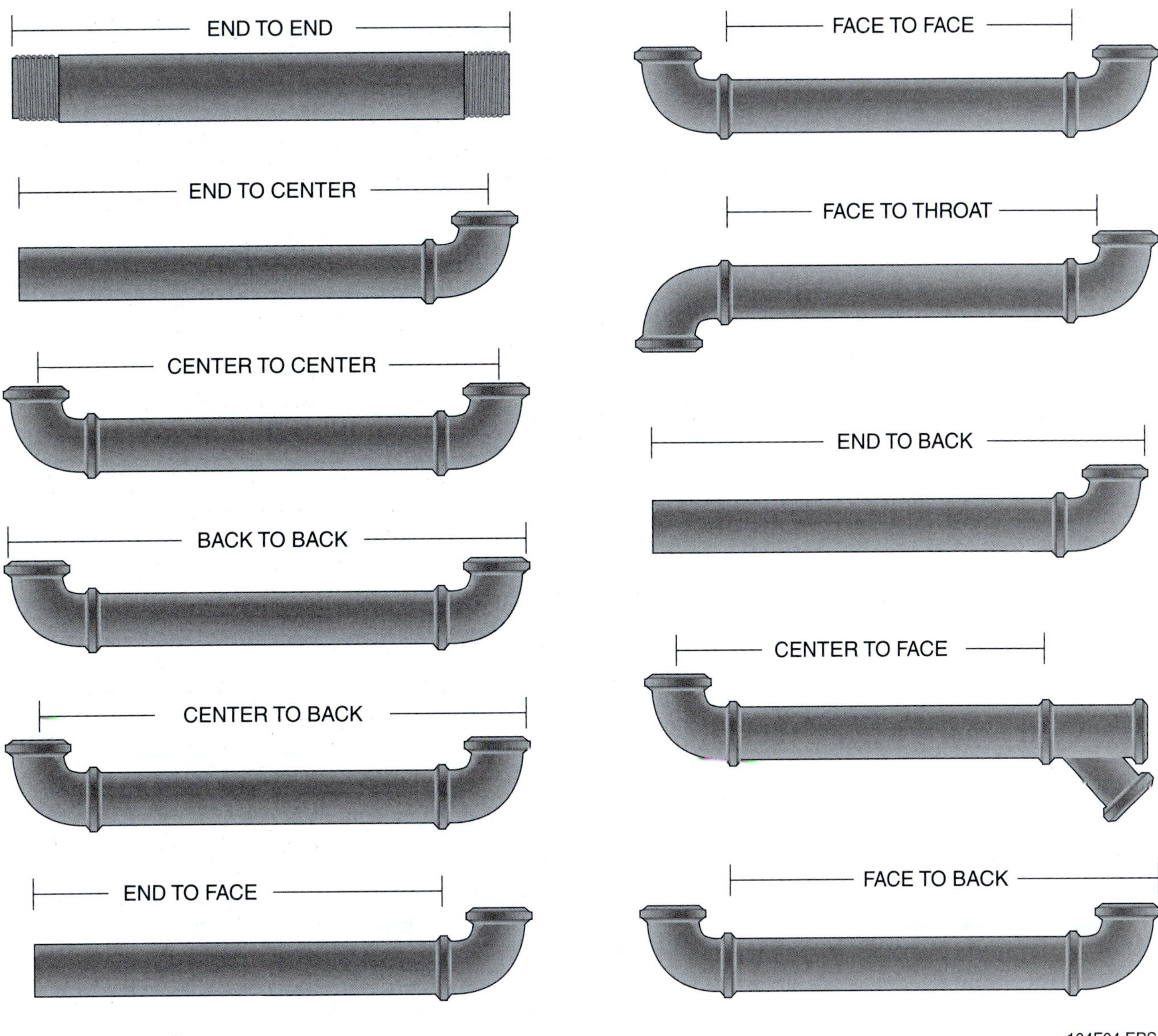

Figure 4 ◆ Measuring techniques.

Figure 5 shows the same information in schematic form. **Schematic drawings** are simple, single-line representations of pipe and fittings. It is more convenient to draw this way than to draw the whole pipe or fitting. Compare *Figures 4* and *5* and note the differences.

The dimensions of the pipe assemblies shown in *Figures 4* and *5* can also be expressed verbally, as *Table 4* shows. Speaking this way may seem strange at first, but this is part of the language of plumbing. To be a plumber, you must be able to understand, speak, and write this language. To get accustomed to this language, read each statement out loud and relate each to one of the sketches in *Figure 5*, where possible. (Note that "ell" is short for "elbow.") For instance, "8 inches end to end" corresponds to the first drawing in *Figure 5*.

3.2.0 Makeup

Thread makeup, also known as thread-in or thread engagement, is the distance that the pipe screws into the threaded fitting. The most convenient way to determine thread makeup is simply to measure the depth of the fitting's socket or thread.

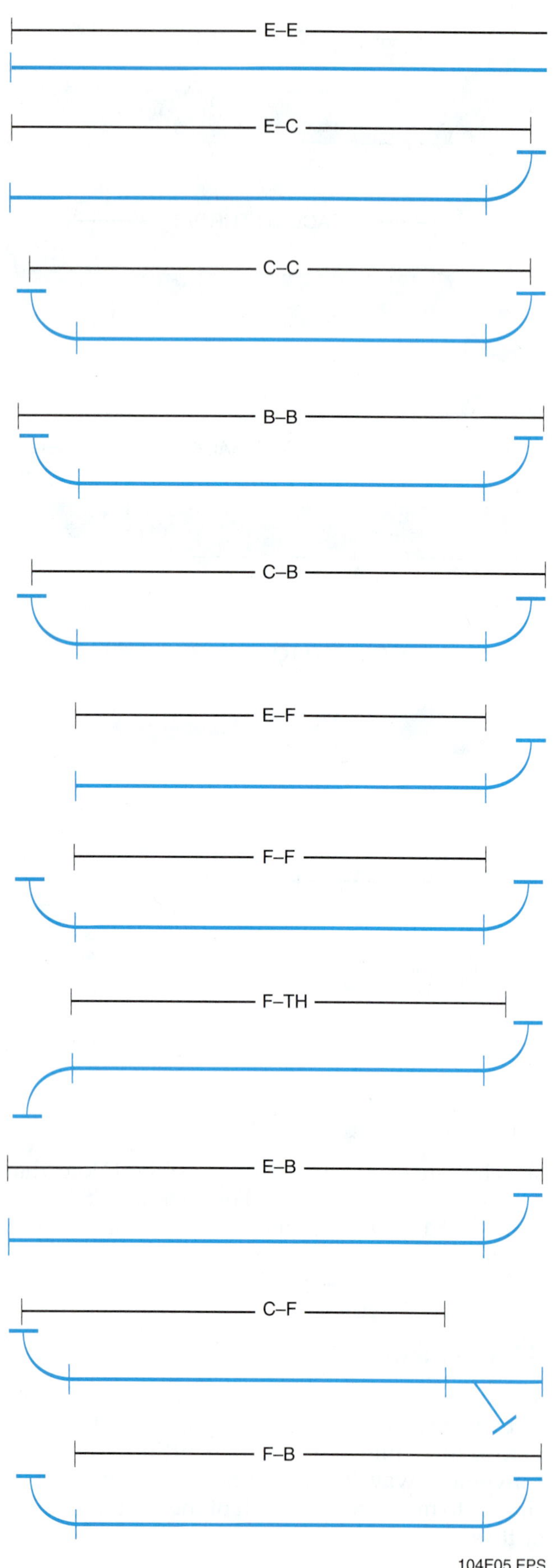

Figure 5 ◆ Schematic views.

Table 4 Verbal Expression of Pipe

Pipe Size	Statement	Abbreviation
1/2"	8 inches end to end	1/2" 8" E–E
3/4"	10 inches end to center of ell	3/4" 10" E–C ell
1"	8 inches end to face of ell	1" 8" E–F ell
3"	17 inches end to back of ell	3" 17" E–B ell
1"	11 inches center of ell to center of tee	1" 11" C ell–C tee
2"	8 1/2 inches face of ell to throat of ell	2" 8 1/2" F ell–TH ell
4"	8 inches face of ell to face of wye	4" 8" F ell–F wye
3/4"	42 inches center of wye to face of ell	3/4" 42" C wye–F ell
6"	14 inches center of tee to back of ell	6" 14" C tee–B ell
1 1/2"	69 inches face of ell to back of ell	1 1/2" 69" F ell–B ell

3.3.0 Takeoff

Fitting allowance, or takeoff, is the distance from the end of the pipe that goes into the fitting (face of fitting) to the center of the fitting. Some joints, such as plastic glued or threaded joints, have takeoffs. Others, including cast-iron no-hub and face-to-face flanged joints, do not have takeoffs.

To calculate fitting allowance, measure the fitting from its center to its face. Then subtract the thread-in measurement.

3.4.0 Using Manufacturer Tables

You can also determine fitting allowances by referring to tables published by fitting manufacturers. *Figure 6* shows fitting dimensions for threaded steel wyes (45-degree angle), tees, and elbows. A capital letter on the drawing of the fitting is keyed to a column on the table. To use the tables, find the pipe size in the first column and then find the dimensions in the columns to the right. For example, the takeoff for a threaded steel 1-inch 90-degree elbow fitting is 1 inch because C = 1½" and T = ½", so 1½" – ½" = 1".

THREADED WYE-45-DEGREE ANGLE

L IS FACE TO FACE
D IS CENTER TO FACE
C IS CENTER TO FACE
T IS THREAD-IN OR MAKEUP

NOMINAL PIPE SIZE	L	C	D	T
1½"	4¾"	3 5/16"	1 7/16"	½"
2"	5⅞"	4 1/16"	1 13/16"	½"
2" × 1½"	5⅞"	4¼"	1⅝"	½"
2½" × 2"	6¼"	4⅝"	1⅝"	¾" × ½"

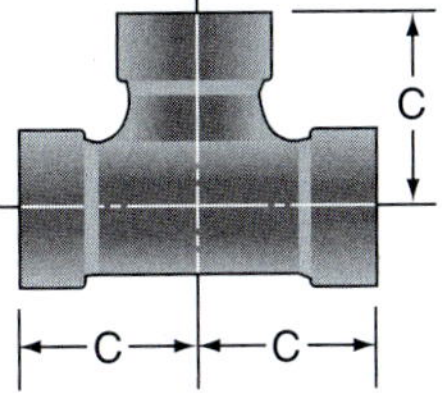

THREADED TEES

C IS CENTER TO FACE
T IS THREAD-IN OR MAKEUP

NOMINAL PIPE SIZE	C	T
⅜"	1"	⅜"
½"	1⅛"	½"
¾"	1⅜"	½"
1"	1½"	½"
1¼"	1¾"	½"
1½"	1⅞"	½"
2"	2¼"	½"
2" × 1½"	2¼"	½"
2½"	2¾"	¾"

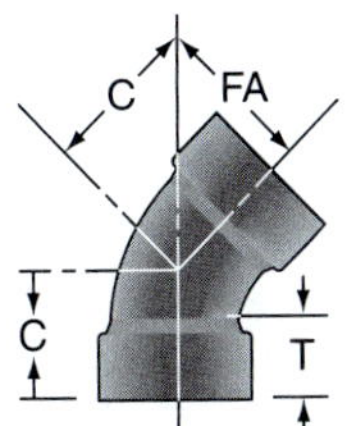

THREADED ELBOWS

FA IS FITTING ANGLE
C IS CENTER TO FACE
T IS THREAD-IN OR MAKEUP

NOMINAL PIPE SIZE	C					T
	FITTING ANGLES					
	90°	60°	45°	22½°	11¼°	
⅜"	1"		¾"			⅜"
½"	1⅛"		¾"			½"
¾"	1⅜"		1"			½"
1"	1½"		1⅛"			½"
1¼"	1¾"	1¼"	1¼"	1⅛"	1"	½"
1½"	1⅞"	1¾"	1½"	1¼"	1⅛"	½"
2"	2¼"	2¼"	1⅝"	1⅜"	1¼"	½"
2½"	2¾"	2½"	1¾"	1½"	1⅜"	¾"

104F06.EPS

Figure 6 ◆ Fitting dimensions for threaded steel pipe.

DID YOU KNOW?

Reading Charts

At first glance, the charts of fitting dimensions in this module may be confusing. Once you get more familiar with them, you will be able to read them easily.

To start, read the title of the chart to ensure that you are looking at the right one. For instance, if you wanted information on threaded tees, you would need to look at the second table in *Figure 6*.

The other information in the top box of each table explains the abbreviations used for the various measurements. For example, in the second table, C refers to center-to-face measure (see the drawing as well) and T is the thread-in.

Now that you know what the abbreviations mean, try reading the table. A table is made up of rows (going across) and columns (going down). The column headings (Nominal Pipe Size, C, and T) tell you what is in each column. Suppose you have a nominal pipe size of 2 inches. To find the corresponding measurements, locate 2 inches in the first column. Now go across that row to find C and T. In long tables, it is helpful to place a ruler right under the row you are reading; this lessens the chance that you will accidentally read another row.

Fitting dimensions vary for different types of fittings made by different manufacturers. For instance, *Figure 7* shows fitting dimensions for the following PVC (polyvinyl chloride) fittings: elbows, tees, and wyes. Cast-iron (soil) pipe, which is not threaded, comes in two types: cast-iron pipe with hub-and-spigot fittings and cast-iron pipe with no-hub fittings. *Figure 8* shows fitting dimensions for both types of cast-iron pipe. *Figure 9* shows fitting dimensions for copper DWV (drain, waste, and vent) pipe, which also is not threaded.

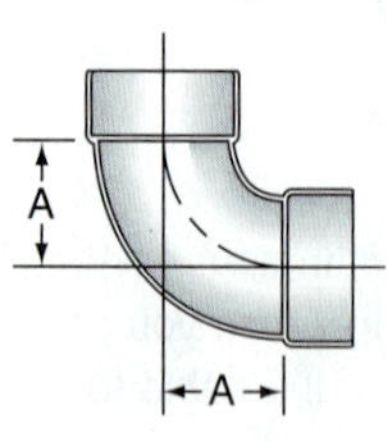

BEND (SANITARY 90° ELL) ALL HUB

SIZE	A	B	C
1½"	1¾"		
2"	2⁵⁄₁₆"		
3"	3¹⁄₁₆"		
4"	3⅞"		
6"	5"		
8"	6"		

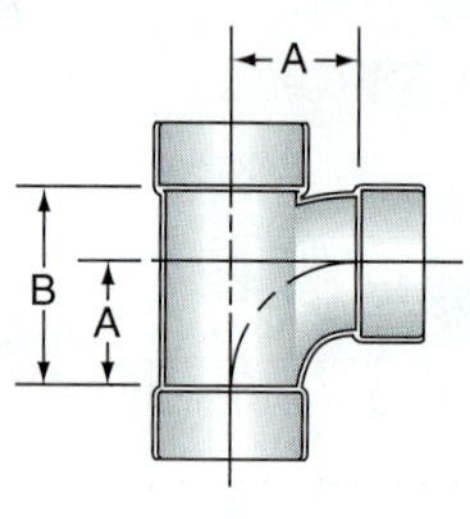

SANITARY TEE ALL HUB

SIZE	A	B	C
1½"	1¾"	2¾"	
2"	2⁵⁄₁₆"	3¹¹⁄₁₆"	
3"	3¹⁄₁₆"	4⅞"	
4"	3⅞"	6⅛"	
6"	5"	8½"	
8"	6"	10½"	

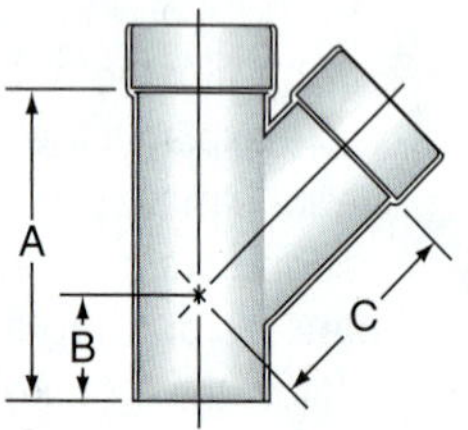

WYE, STREET (45° WYE) SPIGOT × HUB × HUB

SIZE	A	B	C
1½"	4¾"	1⅞"	2⅞"
2"	5⅞"	2¼"	3⅝"
3"	8⅛"	3⅛"	5"
4"	10"	3⅝"	6⅜"

AVAILABLE IN PVC ONLY

104F07.EPS

Figure 7 ◆ Fitting dimensions for PVC pipe.

3.5.0 Calculating Pipe Length

The measuring technique you use determines how you calculate the length of pipe to cut. You will use fitting dimensions as part of your calculations in most pipe-measuring techniques.

3.5.1 Center-to-Center Method

To determine the length of pipe between two fittings using the center-to-center method, subtract the fitting allowances from the center-to-center measurement (see *Figure 10*).

For example, say you want to know the length of a threaded steel pipe to cut to fill a space between two threaded tees. The center-to-center measurement between the two fittings is 50 inches. You are using 1¼-inch threaded steel pipe, which, as you can see by referring to *Figure 6*, has a ½-inch thread-in (T) and a 1¾-inch center-to-face measurement (C).

Step 1 Determine the fitting allowance (or take-off) by subtracting T from C.

$$C - T = \text{fitting allowance}$$

$$1\tfrac{3}{4}" - \tfrac{1}{2}" = 1\tfrac{1}{4}"$$

Step 2 Because you have two fittings, you must add the allowances together to get the total fitting allowance.

$$\text{Total fitting allowance} = 1\tfrac{1}{4}" + 1\tfrac{1}{4}" = 2\tfrac{1}{2}"$$

Step 3 Subtract the total fitting allowance from the 50-inch center-to-center measurement.

$$50" - 2\tfrac{1}{2}" = 47\tfrac{1}{2}"$$

You need to cut a pipe that is 47½ inches long.

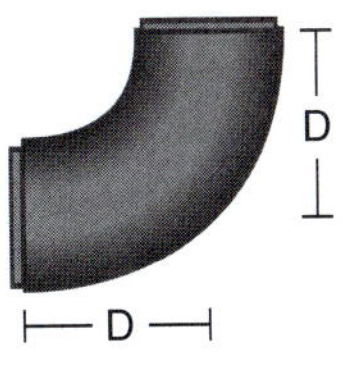

QUARTER BEND		
SIZE (IN)	D	WEIGHT (LB)
1½	4¼	1.9
2	4½	2.4
3	5	3.9
4	5½	6.0
5	6½	10.0
6	7	12.0
8	8½	25.0
4 × 3	5½	6.0

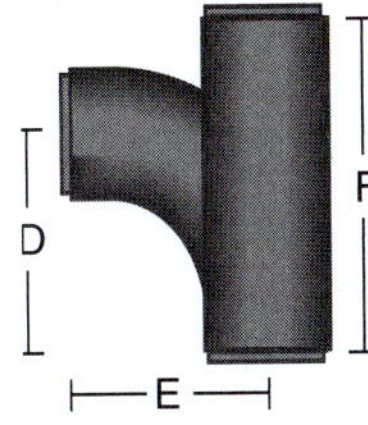

SANITARY TEE				
SIZE (IN)	E	F	D	WEIGHT (LB)
1½ × 1½	4¼	6½	4¼	2.3
2 × 1½	4½	6⅝	4¼	3.0
2 × 2	4½	6⅞	4¼	2.9
3 × 1½	5	6	4¼	5.0
3 × 2	5	6⅞	4½	4.0
3	5	8	5	4.7
4 × 2	5½	6⅞	4½	6.0
4 × 3	5½	8	5	7.2
4	5½	9⅛	5½	8.0
5 × 2	6½	8½	5	8.7
5 × 3	6	9 5/16	5½	10.5
5 × 4	6	10 13/32	6	11.5
5	6½	11 7/16	6½	15.0
6 × 2	6½	8 3/16	5	12.0
6 × 3	6½	9 3/16	5½	23.0
6 × 4	6½	10 1/16	6	11.5
6 × 5	7	11½	6½	27.0
6	7	12½	7	15.0
8 × 4	7½	11½	6½	21.0
8 × 5	8	12½	7	24.0
8 × 6	8	13½	7½	24.0
8	8	15½	9½	28.0

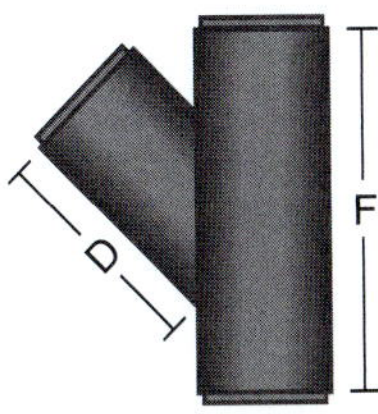

WYE			
SIZE (IN)	D	F	WEIGHT (LB)
1½ × 1½	4	6	2.2
2 × 2	4⅝	6⅝	2.6
3 × 1½	4⅝	6⅝	3.0
3 × 2	5 5/16	6⅝	3.7
3	5¾	8	4.9
4 × 2	6	6⅝	5.5
4 × 3	6½	8	7.2
4	7 1/16	9½	8.8
5 × 2	7½	8 1/16	7.2
5 × 3	8	9 11/16	10.5
5 × 4	8½	11 3/16	12.0
5	9½	12⅝	14.5
6 × 2	8¼	8 5/16	9.5
6 × 3	8¾	9¾	11.0
6 × 4	9¼	11 3/16	13.0
6 × 5	10¼	12½	16.0
6	10¾	14 1/16	18.0
8 × 3	9⅞	10	19.0
8 × 4	10⅜	11 7/16	22.0
8 × 5	11⅜	12⅞	25.0
8 × 6	11 13/16	14 13/16	28.0
8	13⅜	17⅛	38.5
10 × 4	11 11/16	12⅝	37.0
10 × 6	13⅛	15 7/16	46.0
10 × 8	14 11/16	18⅜	52.0
10	16½	21½	72.0

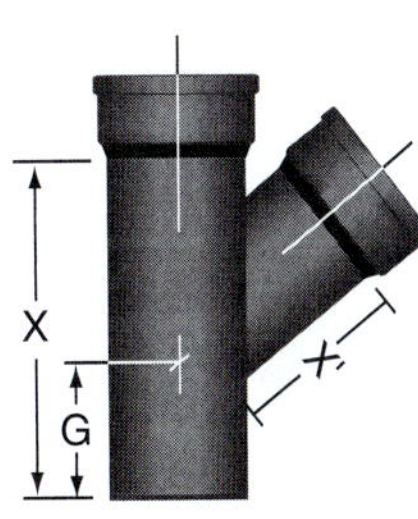

WYE					
				WEIGHT (LB)	
SIZE (IN)	G	X	X'	SV	XH
2 × 2	4	8	4	7	8
3 × 2	4 3/16	9	5	10	14
3 × 3	5	10½	5½	13	17
4 × 2	3⅝	9	5¾	12	17
4 × 3	4 7/16	10½	6¼	14	20
4 × 4	5¼	12	6¾	17	24
5 × 2	3⅛	9	6½	14	–
5 × 3	3⅞	10½	7	17	24
5 × 4	4 11/16	12	7½	19	27
5 × 5	5½	13½	8	22	32
6 × 2	2 9/16	9	7¼	17	–
6 × 3	3⅜	10½	7¾	19	27
6 × 4	4 3/16	12	8¼	22	31
6 × 5	4 15/16	13½	8¾	24	35
6 × 6	5¾	15	9¼	28	40
8 × 2	3⅛	10½	8½	29	–
8 × 3	3 15/16	12	9	32	–
8 × 4	4¾	13½	9½	36	52
8 × 5	5½	15	10	39	–
8 × 6	6 5/16	16½	10½	44	63
8 × 8	7 11/16	19½	11 13/16	55	82
10 × 3	2 9/16	12	11	50	–
10 × 4	3 9/16	13½	11⅛	53	74
10 × 5	4 5/16	15	11⅝	57	–
10 × 6	5⅛	16½	12⅛	61	86
10 × 8	6½	19½	13 7/16	77	110
10 × 10	8	22½	14½	94	133
12 × 4	4⅛	15	12 7/16	70	–
12 × 5	4⅞	16½	12 15/16	74	–
12 × 6	5 11/16	18	13 7/16	80	111
12 × 8	7 1/16	21	14¾	96	136
12 × 10	8 9/16	24	15 13/16	115	160
12 × 12	10⅛	27	16⅞	135	186
15 × 4	2¼	15	15	130	130
15 × 6	4	18	15¾	109	–
15 × 8	5⅜	21	17 1/16	127	182
15 × 10	6⅞	24	18⅛	152	213
15 × 12	8 7/16	27	19 3/16	242	242
15 × 15	10¾	31½	20¾	338	338

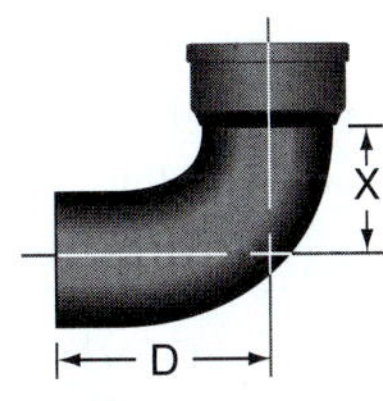

QUARTER BEND				
			WEIGHT (LB)	
SIZE (IN)	D	X	SV	XH
2	6	3¼	4	5
3	7	4	7	10
4	8	4½	11	15
5	8½	5	13	19
6	9	5½	17	24
8	11½	6⅝	34	51
10	12½	7⅝	55	78
12	15	8¾	80	111
15	16½	10¼	169	169

104F08.EPS

Figure 8 ◆ Fitting dimensions for cast-iron pipe.

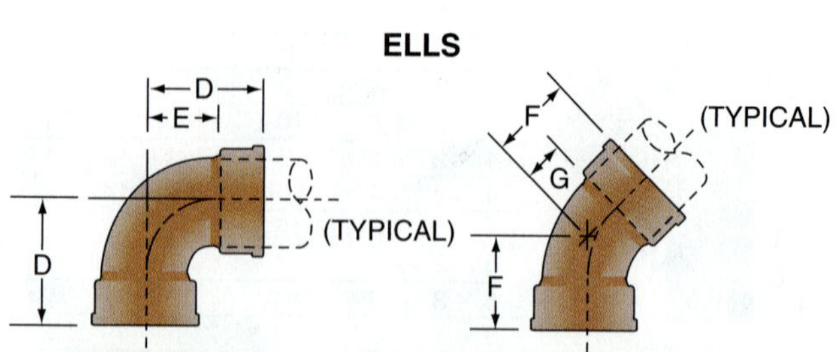

TEE

M
L
L
M

NOTES:
1. DIMENSIONS D AND F ARE CENTER TO END OF FITTING.
2. DIMENSIONS E AND G ARE CENTER OF FITTING TO END OF PIPE.
3. DIMENSIONS ARE IN INCHES.

NOMINAL PIPE SIZE (IN)	90° ELBOWS			45° ELBOWS		
	D	E	WT. 90° (LB)	F	G	WT. 45° (LB)
¼	1.88	1.56	0.13	1.02	0.70	0.09
⅜	1.94	1.62	0.17	1.20	0.89	0.11
½	2.25	1.94	0.25	1.38	1.06	0.17
¾	1.91	2.59	0.34	1.25	0.94	0.24
1	2.38	2.00	0.53	1.50	1.12	0.35
1¼	2.88	2.44	0.76	1.78	1.34	0.51
1½	3.50	2.94	0.96	2.19	1.62	0.67
2	4.28	3.72	1.81	2.53	1.97	1.21
2½	5.38	4.62	2.65	3.19	2.44	1.77
3	6.17	5.41	4.54	3.52	2.76	2.68
4	7.78	6.97	8.55	4.27	3.45	4.88
5	9.53	8.59	13.32	5.14	4.20	6.87
6	11.17	10.17	19.90	5.91	4.91	11.61
8	14.62	13.38	38.50	7.59	6.34	28.08
10	18.00	16.56		9.22	7.78	
12	21.25	19.69		10.70	9.14	

NOTES:
1. DIMENSION M IS THE CENTER TO END OF THE FITTING.
2. DIMENSION L IS THE CENTER OF FITTING TO END OF PIPE.
3. DIMENSIONS ARE IN INCHES.

NOMINAL PIPE SIZE (IN)	STRAIGHT TEES		
	M	L	WT. (LB)
¼	0.69	0.38	0.13
⅜	0.91	0.59	0.17
½	1.09	0.78	0.23
¾	1.25	0.94	0.34
1	1.50	1.12	0.53
1¼	1.88	1.44	0.87
1½	2.25	1.69	1.21
2	2.50	1.94	1.87
2½	3.00	2.25	2.80
3	3.38	2.62	4.33
4	4.12	3.31	8.09
5	4.88	3.94	11.17
6	5.62	4.62	16.58
8	7.00	5.75	27.27
10	8.50	7.06	
12	10.00	8.43	

104F09.EPS

Figure 9 ◆ Fitting dimensions for copper DWV pipe.

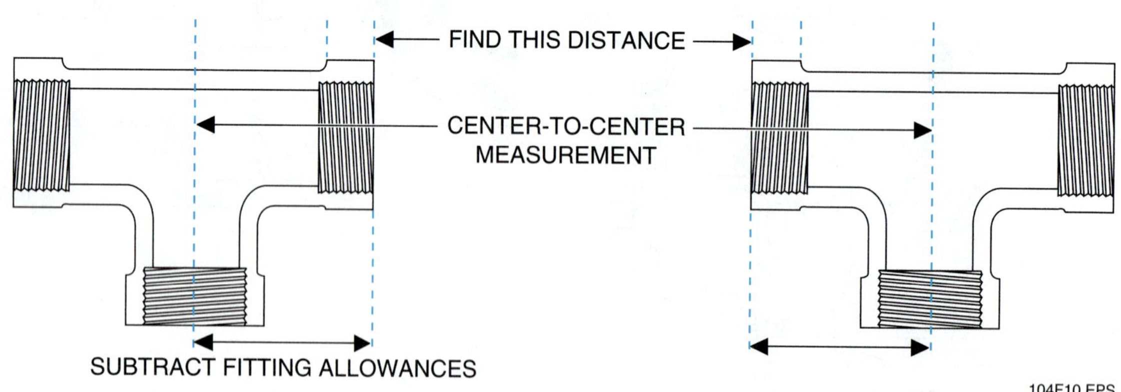

Figure 10 ◆ Center-to-center method.

3.5.2 End-to-Center Method

To calculate the end-to-end dimension when you know the end-to-center measurement, subtract one fitting allowance, because there is just one fitting (see *Figure 11*).

Suppose you need to connect two ¾-inch threaded steel tees. The end-to-center measurement is 17½ inches. Calculate the length of pipe you need to cut.

Step 1 Determine the fitting allowance by subtracting the thread-in from the center-to-face measurement (see *Figure 11*):

Fitting Allowance = 1⅜" − ½" = 1⅛" − 2⁄8" = ⅞"

Step 2 Subtract the fitting allowance from the end-to-center measurement.

17½" − ⅞" = 16 12⁄8" − ⅞" = 16⅝"

You need 16⅝ inches of ¾-inch steel pipe.

3.5.3 Face-to-Face Method

To determine the required end-to-end dimension given a face-to-face measurement, you need to add the thread makeup for each fitting (see *Figure 12*).

Example: Calculate the length of steel pipe required to connect ½-inch threaded steel tees if the face-to-face measurement is 6¾ inches.

6¾" + ½" + ½" = 7¾"

You need 7¾ inches of ½-inch steel pipe.

3.5.4 General Principles of Measurement

As shown in *Figure 4*, there are many measuring techniques, depending on the information you have and the type of installation. You may have noticed the following general principles underlying all of these methods of measurement:

- When you have a center dimension, subtract the fitting allowance for that end.
- When you have a face dimension, add the thread makeup for that end.
- When you have an end dimension, you do not need to add or subtract anything.

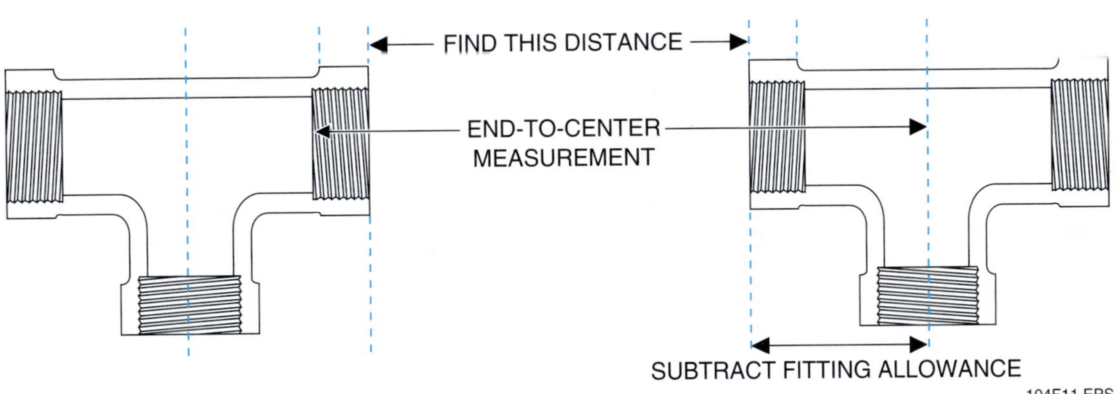

Figure 11 ◆ End-to-center method.

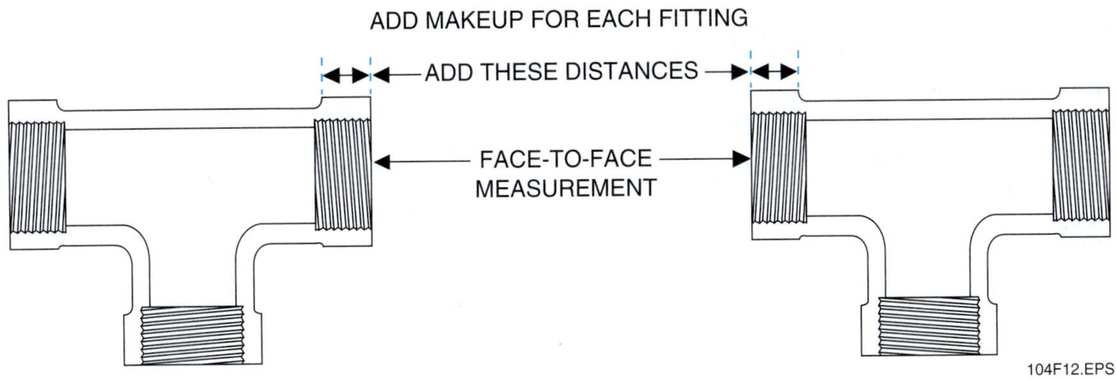

Figure 12 ◆ Face-to-face method.

Review Questions

Section 3.0.0

1. The extensions of the centerline of the pipe and the centerline of the fitting intersect inside the fitting to create a(n) _____.
   a. centerline
   b. single line
   c. center point
   d. end-to-end dimension

2. The distance that a pipe screws into a fitting is called the _____.
   a. fitting allowance
   b. thread makeup
   c. center point
   d. center-to-face dimension

3. A nominal ½-inch pipe that is 10 inches long from the end to the center of an ell is abbreviated as _____.
   a. ½–10E–C
   b. ½" C–Ell
   c. ½" 10" E–C ell
   d. 10" E–Ell C

4. The most convenient way to determine thread makeup is to measure the depth of the fitting's socket or thread.
   a. True
   b. False

5. _____ is the distance from the end of the pipe that goes into the fitting to the center of the fitting.
   a. Center-to-center
   b. Thread-in
   c. Makeup
   d. Fitting allowance

6. You need to cut a threaded steel pipe to fill a space between two ⅜-inch threaded tees. The center-to-center measurement between the two fittings is 15¾ inches. The end-to-end dimension required for this pipe is _____ inches.
   a. ⅝
   b. 14⅜
   c. 14½
   d. 15¾

7. You need to cut a threaded steel pipe to connect two ¾-inch threaded tees. The center-to-center measurement is 41⅞ inches. The end-to-end dimension required for this pipe is _____ inches.
   a. ⅞
   b. 40⅛
   c. 41⅞
   d. 41

8. In an installation, two 1-inch threaded tees need to be connected. The center-to-center measurement is 42 inches. The end-to-end dimension required for this pipe is _____ inches.
   a. 1½
   b. 40
   c. 42
   d. 44

9. The end-to-end dimension required for a ½-inch threaded pipe between two tees with an end-to-center length of 29 inches is _____ inches.
   a. ⅝
   b. 1⅛
   c. 23⅜
   d. 29

10. The end-to-end dimension required for a 1½-inch threaded pipe between two tees with a face-to-face length of 18⅞ inches is _____ inches.
    a. ½
    b. 18⅞
    c. 19⅞
    d. 21⅞

Summary

Skilled, productive plumbers know mathematics. Being able to use math easily and accurately in the field leads to more efficiency and greater productivity. This helps make you a more valued employee, and your career will benefit as a result.

You have been introduced to methods of measuring pipe and have learned how to find the cut length of pipe that runs between fittings. With this groundwork, you can progress in your math skills and your trade skills. The more you know, the more you can take pride in your abilities.

Notes

Trade Terms Quiz

Fill in the blank with the correct trade term that you learned from your study of this module.

1. The _______________, _______________, and _______________ describe the basic parts of a fitting; they define the beginning and ending points of the measurement.
2. The lengths of pipe and fittings are measured along _______________.
3. The _______________ is where the extension of the centerline of the pipe and the centerline of the fitting meet inside the fitting.
4. To calculate _______________, subtract the thread-in measurement from the center-to-face measurement.
5. 89 ÷ 3 = 29 with a(n) _______________ of 2.
6. _______________ are more convenient than drawings of the whole pipe or fitting.
7. For threaded fittings, the _______________ is the distance that the pipe screws into the fitting.
8. The _______________ of a fitting is not used often in measuring pipe.

Trade Terms

Back	Center point	Remainder	Throat
Center	Face	Schematic drawings	
Centerline	Fitting allowance	Thread makeup	

Profile in Success

Tom LeDuc

President/CEO
LeDuc & Dexter, Inc.
Santa Rosa, California

Tom LeDuc was born in Santa Rosa, California, and attended Montgomery High School there. At the age of 12, he began working at his father's plumbing warehouse, sweeping the floor. He attended a junior college for one year, where he studied business administration. When he was 18 years old, he had the opportunity to get into a plumbing apprenticeship program for Local Union 38 out of San Francisco and has been working in the industry ever since. Today, he is president and chief executive officer of LeDuc & Dexter, Inc.

How did you become interested in this industry?
I grew up in a plumbing family. When I was 10 years old, my father went into the plumbing business. He has a wholesale plumbing business. I had exposure to plumbing at a very young age, and I was always involved. From 12 years old on, I spent vacations and time off working at the shop. I served the first half of my apprenticeship at my father's company. I went to work for another company to get experience working for someone else. I worked for three companies while I was an apprentice.

What path did you take to your current position?
I moved from journeyman to foreman to superintendent to project manager. I went from journeyman to foreman in the field and then I moved into the office as a project manager for Empire Swift Plumbing, which was the company I worked for from 1971 to 1982. I started my own business in 1982 with my partner, Art Dexter. We worked together at Empire and have been partners for 21 years. I started by crawling under houses, plus bidding on the work, buying the material, and selling the material. We started with two employees and now we have 75.

In California, the company is licensed, not the individual. The business owner holds the licenses. You can use work experience or training to qualify for the licenses. I first earned my plumbing contractor's license, and then I got a fire protection license. I also completed fire protection work and went to school through the Plumbing-Heating-Cooling Contractors—National Association (PHCC) for three two-day sessions over a six-month period. I received a general contractor's license, a heating and air conditioning license, a boiler and hot water license, and a sheet metal license.

What are some things you do on the job?
I'm the head administrator, the person who negotiates insurance and contracts. I do all the banking, legal work, and licensing. I also spend a huge amount of time working with the trade associations. I am president of the North Coast Builders Exchange, a multi-trade association for the construction industry. With 1,800 members, it is the largest builders' exchange in California and the third largest in the United States. I am also on the State Apprenticeship Master Committee for the PHCC of California.

We handle new home construction, including single-family homes, subdivisions, apartments, and a lot of custom homes. On the commercial end, we do a tremendous amount of winery work, institutional facilities, and high-tech industry work. We even did a huge racetrack facility. I believe that you should service what you do, so we started a service company, called Super Service Plumbing.

What does it take to be successful in your trade?
To be successful in our trade—as a person working in the trade, not just a business owner—it takes passion, dedication, and commitment. You need the willingness to work hard. The number one thing is passion. You can always learn if you have passion.

Your work is something to be proud of. There are real opportunities for a great career, because you can start out sweeping floors and end up owning the company. All of those options are available. You can

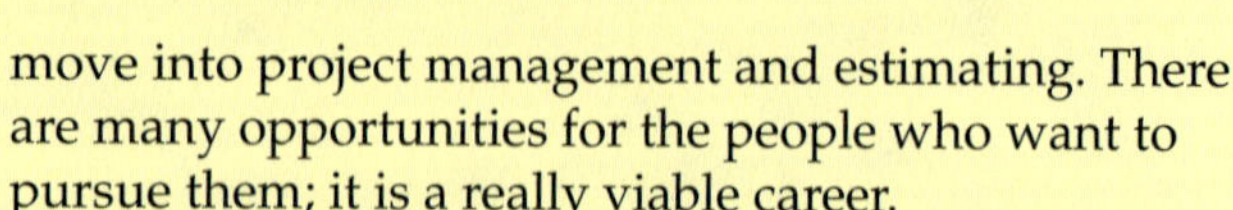

move into project management and estimating. There are many opportunities for the people who want to pursue them; it is a really viable career.

What is the most interesting aspect of your profession? What makes your trade stand out from others?
I like dealing with people: the employees and the customers. That's the most fun. Plumbing stands out because you have to know so many different things. Our trade protects the health and safety of the nation.

What do you like most—and least—about your job?
I most like working with people, meeting a lot of people. I least like collecting money.

What would you say to someone entering the trade today?
This is a very honorable profession. You can be proud of doing a good job for the customer and you are really helping people. It is a great way to make a good living!

What can an apprentice expect to earn during the first years on the job? What can that person expect to earn after 10-plus years in the industry?
Open-shop apprentice rates have a starting wage of $10 per hour. There are percentage increases each year. The state of California has published minimum wages. Apprentices build up through the end of the program to $18 per hour. Apprentices receive raises every six months or 1,000 hours of work (plus hours of school). We always pay above the scale. Journeymen make up to $30 per hour, depending on their capabilities.

Trade Terms Introduced in This Module

Back: Part of the fitting that is opposite to the side with an opening or face.

Center: A point exactly halfway between two other points or surfaces.

Centerline: On a drawing, a line that shows the center of an object.

Center point: Point created where the centerline of the pipe and the centerline of the fitting meet within the fitting. The center point is used to determine the correct length of a pipe.

Face: The open end of a fitting where a pipe is joined to the fitting, such as the opening of the inlet or either end.

Fitting allowance: The distance from the end of the pipe that goes into a fitting to the center of the fitting. Also called fitting takeoff.

Remainder: The leftover amount in a division problem. For example, in the problem $34 \div 8$, 8 goes into 34 four times ($8 \times 4 = 32$) and 2 is left over or, in other words, it is the remainder.

Schematic drawings: Simple, single-line representations (drawings) of pipe and fittings.

Thread makeup: The distance that a pipe screws into a fitting. Also called thread engagement or thread-in.

Throat: The part of the fitting where you thread in another pipe or fitting.

Additional Resources and References

This module is intended to present thorough resources for task training. The following reference works are suggested for further study. These are optional materials for continued education rather than for task training.

Plumber's and Pipefitter's Calculations Manual, 1999. R. Dodge Woodson. McGraw-Hill Professional.

American Society of Civil Engineers. "Metrication—The Basics." Accessed Sept. 25, 2003. http://users.vnet.net/cmstone/metric/basics.htm.

NCCER CURRICULA — USER UPDATE

NCCER makes every effort to keep its textbooks up-to-date and free of technical errors. We appreciate your help in this process. If you find an error, a typographical mistake, or an inaccuracy in NCCER's curricula, please fill out this form (or a photocopy), or complete the online form at **www.nccer.org/olf**. Be sure to include the exact module ID number, page number, a detailed description, and your recommended correction. Your input will be brought to the attention of the Authoring Team. Thank you for your assistance.

Instructors – If you have an idea for improving this textbook, or have found that additional materials were necessary to teach this module effectively, please let us know so that we may present your suggestions to the Authoring Team.

NCCER Product Development and Revision

13614 Progress Blvd., Alachua, FL 32615

Email: curriculum@nccer.org
Online: www.nccer.org/olf

❑ Trainee Guide ❑ AIG ❑ Exam ❑ PowerPoints Other ______________________________

Craft / Level: Copyright Date:

Module ID Number / Title:

Section Number(s):

Description:

Recommended Correction:

Your Name:

Address:

Email: Phone:

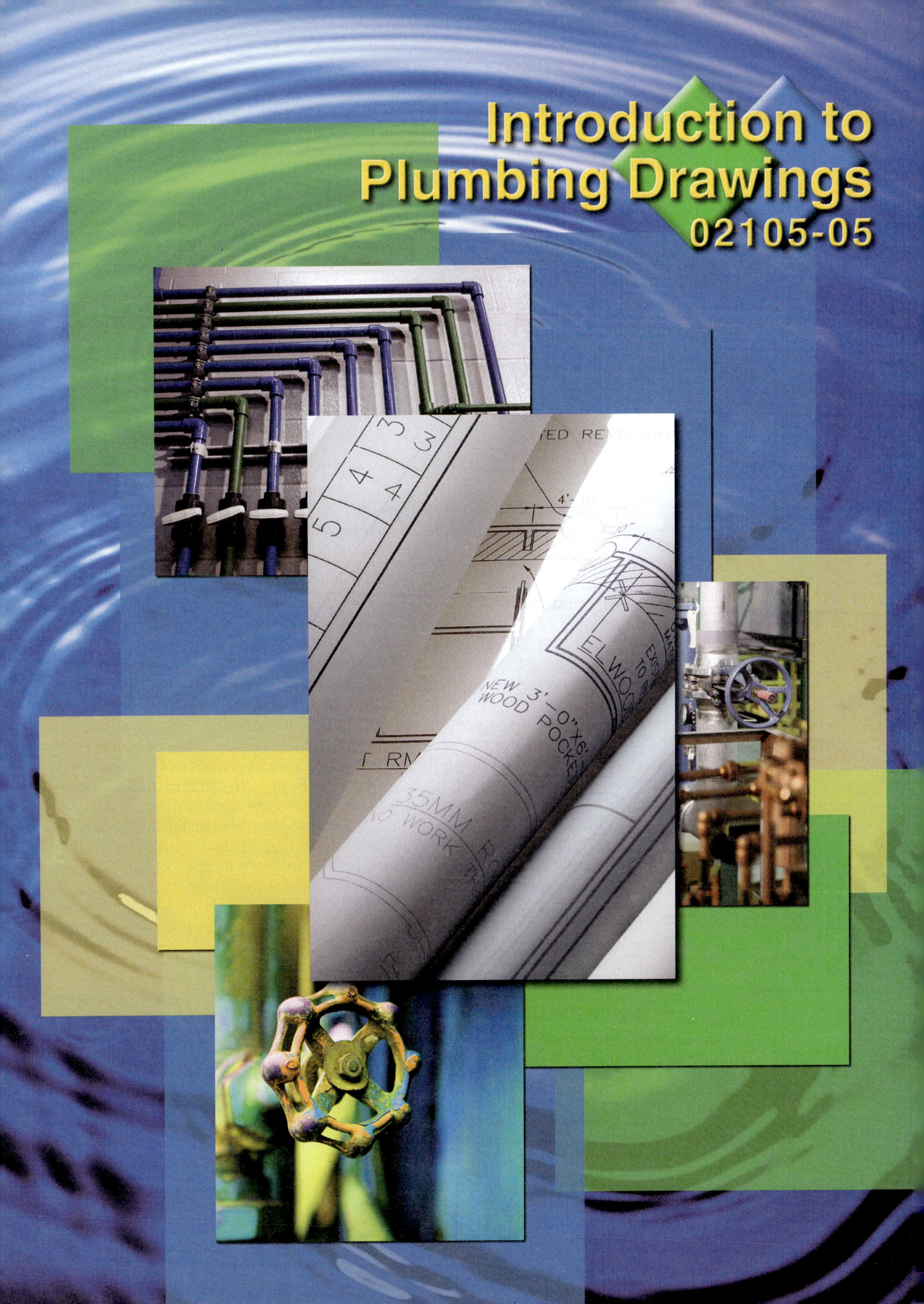
Introduction to
Plumbing Drawings
02105-05

02105-05

Introduction to Plumbing Drawings

Topics to be presented in this module include:

Overview

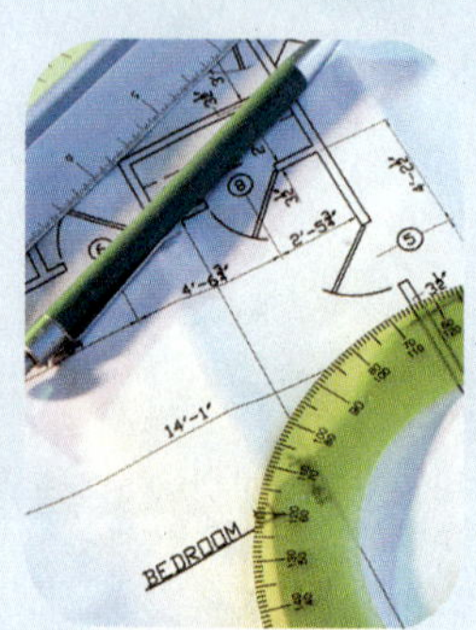

Every building's design begins on paper. Construction drawings (or blueprints) illustrate design, location, and dimensions. Plumbers must learn to work with and interpret the complete set of construction drawings. Being able to do so enables plumbers to direct work and follow written instructions included on the drawings.

Understanding the components of a drawing is essential to being able to interpret it accurately. Basic information about each drawing is included in the title block. Scale and dimension indicate actual sizes and distances. Symbols for pipes and fittings represent specific objects and materials. Notes and specifications include information about the quality of the work, the materials to be used, and the method of installation. If a set of drawings does not include specifications, plumbers follow applicable codes. Every jurisdiction adopts a specific plumbing code to ensure that plumbing systems are installed properly.

In addition to interpreting existing drawings, plumbers also must be able to create simple sketches and isometric drawings. These drawings enable plumbers to communicate with other workers and convey specific information about an installation. Manufacturers also produce drawings that contain detailed product information. The ability to read these drawings to determine specifications and interpret details about piping assemblies is essential to the plumbing trade.

Focus Statement

The goal of the plumber is to protect the health, safety, and comfort of the nation job by job.

Code Note

Codes vary among jurisdictions. Because of the variations in code, consult the applicable code whenever regulations are in question. Referencing an incorrect set of codes can cause as much trouble as failing to reference codes altogether. Obtain, review, and familiarize yourself with your local adopted code.

Objectives

When you have completed this module, you will be able to do the following:

1. Identify pictorial (isometric and oblique), schematic, and orthographic drawings, and discuss how different views are used to depict information about objects.
2. Identify the basic symbols used in schematic drawings of pipe assemblies.
3. Explain the types of drawings that may be included in a set of plumbing drawings and the relationship among the different drawings.
4. Interpret plumbing-related information from a set of plumbing drawings.
5. Sketch orthographic and schematic drawings.
6. Use an architect's scale to draw lines to scale and to measure lines drawn to scale.
7. Discuss how code requirements apply to certain drawings.

Key Trade Terms

Approved submittal data
Architect's scale
Catalog drawing
Computer-aided drafting
Construction drawing
Coordination drawing
Cutaway drawing
Details
Dimension line
Easement
Electrical drawing
Elevation drawing
Exploded drawing
Extension line
Fixture drawing
Floor plan
Foundation plan
HVAC
Isometric drawing
Oblique drawing
Orthographic drawing
Pictorial drawing
Plot plan
Plumbing drawing
Riser diagram
Scale
Schematic drawing
Sepia
Setback
Side yards
Single-line drawing
Site plan
Specifications
Specs
Submittal data
Symbol
Takeoff

Required Trainee Materials

1. Appropriate personal protective equipment
2. Pencils, mechanical pencils, and paper
3. Eraser
4. Calculator
5. Architect's scale
6. T square
7. 30-60 degree triangle or angle
8. Graph paper or isometric paper
9. Copy of local adopted code

Prerequisites

Before you begin this module, it is recommended that you successfully complete *Core Curriculum; Plumbing Level One,* Modules 02101-05 through 02104-05.

This course map shows all of the modules in the first level of the *Plumbing* curriculum. The suggested training order begins at the bottom and proceeds up. Skill levels increase as you advance on the course map. The local Training Program Sponsor may adjust the training order.

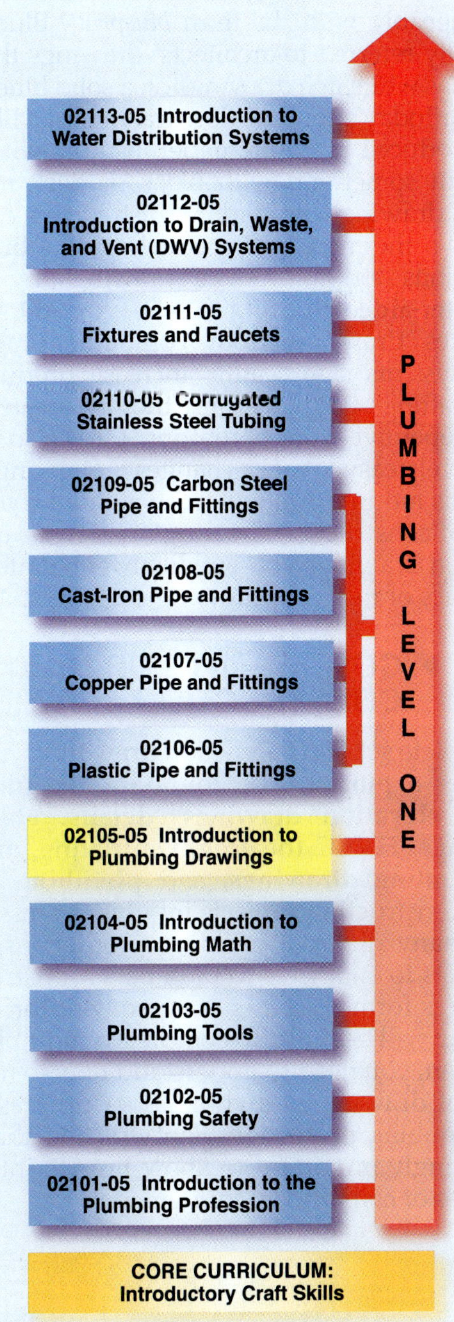

105CMAP.EPS

1.0.0 ◆ INTRODUCTION

Before the first shovel of earth is moved, a building has already had a long life on paper. Every building, regardless of its size, begins life on a designer's drawing board or computer terminal. These drawings illustrate the elements of a building, including its design, location, and dimensions. The ability to read drawings is an essential skill. Drawings allow you to visualize a finished project before starting. By visualizing, you can plan ahead and anticipate potential problems, which saves time and money on the job. Developing this skill takes time and experience.

The term ***construction drawing*** is often used interchangeably with the term *blueprint.* Blueprints originally referred to architects' drawings that appeared as white line drawings on a solid blue background. Today's construction drawings are likely to be reproduced on white paper, usually with blue lines but sometimes with black or dark reddish lines. A drawing with dark reddish lines is called a **sepia.** Furthermore, most people in the building professions today produce drawings by using **computer-aided drafting (CAD).** A CAD system generates drawings from computer programs. CAD systems significantly increase productivity over drafting by hand because they automate much of the repetitive work of drafting. CAD also makes it relatively easy to make changes to drawings.

In this module you will learn about the components of construction drawings, the types of construction drawings you will likely encounter, and the basics of reading plumbing drawings.

2.0.0 ◆ COMPONENTS OF CONSTRUCTION DRAWINGS

A complete set of construction drawings typically includes a **plot plan,** a **foundation plan, floor plans, elevation drawings, details, electrical drawings, HVAC (heating, ventilating, and air conditioning) drawings,** and **plumbing drawings.** Construction projects involve professionals from many different building professions, from architects to engineers to plumbers. Construction drawings form the basis of agreement for how a building will be built. To complete a plumbing installation, you need to understand how to read all types of drawings. As your career progresses in the plumbing profession, you will find that it is increasingly important to know how to interpret all types of construction drawings.

2.1.0 Plot Plan

The plot plan, also known as a **site plan,** shows the location of the building on the property and shows where utilities, such as the water, storm sewer, sanitary sewer, and gas mains, are installed (see the large-scale example in *Figure A-1* in the *Appendix*). The plot plan also indicates where you may not build on certain parts of the property. For example, **setback** requirements establish the minimum distance that must be maintained between the property line and the structure. **Easements** specify places on the property where utilities can be installed (see *Figure 1*). Codes may set a minimum width for **side yards** to provide access to rear yards, to reduce the possibility of a fire jumping from one building to another, or to promote ventilation.

2.2.0 Foundation Plan

The foundation plan shows the lowest level of a building, including concrete footings, slabs, and foundation walls. It may also show the drainage, waste, and vent (DWV) piping and the water supply piping that must be installed below the first floor. A foundation plan may also indicate the positions of floor drains and other piping that must be installed beneath the basement floor.

2.3.0 Floor Plan

The floor plan shows an aerial view of the layout of rooms on each level of a building and the shape and size of each room (see the large-scale example in *Figure A-2* in the *Appendix*). Think of the floor plans as showing a view of a building from above if the roof were removed. You will use floor plan drawings to locate the position of water supply and waste piping.

It is standard practice to install the cold water supply at the right of each fixture. Having the cold water controlled by the right-hand faucet and the hot controlled by the left-hand faucet prevents burns, because people are familiar with this setup.

2.4.0 Elevation Drawings

Elevation drawings show the vertical elements of a building. Elevation drawings and approved submittal data should relate to the dimension sheets and match up to the architectural elevations and details. This ensures a proper fit and accurate installation. They show the height and width of interior or exterior walls. You will find the elevation drawings in the overall set of construction drawings for a project. See an example of an interior elevation drawing in *Figure 2.*

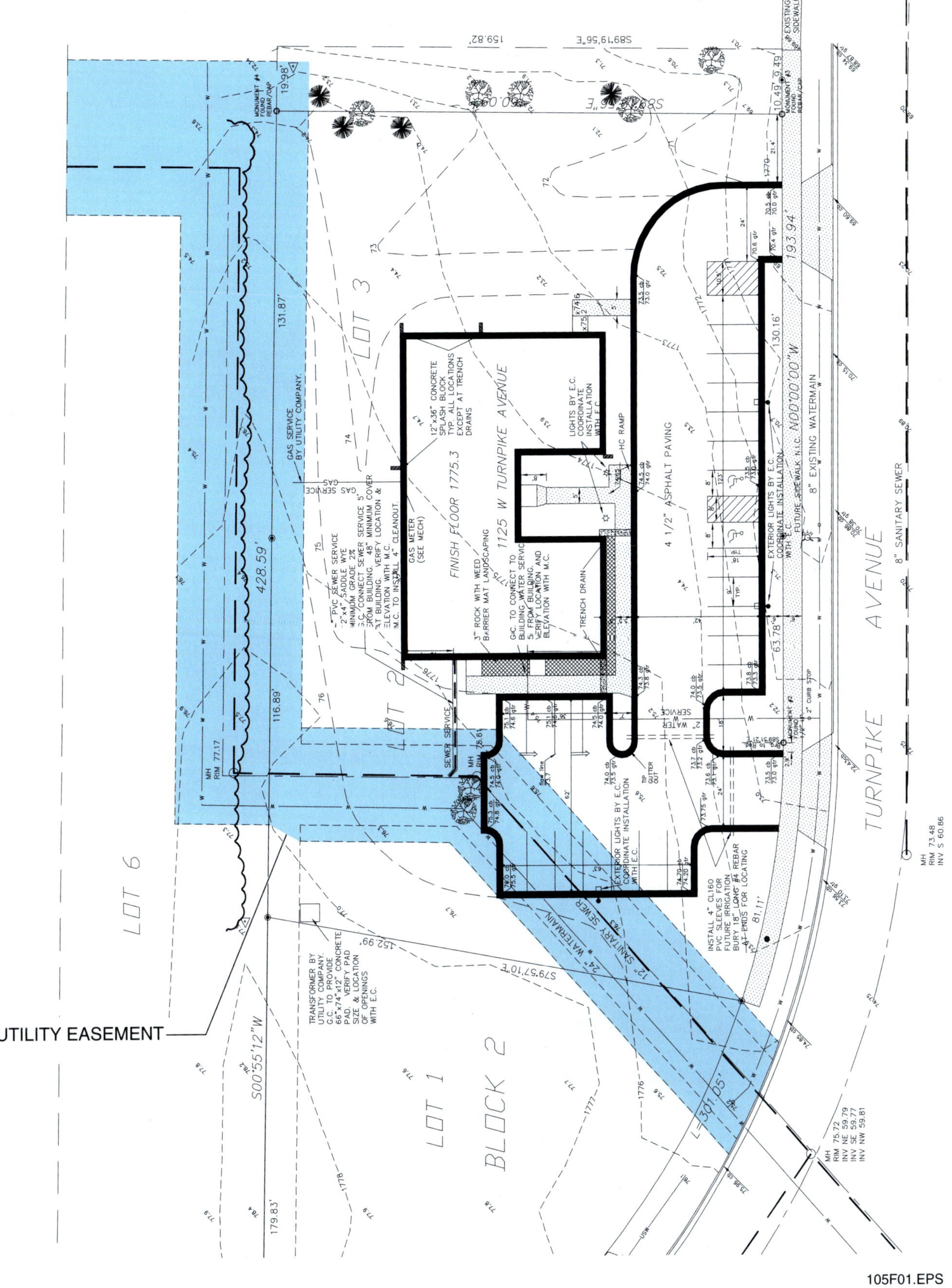

Figure 1 ◆ Utility easement.

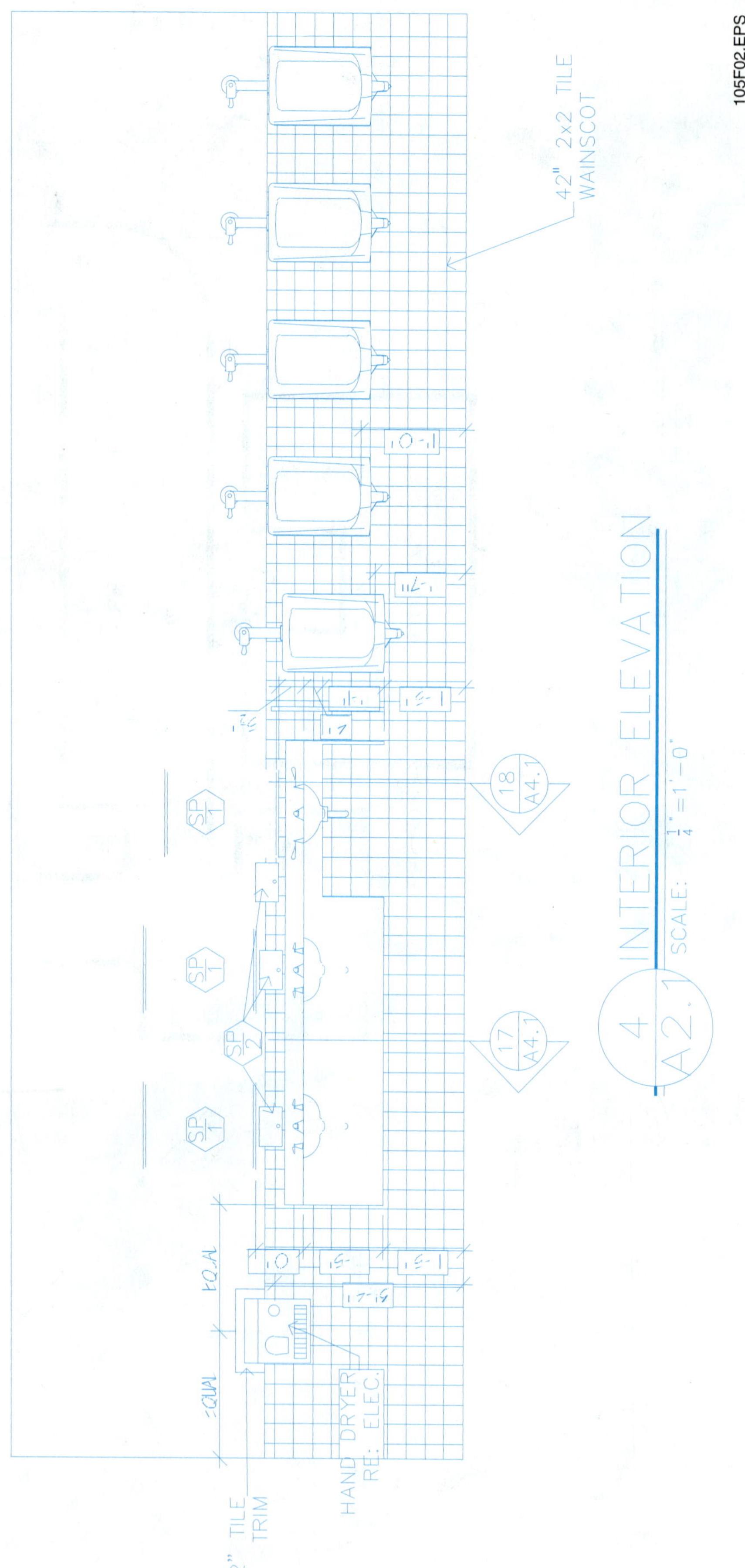

Figure 2 ◆ Interior elevation drawing.

For another example, find the interior elevations in *Figure A-3* in the *Appendix*. You can use these drawings to find information about where exterior hose bibbs go, how storm drainage should be installed, and how vent stacks should be installed through the roof.

2.5.0 Details

Details are enlarged drawings, usually presented as part of the main drawing, of important structural elements or special features of the building such as cabinets or staircases. Because they are enlarged, detail drawings let you see an area of the drawing more clearly. For an example of a detail drawing, see the Janitor's Sink Detail in *Figure A-4* in the *Appendix*.

2.6.0 Electrical Drawings

Electrical drawings are sometimes superimposed on the floor plan. They show how the lighting and power layout will appear on the plan. They also show the location of outlets, switches, and electrical fixtures, including electrical controls and programmable controllers. Electrical symbols illustrate how a specific part of the system works. These drawings also explain how to execute the electrical plan and include electrical specifications.

2.7.0 HVAC Drawings

HVAC drawings are also superimposed on floor plans. These drawings show the location of the furnace and air conditioning equipment, as well as the location of ducts and registers or piping and radiators. HVAC drawings include an electrical diagram that shows the electrical circuitry for the HVAC system. HVAC plans include both mechanical and electrical drawings. Architects may redraw HVAC drawings to scale to illustrate components using their true shapes.

2.8.0 Plumbing Drawings

Like electrical and HVAC drawings, plumbing drawings are superimposed on floor plans. Plumbing drawings show the piping system, with the location of fixtures and pipe runs and the size and type of pipe to be installed (see the large-scale example in *Figure A-4* in the *Appendix*). They show the diameter of each run of pipe and the fittings required to make the installation. You can estimate the length of each run of pipe by looking at the measurements on the floor plans, but you will not know the exact length of pipe until you are actually installing it.

2.9.0 Coordination Drawings

Coordination drawings are dimensioned drawings with elevations and sections, if needed, that indicate proposed routing of system components. Based as nearly as possible on contract drawings, the intent of a coordination drawing is to identify and resolve conflicts that may occur with building elements and other trade contractors. This drawing should also include the actual equipment that will be provided as well as clearances required by that equipment.

Review Questions

Sections 1.0.0–2.0.0

1. The term *blueprints* is often used interchangeably with the term _____.
   a. sepias
   b. CADs
   c. construction drawings
   d. specifications

2. The _____ plan shows the location of the sewer and water mains and shows the location of the building on the property.
   a. foundation
   b. plot
   c. elevation
   d. plumbing

Match the following components of construction drawings with their corresponding definitions.

3. plot plans	a. show important elements more clearly
4. floor plans	b. show furnishings and paint colors
5. elevation drawings	c. show the location of the building on the property
6. details	d. show the vertical elements of a building
	e. show the layout of rooms in a building

7. To identify where floor drains and DWV pipes will be installed beneath the basement floor, refer to the _____ plan.
   a. floor
   b. details
   c. foundation
   d. plot

8. A plumber should consult _____ plans to make sure there are no conflicts with furnace and air conditioning equipment.
   a. floor
   b. HVAC
   c. foundation
   d. elevation

3.0.0 ◆ READING PLUMBING DRAWINGS

To be a successful plumber, you must be able to read plumbing drawings. If you can read drawings, you can direct others, which enables you to earn more money. To do so, you need to understand the relationship between the dimensions on the drawing and the dimensions of the structure itself and know how to read the dimensions. You also need to understand the written requirements that may be included with the drawings. You must follow all applicable plumbing codes in your area.

3.1.0 Title Block

Each drawing contains a title block, which includes information such as the title of the project, the date of the drawing, the logo of the company that created the drawing, the name or initials of the person who drafted the drawing, and the owners and address of the property. When you make any construction drawings, always put a title block on that drawing. The title block usually appears in the lower right-hand corner of the drawing, but the location can vary depending on the company's placement system. The title block is the first thing you should read on construction drawings.

3.2.0 Scale

Construction drawings are usually drawn to **scale.** The scale indicates the size relationship between an object in the drawing and the object's actual size. The type of scale used on a drawing depends on the size of the objects being shown, the space available on the paper, and the type of plan. Scale drawings show objects such as buildings, rooms, doors and windows, or piping assemblies reduced to a smaller size.

For example, a wall that measures 8 feet high and 14 feet wide obviously cannot be drawn full size. By using scale, an architect, designer, or plumber can represent the wall in a smaller size while keeping the same proportions of height and width as the full-size wall. The proportional reduction allows you to determine the actual size of the object from the drawing.

Typical scales for plumbing drawings are ⅛ = 1' or ¼ = 1'. If the scale is ¼ = 1', this means that for every ¼ inch on the drawing, the real object takes up 1 foot. In other words, the drawing is 1/48 the size of the real object. Scale makes drawings convenient to handle. The scale of the drawing usually appears in the title block of the drawing. Be aware that some drawings are not drawn to scale. A note on such drawings will read *not to scale* or *NTS*.

CAUTION

When a plan is marked *NTS*, you cannot measure dimensions on the drawing and use them to build the project. Not-to-scale drawings give relative positions and sizes. These measurements are approximate and are not accurate enough for construction.

The term *scale* is also used to describe the three measuring tools that are used to draw or measure the lines of a construction drawing: **architect's scale,** engineer's scale, and metric scale.

To draw and measure scaled drawings, you use an architect's scale (see *Figure 3*). The architect's scale is used on all plans other than site plans. The architect's scale is a ruler. It may be either flat or triangular in shape. If it is flat, it contains four different scales; if it is triangular, it contains 10 scales. The triangular form is commonly used because it contains a variety of scales on a single tool. You can read an architect's scale from left to right or from right to left, depending on which scale you are reading. For example, you would read the ¼ scale from right to left and the ⅛ scale from left to right. *Table 1* lists the scales on a triangular architect's scale. The full size appears in the first row of the table. Notice that all scales are given in reference to 1 foot.

Table 1 The Scales on a Triangular Architect's Scale and Corresponding Reduction

Designation	Scale	Reduction
12	12 = 1'	Full size
3	3 = 1'	One-quarter size
1½	1½ = 1'	One-eighth size
1	1 = 1'	One-twelfth size
¾	¾ = 1'	One-sixteenth size
½	½ = 1'	One twenty-fourth size
⅜	⅜ = 1'	One thirty-second size
¼	¼ = 1'	One forty-eighth size
3/16	3/16 = 1'	One sixty-fourth size
⅛	⅛ = 1'	One ninety-sixth size
3/32	3/32 = 1'	One one-hundred twenty-eighth size

Each scale is set up the same way. First, you see the designation of the scale itself, such as ¼ or 3/32. The designation is followed by a series of lines, called graduations. These graduations represent exactly 1 foot as it is drawn with that particular scale.

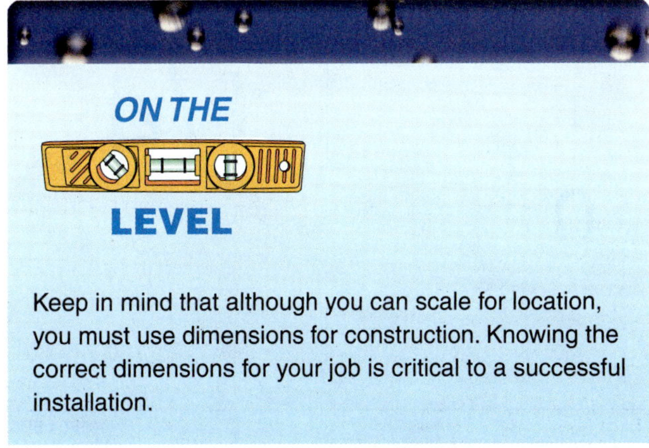

Keep in mind that although you can scale for location, you must use dimensions for construction. Knowing the correct dimensions for your job is critical to a successful installation.

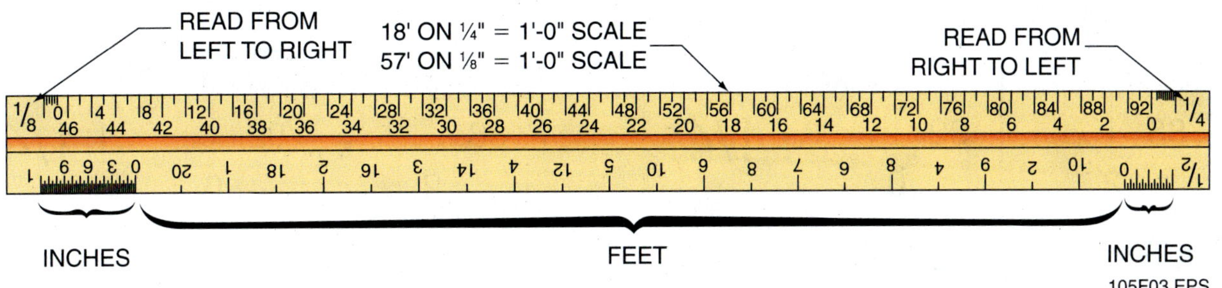

Figure 3 ◆ Architect's scale.

Notice that the number 0 appears at one end of the 1-foot representation on every scale. This is where you start measuring. On the ¼ scale, for example, graduations to the left of the zero signify feet, and graduations to the right signify inches (see *Figure 4*).

Each scale has a different number of graduations to the right of zero. Not all of them signify inches. As you look at the architect's scale, the numbers may confuse you. Refer back to the ¼ scale in *Figure 4*. As you move to the left you will see a line of numbers beginning with 0 and ending with 46, as well as a line of numbers beginning with 92 and ending with 0. Remember, you read the ¼ scale from right to left. Only the numbers from 0 to 46 are used on the ¼ scale. The other numbers are for the ⅛ scale, which is read from left to right.

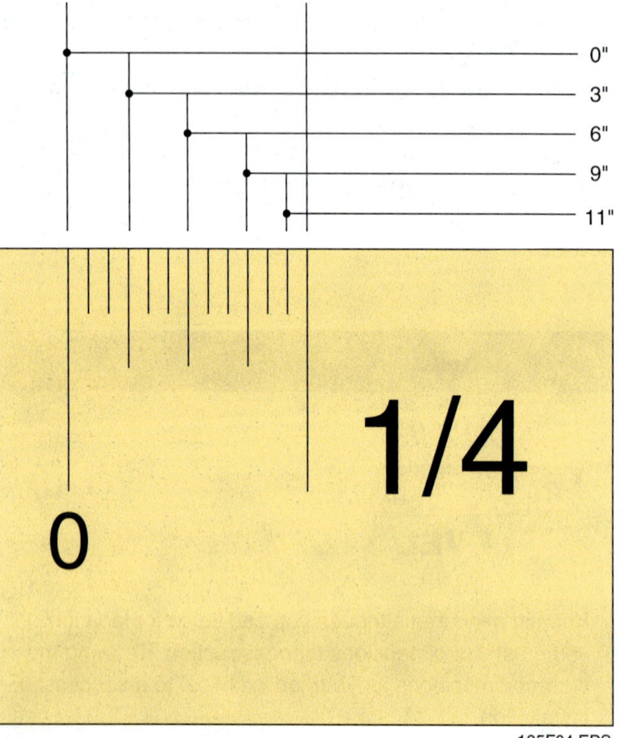

Figure 4 ◆ The ¼ scale.

105F05.EPS

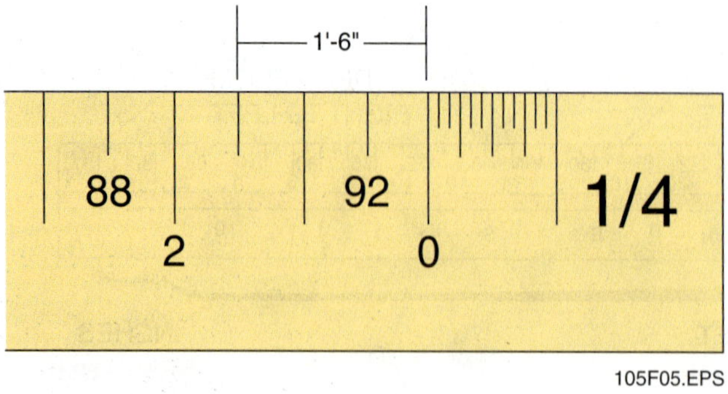

Figure 5 ◆ 1 foot, 6 inches on the ¼ scale.

The system is the same for every face of the architect's scale. Rotate your scale until you see the 1 and ½ scales. Reading the ½ scale from left to right, you see a row of numbers beginning with 0 and ending with 20. Reading the 1 scale from right to left, you see a row of numbers beginning with 0 and ending with 10. You will see the same system on the 3/32 and 3/16 scales, the 1½ and 3 scales, and the ⅜ and ¾ scales.

Refer to the ¼ scale in *Figure 5*. Between 0 and 2 feet on this scale, you see four graduations. (The shorter lines along the top belong to the ⅛ scale, but you also use them when you read the ¼ scale.) Each graduation along the ¼ scale represents 6 inches in the length of an actual object. Thus, a line drawn from 0 to the graduation before the 2 would be 18 inches, or 1 foot, 6 inches long. You can see how 1 foot, 10 inches is represented on the ¼ scale in *Figure 6*.

The engineer's scale is divided into decimal graduations (10, 20, 30, 40, 50, and 60 divisions to the inch). The engineer's scale is used when an area is too large to be represented by the usual scale. For example, the usual scale has a single foot represented by a portion of an inch, and an engineer's scale might be used to represent a larger number of feet per inch. The engineer's scale is used for plotting and map drawing and in the graphic solution of problems such as survey and site plans.

The metric scale is divided into centimeters (cm), with each centimeter divided into 10 millimeters (mm) or 20 half-millimeters. Some scales are made with metric divisions on one edge and inch divisions on the opposite edge. Many companies express measurements in both metric and English units.

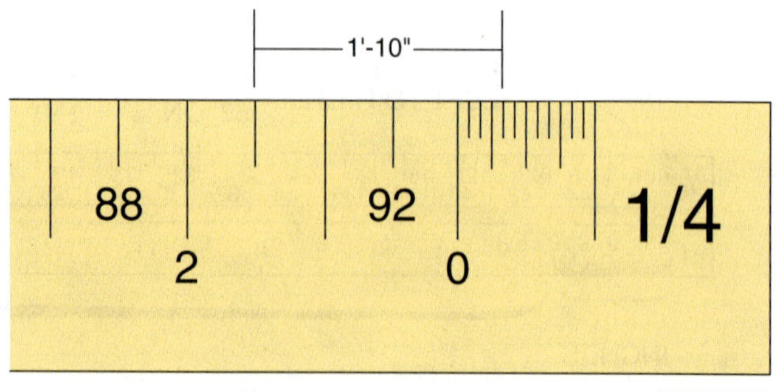

Figure 6 ◆ 1 foot, 10 inches on the ¼ scale.

3.3.0 Dimensioning

Dimensions that appear on a drawing show actual distances. To indicate the dimension of a distance on a drawing, you use **dimension lines.** You can create these lines with arrows or slashes at a termination line drawn perpendicular to the dimension line. Because of space constraints on the drawing, you may need to use **extension lines** to indicate the limits of the dimension away from the actual points of the dimension (see *Figure 7*). Be sure to note the point from which the dimension is taken. Some dimensions are shown from end to end. Others are shown from center lines.

3.4.0 Symbols for Fittings, Pipe, and Fixtures

Plumbers often use **symbols** when they are drawing plans for pipe and fittings. A symbol is a mark or drawing that represents a specific object or material. Using symbols allows you to produce drawings quickly and make plans easy for others to understand.

Be aware that people do not always use the same symbols to represent the same objects. Most drawings include a legend that explains the meaning of the various symbols. Always consult the legend before reading the drawing.

See the symbols used for fittings in *Figure 8*. Seemingly small differences among symbols actually make a huge difference in identifying various fittings. Note the difference in the symbols for fittings that are facing the observer and fittings with the opening facing away. For example, observe the difference in tees where the outlet is up and where it is down. Also, note how the symbols differ for each type of pipe joint listed across the top of the table.

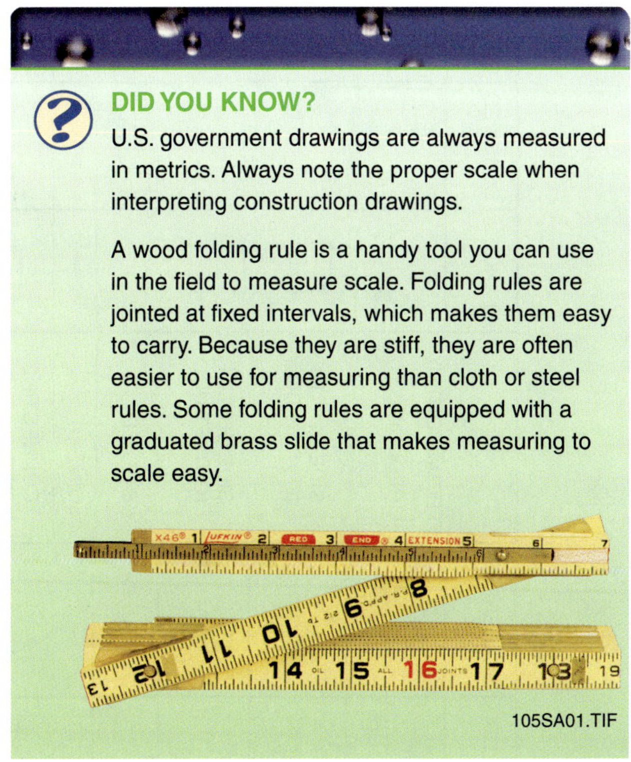

DID YOU KNOW?

U.S. government drawings are always measured in metrics. Always note the proper scale when interpreting construction drawings.

A wood folding rule is a handy tool you can use in the field to measure scale. Folding rules are jointed at fixed intervals, which makes them easy to carry. Because they are stiff, they are often easier to use for measuring than cloth or steel rules. Some folding rules are equipped with a graduated brass slide that makes measuring to scale easy.

105SA01.TIF

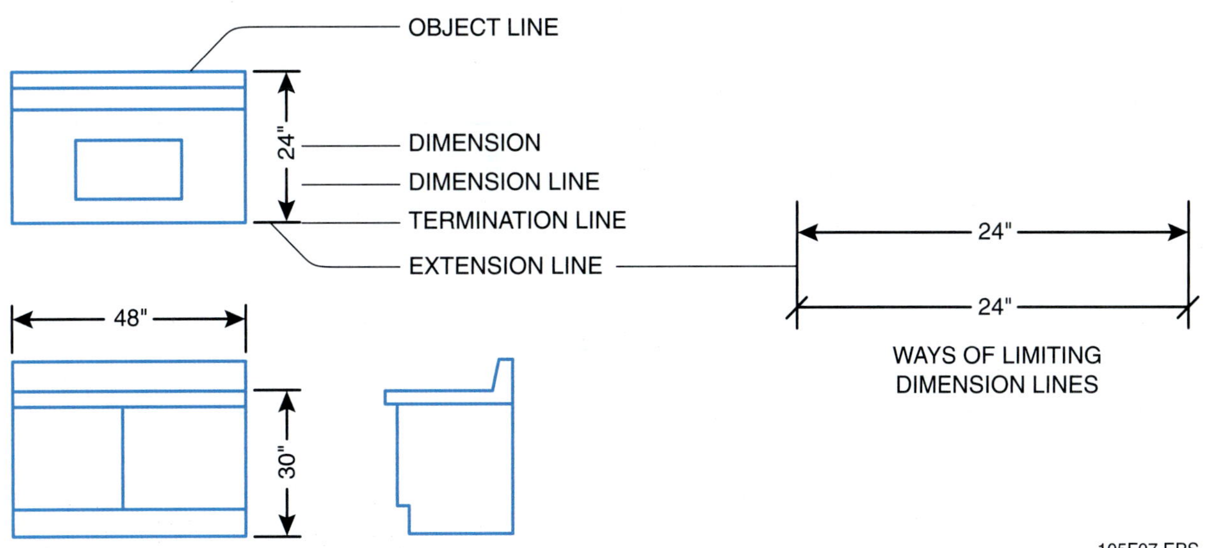

Figure 7 ◆ Extension lines and dimension lines.

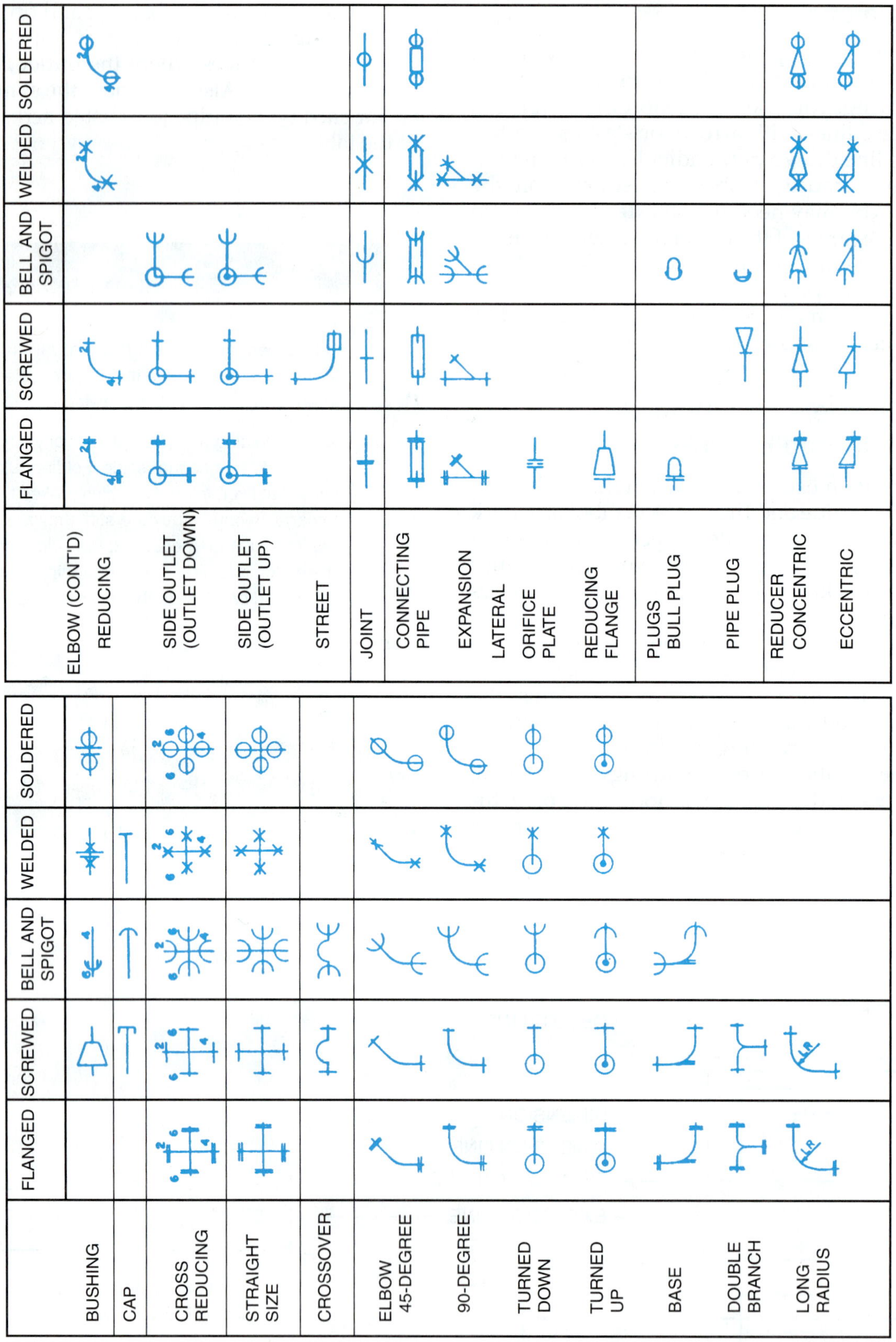

	FLANGED	SCREWED	BELL AND SPIGOT	WELDED	SOLDERED
BUSHING					
CAP					
CROSS REDUCING					
STRAIGHT SIZE					
CROSSOVER					
ELBOW 45-DEGREE					
90-DEGREE					
TURNED DOWN					
TURNED UP					
BASE					
DOUBLE BRANCH					
LONG RADIUS					

	FLANGED	SCREWED	BELL AND SPIGOT	WELDED	SOLDERED
ELBOW (CONT'D) REDUCING					
SIDE OUTLET (OUTLET DOWN)					
SIDE OUTLET (OUTLET UP)					
STREET					
JOINT					
CONNECTING PIPE					
EXPANSION					
LATERAL					
ORIFICE PLATE					
REDUCING FLANGE					
PLUGS BULL PLUG					
PIPE PLUG					
REDUCER CONCENTRIC					
ECCENTRIC					

105F08A.EPS

Figure 8 ◆ Fitting symbols. (1 of 3)

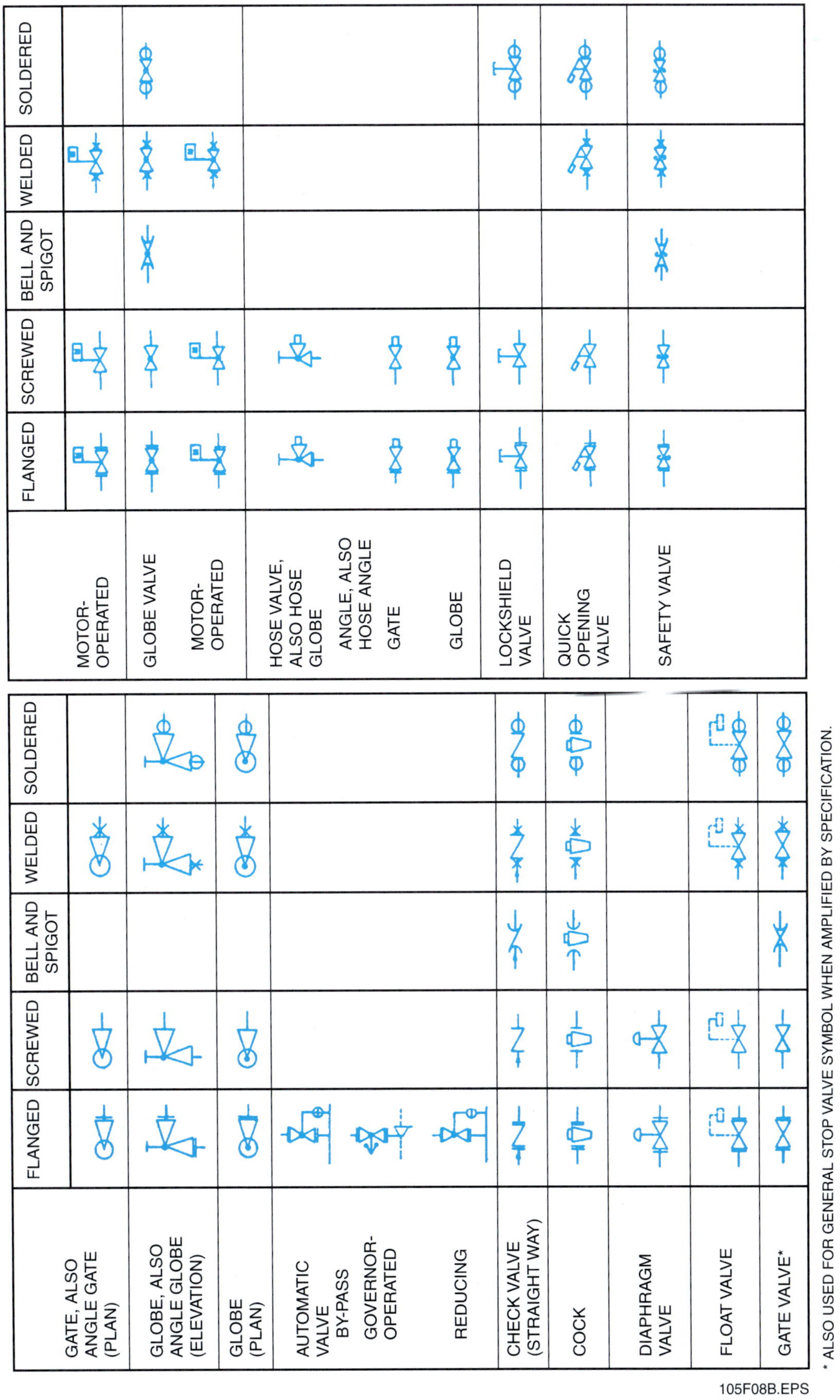

	FLANGED	SCREWED	BELL AND SPIGOT	WELDED	SOLDERED
GATE, ALSO ANGLE GATE (PLAN)					
GLOBE, ALSO ANGLE GLOBE (ELEVATION)					
GLOBE (PLAN)					
AUTOMATIC VALVE					
BY-PASS					
GOVERNOR-OPERATED					
REDUCING					
CHECK VALVE (STRAIGHT WAY)					
COCK					
DIAPHRAGM VALVE					
FLOAT VALVE					
GATE VALVE*					

	FLANGED	SCREWED	BELL AND SPIGOT	WELDED	SOLDERED
MOTOR-OPERATED					
GLOBE VALVE					
MOTOR-OPERATED					
HOSE VALVE, ALSO HOSE GLOBE					
ANGLE, ALSO HOSE ANGLE					
GATE					
GLOBE					
LOCKSHIELD VALVE					
QUICK OPENING VALVE					
SAFETY VALVE					

* ALSO USED FOR GENERAL STOP VALVE SYMBOL WHEN AMPLIFIED BY SPECIFICATION.

105F08B.EPS

Figure 8 ◆ Fitting symbols. (2 of 3)

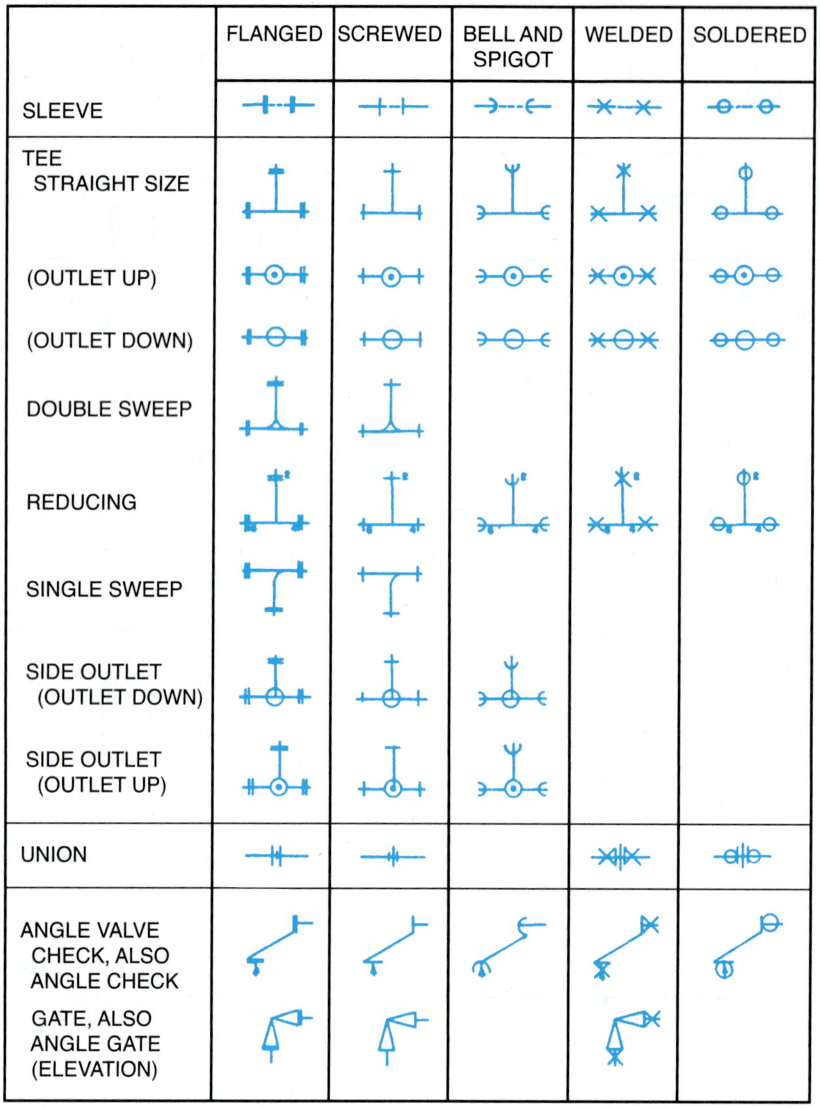

	FLANGED	SCREWED	BELL AND SPIGOT	WELDED	SOLDERED
SLEEVE					
TEE STRAIGHT SIZE					
(OUTLET UP)					
(OUTLET DOWN)					
DOUBLE SWEEP					
REDUCING					
SINGLE SWEEP					
SIDE OUTLET (OUTLET DOWN)					
SIDE OUTLET (OUTLET UP)					
UNION					
ANGLE VALVE CHECK, ALSO ANGLE CHECK					
GATE, ALSO ANGLE GATE (ELEVATION)					

105F08C.EPS

Figure 8 ◆ Fitting symbols. (3 of 3)

See the typical symbols for pipe in *Figure 9*. Note that, generally, drain lines are represented by solid lines and vent lines are represented by broken lines. If you combine the pipe symbols with the fitting symbols, you can illustrate almost any piping system. You can also use symbols for plumbing fixtures to show what fixtures are connected to various runs of pipe (see *Figure 10*).

3.5.0 Notes

Some of the information on a set of drawings is found under the heading of *Notes*. Notes may correspond to numbers on the drawing. They provide more detail about the materials you will need and special installation requirements (see *Figure 11*). Always look for the *Notes* section on drawings.

NOTE

Remember that people use different symbols to represent the same objects. To explain what the various symbols mean, most drawings include a legend. Always consult the legend before reading the drawing.

Symbols for Other Trades

As you work with construction drawings, you will see many types of building symbols that different building professions use. In addition to plumbing symbols, you may see architectural symbols, engineering symbols, mechanical symbols, and electrical symbols. Become familiar with the symbols that electricians and carpenters use, for example, so that when you see them on drawings you will understand how they affect your work. Knowing that an electrical conduit is located close to your plumbing pipes, for example, may allow you to prevent costly errors during installation.

WASTE WATER	
DRAIN OR WASTE–ABOVE GRADE	————
VENT	- - - - - - - -
COMBINATION WASTE AND VENT	—— CWV ——
ACID WASTE	—— AW ——
ACID VENT	- - - AV - - -
INDIRECT DRAIN	—— D ——
STORM DRAIN	—— SD ——
SEWER-CAST IRON	S-CI
WATER SUPPLY	
CIRC. HOT CITY WATER	—··—··—
CHILLED DRINK. WATER	—···—···—
COLD INDUSTRIAL WATER	—··—··—
HOT INDUSTRIAL WATER	—···—···—
CIRC. HOT INDUS. WATER	—····—····—
COLD CITY WATER	—·—·—·
HOT CITY WATER	—--—--—
OTHER PIPING	
GAS-LOW PRESSURE	— G — G —
GAS-MEDIUM PRESSURE	—— MG ——
GAS-HIGH PRESSURE	—— HG ——
COMPRESSED AIR	—— A ——
VACUUM	—— V ——
VACUUM CLEANING	—— VC ——
OXYGEN	—— O ——
LIQUID OXYGEN	—— LOX ——

105F09.EPS

Figure 9 ◆ Typical pipe symbols.

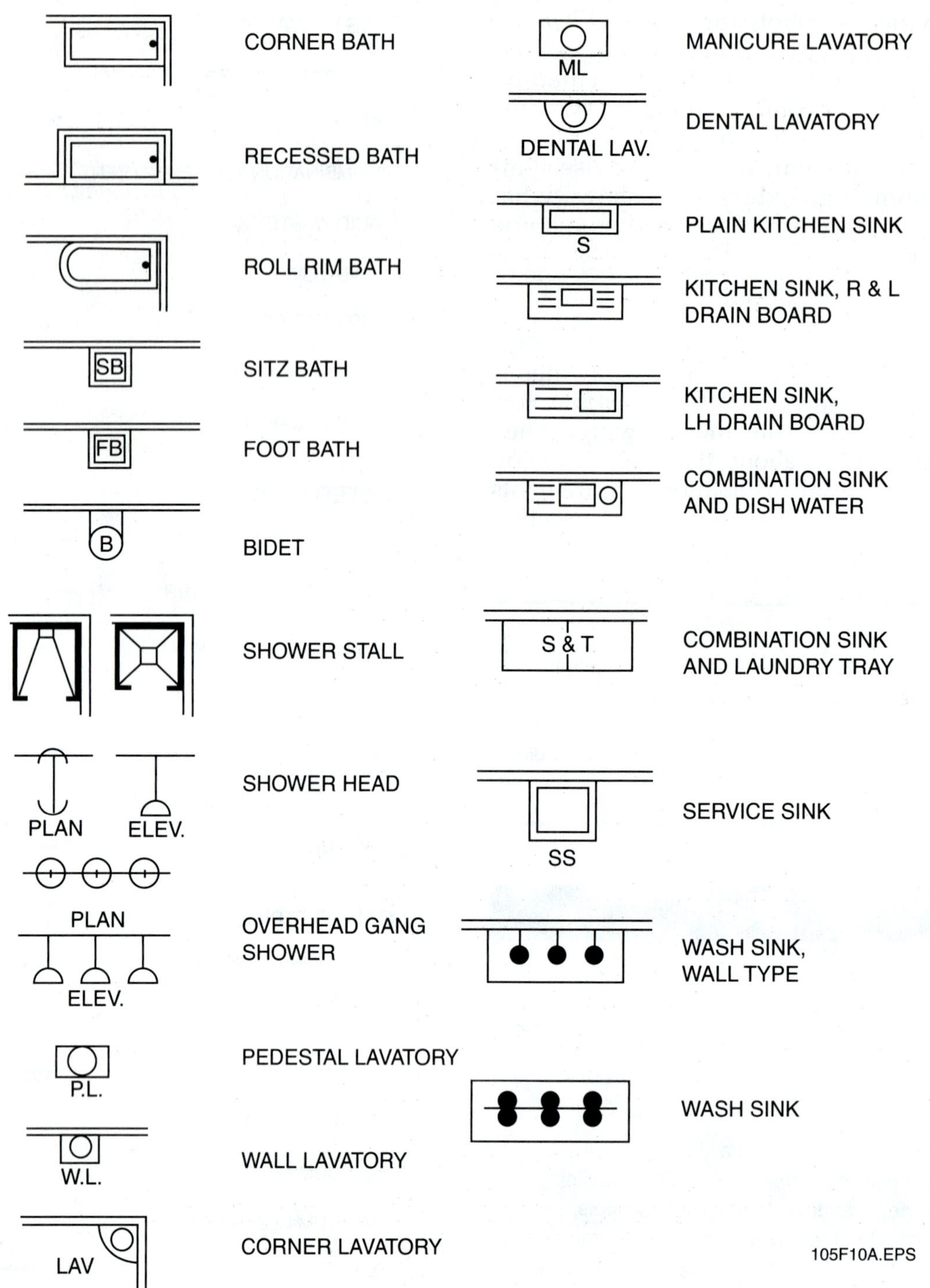

Figure 10 ◆ Fixture symbols. (1 of 2)

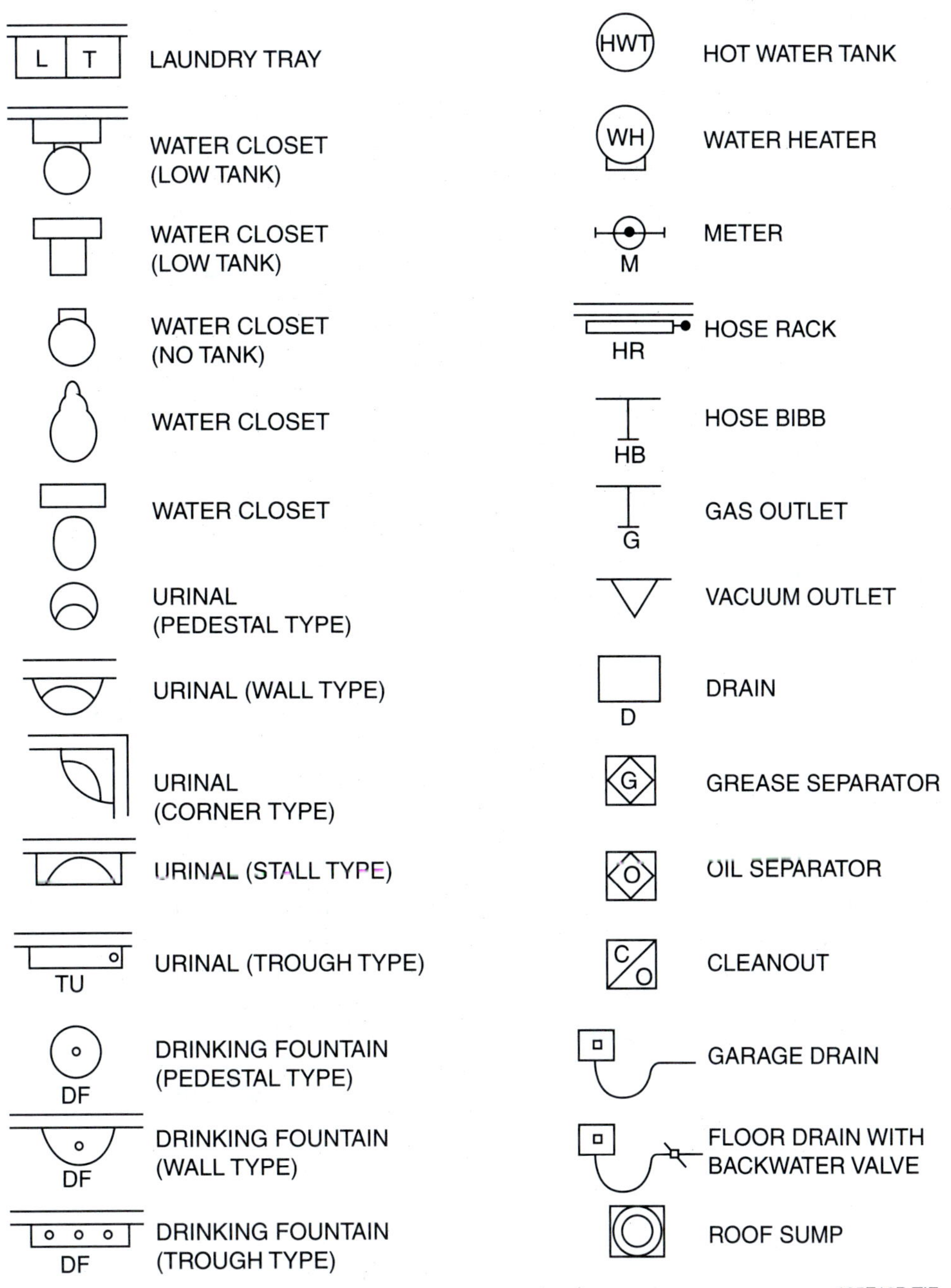

Figure 10 ◆ Fixture symbols. (2 of 2)

NOTES BY SYMBOL "○"

① CAP EXISTING INLETS AND OUTLETS TO EXISTING ACID DILUTION BASIN BELOW FLOOR. REWORK EXISTING ACID WASTE LINE TO CONNECT TO NEW 4" ACID WASTE FROM ABOVE.

② REWORK EXISTING 3" AW TO CONNECT TO NEW 4" AW DOWN.

③ REFER TO ARCH. DRAWINGS FOR FURRINGS.

④ 2½" CW, 1½" HW–UP TO 1ST FLOOR CLG. ¾" HWR DN. TO BELOW FLOOR RE: ARCH. DRAWINGS FOR FURRING.

⑤ SLEEVE STRUCTURAL WALL IN CRAWL SPACE. REFER TO STRUCTURAL DRAWINGS.

⑥ 2" WASTE & 4" AW DOWN IN WALL/CHASE. COORDINATE WITH HVAC PIPING.

⑦ REFER TO ARCHITECTURAL DRAWINGS FOR WALL FURRING.

105F11.EPS

Figure 11 ◆ Sample notes.

3.6.0 Specifications

Specifications, also called **specs,** provide detailed information about the quality of workmanship required on a project as well as the specific materials and method of installation you must use. The person on the project who has been designated responsible for overseeing the project will have a set of specifications. Specs may also indicate that you must follow certain practices in testing the plumbing system. Not all drawings contain a separate set of specs. In that case, you should follow the applicable plumbing codes for the type and quality of materials and installation practices that must be used. Often you will find that the plans and specifications for a building are not complete. For example, plumbing drawings may not be included beyond the location of fixtures on the floor plan. When this happens, it is the plumber's responsibility to design the piping system so that it meets all applicable requirements. You need much more knowledge about plumbing codes and sizing of pipe before you can undertake this responsibility.

3.7.0 Role of Plumbing Codes

The basic purpose of codes is to protect the health and safety of the people who use the system by preventing improper installations. Codes are the law in a community and are enforceable through the courts. Each state, county, city, or other political entity establishes an authority having jurisdiction (AHJ) that is authorized to enforce the plumbing codes adopted by that community. The AHJ may be an agency, department, or board.

Codes also define the quality required for all plumbing installations. The plans and specifications for a particular building may call for installation that is of higher quality than that required by applicable codes, which is perfectly acceptable. But any plan or specification that calls for a plumbing installation of lower quality than that required by the applicable codes is in violation and should be addressed through proper channels. Procedures are usually included in the architectural drawings.

Review Questions

Section 3.0.0

1. The title block on a drawing includes information such as _____.
   a. the project specifications
   b. the initials of the person who drafted the drawing
   c. related easements
   d. contact information for the AHJ

2. To _____ a drawing means to show the actual building, room, or object reduced in proportion from full size to a smaller size.
   a. dimension
   b. spec
   c. measure
   d. scale

3. If the scale on a drawing reads SCALE: ⅛ = 1', then every ⅛ inch on the drawing represents _____.
   a. ⅛ inch
   b. 1 foot
   c. ⅛ foot
   d. 8 feet

4. The instrument used to draw and measure scaled drawings is a(n) _____.
   a. dimension scale
   b. architect's scale
   c. orthographic scale
   d. plumber's scale

5. The number _____ is found at one end of the 1' representation on every scale and is the point where you start measuring.
   a. 0
   b. ¼
   c. 1
   d. 2

6. Use _____ to establish the limits of a measurement on a drawing.
   a. specifications
   b. scales
   c. extension lines
   d. notes

7. Using _____ correctly makes it possible to produce a very understandable drawing quickly.
   a. symbols
   b. 30-degree angles
   c. exploded drawings
   d. isometrics

8. To check for certain practices that must be used in installing and testing a plumbing system, a plumber refers to the _____.
   a. title block
   b. scale
   c. dimensions
   d. specifications

9. If the plumbing specifications for a project are incomplete, it is the _____ responsibility to design the piping system so that it meets all applicable requirements.
   a. plumber's
   b. project manager's
   c. building inspector's
   d. general contractor's

10. It is acceptable for the plans and specifications for a particular building to call for an installation that is of _____ than that required by applicable codes.
    a. lower quality
    b. higher quality
    c. lower cost
    d. higher cost

4.0.0 ◆ TYPES OF DRAWINGS

As a plumber, you will work with many different types of drawings. You must be able to read and create simple sketches as well as complex **isometric drawings.** Even though each type of drawing may use a different technique to convey information, drawings have many similarities.

4.1.0 Sketches

Every plumber must be able to communicate with other workers. For example, you must be able to show another plumber how pipe is to be installed or how the building structure must be modified to install plumbing. Plumbers often make sketches on any convenient surface: scraps of paper, wood, the floor, or the wall. These simple sketches can communicate very effectively. Sketches, like all construction drawings, should include a title block.

4.2.0 Pictorial Drawings

Pictorial drawings show a three-dimensional view of an object. Much like a photograph, pictorial drawings present a picture of the object that is very similar to the actual object. Plumbers use two types of pictorial drawings: isometric and oblique.

4.2.1 Isometric Drawings

An isometric drawing is a pictorial drawing in which all vertical lines are shown vertically but all horizontal lines are projected at a 30-degree angle and appear to go back into the horizon (see *Figure 12*). Isometrics create an illusion of three dimensions. You will encounter isometric drawings often.

When converting floor plan drawings to isometrics, remember that any horizontal lines from the floor plan convert to 30-degree angled lines on an isometric drawing. A drawing of a simple box may be the easiest way to explain this process because it relates directly to the shape of most buildings and rooms (see *Figure 13*). Note that the vertical edges on the box remain vertical in the sketch. Lines that are horizontal in the front view are drawn 30 degrees from horizontal. Lines that are horizontal in the right-side view are also drawn at 30 degrees from horizontal. The dashed lines indicate the hidden back and bottom edges of the box.

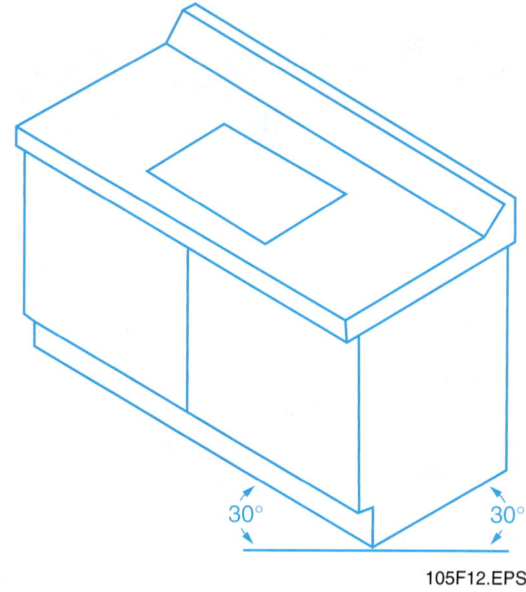

Figure 12 ◆ Isometric drawing.

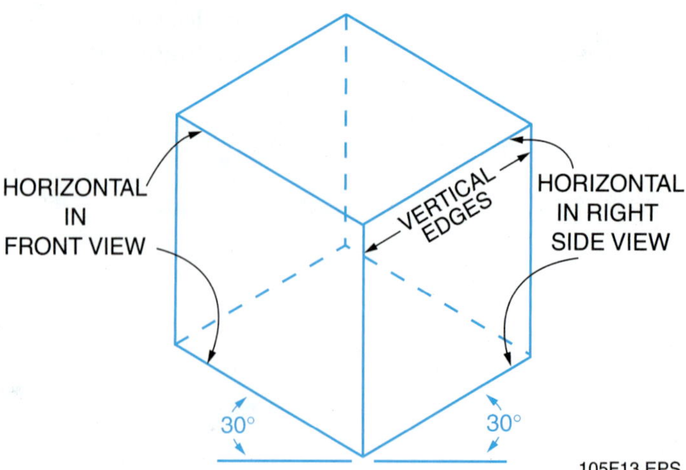

Figure 13 ◆ Isometric sketch of a simple box.

Practice sketching several different boxes. Draw some cubes and draw some rectangles that are long and thin. Do all of your sketches produce an illusion of three dimensions?

As a plumber, you must be able to read and create isometric sketches of piping assemblies. The piping system typically needs to fit within or near the walls. Sketches can be made without including the walls as part of the sketch. However, you must know where the walls are located and position the fitting to make the required bends in the piping system. See the steps for making a sketch in *Figure 14*.

Dimensions on isometric drawings are usually not to scale. Distortion is permitted on these drawings so that they can illustrate three dimensions, and this distortion makes scaling difficult. You can use isometric paper to draft isometric drawings (see *Figure 15*).

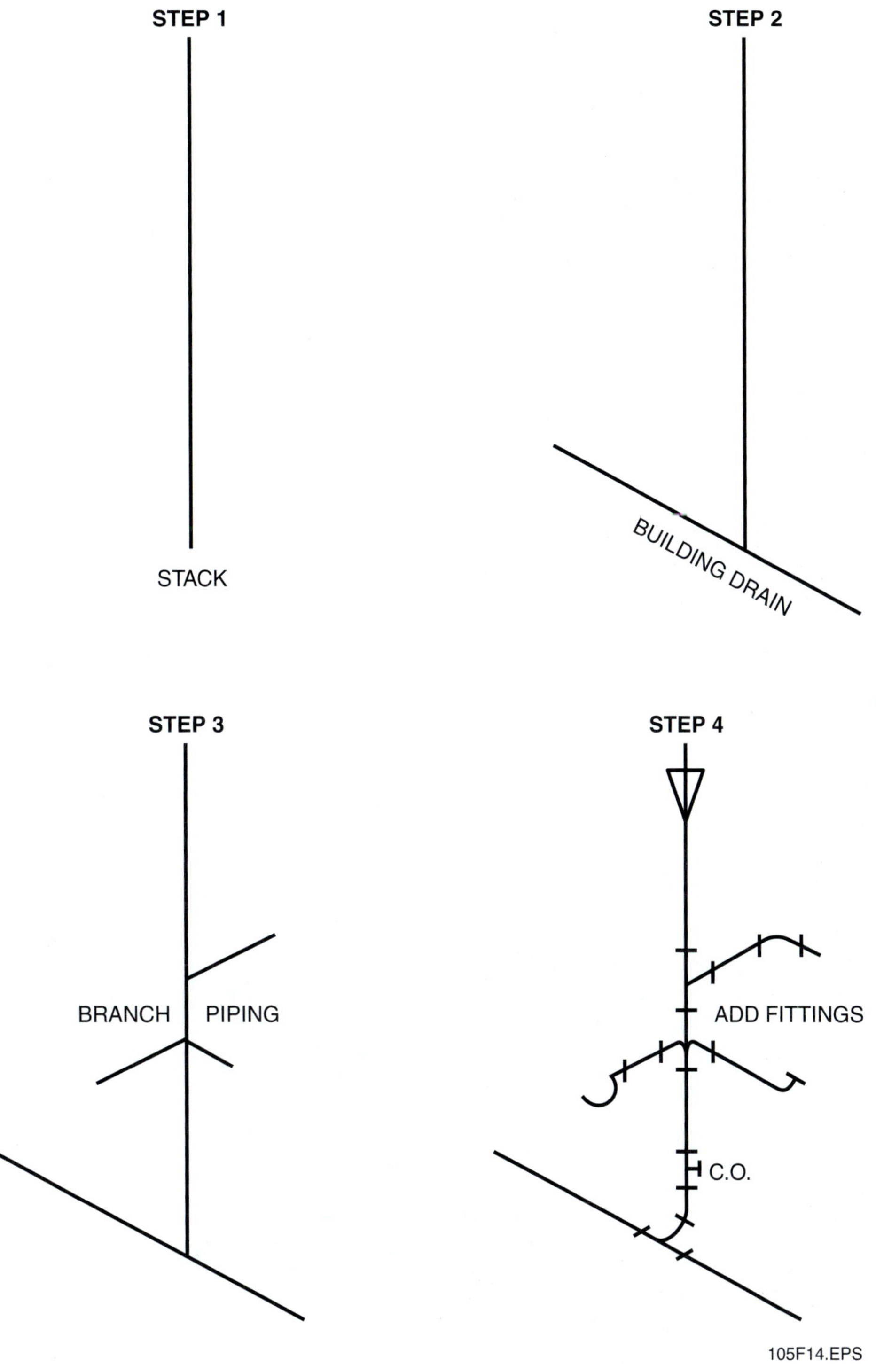

Figure 14 ◆ Steps for making sketches.

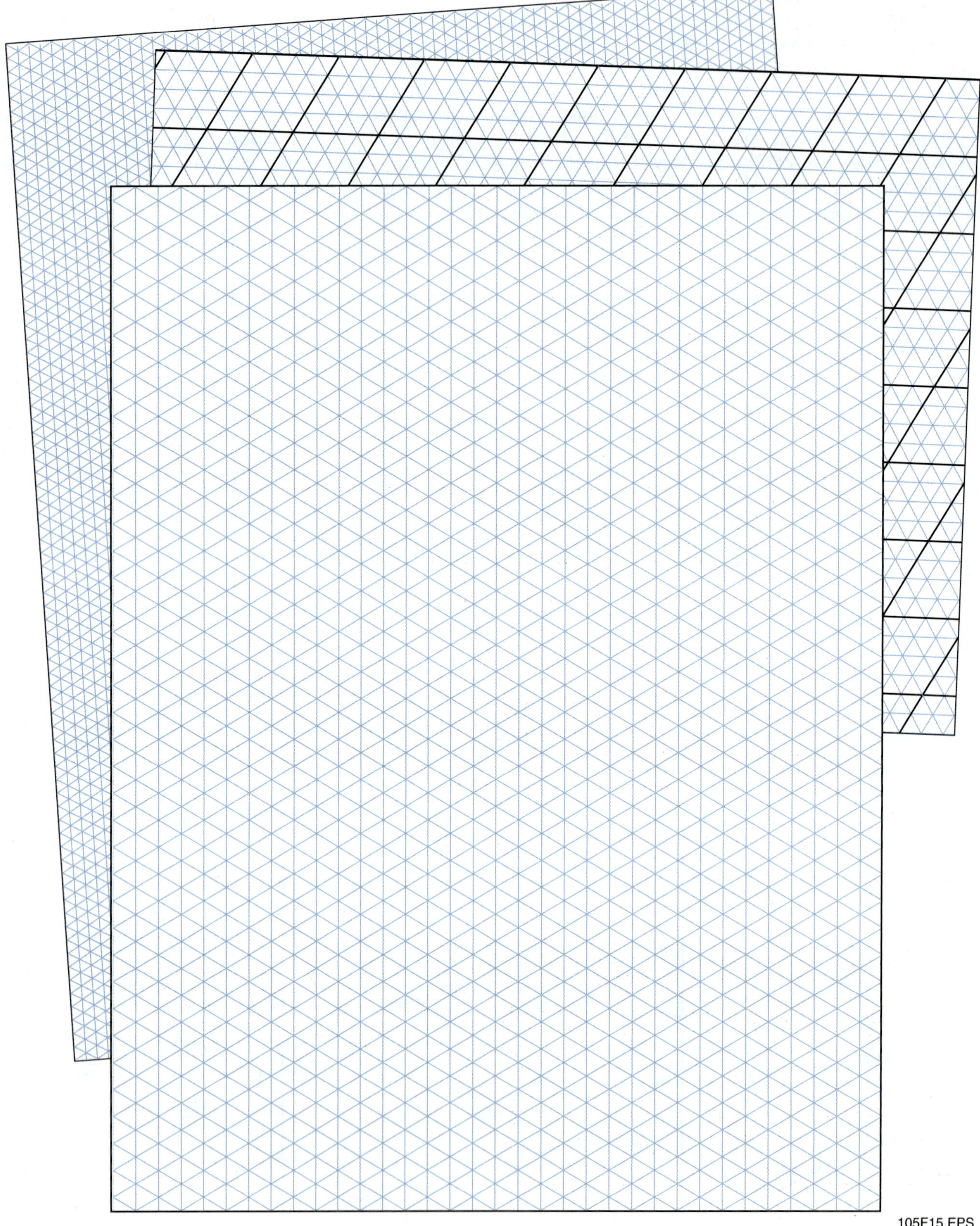

105F15.EPS

Figure 15 ◆ Isometric paper.

See the isometric representation of a stack vent in *Figure 16*. A full circle drawn at the end of an angled line indicates that the line continues up from that end of the horizontal line. A half-circle indicates that the pipe continues downward from the end of the horizontal line. These lines refer to vertical lines on the floor plan. To practice creating isometric drawings, draw what you see in *Figure 16* by following the steps shown in *Figure 14*.

Riser diagrams are more complex isometric drawings of all the plumbing in the building (see the large-scale example in *Figure A-5*). Riser diagrams show vertical and horizontal piping along with sizes and a riser number that refers back to the full set of plumbing drawings. These diagrams typically include a domestic water diagram and a DWV diagram.

Plumbers use riser diagrams to do material **takeoffs.** Takeoffs are detailed lists of all the material and equipment necessary to construct a project. Plumbers use them as checklists when ordering the items required for the job.

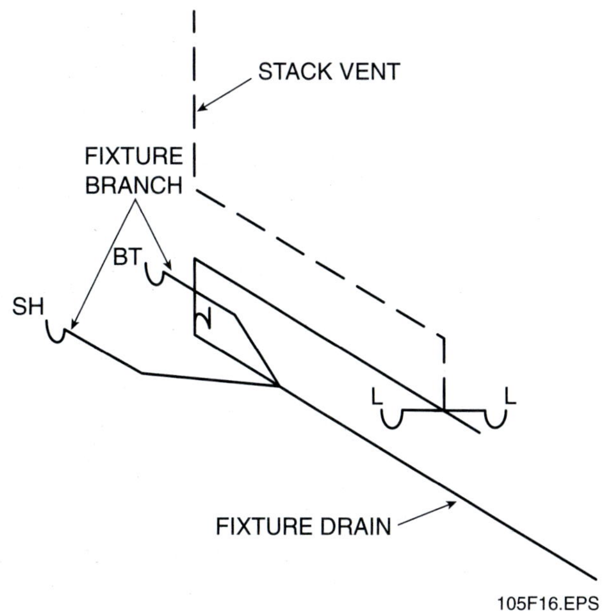

Figure 16 ◆ Isometric sketch of a stack vent.

4.2.2 Oblique Drawings

An **oblique drawing** is a type of pictorial drawing that creates the illusion of depth by using lines that are drawn at 45 or 60 degrees to the horizontal (see *Figure 17*).

For comparison, refer back to the isometric drawing in *Figure 12*. Note that *Figure 12* and *Figure 17* are similar in that the sides and top of the object presented are not rectangular. The corners are not drawn at 90-degree angles even though the object itself has 90-degree corners.

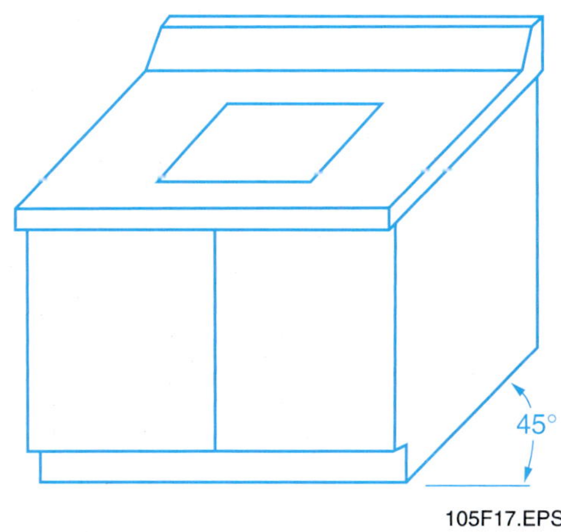

Figure 17 ◆ Oblique drawing.

4.3.0 Schematic Drawings

Most plumbing drawings are **schematic drawings,** also called **single-line drawings.** In a single-line drawing, the line represents the centerline of the pipe, so these drawings can be used to represent pipe of any diameter. See the schematic sketch of a piping stack in *Figure 18*. For comparison, refer back to the isometric representation of a stack vent in *Figure 16*. Note the similarities and differences between the two.

4.4.0 Orthographic Drawings

To communicate information about the exact size and shape of an object, designers often use **orthographic drawings,** which show dimensions that are proportional to the actual dimensions of the object. In orthographic drawings, designers draw lines that are scaled-down representations of real dimensions. Every 12 inches, for example, may be represented by ¼ inch on the drawing. An orthographic drawing is created by viewing the object from one or more of six possible directions and recording how the object looks in each of the views (see *Figure 19*). You can create more views of the object by looking at it from other angles (see *Figure 20*). Although you can present as many as six different views, generally you will not need more than three to show the object completely.

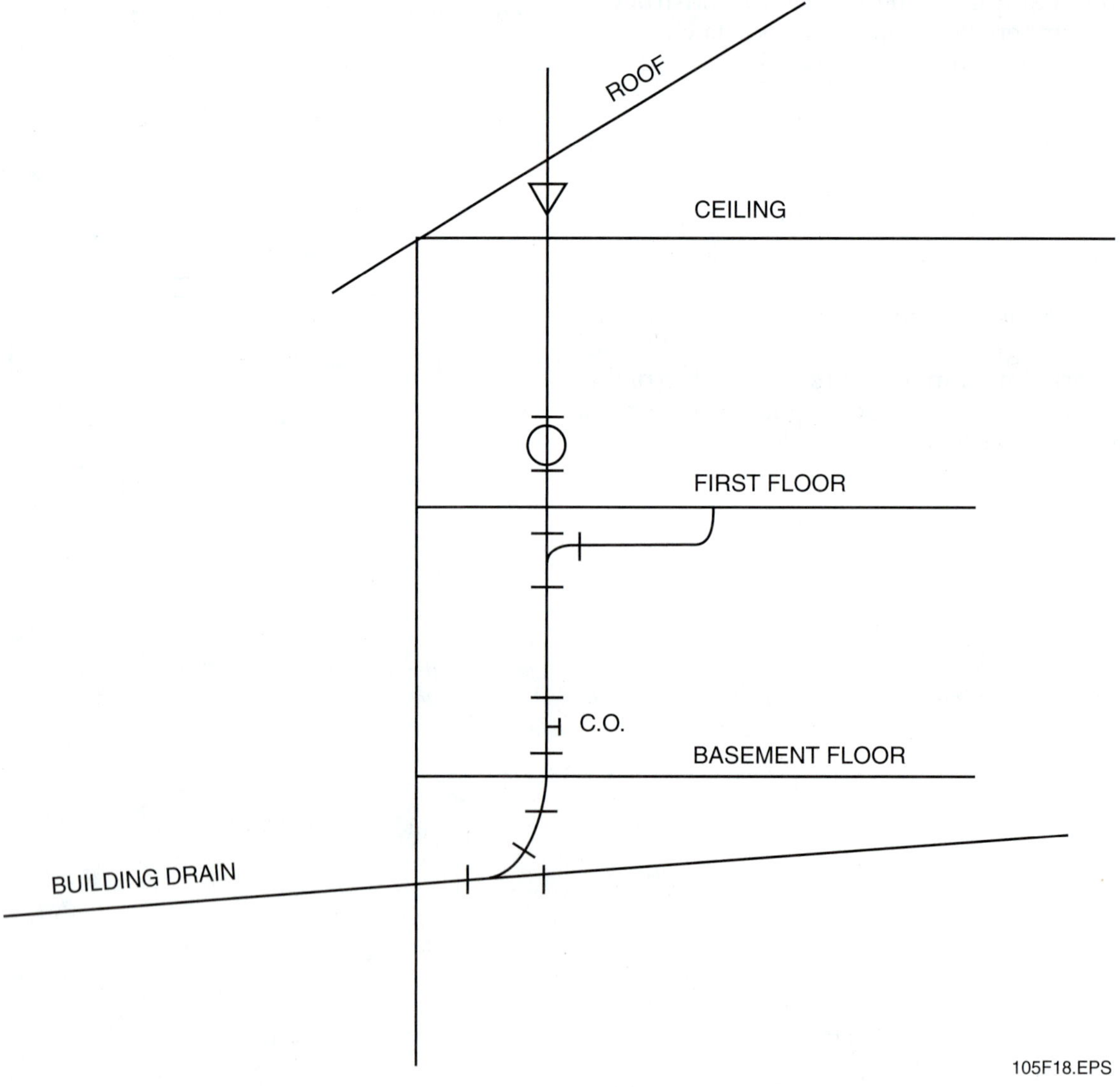

Figure 18 ◆ Schematic sketch of a piping stack.

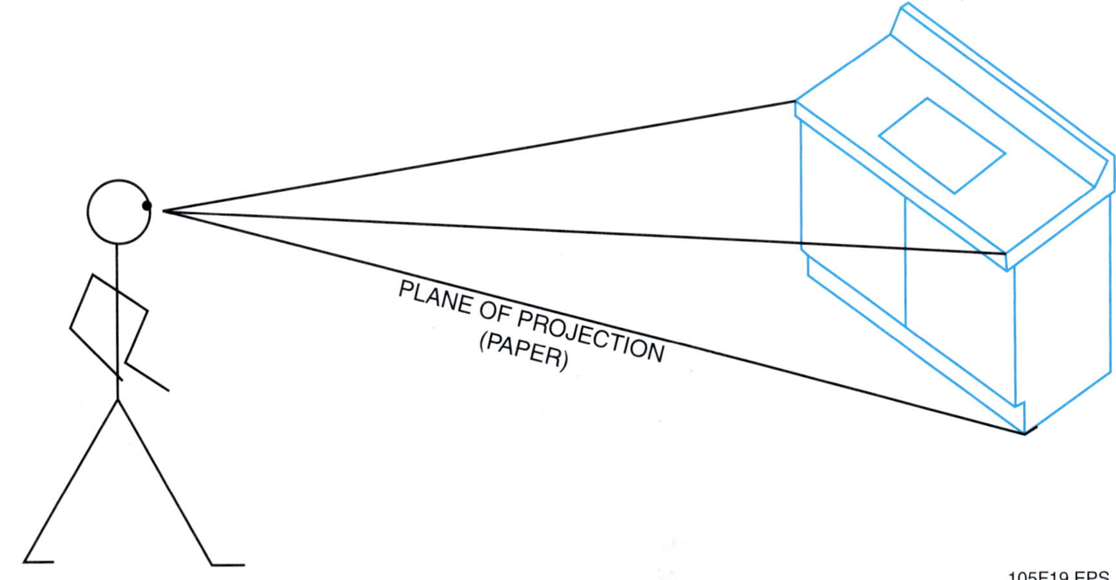

Figure 19 ◆ Orthographic projection.

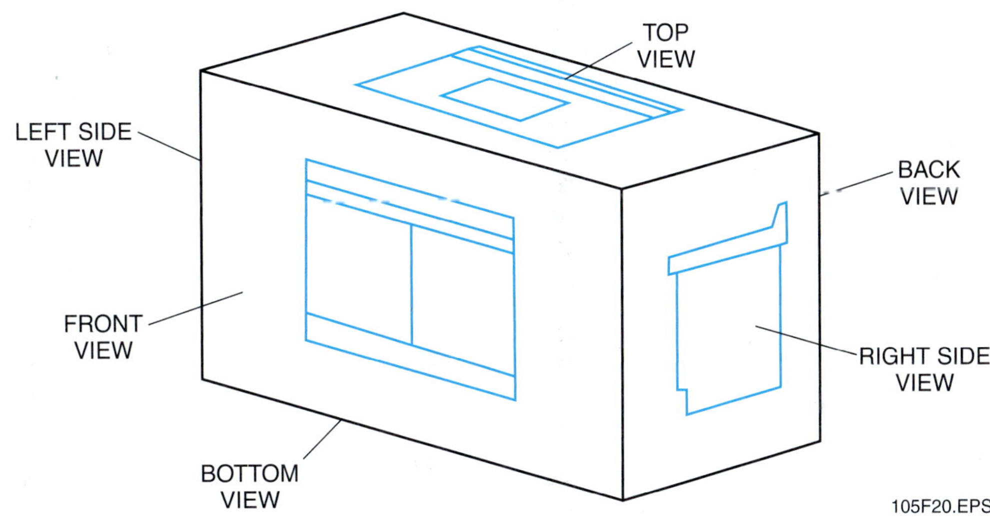

Figure 20 ◆ Additional planes for orthographic projection.

To make orthographic drawings understandable, designers often draw the views in the proper relationship to one another. Notice in *Figure 21* how the views from different angles are related to the finished drawing.

To read an orthographic drawing, you have to learn to visualize what the finished object looks like from the views given. Be patient because this skill requires a lot of practice. Examine the orthographic drawing and pictorial drawing of the same object in *Figure 22*. Study and compare these drawings carefully. Also look at the two drawings in *Figure 23*. One is a schematic representation of a piping assembly; the other is an orthographic drawing of the same thing. Note the differences between these two drawings.

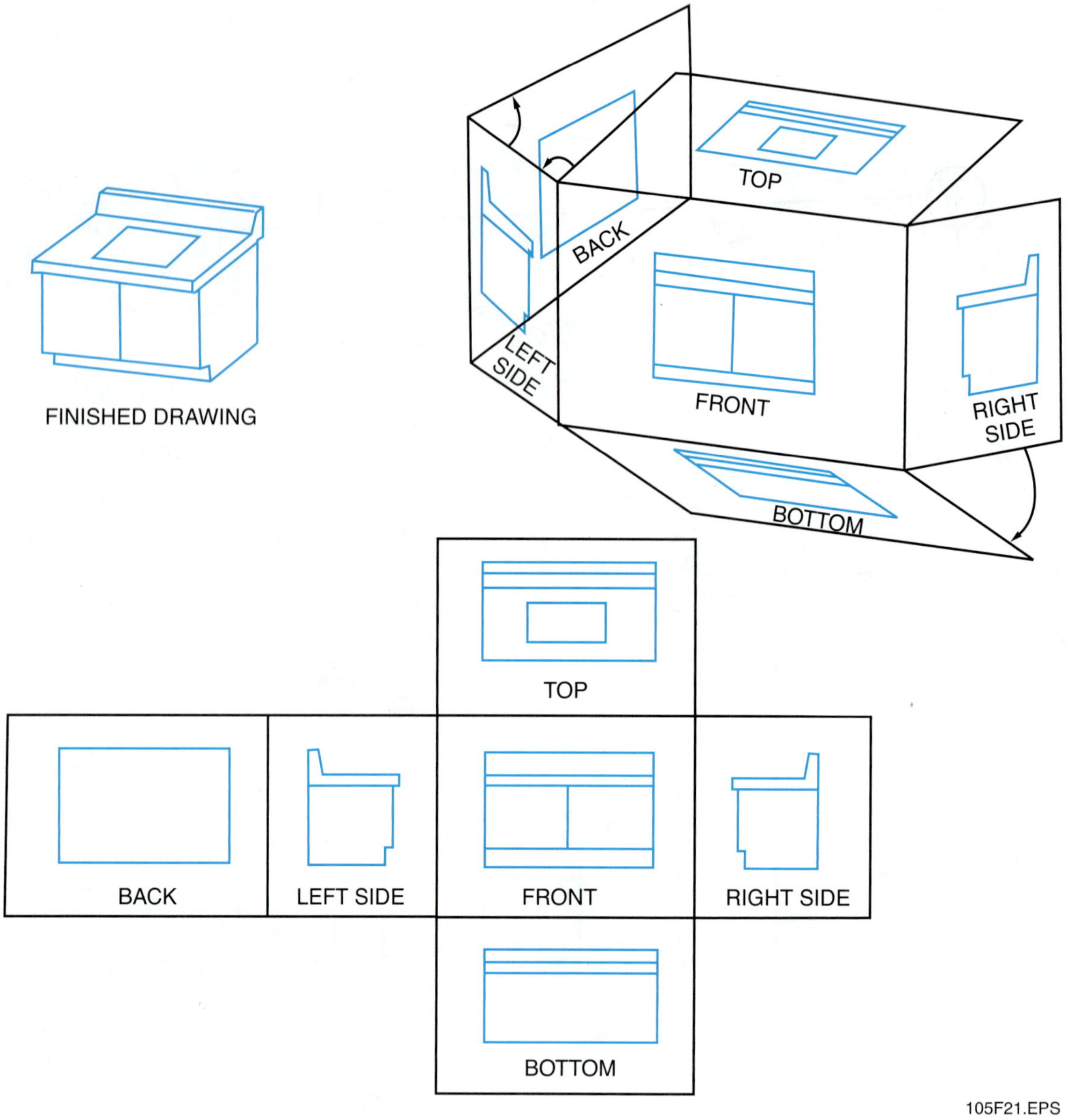

FINISHED DRAWING
BACK
TOP
LEFT
SIDE
FRONT
RIGHT
SIDE
BOTTOM
TOP
BACK
LEFT SIDE
FRONT
RIGHT SIDE
BOTTOM
105F21.EPS

Figure 21 ◆ Relationship of orthographic views.

ORTHOGRAPHIC
PICTORIAL
105F22.EPS

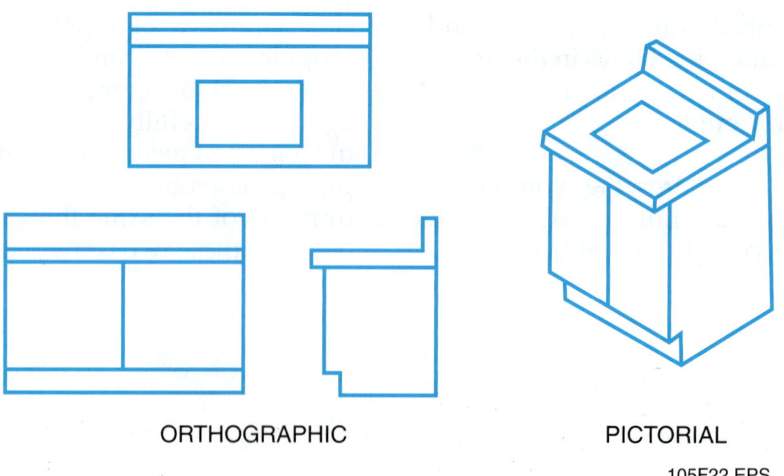

Figure 22 ◆ Orthographic and pictorial views of the same object.

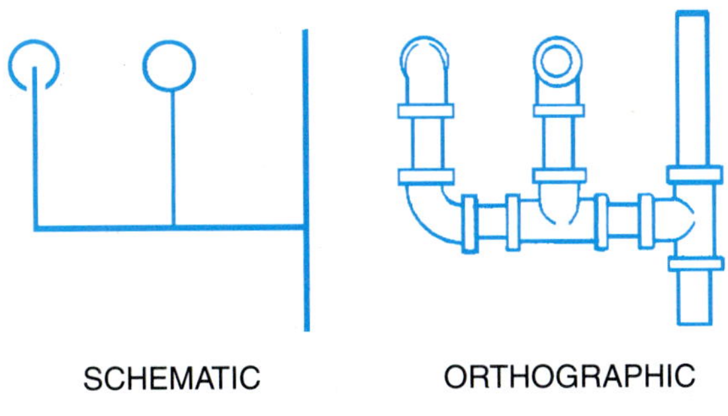

Figure 23 ◆ Schematic and orthographic drawings of the same piping assembly.

4.5.0 Types of Plumbing Drawings

Most plumbing drawings include riser diagrams, floor plans, isometric views, and details. You will work with many different types of drawings. For example, you must be able to find the size and dimensions of fittings from manufacturers' catalogs, and you must also be able to interpret details about assemblies. Study each drawing in this section carefully, and think about what information it contains and how it shows that information.

4.5.1 Approved Submittal Data

Manufacturers of pipe and fittings publish catalogs that contain information about their products. Many of them publish dimensional catalogs that show each of their fittings along with overall sizes and, in some cases, weights. Plumbers use **submittal data,** also called **catalog drawings,** to determine such factors as overall dimensions, actual inside dimensions, fitting depth, and laying length of the fittings. When the architect or engineer approves the fixture or fitting in a catalog drawing, it becomes **approved submittal data.**

See *Figure 24* for an example of a plastic DWV system P-trap with a solvent-welded joint from a catalog. The method used in this drawing to show the dimensions of the fitting is typical. The drawing uses capital letters (A through E) that are keyed to a table. The table provides information about available sizes. This method is efficient because manufacturers can use the same drawing to represent fittings of different sizes.

Figure 25 uses a similar technique to give information about a cast-iron wye. The table provides information for 39 separate fittings. By checking the various combinations of pipe sizes (G, X, and X'), you can find a combination that will work for your installation. (Note that the SV and XH in the table stand for service weight and extra-heavy pipes, respectively.)

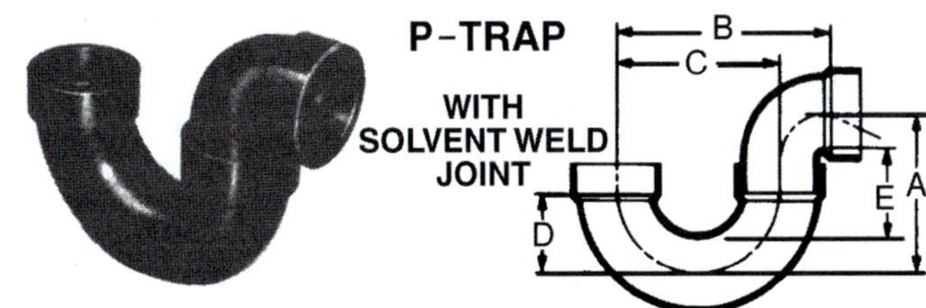

SIZE	DESCRIPTION H x H	A	B	C	D	E
1½		3¾	5 7/32	4	1 13/16	2⅛
2		4 9/16	7½	5¼	2¼	2½
3		6 5/16	8 7/16	6 5/16	2 11/16	3¼
4		7⅞	10 13/16	8⅛	3½	3⅞

105F24.TIF

Figure 24 ◆ Catalog drawing of a P-trap.

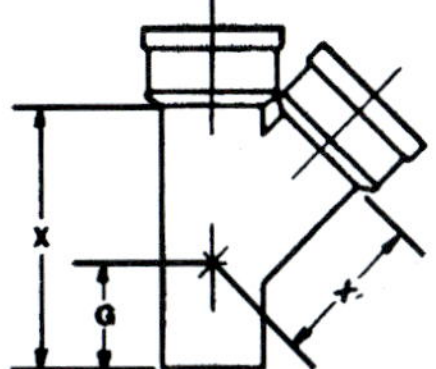

Wye					
				Weights	
Size	G	X	X'	SV	XH
2 × 2	4	8	4	7	8
3 × 2	4 3/16	9	5	10	14
3 × 3	5	10½	5½	13	17
4 × 2	3⅝	9	5¾	12	17
4 × 3	4 7/16	10½	6¼	14	20
4 × 4	5¼	12	6¾	17	24
5 × 2	3⅛	9	6½	14	20
5 × 3	3⅞	10½	7	17	24
5 × 4	4 11/16	12	7½	19	27
5 × 5	5½	13½	8	22	32
6 × 2	2 9/16	9	7¼	17	23
6 × 3	3⅜	10½	7¾	19	27
6 × 4	4 3/16	12	8¼	22	31
6 × 5	4 15/16	13½	8¾	24	35
6 × 6	5¾	15	9¼	28	40
8 × 2	3⅛	10½	8½	29	42
8 × 3	3 15/16	12	9	32	47
8 × 4	4¾	13½	9½	36	52
8 × 5	5½	15	10	39	57
8 × 6	6 5/16	16½	10½	44	63
8 × 8	7 11/16	19½	11 13/16	55	82
10 × 3	2 9/16	12	11	50	71
10 × 4	3 9/16	13½	11⅛	53	74
10 × 5	4 5/16	15	11⅝	57	80
10 × 6	5⅛	16½	12⅛	61	86
10 × 8	6½	19½	13 7/16	77	110
10 × 10	8	22½	14½	94	133
12 × 4	4⅛	15	12 7/16	70	97
12 × 5	4⅞	16½	12 15/16	74	104
12 × 6	5 11/16	18	13 7/16	80	111
12 × 8	7 1/16	21	14¾	96	136
12 × 10	8 9/16	24	15 13/16	115	160
12 × 12	10⅛	27	16⅞	135	186
15 × 4	2¼	15	15	130	130
15 × 6	4	18	15¾	109	152
15 × 8	5⅜	21	17 1/16	127	182
15 × 10	6⅞	24	18⅛	152	213
15 × 12	8 7/16	27	19 3/16	170	242
15 × 15	10¾	31½	20¾	204	290

105F25.EPS

Figure 25 ◆ Catalog drawing of a wye.

4.5.2 Fixture Drawings

Plumbers often use **fixture drawings** to determine whether a certain fixture will fit in the available space. See the fixture drawing of a typical water closet in *Figure 26*. Notice how the drawing shows the overall dimensions and the location of the water supply piping.

Figure 27 is a drawing of a typical lavatory. Two views are required to show the size and shape of this lavatory. The dimensions for roughing-in the water supply pipes and the drains vary depending on the type of drain installed. Study this drawing and answer the following questions:

1. What is the diameter of the opening that must be cut out of the countertop?
2. What is the overall depth of the lavatory?
3. If a pop-up drain is used, how far from the floor should the drainpipe be located?
4. How far apart must the cold and hot water connections be on the faucet that is installed in this lavatory?

NOTE

Fixture drawings are specifically used to select appropriate fixtures for available space. These drawings are not to be used for construction.

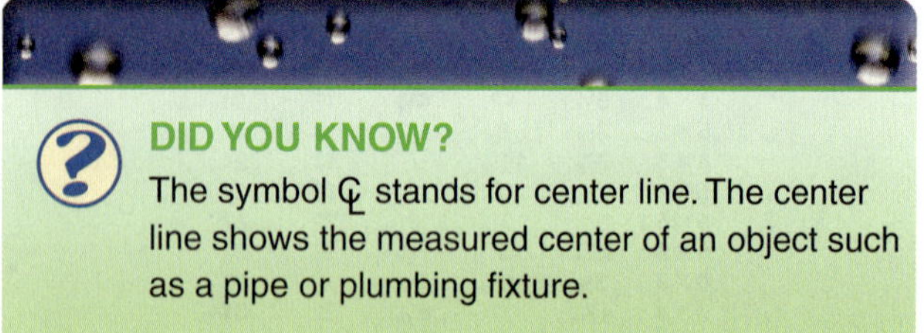

DID YOU KNOW?

The symbol ℄ stands for center line. The center line shows the measured center of an object such as a pipe or plumbing fixture.

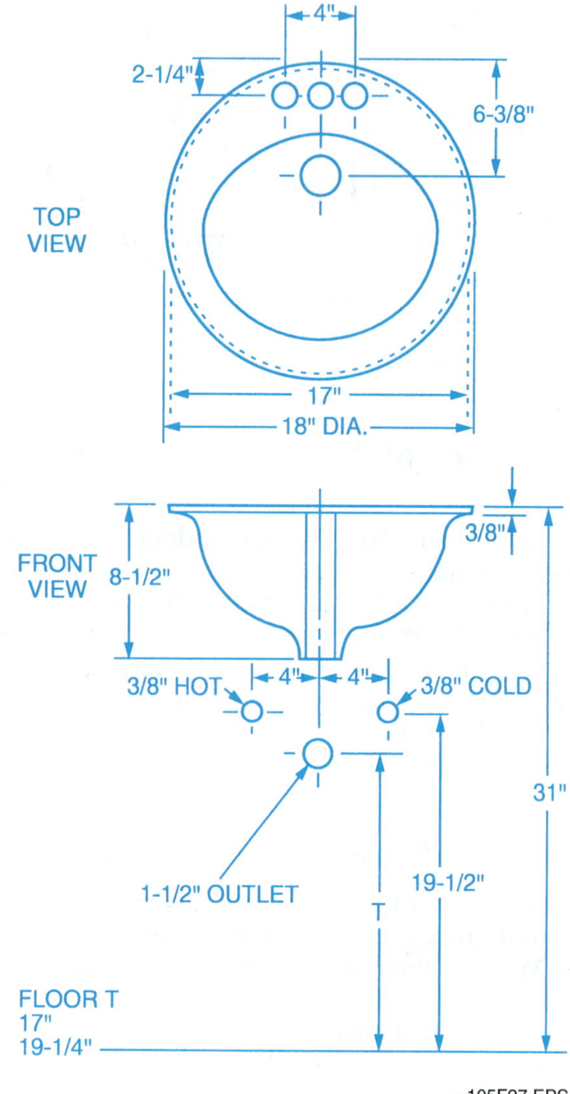

Figure 27 ◆ Fixture rough-in drawing of a lavatory.

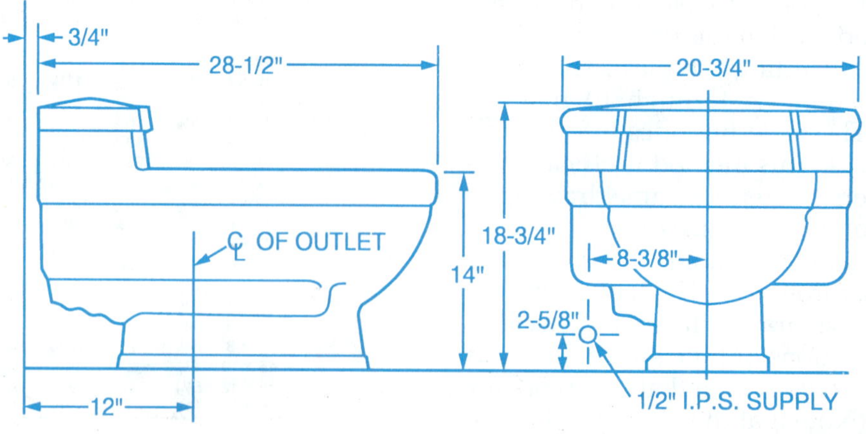

Figure 26 ◆ Fixture rough-in drawing of a water closet.

DID YOU KNOW?

Fixture drawings such as the ones found in *Figures 26* and *27* are also called fixture rough-in sheets, manufacturers' fixture spec sheets, and cut sheets. Changes in the building structure, such as wall placement and wall covering, could affect your work. You must remeasure all fixtures before installing them to ensure proper fit.

Similar drawings are used to show other types of fixtures, such as bathtubs (see *Figure 28*). The top view is typical of views shown in orthographic drawings, but the side and end views are drawn as section views. Section views illustrate the internal shape of objects. Also, part of the end view has been eliminated, and the side view has been moved to the left of its normal location. Study this drawing carefully and then answer the following questions:

1. In how many different sizes is this bathtub manufactured?
2. What is the overall length of the smallest bathtub available?
3. How high above the subfloor will the side of the bathtub be?
4. Will an opening need to be cut in the subfloor to permit installation of the bathtub?
5. How far above the subfloor will the stub-out for the showerhead be located?

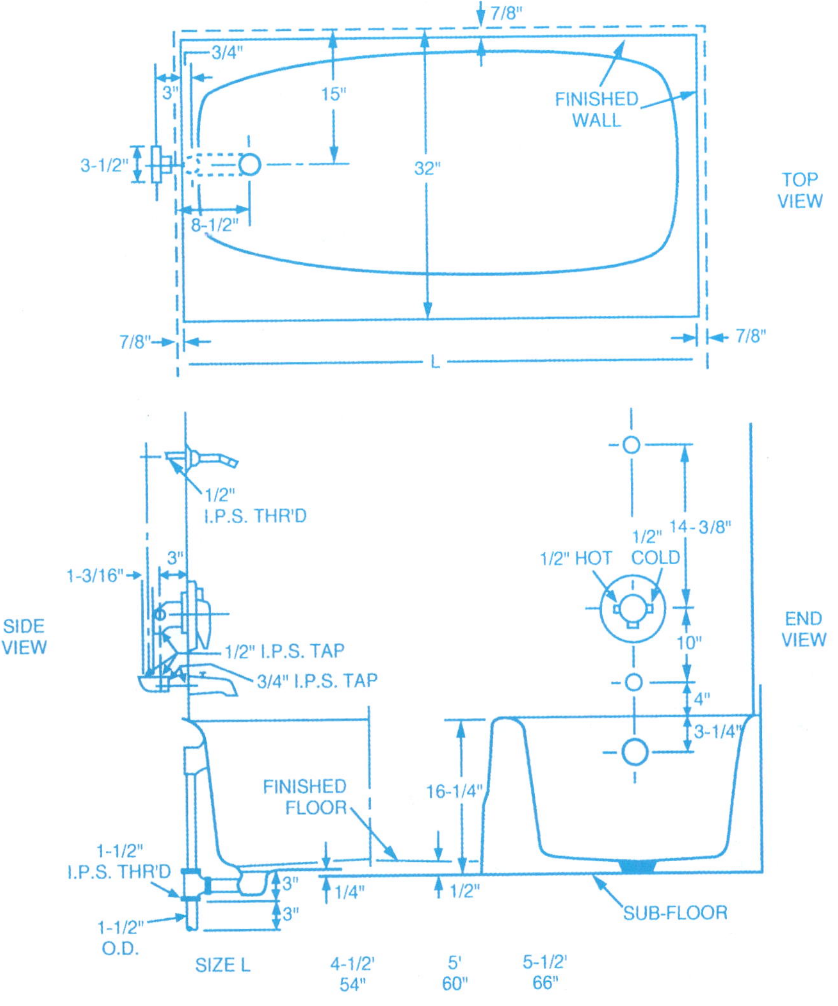

Figure 28 ◆ Fixture rough-in drawing of a bathtub.

4.5.3 Exploded Drawings

Exploded drawings, also called assembly drawings, illustrate how to assemble complex objects. They show the relationship of the individual parts to the whole object. This type of drawing may also show the illustrated parts breakdown (IPB). *Figure 29* shows an exploded view of a pop-up assembly with all its parts. Note how the parts are shown so that you can easily visualize how to assemble the pop-up.

Figure 30 is a more complicated type of exploded drawing than *Figure 29*. Each part is numbered to correspond to a number in a parts list (not shown). In this image, both a handle version and a push-button version of the flush valve are shown on the same drawing. Additionally, two types of control stops may be installed. An exploded drawing of a kitchen faucet appears in *Figure 31*. Using these types of drawings, you should be able to disassemble and reassemble any plumbing product and identify needed repair parts.

4.5.4 Cutaway Drawings

For many products, a **cutaway drawing** provides enough detail to show how the product is constructed (see *Figure 32*). Note in the drawing that each part is numbered to correspond to the parts list. Also note that the dimensions of many sizes of this valve are given in the specifications table. This method eliminates the need to make a separate drawing for each size valve.

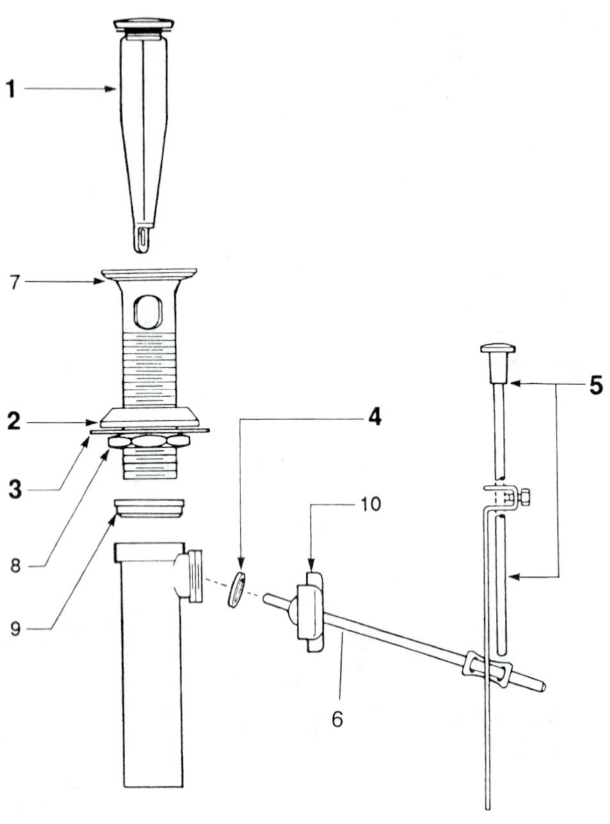

MOEN

METALLIC/NON-METALLIC LAVATORY WASTE ASSEMBLIES MODELS 96497 & 96107

ITEM	CAT. NO.	MFG.	DESCRIPTION
1	52403 ✓	12314	Waste Plug Assy.
2	52102 ✓	11987	Gasket
3	58689	11990	Washer
4	52249 ✓	12312	Pivot Seat
5	58579 ✓	11986	Rod & Knob (bulk)
	58700 ✓		Rod & Knob (carded)
6	51910 ✓	11985	Pivot Rod
10	58675 ✓	11998	Pivot Nut

PARTS AVAILABLE THROUGH SPECIAL ORDER

ITEM	CAT. NO.	MFG.	DESCRIPTION
7	----	11984	Waste Seat Kit
8	----	12313	Mounting Nut
9	----	11997	Seal

105F29.TIF

Figure 29 ◆ Exploded view of a pop-up assembly.

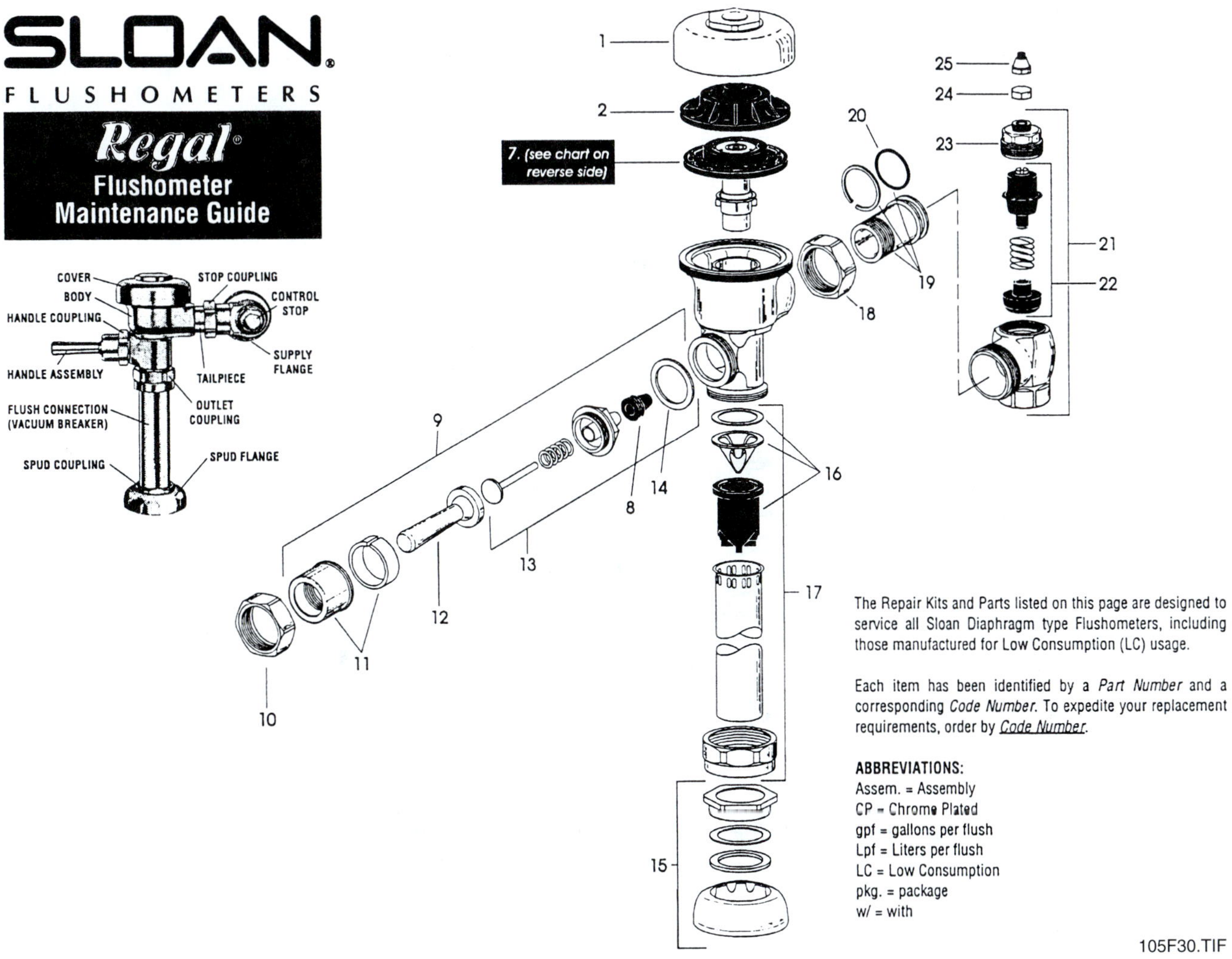

Figure 30 ◆ Exploded drawing with IPB of a flush valve.

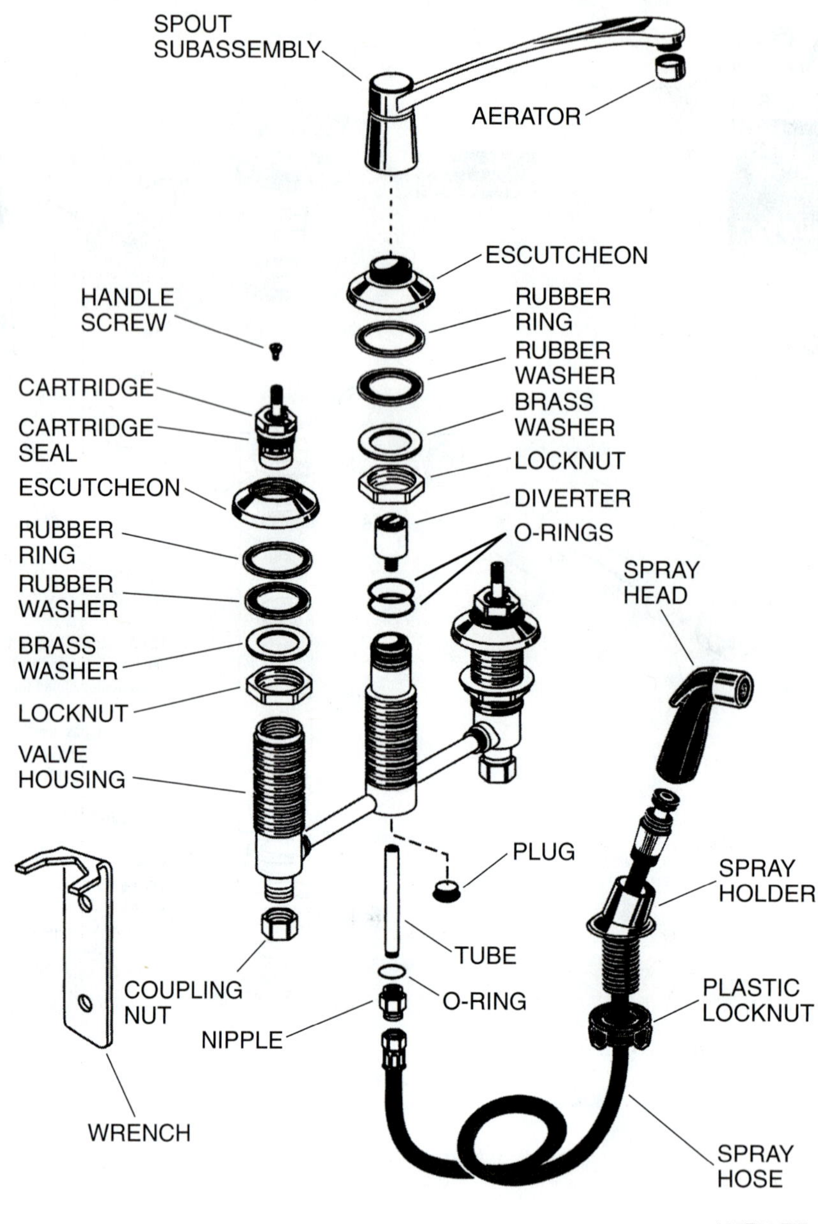

Figure 31 ◆ Exploded drawing of a kitchen faucet.

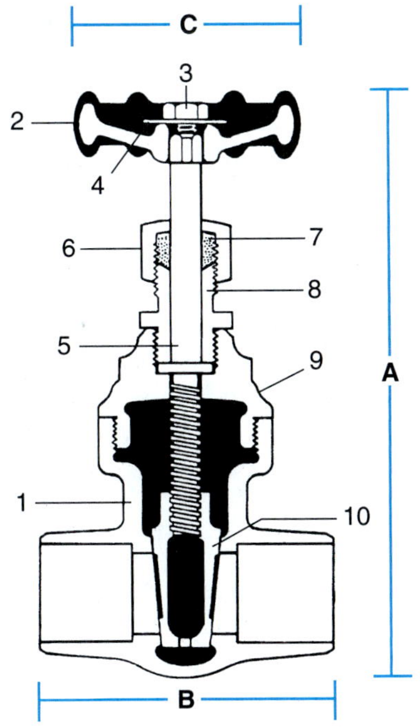

PARTS AND MATERIALS		
Item	**Description**	**Material**
1	Body	ASTM B-62
2	Handwheel	Malleable iron
3	Handwheel nut	ASTM B-16
4	Identification plate	Brass
*5	Stem	ASTM B-16, B-62
*6	Packing nut	ASTM B-16, B-62
7	Packing	Teflon®-Asbestos
*8	Stuffing box	ASTM B-16, B-62
9	Centerpiece	ASTM B-62
*10	Disc	ASTM B-62

*B-62 used in larger sizes

SPECIFICATIONS				
Valve Size	**Dim. A**	**Dim. B**	**Dim. C**	**Weight Pounds**
3/8"	4 1/8"	1 5/8"	2 1/16"	0.8
1/2"	4 1/4"	2 3/16"	2 1/16"	0.9
3/4"	4 7/16"	2 7/8"	2 1/2"	1.4
1"	5 7/16"	3 3/8"	2 1/2"	2.1
1 1/4"	6 1/4"	3 7/16"	2 1/2"	2.9
1 1/2"	7 1/16"	3 3/4"	3"	3.8
2"	8 1/4"	4 3/8"	3 1/2"	6.3
2 1/2"	9 3/10"	5 3/8"	4 1/8"	10.9
3"	11 1/8"	6 1/4"	5 1/4"	16.4

105F32.EPS

Figure 32 ◆ Cutaway drawing of a gate valve.

Review Questions

Section 4.0.0

1. Unlike other construction drawings, sketches do not need to have a title block.
   a. True
   b. False

2. A(n) _____ drawing gives the impression of three dimensions and provides a picture that resembles the physical object.
   a. orthographic
   b. schematic
   c. isometric
   d. linear

3. When converting floor plan drawings to isometrics, remember that any horizontal lines from the floor plan convert to _____ angled lines on an isometric drawing.
   a. 15-degree
   b. 30-degree
   c. 45-degree
   d. 60-degree

4. _____ diagrams show vertical and horizontal piping along with sizes and a riser number that refers back to the full set of plumbing drawings.
   a. Riser
   b. Water
   c. Electrical
   d. DWV

5. Another term for a schematic drawing is a(n) _____.
   a. oblique drawing
   b. pictorial drawing
   c. single-line drawing
   d. riser diagram

6. A(n) _____ drawing is made by looking at the object from one or more of six possible angles.
   a. projection
   b. orthographic
   c. oblique
   d. pictorial

7. Plumbers use _____ drawings to determine such factors as overall dimensions, actual inside dimensions, fitting depth, and laying length of the fittings.
   a. schematic
   b. isometric
   c. cutaway
   d. catalog

8. Plumbers often use _____ drawings to determine whether a certain fixture will fit in the available space.
   a. fixture
   b. schematic
   c. orthographic
   d. linear

9. A(n) _____ drawing shows the relationship of the individual parts to the object as a whole.
   a. oblique
   b. isometric
   c. exploded
   d. schematic

10. A(n) _____ drawing provides enough detail to show how the product is constructed.
    a. schematic
    b. cutaway
    c. exploded
    d. catalog

Summary

Information found on plumbing drawings is critical to performing quality work. Remember that it takes many different types of drawings, along with the specifications for the project, to get a complete picture of the plumbing requirements. Pictorial drawings show depth. Orthographic drawings show objects as if you were looking at them head-on from certain angles. The ability to read and understand different kinds of drawings is an essential plumbing skill, as is the ability to make orthographic or isometric sketches and to draw plumbing assemblies. Learning the basics of reading and interpreting drawings is a building block for future study. You must practice construction-drawing reading, like all other plumbing skills, and study throughout your career.

Notes

Trade Terms Quiz

Fill in the blank with the correct trade term that you learned from your study of this module.

1. You will find the location of the furnace and air conditioning equipment, as well as the location of ducts and registers or piping and radiators, in _______________.
2. A(n) _______________ is created by viewing an object from one or more of six possible directions and recording how the object looks in each of the views.
3. The _______________ may include information about the drainage, waste, and vent (DWV) piping and the water supply piping that must be installed below the first floor.
4. _______________ allow you to determine whether a certain fixture will fit in the available space.
5. _______________ are lines with arrows or slashes at a termination line used to indicate the dimension of a distance on a drawing.
6. The _______________ is the minimum distance that must be maintained between a structure and the structure's property line, such as a street.
7. In _______________, also called _______________, the line represents the centerline of the pipe, so these drawings can be used to represent pipe of any diameter.
8. _______________ are marks or drawings that represent specific objects or materials.
9. The reasons that codes may set a minimum width for _______________ include providing access to rear yards and reducing the possibility of a fire jumping from one building to another.
10. _______________ illustrate how to assemble complex objects and show the relationship of the individual parts to the whole object.
11. Pictorial drawings that create the illusion of depth by using lines that are drawn at 45 or 60 degrees to the horizontal are _______________.
12. _______________ indicates the size relationship between an object in a drawing and the completed object itself.
13. _______________ specify places on the property where utilities can be installed.
14. A drawing with dark reddish lines is called a(n)_______________.
15. A section drawing that shows the details of how a product is constructed is called a(n) _______________.
16. _______________ are detailed lists of all the material and equipment necessary to construct a project.
17. A(n) _______________, also called _______________, provides information about where you may not build on certain parts of the property.
18. After the architect or engineer approves the fixture or fitting in a catalog drawing, it becomes _______________.
19. _____ aids productivity by automating the repetitive work of drafting.
20. _______________ present a three-dimensional picture of an object that is very similar to the actual object.
21. In _______________, all vertical lines are shown vertically but all horizontal lines are projected at a 30-degree angle and appear to go back into the horizon.
22. Limited space may require you to use _______________ to indicate the limits of a dimension on a drawing away from the actual points of the dimension.
23. Sections of the main drawing that are enlarged to show important structural elements more clearly are called _______________.
24. The _______________ shows an aerial view of the layout of rooms on each level of a building and the shape and size of each room.

25. The term *blueprint* is used interchangeably with _______________. It originally referred to drawings that appeared as white line drawings on a solid blue background.

26. A(n) _______________ is a dimensioned drawing with elevations and sections that indicate a proposed routing of system components.

27. _______________ show the location of outlets, switches, and electrical fixtures in a structure.

28. _______________ are complex isometric drawings that show vertical and horizontal piping along with sizes and a riser number that refers back to the full set of plumbing drawings.

29. Drawings of plumbing fixtures that are found in manufacturers' catalogs are called either _______________ or _______________.

30. _______________ show the piping system with the location of fixtures and pipe runs and the size and type of pipe to be installed.

31. The requirements that provide detailed information about the quality of workmanship required on a project, as well as the specific materials and method of installation you must use, are called _______________.

32. The type of measuring device used to draw and measure scaled drawings is a(n) _______________.

33. _______________ show the vertical elements of a building, such as the height and width of interior or exterior walls.

Trade Terms

Approved submittal data
Architect's scale
Catalog drawing
Computer-aided drafting (CAD)
Construction drawing
Coordination drawing
Cutaway drawing
Details
Dimension line
Easement
Electrical drawing
Elevation drawing
Exploded drawing
Extension line
Fixture drawing
Floor plan
Foundation plan
HVAC (heating, ventilating, and air conditioning) drawing
Isometric drawing
Oblique drawing
Orthographic drawing
Pictorial drawing
Plot plan
Plumbing drawing
Riser diagram
Scale
Schematic drawing
Sepia
Setback
Side yards
Single-line drawing
Site plan
Specifications
Specs
Submittal data
Symbol
Takeoff

Profile in Success

Radford (Rad) Mitchell, Sr.

Virginia Registered Plumbing Apprenticeship Program
Norfolk Technical Vocational Center
Fourth-Year Plumbing Instructor
Virginia Beach, Virginia

Radford Mitchell was born in Richmond, Virginia, and attended Granby High School in Norfolk. Following high school, he joined the U.S. Army, where he was a communications specialist in Vietnam. Early in his career, Rad took science classes at Tidewater Community College, working toward an associate's degree in natural science (much of the science was related to liquid flow). Later, Rad began working as an evening plumbing apprenticeship instructor with the Virginia Registered Plumbing Apprenticeship Program at the Norfolk Technical Vocational Center. After 27 years of work in the plumbing trade, Rad decided to continue his education full time. He attended Old Dominion University and received a bachelor's degree in geology with a minor in education. Inspired by his success as an apprenticeship teacher, Rad put his degrees to good use and became a public school science teacher while also continuing his evening work in the plumbing apprenticeship.

How did you become interested in this industry?
My first job in construction was in masonry. I laid brick from age 11 on. I would visit relatives who were all in construction, so I was brought up working for different uncles, working in tile and framing and brick and plastering. I have always been in construction. Out of high school I went to work as a plumber's helper and I thought, "Who knows? I may learn a trade that I can fall back on." My buddy had a job in plumbing, and one thing led to another.

What path did you take to your current position?
I went to work as a plumber's helper, working for E.B. Sam's Plumbing and Heating. While going through the plumbing apprenticeship program, I received a journeyman's license. After another year working as a journeyman mechanic, I passed the City of Virginia Beach master's plumbing exam. After an additional year as a master plumber with E.B. Sam's, working mostly with new commercial installations, I decided I wanted to be a plumbing inspector for the City of Virginia Beach. As an inspector, I obtained a Virginia state master's license in plumbing, HVAC, and gas fitting. Gradually, over 20 years, I passed other state exams to earn certifications in plumbing inspections, mechanical inspections, and as a plumbing/mechanical plan reviewer.

In addition to plumbing and mechanical inspections, I was responsible for the development and implementation of a cross-connection backflow prevention program for Virginia Beach. To develop the program, I focused on existing multistory hospitals, funeral homes, and other high-hazard commercial buildings first, and then went on to inspect schools and other commercial buildings for any needed backflow prevention devices.

I moved into education and was teaching 8th grade Earth science full time at a magnet school and then went on to teach 11th and 12th grade high school oceanography. I moved from the magnet school to a regular school because with the gifted students, I couldn't mention vocational school as an alternative to college. In the regular high school setting, this was not a problem.

Now I'm looking forward to leaving full-time teaching and going back to government code enforcement, possibly as a combination code inspector. I'd also like to start up a building trades academy where students could benefit the Habitat for Humanities Program as part of their training. I will never leave the evening apprenticeship work, however, as it is what motivated me to teach full time in the first place.

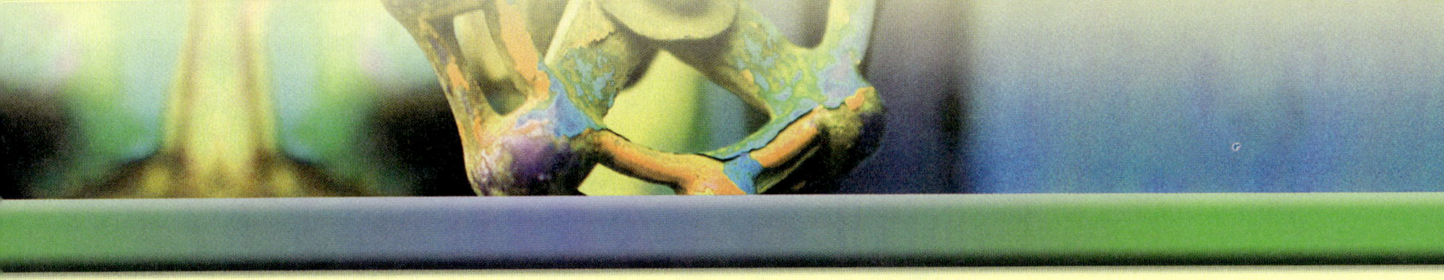

What does it take to be successful in your trade?
Attitude is the most important thing, as well as a desire to become a productive part of society. Even if you are meandering along and you can't see the vocational light at the end of the educational tunnel, try to have some direction. You need to have something to fall back on. You have to pay attention, be knowledgeable, maintain some form of education as you are learning, have mechanical talent, be good at assembling things, and work well with other people.

It is important to be willing to work well with other trades and with people on the job. You should have a certain amount of caring. Your role in society is to help the safety and welfare of the general public. To maintain indoor plumbing, to keep it up and running, the plumber is a godsend in some cases. Care about your customers, wipe your feet, and understand that they need your help, professionalism, and expertise. This job requires a balance between productivity and quality. Do things right, meet the minimum code, be thorough, and have tenacity.

What are some things you do on the job?
I provide career counseling and guide trainees through the development of their résumés. I first provide the men and women with a list of the top 10 things needed for the résumés. The trainees need to have this historical information when we meet later on to start typing the résumés. I also help them to recognize their skills and abilities. Until making a list, many do not realize just how much they have to offer an employer!

I introduce trainees to the codes. In the apprentices' fourth year, I apply what they have learned to real-life situations. The trainees apply what they have learned to code interpretation, such as finding the code violations in special isometric drawings that I have drawn and labeled with code discrepancies.

I help trainees understand code theory and design so that, on the job, they are not installing the system a certain way out of fear of the inspector, but because they understand why they should do it that way. I want them to have the confidence to make judgment calls on the job. I help them find the balance between productivity and quality through their knowledge of the code.

I am on the PHCC curriculum revision committee. Additionally, I focus on student retention in all four levels of the apprenticeship, and I provide a lot of trade-related career information about spin-off careers that are available in the plumbing industry.

What is the most interesting aspect of your profession? What makes your trade stand out from others?
The most interesting aspect is being able to see clearly that you have a valuable role in society. With plumbing, your work is eliminating disease, reducing potential for injury, reducing unsanitary conditions, providing conveniences such as running water and indoor plumbing, and eliminating the introduction of vermin. We have a powerful role in society to protect health.

It's challenging to craft a system that will pass inspection. You never know how an inspector will interpret code because it changes from city to city. A multicode inspector may be an expert in electrical work but not in plumbing. It's a challenging part of the trade.

What do you like most—and least—about your job?
I most like putting productive young people into the workforce in an occupation that they can take pride in and that makes them feel good about themselves.

I least like not being able to reach everybody about the importance of remaining in education and completing something. It concerns me when people drop out of the apprenticeship program but haven't left the trade. They develop a pattern of not finishing things and end up playing catch-up throughout their lives.

What would you say to someone entering the trade today?
Be flexible, especially with the code. Interpretation of the code is so important. Get used to change. Remember that you have a major role in society—in helping the general public. Continue your education so that doors don't shut in your face. Enjoy your job and find the right profession. With careers, don't feel

guilty about starting one thing and moving to another.

You don't have to remain in repair work or in new installations for your whole career. There are many related professions. Develop a foundation of knowledge and experience that you can draw on later as you fine-tune your skills. You can work in sales, in a supply house, or in any number of things that are related to the plumbing profession. You can hook up the equipment for such things as sprinklers, filters, and specialized photography or medical equipment. You can begin in installation and end up in management.

What can an apprentice expect to earn in his/her first years on the job in your area? What can he/she expect to earn after 10-plus years in the industry?
Apprentices in my area earn between $9.50 to $10.00 per hour. In the first years, trainees get a raise of $1.00 or so every six months to a year up to $14.00 or so per hour by their fourth year. You'll make steady increases. With the right company and your boss's encouragement, you could be a mechanic making $17.00 to $19.00 per hour after graduation! With extra training in a specialized area (such as working as a plumber in a hospital or for a hotel chain), earnings increase. After 10 years, hourly rates move close to $22.00. Foreman, supervisor, and estimator salaries go up even higher.

Trade Terms Introduced in This Module

Approved submittal data: The fixtures and fittings in catalog drawings (or submittal data) that have been approved by the architect or engineer.

Architect's scale: A measuring device that uses smaller units, such as ½ inch or ¼ inch, to represent 1 foot. Architect's scales are also issued in metric units. The units are used so that all building measurements are in proportion to their actual measurements but able to fit in the drawings.

Catalog drawing: A drawing of a plumbing fixture that is found in manufacturers' catalogs. Also referred to as submittal data.

Computer-aided drafting (CAD): A sophisticated design program used on computers. It allows designers to create drawings in two or three dimensions.

Construction drawing: A drawing that shows the design, location, and dimensions of a building and its various components.

Coordination drawing: A dimensioned drawing with elevations and sections that indicate the proposed routing of system components. The dimensioned coordination drawing includes the actual dimensioned equipment as well as the access space this equipment will need.

Cutaway drawing: A section drawing that shows the construction elements of a particular part of a building or fixture.

Details: Sections of construction drawings that are enlarged to make them clearer.

Dimension line: A line on a drawing with a measurement indicating actual length.

Easement: A designated right-of-way, such as the access guaranteed to utility companies for repair of utilities that are located on, or cross over, private land, or for vehicles to cross private property.

Electrical drawing: A drawing that shows the location of outlets, switches, and electrical fixtures. An electrical drawing may be superimposed on the floor plan.

Elevation drawing: A drawing of a structure showing a side, front, or back view.

Exploded drawing: A drawing that shows how to assemble a complex product. It also shows the relationship of the individual parts to the object as a whole. Also referred to as an assembly drawing.

Extension line: A line used on a drawing to locate a dimension away from the actual points of the dimension. This method is used when a drawing would be too crowded or cluttered if the dimension were shown within the two points.

Fixture drawing: A drawing that shows the components of a fixture in detail.

Floor plan: A construction drawing of a building looking down at the floor from above (bird's-eye view). The plan shows at least the outline of the wall locations and lengths to scale. Normally, this drawing is oriented such that the top of the page represents north.

Foundation plan: A construction drawing showing the placement and dimensions of a building foundation.

HVAC (heating, ventilating, and air conditioning) drawing: A construction drawing that shows the placement of the furnace and air conditioning equipment and the location of ducts and registers or pipes and radiators.

Isometric drawing: A pictorial drawing that creates the illusion of a three-dimensional object. All horizontal lines are projected at a 30-degree angle.

Oblique drawing: A pictorial drawing that shows the shape of an object. It shows the front of the object with the body of the object at a slight angle.

Orthographic drawing: A construction drawing that shows straight-on views of the different sides of an object. Orthographic drawings are used for elevation drawings.

Pictorial drawing: A drawing that shows a three-dimensional view of an object.

Plot plan: A drawing of a structure that includes the dimensions of the building site, location of the structure in relation to the property boundaries, elevation of key points, existing and finish contour lines, utility services, and compass directions. Also referred to as a site plan.

Plumbing drawing: A construction drawing that shows the location of fixtures and pipe runs, and gives the size and type of pipe to be installed.

Riser diagram: A drawing that shows vertical and horizontal piping along with sizes and a riser number that refers back to the full set of plumbing drawings.

Scale: The relationship of the dimensions on a drawing to the actual dimensions of the structure. For example, in a ¼ scale, 0.25 inches represent 1 foot. Scale is often provided in both English and metric units.

Schematic drawing: A single-line drawing of a plumbing system, electrical wiring routing, or circuit.

Sepia: A print or construction drawing with dark reddish-brown lines on a light background.

Setback: The distance a code requires between a building and a property line, such as the street.

Side yards: The spaces along the sides of a structure that provide access to rear yards, reduce the possibility of fire jumping from one building to the next, and promote ventilation around the structure.

Single-line drawing: A plumbing drawing that uses a single line to represent the centerline of a pipe. Single-line drawings can be used to represent pipe of any diameter.

Site plan: Another term for plot plan.

Specifications: Written requirements included with the drawings or blueprints of a construction project. They provide more details or descriptions of the technical standards that must be met during construction. Specifications usually override drawings, but are overridden by the contract. Also referred to as specs.

Specs: Another term for specifications.

Submittal data: Another term for catalog drawings.

Symbol: A mark or drawing used to indicate a specific object, material, class, or entity. A legend shows the symbols used on a drawing and their meanings.

Takeoff: The process by which detailed lists are compiled, based on drawings and specifications, of all the material and equipment necessary to construct a project. Such a list is also called a material takeoff.

Appendix

Plans and Diagrams

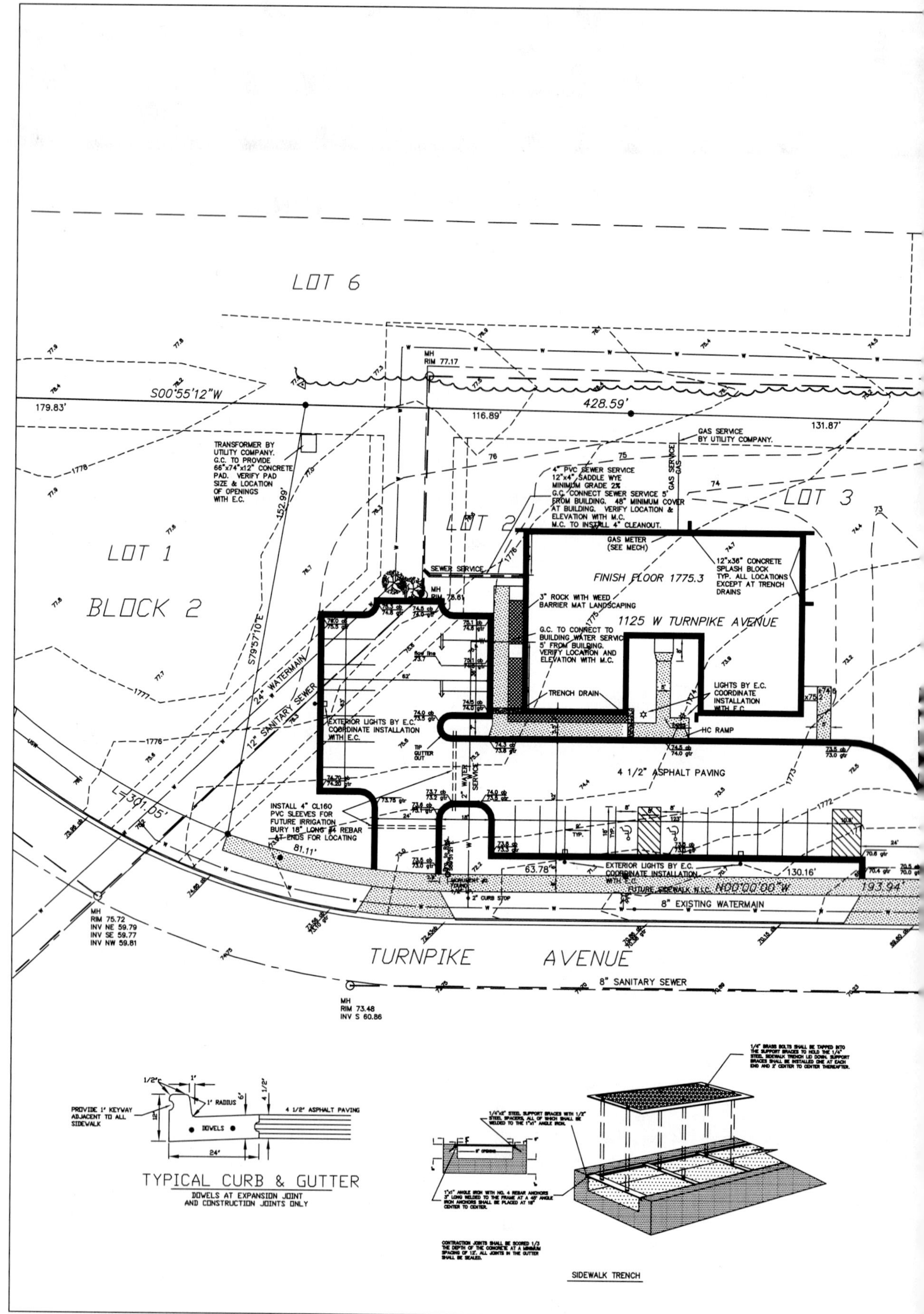

Figure A-1 ◆ Site plan showing sewer, water main, and gas lines.

HORIZONTAL CONTROL
MONUMENT #1 N 5413.7450 E 5100.9890
MONUMENT #3 N 4987.0740 E 4934.0270
MONUMENT #2 N 5181.0332 E 4934.0005
MONUMENT #4 N 4985.2186 E 5094.0956
LEGEND
EXISTING SANITARY SEWER
EXISTING WATERMAIN
GAS SERVICE
TREE OR SPRUCE
TELEPHONE OR CATV RISER
HYDRANT
GATE VALVE OR CURB STOP
MONUMENT FOUND OR SET
EXISTING GRADE
EXISTING CONTOURS
PROPOSED CONTOURS
SANITARY SEWER SERVICE
WATER SERVICE
PROPOSED GRADE
BENCH MARKS
TOP OF HYDRANT #1675 TURNPIKE AVE AND 3RD NORTH OF CAPITOL AVE ELEVATION 1779.26
TOP OF HYDRANT #1674 TURNPIKE AVE AND 2ND NORTH OF CAPITOL AVE ELEVATION 1773.08
TOP OF HYDRANT #1673 TURNPIKE AVE AND 1ST NORTH OF CAPITOL AVE ELEVATION 1770.94
NOTES
SEWER AND WATER INFORMATION WAS TAKEN FROM PLANS.
19.98'
159.82'
S89°19'56"E
10.49'
9.49'
EXISTING SIDEWALK
CONSTRUCTION NOTES
GENERAL
G1. ALL CONSTRUCTION SHALL CONFORM TO APPLICABLE CITY, STATE AND FEDERAL CODES.
G2. ALL SITE CONSTRUCTION IN CITY RIGHT OF WAY SHALL CONFORM TO THE MOST RECENT EDITION OF THE CITY OF BISMARCK CONSTRUCTION SPECIFICATIONS FOR MUNICIPAL PUBLIC WORKS IMPROVEMENTS AS MODIFIED BY THE ATTACHED PLANS.
G3. CONTRACTOR MUST BE BONDED BY THE CITY OF BISMARCK FOR ALL CONSTRUCTION IN CITY RIGHT OF WAY OR WHICH CONNECTS TO A CITY UTILITY.
G4. CONTRACTOR WILL BE REQUIRED TO OBTAIN ALL NECESSARY PERMITS. OWNER WILL OBTAIN CITY OF BISMARCK STORM WATER PERMIT, CONTRACTOR SHALL COMPLY WITH PERMIT REQUIREMENTS.
G5. IT WILL BE NECESSARY TO HAVE A PRIVATE ENGINEERING AND/OR TESTING FIRM SET GRADES, DO COMPACTION TESTS AND PREPARE RECORD DRAWINGS FOR THE SEWER LINES. THE CITY ENGINEERING DEPARTMENT WILL REQUIRE TEST REPORTS AND RECORD DRAWINGS PRIOR TO ACCEPTING THE PROJECT AFTER COMPLETION.
G6. PROTECT IMPROVEMENTS ON SITE AND ON ADJOINING PROPERTIES. PROVIDE BARRICADES, COVERINGS, OR OTHER TYPES OF PROTECTION AS NECESSARY TO PREVENT DAMAGE AND TO SAFEGUARD AGAINST INJURY. RESTORE TO ORIGINAL CONDITION IMPROVEMENTS DAMAGED BY THE WORK OR IMPROVEMENTS WHICH REQUIRED TEMPORARY REMOVAL DURING CONSTRUCTION.
G7. VERIFY GRADES, LINES, LEVELS, LOCATIONS, AND DIMENSIONS AS SHOWN AND INSPECT SURFACES THAT ARE TO RECEIVE WORK BEFORE PROCEEDING. CONTRACTOR SHALL BE SOLELY RESPONSIBLE FOR ACCURACY OF MEASUREMENTS AND LAYING OUT OF HIS WORK. NOTIFY OWNER IN WRITING IN CASE OF UNSUITABLE CONDITIONS, DEFECTIVE SUBSTRATES, OR DISCREPANCIES IN CONTRACT DOCUMENTS. STARTING OF WORK SHALL IMPLY ACCEPTANCE OF CONDITIONS.
WATER SERVICE
W1. MAINTAIN MINIMUM 7.5 FEET OF COVER OVER WATER SERVICE LINE.
SITE WORK
S1. ALL FILL MATERIAL PLACED TO BRING THE SITE TO FINISH GRADE SHALL BE AS SPECIFIED IN THE GEOTECHNICAL REPORT. FILL MATERIAL PLACED SHALL BE COMPACTED AS SPECIFIED IN THE GEOTECHNICAL REPORT.
S2. ALL EXCESS EXCAVATION SHALL BE REMOVED FROM THE PROPERTY AND DISPOSED OF BY THE CONTRACTOR.
S3. CONTRACTOR SHALL TOPSOIL ALL DISTURBED AREAS. PROVIDE MINIMUM 6" TOPSOIL OVER ALL AREAS TO BE SEEDED OR SODDED. SITE TOPSOIL MAY BE USED. ALL SHORTAGES OF TOPSOIL SHALL BE MADE UP BY CONTRACTOR. THE CONTRACTOR SHALL PROVIDE NATURAL, FRIABLE, FERTILE FINE SANDY LOAM TOPSOIL. THE TOPSOIL SHALL BE FREE FROM SUBSOIL, NOXIOUS WEED, STONES, LIME, CEMENT, ASHES, SLAG, OR OTHER DELETERIOUS MATTER.
ASPHALT
A1. ASPHALT PAVEMENT IN PARKING AREAS SHALL CONFORM TO THE CITY OF BISMARCK SPECIFICATIONS FOR TYPE "C" PAVING.
A2. THE TOP 6 INCHES OF SUBGRADE SHALL BE SCARIFIED AND RECOMPACTED AS REQUIRED BEFORE PLACEMENT OF ALL PAVEMENT SECTIONS.
SANITARY SEWER SERVICE
SS1 PVC SEWER SERVICE PIPE SHALL BE GASKETED JOINT PVC SDR 35 OR 26 MEETING THE REQUIREMENTS OF ASTM D3034
TESTING
T1. CONTRACTOR SHALL COORDINATE ALL TESTING AS REQUIRED BY THE CONTRACT DOCUMENTS AND AS FOLLOWS:
T2. A MINIMUM OF TWO (2) COMPACTION TESTS WILL BE REQUIRED IN PAVED AREAS. THE AREA REPRESENTED BY FAILING TESTS SHALL BE REMOVED OR REWORKED UNTIL A PASSING TEST IS OBTAINED. SEE GEOTECHNICAL REPORT FOR COMPACTION REQUIREMENTS.
T3. FOUR (4) ASPHALT DENSITY TESTS WILL BE REQUIRED. TWO (2) ON EACH LIFT. DENSITY REQUIRED SHALL BE 92% ON BASE COURSE AND 95% ON SURFACE COURSE. TESTS MAY BE OBTAINED BY A NUCLEAR DENSITY DEVICE.
T4. EXTERIOR CONCRETE TESTING. 1 SET OF 3 STANDARD CYLINDERS SHALL BE TAKEN FOR EACH DAYS POUR OF UP TO 20 CY. AND ADDITIONAL SET SHALL BE TAKEN FOR EVERY 25 CY THEREAFTER. A SLUMP TEST AND AIR ENTRAINMENT TEST WILL BE REQUIRED FOR EACH SET OF CYLINDERS.
CONCRETE
C1. ALL EXTERIOR CONCRETE SHALL HAVE A MINIMUM STRENGTH OF 4000 PSI MEASURED AT 28 DAYS. MAXIMUM ALLOWABLE SLUMP SHALL BE 4 INCHES. THE AIR CONTENT SHALL FALL IN THE RANGE OF 5% TO 7% ENTRAINED AIR.
C2. EXPANSION JOINTS SHALL BE SEALED USING A ONE PART POLYURETHANE TYPE, "SONOLASTIC N.P. 1" SEALANT AS MADE BY SONNELLORN BUILDING PRODUCTS DIVISION OR "SIKAFLEX-1A" AS MADE BY SIKA CHEMICAL CORP., OR "DYNATROL 1" AS MADE BY PERCORA CORP. OR EQUAL.
SURVEYING
SU1. CONTRACTOR SHALL PROVIDE ALL SURVEYS STAKING AND LAYOUT AS REQUIRED TO COMPLETE THE WORK.
STRIPING
ST1. ALL STRIPES SHALL BE 4" WIDE YELLOW NON-REFLECTORIZED PAINT MEETING THE REQUIREMENTS OF ASSHTO M248 APPLIED AT A RATE OF 110 SQUARE FEET PER GALLON.
WARRANTY / DISCLAIMER
The designs represented in these plans are in accordance with established practices of civil engineering. However, neither Swenson, Hagen & Co. nor its personnel can or do warranty these plans as constructed except in the specific cases where Swenson, Hagen & Co personnel inspect and control the physical construction on a full time basis.
SAFETY NOTICE TO CONTRACTOR
In accordance with generally accepted construction practices the Contractor will be solely and completely responsible for conditions of the job site, including safety of all persons and property during performance of the work. This requirement will apply continuously and not be limited to normal business hours.
UTILITY NOTICE TO CONTRACTOR
The location of existing underground utilities are shown in an approximate way only and have not been independently verified by the Owner or its representatives. The Contractor shall determine the exact location of all existing utilities before commencing work, and agrees to be fully responsible for any and all damages which might be occasioned by the Contractor's failure to exactly locate and preserve any and all underground utilities. Underground locates can be made by using the "One Call" number 1-800-795-0555.
Z
20' 10' 0 10' 20'
DRAWING DESCRIPTION:
SITE PLAN
PLOT SCALE:
FILE NAME:
© 1999 Ritterbush - Ellig - Hulsing P.C. Architects - Planners
REH
RITTERBUSH-ELLIG-HULSING P.C.
RITTERBUSH - ELLIG - HULSING P.C.
ARCHITECTS - PLANNERS
209 NORTH SEVENTH STREET
BISMARCK, NORTH DAKOTA 58501-4495
TELE: 701-223-7780 FAX: 701-258-6564
GERMANS FROM RUSSIA HERITAGE SOCIETY
BISMARCK, NORTH DAKOTA
REVISIONS
DRAWN BY
DATE
MAY 30, 2000
PROJECT NO.
9918
C-2

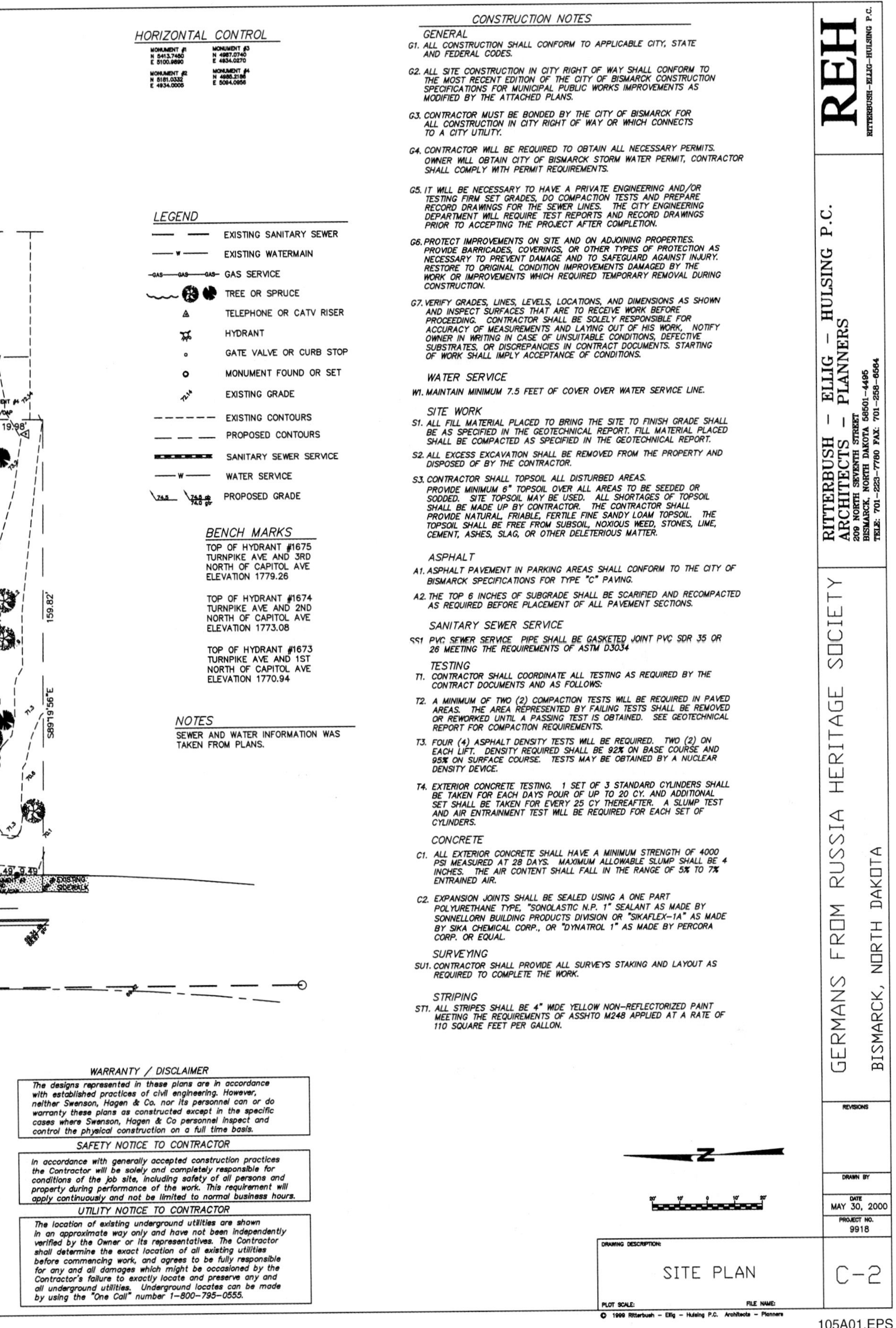

105A01.EPS

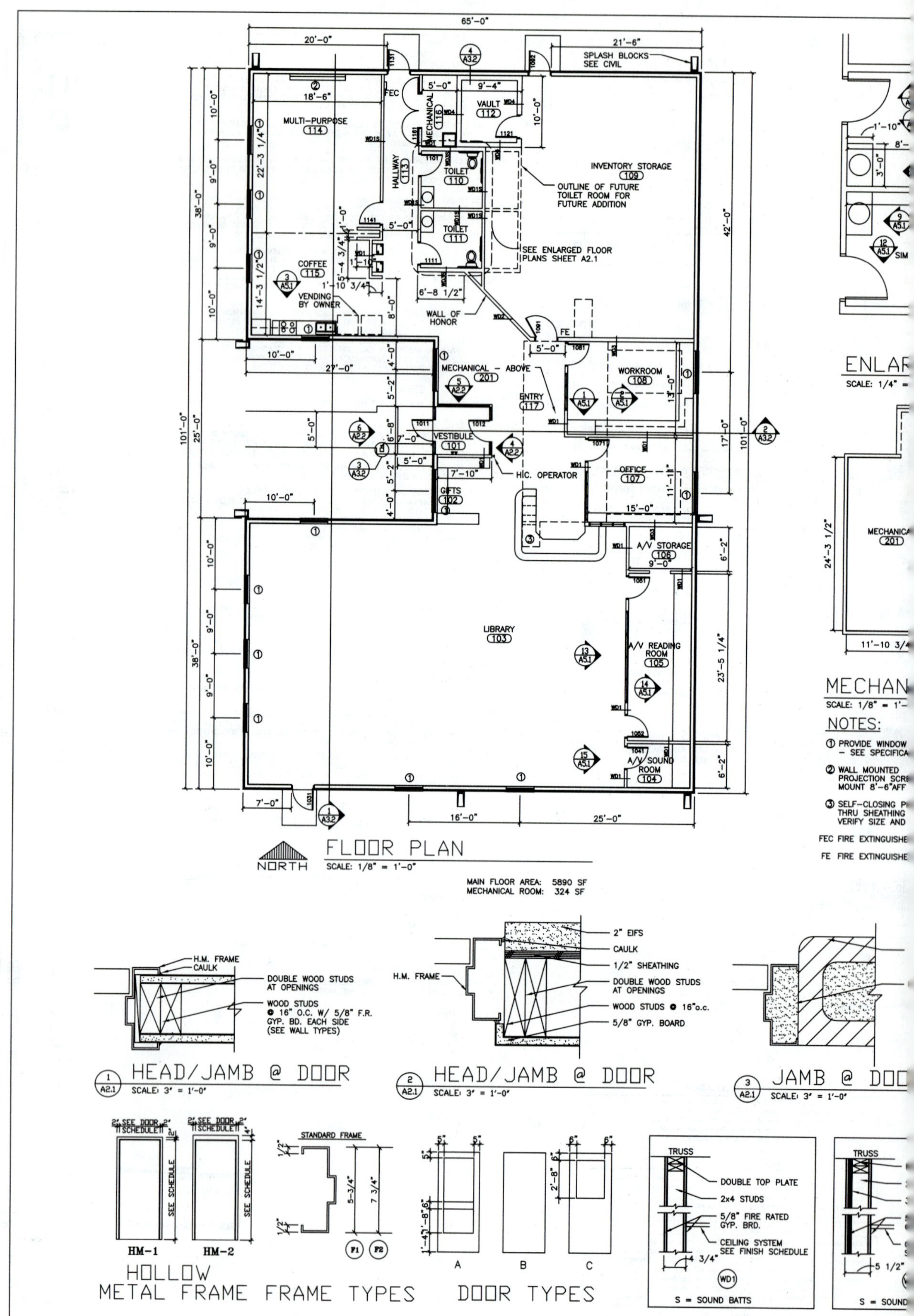

Figure A-2 ◆ Floor plan.

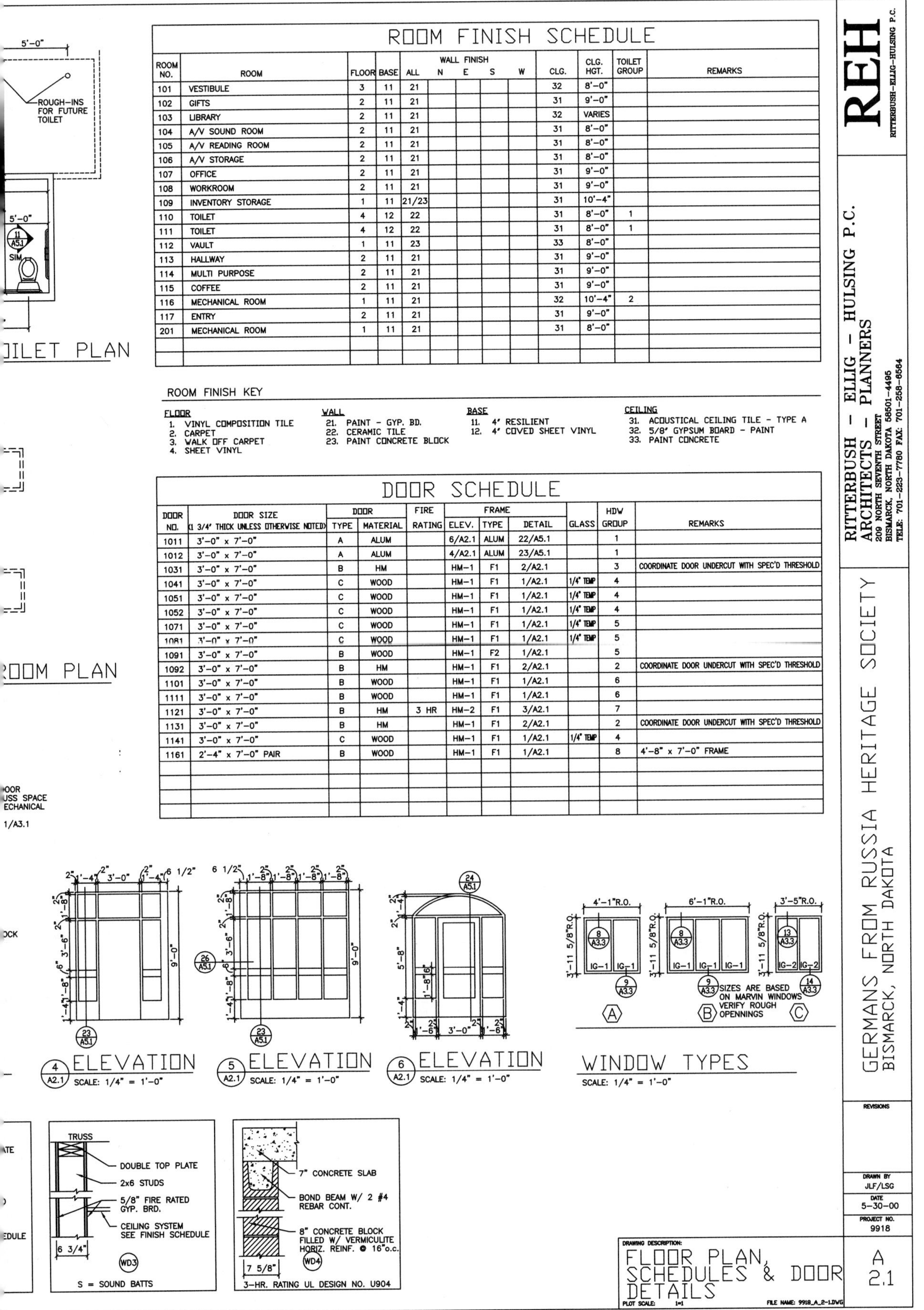

ROOM FINISH SCHEDULE

ROOM NO.	ROOM	FLOOR	BASE	WALL FINISH ALL	N	E	S	W	CLG.	CLG. HGT.	TOILET GROUP	REMARKS
101	VESTIBULE	3	11	21					32	8'-0"		
102	GIFTS	2	11	21					31	9'-0"		
103	LIBRARY	2	11	21					32	VARIES		
104	A/V SOUND ROOM	2	11	21					31	8'-0"		
105	A/V READING ROOM	2	11	21					31	8'-0"		
106	A/V STORAGE	2	11	21					31	8'-0"		
107	OFFICE	2	11	21					31	9'-0"		
108	WORKROOM	2	11	21					31	9'-0"		
109	INVENTORY STORAGE	1	11	21/23					31	10'-4"		
110	TOILET	4	12	22					31	8'-0"	1	
111	TOILET	4	12	22					31	8'-0"	1	
112	VAULT	1	11	23					33	8'-0"		
113	HALLWAY	2	11	21					31	9'-0"		
114	MULTI PURPOSE	2	11	21					31	9'-0"		
115	COFFEE	2	11	21					31	9'-0"		
116	MECHANICAL ROOM	1	11	21					32	10'-4"	2	
117	ENTRY	2	11	21					31	9'-0"		
201	MECHANICAL ROOM	1	11	21					31	8'-0"		

ROOM FINISH KEY

FLOOR
1. VINYL COMPOSITION TILE
2. CARPET
3. WALK OFF CARPET
4. SHEET VINYL

WALL
21. PAINT - GYP. BD.
22. CERAMIC TILE
23. PAINT CONCRETE BLOCK

BASE
11. 4" RESILIENT
12. 4" COVED SHEET VINYL

CEILING
31. ACOUSTICAL CEILING TILE - TYPE A
32. 5/8" GYPSUM BOARD - PAINT
33. PAINT CONCRETE

DOOR SCHEDULE

DOOR NO.	DOOR SIZE (1 3/4" THICK UNLESS OTHERWISE NOTED)	DOOR TYPE	DOOR MATERIAL	FIRE RATING	FRAME ELEV.	FRAME TYPE	FRAME DETAIL	GLASS	HDW GROUP	REMARKS
1011	3'-0" x 7'-0"	A	ALUM		6/A2.1	ALUM	22/A5.1		1	
1012	3'-0" x 7'-0"	A	ALUM		4/A2.1	ALUM	23/A5.1		1	
1031	3'-0" x 7'-0"	B	HM		HM-1	F1	2/A2.1		3	COORDINATE DOOR UNDERCUT WITH SPEC'D THRESHOLD
1041	3'-0" x 7'-0"	C	WOOD		HM-1	F1	1/A2.1	1/4" TEMP	4	
1051	3'-0" x 7'-0"	C	WOOD		HM-1	F1	1/A2.1	1/4" TEMP	4	
1052	3'-0" x 7'-0"	C	WOOD		HM-1	F1	1/A2.1	1/4" TEMP	4	
1071	3'-0" x 7'-0"	C	WOOD		HM-1	F1	1/A2.1	1/4" TEMP	5	
1081	3'-0" x 7'-0"	C	WOOD		HM-1	F1	1/A2.1	1/4" TEMP	5	
1091	3'-0" x 7'-0"	B	WOOD		HM-1	F2	1/A2.1		5	
1092	3'-0" x 7'-0"	B	HM		HM-1	F1	2/A2.1		2	COORDINATE DOOR UNDERCUT WITH SPEC'D THRESHOLD
1101	3'-0" x 7'-0"	B	WOOD		HM-1	F1	1/A2.1		6	
1111	3'-0" x 7'-0"	B	WOOD		HM-1	F1	1/A2.1		6	
1121	3'-0" x 7'-0"	B	HM	3 HR	HM-2	F1	3/A2.1		7	
1131	3'-0" x 7'-0"	B	HM		HM-1	F1	2/A2.1		2	COORDINATE DOOR UNDERCUT WITH SPEC'D THRESHOLD
1141	3'-0" x 7'-0"	C	WOOD		HM-1	F1	1/A2.1	1/4" TEMP	4	
1161	2'-4" x 7'-0" PAIR	B	WOOD		HM-1	F1	1/A2.1		8	4'-8" x 7'-0" FRAME

105A02.EPS

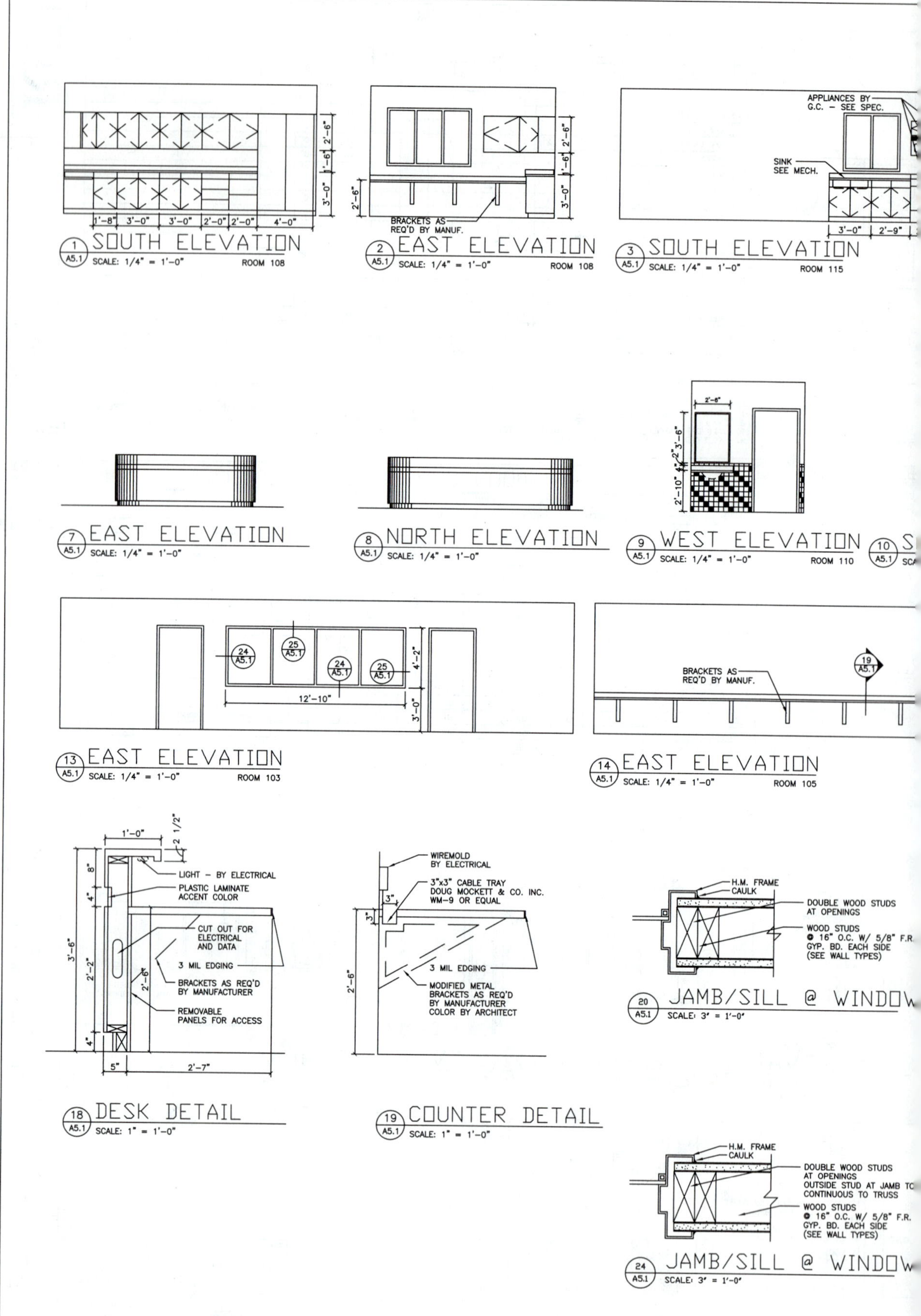

Figure A-3 ◆ Elevation drawings.

DESK PLAN
SCALE: 1/4" = 1'-0"
WEST ELEVATION
SCALE: 1/4" = 1'-0"
PLASTIC LAMINATE END PANEL
PROVIDE CASTORS & FINISHED SIDES @ ALL UNITS
LOCK FILE
SOUTH ELEVATION
SCALE: 1/4" = 1'-0"
BRACKETS AS REQ'D BY MANUF.
ACCENT STRIP (ALL WALLS)
ELEVATION
ROOM 110
EAST ELEVATION
SCALE: 1/4" = 1'-0"
ROOM 110
NORTH ELEVATION
SCALE: 1/4" = 1'-0"
ROOM 110
BRACKETS AS REQ'D BY MANUF.
EAST ELEVATION
SCALE: 1/4" = 1'-0"
ROOM 104
EAST ELEVATION
SCALE: 1/4" = 1'-0"
ROOM 117
NORTH ELEVATION
SCALE: 1/4" = 1'-0"
ROOM 103
WOOD STUDS @ 16" O.C. W/ 5/8" F.R. GYP. BD. EACH SIDE (SEE WALL TYPES)
DOUBLE WOOD HEADER
CAULK
H.M. FRAME
HEAD @ WINDOW
SCALE: 3" = 1'-0"
WOOD STUDS @ 16" O.C. W/ 5/8" F.R. GYP. BD. EACH SIDE (SEE WALL TYPES)
2 2x4s
DOUBLE 2x6 HEADER
2 2x4s
CAULK
H.M. FRAME
HEAD @ WINDOW
SCALE: 3" = 1'-0"
2" E.I.F.S. W/ BACKSTOP & REINFORCING MESH AROUND DOOR OPENNING & @ CORNERS
HEADER – SEE STRUCTURAL
5/8" GYPSUM BOARD ON 2x4s @ 16"o.c.
CONTINUOUS FLASHING WITH SEALANT, BACKER ROD, & SPANDREL WINDOW
CAULK & FOAM ROD BACK-UP CONTINUOUS BOTH SIDES
1" INSULATED GLASS
ALUMINUM FRAME
ALUMINUM DOOR
1" INSULATED GLASS
DETAIL
SCALE: 1 1/2" = 1'-0"
2x4 TO TRUSS, BRACE 4'-0"o.c.
5/8" GYPSUM ON 2x4s @ 16"o.c.
ACOUSTICAL CEILING
CAULK & FOAM ROD BACK-UP CONT. BOTH SIDES
ALUM. FLAT STOCK COVER PLATES FASTENED TO FRAME WITH CONT. 1"x1" ALUM. CLIPS BY WINDOW MANUF.
CAULK & FOAM ROD BACK-UP CONT. BOTH SIDES
DETAIL
SCALE: 1 1/2" = 1'-0"
ALUMINUM FRAME
4 1/2"x4 1/2" ALUM. TUBE
ALUMINUM FRAME
DETAIL
SCALE: 1 1/2" = 1'-0"
REH
RITTERBUSH-ELLIG-HULSING P.C.
RITTERBUSH – ELLIG – HULSING P.C.
ARCHITECTS – PLANNERS
209 NORTH SEVENTH STREET
BISMARCK, NORTH DAKOTA 58501-4495
TELE: 701-223-7780 FAX: 701-258-6564
GERMANS FROM RUSSIA HERITAGE SOCIETY
BISMARCK, NORTH DAKOTA
REVISIONS
DRAWN BY
JLF/LSG
DATE
5-30-00
PROJECT NO.
9918
DRAWING DESCRIPTION:
INTERIOR ELEVATIONS, CASEWORK & DETAILS
A 5.1
PLOT SCALE: 1=1
FILE NAME: 9918_A_5-1.DWG
© 2000 Ritterbush – Ellig – Hulsing P.C. Architects – Planners

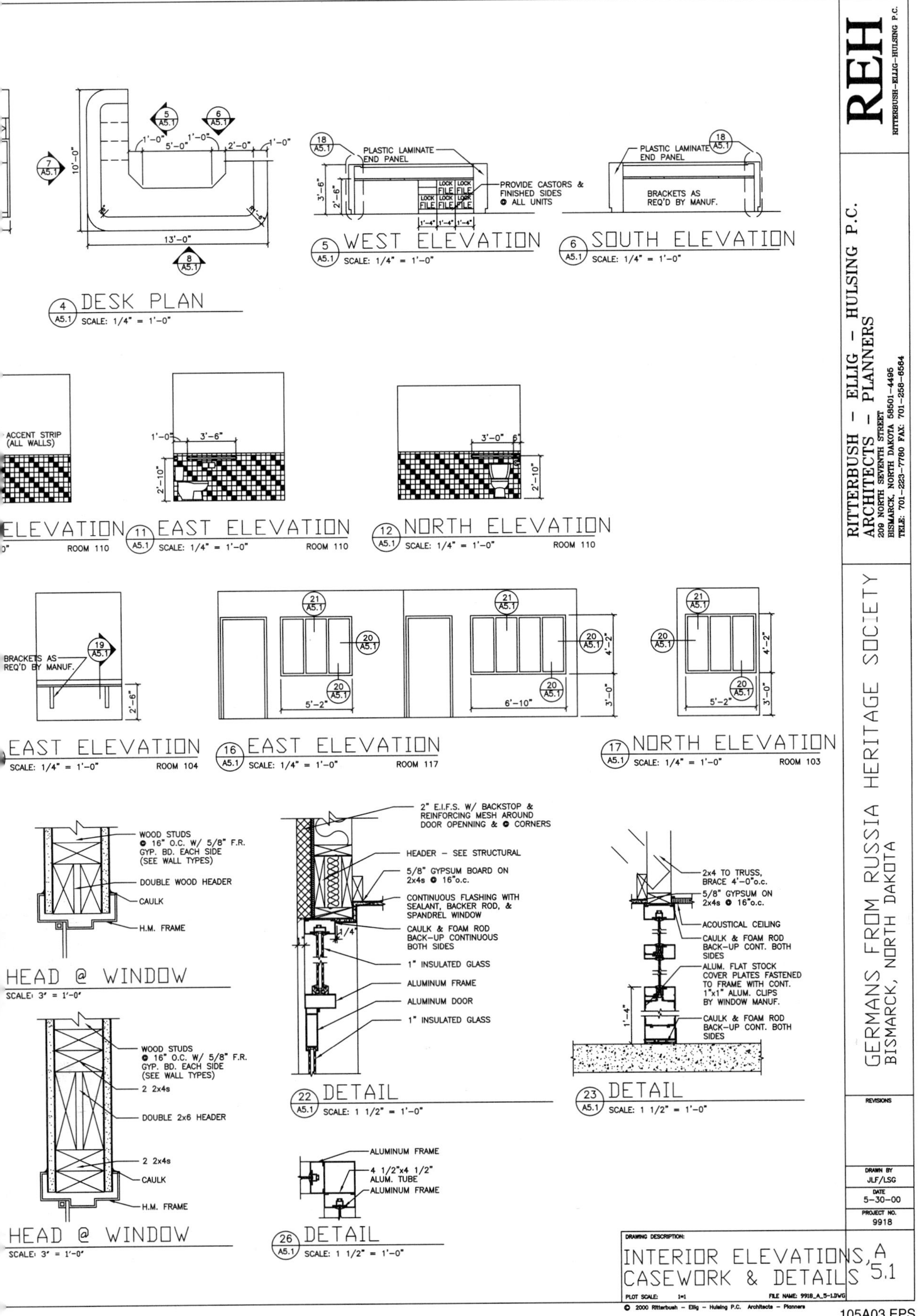

105A03.EPS

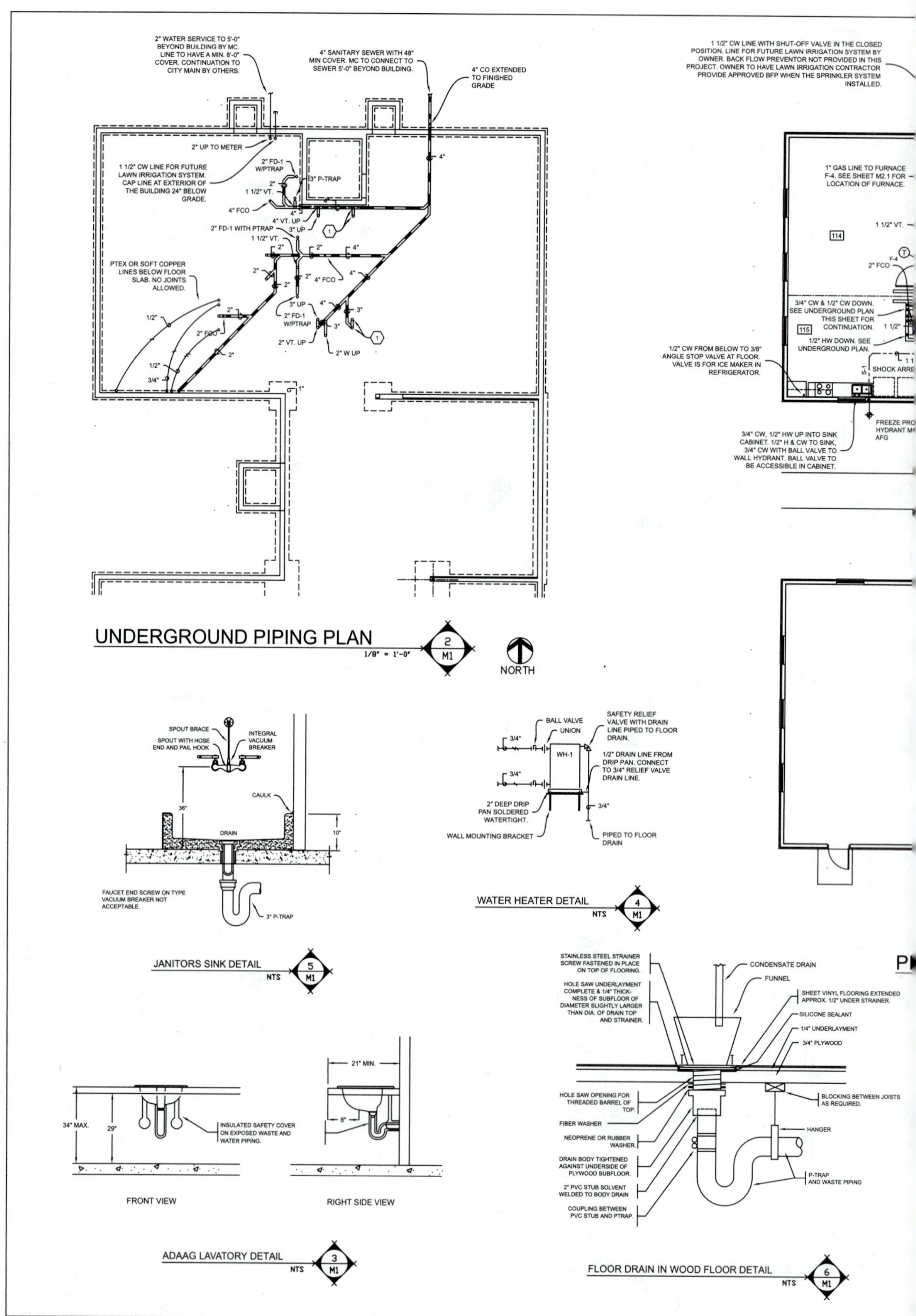

Figure A-4 ◆ Piping plan.

PITCH BACK TO HUMIDIFIER
DUCT
CONDENSATE SEPARATOR
1/2" CW
H-1
STEAM DISPERSION TUBE
1/2"
1" COPPER CABINET DRAIN
2" TO 5" TRAP HEIGHT
FOLLOW MANUFACTURERS INSTRUCTIONS FOR STEAM DISPERSION TUBE PLACEMENT AND STEAM HOSE LAYOUTS.
1/2" COPPER TO FD-2
1" COPPER OVERFLOW AND UNIT DRAIN TO FD-2.
HUMIDIFIER H-1 DETAIL
NTS
8
M1
KEYNOTES
1 3" WASTE AND 2" VENT CAPPED 2" BELOW FLOOR (VERIFY DEPTH WITH GC) FOR FUTURE WATER CLOSET. MARK FLOOR AT WASTE AND VENT LOCATIONS.
2" FD-1 IN ROOMS 110 AND 111
4" VT., 4" VTR
1/2" CW LINE CAPPED FOR FUTURE
4" FCO
1 1/2" VT.
1/2" CW LINE CAPPED FOR FUTURE
FURNACE F-3
HUMIDISTAT H-2
FREEZE PROOF WALL HYDRANT MOUNTED 24" AFG
1/2" CW UP TO HUMIDIFIERS
GAS METER
1 1/2" GAS LINE UP. SEE SHEET M2.1 FOR CONTINUATION.
UNDERGROUND GAS SERVICE AND METER BY MDU. VERIFY ROUTING OF UNDERGROUND LINE WITH FUTURE EXPANSION. SEE ARCHITECTURAL PLANS.
FURNACE F-1a AND F-1b
HUMIDISTAT H-1
FREEZE PROOF WALL HYDRANT MOUNTED 24" AFG
PLAN
1/8" = 1'-0"
1
M1
NORTH
PIPE LEGEND
UNDER GROUND WASTE
ABOVE GROUND WASTE
ABOVE GROUND VENT
DOMESTIC COLD WATER
DOMESTIC HOT WATER
NATURAL GAS
BALL VALVE
SHUTOFF VALVE IN VERTICAL
STANDARD ABBREVIATIONS
VT - VENT
VTR - VENT THRU ROOF
W - WASTE
CW - COLD WATER
HW - HOT WATER
RHW - RECIRCULATING HOT WATER
RD - ROOF DRAIN
FD - FLOOR DRAIN
CO - CLEAN OUT
FCO - FLOOR CLEAN OUT
WCO - WALL CLEAN OUT
AFF - ABOVE FINISHED FLOOR
AFG - ABOVE FINISHED GRADE
MC - MECHANICAL CONTRACTOR
EC - ELECTRICAL CONTRACTOR
GC - GENERAL CONTRACTOR
ATC - AUTOMATIC TEMPERATURE CONTROLS
FA - FRESH AIR
CA - COMBUSTION AIR
SA - SUPPLY AIR
RA - RETURN AIR
PRV - POWER ROOF VENTILATOR
BDD - BACK DRAFT DAMPER
OBD - OPPOSED BLADE DAMPER
VD - VOLUME DAMPER
FDR - FIRE DAMPER
AD - ACCESS DOOR
WH - WALL HYDRANT
HVAC LEGEND
POSITIVE PRESSURE AIR DUCT
NEGATIVE PRESSURE AIR DUCT
MANUAL VOLUME DAMPER
ELBOW WITH TURNING VANE
COMB. SMOKE/FIRE DAMPER
MOTORIZED DAMPER
FIRE DAMPER
REHEAT COIL
ACCESS DOOR
HI-EFFICIENCY TAKEOFF (W/ VOLUME DAMPER)
LINE FOR FUTURE LAWN IRRIGATION SYSTEM. A BACK FLOW PREVENTOR WILL NEED TO BE INSTALLED OUTSIDE BUILDING ON THIS LINE WHEN THE IRRIGATION SYSTEM IS INSTALLED. BACK FLOW PREVENTOR IS NOT INCLUDED IN THIS PROJECT.
UNTIL IRRIGATION SYSTEM IS INSTALLED, THIS VALVE SHALL REMAIN CLOSED.
2" BYPASS LINE WITH BALL VALVE NORMALLY CLOSED
DRAIN VALVE
CAP LINE BELOW GRADE. WHEN THE IRRIGATION SYSTEM IS INSTALLED IN THE FUTURE, THE INSTALLER SHALL PROVIDE BACK FLOW PREVENTOR ON LINE OUTSIDE BUILDING.
STAND
WATER METER DETAIL
NTS
7
M1
ROOM SCHEDULE
ROOM NO. ROOM NAME ROOM NO. ROOM NAME
101 VESTIBULE 110 TOILET
102 GIFTS 111 TOILET
103 LIBRARY 112 VAULT
104 A/V SOUND ROOM 113 HALLWAY
105 A/V READING ROOM 114 MULTI-PURPOSE
106 A/V STORAGE 115 COFFEE
107 OFFICE 116 MECHANICAL
108 WORKROOM 117 ENTRY
109 INVENTORY STORAGE 201 MECHANICAL
DRAWING DESCRIPTION:
PIPING PLAN
PLOT SCALE: AS NOTED
FILE NAME: GRHS-P-6.DWG
© 1998 Ritterbush – Ellig – Hulsing P.C. Architects – Planners
REH
RITTERBUSH-ELLIG-HULSING P.C.
RITTERBUSH – ELLIG – HULSING P.C.
ARCHITECTS – PLANNERS
209 NORTH SEVENTH STREET
BISMARCK, NORTH DAKOTA 58501-4495
TELE: 701-223-7780 FAX: 701-258-6564
GERMANS FROM RUSSIA HERITAGE SOCIETY
BISMARCK, NORTH DAKOTA
REVISIONS
DRAWN BY SSB
DATE 5-30-00
PROJECT NO. B99548
M 1

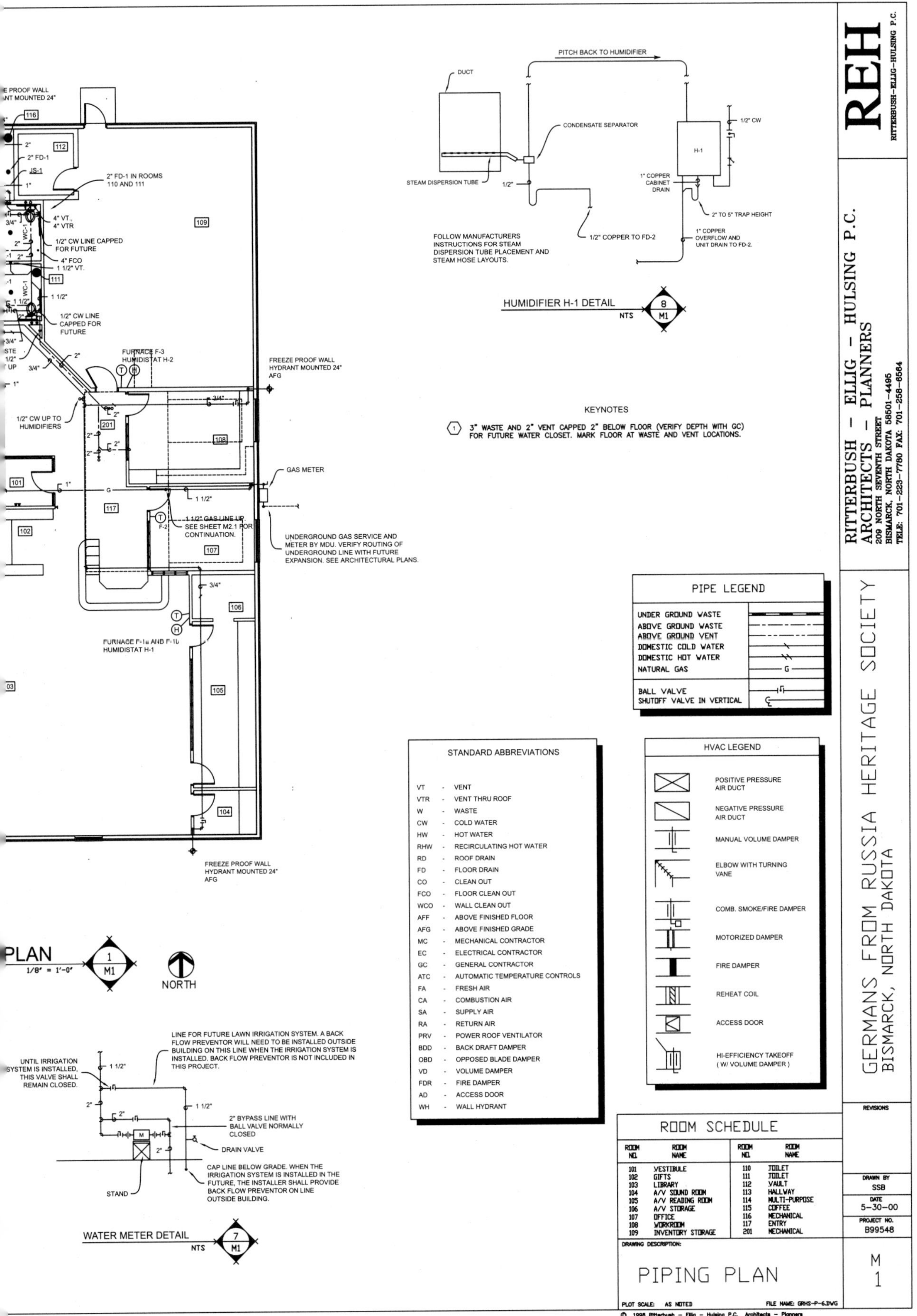

105A04.EPS

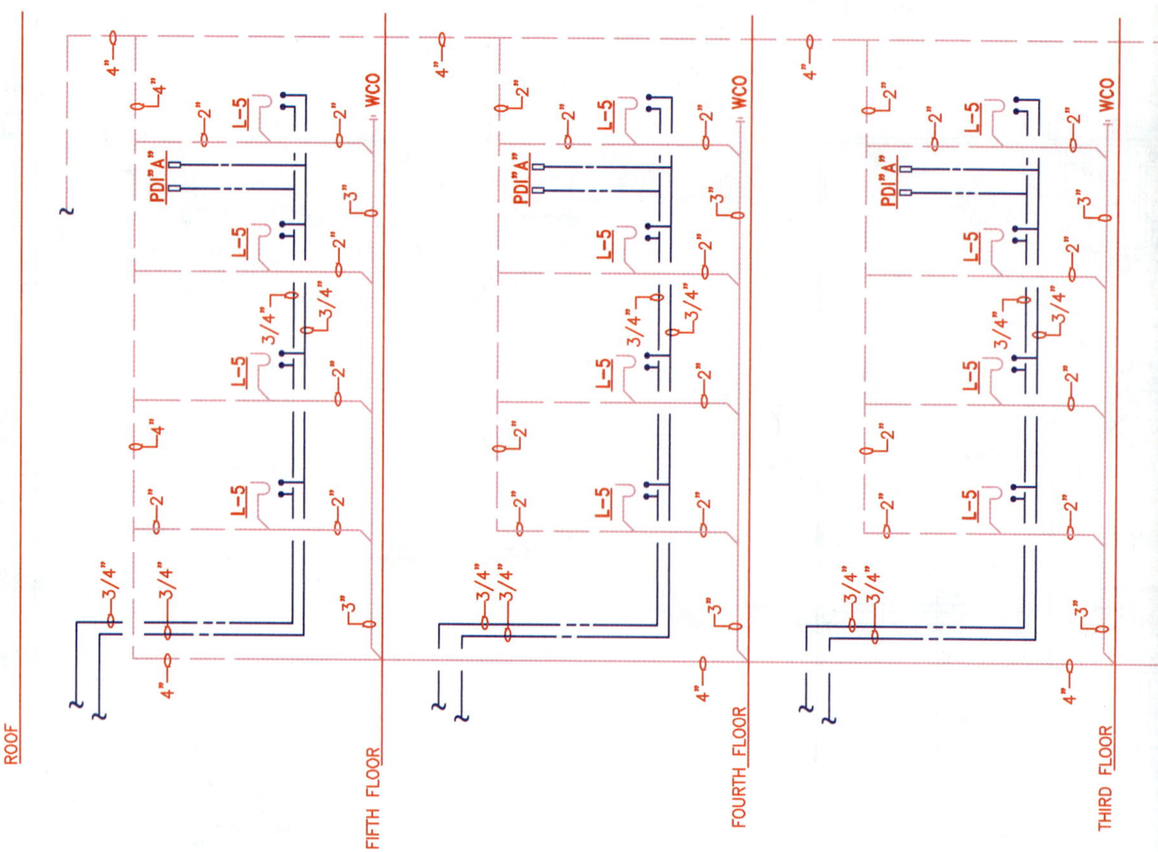

Figure A-5 ◆ Riser diagram.

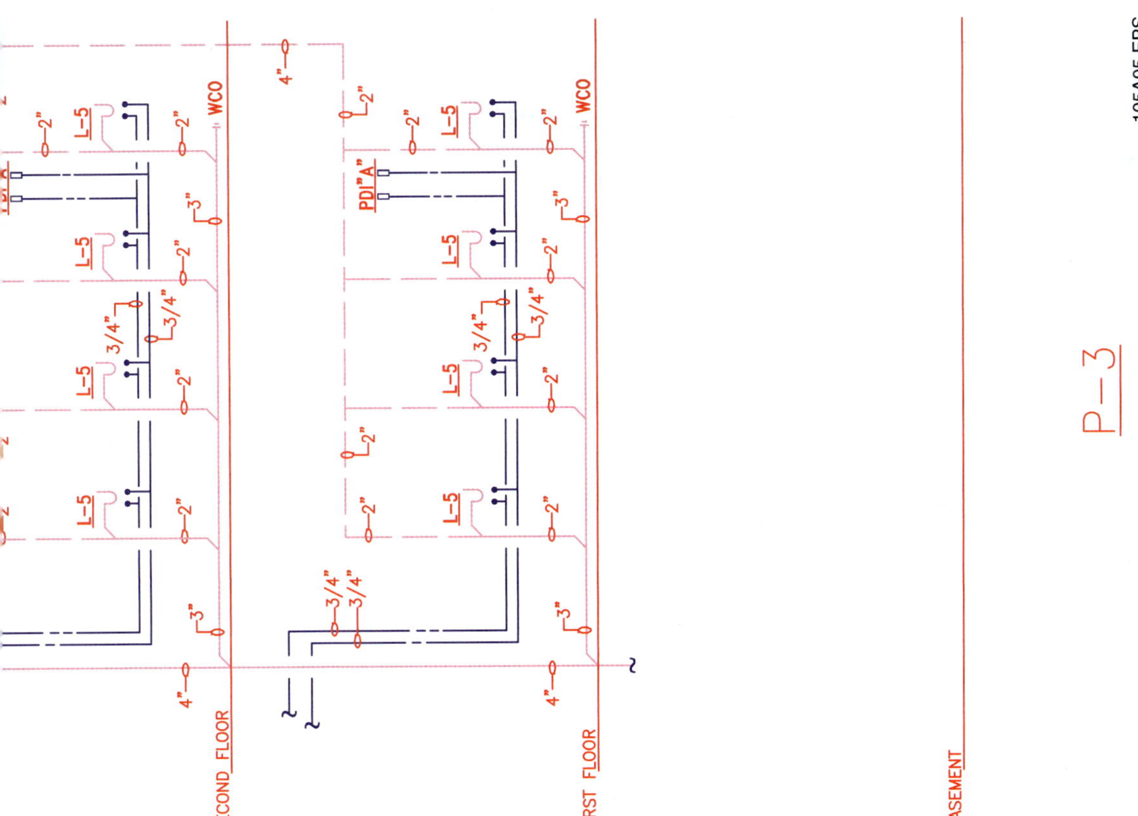
SECOND FLOOR
FIRST FLOOR
BASEMENT
WCO
PDI "A"
L-5
P-3
105A05.EPS

Additional Resources and References

ADDITIONAL RESOURCES

This module is intended to present thorough resources for task training. The following reference works are suggested for further study. These are optional materials for continued education rather than for task training.

Blueprint Reading Basics, 1996. Warren Hammer. New York: Industrial Press.

Blueprint Reading for the Building Trades, 1989. John Traister. Carlsbad, CA: Craftsman Book Co.

Core Curriculum, 2004. NCCER. Upper Saddle River, NJ: Prentice Hall.

Technical Drawing, 2003. Frederick E. Giesecke et al. Upper Saddle River, NJ: Prentice Hall/ Pearson Education.

REFERENCES

Cast Iron Soil Pipe Institute website, www.cispi.org, *Cast Iron Soil Pipe and Fittings Handbook,* www.cispi.org/handbook.htm, reviewed August 2003.

Dictionary of Architecture and Construction, 1993. Cyril M. Harris, ed. New York: McGraw-Hill.

Exploring Drafting: Basic Fundamentals, 1991. John R. Walker. South Holland, IL: Goodheart-Willcox.

Illustrated Dictionary of Building Materials and Techniques, 1993. Paul Bianchina. New York: John Wiley & Sons, Inc.

NCCER CURRICULA — USER UPDATE

NCCER makes every effort to keep its textbooks up-to-date and free of technical errors. We appreciate your help in this process. If you find an error, a typographical mistake, or an inaccuracy in NCCER's curricula, please fill out this form (or a photocopy), or complete the online form at **www.nccer.org/olf**. Be sure to include the exact module ID number, page number, a detailed description, and your recommended correction. Your input will be brought to the attention of the Authoring Team. Thank you for your assistance.

Instructors – If you have an idea for improving this textbook, or have found that additional materials were necessary to teach this module effectively, please let us know so that we may present your suggestions to the Authoring Team.

NCCER Product Development and Revision

13614 Progress Blvd., Alachua, FL 32615

Email: curriculum@nccer.org
Online: www.nccer.org/olf

❑ Trainee Guide ❑ AIG ❑ Exam ❑ PowerPoints Other ______________________________

Craft / Level: Copyright Date:

Module ID Number / Title:

Section Number(s):

Description:

Recommended Correction:

Your Name:

Address:

Email: Phone:

Plastic Pipe and Fittings

02106-05

02106-05

Plastic Pipe and Fittings

Topics to be presented in this module include:

Overview

Plastic has changed the way plumbers work. Plastic pipe is strong, durable, and requires little maintenance. Although it is easier to use than metal pipe, it does have some disadvantages. It can be affected by temperatures, give off toxic fumes, and become flammable. Plumbers must learn how to safely and properly handle plastic pipe and fittings.

Plumbers must be familiar with several types and sizes of plastic pipe, each with its own properties. Each type of plastic pipe is designed for different applications and has different installation requirements. All plastic pipe must be handled and stored properly. A pipe's application determines the type of fittings and joints that are compatible. Specific fittings are used for water supply and drain, waste, and vent (DWV) systems. Plumbers use special tools to measure, cut, and join plastic pipe and fittings. While measuring and cutting techniques are standard, different types of plastic pipe have their own methods for joining.

Plastic pipe can be supported using a variety of methods. The material, size, applications, and installation affect the type of support. Hangers are used for horizontal support; pipe riser clamps provide vertical support. Plans and specifications dictate codes that must be followed. After completing an installation, the system must be hydrostatically pressure-tested.

Focus Statement

The goal of the plumber is to protect the health, safety, and comfort of the nation job by job.

Code Note

Codes vary among jurisdictions. Because of the variations in code, consult the applicable code whenever regulations are in question. Referencing an incorrect set of codes can cause as much trouble as failing to reference codes altogether. Obtain, review, and familiarize yourself with your local adopted code.

Objectives

When you have completed this module, you will be able to do the following:

1. Identify types of materials and schedules of plastic piping.
2. Identify proper and improper applications of plastic piping.
3. Identify types of fittings and valves used with plastic piping.
4. Identify and determine the kinds of hangers and supports needed for plastic piping.
5. Identify the various techniques used in hanging and supporting plastic piping.
6. Properly measure, cut, and join plastic piping.
7. Explain proper procedures for the handling, storage, and protection of plastic pipes.

Key Trade Terms

ABS
Bell-and-spigot pipe
Cellular core wall
Chamfer
Chemically inert
Compression collar
CPVC
Elastomeric
Fusion fitting
Hydronic
Hydrostatically pressure-test
Inside diameter
Interference fit
Outside diameter
PB
PE
PEX
Pipe riser clamp
Pressure rating
psi
PVC
Ring-tight gasket fitting
Schedule
Size dimension ratio
Solid wall
Solvent weld
Thermoplastic pipe
Thermosetting pipe
Transition fitting
Water hammer

Required Trainee Materials

1. Appropriate personal protective equipment
2. Pencils and paper
3. Copy of local adopted code

Prerequisites

Before you begin this module, it is recommended that you successfully complete *Core Curriculum; Plumbing Level One,* Modules 02101-05 through 02105-05.

This course map shows all of the modules in the first level of the *Plumbing* curriculum. The suggested training order begins at the bottom and proceeds up. Skill levels increase as you advance on the course map. The local Training Program Sponsor may adjust the training order.

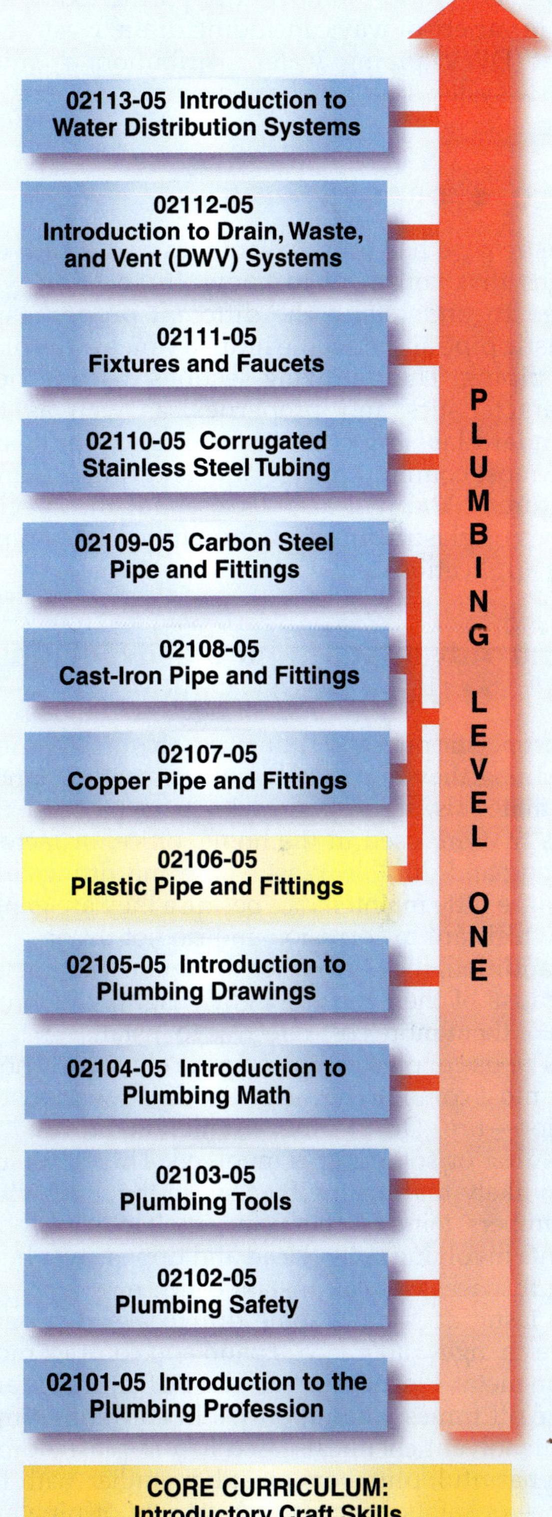

106CMAP.EPS

1.0.0 ◆ INTRODUCTION

Plastic has revolutionized plumbing, a trade that was based for hundreds of years on metal pipe and fittings. Not widely used by the plumbing industry until the 1950s, plastic is the newest of plumbing materials and has changed the way plumbers work. Plumbers use plastic today in a wide variety of ways, including drain, waste, and vent (DWV) piping; water distribution; chemical waste systems; and fuel gas piping.

2.0.0 ◆ PLASTIC PIPE

Plastic pipe has both advantages and disadvantages that you need to know. In addition, you need to understand the different properties of plastic pipe and how the industry measures plastic piping. The following sections discuss these characteristics and properties, as well as the standard labeling practice for plastic piping and the most common plastic pipe and fitting manufacturers. You will also learn about the specific types of plastic pipe that you will use in plumbing applications.

2.1.0 Advantages and Disadvantages of Using Plastic Pipe

Most plastic pipe and fittings are **chemically inert,** meaning they do not react with other substances or materials, so they can withstand most chemicals that are used in the home, office, or factory. Strong and durable, plastic pipe and fittings also require little maintenance because they are generally resistant to corrosion and do not pit or scale. In addition, they are easier to handle than metals because of their relative light weight, which can make them more cost-effective to install.

The use of plastic pipe helps to eliminate one of the most common hazards associated with installation—fire. During the installation process, no welding or soldering is involved. Therefore, it is less likely that fires will occur. In the past, when plumbers joined pipe using molten lead, accidents involving spilled lead and tipped-over heating furnaces were a constant concern.

Plastic does have some drawbacks. Temperature changes cause it to expand and contract more than metal. Plastic pipe can also be flammable and give off fumes when it burns. In addition, fumes from solvent chemicals used to join plastic pipes are harmful; plumbers must be familiar with the types of safety hazards each variety of pipe and chemical solvent presents. It is your responsibility to know how to protect yourself on the job site. Always consult the manufacturer's specifications regarding the properties of the pipe you are using.

All plastic pipe is affected by ultraviolet (UV) radiation to some degree. That means that you should store plastic piping under cover during a construction project to protect it from the sun. In general, short-term sun exposure will not harm the piping, but long-term exposure will degrade the plastic material. Always consult the manufacturer's instructions regarding the proper way to select and protect the particular piping you are using.

2.2.0 Properties of Plastic Pipe

Plastics contain polymers, long chains of molecules that manufacturers can mold, cast, or force through a die into desired shapes. Manufacturers can heat and shape resins in the form of pellets, powders, and solutions in various combinations to give the finished product the desired properties.

Each type of plastic pipe has its own unique properties. These properties determine where and how you can install it, how you can join it to other piping, and how you should support it. There are two general categories of plastic: thermoplastic and thermosetting. **Thermoplastic pipe** can be repeatedly softened by heating and hardened by cooling. When softened, thermoplastics can be molded into desired shapes. **Thermosetting pipe** changes chemically when heated, so that once hardened by heat or chemicals, it is hardened permanently.

In addition, plastic pipe can be either rigid or flexible. Rigid pipe is straight and maintains its shape. Flexible pipe bends, so although it requires fewer fittings and joints, it requires more support. Flexible pipe is often referred to as tubing. Manufacturers sell flexible pipe in coils.

The construction of plastic piping can vary as well. Pipes can be **solid wall** or **cellular core wall** construction. Solid wall does not contain trapped air. A pipe with cellular core construction, also called foam core construction, has walls that contain trapped air, so it is lighter weight. Cellular core wall piping consists of three layers of plastic: solid inner and outer layers and a foam middle layer. Because it consists of less material, it is also less expensive than solid wall pipe. Plumbers use solid wall and cellular core wall pipe interchangeably for DWV applications.

NOTE

When choosing plastic piping, always consult the specifications for your project. Always adhere to applicable local codes.

2.3.0 Plastic Pipe Sizing

Plastic pipe is manufactured with various wall thicknesses, commonly called **schedules** or **size dimension ratios (SDRs).** SDR relates wall thickness to the diameter of the pipe, using a set ratio of wall thickness to diameter. As the pipe diameter gets bigger, the wall gets thicker. An SDR number, such as SDR 35, designates this ratio. The ratio between a pipe's **outside diameter (OD),** or the distance between the outer walls of a pipe, and its wall thickness is constant for each pipe size. The larger the SDR number, the thinner the wall versus the diameter. For example, a 2-inch pipe (OD of 2.375 inches) with an SDR of 11 has a wall thickness of 0.216 inches (see *Table 1*). The same pipe with an SDR of 17 has a wall thickness of 0.14 inches. Be aware that pipe can also be measured by its **inside diameter (ID),** the distance between its inner walls.

In an effort to simplify and standardize the use of plastic pipe and fittings, manufacturers designated two SDR ratios as Schedule 40 (thin wall), which is the most commonly used, and Schedule 80 (thick wall). Most plumbing codes designate Schedule 40 as the minimum size for use in or under buildings.

2.4.0 Labeling (Markings)

Plumbers must use plastic pipe and fittings that are labeled and approved for use in plumbing systems. The label on a plastic pipe will include the following:

- Nominal pipe diameter, copper tube size (CTS) or iron pipe size (IPS)
- Schedule or SDR
- Type of material, **PVC (polyvinyl chloride), CPVC (chlorinated polyvinyl chloride),** and so on
- **Pressure rating,** usually expressed as **psi (pounds per square inch)** at a certain temperature
- Relevant standard for the pipe, for example, *American Society for Testing and Materials (ASTM) D-3309*
- Listing body or laboratory, for example, the National Sanitation Foundation
- Name of manufacturer or brand name
- Country of origin

This label ensures that a recognized third party—the listing body or laboratory—has tested the material and approved it for the use intended (see *Figure 1*). For example, the ASTM numbers

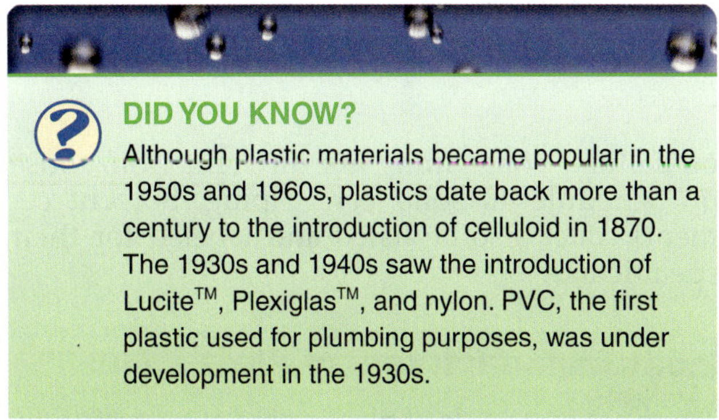

Table 1 Minimum Wall Thickness of 2-Inch Pipe

Plastic Pipe and Fittings Association
Minimum Wall Thicknesses of 2-Inch Pipe
Based on SDR/SIDR

IPS-OD SDR (Outside Diameter = 2.375 inches) SDR (D 3035)	Wall	IPS-ID SIDR (Inside Diameter = 2.067 inches) SIDR (D 2239)	Wall
7	0.339	5.3	0.390
9	0.264	7	0.295
11	0.216	9	0.230
13.5	0.176	11.5	0.180
17	0.140	15	0.138
21	0.113	19	0.109
26	0.091	24	0.089
32.5	0.073	30.5	0.071

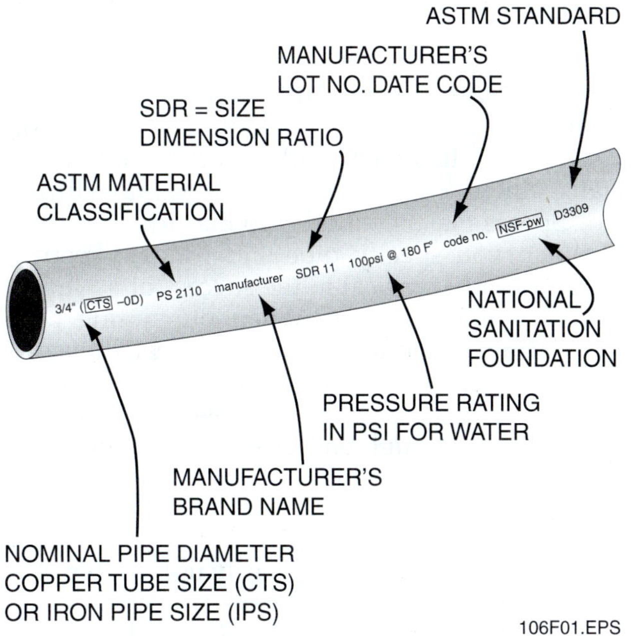

Figure 1 ◆ Typical pipe label.

DID YOU KNOW?

ASTM was organized in 1898 and has grown into one of the largest voluntary standards development systems in the world.

ASTM is a not-for-profit organization that provides a forum for producers, users, consumers, and representatives of government and academia to meet and write standards for materials, products, systems, and services. ASTM publishes standard test methods, specifications, practices, guides, classifications, and terminology. ASTM develops standards that cover metals, paints, plastics, textiles, petroleum, construction, energy, the environment, consumer products, medical services and devices, computerized systems, electronics, and many other areas.

More than 10,000 ASTM standards are published each year in the 72 volumes of the *Annual Book of ASTM Standards.*

Restrictions on Plastics

Consult your local codes for restrictions on the use of plastics in commercial fire-rated buildings and in multistory applications.

Table 2 Types of Plastic Pipe and Their Applications

Pipe	Applications
ABS	DWV sanitary systems, corrosive waste
PVC	DWV sanitary systems, cold water service
CPVC	Hot and cold water distribution
PE	Cold water service, corrosive waste, gas service
PEX	Hot and cold water distribution
PB	Hot and cold water distribution

on cellular core piping are different from those on Schedule 40 solid core piping. Solvent cements must also be listed and labeled for their specific use.

2.5.0 Manufacturers of Plastic Pipe

There are hundreds of plastic pipe manufacturers, and different manufacturers make different pipes for different applications. For example, some specialize in pipes for hot or cold water systems, DWV systems, gas, or radiant heating. Manufacturers of plastic pipe include Charlotte Pipe, Vanguard, Bristol Pipe, J-M Manufacturing, and Cresline Plastic Pipe Co. Manufacturers that specialize in fittings include Nibco Manufacturing Co., Spears Plastics, and Lasco Plastics.

When you specify pipe and fittings, keep in mind that fit is critical to the success of a plumbing system installation. When purchasing pipe and fittings, ensure that they all meet the same standard for fit and materials. One manufacturer's pipe may not be compatible with another's fittings, even though both meet code standards.

2.6.0 Types of Plastic Pipe

The plumbing industry primarily uses six types of plastic pipe, each with different applications and installation requirements (see *Table 2*). This section reviews commonly used types of plastic pipe.

2.6.1 ABS

ABS (acrylonitrile-butadiene-styrene) pipe and fittings (*Figure 2*) are made from a thermoplastic resin. ABS is the standard material for many types of DWV systems. ABS pipe is available in diameters ranging from 1½ inches to 6 inches. ABS pipe wall comes in both solid wall and cellular core wall construction, which are interchangeable in plumbing applications.

Because ABS is light, it is easy to handle and install. A 3-inch-diameter, 10-foot-long section weighs less than 10 pounds. ABS performs well at extreme temperatures because it absorbs heat and cold slowly, an important feature for a system that handles both hot and cold wastes.

ABS is highly resistant to household chemicals. In tests, it showed no effect from such common products as detergent, bleach, and household drain cleaners. Sewage treatment plants use ABS because it stands up to the highly corrosive and abrasive liquids commonly found in such systems.

ABS is strong and long-lasting. In 1959, ABS pipe was used in an experimental residence. Twenty-five years later, an independent research firm dug up and analyzed a section of the pipe and found no evidence of rot, rust, or corrosion. The pipe also withstands earth loads, slab foundations, and high surface loads without collapsing.

106F02.TIF

Figure 2 ◆ ABS pipe and fittings.

2.6.2 PVC

PVC is a rigid pipe with high-impact strength that is manufactured from a thermoplastic material (see *Figure 3*). The material has an indefinite life span under most conditions. Plumbers frequently use PVC in cold water systems. PVC is interchangeable with ABS pipe for DWV systems. It is also used to transport many chemicals because of its chemical-resistant properties. PVC is available in solid wall and cellular core wall construction.

Solid core PVC pipe can be used in high-pressure systems, but only to carry low-temperature water. You must protect it from sunlight because ultraviolet light degrades the thermoplastic materials. It is lightweight, easy to handle and install, has joint flexibility that handles ground movement without leaking, and lasts a long time with no maintenance as long as it is protected from sunlight.

2.6.3 CPVC

CPVC pipe and fittings (*Figure 4*) are made from an engineered vinyl polymer. Plumbers use CPVC in hot and cold water distribution systems. Improvements made to its parent polymer, polyvinyl chloride, added high-temperature performance and improved impact resistance to this material. CPVC is produced in standard CTSs from ½ inch to 2 inches, with a full line of fittings.

CPVC is acceptable under many model codes for indoor use. Its molecular structure practically eliminates condensation in the summer and heat loss in the winter, decreasing the likelihood of costly drip damage to walls or structures. CPVC pipe's smooth, friction-free interior surfaces result in lower pressure loss and higher flow rates and provide less opportunity for bacteria growth. CPVC does not break down in the presence of aggressive, or chemically reactive, water. Like other types of plastic, it does not rust, pit, or scale.

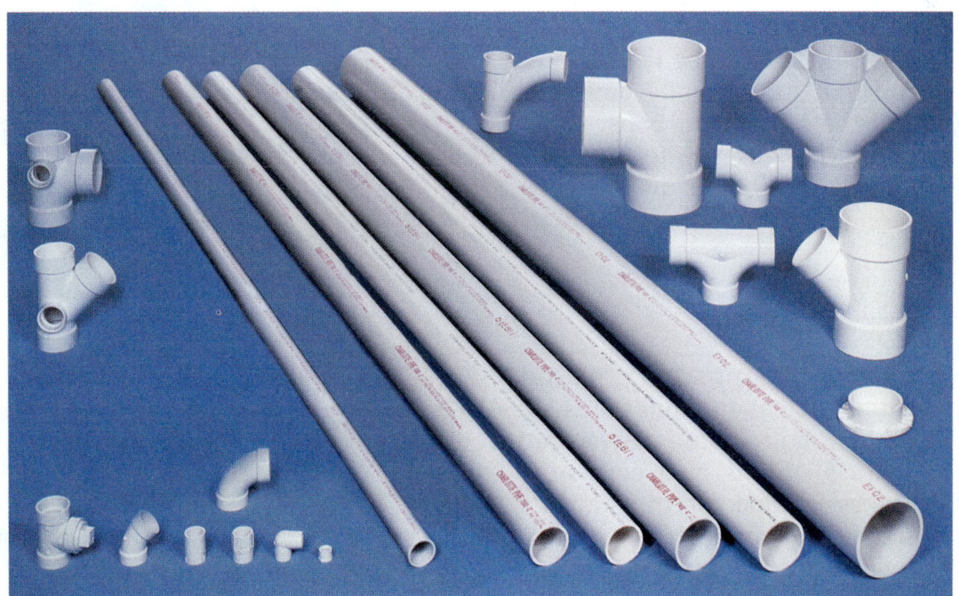

106F03.TIF

Figure 3 ◆ PVC pipe and fittings.

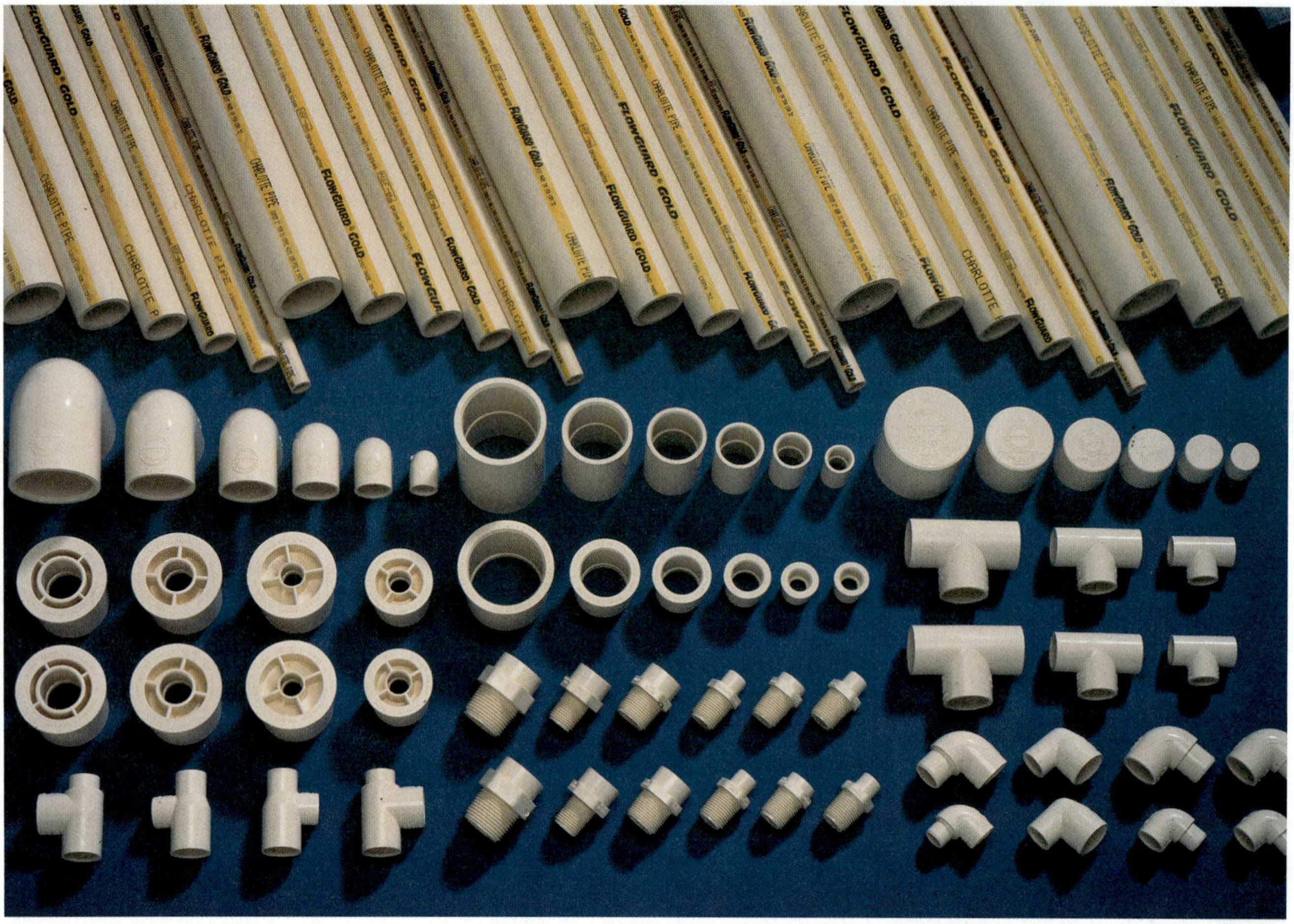

Figure 4 ◆ CPVC pipe and fittings.

106F04.TIF

CPVC is lightweight and easy to install. Recent improvements to CPVC have made the pipe stronger and more durable during installation. The strength of CPVC is a clear advantage for plumbers working in cold-weather states. In laboratory tests down to 20°F, CPVC pipe withstood a **water hammer** drop that substantially damaged copper piping under the same conditions. Water hammer is an extreme change in water pressure within a pipe that can cause a loud, banging sound.

2.6.4 PE

PE (polyethylene) is a thermosetting plastic. Plumbers commonly use PE as tubing because of its strength, flexibility, and chemical-resistant properties (see *Figure 5*). PE is also corrosion-resistant, which makes it ideal for transporting chemical compounds. It will not deteriorate when exposed to ultraviolet light, so you can install it outdoors without a protective coating. PE is used for cold water and underground gas service lines (outside buildings).

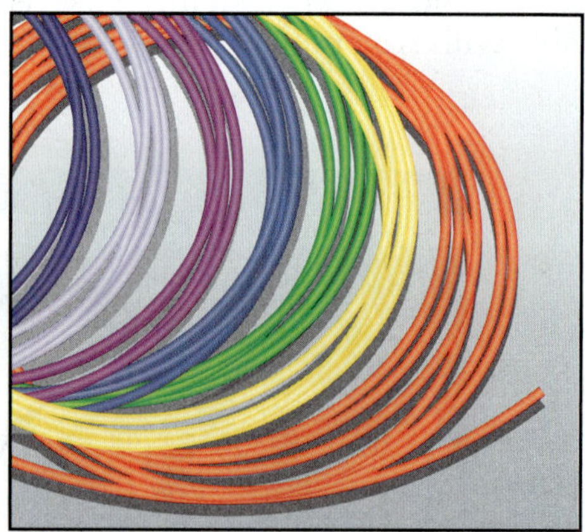

106F05.EPS

Figure 5 ◆ PE tubing.

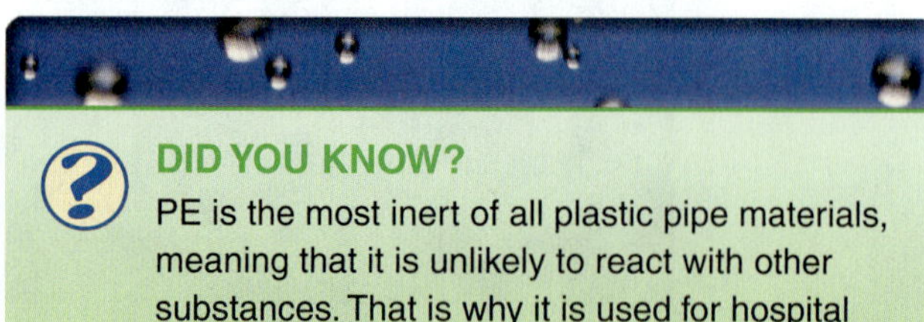

DID YOU KNOW?

PE is the most inert of all plastic pipe materials, meaning that it is unlikely to react with other substances. That is why it is used for hospital tubing and in soda fountains.

2.6.5 PEX

PEX (cross-linked polyethylene) tubing (*Figure 6*) is formed when high-density polyethylene is subjected to heat and high pressure. Because it resists high temperatures, pressure, and chemicals, it is ideal for potable (free from impurities in amounts sufficient to cause disease or harmful effects) hot and cold water systems, **hydronic** radiant floor systems, baseboard and radiator connections, and snow-melt systems. Plumbers commonly use PEX for manifold plumbing distribution systems because of its flexibility.

2.6.6 PB

PB (polybutylene) is a thermoplastic pipe that plumbers used extensively for water supply piping from the late 1970s to the mid-1990s. In many cases, PB piping has become weak and failed without warning. Experts believe this may be because of exposure to oxidants, such as chlorine, in public water systems. For this reason, PB is no longer available, so it is very unlikely that you will install PB piping. Many buildings still contain PB piping, though, and repair fittings are available, so you need to be able to identify it (see *Figure 7*).

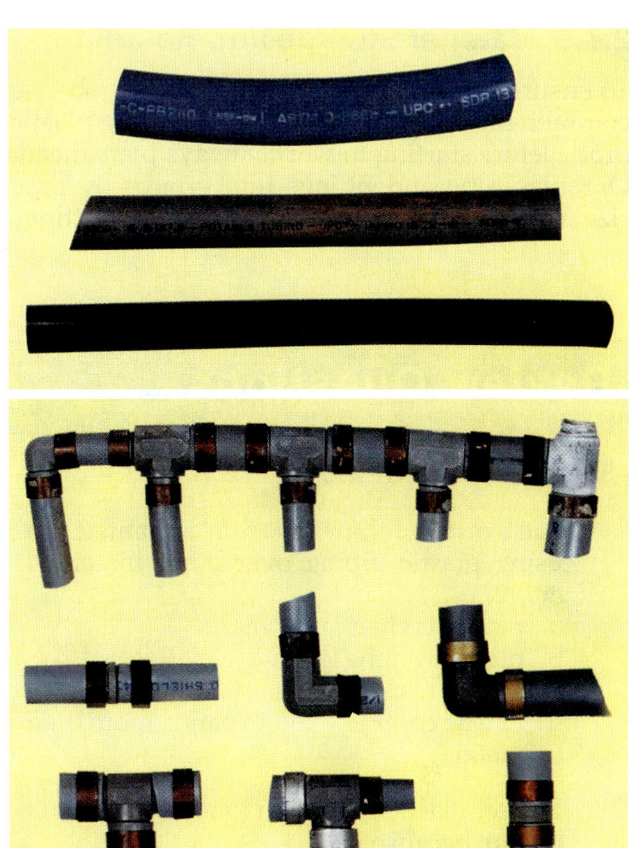

106F07.EPS

Figure 7 ◆ PB piping and fittings.

106F06.TIF

Figure 6 ◆ PEX tubing.

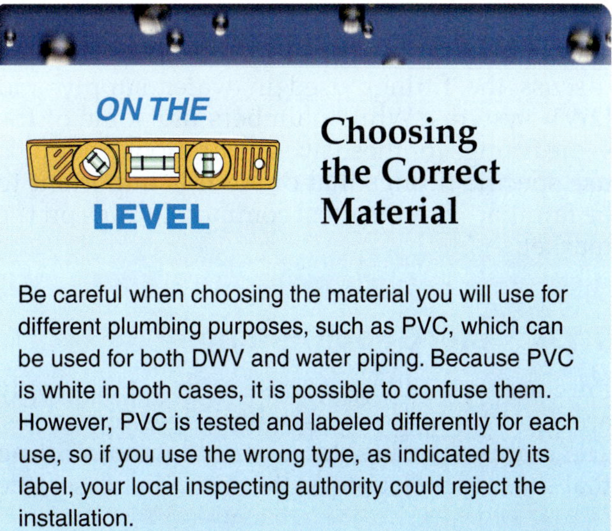

Choosing the Correct Material

Be careful when choosing the material you will use for different plumbing purposes, such as PVC, which can be used for both DWV and water piping. Because PVC is white in both cases, it is possible to confuse them. However, PVC is tested and labeled differently for each use, so if you use the wrong type, as indicated by its label, your local inspecting authority could reject the installation.

2.7.0 Material Storage and Handling

To ensure maximum productivity on the job, use common sense when storing and handling plastic pipe. Before starting to work, always plan ahead. Organize pipe and fittings into groups by pipe size and type. In addition, store pipe and fittings close to where you will be working so they are easy to access. Remember that you must protect many types of plastic pipe from ultraviolet light. When handling pipe and fittings, be careful not to bend or damage them. Pipe that is bent or otherwise damaged not only costs money, but also makes you less productive.

Review Questions

Sections 1.0.0–2.0.0

1. Each of the following is an advantage to using plastic piping over other materials *except* _____.
   a. plastic is chemically inert
   b. plastic weighs less
   c. plastic resists corrosion
   d. plastic contracts and expands more than metal

2. Protect plastic pipe from long-term sun exposure because it _____.
   a. will melt
   b. will darken in color
   c. will degrade
   d. is flammable

3. With SDR pipe, the ratio ensures that the pipe wall gets thicker as the diameter gets smaller.
   a. True
   b. False

4. Each of the following is included on pipe labels *except* _____.
   a. UV exposure rating
   b. pressure rating
   c. name of manufacturer
   d. country of origin

5. _____ is an example of a thermosetting plastic.
   a. ABS
   b. PE
   c. PEX
   d. PVC

3.0.0 ◆ FITTINGS

A pipe's use determines what kinds of fittings and joints are needed. The following sections discuss the fittings used in water supply and DWV systems. While plumbers use some of the same fittings for these two applications, they also use specific fittings that differ. It is important to be familiar with the most common fittings on the market.

3.1.0 Water Supply Fittings

Pressure-type fittings for use with water supply are short turn, or radius, with ledges or shelves. The radius describes the curve or bend of a fitting that changes direction. As the radius increases, the change in flow direction becomes smoother. This is particularly important for waste drain systems, where the smoother flow keeps organic waste from being deposited along the pipeline.

The most common types of water supply fittings include those listed in *Table 3* and shown in *Figure 8*.

3.2.0 DWV Fittings

DWV fittings have smooth interior passages with no ledges. They also have a longer radius that makes directional changes smoother and less likely to collect solids (see *Table 4* and *Figure 9*). When selecting DWV fittings, always consult all applicable codes and manufacturers' installation recommendations.

Table 3 Water Supply Fittings and Their Uses

Fitting	Use
Union	Mechanically connects two pipes, usually at a termination downstream of a service valve
Reducer	Connects pipes of different sizes
Elbow	Changes the direction of rigid pipe by either 90 degrees or 45 degrees
Tee	Provides an opening to connect a branch pipe at 90 degrees to the main pipe run
Coupling	Joins two lengths of the same pipe size when making a straight run
Cap	Plugs water outlets when testing the system or creates an air chamber to eliminate water hammer
Plug	Closes openings in other fittings or seals the end of a pipe
Manifold	Runs several water supply lines from the main supply to different fixtures

Table 4 DWV Fittings and Their Uses

2003 International Plumbing Code®
Table 706.3, Fittings for Change in Direction

Type of Fitting Pattern	Change in Direction: Horizontal to Vertical	Vertical to Horizontal	Horizontal to Horizontal
Sixteenth bend	X	X	X
Eighth bend	X	X	X
Sixth bend	X	X	X
Quarter bend	X	X[a]	X[a]
Short sweep	X	X[a,b]	X[a]
Long sweep	X	X	X
Sanitary tee	X[c]		
Wye	X	X	X
Combination wye and eighth bend	X	X	X

For SI: 1 inch = 25.4 mm

a. The fittings shall only be permitted for a 2-inch or smaller fixture drain.

b. Three inches or larger.

c. For a limitation on double sanitary tees, see *Section 706.3.*

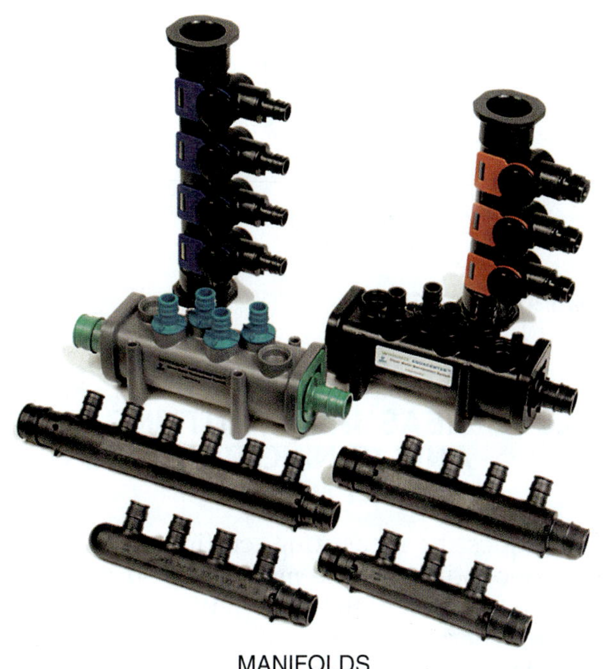

Figure 8 ◆ Water supply fittings.

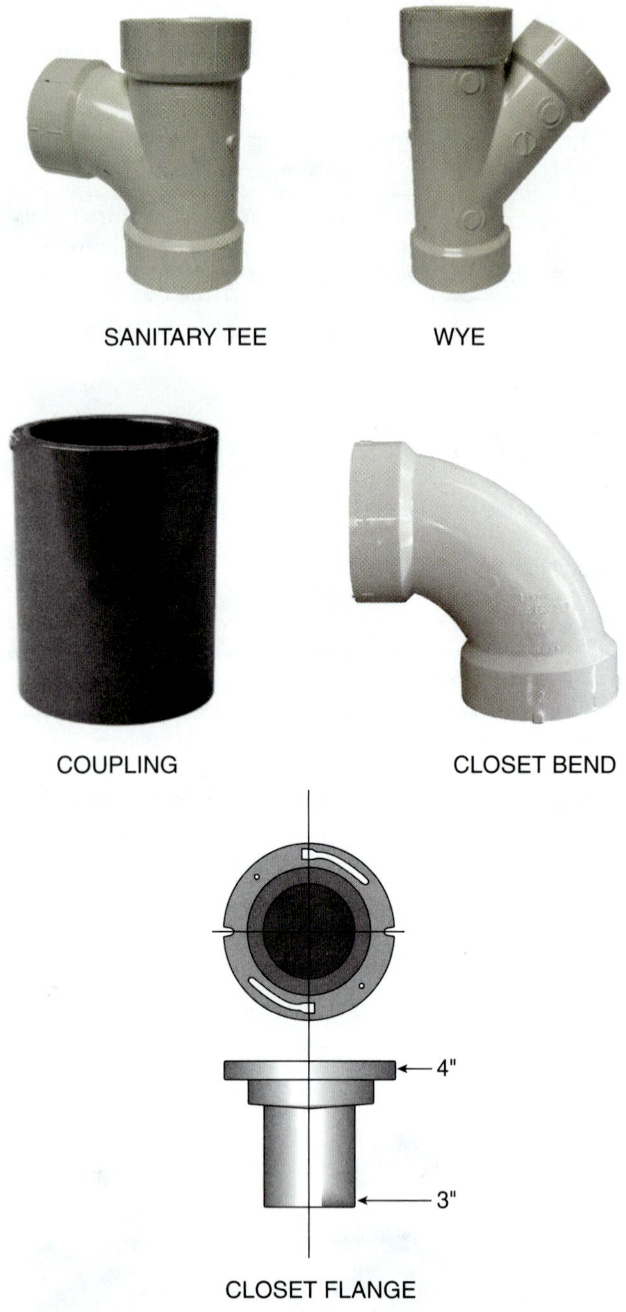

Figure 9 ◆ DWV fittings.

4.0.0 ◆ MEASURING, CUTTING, AND JOINING PLASTIC PIPE AND FITTINGS

Techniques for measuring, cutting, and joining vary depending on the materials you are using and the function of the pipe (DWV or gas, for instance). There are many tools you will need to become familiar with in order to measure, cut, and join materials properly. The following sections explain methods for working with ABS, PVC, CPVC, PEX, and PE pipe used in DWV and water systems.

4.1.0 Measuring

It is important to plan ahead when you are installing ABS and PVC pipe and fittings in DWV systems. These systems have a built-in slope, also called a pitch or fall, that you will learn more about elsewhere in this curriculum, and you must lay them accurately, with pipes cut to exact lengths. You cannot fix mistakes later with heat or hammers.

When you are measuring any pipe, be sure to allow for depth of joints. Take measurements to the full depth of the socket, not with pipe partly inserted into the socket. This is especially important when you use solvent cement to join piping. With this method, you must dry fit the installation, then mark alignment for fittings before you make the joint.

4.2.0 Cutting

Plastic pipe requires a square cut for good joint integrity. Cut tubing as squarely as possible to create the best bonding area within a joint. If you see any indication of damage or cracking at the tubing end, cut off at least 2 inches beyond any visible crack.

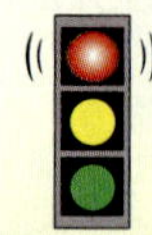

WARNING!

Follow all manufacturer-recommended precautions when cutting or sawing pipe or when using any flame, heat, or power tools. Always wear appropriate PPE.

After cutting, you should ream the pipe. Reaming removes the small burr that results when you cut pipe. Burrs left on piping can cause it to corrode and can prevent proper contact between tube and fitting during assembly. A pipe that is reamed correctly provides a smooth inner surface for better flow. You must also remove burrs on the outside of the pipe to ensure a good fit. Create a slight bevel on the end of the tube to make it easier for the tube to fit into the fitting socket and lessen the chances of pushing solvent cement to the bottom of the joint. Tools used to ream pipe ends include the reaming blade on the tubing cutter, files, pocketknives, and deburring tools (*Figure 10*).

4.2.1 Cutting PVC and ABS Pipe

You can cut PVC and ABS pipe with appropriate pipe cutters, a handsaw, or a power saw

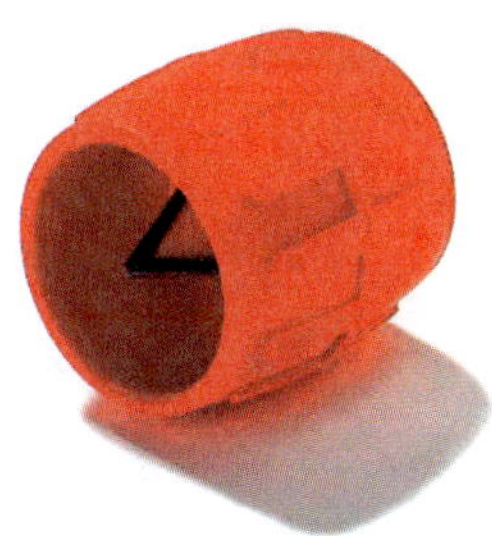

Figure 10 ◆ Deburring tools.

equipped with a carbide tip or abrasive blade. Plastic pipe cutters have one to four cutting wheels that can be rotated around a pipe to cut it. Wheels are available to fit standard cutters. You can also use ratchet shears or lightweight, quick-adjusting cutters designed exclusively for plastic piping (*Figure 11*).

To make sure you get a square cut, use a power saw on large jobs and a miter box on small jobs (see *Figure 12*). Ensure that you use the proper plastic saw for cutting PVC pipe. If these are not

Use Sharp Wheels

Do not use pipe cutters with dull wheels, especially wheels that have been used to cut metal, because they will exert too much pressure, causing larger shoulders and burrs that you will need to remove. Always use blade-type wheels, which actually slice through the plastic.

available, scribe the pipe and cut to the mark. After cutting the pipe, ream it inside and **chamfer** the edge to remove burrs, shoulders, and ragged spots. Chamfering involves beveling the edge of the pipe to a 45-degree angle. Chamfering is good practice because it provides for a more secure joint. When the socket and pipe fit tighter, you reduce the possibility of leakage from the pipes.

CAUTION

Mark carefully! Proper alignment in the final assembly is critical. To ensure alignment, carefully mark the positions of any fittings that will be rolled or otherwise aligned.

4.3.0 Joining

Typically, water supply and DWV fittings are joined using the same installation techniques.

Figure 11 ◆ Cutting tools used for plastic pipe.

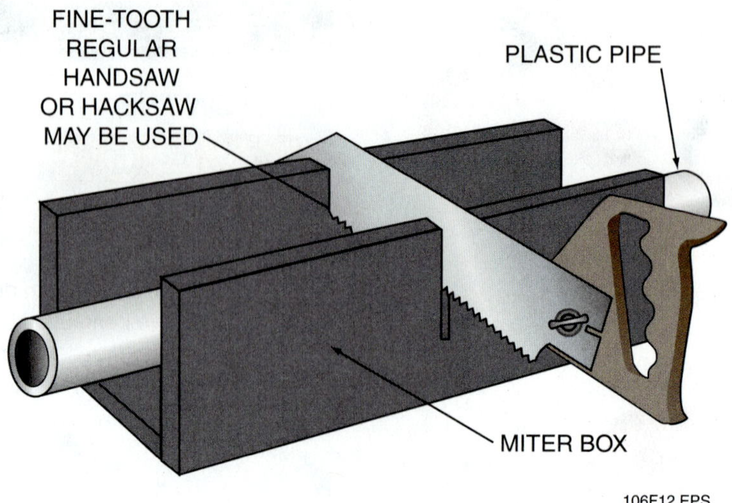

Figure 12 ◆ Using a miter box on small jobs.

Among the installation methods are mechanical fittings, such as **ring-tight gaskets, bell-and-spigot,** threaded joints, heat fusion, and **solvent weld.**

Ring-tight gasket fittings have a rubber O-ring or gasket in the socket. You must bevel the pipe at the leading edge to allow it to pass the gasket that forms the seal (see *Figure 13*). The bell-and-spigot pipe has a bell on one end with an internal **elastomeric** seal. The spigot (straight end) of the next pipe is fitted into the bell end to form a fluid-tight joint.

Plumbers can also connect some plastic pipe to dissimilar pipe with a **transition fitting.** Refer to your local applicable code and manufacturer's specifications to ensure that you are using the proper transition fittings.

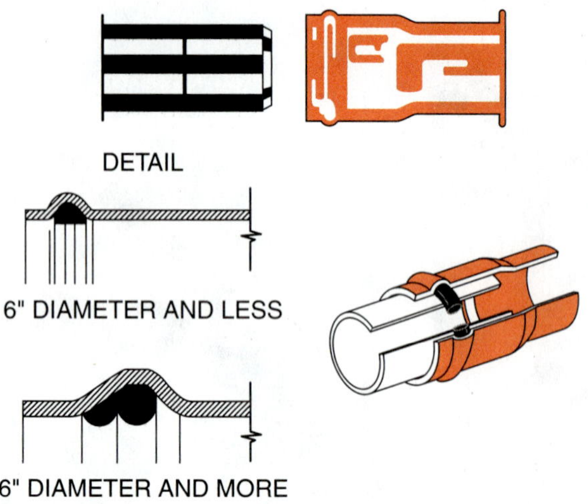

Figure 13 ◆ Ring-tight gasket fitting.

There are many different types of **fusion fittings,** so always consult your local applicable code and the manufacturer's specifications for proper joining and installation methods (see *Figure 14*). You will learn more about fusion fittings and how to create a fusion weld elsewhere in this curriculum.

4.3.1 Solvent Welding (Solvent Cementing)

Plumbers often join ABS, PVC, and CPVC plastic pipe and fittings with solvent cement to form a solvent weld. The process of solvent welding is also referred to as solvent cementing. Solvent-weld fittings have sockets that the pipe fits into (see *Figure 15*). Plumbers apply the solvent cement to the pipe end and the inside of the fitting end, which temporarily softens the joining surfaces. This brief softening period lets you seat the pipe into the socket's **interference fit**—that is, the fit gets tighter as the pipe is pushed into the socket. The softened surfaces then fuse together, and joint strength develops as the solvents evaporate. The resulting joint is stronger than the pipe itself. A major advantage of the solvent-weld process is that it eliminates the need for torches or lead pots, which helps prevent the risk of fire during plumbing installations.

The solvent cements you use will depend on the materials you are solvent welding. Each type of plastic has its own cement, and the cement may vary based on the size of the pipe and fittings. Be careful to choose the right cement for the job. Most codes require that you use primer before solvent cementing pipe. Always refer to all applicable codes and manufacturer's instructions, and ask

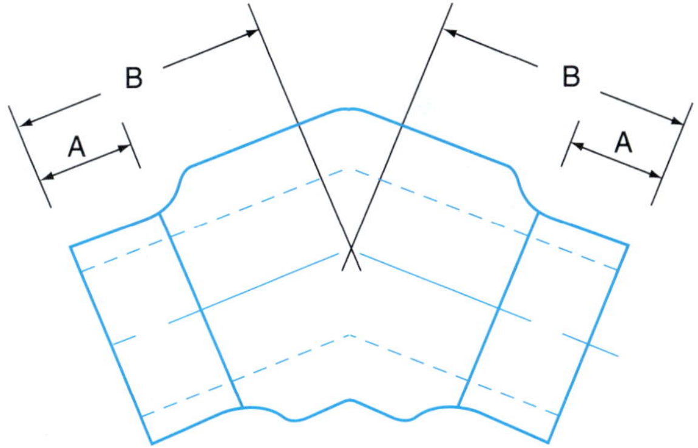

45-DEGREE ELBOW — BUTT FUSION — MOLDED

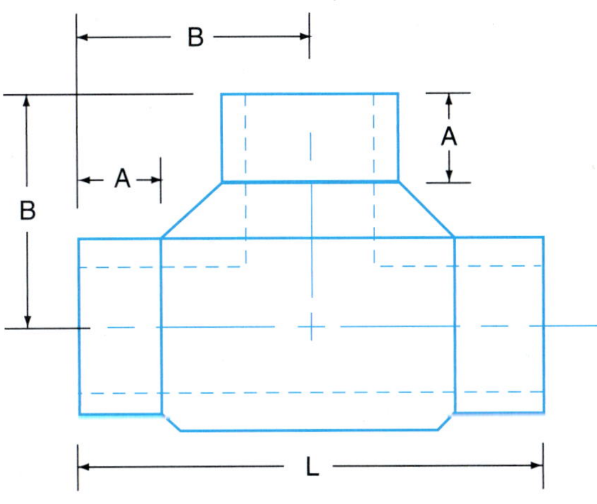

TEE — BUTT FUSION — MOLDED

90-DEGREE ELBOW — BUTT FUSION — MOLDED

106F14.EPS

Figure 14 ◆ Fusion fittings.

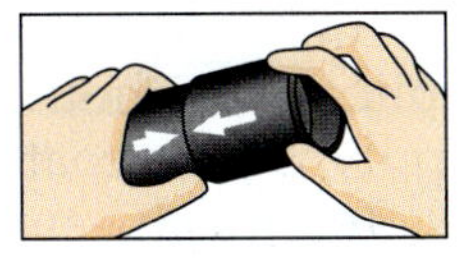

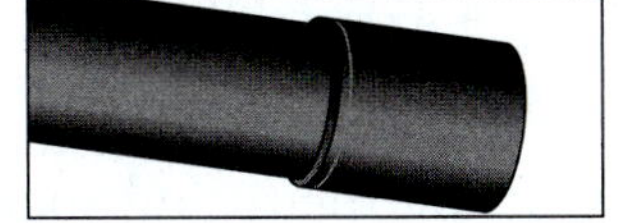

106F15.EPS

Figure 15 ◆ Solvent-welded fitting.

Purple versus Clear Primer

Two types of plastic pipe primer are available: purple and clear. Use purple primer; code inspectors will look specifically for the purple stain as proof that the pipes were primed before the solvent cement was applied. In some cases, with permission, you may use a clear primer on trim-out work.

Joining Pipe with Solvent Cement

Before you cement a joint, make sure the pipe and fitting are free of dust, dirt, water, and oil. Solvent cement acts fast. It is important to move quickly and efficiently when you are joining pipe using solvent welding.

your supervisor if you are unsure which cement to choose.

Because the cement hardens fast, you must move quickly and efficiently when joining pipe. The manufacturer's instructions will show minimum cure times for different size tubes at different temperatures before you can pressure-test the joint. Solvent cement and cure times also depend on relative humidity. Cure time is shorter for drier environments, smaller sizes, and higher temperatures.

WARNING!

Avoid unnecessary skin or eye contact with primers and cements. If contact occurs, wash immediately. Use protective eyewear and gloves during any solvent-weld procedure and consult the MSDS for the cement you are using. An MSDS should be available to anyone on the job site.

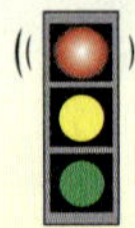

WARNING!

Keep primers and cements away from any sources of ignition, such as heat and open flames, and always apply these products in a well-ventilated area. Refer to the manufacturer's instructions for specific safety precautions.

4.3.2 Joining CPVC or PVC Pipe and Fittings

When you join CPVC or PVC pipe and fittings, use only CPVC or PVC cement or all-purpose cement conforming to *ASTM F-493* standards, or the joint may fail.

Use special care when assembling CPVC and PVC systems in extremely low temperatures (below 40°F) or extremely high temperatures (above 100°F). In extremely hot environments, make sure both surfaces to be joined are still wet with cement when you put them together. If the cement has dried, the two surfaces may not adhere. Adapters are available to connect CTS CPVC pipe to CPVC Schedule 40 and 80 pipe for systems requiring piping diameters larger than 2 inches.

To join CPVC or PVC pipe and fittings with solvent cement, follow these steps (see *Figure 16*). Note that these steps illustrate typical instructions. When joining pipe, always follow the cement manufacturer's instructions.

Step 1 Cut the pipes to the lengths you need, then test fit the pipes and fittings. The pipes should fit tightly against the bottom of the hubbed sockets on the fittings.

Step 2 Clean the surfaces you are joining by lightly scouring the ends of the pipe with emery paper, then wiping with a clean cloth. This ensures that the primer you apply later will soften the plastic before you apply solvent cement.

Step 3 Clean the socket interior and the spigot area of all dirt with a rag or brush.

Step 4 Mark the pipe and fitting with a felt-tipped pen to show the proper position for alignment. Also mark the depth of the fitting sockets on the pipes to make sure that you fit them back in completely when joining.

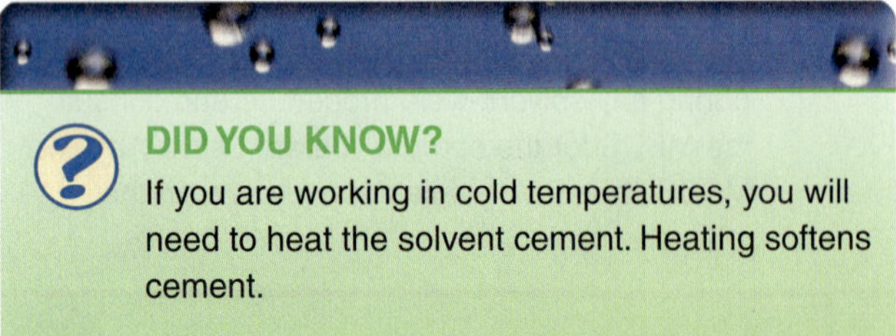

DID YOU KNOW?

If you are working in cold temperatures, you will need to heat the solvent cement. Heating softens cement.

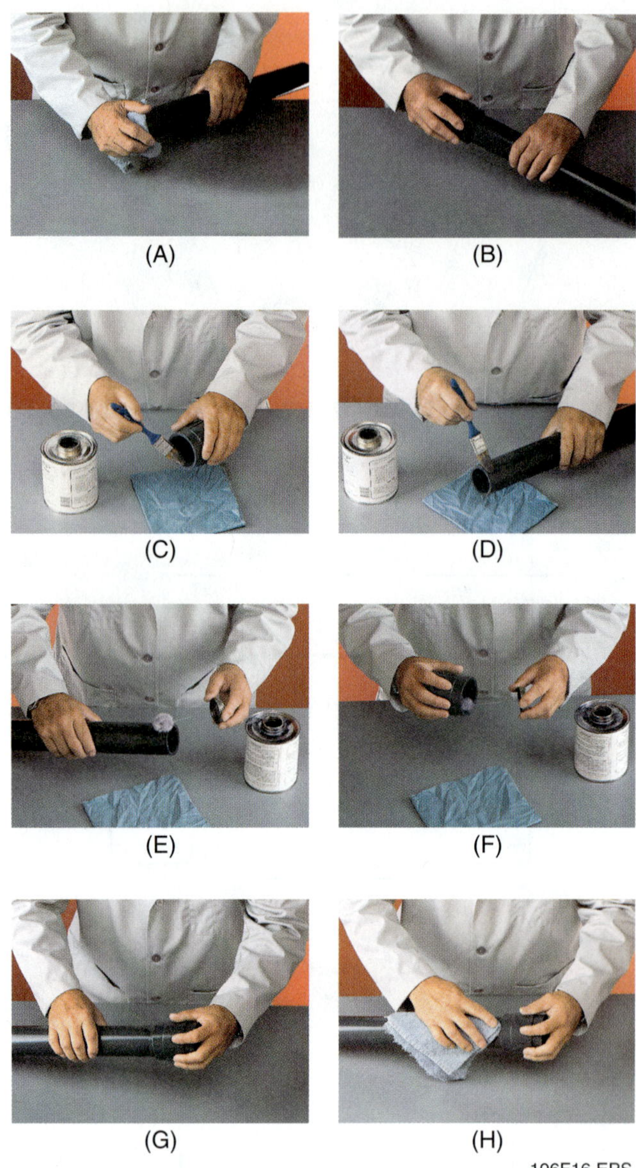

Figure 16 ◆ Joining CPVC or PVC pipe and fittings with solvent cement.

Step 5 Apply the primer to the surfaces you are joining. The primer will soften the plastic in preparation for the solvent-weld process.

Step 6 While the primer is still wet, apply a heavy, even coat of cement to the pipe end. Use the same applicator without additional cement to apply a thin coat inside the fitting socket. Too much cement can clog waterways. Do not allow excess cement to puddle in the fitting and pipe assembly.

Step 7 Immediately insert the tubing into the fitting socket, rotating the tube one-quarter to one-half turn while inserting. This motion ensures an even distribution of cement in the joint. Properly align the fitting.

Step 8 Hold the assembly firmly until the joint sets up. An even bead of cement should be visible around the joint. If this bead does not appear all the way around the socket edge, you may not have applied enough cement. In this case, remake the joint to avoid the possibility of leaks. Wipe excess cement from the tubing and fitting surfaces. Always follow the cement manufacturer's instructions.

4.3.3 Installing PVC Bell-and-Spigot Pipe

PVC bell-and-spigot pipe is generally used outdoors for gravity sewers. These outdoor systems are typically installed to connect with municipal utilities. To install PVC bell-and-spigot pipe, follow these steps (see *Figure 17*):

Step 1 Prepare the inner surface of the bell according to the manufacturer's instructions. Ensure that the groove is free of dirt and other particles.

Step 2 Fold the gasket into a heart shape with the nose or rounded part of its cross section facing out of the mouth of the bell.

Step 3 Insert the gasket into the bell and work it into its groove until it is smooth and free from waves. You may have to snap the gasket, or wet it with clean water or a wet rag, to make sure it goes in place completely. Mark the pipe, creating a memory mark to show the proper position for alignment.

Step 4 After you place the gasket in the bell, thoroughly coat its exposed surface with lubricant. Then apply the lubricant to the entire surface of the spigot end up to the memory mark. Make especially sure that the tapered portion of the spigot is thoroughly coated. When you have finished lubricating, the pipe is ready to be joined.

Step 5 Line up the spigot with the bell and insert it straight into the bell. The spigot end of the pipe has a mark to indicate the proper depth of insertion. This mark will be about flush with the end of the bell when the joint is fully assembled. The memory mark must never be more than ⅜ inch from the end of the bell after assembly.

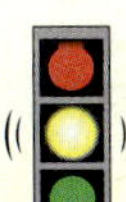

CAUTION

When you are installing a ring-tight PVC gasket, you must assemble the pipe either by hand or by using a bar or block. Never swing or stab the pipe to join it.

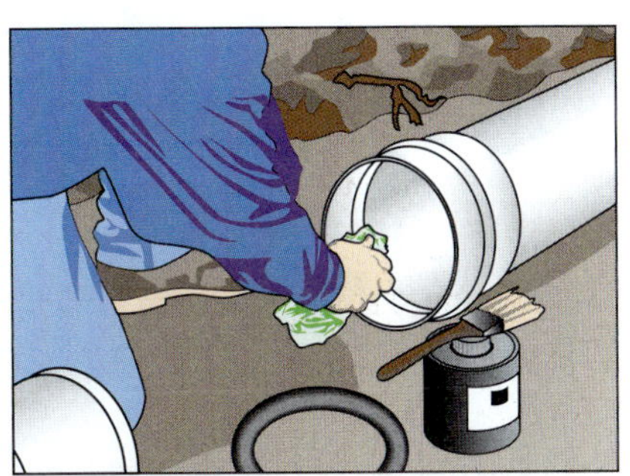

106F17.EPS

Figure 17 ◆ Installing PVC gravity sewer pipe.

4.3.4 Joining PEX Tubing

Because PEX tubing resists high temperature and chemicals, you cannot join it with solvent cement or heat fusion. The most common method is to use an insert and a crimp-ring system. The other tools used to join PEX tubing include the tubing cutter, hand-crimping tool, and go-no-go gauge (see *Figure 18*).

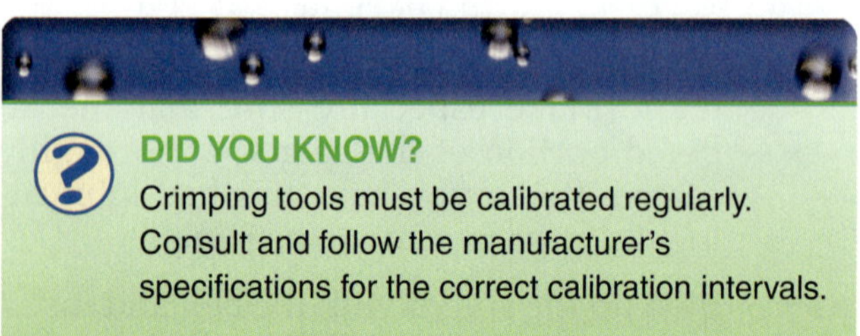

DID YOU KNOW?
Crimping tools must be calibrated regularly. Consult and follow the manufacturer's specifications for the correct calibration intervals.

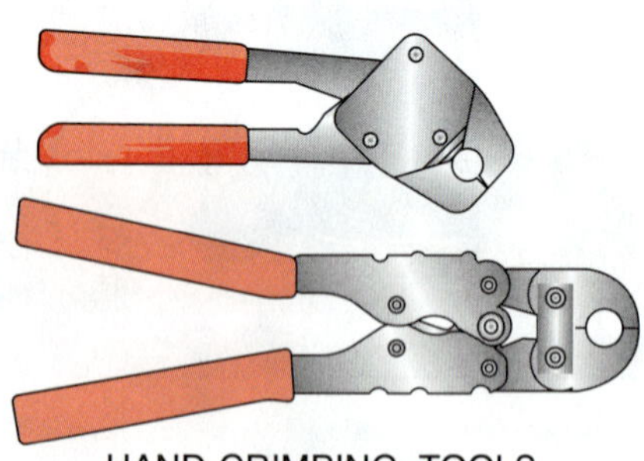

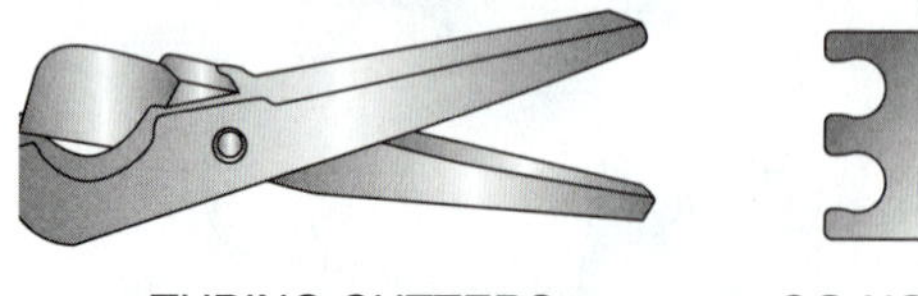

Figure 18 ◆ Tools for joining PEX tubing.

To join PEX tubing, follow these steps (see *Figure 19*):

Step 1 Square cut the tubing perpendicular to the length of the tubing, using a cutter designed for plastic tubing. Remove all excess material or burrs that might affect the fitting connection.

Step 2 Slide a PEX ring over the end of the tube, and extend it no more than 1/16 inch.

Step 3 Open the handles of an expander tool and insert the tool's expansion head into the end of the tubing until it stops. Be sure you

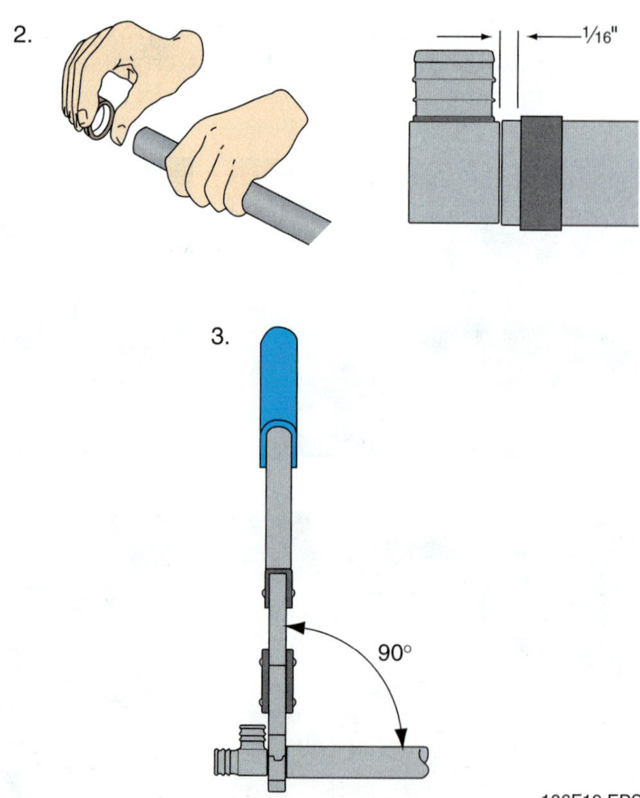

Figure 19 ◆ Joining PEX tubing.

have the correct size expander head in the tool. Place the free handle of the tool against your hip, or place one hand on each handle when necessary.

Step 4 Fully separate the handles and bring them together. Repeat this process until the tubing and ring are snug against the shoulder on the expansion head. Before the final expansion, withdraw the tubing from the tool and rotate the tool one-eighth of a turn. This prevents the tool from forming ridges in the tubing.

Step 5 Expand the tubing one final time. Immediately remove the tool and slide the expanded tubing over the fitting until the tubing reaches the stop on the fitting. Hold the fitting in place for two or three seconds until the tubing shrinks onto the fitting so that it holds the fitting firmly. For a proper connection, the tubing and PEX ring must be snug against the stop of the shoulder fitting. If there is more than 1⁄16 inch between the ring and the fitting, square cut the tubing 2 inches away from the fitting and make another connection using a new PEX ring.

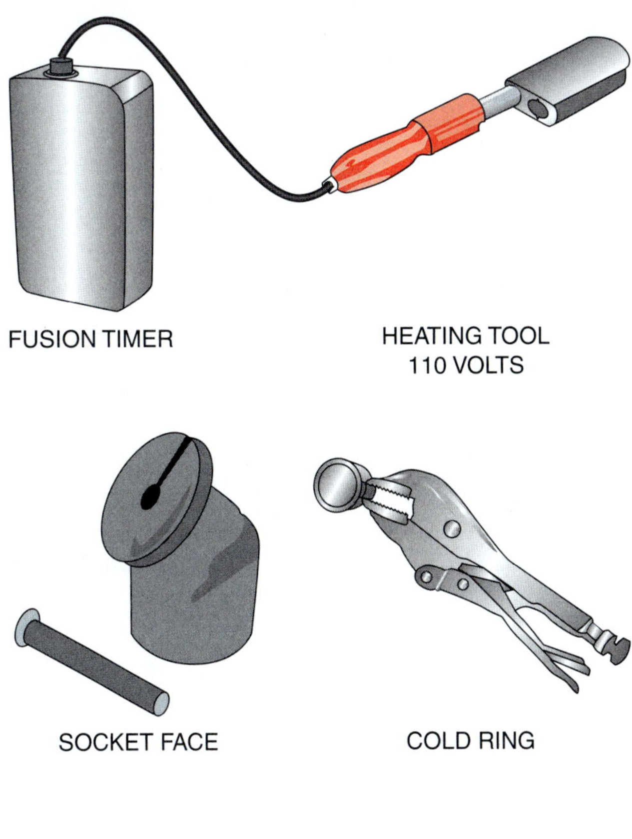

Figure 20 ◆ PE fusion tools.

4.3.5 Joining PE Tubing

Because PE tubing is resistant to chemicals, you must join it by heat fusion or with mechanical joints and clamps. PE joined by fusion is similar to a weld on steel—the materials of the joined parts merge so they are indistinguishable from each other. This process gives the joint the same positive characteristics as the pipe itself.

PE fusion often requires special training and certification. Manufacturers of PE products and joining equipment often provide this training and certification free of charge. New techniques that involve **compression collars** for joining PE are becoming popular because they require less training.

Some of the tools that plumbers use in the fusion process include a temperature indicator stick, a heating tool, a fusion timer, a socket face, and a cold ring (see *Figure 20*). Plumbers use the temperature indicator stick to make sure that piping has reached the required temperature for successful fusion. They use the stick to mark a particular area on the pipe. When the pipe reaches the desired temperature, the mark will melt. During heat fusion, the surface of the socket face comes in direct contact with the pipe or fitting.

One of the most common methods for joining PE tubing is the butt-fusion method. To join PE tubing using the butt-fusion method, follow these steps:

Step 1 Cut the ends of the tubing square with a tubing cutter.

Step 2 Mark the tubing with the proper temperature indicator stick and heat the tubing ends with a heating tool.

Step 3 When the tubing reaches the required temperature, remove the heating tool.

Step 4 Press the tubing ends together to form a tight seal at the joint.

Step 5 Allow the joint to cool before applying force.

Review Questions

Sections 3.0.0–4.0.0

1. A pipe's _____ determines what kinds of fittings and joints are needed.
   a. length
   b. use
   c. inside diameter
   d. outside diameter

Match the following water supply fittings with their intended use.

2. coupling	a. to run several water supply lines from the main supply to different fixtures
3. elbow	b. to join two lengths of the same pipe size when making a straight run
4. manifold	c. to change direction of pipe by 22.5 or 60 degrees
5. reducer	d. to change direction of pipe by 45 or 90 degrees
	e. to connect pipes of different sizes

6. DWV fittings have a _____ that makes directional changes smoother and less likely to collect solids.
   a. smaller circumference
   b. longer radius
   c. shorter diameter
   d. greater perimeter

7. In DWV systems, a sanitary tee can be used to change the direction of a pipe from _____.
   a. horizontal to vertical
   b. vertical to horizontal
   c. horizontal to horizontal
   d. vertical to vertical

8. When you are measuring any pipe that needs to be cut and joined, be sure to calculate the depth of the _____.
   a. weld
   b. fitting
   c. joint
   d. socket

9. If you see any indication of damage or cracking at the tubing end, cut off at least _____ inch(es) beyond any visible crack.
   a. 1
   b. 2
   c. 3
   d. 4

10. The bell-and-spigot pipe has a bell on one end with an internal _____ seal.
    a. pressure
    b. air
    c. water
    d. elastomeric

11. Plumbers can connect some plastic pipe to dissimilar pipe with a _____ fitting.
    a. solvent weld
    b. heat fusion
    c. transition
    d. threaded joint

12. A(n) _____ temporarily softens the pipe and the fitting materials, allowing them to be fitted together and fused.
    a. expander tool
    b. adapter
    c. solvent cement
    d. polymer

13. Because the cement hardens fast, you must move slowly and methodically when joining pipe to avoid making mistakes.
    a. True
    b. False

Review Questions

14. Use special care when assembling CPVC and PVC systems in temperatures _____.
    a. below 40°F
    b. below 40°F and above 100°F
    c. between 40°F and 100°F
    d. above 100°F

15. _____ bell-and-spigot pipe is generally used outdoors for gravity sewers, which are typically installed to connect with municipal utilities.
    a. PVC
    b. PEX
    c. CPVC
    d. PE

16. Because it resists high temperature and chemicals, you cannot join _____ tubing with solvent cement or heat fusion.
    a. ABS
    b. PEX
    c. CPVC
    d. PE

17. Because _____ is resistant to chemicals, you must join it by heat fusion or with mechanical joints and clamps.
    a. ABS
    b. PEX
    c. CPVC
    d. PE

18. Plumbers use a _____ to make sure that piping has reached the required temperature for successful fusion.
    a. mercury thermometer
    b. temperature indicator stick
    c. heating tool
    d. fusion timer

5.0.0 ◆ PIPE SUPPORTS

Plastic pipe can be supported using several different methods. The type of support you use depends on the pipe material, its size, its use, and whether the pipe is installed in a horizontal or vertical position, as well as the system specifications and applicable plumbing codes.

When architects and engineering consultants design plumbing installations, they provide plans and specifications that completely describe the proposed system. Specifications are based on codes or ordinances and must be followed. A specification for pipe hangers, for example, may read, "All piping shall be supported with hangers spaced no more than 10 feet apart (on center). Hangers shall be the malleable iron split-ring type and shall be as manufactured by XYZ Hangers, Inc., or other approved vendor." You should always refer to applicable codes and the manufacturer's installation instructions for specific types and intervals appropriate for your particular installation.

If you are installing pipe in a seismically active area—that is, where earthquakes are a possibility—local codes will require seismic restraints. The purpose of these additional requirements is to ensure that the pipe is securely fastened to the structure in the event of excessive vibration. For example, some codes require hangers and supports to be used at closer intervals than in nonseismically active areas. In addition, they may require that you leave extra spacing for pipes where they meet walls and floors to allow for anticipated movement.

DID YOU KNOW?

Local codes often include spacing requirements for hangers for different types and sizes of plastic pipe. This is an example of a typical local code.

Table S–1

2003 International Plumbing Code®
Table 308.5, Hanger Spacing

Piping Material	Maximum Horizontal Spacing (feet)	Maximum Vertical Spacing (feet)
ABS pipe	4	10[b]
Aluminum tubing	10	15
Brass pipe	10	10
Cast-iron pipe	5[a]	15
Copper or copper-alloy pipe	12	10
Copper or copper-alloy tubing, 1¼-inch diameter and smaller	6	10
Copper or copper-alloy tubing, 1½-inch diameter and larger	10	10
PEX pipe	2.67 (32 in)	10[b]
PEX/aluminum/PEX (PEX-AL-PEX) pipe	2.67 (32 in)	4[b]
CPVC pipe or tubing, 1 inch or smaller	3	10[b]
CPVC pipe or tubing, 1¼ inches or larger	4	10[b]
Steel pipe	12	15
Lead pipe	Continuous	4
PB pipe or tubing	2.67 (32 in)	4
PE/aluminum/PE (PE-AL-PE) pipe	2.67 (32 in)	4[b]
PVC pipe	4	10[b]
Stainless steel drainage systems	10	10[b]

For SI: 1 inch = 25.4 mm, 1 foot = 304.8 mm

a. The maximum horizontal spacing of cast-iron pipe hangers shall be increased to 10 feet where 10-foot lengths of pipe are installed.

b. Mid-story guide for sizes 2 inches and smaller.

5.1.0 Hangers

Plumbers use hangers (*Figure 21*) for horizontal support of pipes and piping. The main purpose of hangers and brackets is to keep the piping in alignment and to prevent it from bending or distorting, but they can also prevent pipes from vibrating. You can attach horizontal hangers to wooden structures with lag screws or large nails. If vibration is a concern, follow the manufacturer's specifications to determine what type of hanger to use for your installation.

Hangers should be strong enough to support the weight of the pipe and its contents and maintain its alignment without sagging. In general, you should support piping at intervals of 4 feet or less, as well as at branches and changes of direction and when using large fittings. Although supports should provide free movement, they must prevent lateral runs from moving up, which could create a reverse grade (also known as slope) on branch piping and back up the system. Avoid hangers that may cut or squeeze pipe and tight clamps or straps that prevent pipe from moving or expanding. Size any holes made for pipe through framing members to allow for free movement. When working with piping in the ground, lay it on a firm bed for its entire length. Always consult applicable codes for specific requirements.

5.2.0 Fasteners

Plumbers often use beam clamps, C-clamps, and suspension clamps to fasten pipes to beams and other metal structures (see *Figure 22*). *Figure 23* shows other examples of horizontal support clamps and brackets. Vertical hangers, also called **pipe riser clamps,** provide vertical support for pipes and tubing (see *Figure 24*). These hangers consist of a friction clamp that you can attach to structural site components to support the vertical load of the pipe. You must use specific fasteners to attach hangers to masonry, concrete, or steel.

CAUTION

The area where the support is to be fastened should be smooth so that the item will have solid footing. Uneven footing might cause the support being fastened to twist, warp, or not tighten properly.

CAUTION

Make sure the fastener is straight after working it around in the hole and installing the washer. The washer centers the fastener and holds it in place until the grout or epoxy hardens. If the grout or epoxy sets and the fastener is not straight, the fastener will be unusable and will have to be removed and the installation repeated.

Use supports for vertical piping at each floor level or as required by the installation design. Mid-story guides can provide greater stability for vertical pipes that run up through the building.

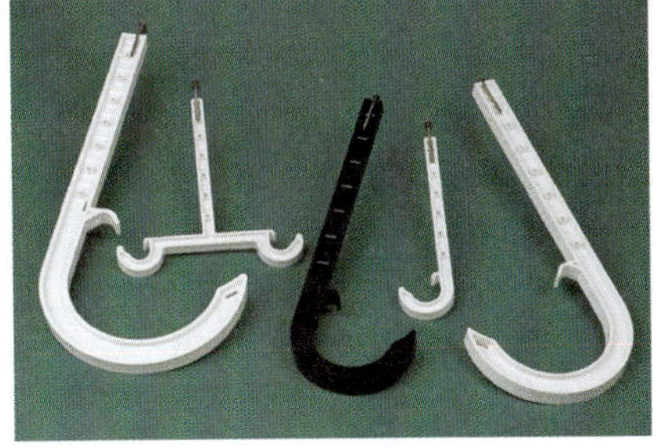

J-HOOKS

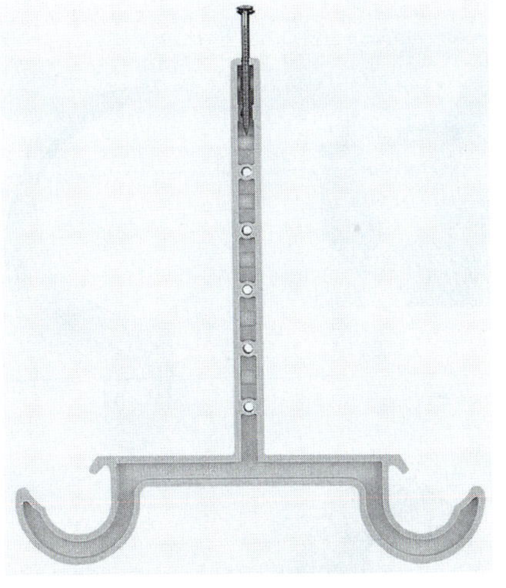

DOUBLE-J HOOKS

DOUBLE-J HOOKS INSTALLED

LOCKING TUBE STRAPS

PIPE HOOKS

106F21.EPS

Figure 21 ◆ Pipe hangers.

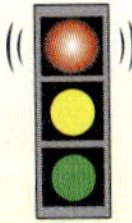

WARNING!

Do not use metal hangers on any plastic pipe. Always use hangers that are made of the same material as the pipe itself. For more information, refer to the *MSS40* hanger standards established by the Manufacturers Standardization Society.

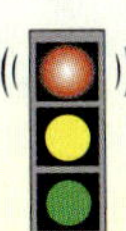

WARNING!

Testing PVC pipe with air can increase the chance of an explosion. Serious injury or death can result if you use too much air, fail to vent trapped air, or fail to depressurize the system.

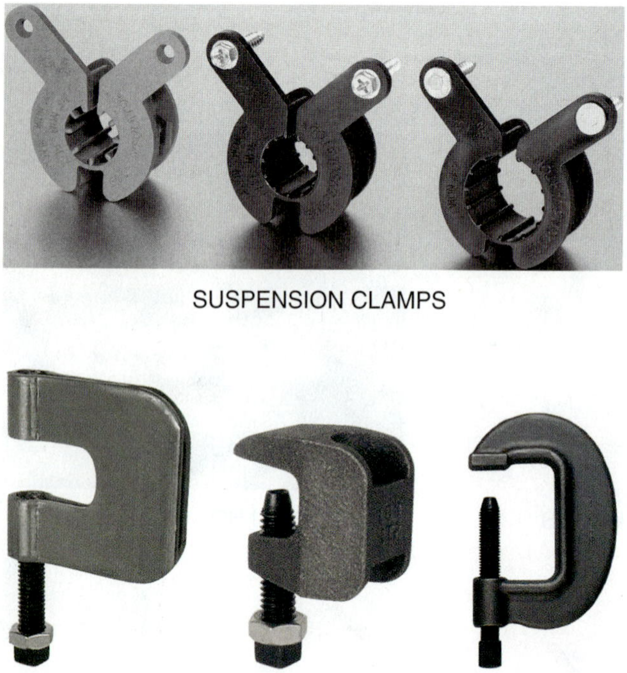
SUSPENSION CLAMPS
I-BEAM CLAMP
BEAM CLAMP
C-CLAMP
106F22.EPS

Figure 22 ◆ Clamps.

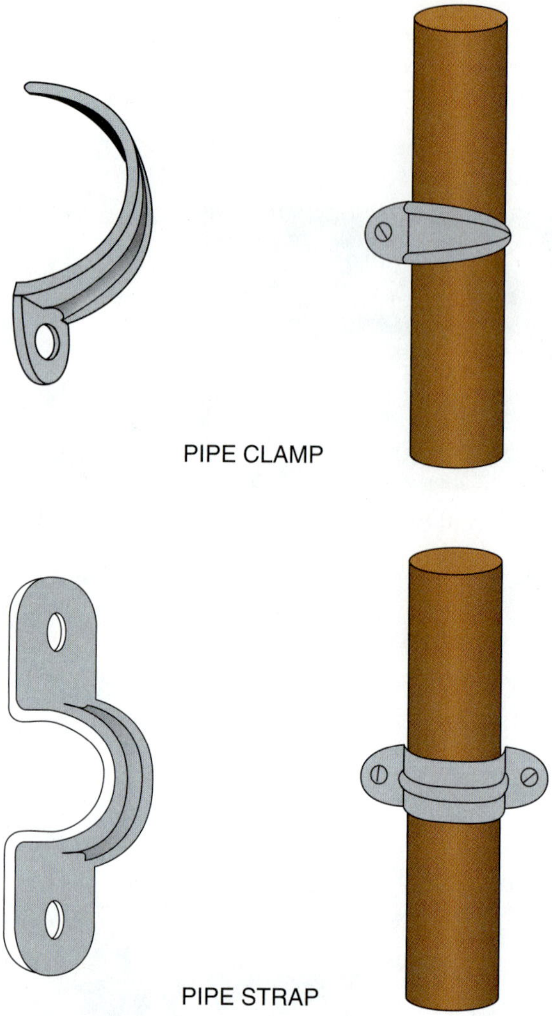
PIPE CLAMP
PIPE STRAP
106F24.EPS

Figure 24 ◆ Vertical hangers.

PIPE SUPPORT
SUPPORT BRACKET
SUPPORT BRACKET INSTALLED
106F23.EPS

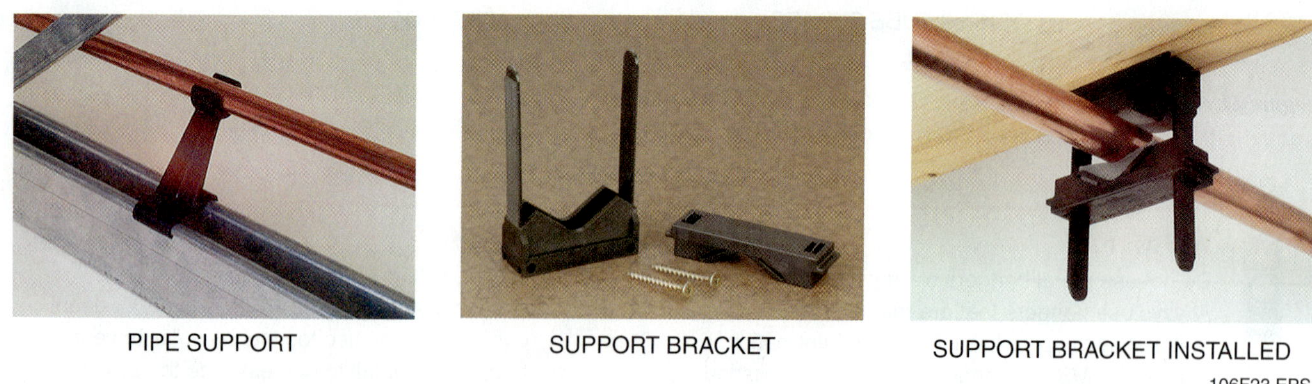

Figure 23 ◆ Horizontal support clamps and brackets.

6.0.0 ◆ PRESSURE TESTING

Once you have completed and cured an installation, you must **hydrostatically pressure-test** the system in accordance with applicable code requirements. This process involves filling the system with water and bleeding all air out from the highest and farthest points in the run. If you find a leak, you must remove and replace it. You can install a new section using couplings. During subfreezing temperatures, you should blow water out of the lines after testing to avoid possible damage to the pipes from freezing.

CAUTION

Never pressure-test a connection until the manufacturer's recommended cure times have been met. After testing a connection, thoroughly flush the system for at least 10 minutes to remove any remaining trace amounts of solvent cement.

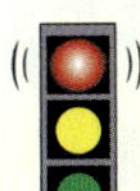

WARNING!

Air testing is not recommended for plastic pipe, although it may be necessary under certain temperatures. Never use pure oxygen for pressure tests. Never test PVC pipe with air, as it will explode. Use air testing for other plastic pipe only when hydrostatic testing is not practical. Always use extreme caution, because air under pressure is explosive. Never test with more than 5 pounds of air pressure. If you perform an air test, you must notify all site personnel of the test, use protective eyewear, and take precautions to prevent impact damage to the system during the test.

Review Questions

Sections 5.0.0–6.0.0

1. The main purpose of pipe hangers is to _____.
   a. ensure adequate pipe vibration
   b. allow pipes to bend easily
   c. maintain horizontal alignment of pipes
   d. maintain vertical alignment of pipes

2. In general, you should support piping at intervals of _____ feet or less, as well as at branches and changes of direction and when using large fittings.
   a. 3
   b. 4
   c. 5
   d. 8

3. _____ are used as vertical support for pipes and tubing.
   a. Beam clamps
   b. C-clamps
   c. Pipe riser clamps
   d. Special fasteners

4. If you find a leak when testing a connection, you must _____.
   a. solvent weld the connection
   b. remove and replace it
   c. perform the hydrostatic test again
   d. air test the connection

5. During subfreezing temperatures, you should blow water out of pipe lines _____ to avoid possible damage to the pipes from freezing.
   a. before installation
   b. during installation
   c. before testing
   d. after testing

Summary

The use of plastics in the plumbing trade has greatly increased the effectiveness and safety of indoor and outdoor plumbing systems. The introduction of different types of plastic pipe and tubing means that today's plumbers must know about new techniques, tools, and applications. Plastic may seem like the answer to all plumbing problems, but exactly which type of plastic you use makes a big difference. You must learn the special techniques for cutting and joining a variety of plastic pipe types. Recognizing the common types of materials used on a job site is a basic skill.

Developing technology affects plumbers now more than ever. Manufacturers are constantly researching and producing new products. To be competitive in the modern plumbing industry, you must keep up with innovations and improvements in installation techniques and materials.

Notes

Trade Terms Quiz

Fill in the blank with the correct trade term that you learned from your study of this module.

1. A(n) _______________ allows threaded plastic pipe to be connected to steel pipe.
2. The _______________ is a measurement of a pipe's wall thickness.
3. Plastic is _______________; it does not react with other substances or materials.
4. A(n) _______________ means that the fit tightens as the pipe is pushed into the socket.
5. The _______________ relates wall thickness to the diameter of the pipe, using a set ratio of wall thickness to diameter.
6. Plumbers commonly use _______________ for manifold plumbing distribution systems because of its flexibility.
7. _______________ changes chemically when heated, so that once hardened by heat or chemicals, it is hardened permanently.
8. After reaming cut pipe, _______________ the edge to remove burrs, shoulders, and ragged spots.
9. The pressure that a pipe can withstand continuously, called the _______________, is required on the pipe's label.
10. When softened, _______________ can be molded into desired shapes.
11. Vertical hangers, also called _______________, provide vertical support for pipes and tubing.
12. Pipe with _______________ construction consists of a single layer of plastic that does not contain trapped air.
13. _______________ pipe will not deteriorate when exposed to ultraviolet light, so you can install it outdoors without a protective coating.
14. Techniques for joining PE that use _______________ are becoming popular because they require less training than heat fusion.
15. A(n) _______________ system heats and cools by circulating water or steam through a closed piping system.
16. After completing an installation, you must _______________ the system by filling the system with water and bleeding all air out from the highest and farthest points in the pipe run.
17. An extreme change in water pressure that can cause a loud, banging sound within a pipe is called _______________.
18. Because it has been known to become weak and fail without warning in previous installations, _______________ pipe is no longer available for installation today.
19. _______________ pipe has a bell on one end with an internal _______________ seal.
20. Pipe can be measured by its _______________, which is the distance between its inner walls.
21. Pipe with _______________ construction is more lightweight and less expensive than solid wall pipe.
22. _______________ fittings have a rubber O-ring or gasket in the socket.
23. The ratio between a pipe's _______________ and its wall thickness is constant for each pipe size.
24. _______________ pipe performs well at extreme temperatures because it absorbs heat and cold slowly.
25. The molecular structure of _______________ pipe practically eliminates condensation in the summer and heat loss in the winter, decreasing the likelihood of costly drip damage to walls or structure.
26. To _______________ joints, plumbers apply solvent cement to the pipe end and the inside of the fitting end, which temporarily softens the joining surfaces.

27. A pipe's pressure rating is measured in _______________.

28. To form a joint using a(n) _______________, you must heat the accessories to the manufacturer's specifications and press them together.

29. _______________ is a rigid pipe with high-impact strength that is manufactured from a thermoplastic material and has an indefinite life span under most conditions.

Trade Terms

ABS (acrylonitrile-butadiene-styrene)
Bell-and-spigot pipe
Cellular core wall
Chamfer
Chemically inert
Compression collar
CPVC (chlorinated polyvinyl chloride)
Elastomeric
Fusion fitting
Hydronic
Hydrostatically pressure-test
Inside diameter (ID)
Interference fit
Outside diameter (OD)
PB (polybutylene)
PE (polyethylene)
PEX (cross-linked polyethylene)
Pipe riser clamp
Pressure rating
psi (pounds per square inch)
PVC (polyvinyl chloride)
Ring-tight gasket fitting
Schedule
Size dimension ratio (SDR)
Solid wall
Solvent weld
Thermoplastic pipe
Thermosetting pipe
Transition fitting
Water hammer

Profile in Success

Bob Muller

John J. Muller Plumbing & Heating Inc.
President/Chief Executive Officer
Matawan, New Jersey

Bob Muller was born in Matawan, New Jersey, and attended Matawan Regional High School. He joined the U.S. Air Force and attended technical school at the Air Force School of Applied Sciences, where he studied aerospace ground equipment, which, along with weapons loading, covered hydraulics, pneumatics, electrical generation and control, and air conditioning. After he returned from Vietnam, Bob left the service and took extension courses in math and science at the University of Southern California, Victor Valley College. He spent three years in college and worked for several companies in the industry but always felt that something was missing. He finally found the missing piece when he returned to his family's business. Today, he is president and CEO of John J. Muller Plumbing & Heating Inc., a third-generation business his grandfather started in 1927.

How did you become interested in this industry?
I always had an interest in mechanical things; I enjoyed the challenge of using my brain and hands and figuring out how systems worked. During high school, I worked as a plumber's helper, digging, threading pipe, running for material, and helping the other mechanics during summers. I had several different jobs before going back to the family business, but there was always something missing. I went back to the family business in 1979 when my grandfather became ill. The tugging feeling was gone once I went back to my family business. I love the industry; it's a great deal of fun.

What path did you take to your current position?
I came from an old German family. There was no free ride, no allowance without work. My family taught me a good work ethic. My father was harder on me than on most of the employees, so I worked my tail off. I had to prove myself. While I was going to school, I worked as a plumber's helper, and then I was an apprentice, journeyman, and master plumber. I worked in the family business until my father retired, and then I was thrown into the business of management.

I would not take the position without proper training, so I attended vocational school. I completed a four-year apprenticeship program and one year of journeyman training. Then I applied for my master plumber's license and hit it on the first try. I now hold certifications for the International Boiler and Radiator Manufacturers' Association and for various manufacturer-sponsored training programs. I renew my licenses every two years. As part of the renewal process, our state requires that every plumber take at least five hours of continuing education prior to renewal. In addition, I continue my education because there are many changes in our industry in material, code issues, and installation practices, which occur almost daily. As the saying goes, "If ya' snooze, ya' lose."

What does it take to be successful in your trade?
If it's one thing, it's that you have to have a good work ethic. And you have to develop this work ethic; it's not God-given. You have to study math and the sciences. With geometry, for example, you must be able to do the numbers, then apply them to something tangible. Reading and retention are also important. You have to be able to read and understand plans and apply what you have read. Written and oral communication is essential. You have to be able to communicate both in writing and speech. Understanding the latest technology, such as computers, is a necessity.

What are some things you do on the job?
It's not just a nine-to-five job. Sometimes it's 24 hours a day, seven days a week. You may try to shut off, but it's hard not to take work with you. So manage your time, stick to deadlines, set short- and long-term goals, and prioritize.

You wear a lot of hats in this business. I came from a military background, but nothing compared to what I am doing now. I took college courses in management and made connections with the Plumbing-Heating-Cooling Contractors—National Association (PHCC—NA). The PHCC—NA provided a lot of help in offering seminars in management, profitability, safety, employee training, and apprenticeship training. But even more important, my affiliation with the PHCC—NA has allowed me a connection with plumbers from all over the country and around the world.

My father was standing by for help, and he is still always ready to help. As a manager, I still go into the field with my employees. I start digging just like everyone else. We have seven employees, and we are adding and growing all the time. We currently affiliate ourselves with the Local Union #9, and we are working with their training programs. Our company offers residential, commercial, and light industrial work.

Now I am also an instructor for continuing education. After I received my master plumbing license, I joined the New Jersey PHCC—NA. I worked through the committees and was president of the association from 1995 to 1996. I am a certified instructor in the apprenticeship program for the Middlesex County Vocational and Technical School. I teach all four years; currently, I'm teaching fourth-year plumbing. I've been doing this for 15 to 20 years now. I've been recognized by the New Jersey General Assembly and the Board of Education of Middlesex County for my contributions to the Apprenticeship Training Program.

What is the most interesting aspect of your profession? What makes your trade stand out from others?
We are in the business of health. We live by the credo, "The plumber protects the health and safety of the nation." Basic sanitation cannot be taken for granted; it is our job to maintain it and improve upon it daily. All you have to do is look at the news and you'll see that, without basic sanitation, thousands of people die daily.

We all have a role. Everyone has a place in the community and we all have a part to make it work. I am not sure I can answer specifically what makes our trade better than another. All the trades are important to the community; plumbing is important, indeed, but what's the point if the carpenter didn't build the structure and the electrician didn't install the lighting? We are part of the overall big picture.

What do you like most—and least—about your job?
I love it all! Bringing young individuals in and making sure that they achieve what they set out to achieve is rewarding. I reached my goals and I enjoy helping others reach theirs. Nothing, absolutely nothing, pleases me more than when a student, former employee, colleague, or friend in the business comes up to me and thanks me for some success they have had in the business.

Probably what I like the least is when I hear of a business or personal failure because somebody didn't apply the education they have received.

What would you say to someone entering the trade today?
You have to be physically strong and mentally focused. Education, education, education. Then, application. Keep yourself in good physical condition. It is an industry where physical and mental conditioning are important. Stress is a killer in this business. A lot of good people have ruined their lives because they turned to drugs and alcohol to relieve the stress. This job requires a positive mental attitude and the ability to communicate with others. Develop a good mental attitude and a good work ethic. If you don't know what that is, just find a good plumber!

What can an apprentice expect to earn in his/her first years on the job in your area? What can he/she expect to earn after 10-plus years in the industry?
New Jersey has higher wages but also high overhead and living expenses. The base wage for plumber pipe fitter Local #9 journeymen is $22.08. This includes the surety fund, welfare fund, and almost 100 percent pension, as well as two weeks of vacation time. The

overtime rate is time-and-a-half for weekends, and holidays get double time. There are a lot of opportunities for people to work the weekends and get those rates if they can withstand the work.

The apprentice salary is based on a percentage of the journeyman rate: 50 percent of the journeyman wage, plus full benefit package, pension fund, and so on, which is managed by the union. Apprentices make 60 percent of the wage in their second year, 70 percent in the third, and 80 percent in the fourth. In their fifth year they begin to earn the full journeyman wage. It's a nonbinding contract between employer and apprentice. Under the bureau of apprenticeships, the formula is the same, but the rate is negotiated.

Trade Terms Introduced in This Module

ABS (acrylonitrile-butadiene-styrene): Plastic pipe and fittings used extensively in drain, waste, and vent (DWV) systems.

Bell-and-spigot pipe: Pipe that has a bell, or enlargement, also called a hub, at one end of the pipe and a spigot, or smooth end, at the other end. The bell and spigot of two different pipes slide together to form a joint. Also called hub-and-spigot pipe.

Cellular core wall: Plastic pipe wall that is low-density, lightweight plastic containing entrained (trapped) air.

Chamfer: To bevel the edge of construction material to a 45-degree angle.

Chemically inert: Does not react with other chemicals.

Compression collar: A piece of hardware that uses compression force to connect sections of polyethylene piping.

CPVC (chlorinated polyvinyl chloride): Plastic pipe and fittings used extensively in hot and cold water distribution systems.

Elastomeric: Rubberized. Made of an elastic substance such as a polyvinyl elastomer.

Fusion fitting: A fitting with a butt that has the same outside diameter and inside diameter as the pipe. It is usually joined to a pipe by heat.

Hydronic: A system that heats and cools by circulating water or steam through a closed piping system.

Hydrostatically pressure-test: To fill a pipe with water and bleed all air out from the highest and farthest points in the run.

Inside diameter (ID): The distance between the inner walls of a pipe; the standard measure of piping used in heating and plumbing.

Interference fit: Fit that tightens as the pipe is pushed into the socket.

Outside diameter (OD): The distance between the outer walls of a pipe.

PB (polybutylene): Plastic piping that was formerly used for plumbing pipe; it is no longer used but is still found in some residences.

PE (polyethylene): Flexible plastic pipe, tubing, and fittings, usually used for water distribution, that do not deteriorate when exposed to sunlight.

PEX (cross-linked polyethylene): Tubing and fittings made with heat and high pressure that resist high temperatures, pressure, and chemicals.

Pipe riser clamp: A vertical extension of pipe hanger that provides support for pipe and tubing.

Pressure rating: The maximum pressure at which a component or system may be operated continuously.

psi (pounds per square inch): A measurement of pressure.

PVC (polyvinyl chloride): Plastic pipe and fittings used for cold water distribution and for industrial water and chemicals, as well as for drain, waste, and vent (DWV) systems.

Ring-tight gasket fitting: Fitting with a rubber O-ring or gasket in the socket.

Schedule: A measurement that describes pipe wall thickness.

Size dimension ratio (SDR): A measurement of pipe size that relates pipe wall thickness to pipe diameter.

Solid wall: Plastic pipe wall that does not contain trapped air.

Solvent weld: A joint created by joining two pipes using solvent cement that softens the material's surface.

Thermoplastic pipe: Pipe that can be repeatedly softened by heating and hardened by cooling. When softened, thermoplastic pipe can be molded into desired shapes.

Thermosetting pipe: Pipe that changes chemically when heated, so that once hardened by heat or chemicals, it is hardened permanently.

Transition fitting: A special fitting used to connect plastic pipe to pipe of a dissimilar material, as specified by applicable code.

Water hammer: An extreme change in water pressure within a pipe that can cause a loud, banging sound and even damage the system.

Additional Resources and References

ADDITIONAL RESOURCES

This module is intended to present thorough resources for task training. The following reference works are suggested for further study. These are optional materials for continued education rather than for task training.

Basic Plumbing with Illustrations, Revised, 1994. Howard C. Massey. Carlsbad, CA: Craftsman Book Company.

Pipefitter Level 2, 1998. NCCER. Upper Saddle River, NJ: Prentice Hall.

Plumber's Handbook, Revised Edition, 1998. Howard C. Massey. Carlsbad, CA: Craftsman Book Company.

Plumbing: Design and Installation, Second Edition, 2002. L. V. Ripka. Homewood, IL: American Technical Publishers.

REFERENCES

J&L Supply website, www.jlsupply.com, *J&L Supply Instruments,* www.jlsupply.com/testing/testinginstr.htm#TEMPILSTIK_TEMPERATURE_INDICATORS, reviewed July 2003.

National Standard Plumbing Code, 2003. Falls Church, VA: Plumbing-Heating-Cooling Contractors—National Association.

Plastic Pipe and Fittings Association. Table 3-c Minimum Wall Thicknesses of 2-Inch Pipe Based on SDR/SIDR (page 20).

Plumbing Apprentice Training Manual for Plastic Piping Systems, 2002. Glen Ellyn, IL: Plastic Pipe and Fittings Association and Plastic Piping Educational Foundation.

Ridgid Tool Company website, www.ridgid.com/Tools/Reamer-and-Deburring-Tools/, reviewed March 8, 2004.

2003 International Plumbing Code. Table 706.3 Fittings for Change in Direction. Falls Church, VA: International Code Council.

2003 International Plumbing Code. Table 308.5 Hanger Spacing. Falls Church, VA: International Code Council.

Wirsbo Systems website, www.wirsbo.com, *Comfort Heating Frequently Asked Questions,* www.wirsbo.com/main.php?pm=1&mm=1&sm=4&pc=homeowner/ho_mm1sm4.php, reviewed July 2003.

NCCER CURRICULA — USER UPDATE

NCCER makes every effort to keep its textbooks up-to-date and free of technical errors. We appreciate your help in this process. If you find an error, a typographical mistake, or an inaccuracy in NCCER's curricula, please fill out this form (or a photocopy), or complete the online form at **www.nccer.org/olf**. Be sure to include the exact module ID number, page number, a detailed description, and your recommended correction. Your input will be brought to the attention of the Authoring Team. Thank you for your assistance.

Instructors – If you have an idea for improving this textbook, or have found that additional materials were necessary to teach this module effectively, please let us know so that we may present your suggestions to the Authoring Team.

NCCER Product Development and Revision

13614 Progress Blvd., Alachua, FL 32615

Email: curriculum@nccer.org
Online: www.nccer.org/olf

❑ Trainee Guide ❑ AIG ❑ Exam ❑ PowerPoints Other ______________________________

Craft / Level: Copyright Date:

Module ID Number / Title:

Section Number(s):

Description:

____

____

____

____

Recommended Correction:

____

____

____

____

Your Name:

Address:

____

Email: Phone:

Copper Pipe and Fittings
02107-05

02107-05
Copper Pipe and Fittings

Topics to be presented in this module include:

Overview

Copper pipe and fittings are used in a variety of plumbing applications. Copper is expensive, but because it is reliable and easy to use, it is also cost-effective. Copper pipe comes in a series of sizes and different wall thicknesses. Annealed copper is soft and flexible, while drawn copper is rigid and hard. Plumbers must use copper pipe and fittings that have been labeled and approved by the manufacturer.

Based on a pipe's application, plumbers select appropriate fittings, which change the direction or size of a pipe run, and valves, which control flow. Plumbers use specific methods to measure, cut, bend, join, and groove copper pipe. These methods depend on the type of pipe and its function, and the various special tools that are needed. Copper pipe can be joined by soldering a sweat joint, creating a compression joint, or making a flare joint. Plumbers must learn to perform each method safely and use all tools properly.

Hangers and supports secure horizontal and vertical runs of copper pipe and are used to prevent leaks and damage. Plumbers select the type of hanger depending on job specifications. Applicable codes dictate appropriate pipe attachments and connectors for each installation, as well as whether the pipe needs to be insulated. After completing an installation, plumbers pressure-test the system for leaks and secure connections.

Focus Statement

The goal of the plumber is to protect the health, safety, and comfort of the nation job by job.

Code Note

Codes vary among jurisdictions. Because of the variations in code, consult the applicable code whenever regulations are in question. Referencing an incorrect set of codes can cause as much trouble as failing to reference codes altogether. Obtain, review, and familiarize yourself with your local adopted code.

Objectives

When you have completed this module, you will be able to do the following:

1. Identify the types of materials and schedules used with copper piping.
2. Identify the material properties, storage, and handling requirements of copper piping.
3. Identify the types of fittings and valves used with copper piping.
4. Identify the techniques used in hanging and supporting copper piping.
5. Properly measure, ream, cut, and join copper piping.
6. Identify the hazards and safety precautions associated with copper piping.

Key Trade Terms

ACR
ACR tubing
Annealing
Bullhead tee
Capillary action
Clevis
Compression joint
Drawn copper
Drop forged
Ferrule
Flare joint
Formability
Head
Insulation
Nominal size
Pressure drop
Roll grooved
Sizing tool
Sweat joint

Required Trainee Materials

1. Appropriate personal protective equipment
2. Pencils and paper
3. Copy of local adopted code

Prerequisites

Before you begin this module, it is recommended that you successfully complete *Core Curriculum; Plumbing Level One,* Modules 02101-05 through 02105-05.

This course map shows all of the modules in the first level of the *Plumbing* curriculum. The suggested training order begins at the bottom and proceeds up. Skill levels increase as you advance on the course map. The local Training Program Sponsor may adjust the training order.

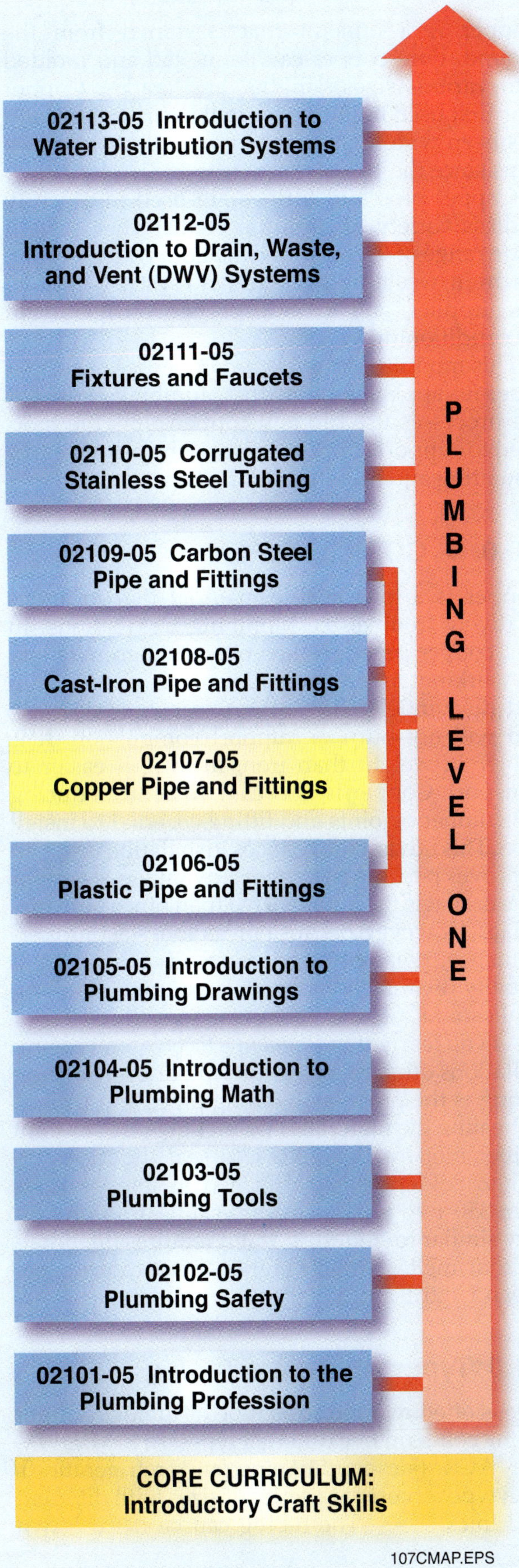

107CMAP.EPS

1.0.0 ◆ INTRODUCTION

Copper is a mineral that is mined from the ground. Pure copper can be melted and molded into various sizes, lengths, and angles. Copper was first used in plumbing in the early 1800s, and has been widely used in hot and cold water systems since the 1930s.

Copper pipe and fittings are used in a nearly endless variety of piping systems. They have a wide range of uses: for hot and cold water supply; for drain, waste, and vent (DWV) systems; for fuel gas supplies; and for transporting refrigerant in air conditioning systems.

This module discusses the properties of copper pipe, its applications in the plumbing industry, the processes used to join copper pipe, the tools used to support copper pipe, and the processes for insulating and pressure testing copper pipe.

2.0.0 ◆ COPPER PIPE

Copper is a long-lasting material that provides relatively trouble-free plumbing installations. In fact, copper pipe installed 60 years ago may still be working successfully. Copper is easy to join and dismantle, resists corrosion extremely well, and will not burn or support combustion. It is lighter in weight than iron, making it easier to transport. Copper is also easy to bend, reducing the number of joints and fittings needed to install pipe. This advantage reduces installation cost and improves performance. A disadvantage of copper is its cost; it is more expensive than other pipe materials. However, because of its long-lasting performance, reliability, and other advantages, it is actually extremely cost-effective for plumbing applications.

As you learned in *Plumbing Tools,* plumbing installations often require soldering or brazing. Soldering is the process of joining pipes and fittings by heating the materials and melting into the joint a filler metal called solder to seal the pipes together. Soldering is performed at temperatures from 350°F to 550°F. Brazing is a heating process very similar to soldering, but it requires filler metals that melt at much higher temperatures (between 1,100°F and 1,500°F) than solders.

2.1.0 Types of Copper Tubing

Pipe is often referred to as tube or tubing. Copper tubing comes in five different types: K, L, M, DWV, and **ACR** (air conditioning and refrigeration). Each type represents a series of sizes with different wall thicknesses. The tubing can be hard **drawn copper,** or soft and flexible annealed copper. Drawn copper is produced by pulling the tube through dies to reduce its diameter. Drawing hardens the copper and makes it very rigid. Annealed copper is produced by the **annealing** process, in which the material is heated and slowly cooled to relieve internal stress. This process reduces brittleness and increases toughness. All types of tube are available in a hard form. Hard forms come in 12- to 20-foot lengths and in diameters ranging from ¼ to 12 inches. Drawn copper tubing is widely used in commercial refrigeration and air conditioning systems. Types K, L, and ACR also are available in soft coils in 40- to 100-foot lengths in diameters ranging from ⅛ to 2 inches.

Manufacturers fill lengths of drawn tubing with nitrogen and plug them at each end to maintain a clean, moisture-free internal condition. This tubing is intended for use with formed fittings to make the necessary bends or changes in direction. It is more self-supporting than annealed copper tubing, such as tubing for air conditioning and refrigeration systems (**ACR tubing**); therefore, it needs fewer supports.

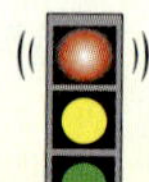

WARNING!

Always check for electrical grounding when working with copper pipe. Copper water supply lines are often used as the electrical grounding. Always shut off electrical power if you break the grounding. When you shut off electrical power, always use approved lockout/tagout procedures to avoid electrical shock. Check with your immediate supervisor before proceeding whenever electrical power is applied.

2.2.0 Copper Pipe Sizing

Pipe sizing varies depending on the type of copper pipe. Types K, L, M, and DWV use nominal, or standard, sizing. This means that the outside diameter (OD) is always ⅛ inch larger than the **nominal size** (see *Figure 1*). The nominal size is the approximate measurement in inches of the inside diameter (ID) for most copper pipes. For example, the OD of a ¾-inch nominal type M pipe measures ⅞ inch. This allows the same fittings to be used with the different wall thicknesses and IDs of different types of copper pipe. The ID is usually close to the stated size. A pipe with a ½-inch ID has approximately ½ inch between the inside walls of the pipe.

Type ACR pipe uses actual OD sizing. A ⅞-inch OD ACR copper pipe is actually the same OD as

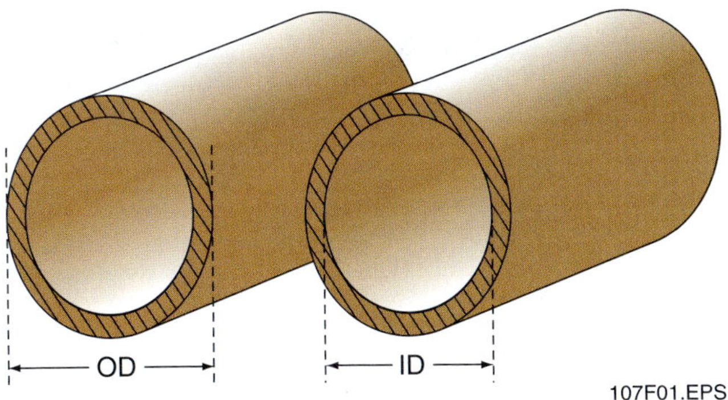

Figure 1 ◆ Copper pipe sizing.

DID YOU KNOW?

Copper is one of the most plentiful metals. It has been in use for thousands of years. In fact, a piece of copper pipe used by the Egyptians more than 5,000 years ago is still in good condition. When the first Europeans arrived in the New World, they found the Native Americans using copper for jewelry and decoration. Much of this copper was from the region around Lake Superior, which separates northern Michigan and Canada.

One of the first copper mines in North America was established in 1664 in Massachusetts. During the mid-1800s, new deposits of copper were found in Michigan. Later, when miners went west in search of gold, they uncovered some of the richest veins of copper in the United States.

¾-inch K, L, or M copper pipe. This means that you can use the same fittings with these pipes.

2.3.0 Labeling (Markings)

Plumbers must use copper pipe and fittings that are labeled and approved for use in plumbing systems. The labels on copper pipe contain very important information. Manufacturers must permanently mark Types K, L, M, and DWV to show the tube type, the name or trademark of the manufacturer, and the country of origin. Manufacturers also print this information on the hard tubes in a color that identifies the tube type. Manufacturers label most soft copper by stamping the information into the product. *Table 1* shows five types of copper tubing and the corresponding color code for labeling.

Table 1 Copper Tubing Color Codes

Type	Color Code	Application
K	Green	
L	Blue	
M	Red	
DWV	Yellow	
ACR	Blue	

Exercise: Using *Table 1*, fill in the blanks to indicate the applications that correspond to each type of copper tubing based on applicable codes in your area.

2.4.0 Copper Pipe Applications

Consider strength and **formability** (ease of bending) when selecting copper pipe. Consult your local plumbing and mechanical codes when selecting pipe; these codes govern which types of pipe you can use in particular applications.

2.5.0 Material Storage and Handling

To ensure maximum productivity on the job, it is important to use common sense when storing and handling copper pipe. Before starting to work, always plan ahead. Always store copper pipe in a secure area. Store pipe and fittings near where you will be working so they are easy to access. Organize pipe and fittings into groups by pipe size and type. When handling pipe and fittings, be careful not to bend or damage them. Pipe that is bent or otherwise damaged not only costs money, but also makes you less productive. Take extra care when working in cold temperatures. Cold copper pipe can stick to bare hands.

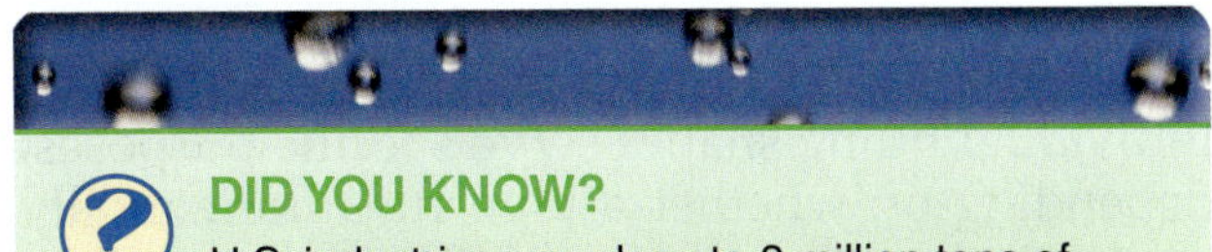

DID YOU KNOW?

U.S. industries use close to 3 million tons of copper each year. About two-thirds of the copper is mined; the rest comes from recycled copper. Discarded objects, dismantled buildings, and worn-out machinery are sources of scrap copper. This scrap copper can be melted down and reused. Many recycling stations pay for scrap copper.

Review Questions

Sections 1.0.0–2.0.0

1. Copper pipe is generally less expensive than other types of pipe.
   a. True
   b. False

2. Pipes that are measured in nominal sizing have an OD that is _____ larger than the nominal size.
   a. ⅛ inch
   b. ¼ inch
   c. ½ inch
   d. 1 inch

3. Types K, L, M, and DWV copper tubing are measured using _____.
   a. ID sizing
   b. OD sizing
   c. nominal sizing
   d. copper tube sizing

Match the following types of copper pipe with their corresponding color code.

4. DWV	a. Orange
5. K	b. Blue
6. M	c. Yellow
7. L	d. Green
	e. Red

8. Take extra care when working with copper pipe in cold temperatures because it can _____.
   a. break
   b. shrink
   c. harm your hands
   d. corrode

3.0.0 ◆ FITTINGS AND VALVES

The types of fittings and valves used with copper pipe depend on how the pipe is used. Fittings allow you to join pipes to each other and change the direction or size of a pipe run, for example. Fittings are designed so they do not block or slow the flow of materials in the pipe. This section discusses water supply fittings, water supply valves, and DWV fittings, as well as relatively new alternatives.

CAUTION

Never use dielectric unions in hot water systems. The unions cannot handle the temperature changes and they may leak.

Source: University of California–Santa Barbara Design, Construction, and Physical Facilities website. http://facilities.ucsb.edu/Standards/15000Plumbing.pdf, reviewed March 9, 2004.

3.1.0 Water Supply Fittings

Individual fitting shapes serve specific purposes, depending on whether a pipe runs horizontally or vertically. Use as few fittings as possible. Fewer fittings mean fewer chances for leaks and **pressure drops,** which are decreases in pressure from one point to another caused by friction losses in a water system. Common copper fittings for use with water supply systems include those listed in *Table 2* and shown in *Figure 2*.

When brazing, use brazing fittings, not standard fittings. Brazing fittings have sockets that are half the depth of standard fittings.

Solder fittings, also called sweat fittings, are special copper or brass fittings that are used for soldering or brazing copper pipe. Solder fittings are made slightly larger than the pipes to be joined, leaving only enough room for solder to flow into the joint. Adapter fittings allow copper tubing to be joined with threaded pipe on one end while the other end is soldered.

Table 2 Common Copper Fittings and Their Descriptions

Fitting	Description
90-degree ell and 45-degree ell	Elbow used to change the direction of the pipeline by 90 or 45 degrees.
Drop ear ell	An elbow that allows you to attach the pipeline to the building frame; frequently used at the last joint before the pipe comes through the wall to be attached to a fixture.
Street 90	An elbow with a male end and a female end that is used to change the direction of the pipeline by 90 degrees.
Street 45	An elbow with a male end and a female end that is used to change the direction of the pipeline by 45 degrees.
Tee	A fitting with three openings used to make branches at 90-degree angles to the main pipe. Reducing tees are used to make branches at 90-degree angles from the main pipe to smaller outlet pipes.
Coupling	A fitting used to connect lengths of pipe on straight runs.
Sweat cap	A fitting used to close the end of a copper pipe.
Female and male adapters	Fittings soldered onto the end of a length of copper tubing to provide a threaded end for attaching a pipe to another threaded pipe.
Sweat-to-compression adapter	A fitting used to adapt a soldered copper tube to a **compression joint** by means of a **ferrule,** which is a brass compression ring used for joining. A compression joint consists of a threaded nut that has been squeezed over a compression ring to seal the joint.
Sweat flange	A fitting used to adapt soldered copper tube to iron pipe flange.
Reducer coupling and reducer bushing	A fitting used to connect pipe to pipe or pipe to fitting with different sizes of pipe.
Sweat union	A fitting soldered to copper tubes, allowing the tubes to be joined to male and female half unions by a threaded shell nut.
Flare fittings	Fittings with a flared end that can be joined with a male cone-shaped tubing end or union. Flare fittings used in refrigeration and air conditioning include a variety of elbows, tees, and unions. They are **drop-forged** brass and are accurately machined to form the 45-degree flare face. The fittings used are based on the size of the tubing. Flare nuts are hexagon-shaped—they have six sides. An adjustable, or open-end, wrench is used with these fittings.
Grooved copper	A mechanical coupling material for rigidly connecting copper tubing that has been **roll-grooved.** Grooved copper fittings are made for connecting copper tubing in sizes from 2 to 6 inches. These fittings have grooved ends, so they can be installed using a wrench. This eliminates the need for soldering or brazing.

Dielectric Unions

If you attach unlike metals such as copper and galvanized pipe, it results in a process called electrolysis. Transition fittings, such as dielectric unions, prevent electrolysis. Dielectric unions isolate the different materials by using a mixture of brass, plastic, or rubber. However, they can cause other problems. For example, calcium deposits can build up, which eventually will lead to blockages. Therefore, you should never use dielectric unions unless they are plastic lined.

Source: Peace of Mind Home Inspection Services, LLC, website. www.getpeace.com/plumbing.htm, reviewed March 9, 2004.

DID YOU KNOW?

Two kinds of solder fittings are available with copper tubing. The first is a wrought copper fitting, which is made from copper tubing that is shaped into different types of fittings. Wrought copper fittings are generally lightweight, are smooth on the outside, and have thin walls. The second type is a cast solder fitting. This type of fitting is made using a mold. The first cast fittings had holes in the sockets to put solder in. Today, the heated copper is poured into the mold and allowed to cool. Cast solder fittings have a rough exterior and come in a wide variety of shapes. They are heavier than wrought solder fittings.

107F02A.EPS

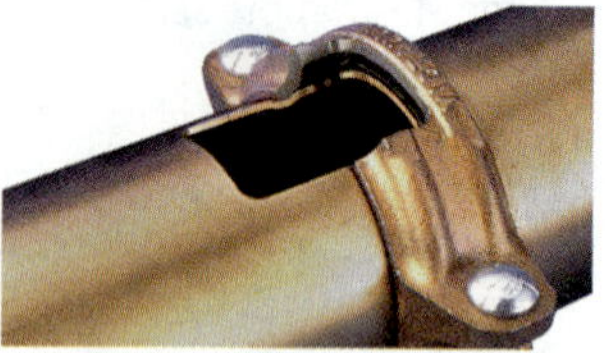

107F02B.TIF

Figure 2 ◆ Fittings for copper tubing.

Plumbing Codes

Plumbing codes have been developed to protect the water supply and public health and safety. Most cities, counties, and states have adopted model codes that are based on suggested national standards. These model codes may be subject to local interpretation and usually reflect local conditions. For example, an area that often has earthquakes or floods will have special requirements to deal with these conditions. The specifications for each job should detail these requirements. Installers take responsibility for correcting code violations, which can be expensive. While standards are guidelines to improve performance or reliability, codes and ordinances are mandatory, and you must follow them.

3.2.0 Water Supply Valves

Valves regulate the flow of liquid. They may provide on/off service or prevent flow reversal through a line. Valves use one or more of the following methods to control flow through a piping system:

- Move a disc or plug into or against a passageway.
- Slide a flat cylindrical or spherical surface across a passageway.
- Rotate a disc or ellipse around a shaft extending across the diameter of a pipe.
- Move a flexible material into the passageway.

Common valves for water supply systems include the following (see *Figure 3*):

- *Gate* – A valve with a wedge-shaped or tapered metal disc that fits into a smooth-ground surface or seat with the same shape, allowing a straight-line flow with little obstruction. It is a good choice for lines that will remain either completely open or completely closed most of the time.
- *Globe* – A valve that controls the flow of liquid with a movable spindle, which lowers to restrict flow through the valve opening. Because this valve is reliable and easy to repair, it is often used in water supply lines inside buildings.
- *Ball* – A valve that consists of a ball with a hole bored through its diameter, mounted on a spindle. When the valve is closed, the hole is at 90 degrees to the valve body so no flow can take place. When the valve is turned a quarter turn

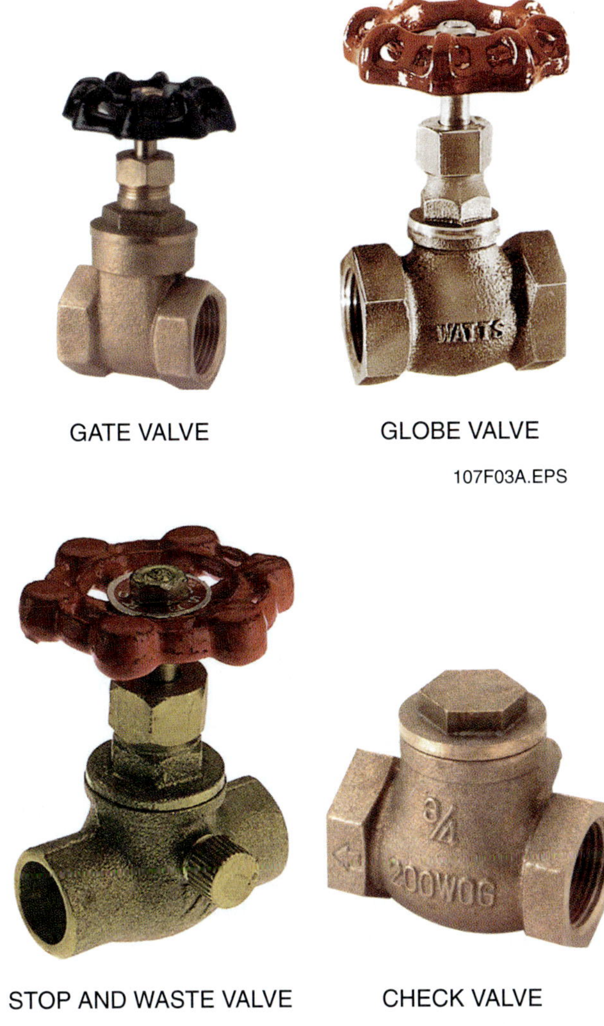

Figure 3 ◆ Common valves for water supply systems.

or opened completely, water flows through the hole. This valve is commonly used at the inlets and outlets of heat exchangers in HVAC (heating, ventilating, and air conditioning) systems and in systems where quick shutoffs may be necessary for in-line maintenance.

- *Compression, stop, and waste* – A valve that is opened or closed by raising or lowering a horizontal disc using a threaded stem. An elastomeric, or rubberized, washer on the end of the stem seals the valve seat, closing off water flow. This valve is most commonly used for draining and freeze protection above ground.
- *Check* – An automatic valve that permits the flow of liquid in one direction only. It prevents reverse flow. This valve is commonly used on domestic and irrigation wells.
- *Stop* – A valve that controls flow of liquids or gases between a building and supply source. It is also called a ground-key valve.

3.3.0 DWV Fittings

DWV fittings are designed to allow liquids and other materials to flow smoothly through them. Common fittings used in DWV systems are elbows (90 degrees, 45 degrees, 22½ degrees), male adapters, ferrule adapters, sanitary tees, cleanout tees, reducing tees, wyes, and reducers.

3.4.0 Alternative Fittings

In addition to the common fittings discussed above, you can use some alternative fittings to join copper pipe. These alternatives include press fittings and mechanically formed tee connections.

Press fittings, also called press-connect fittings, are mechanical fittings that connect pipe by means of a cold press fit system. First introduced in the 1970s in Germany and more recently in the U.S., they are used primarily for potable water systems. Press fittings are created by an electric press tool or a hand pressing tool (*Figure 4*). The tool crimps the fitting around the pipe against an O-ring inside the fitting. This ensures that the connection is strong and leakproof.

A **bullhead tee** is a tee fitting used on branches that are longer than the main line. The term is also used to describe a tee fitting in which the outlet is larger than the opening of the straight run. If tee fittings are not properly installed, they can cause a condition known as bullheading, which results from the larger outlet opening. When the flow of liquid hits the back wall of the tee, it causes turbulence. This adds to the pressure drop caused by liquid moving from a smaller pipe to a larger pipe. This turbulence may also cause a banging in the line. If more than one tee is installed in the line, a straight piece of pipe with a length between tees of 10 times the pipe's diameter is recommended to reduce turbulence. For example, a pipe that is 4 inches in diameter should have 40 inches of pipe between each tee (10 inches × 4 inches = 40 inches).

Mechanically formed tee connections are joints created by a tee-pulling tool. The tool allows you to drill into a section of pipe and create a tee connection

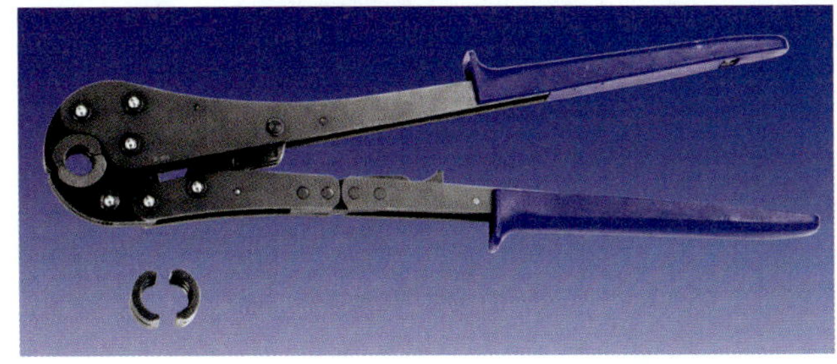

Figure 4 ◆ Hand pressing tool.

(see *Figure 5*). You must use brazing to join a branch line to the pipe. This method increases productivity because you create only one brazed joint rather than three soldered joints to form the tee connection (see *Figure 6*). This method is commonly used to create manifolds as well as copper fire sprinkler installations. Some codes do not permit tee-pulling tools for drainage because of the possibility of leakage. Follow the manufacturer's instructions and always consult applicable codes.

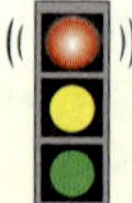

WARNING!

Always work under the direct supervision of your instructor or foreman, and always practice safety precautions to protect yourself and your co-workers from injury and to prevent equipment damage. Do not try to do any work that you have not been specifically trained to do. For example, do not operate valves unless you know exactly what the result will be. Before you work independently on a system, you must know the temperature and pressure conditions that exist at every point in the system, and how those conditions can be affected by malfunctions or by changes in valve positions.

107F05.TIF

Figure 5 ◆ Tee-pulling tool.

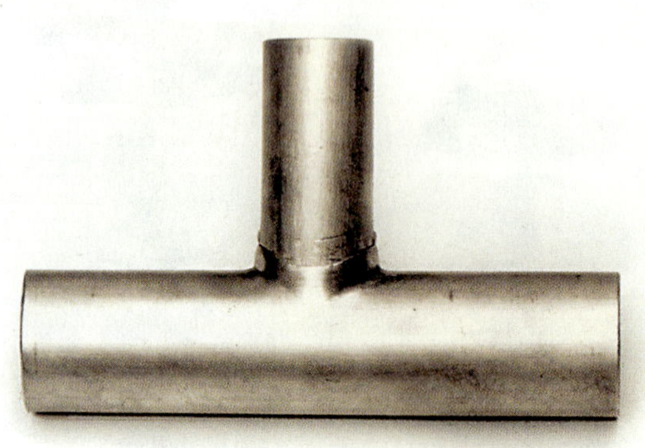

107F06.TIF

Figure 6 ◆ Mechanical tee connection.

4.0.0 ◆ MEASURING, CUTTING, BENDING, JOINING, AND GROOVING

Some measuring, cutting, reaming, bending, joining, and grooving techniques are related specifically to copper. Different techniques may be used depending on the type and function of the copper pipe you are using. You will need to become familiar with many tools to install copper tubing properly. The following sections explain methods for measuring, cutting, bending, joining, and grooving copper tubing.

WARNING!

Before you begin a job, think through the potential hazards and wear the appropriate PPE, including, but not limited to, safety glasses, a hard hat, and gloves. In addition, beware of hair creams and shave creams; they can be flammable. Always make sure you have quick access to a fire extinguisher on the job.

4.1.0 Measuring

It is extremely important to measure pipe carefully. Several methods of measuring copper pipe are described in the following list (see *Figure 7*):

- *End-to-end* – Measure the full length of the pipe.
- *End-to-center* – Use for pipe that has a fitting joined on one end only; pipe length is equal to the measurement minus the end-to-center dimension of the fitting.
- *Center-to-center* – Use with a length of pipe that has fittings joined on both ends; pipe length is equal to the measurement minus the sum of the end-to-center dimensions of the fittings.

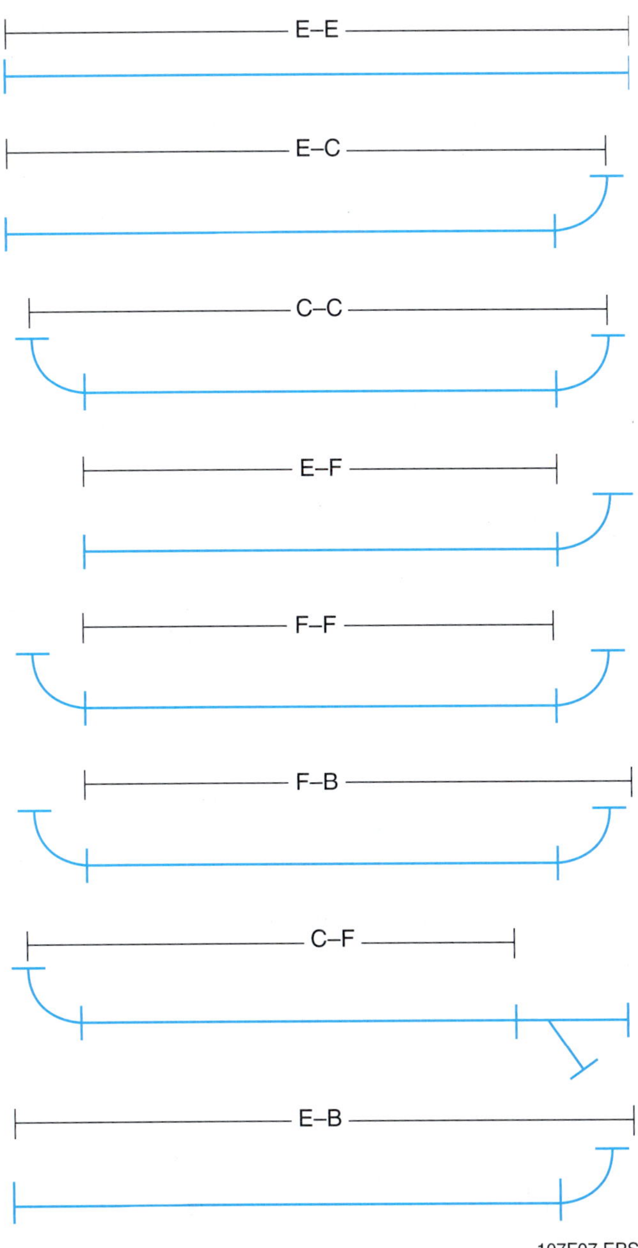

Figure 7 ◆ Measuring copper pipe.

- *End-to-face* – Use for pipe that has a fitting joined on one end only; pipe length is equal to the measurement.
- *Face-to-face* – Use for same situation as center-to-center measurement; pipe length is equal to the measurement.
- *Face-to-back* – Use with a length of pipe that has fittings joined on both ends; pipe length is equal to the measurement plus the distance from the face to the back of one sweated-on fitting.
- *Center-to-face* – Use with a length of pipe that has fittings on both ends; pipe length is equal to the measurement from the center of one of the fittings to the face of the opposite fitting, plus twice the insertion length.
- *End-to-back* – Use for pipe that has a fitting joined on one end only; pipe length is equal to the measurement plus the length of the sweated-on fitting.

4.2.0 Cutting

You can cut copper tubing with a handheld tube cutter, a hacksaw, or a midget cutter. The handheld tube cutter (*Figure 8*) is preferred because it makes a cleaner joint and leaves no metal particles. Use a tube cutter that is the right size for the copper you are cutting, and make sure that the proper cutting wheel is in place. Plumbers use internal tube cutters (*Figure 9*) for trimming extended ends of installed water closet bowl and shower waste lines below the level of the flange.

Figure 8 ◆ Handheld tube cutter.

Figure 9 ◆ Internal tube cutter.

After cutting, ream all cut tube ends to the full inside diameter of the tube. Reaming removes the small burr (rough inside edge) created when you cut the pipe. Burrs left on tubing can cause the pipe to corrode. A tube that is reamed correctly provides a smooth inner surface for better flow. You must also remove burrs on the outside of the tube to ensure a good fit. Tools used to ream tube ends include the reaming blade on the tube cutter, files (round or half-round), a pocketknife, and a deburring tool. If your pipe becomes deformed, you can use a **sizing tool,** which consists of a plug and a sizing ring, to bring the pipe back to roundness. Refer to *Figure 8* for an example of a deburring blade on a tube cutter. A variety of models are available, and the cutting sizes range from ⅛- to 4⅛-inches OD.

To use the handheld tube cutter, follow these steps (see *Figure 10*):

Step 1 Place the tube cutter on the tube at the point where you want to cut. Tighten the knob, forcing the cutting wheel against the tube.

Step 2 Make the cut by rotating the cutter around the tube under constant pressure.

Step 3 Use the built-in deburring blade to remove any burrs from inside the tube.

To cut larger size drawn tubing, you can use a hacksaw and a vise. A vise is a gripping tool that secures an object while you work on it. Using a vise helps you square the ends and allows more accurate cuts. The blade of the hacksaw should have at least 32 teeth per inch. Avoid leaving saw cuttings inside the tubing. File the end to produce a smooth surface, and carefully clean the inside of the tubing with a cloth. To cut large quantities of tubing, use portable power tools with abrasive wheels to cut, clean, and buff the tubing.

4.3.0 Bending

Unlike other types of rigid metal pipe, such as cast iron or steel, copper pipe can be bent. In fact, tests have shown that the bursting strength of bent pipe is greater than that of regular tubing. Bending reduces the number of joints and fittings in a plumbing system. This can reduce leaks, and ultimately installation time and costs as well. When you are using soft tubing, it is best to bend rather than cut it. You can easily bend small-diameter tubing by hand. Take care not to flatten the tube with a bend that is too sharp. For large tubing, the minimum bending radius to which a tube may be curved is up to 10 times the tube's diameter. For smaller tubes, it may be up to five times the tube's diameter.

You can place tube-bending springs on the outside of the tube to prevent it from collapsing while you are bending it by hand (see *Figure 11*). You can use tube-bending equipment to make accurate and reliable bends in both soft and hard tubing. This equipment has various sizes of forming attachments for use in bending tubing with diameters up to ⅞ inch at any angle up to 180 degrees (see *Figure 12*).

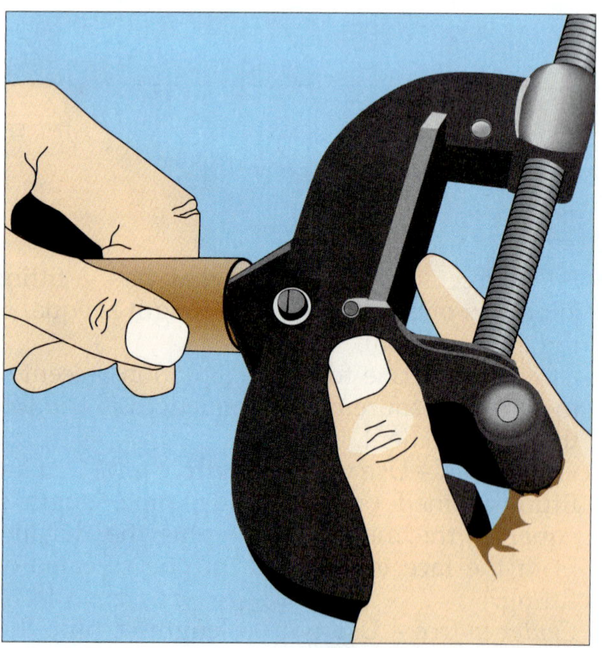

107F10.EPS

Figure 10 ◆ Using a handheld tube cutter.

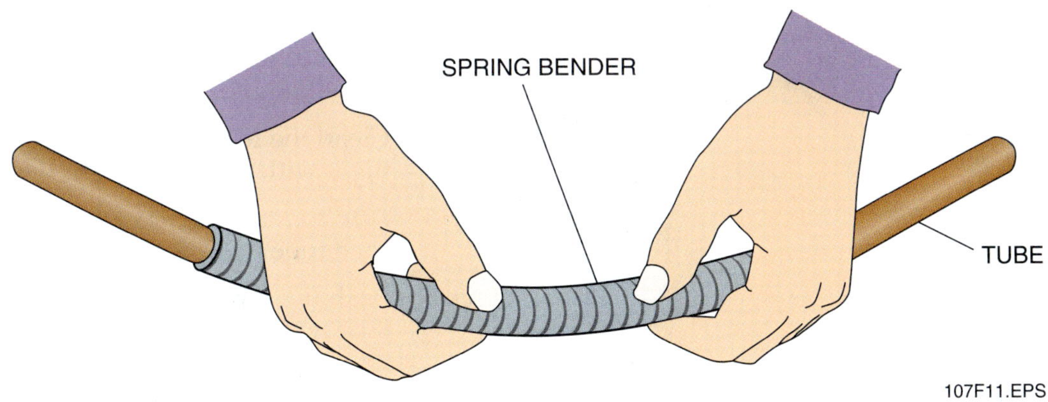

Figure 11 ◆ Tube-bending spring.

Figure 12 ◆ Tube-bending equipment.

4.4.0 Joining

Copper can be joined in many ways, including by a **sweat joint,** a compression joint, or a **flare joint.** The method depends on the plumbing application and environmental factors. Sweat joints, for example, require a heating process. You should not use this method when a fire hazard exists at the job site. As always, follow all applicable codes and manufacturer's instructions when joining copper pipe.

4.4.1 Soldering

Soldering is a type of heat bonding in which copper pipe is joined when a soft filler metal is melted in the joint between the two pipes. Solder fittings are made slightly larger than the pipes to be joined, leaving only enough space for solder to flow into the joint. This space is called the capillary action gap. Solder melts and flows into the gap in a process known as **capillary action.** Capillary action occurs regardless of whether the molten solder is flowing up, down, or horizontally. Adapter fittings allow copper tubing to be joined with threaded pipe on one end while the other end is soldered.

A sweat joint is made by measuring, cutting, reaming, cleaning, and applying flux to the copper pipe, then adding solder and heating to a certain temperature until the solder flows into the joint. Always use lead-free solder. Flux is a paste that acts as a wetting and cleaning agent and aids the soldering process by preventing oxidation of the joint, which would damage the copper. Flux should be lead-free and water-soluble. The heat required for soldering (350°F to 550°F) can be generated by electricity or various kinds of gases. Use sweat joints with hard or soft copper with nominal sizes from ⅛ to 4 inches.

Tools used to solder copper tubing to fittings include a tube cutter, fitting brush, solder, flux brush, and soldering torch (see *Figure 13*). In place

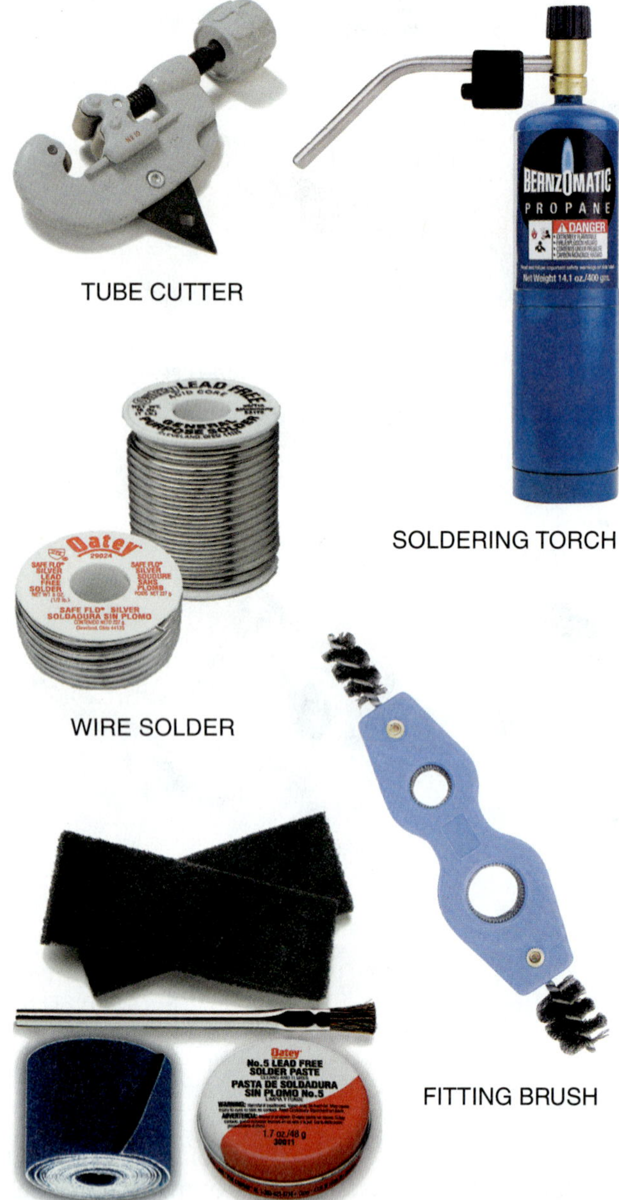

Figure 13 ◆ Tools for soldering copper tubing to a fitting.

of a fitting brush, you can use any abrasive other than steel wool. Steel wool contains oil, which contaminates the pipe.

To solder copper tubing to a fitting, follow these steps (see *Figure 14*):

Step 1 Measure and mark tubing to length, including the fitting allowance for the portion of the tube that will extend inside fittings.

Step 2 Use a tube cutter to cut the tubing.

Step 3 Use a reaming tool on the end of the tube cutter to remove the burr from the inside of the tubing.

Step 4 Clean the inside of each fitting by scouring it with a fitting brush.

Step 5 Apply a thin layer of flux to the end of each tube using a flux brush.

Step 6 Push the tubing into the fitting, and turn it a few times to spread the flux evenly.

Step 7 Hold the tip of a soldering torch flame against the fitting. When the flux starts to sizzle, move the flame around to the other side of the fitting to heat it evenly. Some plumbers use a special soldering torch fired from a 20-pound tank of gas called a B-tank.

Step 8 When the flux starts to bubble, remove the torch, and touch the wire solder to the point where the tubing enters the fitting. The solder will melt and be drawn into the joint. Once a line of solder shows completely around the joint, the connection is filled with solder.

Step 9 Allow the joint to cool until the solder solidifies.

Step 10 Once the solder has solidified, wipe it clean with a soft, wet cloth to remove any flux.

Brazing is a heating process very similar to soldering, but it requires filler metals that melt at much higher temperatures (between 1,100°F and 1,500°F) than solders. Plumbers use brazing when joints must be very strong. For example, brazing is often required to join refrigeration pipe.

4.4.2 Creating Compression Joints

A compression joint (*Figure 15*) is a mechanical joint that is made by measuring, cutting, and reaming the pipes and using compression fittings. With this method, you tighten a threaded nut to squeeze a compression ring to seal the joint. This

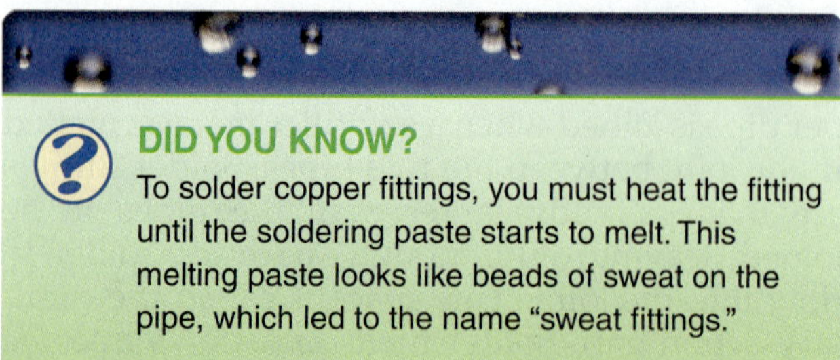

DID YOU KNOW?

To solder copper fittings, you must heat the fitting until the soldering paste starts to melt. This melting paste looks like beads of sweat on the pipe, which led to the name "sweat fittings."

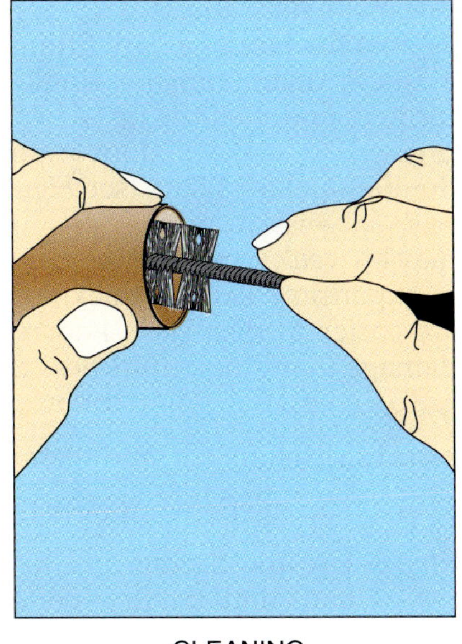

CLEANING

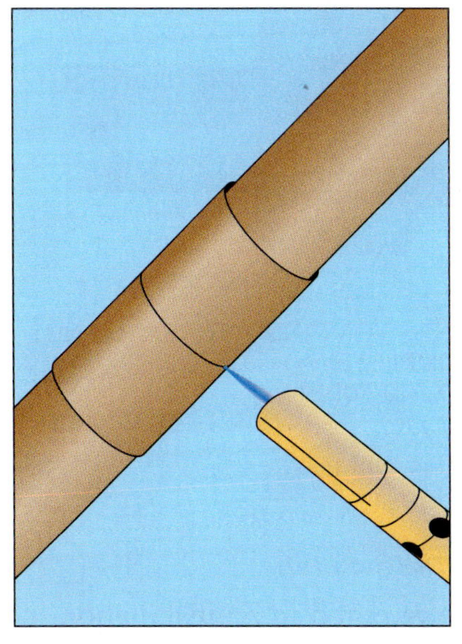

HEATING

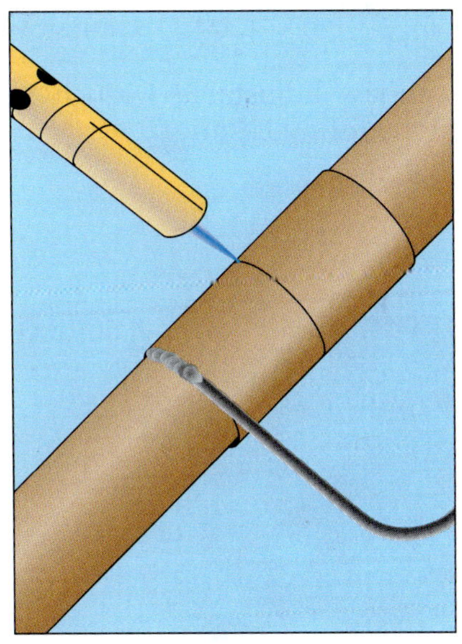

MELTING THE SOLDER

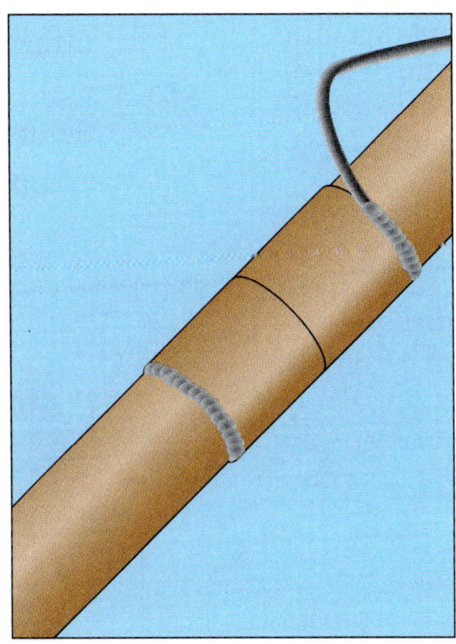

FILLING THE JOINTS

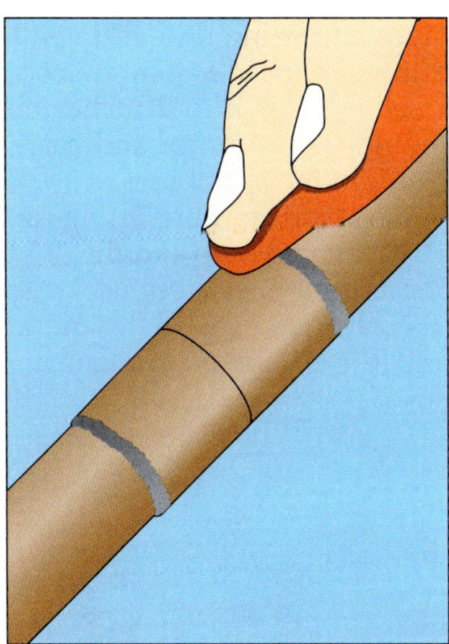

WIPING REMAINING FLUX AFTER SOLDER COOLS

107F14.EPS

Figure 14 ◆ Soldering copper tubing to a fitting.

kind of joint is often used for joining refrigerant tubing. It is a popular method because it uses a threaded fitting and takes less time than making soldered or flared joints.

To create a compression joint, follow these steps:

Step 1 Measure, cut, and ream the tube. Make sure the tube is cut square and is free of burrs.

Step 2 Slip the nut over the tube, and slide the compression ring on the tube with the teeth facing the tube's end.

Step 3 Install the cone with the convex surface toward the end of the tube. To be sure the fitting goes together completely, make sure that ¼ inch of the tube extends beyond the cone when working with a ½-inch tube, and that ½ inch of the tube extends beyond the cone when working with a ¾-inch tube.

Step 4 Push the nut onto the fitting, and tighten. When the fitting squeaks, turn the nut one more full turn.

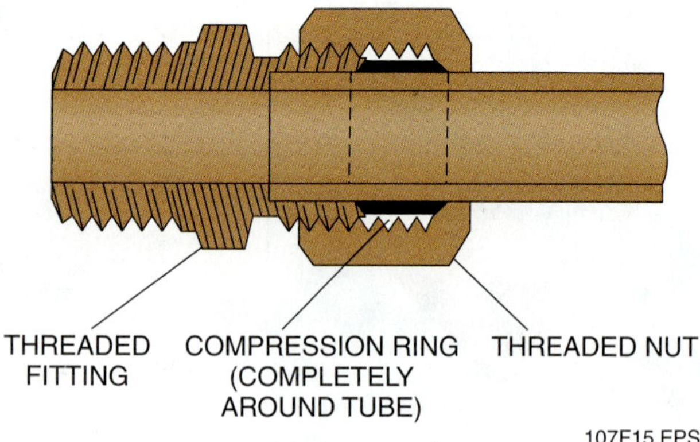

Figure 15 ◆ Compression joint.

4.4.3 Creating Flare Joints

Flare joints may be required in an installation in which a fire hazard exists and a torch for soldering or brazing is not permitted. A flare joint is made by measuring, cutting, and reaming the pipes and using flare fittings. This kind of joint is commonly used to join soft copper tubing with diameters from ¼ to 2 inches. Soft copper fittings should be leakproof and easily dismantled with the right tools.

Two kinds of flare fittings are popular: the single-thickness flare and the double-thickness flare. For both types, you use a special flaring tool (*Figure 16*) to expand the end of the tube outward into the shape of a cone, or flare. The single-thickness flare forms a 45-degree cone that fits against the face of a flare fitting (see *Figure 17*). In a single operation, the single-thickness flare is formed, and then the lip is folded back and compressed to make a double-thickness flare. The double-thickness flare (see *Figure 18*) is preferable with larger-size tubing. A single-thickness flare may be weak when used under excessive pressure or expansion. Double-thickness flare connections are easier to dismantle and reassemble without damage than single-thickness flare connections.

To make a flare joint, follow these steps:

Step 1 Measure, cut, and ream the tubing.

Step 2 Slip the flare nut over the tubing.

Step 3 Use the flaring tool to flare the tubing's ends until the fit is perfect.

Step 4 Slide the nuts onto each end of the flare tubing. Then gently bend or shape the tubing by hand over the male thread of each fitting.

Step 5 Use a smooth-jaw adjustable wrench to tighten until the fitting is snug.

Step 6 Test the joint for leaks.

Figure 16 ◆ Flaring tool.

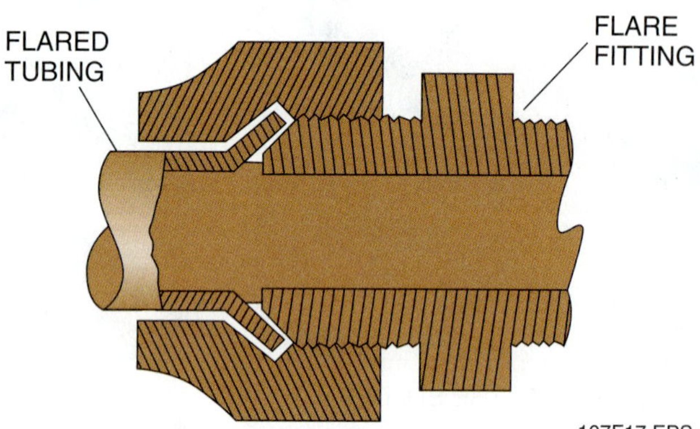

Figure 17 ◆ Single-thickness flare.

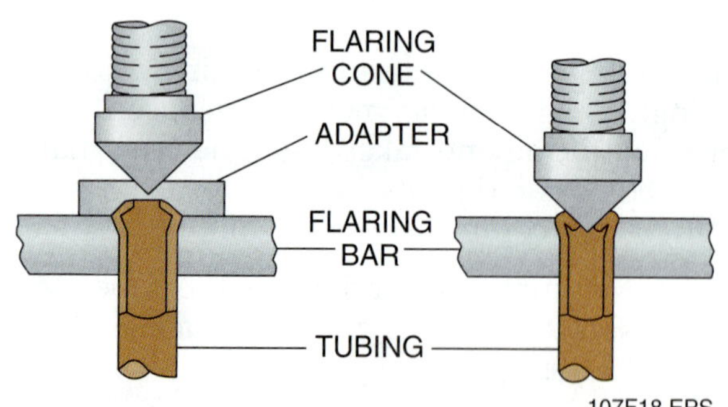

Figure 18 ◆ Double-thickness flare.

4.5.0 Grooving

You can groove copper pipe in two ways: by rolling or by cutting. Roll grooving involves cold forming pipe—it does not remove any metal from the pipe. Cut grooving removes metal from the OD of the pipe. You will learn more about grooving pipe elsewhere in this curriculum. Always follow applicable code and manufacturer's instructions when grooving copper pipe.

Review Questions

Sections 3.0.0–4.0.0

1. _____ are used to make branches in a pipeline.
   a. Tees
   b. Adapters
   c. Couplings
   d. Flanges

2. Grooved copper fittings are made for connecting copper tubing in sizes from _____.
   a. 0 to 4 inches
   b. 2 to 6 inches
   c. 4 to 8 inches
   d. 6 to 10 inches

3. The valve most often used for draining and freeze protection above ground is the _____ valve.
   a. ball
   b. compression, stop, and waste
   c. globe
   d. gate

4. The automatic valve that permits the flow of liquid in one direction only and prevents reverse flow is the _____ valve.
   a. globe
   b. gate
   c. check
   d. stop

5. A _____ tee is a fitting that is used on branches that are longer than the main line.
   a. bullhead
   b. sanitary
   c. reducing
   d. cleanout

6. To measure a length of pipe that has fittings joined on both ends, use the _____ method.
   a. end-to-end
   b. end-to-center
   c. center-to-center
   d. end-to-face

7. When cutting copper tubing, the _____ is preferred because it makes a cleaner joint and leaves no metal particles.
   a. handheld tube cutter
   b. hacksaw
   c. midget cutter
   d. power saw

8. A _____ helps prevent the tube from collapsing while you are bending it by hand.
   a. tube-reaming tool
   b. handheld burring blade
   c. tube-shielding cover
   d. tube-bending spring

9. A(n) _____ joint is commonly used to join soft copper tubing from ¼ to 2 inches.
   a. compression
   b. flare
   c. sweat
   d. angle

10. You can _____ copper pipe in two ways: by rolling or by cutting.
   a. bend
   b. groove
   c. join
   d. measure

5.0.0 ◆ INSTALLING PIPE HANGERS AND SUPPORTS

Hangers and supports are designed to hold and support pipe in either a horizontal or a vertical position. All pipes must be installed and supported so that both the pipe and its joints remain leakproof. Improper support can cause the piping system to sag. This causes stress on the pipe and fittings and, over time, increases the chance of breaks or leaks in the piping system. Without proper support, the drainage pipe can shift from its proper angle and form traps. These traps fill with liquid and solid wastes that block the pipeline. Copper pipe is supported in a variety of ways and with various sizes of hangers and supports. Be sure to follow the manufacturer's instructions for specific installation instructions, and consult applicable codes.

If you are installing pipe in a seismically active area—that is, where earthquakes are a possibility—local codes will require seismic restraints. The purpose of these restraints is to ensure that the pipe is securely fastened to the structure in the event of excessive vibration. For example, some codes require hangers and supports to be used at closer intervals than in non-seismically active areas. In addition, they may require you to leave extra spacing for pipes where they meet walls and floors to allow for anticipated movement.

5.1.0 Types of Pipe Hangers and Supports

Which hanger to use depends on the job specifications, the documents that describe the quality of the materials and work required. Specifications for hangers and supports are determined by the following factors:

- The combined weight of the pipe fittings and valves
- The maximum weight of the contents that the pipe might carry
- The material (wood, concrete, steel) the hanger will be attached to
- The distance from the anchor point to the pipe
- The potential for corrosion between the hanger and the pipes, fittings, or valves
- The expansion and contraction of the piping system
- The vibration of equipment attached to the piping system

Different types of hangers and supports are designed to hold and support pipe in either a horizontal or vertical position. They are manufactured in various materials, including carbon steel, malleable iron, cast iron, and plastics. They are available with different finishes, including copper plate, black, galvanized, chrome, and brass. To reduce corrosion, hangers and supports should be made of the same material as the pipe. If another material is used, the pipe must be shielded.

The basic components of hangers and supports can be placed into the following three major categories:

- Pipe attachments
- Connectors
- Structural attachments

5.1.1 Pipe Attachments

A pipe attachment is the part of the hanger that touches or connects directly to the pipe. It may be designed for either heavy duty or light duty, for covered (insulated) pipe or plain pipe. Applicable fire codes may restrict the use of hangers made of material other than copper. In some cases, you can use plastic-coated hangers. For specific requirements, consult applicable codes. Examples of pipe attachments are shown in *Figure 19*.

Hangers are used for horizontal or vertical support of pipes. The main purpose of hangers and brackets is to keep the piping in alignment and prevent it from bending or distorting. Examples of hangers are shown in *Figure 20*.

Pipe can be supported on wood-frame construction with several styles of pipe attachments. These include pipe hooks, J-hooks, tube straps, plumber's tape (also called strap iron or band

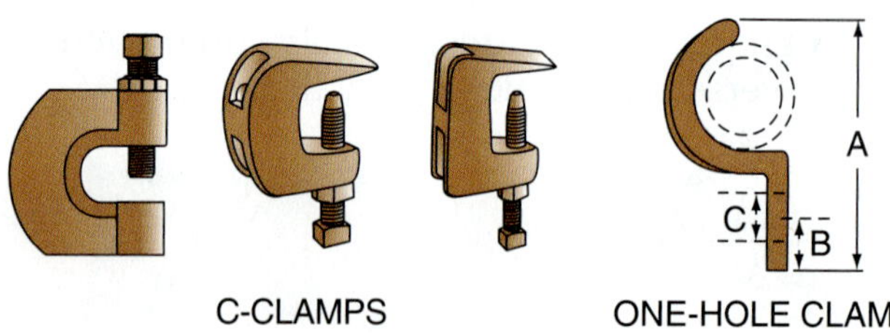

Figure 19 ◆ Pipe attachments.

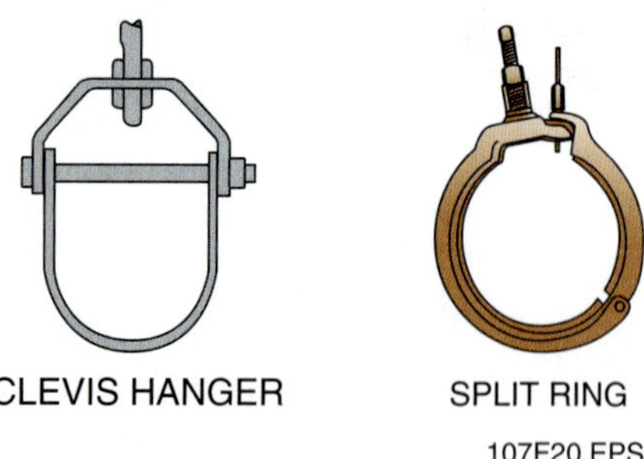

Figure 20 ◆ Pipe hangers.

iron), and pipe straps (see *Figure 21*). In these examples, the pipe attachment, connector, and structural attachment are all one unit. Note that plumber's tape may not be allowed in your area. When choosing hangers of any type, always consult applicable codes.

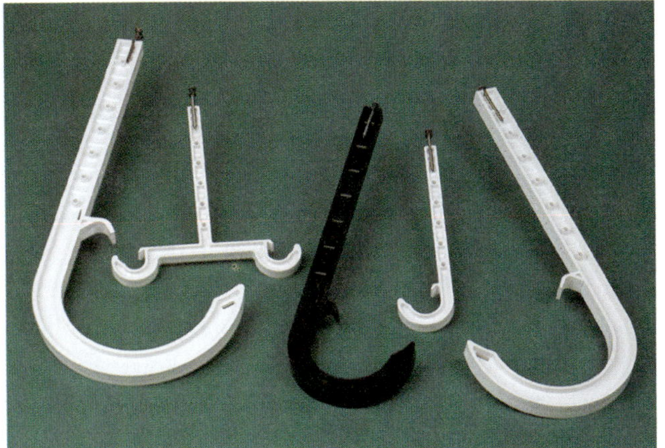

Pipe hangers are used mainly to support pipe, but they can also be used as a support for vibration isolators (see *Figure 22*). If no vibration problems are expected, use commonly accepted plumbing practices. If vibration may be a problem, the specifications should tell you what materials to use.

To fasten pipes to beams and other metal structures, one-hole clamps, steel brackets, beam clamps, or C-clamps are used. Vertical hangers, also called pipe riser clamps, consist of a friction clamp that can be attached to structural site components to support the vertical load of the pipe (see *Figure 23*). Special fasteners are used to attach hangers to masonry, concrete, or steel.

Other pipe attachments include universal pipe clamps and standard 1⅝-inch or 1½-inch channels (see *Figure 24*). The notched steel clamps are inserted by twisting them into position along the slotted side of the channel. The pipes can be aligned as close to one another as the couplings allow.

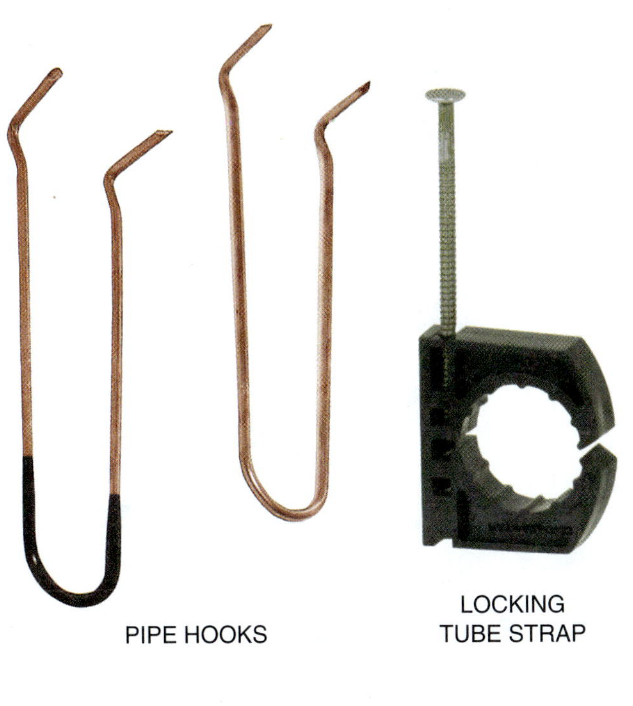

Figure 21 ◆ Pipe hangers for wood-frame construction.

Figure 22 ◆ Vibration isolator installed in hanger.

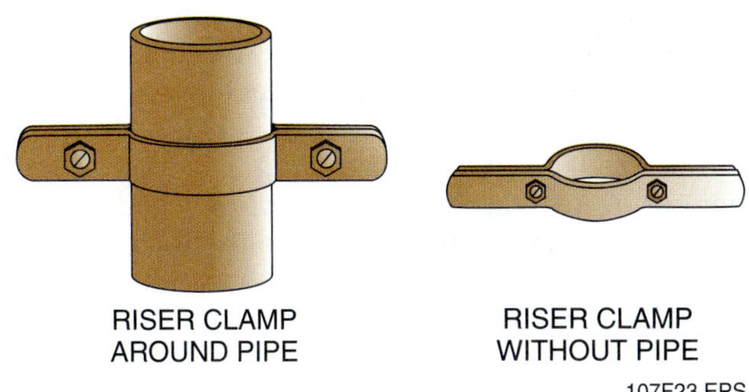

Figure 23 ◆ Vertical hangers.

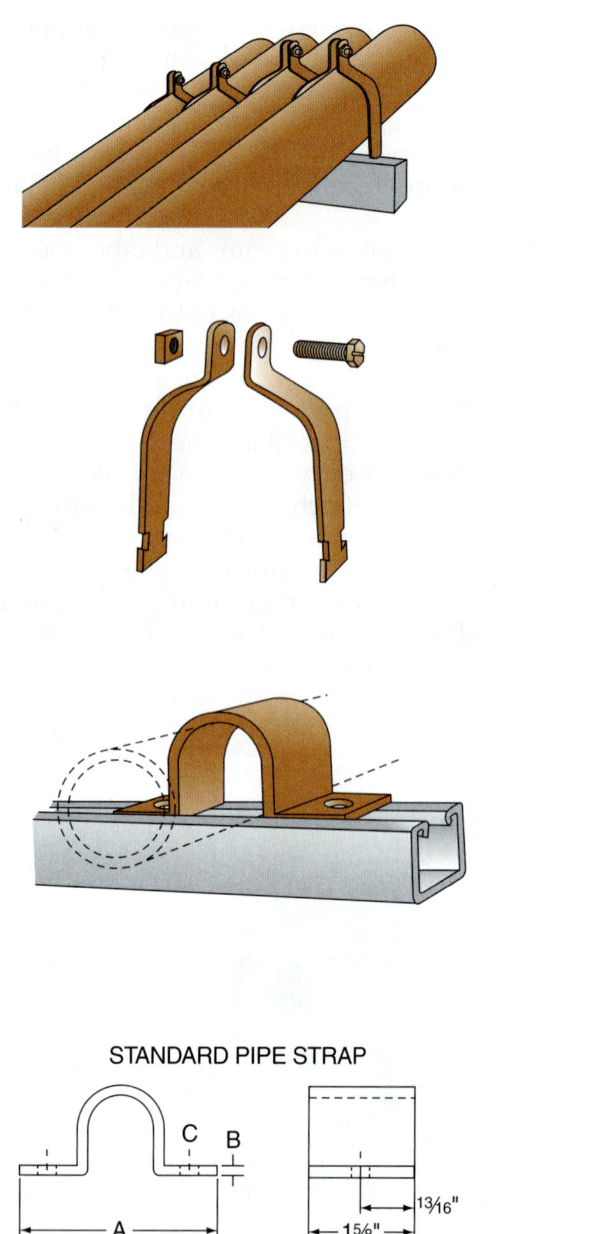

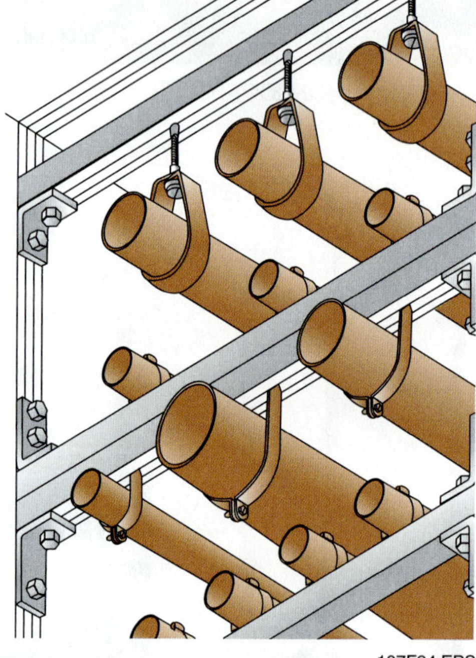

Figure 24 ◆ Universal pipe clamps and channels.

5.1.2 Connectors

The connector section of the hanger is the part that links the pipe attachment to the structural attachment. Connectors can be divided into two groups: rods and bolts, and other rod attachments.

Rod attachments include eye sockets, extension pieces, rod couplings, reducing rod couplings, hanger adjusters, turnbuckles, **clevises,** and eye rods. A clevis is an iron bent into the form of a U, with holes in the ends to receive a bolt or pin. Some of these connectors are shown in *Figure 25*.

5.1.3 Structural Attachments

Structural attachments are used to anchor the pipe hanger assembly securely to the structure. Structural attachments include threaded drop-in anchors, plastic or lead mollies, toggles, C-beam clamps, pound-in nail anchors, wedge anchors, threaded rods, and strut channels.

What you use to install hangers and supports will depend on the item and the type of material to which you are attaching it (see *Table 3* and see *Figure 26*).

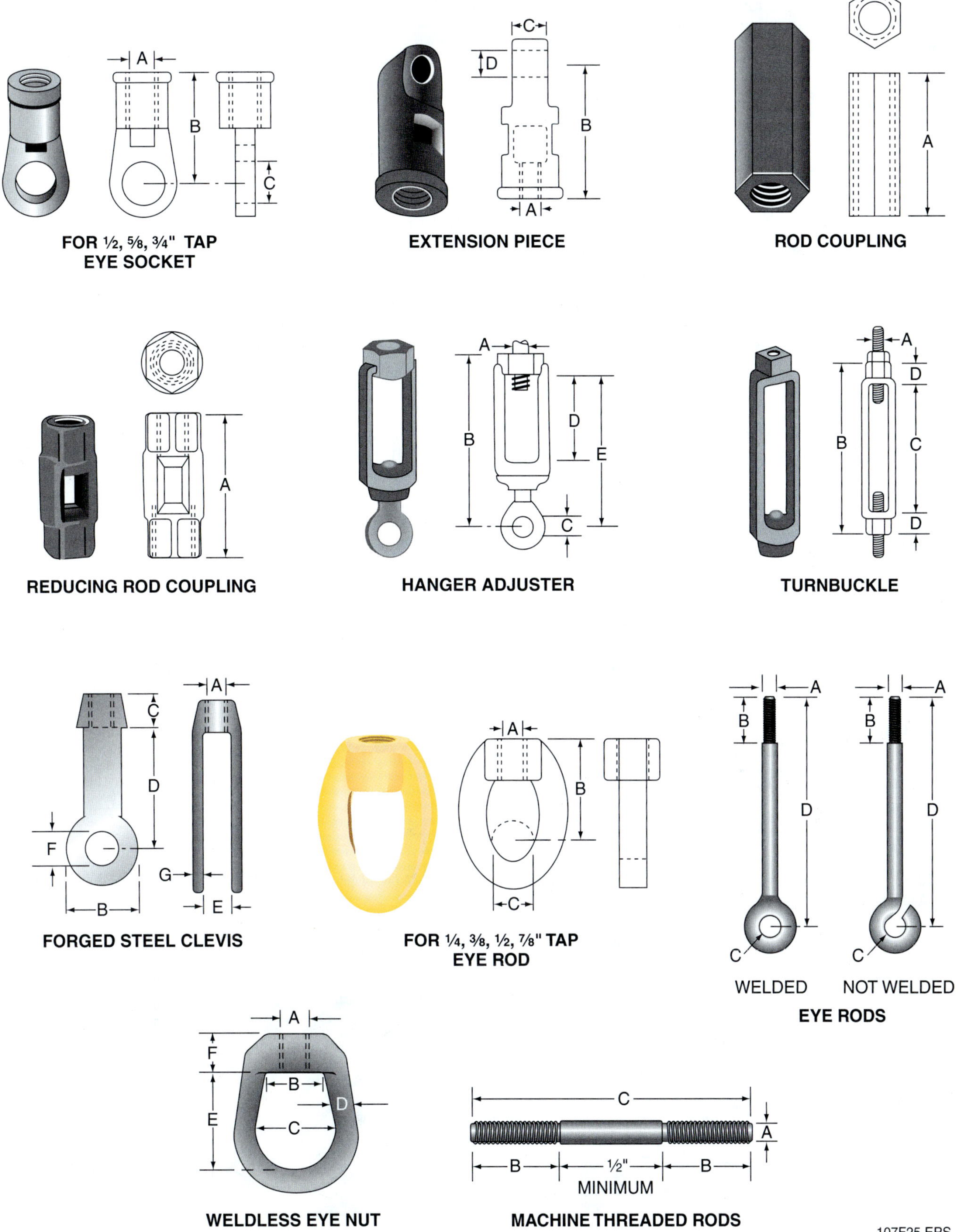

Figure 25 ◆ Rod attachment connectors.

Table 3 Types of Structural Attachments and Their Descriptions

Structural Attachment	Description
Threaded drop-in anchors	Most commonly used in concrete ceilings, walls, or floors. Different sizes of anchors are rated for different weights or spreads between anchors.
Toggles	Used for any hollow wall or ceiling support, including drywall and block.
Pound-in nail anchors	Used in concrete, brick, and stone.
Threaded rod hangers	Used in wood. Various lengths of all-thread rod can be cut to a specified length. Common sizes of all-thread rod are ⅜ inch to ½ inch.
C-clamps or beam clamps	Usually used to hook to iron or to a beam. These include set-bolts that must be tightened to ensure that they will not come loose.
Channel (strut)	Available in various lengths, widths, and heights and can be bolted or welded together. Various floor, wall, and overhead strut brackets are available.

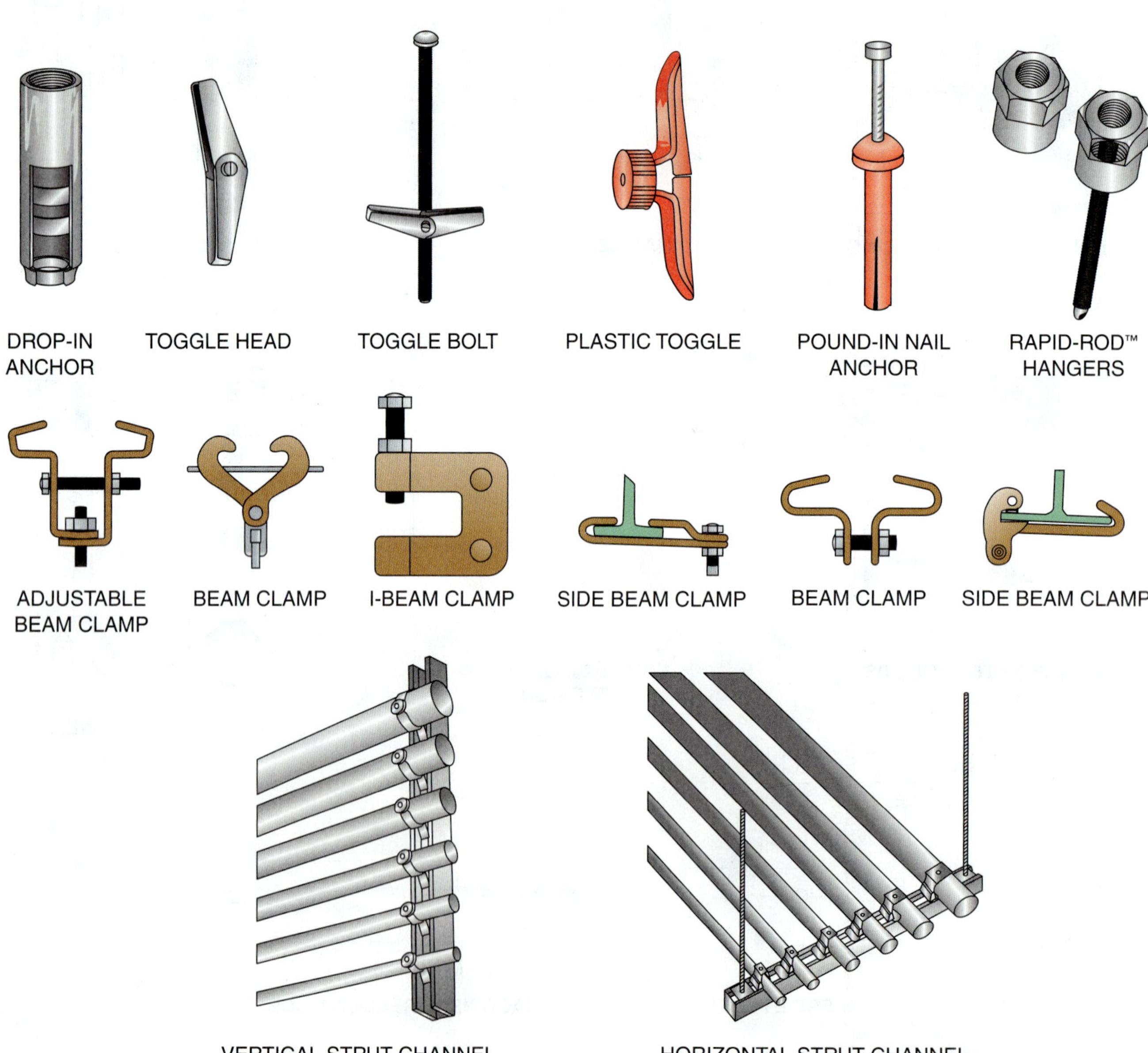

Figure 26 ◆ Structural attachments.

6.0.0 ◆ INSULATING PIPES

Under certain temperature and humidity conditions, condensation will form on cold water pipe and may drip into equipment or occupied areas. Similarly, heat can escape from hot water piping. To prevent these occurrences, some pipe is insulated (*Figure 27*).

Insulation is a material that prevents the transfer of heat. Cork, glass fibers, mineral wool, and polyurethane foams are examples of insulating materials. Insulation should be fire-resistant, moisture-resistant, and vermin-proof.

If the insulation cannot be installed before the tubing is connected, it must be split lengthwise to fit onto the pipe. Split seams and connecting seams must then be sealed. Insulation should not be stretched, because its effectiveness will be reduced. You generally do not put the insulation on the pipe until after pressure testing is performed. However, if you must do so before testing takes place, you must keep the joints exposed for the testers.

Some pipes are always insulated, and others are insulated only under certain conditions. Local building codes and job specifications will describe insulation requirements. You will learn more about which pipes to insulate and under what conditions elsewhere in this curriculum.

7.0.0 ◆ PRESSURE TESTING

Once an installation is complete, you must pressure test the system for leaks to be sure that all connections are secure. The system must also be inspected to make sure that the work has been done according to the job specifications and applicable codes.

You can pressure test copper pipe with water or air. You are required to supply all the equipment needed for testing, including test plugs, caps, plugs, and a source of compressed air (if air testing is required). Water testing can be done with test plugs and a hose. To test, close all outlets to the piping system with test plugs, caps, and plugs. Fill the system with water or air under pressure and look for leaks.

To water test a DWV system, you must have a minimum of 10 feet of water **head.** Consult all applicable codes to determine the appropriate pressure for testing. To do this, fill the vent stack completely with water. Once you have filled the DWV piping, inspect the whole piping system for leaks. This inspection is required before any part of the system can be covered. The plumbing inspector will also check for cross-connections, defective or inferior materials, and poor work.

To water test the water supply system, close all outlets and fill the system with water from the main. Inspect the system for leaks or other problems. Where codes specify water testing at pressures higher than those available from the normal water supply, you can use a hydraulic test pump (*Figure 28*) to produce the required pressure. Refer to code for the correct pressures to use.

To air test a system, the plumber or inspector fills the piping with compressed air. By using a pressure gauge at the test plug, the plumber or inspector can determine if any pressure is lost somewhere in the piping. Another useful method for identifying leaks is to brush soapsuds onto joints. If there are leaks in a joint, bubbles will form there.

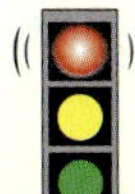

WARNING!

Never use oxygen, acetylene, or other gases to pressure test a refrigerant system. Oxygen will cause an explosion if it comes into contact with refrigerant oil. Acetylene is highly flammable. The only gas other than refrigerant that should be introduced into a system is nitrogen.

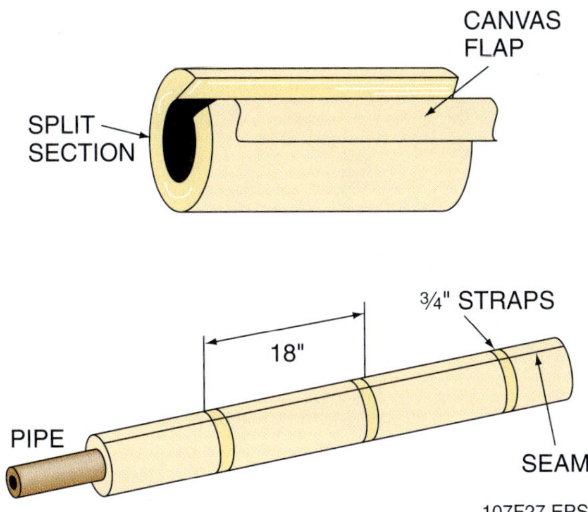

Figure 27 ◆ Insulated pipe.

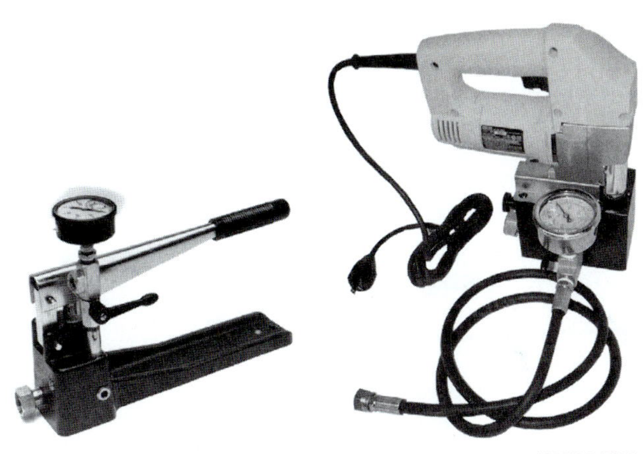

Figure 28 ◆ Hydraulic test pumps.

Review Questions

Sections 5.0.0–7.0.0

1. Pipe hangers are primarily used to support pipe, but they can also be used to support _____.
   a. pressure checks
   b. channel struts
   c. vibration isolators
   d. insulation support

2. _____ can be attached to structural site components to support the vertical load of the pipe.
   a. J-hooks
   b. Eye sockets
   c. Wedge anchors
   d. Pipe riser clamps

3. A connector is the part of the hanger that links the pipe itself to the structural attachment.
   a. True
   b. False

4. Each of the following is an example of an insulating material for copper pipe *except* _____.
   a. cork
   b. polyethylene foam
   c. glass fibers
   d. mineral wool

5. In addition to checking for leaks, a plumbing inspector checks for _____ when water testing the completed installation of a water system.
   a. inadequate water pressure
   b. defective materials
   c. improper insulation
   d. inadequate pipe supports and hangers

Summary

A plumber must be able to work with several kinds of pipe, including copper. This module introduced you to soft and hard copper pipe and its uses. You have also learned about the various fittings and valves that you can use to join copper pipe for water distribution systems and DWV systems. You now know that pipe must be properly supported and insulated, and you know the different kinds of hangers and supports that you can use. The requirements for pipe hangers and insulation are usually specified in the job specifications or by local building codes.

Copper can be joined in many ways, including sweat joints, compression joints, or flare joints. This module has introduced you to all three methods of joining copper pipe, and how to measure, cut, ream, and bend pipe that you are going to join.

A completed piping installation must be inspected and tested. At this point, you will not be doing the testing, but you should be familiar with the procedures. Throughout this module you have read about special safety practices to prevent injury to yourself and others or damage to equipment. Follow these practices as you go into the field and work with copper pipe and fittings.

Notes

Trade Terms Quiz

Fill in the blank with the correct trade term that you learned from your study of this module.

1. Grooved copper is a mechanical coupling material for rigidly connecting copper tubing that has been _______________.
2. _______________ is the approximate measurement in inches of the inside diameter of pipe for most copper pipes. The exception is the measurement for ACR tubing, which is based on the outside diameter.
3. A brass compression ring used for joining is called a(n) _______________.
4. A(n) _______________ is commonly used to join soft copper tubing with diameters from ¼ to 2 inches.
5. Copper tubing that is described as soft and flexible is created through a process called _______________.
6. Flare fittings are made of _______________ brass.
7. If pipe becomes deformed, use a(n) _______________ to work the pipe back to roundness.
8. _______________, or air conditioning and refrigeration systems, is one of many plumbing applications in which copper tubing is used.
9. The use of a(n) _______________, a fixture in which the main line is smaller than the branch, can prevent turbulence in the system.
10. A(n) _______________ is a rod attachment made of iron that is commonly in the form of a U.
11. To prevent the transfer of heat, install _______________. Materials used for this purpose include cork, glass fibers, mineral wool, and polyurethane foams.
12. Annealed copper tubing that is manufactured specifically for use in air conditioning and refrigeration work is called _______________.
13. A(n) _______________ consists of a threaded nut that has been squeezed over a compression ring to join pipes.
14. To water test a DWV system, you must have at least 10 feet of water _______________ in the system.
15. _______________, or hard, copper tubing is widely used in commercial refrigeration and air conditioning systems.
16. A(n) _______________ is created by soldering.
17. Using fewer fittings reduces the chance for _______________, or decreases in pressure from one point to another caused by friction losses in a water system.
18. The process that occurs during soldering in which solder melts and flows into the gap between the pipe and the fitting is known as _______________.
19. You should consider strength and _______________, or ease of bending, when selecting copper pipe.

Trade Terms

ACR
ACR tubing
Annealing
Bullhead tee
Capillary action
Clevis
Compression joint
Drawn copper
Drop forged
Ferrule
Flare joint
Formability
Head
Insulation
Nominal size
Pressure drop
Roll grooved
Sizing tool
Sweat joint

Profile in Success

Charles Robbins

M. Davis and Sons
Training Director
Wilmington, Delaware

Charles Robbins was born in Cook County, Illinois. He attended Upper Darby High School in Pennsylvania and began college not knowing exactly what he wanted to do. Charles began his career in construction as a shipfitter, but a friend gave him the opportunity to get into an accelerated apprenticeship program. When shipping took a downturn, he went to work for a plumbing contractor, where he was offered his plumbing apprenticeship. The blueprint reading, shipbuilding, and basic measuring skills that he learned as a shipfitter served him well in plumbing. For the past 19 years, Charles has been with M. Davis and Sons, the oldest privately owned company in Delaware.

How did you become interested in this industry?
I enjoyed plumbing more than shipfitting. The man who hired me actually had me drive to North Philadelphia just to dig ditches. I worked all day for $15. One day he asked, "Can I call you if I need you?" I thought about it and said yes. He had a one-man shop, so when he needed help, he called me. I started learning more and more, and soon he would leave me alone to do the work on my own.

What path did you take to your current position?
I was working for a man who ran a job shop—bathrooms and additions—but he wasn't keeping me busy. I had the opportunity to go into new construction when I was still an apprentice. At a plumbing class I ran into a man that I knew when I was a Boy Scout. His father owned a plumbing shop and he invited me to come work for them. They remembered me from when I was a kid. I finished the last two years of my apprenticeship with Charles H. Fischer, Inc.

The normal apprenticeship was a five-year program. I finished in four years because of the accelerated program at Sun Ship. I had taken welding and blueprint reading and they credited that amount of time to me. I was a journeyman plumber when I left Charles H. Fischer for M. Davis and Sons.

After I began with M. Davis and Sons, I took a year off from school. I got my master plumber's license and began a heating, ventilating, and air conditioning (HVAC) apprenticeship program. I got my HVAC journeyman's card and then my master's HVAC license. Today I hold master plumber's licenses in New Jersey; Delaware; Maryland; New Castle County, Delaware (upper portion of the state); the city of Wilmington; and the city of Philadelphia. In Delaware I have an HVAC master's license, in Maryland I hold a limited electrical license, and I am a certified member of the Refrigeration, Service, and Engineers Society.

M. Davis and Sons is an industrial contractor. I do a lot of work in the petrochemical arena. I found that very interesting, so that's where I ended up. I started working in the field in industrial maintenance. I took on more responsibility and started estimating. Then I became a foreman and took the route to superintendent and division manager. I oversaw $9 million worth of work and 110 employees.

I went from division manager to training when the CEO asked me what I wanted to do. I always enjoyed training. I was always sent to classes, because I would ask, "Can I take this class, or that class?" I thought that I might as well start training other people. I have enough gray hair! And it's a pleasant opportunity to help others.

What does it take to be successful in your trade?
I started as a plunger-carrying plumber and worked my way up. I took on more responsibility. It takes hard work. Always look for opportunities to learn and stay on top. Be open-minded to new things. I find it exciting to learn something new. I wanted to see new things and see how they work.

What are some things you do on the job?
I conduct small-group training. On some assignments I go to the job site and grab the new guys. I show them basic knot tying, how to pick out the right tips, how to use the right number of teeth for the specific material they are cutting, how to select the right glue. I conduct on-site training and classroom training. I am really hands-on. I teach them how to use equipment and machinery. Adult learners need instant gratification. My trainees feel good about learning new methods.

I also work with people who are switching careers. I focus on an area they are interested in and help them. I do a lot with high schools, talking about careers in construction. A lot of schools invite me to be on panels.

I am active in the Plumbing-Heating-Cooling Contractors–National Association (PHCC–NA). I am past president of PHCC of Delaware, past president of Refrigeration Service Engineers Society (RSES), and treasurer of PHCC–NA. I am also active in Associated Building Contractors and serve as the educational chair. As a master trainer for NCCER, I am currently teaching the second-year apprenticeship program in Delaware.

What is the most interesting aspect of your profession? What makes your trade stand out from others?
A plumber protects the health of the nation. A house isn't livable without plumbing. I truly believe that. I have worked in all aspects of the trade. It's great to go from start to finish, and it is exciting to see something come out—to see the end product. A tradesperson wants to show off his or her accomplishments. Take a bridge builder, for example; some people in the trade are just waiting on quitting time, while another guy will say, "I'm doing this so my family can cross the water and go to the beach."

What do you like most—and least—about your job?
I most like interaction with other people and seeing something completed. I like to see something come together, something get built, the end product completed professionally and in an appealing way.

I least like the people who are along for the ride, but who do not have an interest in what they are doing. They say, "That's good enough," but that's not the right attitude.

What would you say to someone entering the trade today?
Training and education is forever. It is never ending.

What can an apprentice expect to earn in his/her first years on the job in your area? What can he/she expect to earn after 10-plus years in the industry?
Rates are geared toward geography. In this area, apprentices can start at $10.00 per hour. After completing their apprenticeship, they can earn around $18.00 to $19.00 per hour. After about 10 years, they earn around $25.00 per hour.

Trade Terms Introduced in This Module

ACR: Air conditioning and refrigeration system.

ACR tubing: Annealed copper tubing that is manufactured specifically for use in air conditioning and refrigeration work.

Annealing: Heating a material and slowly cooling it to relieve internal stress. This process reduces brittleness and increases toughness.

Bullhead tee: A tee fitting used on a branch that is longer than the main line, or that has one outlet larger than the run openings.

Capillary action: The process during soldering in which the molten solder flows into the narrow gap between the pipe and the joint, regardless of whether the solder is flowing up, down, or horizontally.

Clevis: An iron, or link in a chain, bent into the form of a horseshoe, stirrup, or letter U, with holes in the ends to receive a bolt or pin.

Compression joint: A method of connection in which tightening a threaded nut squeezes a compression ring to seal the joint.

Drawn copper: Tubing produced by pulling the tube through dies to reduce its diameter. The drawing process hardens the copper and makes it very rigid.

Drop forged: A characteristic of a product made when heated metal is pounded or shaped between dies with a drop hammer or press.

Ferrule: A brass compression ring used for joining.

Flare joint: A fitting in which one end of each tube to be joined is flared outward using a special tool. The flared tube ends mate with the threaded flare fitting and are secured to the fitting with flare nuts.

Formability: The ease with which a material can bend.

Head: The height of a water column, measured in feet. One foot of head is equal to 0.433 pounds per square inch (psi).

Insulation: A substance that retards the flow of heat.

Nominal size: Approximate measurement in inches of the inside diameter (ID) of pipe for most copper pipes. However, nominal size of ACR tubing is based on the outside diameter (OD).

Pressure drop: A decrease in pressure from one point to another caused by friction losses in a fluid system, such as a water system.

Roll grooved: A type of copper piping that is compatible with grooved copper fittings.

Sizing tool: A tool consisting of a plug and a sizing ring that is used to reshape a deformed pipe back to roundness.

Sweat joint: A pipe joint made by applying solder to the joint and heating it until it flows into the joint.

Additional Resources and References

ADDITIONAL RESOURCES

This module is intended to present thorough resources for task training. The following reference works are suggested for further study. These are optional materials for continued education rather than for task training.

The Copper Tube Handbook, 1995. New York: Copper Development Association.

Engineering Lab, Inc. website, www.engineeringlab.com, *Flux, Solder, and Cleaning* (Online Course), www.engineeringlab.com/fluxsolder.html?source=goto, reviewed July 2003.

Modern Plumbing, 1997. E. Keith Blankenbaker. Tinley Park, IL: The Goodheart-Willcox Company, Inc.

Modern Refrigeration and Air Conditioning, 1996. Andrew D. Althouse, Carl H. Turnquist, and Alfred F. Bracciano. Tinley Park, IL: The Goodheart-Willcox Company, Inc.

Pipefitter's Handbook, Third Edition, 1967. Forrest R. Lindsey. New York: Industrial Press, Inc.

Plumbing and Mechanical website, www.pmmag.com, *Throw Away Your Torches,* Julius Ballanco, P.E. http://www.pmmag.com/pm/cda/articleinformation/features/bnp__features__item/0,,5020,00+en-uss_01dbc.html, publication date: June 2000, reviewed July 2003.

REFERENCES

Canadian Copper and Brass Development Association website, www.ccbda.org, *Copper Tube and Fittings Manual, Publication No. 28E,* www.ccbda.org/publications/pub28E/28e-publicationp1.html, reviewed July 2003.

The Copper Tube Handbook, 1995. New York: Copper Development Association.

Efficient Building Design, Volume 3: Water and Plumbing, 2000. Ifte Choudhury and J. Trost. Upper Saddle River, NJ: Prentice Hall.

NSF International website, www.nsf.org, *Certification Underway for Press-Connect Fittings,* www.nsf.org/newsletters/plumbing01-1/page4.html, reviewed July 2003.

NCCER CURRICULA — USER UPDATE

NCCER makes every effort to keep its textbooks up-to-date and free of technical errors. We appreciate your help in this process. If you find an error, a typographical mistake, or an inaccuracy in NCCER's curricula, please fill out this form (or a photocopy), or complete the online form at **www.nccer.org/olf**. Be sure to include the exact module ID number, page number, a detailed description, and your recommended correction. Your input will be brought to the attention of the Authoring Team. Thank you for your assistance.

Instructors – If you have an idea for improving this textbook, or have found that additional materials were necessary to teach this module effectively, please let us know so that we may present your suggestions to the Authoring Team.

NCCER Product Development and Revision
13614 Progress Blvd., Alachua, FL 32615

Email: curriculum@nccer.org
Online: www.nccer.org/olf

❑ Trainee Guide ❑ AIG ❑ Exam ❑ PowerPoints Other ______________________________

Craft / Level: Copyright Date:

Module ID Number / Title:

Section Number(s):

Description:

____

____

____

____

Recommended Correction:

____

____

____

____

Your Name:

Address:

____

Email: Phone:

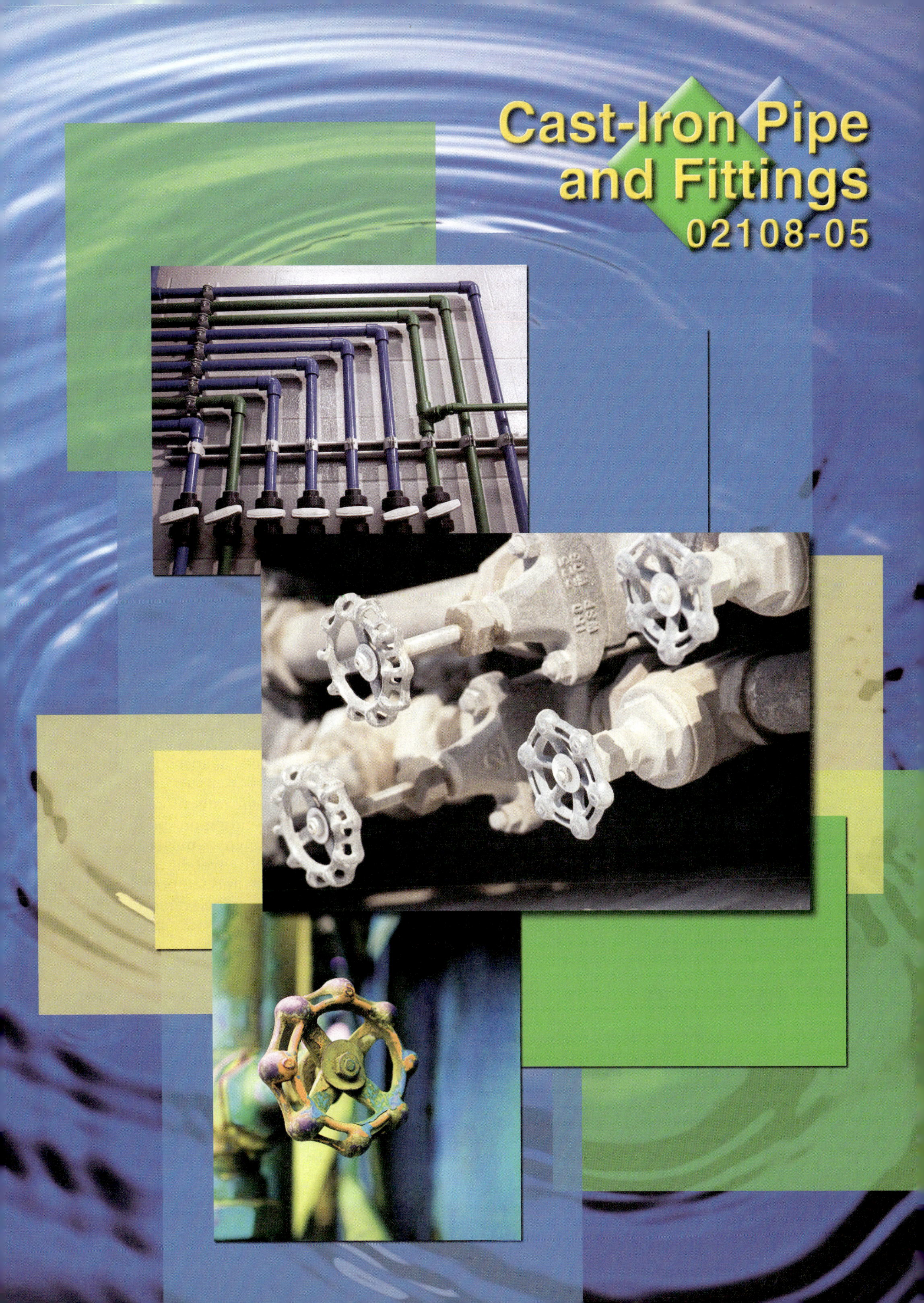
Cast-Iron Pipe
and Fittings
02108-05

02108-05

Cast-Iron Pipe and Fittings

Topics to be presented in this module include:

Overview

Cast-iron pipe is used for drain, waste, and vent (DWV) systems in residential, commercial, and industrial plumbing. Cast iron is strong, durable, and resistant to corrosion. Cast-iron pipe is molded from gray cast iron, which has a controlled carbon content and is brittle, and ductile iron, which is made from molten gray iron and magnesium and is less brittle.

Plumbers work with hub-and-spigot and no-hub cast-iron pipe. Both types are available in various lengths, diameters, and wall thicknesses. Plumbers must be familiar with the materials, schedules, and fittings used with cast-iron piping. Plumbers must be able to measure, cut, and join cast-iron pipe. Compression joints and lead and oakum joints are used to join hub-and-spigot pipe. Pouring these joints requires experience and a variety of special tools. To join no-hub pipe, plumbers use couplings.

Hangers are used to align horizontal and vertical runs of cast-iron pipe. Plumbers select the type of hanger or support depending on the materials to which it is attached. Pipe hangers can be attached to wood, masonry, concrete, or steel using special fasteners designed to handle the weight of the installation. Plumbing codes specify the number of hangers and supports, as well as the required spacing. After installation, inspectors test the system using a hydrostatic, air pressure, smoke, or peppermint test.

Focus Statement

The goal of the plumber is to protect the health, safety, and comfort of the nation job by job.

Code Note

Codes vary among jurisdictions. Because of the variations in code, consult the applicable code whenever regulations are in question. Referencing an incorrect set of codes can cause as much trouble as failing to reference codes altogether. Obtain, review, and familiarize yourself with your local adopted code.

Objectives

When you have completed this module, you will be able to do the following:

1. Recognize proper and improper applications of cast-iron piping.
2. Identify the material properties, storage, and handling requirements of cast-iron piping.
3. Identify the types of materials and schedules used in cast-iron piping.
4. Identify the types of fittings used with cast-iron piping.
5. Identify the various techniques used in hanging and supporting cast-iron piping.
6. Properly measure, cut, and join cast-iron piping.
7. Identify the hazards and safety precautions associated with cast-iron piping.

Key Trade Terms

Bend
Caulking iron
Ceiling caulking iron
Closet bend
Closet flange
Compression joint
Coupling
Double sanitary tee
Double wye
Ductile iron
Heel inlet
Hub-and-spigot cast-iron pipe
Increaser
Insertion length
Inside caulking iron
Joint runner
Ladle
Lead and oakum joint
Lead pot
Long bend
Main stack
No-hub cast-iron pipe
Oakum fiber
Outside caulking iron
Packing iron
Pickout iron
Powder-actuated fastening tool
Propane melting furnace
Rotary hammer drill
Running rope
Sanitary cross
Sanitary tee
Service (SV)
Slope
Soil pipe cutter
Spring clamp
Sway brace
Sweep
Torque wrench
Trap
Wye
Yarning iron

Required Trainee Materials

1. Appropriate personal protective equipment
2. Pencils and paper
3. Soapstone
4. Copy of local adopted code
5. Calculator

Prerequisites

Before you begin this module, it is recommended that you successfully complete *Core Curriculum; Plumbing Level One,* Modules 02101-05 through 02105-05.

This course map shows all of the modules in the first level of the *Plumbing* curriculum. The suggested training order begins at the bottom and proceeds up. Skill levels increase as you advance on the course map. The local Training Program Sponsor may adjust the training order.

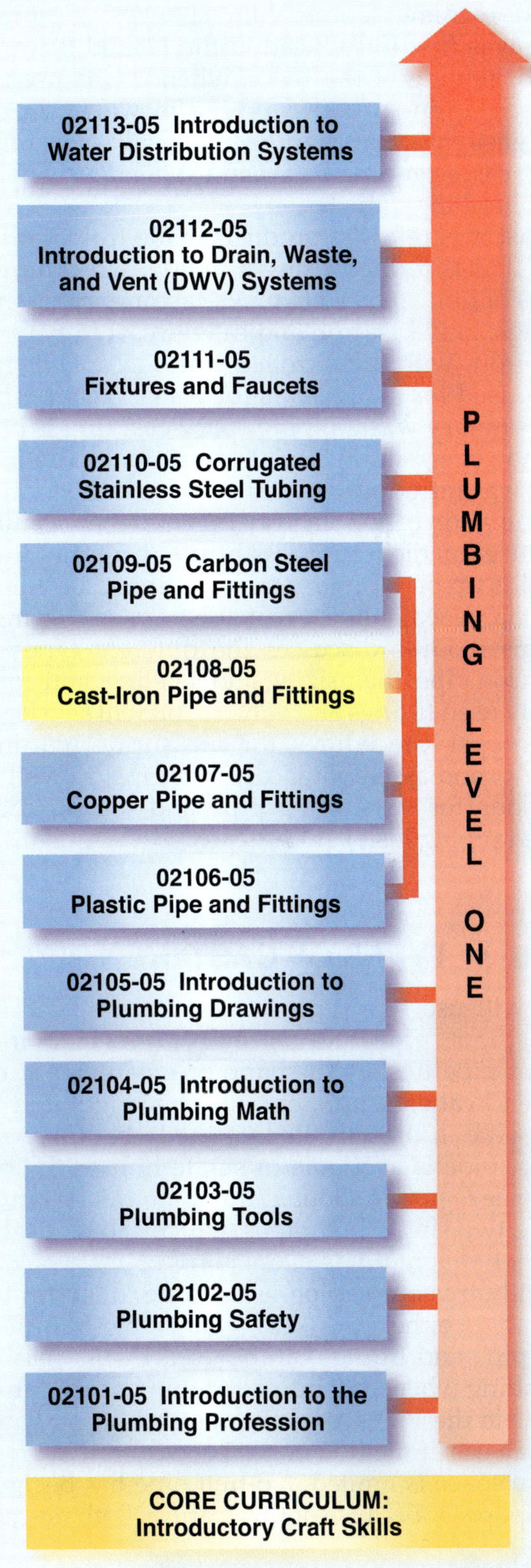

108CMAP.EPS

1.0.0 ◆ INTRODUCTION

Cast-iron pipe, often called soil pipe, is used for drain, waste, and vent (DWV) systems in residential, commercial, and industrial plumbing. Cast iron is a strong and durable piping material that resists corrosion and abrasion. It also can withstand extreme temperature changes. Cast-iron pipe was first introduced in the United States at the beginning of the 1800s, when it was used to replace rotting wooden piping in water supply and gaslight systems in Pennsylvania. Since that time, cast-iron pipe has been used extensively in the plumbing industry.

Cast iron refers to products made from molten iron that is formed using a mold. Two common types of iron are used to make cast-iron pipes and fittings: gray cast iron and ductile cast iron. Gray cast iron, with a controlled carbon content, has been used for many years for water main piping and sanitary waste piping. Gray cast iron is brittle, however, and in 1955, less brittle **ductile iron** pipe was introduced. Ductile iron is made from a combination of molten gray iron and magnesium. Because ductile iron is less brittle than gray iron, it holds up better under the weight of a building. Most large water lines are made from ductile iron.

This module discusses the different types of cast-iron pipe, their sizing and labeling, and commonly used fittings and valves. You will also learn the proper procedures for measuring, cutting, joining, and assembling cast-iron pipe, as well as methods for supporting pipe and testing completed piping installations.

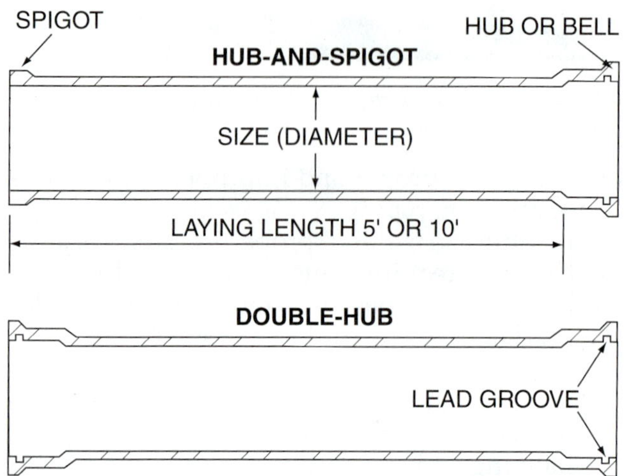

Figure 1 ◆ Hub-and-spigot and double-hub cast-iron pipe.

Figure 2 ◆ No-hub cast-iron pipe.

2.0.0 ◆ TYPES OF CAST-IRON PIPE

You will use two types of cast-iron pipe: **hub-and-spigot pipe** and **no-hub pipe.** Hub-and-spigot pipe has a bell shape or enlargement on one end called a hub, or bell. The spigot end of each pipe in the run fits into the hub. Hub-and-spigot pipe is available in single hub or double hub (see *Figure 1*). Double-hub pipe can be cut to create two single-hub pipes, which can be useful for a job that requires many joints.

No-hub cast-iron pipe, as its name suggests, has no hub on either end (see *Figure 2*). No-hub cast-iron pipe and fittings can be used in all areas of plumbing where cast iron is recommended. Introduced in the 1960s, no-hub pipe and fittings provide an easier method of joining cast iron in areas where space is limited. No-hub pipe has become very popular because of this feature. Both no-hub and hub-and-spigot cast-iron pipe are widely used by plumbers today.

2.1.0 Sizes

Single hub-and-spigot cast-iron pipe is available in lengths of 2½, 3½, 5, and 10 feet and in diameters of 2, 3, 4, 5, 6, 8, 10, 12, and 15 inches. The smaller diameters are used in residential construction. The length of the pipe, referred to as the laying length, is the measurement from the base of the hub to the end of the spigot.

Double-hub pipe is available in 30-inch and 60-inch lengths. Pipe in 30-inch lengths can be cut into two short single-hub pipes. No-hub cast-iron

DID YOU KNOW?

Rock with high concentrations of iron is called iron ore. This ore is mined from beneath the earth's surface. To convert iron ore into iron products, manufacturers refine the ore to remove impurities. Refining iron ore requires extreme heat and a reducing agent to draw oxygen out of the iron ore. This reducing agent, called coke, is a grayish-black substance composed mostly of carbon. When the coke is burned, it gives off intense heat and draws the oxygen out of the iron ore. Limestone is also added to the coke to remove unwanted minerals such as sulfur and phosphorus from the iron. The result is refined iron that manufacturers can use to make iron products.

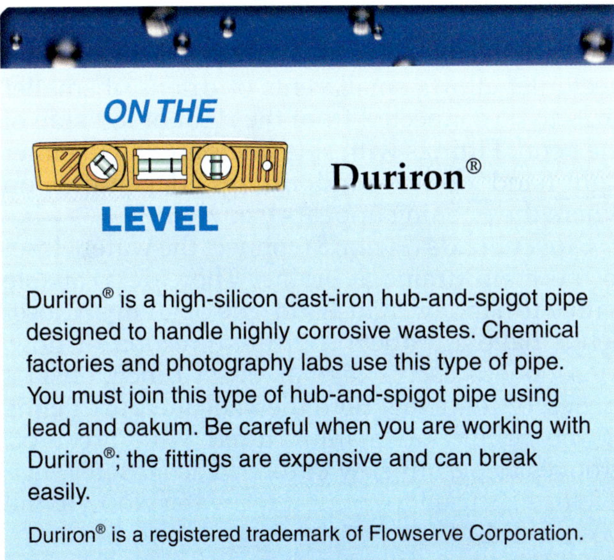

ON THE LEVEL

Duriron®

Duriron® is a high-silicon cast-iron hub-and-spigot pipe designed to handle highly corrosive wastes. Chemical factories and photography labs use this type of pipe. You must join this type of hub-and-spigot pipe using lead and oakum. Be careful when you are working with Duriron®; the fittings are expensive and can break easily.

Duriron® is a registered trademark of Flowserve Corporation.

pipe is manufactured in lengths of 10 feet. It is available in diameters of 1½, 2, 3, 4, 5, 6, 8, 10, 12, and 15 inches. No-hub and hub-and-spigot pipe are generally made to the same thickness.

Fewer types of fittings are available for pipe sizes larger than 8 inches. Generally, when sizes above 15 inches are required under a building, the specifications will require you to use ductile iron pipe, as permitted by applicable codes.

Cast-iron pipe is classified by its wall thickness. Manufacturers make two thicknesses of pipe: **service** and extra-heavy, which is available only for hub-and-spigot pipe. Service is marked SV; the term *service* indicates the thickness of the pipe, which can vary by size. Extra-heavy is marked XH. Service is the most commonly used hub-type pipe. However, if a structure is subject to vibration, settling, or excessive corrosion, the specifications or codes may require extra-heavy pipe and fitting. You should always use fittings that are made to work specifically with the class of pipe you are using.

DID YOU KNOW?

Man has understood how to use iron longer than he has understood how to use any other metal. Prehistoric people used iron from meteorites to make tools, weapons, and other objects. Refined iron tools and weapons marked the Iron Age, which began about 1200 B.C.E. in the Near East. Until the 1800s, iron was the main industrial metal. With the advent of the Industrial Revolution in the mid-1800s, manufacturers discovered an economical way of making steel (a mixture of iron and carbon). Steel soon replaced iron as the basic material for making tools and equipment for industry.

2.2.0 Labeling

All cast-iron pipe and fittings with the "C_I" trademark (C_I®) are manufactured according to the standards set by the Cast Iron Soil Pipe Institute (CISPI). According to these standards, all pipe must be measured before the manufacturer can distribute it. In addition, the manufacturer must label all pipe and fittings with its name or trademark and with the CISPI trademark; fittings are frequently also labeled with the fitting's diameter (see *Figures 3* and *4*). On no-hub cast-iron pipe, the label appears about 1½ inches from the end of the pipe.

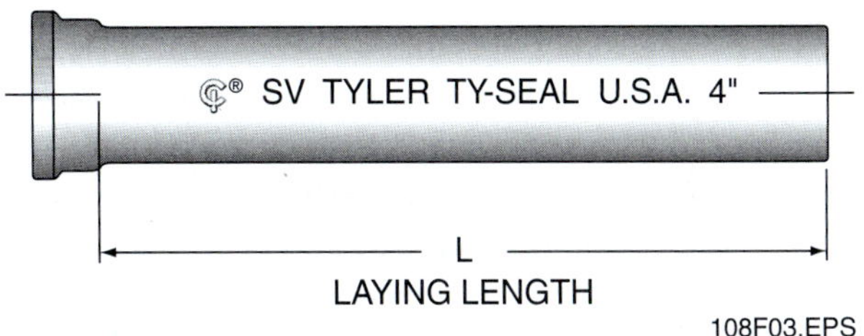

Figure 3 ◆ Pipe label.

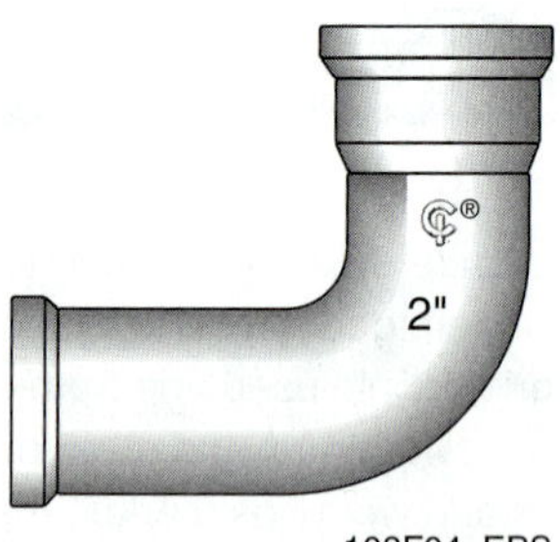

Figure 4 ◆ Fitting label.

CAUTION

If no-hub plumbing materials are not labeled correctly, it may be an indication that they were not manufactured according to the standards of the Cast Iron Soil Pipe Institute. In most cases, plumbing codes do not permit unlabeled connections to be installed.

2.3.0 Material Storage and Handling

To ensure maximum productivity on the job, use common sense when storing and handling cast-iron pipe. Before starting to work, always plan ahead. Organize pipe and fittings into groups by pipe size and type. Store pipe and fittings near where you will be working so they are easy to access. When handling pipe and fittings, be careful not to bend or damage them. Pipe that is bent or otherwise damaged not only wastes money, but also makes you less productive. When storing and handling cast-iron pipe, ensure that dirt, insects, rodents, and animals are prevented from entering the pipe.

3.0.0 ◆ FITTINGS

Fittings are used to change the direction of flow in a system. Almost any kind of pipe assembly is possible with the wide variety of pipe fittings made today. Plumbing codes require that different types of fittings be used in different types of plumbing systems. This section identifies the types of cast-iron fittings.

3.1.0 Bends

Bends are fittings that change the direction of piping. Bend dimensions are expressed as fractions of a complete circle. A circle contains 360 degrees. To determine the number of degrees a given bend turns, multiply the fraction by 360 degrees. For example, a ⅛ bend turns the pipe ⅛ × 360 degrees, or 45 degrees. Bends are available in most standard pipe sizes, the smallest being 1½ inches in diameter. See *Figure 5* for common bends.

Long bends (*Figure 6*) are similar to standard bends, except that long bends consist of one leg that is longer than the other. Long bends are typically sized with a two-number system. The first number indicates the diameter of the bend; the second number indicates the length of the longer leg of the fitting. For example, a 4 × 8, ¼ bend is 4 inches in diameter, and its longer leg is 8 inches in length. Long bends are expensive and should be used only at the base of plumbing stacks.

Sweeps are similar to ¼ bends, but the radius of the curve is much greater (refer to *Figure 5*). The greater radius allows for smoother flow of wastes through the piping system. Two types of sweeps exist: long sweeps and short sweeps. They differ in the radius of the curve made by the fittings and by their laying length. Long sweeps have a longer radius and a longer laying length; short sweeps have a shorter radius and a shorter laying length. Always check applicable codes to determine which sweep you should use for your installation.

A number of bends have side or **heel inlets.** The heel inlet allows small vents or drains of smaller sizes to be connected from the right or left side of the bend. Fittings with side inlets are called either right-hand or left-hand fittings, depending on where the opening appears (see *Figure 7*).

Closet bends (*Figure 8*) connect the water closet to the main drainage piping. They are available with left-hand or right-hand side openings. Closet bends have a reduction in the direction of flow. Closet bends can be regular, offset 1 inch, offset 2 inches, or reducing. Both the inlet and outlet ends of the fitting are manufactured with break-off grooves to permit easy cutting. The **closet flange** (*Figure 9*) is used to connect the water closet to the closet bend.

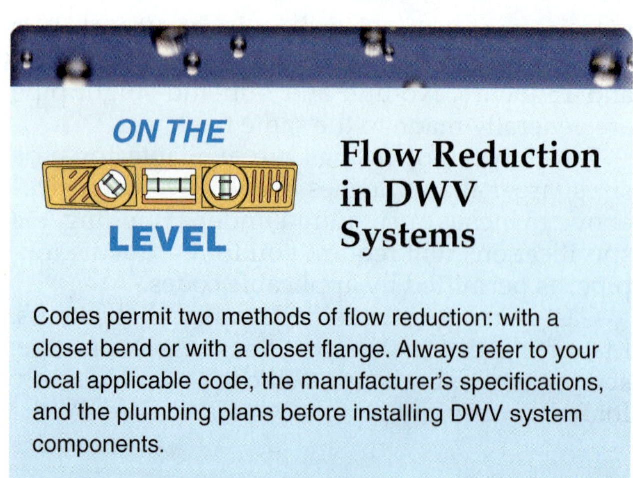

Flow Reduction in DWV Systems

Codes permit two methods of flow reduction: with a closet bend or with a closet flange. Always refer to your local applicable code, the manufacturer's specifications, and the plumbing plans before installing DWV system components.

BEND	DEGREE OF TURN
1⁄4	90
1⁄5	72
1⁄6	60
1⁄8	45
1⁄16	22½

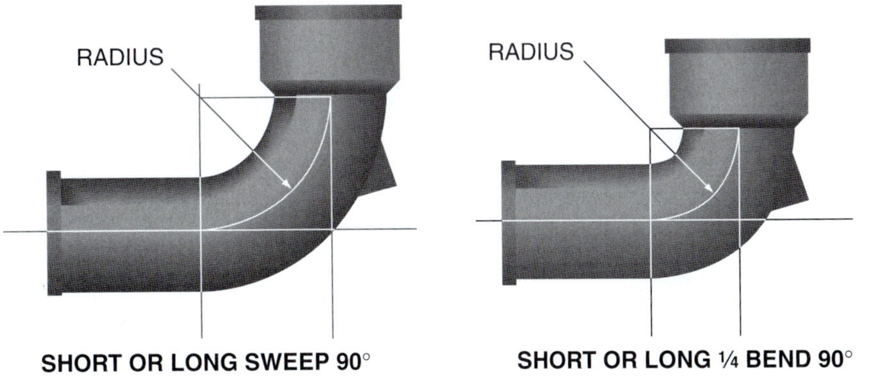

RADIUS
SHORT OR LONG SWEEP 90°
RADIUS
SHORT OR LONG 1⁄4 BEND 90°

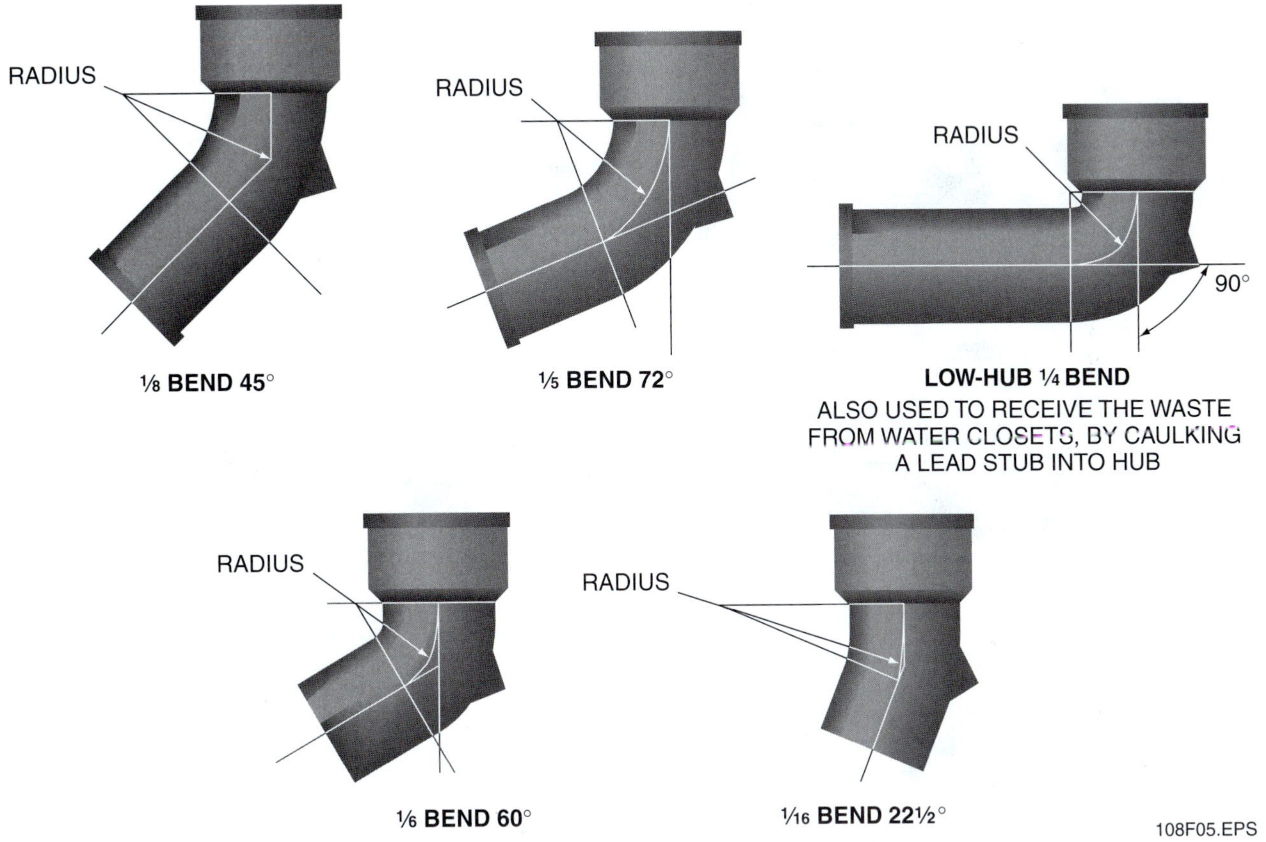

RADIUS
1⁄8 BEND 45°
RADIUS
1⁄5 BEND 72°
RADIUS
90°
LOW-HUB 1⁄4 BEND
ALSO USED TO RECEIVE THE WASTE FROM WATER CLOSETS, BY CAULKING A LEAD STUB INTO HUB
RADIUS
1⁄6 BEND 60°
RADIUS
1⁄16 BEND 22½°
108F05.EPS

Figure 5 ◆ Common bends.

LONG LEG IS 12" IN LENGTH
4" DIAMETER
4 × 12
LONG 1⁄4 BEND
108F06.EPS

Figure 6 ◆ Long bend.

HUB (OR BELL)
RIGHT INLET
SPIGOT
108F07.EPS

Figure 7 ◆ Bend with heel inlet.

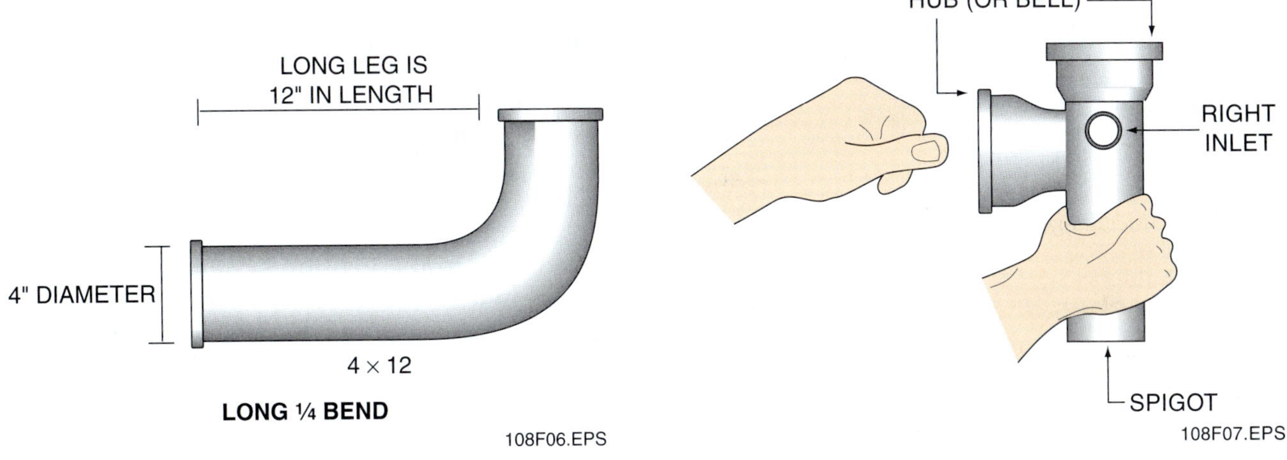

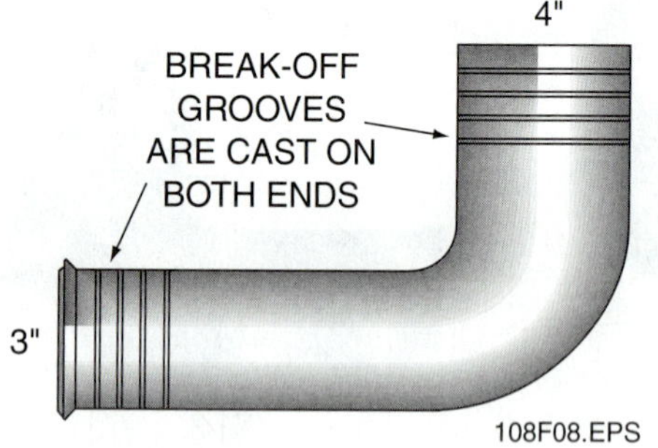

Figure 8 ◆ Closet bend.

Plumbing Codes and Cast-Iron Pipe

Plumbing codes are designed to protect the health and safety of the public. While codes vary from location to location, they are all based on sound principles of sanitation and safety.

Plumbing codes regulate every aspect of cast-iron pipe and fitting selection and installation. Take the time to become familiar with the state, city, or local plumbing codes that apply to the area where you work. Some fittings may be allowed in one area but be strictly forbidden in another area.

Figure 9 ◆ Closet flange.

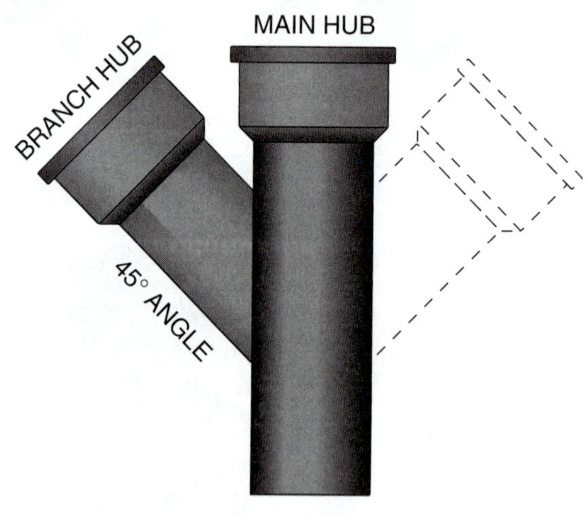

Figure 10 ◆ Wye.

3.2.0 Branches

A variety of fittings allow you to connect pipes to one another to create branches in the piping system. **Wyes** and **double wyes** are used to join two or three drains into one pipe (see *Figure 10*). Wyes connect branches at 45-degree angles to the **main stack,** the vertical main pipe that runs between the building drain and the highest horizontal drain in the system. Sometimes the branch fitting has a threaded or tapped opening in either the main or branch hub. You can connect a cleaned plug (used to close the fitting or pipe) or a threaded pipe from a plumbing fixture to this opening. Use an inverted wye in the venting system. Refer to your local applicable code to determine the correct types of branches for different installations (see *Table 1*).

The **sanitary tee** allows you to connect pipe at right-angle intersections in branches that run from horizontal to vertical (see *Figure 11*). Model

Table 1 DWV Fittings and Their Uses

Type of Fitting Pattern	2003 International Plumbing Code® Table 706.3, Fittings for Change in Direction — Change in Direction: Horizontal to Vertical	Vertical to Horizontal	Horizontal to Horizontal
Sixteenth bend	X	X	X
Eighth bend	X	X	X
Sixth bend	X	X	X
Quarter bend	X	X[a]	X[a]
Short sweep	X	X[a,b]	X[a]
Long sweep	X	X	X
Sanitary tee	X[c]		
Wye	X	X	X
Combination wye and eighth bend	X	X	X

For SI: 1 inch = 25.4 mm

a. The fittings shall only be permitted for a 2-inch or smaller fixture drain.

b. Three inches or larger.

c. For a limitation on double sanitary tees, see *Section 706.3.*

codes restrict the use of sanitary tees to sanitary drainage systems where the flow of materials is from the horizontal to the vertical runs. The **double sanitary tee,** also called a **sanitary cross,** allows you to connect two branch lines entering the system from opposite directions. This fitting allows you to place the horizontal runs and vertical runs centrally. You can also use tees on vent lines. Tees are available with threaded or tapped openings as well as side openings. Like bends, these fittings are designed with right-hand or left-hand openings.

The dimensions of the tee indicate the diameter of the openings. If all openings are the same diameter, only one dimension is given. When the branch openings are smaller than the straight-through openings, the fitting is labeled with two numbers, such as 4 × 3. The first number is the diameter of the openings on the straight-through run; the second number is the diameter of the branch openings.

3.3.0 Increasers

An **increaser** (*Figure 12*) is a fitting that connects two pipes of different diameters. It is used to increase the size of a straight-through line of pipe. Increasers are commonly used to connect a larger pipe to the top of a vent stack. This provides for a larger-sized pipe above the heated area of a building, which keeps frost buildup from closing the opening in the vent stack.

3.4.0 Traps

Traps are fittings designed to provide a liquid seal that prevents sewer gas from escaping into a building. Traps are installed between the drainage line and fixtures or drains. Many traps are designed with cleanout plugs that you can remove to clean out the trap. Common traps include P-traps and S-traps (*Figure 13*). P-traps are commonly used at sinks and lavatories. S-traps are no longer installed but are still found in older buildings. Running traps may be illegal in one area but not in another. Always consult applicable codes before working with running traps.

Traps

The trap is a simple, elegant solution to the plumbing problem of separating a living space from a sewage system. The water seal in a trap prevents gas and fumes from entering the building through the fixtures. Traps also act as a barrier to vermin that may enter a building's living space through the plumbing piping. Nonetheless, waste systems without traps are still found in some parts of the world.

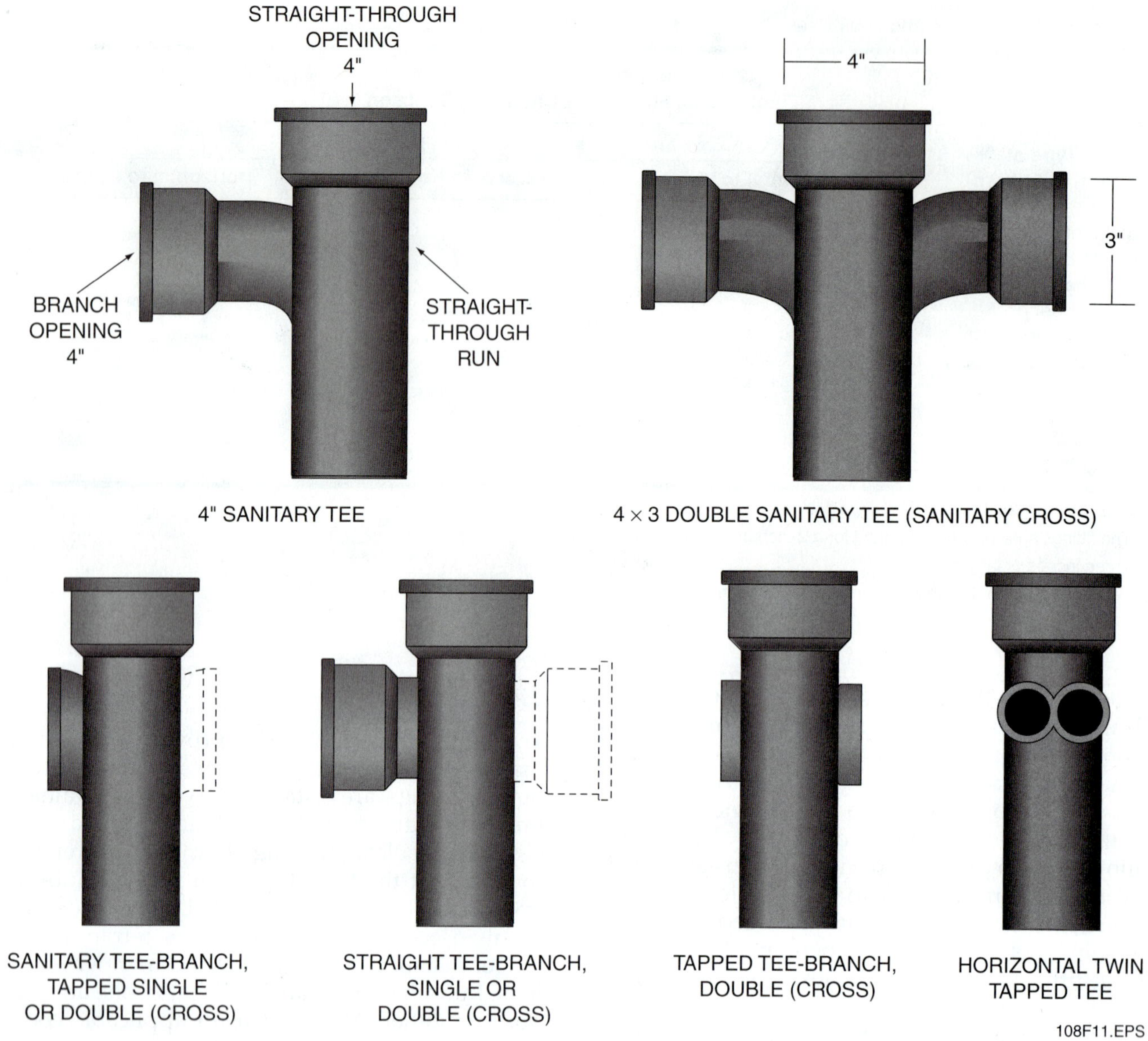

STRAIGHT-THROUGH
OPENING
4"
BRANCH
OPENING
4"
STRAIGHT-
THROUGH
RUN
4" SANITARY TEE
4"
3"
4 × 3 DOUBLE SANITARY TEE (SANITARY CROSS)
SANITARY TEE-BRANCH,
TAPPED SINGLE
OR DOUBLE (CROSS)
STRAIGHT TEE-BRANCH,
SINGLE OR
DOUBLE (CROSS)
TAPPED TEE-BRANCH,
DOUBLE (CROSS)
HORIZONTAL TWIN
TAPPED TEE
108F11.EPS

Figure 11 ◆ Tees.

HUB
SPIGOT END
108F12.EPS

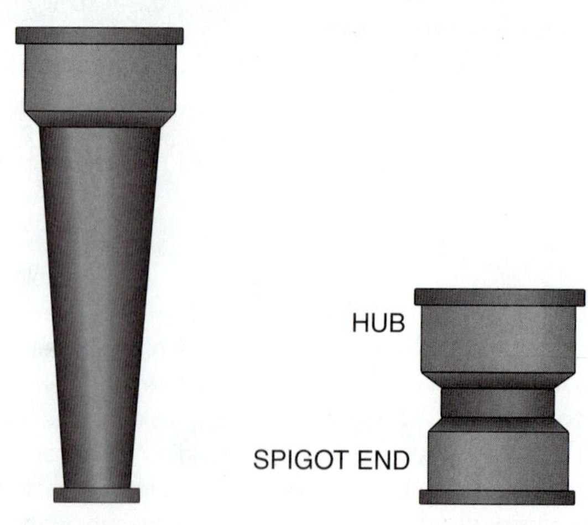

Figure 12 ◆ Increasers.

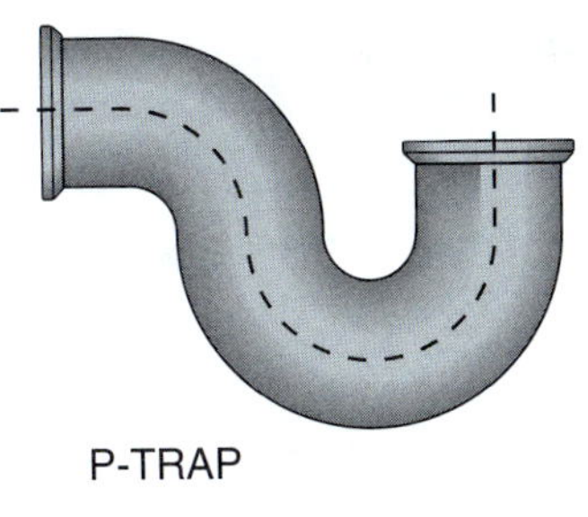

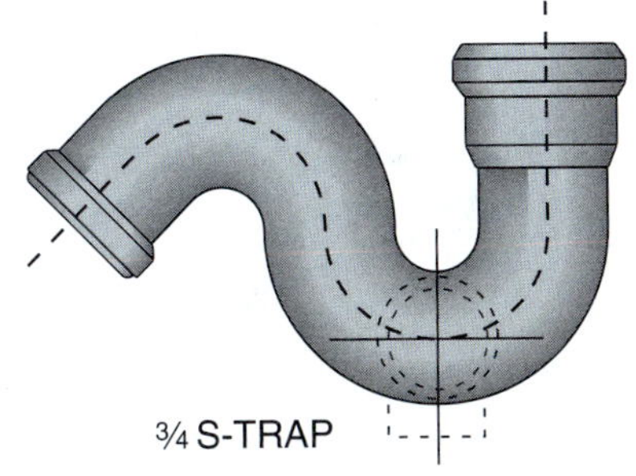

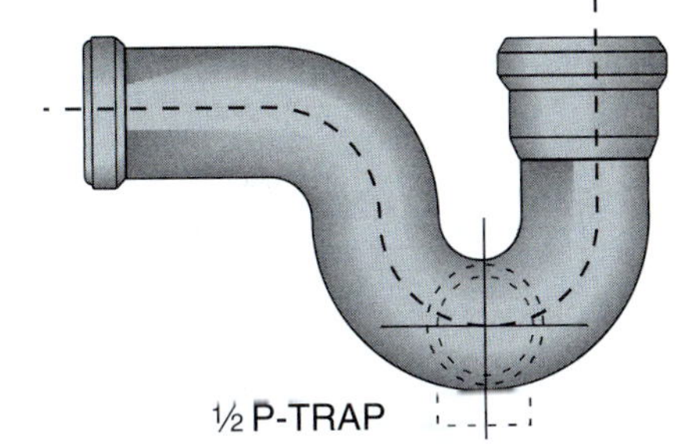

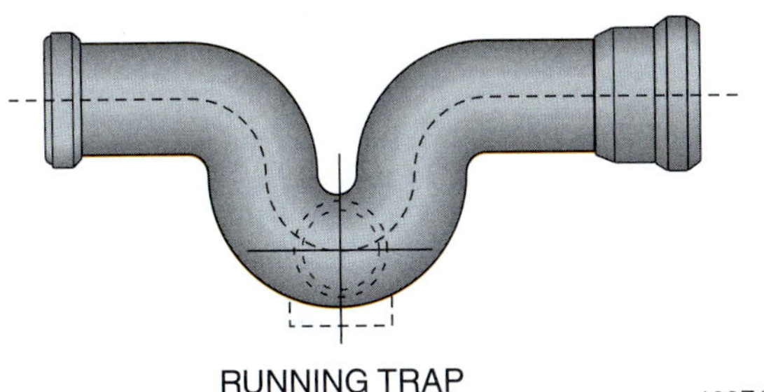

108F13.EPS

Figure 13 ◆ Traps.

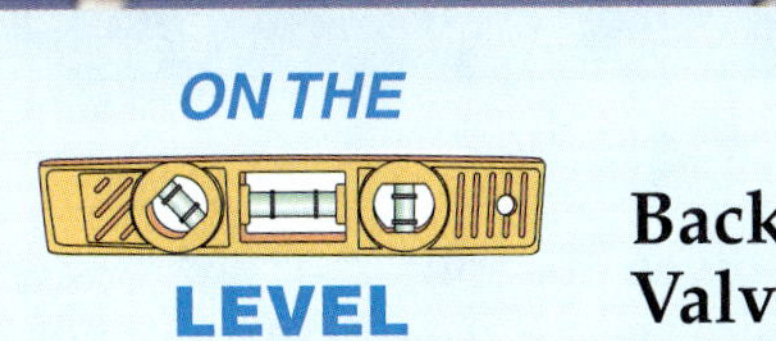

Backwater Valves

The backwater valve is a specially designed cast-iron check valve for drainage pipe systems. A check valve prevents fluid from going backward in the pipe system; the backwater valve prevents the backflow of sewage in the system. The valve is available for both service hub-and-spigot pipe and no-hub pipe connections. When plumbers install backwater valves underground, they usually install them in a valve box or vault so they can access them easily. Backwater valves are for special applications only, and they are the only type of valves that some jurisdictions allow. Refer to applicable codes.

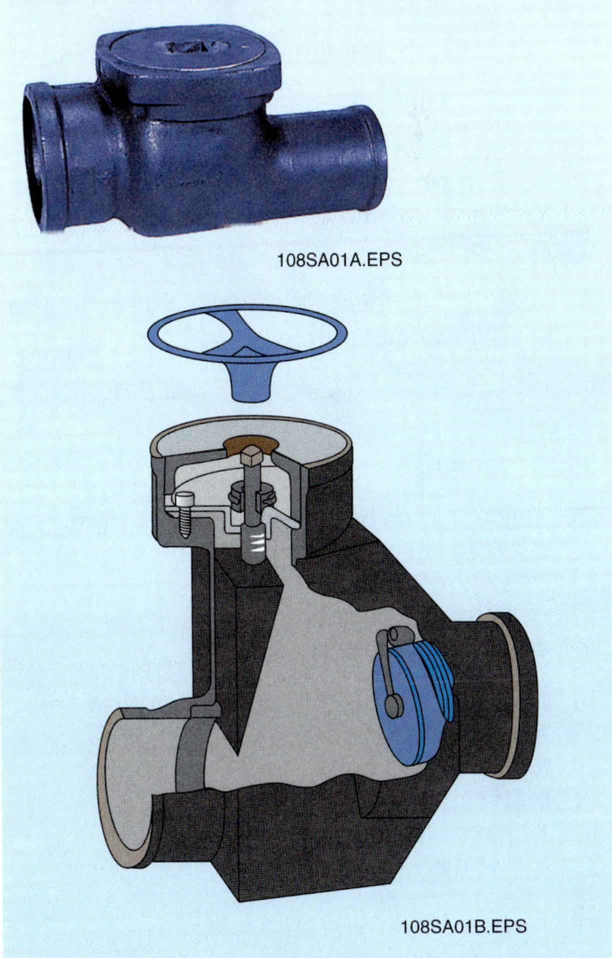

108SA01A.EPS

108SA01B.EPS

Review Questions

Sections 2.0.0–3.0.0

1. _____ pipe gives the plumber an easier method of joining cast iron in areas where space is limited.
   a. Gray iron
   b. Ductile
   c. No-hub
   d. Hub-and-spigot

2. When pipe sizes above 15 inches are required under a building, the specifications will require you to use _____ pipe.
   a. ductile iron
   b. service-weight
   c. extra-heavy
   d. double-hub

3. _____ pipe and fittings are recommended for structures that are subject to vibration, settling, or excessive corrosion.
   a. Standard
   b. Extra-heavy
   c. Service-weight
   d. Ductile

4. A _____ connects the water closet to the main drainage piping.
   a. closet bend
   b. wye
   c. trap
   d. sanitary cross

5. A _____ joins two drains into one pipe.
   a. closet bend
   b. wye
   c. sweep
   d. sanitary cross

6. A _____ changes the direction of flow from horizontal to vertical.
   a. trap
   b. wye
   c. closet flange
   d. sanitary cross

7. A(n) _____ connects one size pipe to another.
   a. closet flange
   b. wye
   c. increaser
   d. sanitary cross

8. A _____ prevents sewer gas from escaping into the building.
   a. closet bend
   b. wye
   c. trap
   d. sweep

Match the following fittings with their corresponding pictures.

9. ⅛ bend _____

10. Long ¼ bend _____

11. Sweep _____

12. Closet bend _____

13. Closet flange _____

14. Wye _____

15. Sanitary tee _____

16. Decreaser _____

17. P-trap _____

18. Tees are typically dimensioned with two numbers, such as 4 × 3. The first number indicates the _____.
   a. length of the straight-through opening
   b. length of the branch
   c. diameter of the straight-through opening
   d. diameter of the branch opening

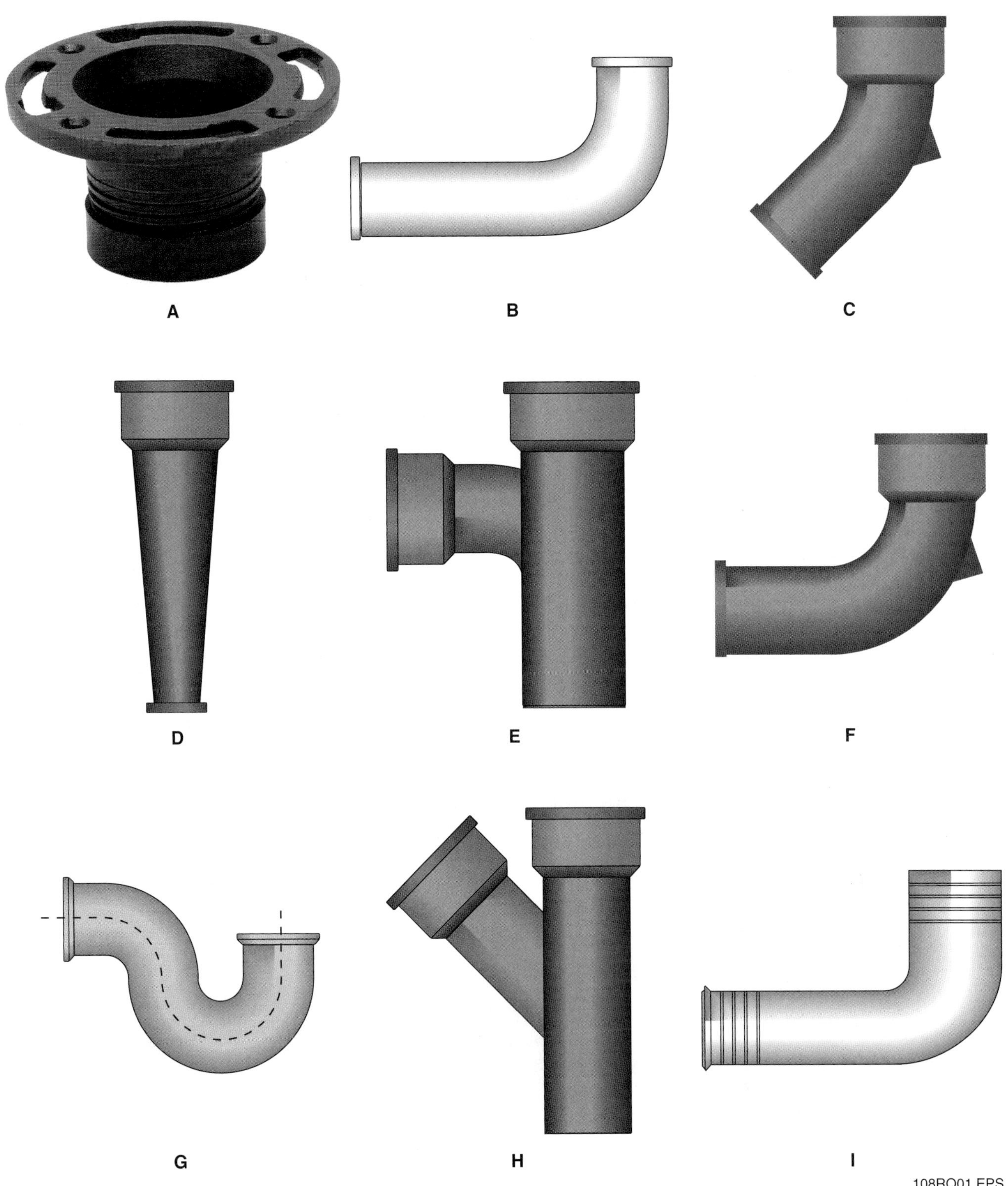
A
B
C
D
E
F
G
H
I
108RQ01.EPS

4.0.0 ◆ MEASURING, CUTTING, AND JOINING

This section describes procedures for measuring, cutting, joining, and assembling cast-iron pipe using **lead and oakum joints.** Be sure to measure and cut pipe carefully. Inaccurate measuring and improper or sloppy cutting cost a business money. If a pipe is cut too short, it is useless. If a pipe is cut improperly, it cannot be used to make a secure joint. Measuring and cutting hub-and-spigot pipe is more involved than working with no-hub pipe.

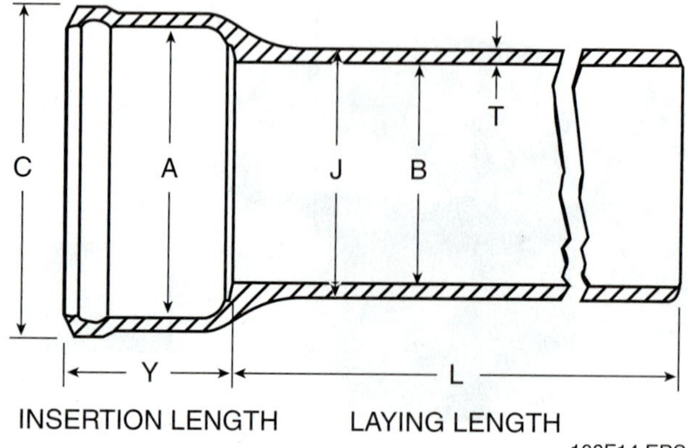

Figure 14 ◆ Measuring hub-and-spigot pipe.

4.1.0 Measuring and Cutting Hub-and-Spigot Pipe

Hub-and-spigot cast-iron pipe has a laying length of 5 feet or 10 feet, measured from the base of the hub to the end of the spigot. In connecting long runs containing many sections of pipe, you must add the **insertion length** to the length of pipe to get a correct measurement. The insertion length is the length of the spigot end that is inserted into the hub (see *Figure 14*). Insertion lengths for the most commonly used pipe sizes are 2½ inches for 2-inch pipe; 2¼ inches for 3-inch pipe; and 3 inches for 4-, 5-, or 6-inch pipe.

To measure hub-and-spigot pipe correctly, you must also know how to locate the center of a fitting. The center of a fitting is the point where the lines of flow through the fitting intersect (see *Figure 15*). For example, in a drainage system, if you are connecting a ¼-bend to a ⅛-bend fitting with a length of soil pipe, and if the center-to-center measurement must be 36 inches, the length of pipe is actually less than 36 inches, because the fittings take up part of the overall measurement (see *Figure 16*). To make a proper measurement, place the fittings together (see *Figure 17*). Then measure the distance between the center marks on the fittings and subtract that number from the overall measurement.

Use a folding rule to measure pipe length. Begin by putting the fittings close together. Then place the 36-inch mark of the rule on the center mark of the right fitting (see *Figure 18*). The mark on the rule that falls over the center mark of the fitting on the left is the length to cut the pipe.

After you measure the pipe, cut it to the desired length with a **soil pipe cutter.** Soil pipe cutters are available in a variety of models (see *Figure 19*). Each cutter can be used to cut a variety of pipe sizes. The soil pipe cutter uses a length of chain, each link of which contains a cutting wheel. When the chain is tightened around the pipe, the cutters press into the metal and break the pipe cleanly.

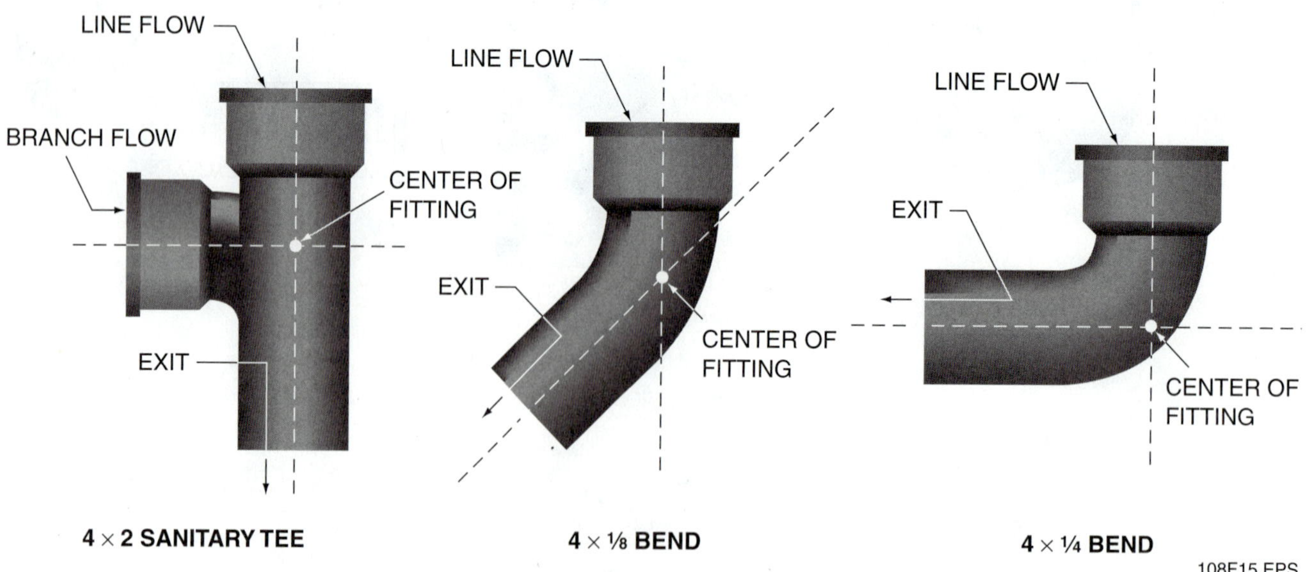

Figure 15 ◆ Finding the center of a fitting.

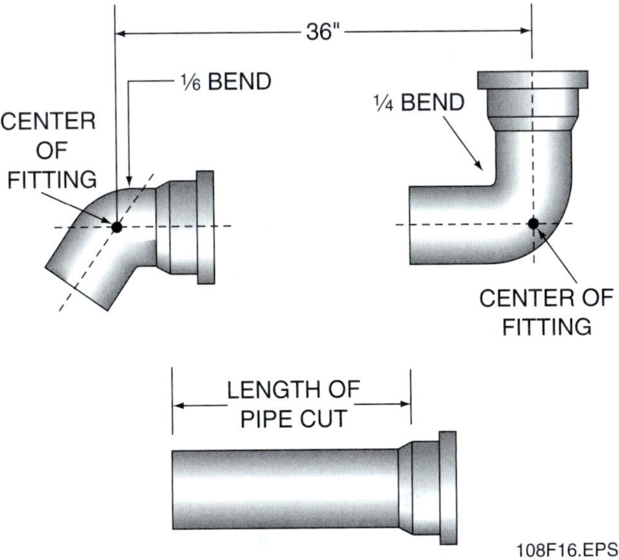

Figure 16 ◆ Determining the cutting length.

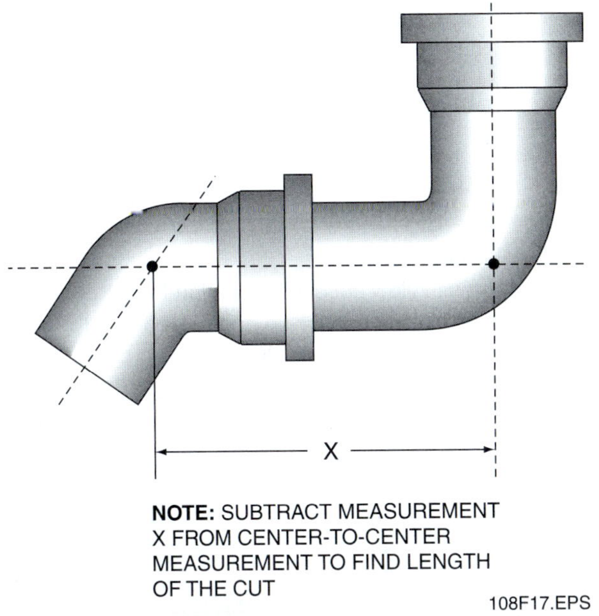

Figure 17 ◆ Placing fittings together.

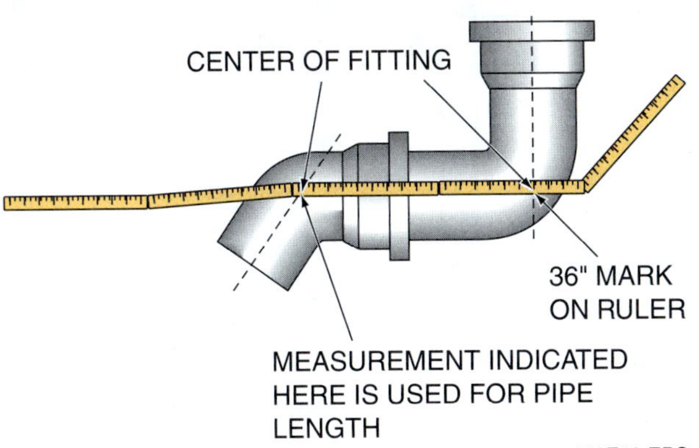

Figure 18 ◆ Using a folding rule.

Figure 19 ◆ Soil pipe cutters.

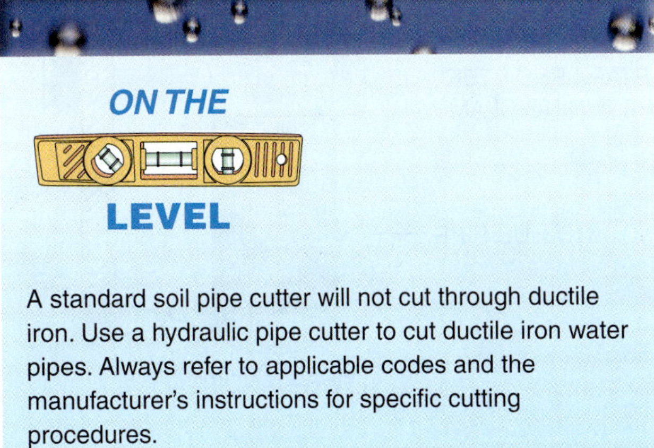

A standard soil pipe cutter will not cut through ductile iron. Use a hydraulic pipe cutter to cut ductile iron water pipes. Always refer to applicable codes and the manufacturer's instructions for specific cutting procedures.

4.2.0 Measuring and Cutting No-Hub Pipe

Measuring no-hub cast-iron pipe is rather simple. Using a folding rule or tape measure, measure the exact distance between the fitting connection and the pipe, or from pipe to pipe, and subtract the allowance for the ridge on the inside of the neoprene gasket (see *Figure 20*).

4.3.0 Joining Hub-and-Spigot Pipe

Hub-and-spigot pipe can be joined by two methods: the **compression joint** or the lead and oakum joint (see *Figures 21* and *22*). Because compression joints are less labor-intensive and less expensive, they are used most often. Whichever method you use, you must ensure proper alignment of the piping to ensure watertight joints. Proper alignment means that the centerline in every straight run of pipe is perfectly straight and that pipe and fittings meet at the correct angle.

4.3.1 Creating Compression Joints

Compression joints are a fast method of joining cast-iron pipe and fittings. They can absorb vibrations and be deflected or bent up to 5 degrees. Compression joints require the use of pipe and fittings that do not have a bead on the spigot end.

To create a compression joint, insert a neoprene gasket into the hub end of the pipe. Use a chain puller, lead hammer, or pushing bar to push or draw the spigot into the hub. This process causes

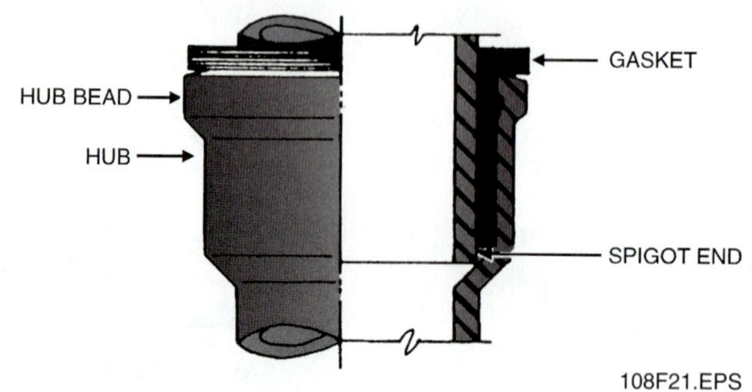

Figure 21 ◆ Compression joint.

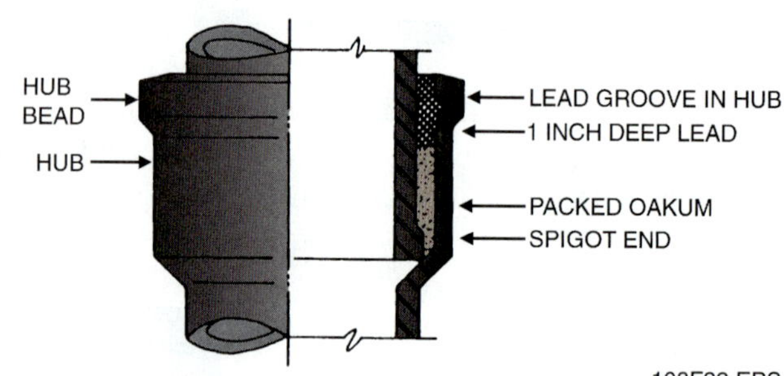

Figure 22 ◆ Lead and oakum joint.

NOTE

Always follow special installation techniques when installing soil pipe above food-handling areas. Consult and follow applicable codes.

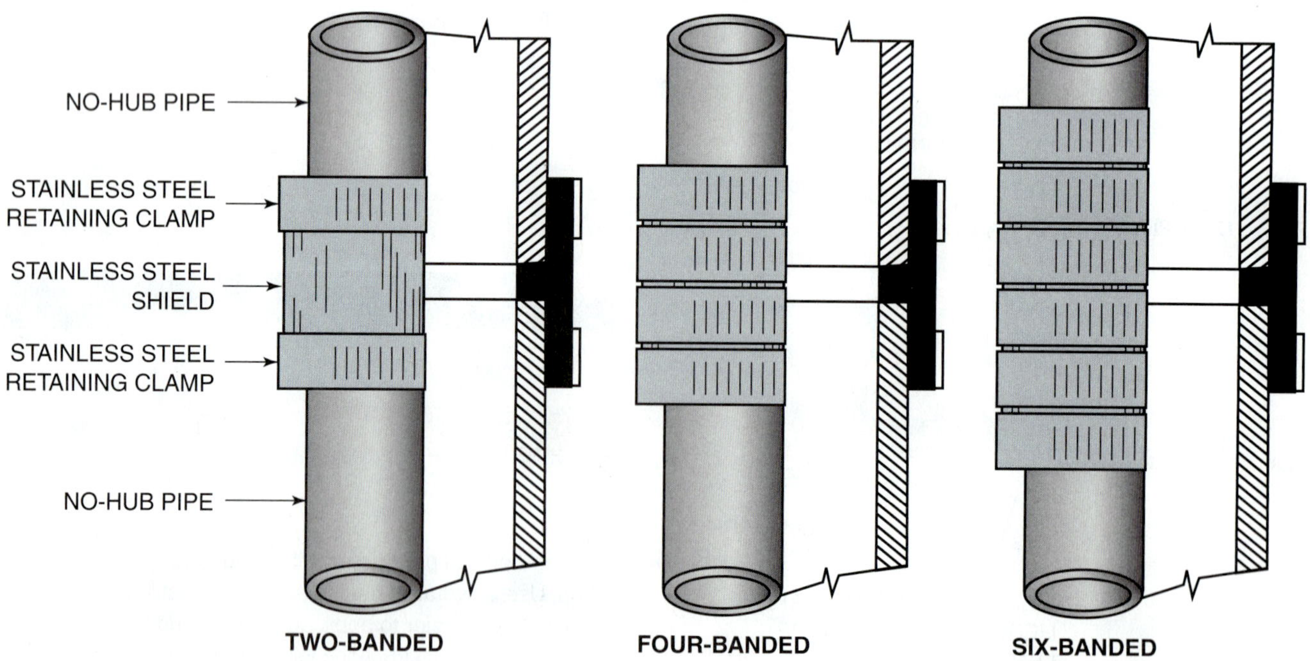

Figure 20 ◆ Gasket allowance.

the neoprene gasket to compress and fill in any air space between the two sections of pipe.

To make a compression joint, follow these steps:

Step 1 Clean the neoprene gasket and pipe ends.

Step 2 Insert the gasket in the hub.

Step 3 Apply a rubber lubricant to the inside of the gasket.

Step 4 Force the spigot end of the pipe into the gasket using a puller tool.

4.3.2 Creating Lead and Oakum Joints

Lead and oakum joints are strong. These joints are made by pouring molten lead over **oakum fiber** to form a watertight seal between the two pieces of pipe, and then caulking the joint after the lead has cooled and solidified. Oakum fiber is a caulking material made of plant fibers that have been treated with tar or a tar derivative. The solidified lead that results from the heating process is called slag.

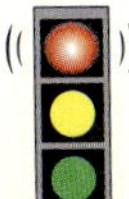

WARNING!

Working with molten lead can be extremely dangerous. For example, moisture in a joint can cause molten lead to explode out of it, causing serious injuries. When you are pouring lead, always wear appropriate PPE, including safety glasses, gloves, high-top shoes or boots, and long pants. Follow all applicable codes regarding specific safety precautions.

For leaded joints there are two types of oakum: white oakum and brown oakum. The entire industry has switched to white oakum because of its uniformity in the packing. Brown oakum, which was a useful packing material in its day, is now used only in certain states as part of their journeyman, master, or contractor license exams. One of the problems with brown oakum is that it comes loose in the packing and prevents a good pour.

Creating lead and oakum joints is a skill that takes much practice. Have patience and realize that you will not learn how to do it overnight. Creating lead and oakum joints also requires many special tools, including the following:

- *Running rope* – This rope is made of fiber and is also referred to as **joint runner.** Running rope is clamped around the pipe to keep the hot, molten lead in the hub during the pouring operation (see *Figure 23*). You must create a gate, or gap, in the rope that allows you to add the lead to the joint. Running rope is used to make horizontal joints (see *Figure 24*).

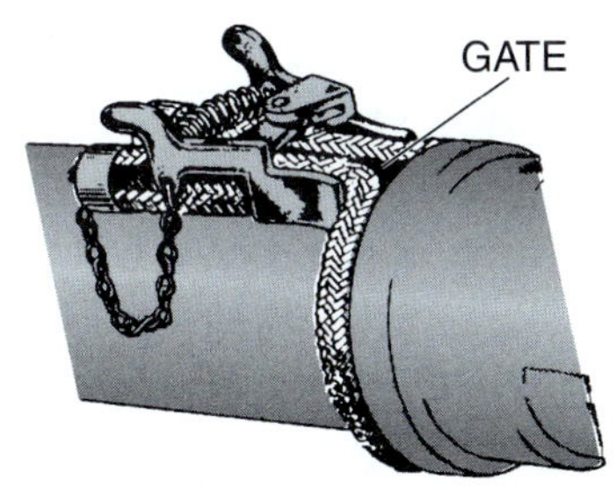

Figure 23 ◆ Running rope and spring clamp.

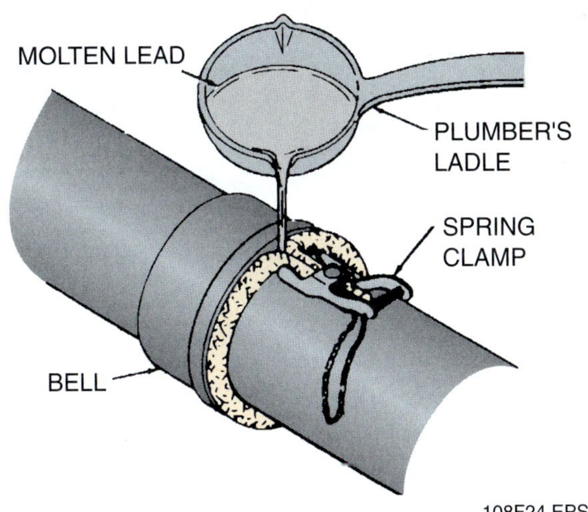

Figure 24 ◆ Horizontal joint.

- *Spring clamp* – This tool is used to clamp the running rope around the pipe next to the hub (refer to *Figure 24*).
- *Propane melting furnace* – This device melts the lead (*Figure 25*). It burns liquefied petroleum (LP) gas. Safety is of the utmost importance in operating a propane furnace. You must follow the manufacturer's operating instructions and maintenance procedures exactly.
- *Lead pot* and *ladle* – The lead pot is the vessel used to hold the molten lead. It is placed on the head of the propane melting furnace during the melting process. The ladle is a spoon-like tool used to dip the molten lead from the pot and pour it into the joint (*Figure 26*).
- *Yarning iron* – This tool is used to pack the oakum fiber into the joint by hand (*Figure 27*). When you use a yarning iron, you do not use a hammer.
- *Packing iron* – This tool has a broad, thick blade and is used to pack oakum into a joint with a hammer (*Figure 28*).
- *Caulking iron* – This tool is used to drive the lead firmly into a joint. Many different types of caulking irons exist (*Figure 29*). Lead shrinks

Figure 25 ◆ Propane melting furnace.

Figure 26 ◆ Lead pot and ladle.

Figure 27 ◆ Yarning iron.

Figure 28 ◆ Packing iron.

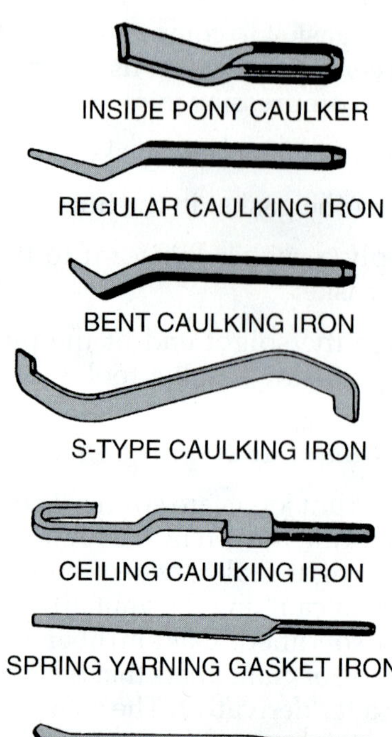

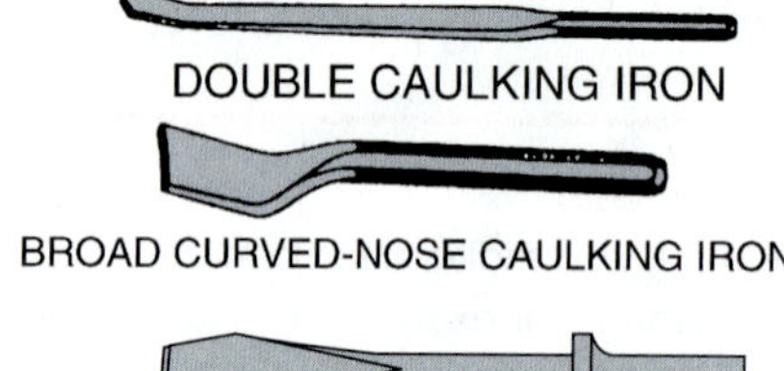

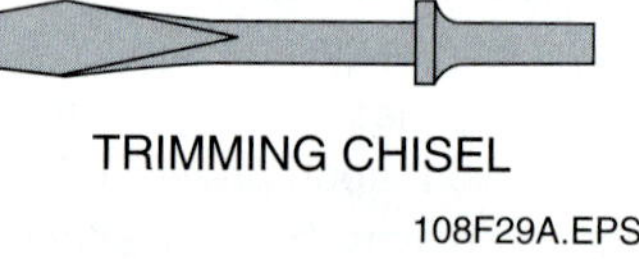

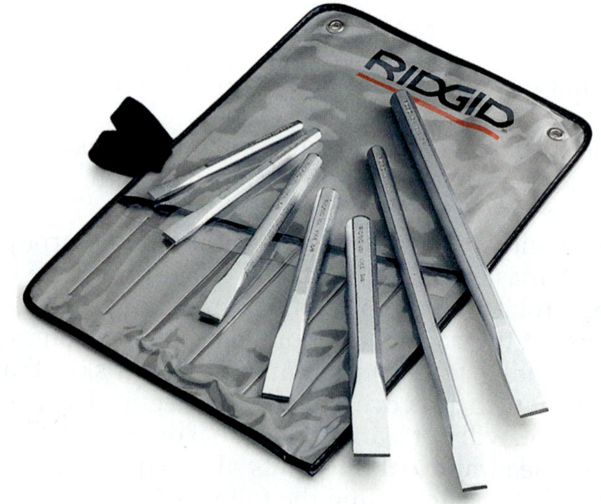

Figure 29 ◆ Caulking irons.

when it cools, so to produce a tight seal, you must caulk the lead into the joint to fill it.

- *Ceiling caulking iron* – This specially shaped tool is used to drive the lead firmly into a joint near the ceiling. Because of its design, this tool is the only caulking iron you can use to caulk a joint next to the ceiling (see *Figure 30*).
- *Inside* and *outside caulking irons* – These tools are used to shape the lead in a joint. The inside caulking iron, which angles to the left from the bottom up, is used to shape the lead near the spigot (see *Figure 31*). The outside caulking iron, which angles to the right, is used to shape the lead to the inside of the hub (see *Figure 32*).
- *Pickout iron* – This tool has a diamond-shaped flat point and is used to remove the lead and oakum from a joint (see *Figure 33*).

Although you may not be required to pour this type of joint, without instruction, the art of pouring brown oakum joints is in danger of being lost. For the sake of plumbing heritage, as well as for practical skill and experience, you should be familiar with the procedure for pouring brown oakum joints. To pour a brown oakum joint, follow these steps:

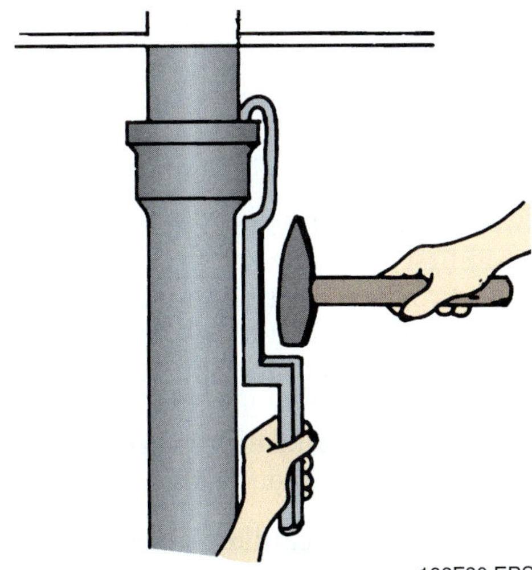

Figure 30 ◆ Using a ceiling caulking iron.

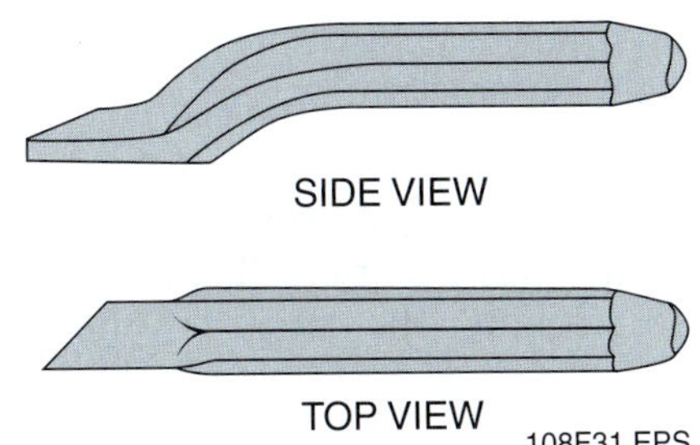

Figure 31 ◆ Inside caulking irons.

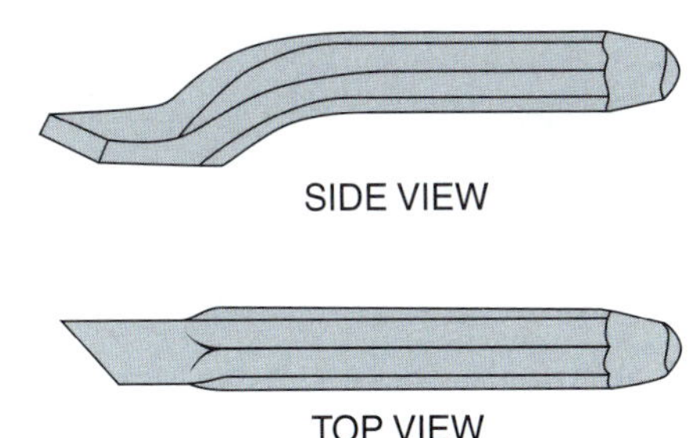

Figure 32 ◆ Outside caulking irons.

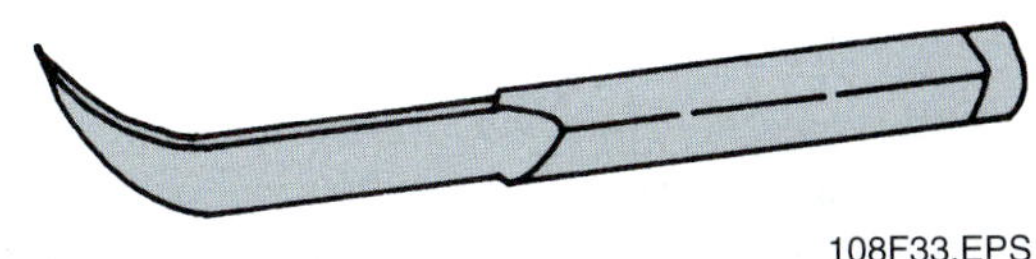

Figure 33 ◆ Pickout iron.

Step 1 Take a string of the brown oakum and, using a yarning iron, make it seat off where the end of the pipe rests against the hub wall.

Step 2 Pack it tightly with a packing iron. This iron is wide and has a flat tip. The packing should be filled up to the top within 1 to 1⅛ inches.

Step 3 After measuring the packing, take the packing iron and heat it up in the lead in the pot.

Step 4 Be sure to wear gloves to protect your hands from the heat in the handle of the iron.

Step 5 Take the hot packing iron and sear the brown oakum, repeating this step as needed to go all the way around the pipe. The searing process will seal the brown oakum down in the pipe and will prevent it from coming loose and ruining the pour.

To ensure a watertight joint, make sure your caulking irons have a good caulking surface. Inspect the surface often and reshape it on a bench grinder as needed to maintain the correct angle. This ensures high-quality, reliable joints. Cleaning the irons and protecting them with a thin film of oil makes them easier to use and makes them last longer.

When you heat the lead in the pot, it will change color. The lead is ready to pour into the joint when it appears to be royal blue. When working with molten lead, you must keep safety in mind at all times. For example, heat the ladle before inserting it into the molten lead; otherwise, it will cause the hot lead to splatter.

Henry Bell's Story

Henry Bell, a plumber in Alabama, lost his sight in one eye while working with lead and oakum joints. This is his story.

"It was 9:00 A.M. Christmas Eve 1967 in Montgomery, Alabama. I was installing hub-and-spigot cast-iron pipe using lead joints. My crew and I were topping out installing vents in the walls and out the roof of a house in a residential subdivision. I had prepared a couple of joints with the oakum and was about to pour them with the molten lead. I dipped the ladle into the lead melting pot and was attempting to turn the burner control valve down because the lead was red hot. It is unclear if there was some moisture on the ladle, or perhaps a few drops of perspiration fell from my face or hands, but the result caused a tragic accident that left me permanently blind in one eye. Local hospitals could not treat my injury. I was transported to the Eye Clinic in Birmingham, where they were able to remove the lead particles from both eyes. The lead particles are on display at the clinic today. After 18 months of surgery and grafting, I regained sight in the other eye. I have had numerous surgeries over the years to keep my sight in that eye."

The Bottom Line: Always wear the appropriate eye protection and keep ladles dry at all times.

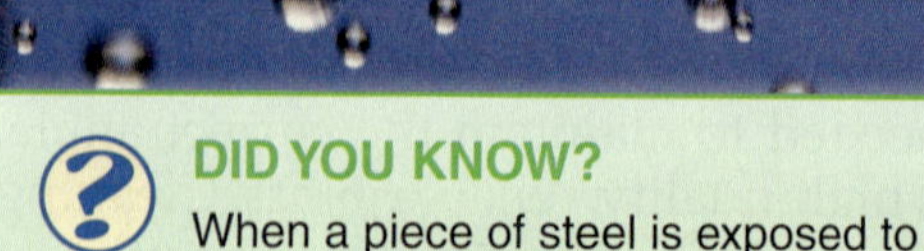

DID YOU KNOW?

When a piece of steel is exposed to heat in the presence of air, it changes colors as the heat increases. The color change is caused by a thin film of oxide that forms as the steel gets hotter. The colors change from a faint straw, to a deeper straw, to dark brown with purple, to dark blue, and, finally, to light blue.

Source: Scale Aero's website. "Tempering Equals Spring/Flex & Strength." www.scaleaero.com/heating_metal.htm, March 15, 2004.

You can use the lead and oakum joining method to create three vertical caulked joints and horizontal caulked joints. With all joints, take care when caulking. If you caulk the lead too tightly, you may create pressures high enough to crack the pipe. If this occurs, you must replace the broken section of pipe. Follow these steps to create each type of joint.

To make vertical caulked joints, follow these steps (see *Figure 34*):

Step 1 Dry the hub and spigot ends, and wipe off any foreign materials. If necessary, dry the ends with a heating torch to eliminate all traces of moisture.

Step 2 Slide the spigot end into the hub of the other pipe, and align the joint. A cut piece of pipe has no spigot bead, so take extra care to center the cut end in the hub.

Step 3 Using a yarning iron, pack the oakum around the pipe. Repeat this operation until the hub is packed to about 1 inch from its top. Pack the oakum with a hammer and packing iron to make a bed for the molten lead.

Step 4 Using the plumber's ladle, carefully pour the molten lead into the joint. Dip enough lead to fill the joint in one pouring. Allow a minute or two for the molten lead to harden and to change in color from royal blue to dull gray. Usually, 1 pound of lead is melted for each inch of pipe size.

Step 5 Caulk the joint using the outside caulking iron first and then the inside caulking iron. Use firm but light hammer blows to drive the lead down on the oakum. Drive the lead into contact with the spigot surface on one edge and with the inner surface of the hub on the other. Strike the first four blows 90 degrees apart around the joint to set the pipe.

To make horizontal caulked joints, follow these steps (see *Figure 35*):

Step 1 Dry the hub-and-spigot ends and wipe off any foreign materials. If necessary, dry the ends with a heating torch to eliminate all traces of moisture.

Step 2 Slide the spigot end into the hub of the other pipe and align the joint. A cut piece of pipe has no spigot bead, so take extra care to center the cut end in the hub.

Step 3 Using a yarning iron, pack the oakum around the pipe. Repeat this operation until the hub is packed to about 1 inch from

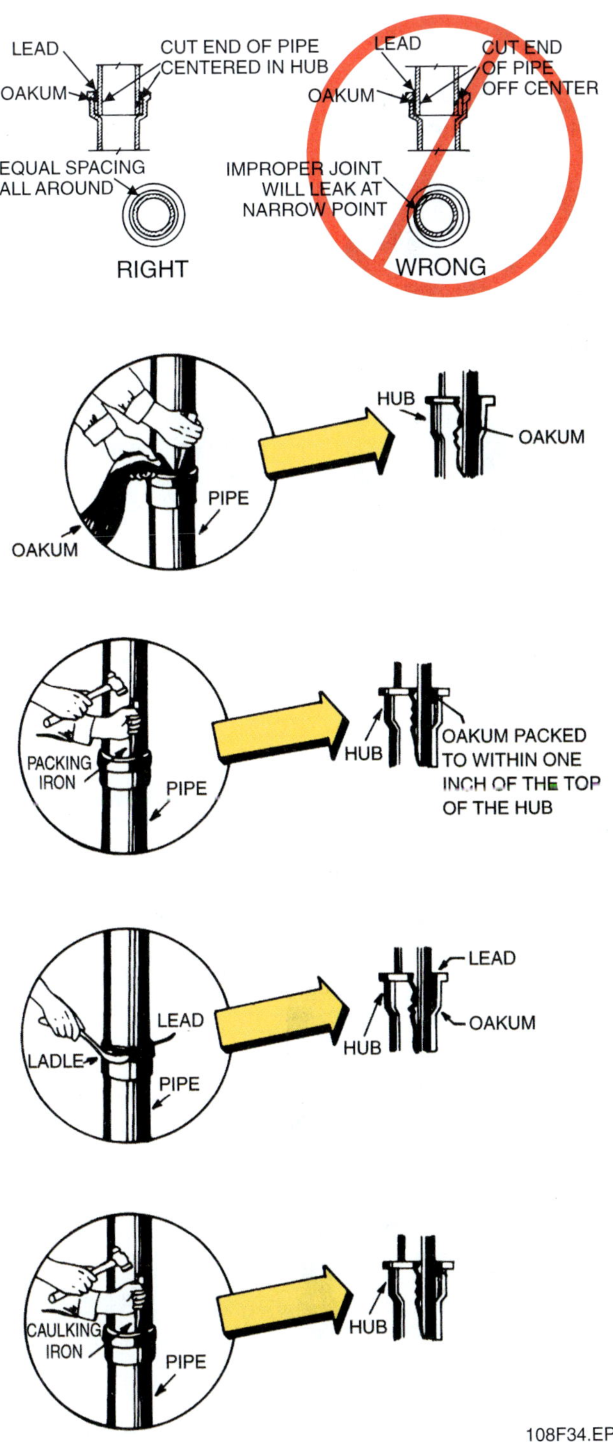

Figure 34 ◆ Making vertical caulked joints.

its top. Pack the oakum with a hammer and packing iron to make a bed for the molten lead.

Step 4 Clamp the running rope in place around the pipe and fill the joint with molten lead.

Figure 35 ◆ Making horizontal caulked joints.

Step 5 After the lead hardens, remove the running rope and trim off the surplus.

Step 6 Use firm but light blows of a ball peen hammer to drive the lead down on the oakum. Strike the first four blows using the caulking iron 90 degrees apart around the joint to set the joint. Use the caulking irons to caulk the joint. Use the inside caulking iron first and then the outside caulking iron. Drive the lead into contact with the spigot surface on one edge and with the inner surface of the hub on the other.

4.4.0 Joining No-Hub Pipe

To join no-hub pipe, use **couplings** (*Figure 36*). Couplings have three components: a gasket, a stainless steel shield, and clamps. Gaskets and clamps are manufactured to conform to the regulations of the CISPI. The gaskets are made of neoprene. They must be labeled as neoprene and

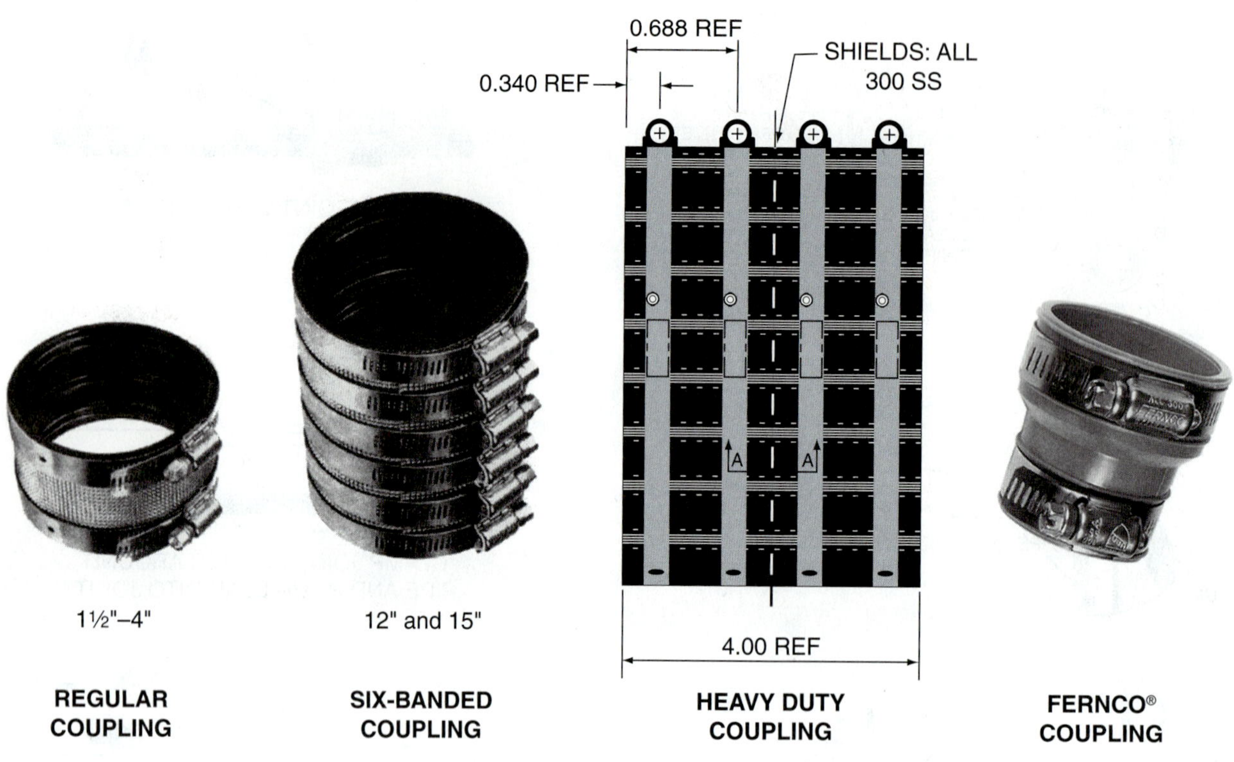

Figure 36 ◆ Couplings.

must have the Institute's trademark and patent number (3, 233, 922).

The gasket is flexible, with a ridge on the inside diameter to control the distance the gasket can be slipped onto the end of no-hub pipe. The shield and clamps are made of stainless steel. They must also be labeled with the Institute's trademark and patent number (3, 233, 922) as well as with the marking "all stainless" or other recognized abbreviations. The corrugated stainless steel shield surrounds the neoprene gasket, with two clamps around the gasket and shield. Transition couplings are available to allow you to join cast-iron pipe to other types of pipe, such as plastic (refer to *Figure 36*). Heavy-duty couplings have four clamps and are available with either a 5⁄16-inch or 3⁄8-inch nut (refer to *Figure 36*).

Before you join the pipe with the coupling, always clean the area of the pipe and fitting that the gasket is to cover. Dirt, mud, sand, or any other material between the gasket and the pipe can cause a leaky connection.

Joining no-hub pipe requires two special tools: a soil pipe cutter and a **torque wrench** (*Figure 37*). The soil pipe cutter is the same tool used to cut hub-and-spigot pipe (refer to *Figure 19*). The torque wrench, which you use to tighten the coupling clamps, has a 5⁄16-inch nut driver. You must tighten each clamp to the correct torque, or approximately 60 inch-pounds of force. Use a torque wrench that is designed to disengage at 60 inch-pounds; you will hear a clicking sound when the wrench disengages.

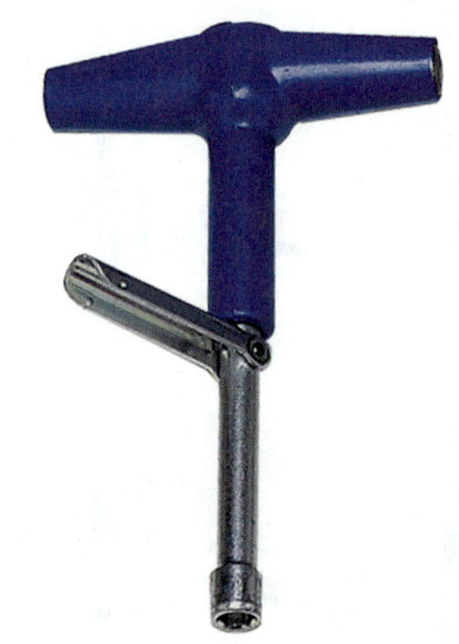

Figure 37 ◆ Torque wrench.

Proper pipe alignment is necessary to guarantee tight joints. Before tightening the clamps, ensure that you have proper alignment and that the coupling is in place. When you are tightening the coupling, be sure that you apply the torque evenly. Switch the wrench from one clamp to the other. Tightening one clamp to 60 inch-pounds before you begin to tighten the other clamp could cause the gasket to become misaligned, and the coupling might leak. One way to be sure you are distributing the clamping pressure evenly between the two clamps is to use an electric drill attachment (*Figure 38*) with the torque wrench. This attachment tightens both clamps at once. Always follow the manufacturer's instructions regarding the appropriate torque for the torque wrench you are using.

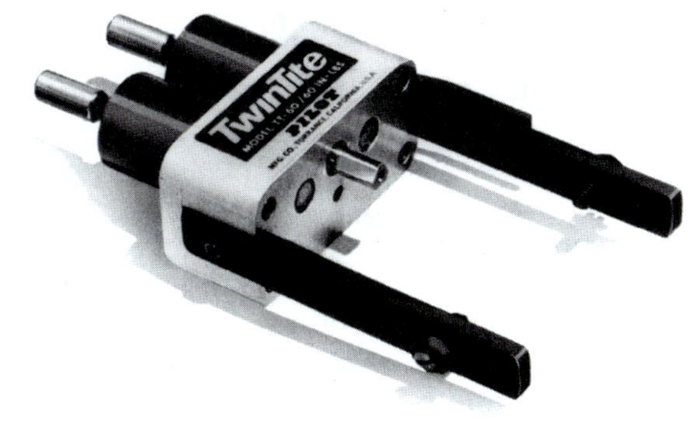

108F38.TIF

Figure 38 ◆ Electric drill attachment.

Review Questions

Section 4.0.0

1. The _____ of a fitting is the point at which the lines of flow through the fitting intersect.
   a. bend
   b. sweep
   c. center
   d. joint

2. All the following tools are used to make a lead and oakum joint *except* a _____.
   a. running rope
   b. gasket
   c. spring clamp
   d. ladle

Match the following tools with their intended use.

3. Yarning iron
4. Packing iron
5. Caulking iron
6. Pickout iron

a. Packs oakum fiber into a joint with a hammer
b. Removes lead and oakum from a joint
c. Removes the gasket from a joint
d. Drives lead firmly into a joint
e. Packs oakum fiber into a joint by hand

7. _____ are used to join no-hub pipe.
   a. Couplings
   b. Bushings
   c. Anchors
   d. Compression joints

8. When using a torque wrench, always ensure the _____ is in place before tightening the clamps.
   a. coupling
   b. bushing
   c. anchor
   d. gasket

5.0.0 ◆ HANGERS AND SUPPORTS

Hangers are used for horizontal or vertical support of pipes and piping. The primary purpose of hangers is to keep the piping in alignment and prevent it from bending, swaying, or distorting. Without proper alignment, waste may not flow through the system properly, and leaks may develop. Plumbing codes require pipe to be securely fastened horizontally and vertically. Never hang one pipe from another or allow one pipe to support another. Always consult applicable codes.

NOTE

The construction drawings and specifications for your job may have unusual pipe support requirements. Always check the construction drawings and specifications before starting your installation.

If you are installing pipe in a seismically active area—that is, where earthquakes are a possibility—local codes will require seismic restraints. The purpose of these requirements is to ensure that the pipe is securely fastened to the structure in the event of excessive vibration. For example, some codes require hangers and supports to be used at closer intervals than in nonseismically active areas. In addition, they may require extra spacing for pipes where they meet walls and floors, to allow for anticipated movement.

5.1.0 Types of Hangers and Supports

The hangers and supports you use will depend on the size of the pipe and the type of material to which the hanger or support is attached. When selecting the types of hangers or supports, you must consider whether the pipe runs horizontally or vertically. All hangers and supports are designed to keep pipe in alignment, but you must also ensure that horizontal pipes do not move back and forth. When selecting supports for vertical pipe runs, you must also take into account the vertical load, or total weight, of the pipe. A variety of hangers are available for horizontal and vertical runs. Proper spacing of hangers and supports is very important.

NOTE

When installing soil pipe above food-handling areas, you must use special installation techniques. As always, consult applicable codes.

5.2.0 Supporting Horizontal Pipe Runs

For horizontal pipe runs, you can use many different types of horizontal support hangers and brackets (*Figure 39*). Your choice will depend on the conditions and the specifications set forth by the contractor, architect, or engineer. Most authorities and plumbing codes specify that pipe must have hangers to support it at each joint. For maximum support, hangers must be no more than 18 inches from each hub. When cast-iron pipe is suspended at certain spacing between nonrigid hangers—generally a spacing of more than 18 inches, but this varies depending on applicable codes—you must also use **sway braces** (see *Figures 40* and *41*) to keep the pipe from swaying. Consult applicable codes when determining the appropriate hanger requirements for your installation.

When your installation involves several pipe runs that are close together and parallel, you can use a properly sized pipe trapeze (*Figure 42*) or brackets made from prepunched channel (*Figure 43*). No matter what kind of support you use on a horizontal run of pipe, be sure to maintain a proper **slope** of ⅛ to ¼ inch per foot. As always, be sure to check all applicable codes.

Horizontal runs of pipe must be installed to provide a slope, which allows the waste to drain away. Slope, also called fall, is the downward

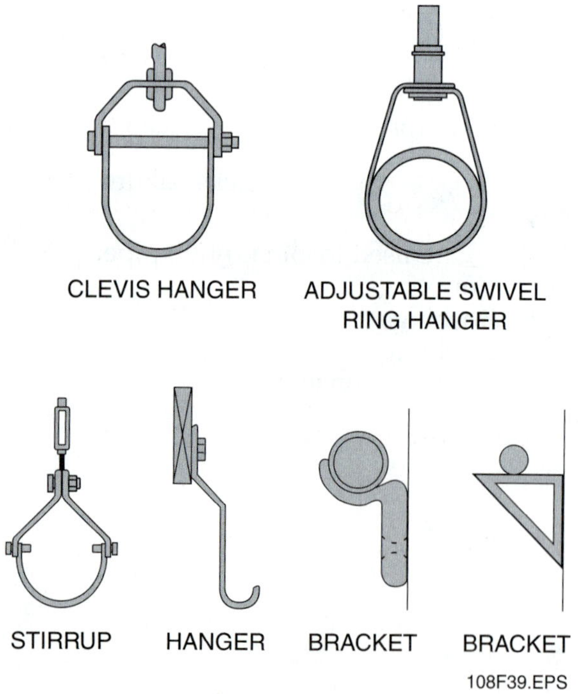

Figure 39 ◆ Horizontal pipe hangers.

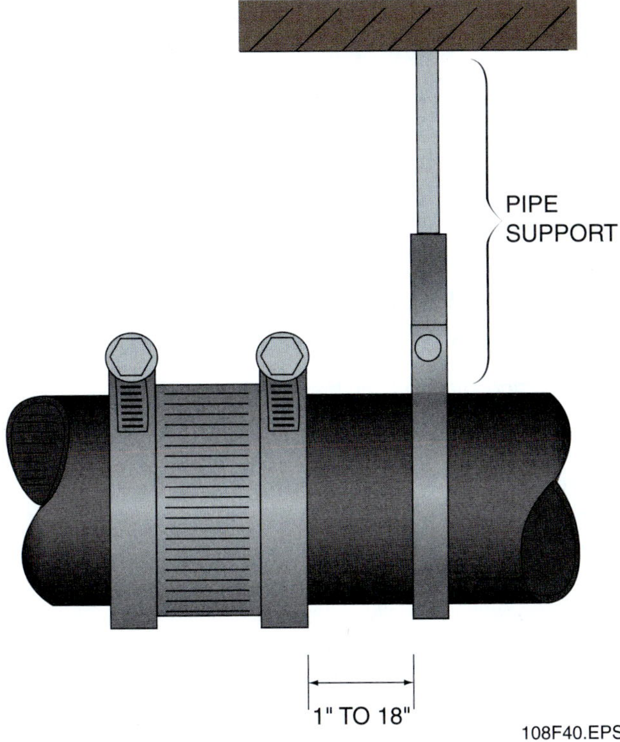

Figure 40 ◆ Hanger distance.

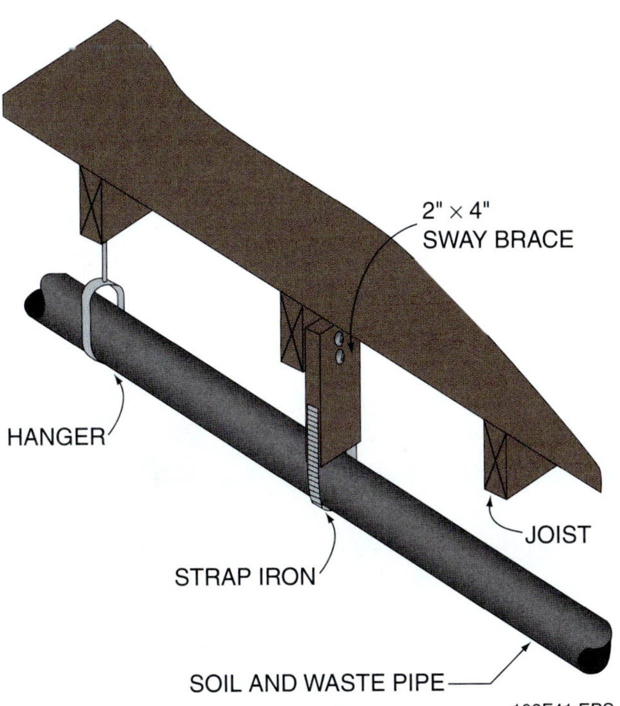

Figure 41 ◆ Sway brace.

slant of the pipe. DWV piping systems depend on gravity to move waste out of the piping system. Without the correct slope, gravity is not effective. Branch pipes generally require a slope of ⅛ inch to ¼ inch per foot on all horizontal runs. At this slope,

Figure 42 ◆ Pipe trapeze.

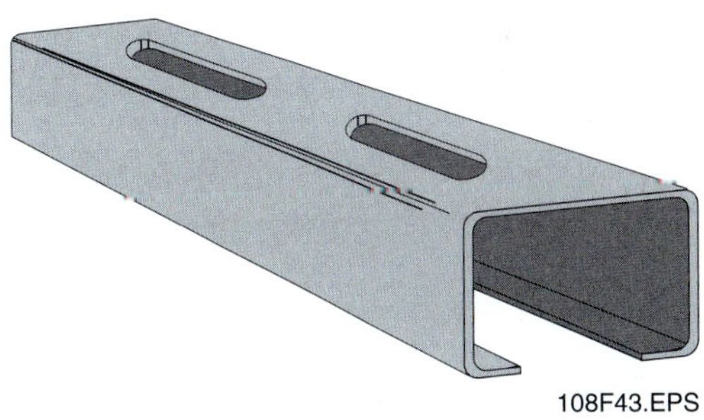

Figure 43 ◆ Prepunched channel.

the waste in the branch moves fast enough to clean the pipe.

If the slope is more than ¼ inch per foot, liquid and solid wastes may separate. When separation occurs, the heavier solid waste is deposited in the bottom of the pipe instead of flowing away with the liquid. If the slope is less than ⅛ inch per foot, the waste flow slows, and the branch tends to become blocked. When the pipe is perfectly aligned, it must be secured to the building structure to ensure that it remains aligned. You will learn more about determining the slope or grade required for DWV pipes elsewhere in this curriculum.

CAUTION

If you are installing pipe in a seismically active area—that is, where earthquakes are a possibility—local codes will require seismic restraints. Consult your local applicable building or plumbing code when installing pipe in such areas.

Seismic Codes

The *2003 International Building Code* (IBC) is considered the leading source for information on seismic code. The 2003 IBC was a combined effort of the Building Officials Code Administration (BOCA), the Southern Building Code Congress International (SBCCI), and the Uniform Building Code (UBC). The 2003 IBC includes seismic hazard maps, seismic design loads, and seismic code criteria.

Source: Kinetics Noise Control website. "Seismic Restraint Capabilities." www.kineticsnoise.com/seismicbrochure.html, reviewed March 15, 2004.

To attach horizontal hangers to wooden structures, use screws, lag screws, or large nails. To fasten pipe to I-beams, bar joists, and other metal structures, use beam clamps or C-clamps (*Figure 44*).

5.3.0 Supporting Vertical Pipe Runs

Vertical hangers are used to support vertical runs of pipe and to maintain pipe alignment. Proper pipe alignment is critical to ensure a leak-proof system. To support the vertical load of the pipe, use a riser clamp (*Figure 45*). Support vertical pipe at each floor level. Mount the riser clamp so the bracket supports the pipe weight directly on the floor.

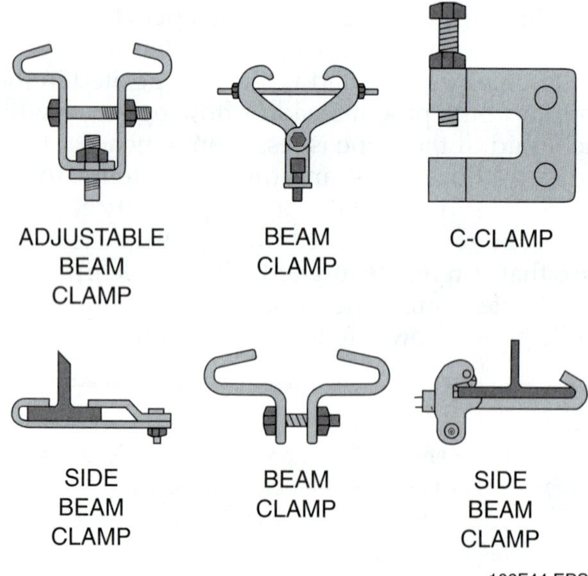

Figure 44 ◆ Beam clamps.

Plumbing Codes for Supports and Fasteners

In selecting supports and fasteners, refer to the applicable code and building specifications. As the plumber, you must select and use the supports and fasteners best suited for the job. You are also responsible for installing hangers or fasteners at the correct intervals along the length of the pipe run. If unusual circumstances exist, the project specifications may direct you to make changes from generally accepted plumbing practices. Never work without knowing the applicable codes and detailed project specifications.

5.4.0 Installation

You can attach pipe hangers or supports to wood, masonry, concrete, and steel. Because of the heavy weight of cast-iron pipe, it is extremely important

Figure 45 ◆ Riser clamp.

for you to choose a fastener with the appropriate holding power for your installation.

5.4.1 Wooden Structures

Use nails, screws, and bolts to fasten pipe to wooden structures. Screws and bolts have more holding power than nails. For both horizontal and vertical pipe runs, use nails only if vibration and strength are not factors in your installation.

5.4.2 Masonry and Concrete Structures

You can use various expansion anchors or threaded masonry fasteners to fasten support hangers to masonry and concrete. Use a **rotary hammer drill** (*Figure 46*) with a special masonry bit to drill into masonry. After drilling the hole, insert a screw anchor, a plastic expansion anchor, or a threaded masonry anchor in the hole (see *Figure 47*). Fasten the screw into the insert to expand the insert, allowing it to hold the screw tightly in the masonry. Use toggle bolts (*Figure 47*) to attach pipe supports to hollow masonry units, such as concrete block.

A special tool that can be used to anchor pipe supports to concrete, masonry, and steel is the **powder-actuated fastening tool** (*Figure 48*). Because this tool requires special training, you must be licensed by the tool manufacturer to use it. This tool shoots a fastener into the structure by exploding a blank 0.22-caliber shell. You must take special care to use the correct size of shell and nail. A shell that is too heavy could drive a nail completely through a steel beam.

108F46.TIF

Figure 46 ◆ Rotary hammer drill.

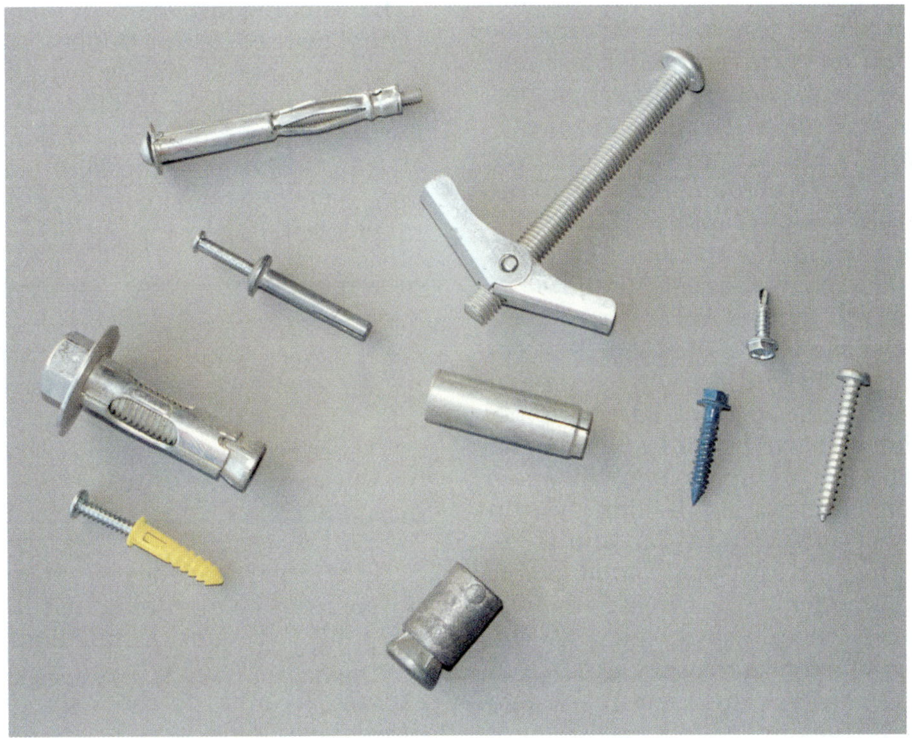

108F47.TIF

Figure 47 ◆ Anchors.

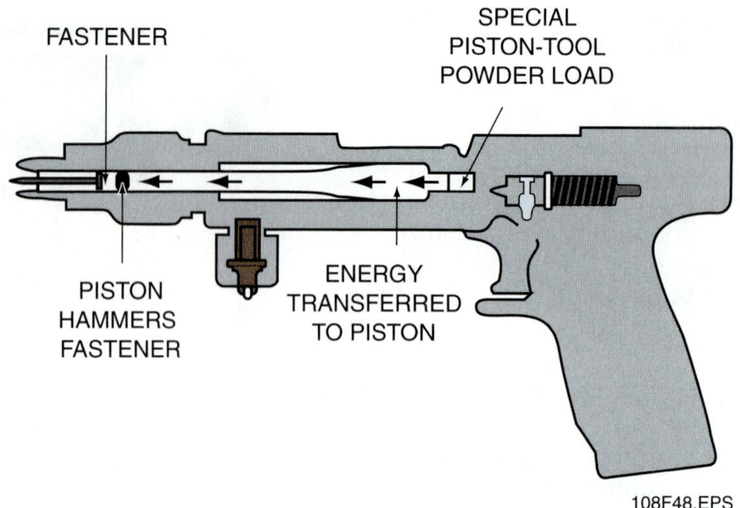

Figure 48 ◆ Powder-actuated fastening tool.

CAUTION

You must be certified by the manufacturer to use a powder-actuated fastening tool. Manufacturers' representatives visit shops and job sites to conduct training in their equipment and to certify people in its use.

6.0.0 ◆ TESTING INSTALLATIONS

Cast-iron piping installations of venting and drainage systems can be tested by a hydrostatic test (water test), air pressure test, smoke test, and peppermint test. Air and water testing can be performed at the rough-in stage. Smoke and peppermint testing must be performed after final installation. Check the local plumbing code to determine which test should be applied. If possible, test only one floor at a time.

Codes have different requirements for the water test. Some require a 10-foot head of water; others a 5-foot head of water. First, plug all openings except for one vertical pipe. Extend the vertical pipe at least 5 feet higher than the horizontal drain piping. If the code requires a 10-foot head of water, extend the pipe at least 10 feet. Fill this pipe with water. This allows you to check for any leaking joints in the system. An inspector may require you to remove cleanout plugs or caps at different points along the system piping to make sure that the water has reached everywhere in the drainage system.

Air testing is similar to water testing except that the pipes are filled with compressed air. The system should generally be pressurized to a maximum of 5 pounds per square inch (psi). Use a pressure gauge at the test plug to determine whether any pressure has been lost. While cast-iron pipe and fittings joined with compression joints or couplings should have a reduction in air pressure during a 15-minute test, a reduction of more than 1 psi indicates failure of the system.

A smoke test involves filling the traps with water and adding a thick smoke into the vent system. The stack should be closed when the smoke appears at the roof opening. A pressure of 1 inch water column is needed throughout the smoke test. The system passes the test if no smoke is visible at any connection in the system.

If plumbing inspectors decide that a smoke test cannot be used, use a peppermint test. To conduct a peppermint test, add 2 ounces of peppermint oil into each roof opening of the system. After adding the peppermint oil, add 10 quarts of hot water (160°F). Seal each terminal. If you can smell peppermint at any point in the system, you have identified a leak.

NOTE

When conducting a peppermint test, it is essential to keep your skin and clothing free of peppermint oil. If you contact peppermint oil, you will need to leave the area until the test is complete.

Adhering to all codes is extremely important to the success of a plumbing installation. An inspector will test your installation to ensure that you have complied with all codes. Inspectors are especially careful to check for cross-connections, defective or inferior materials, or poor craftsmanship. Because it is important to inspect the joints in the piping system visually, the inspection must be done before you cover the pipes.

Review Questions

Sections 5.0.0–6.0.0

1. When selecting supports for vertical pipe run, you must take into account the vertical load, or _____, of the pipe.
   a. total weight
   b. total length
   c. total circumference
   d. total radius

2. Most authorities and plumbing codes require hangers to support horizontal pipe at each _____.
   a. brace
   b. joint
   c. bend
   d. clamp

3. When several pipe runs are close together and parallel, you can use a properly sized _____ to support the pipe.
   a. joint
   b. anchor
   c. sway brace
   d. pipe trapeze

4. A _____ clamp is used to support the vertical load of the pipe.
   a. riser
   b. clevis
   c. spring
   d. C-

5. Nails can be used with wooden structures only if _____ are not factors.
   a. vibration and weight
   b. strength and expansion
   c. vibration and strength
   d. weight and expansion

6. A _____ can be used to anchor pipe supports to concrete, masonry, and steel.
   a. toggle bolt
   b. rotary hammer drill
   c. threaded masonry anchor
   d. powder-actuated fastening tool

7. To test a cast-iron piping installation, you can use any of the following methods *except* a _____.
   a. hydrostatic test
   b. sway test
   c. peppermint test
   d. smoke test

8. In an air test, the system should be pressurized to a maximum of _____ psi.
   a. 1
   b. 3
   c. 5
   d. 7

9. In a smoke test, a pressure of _____ inch water column is needed throughout the smoke test.
   a. 1
   b. 2
   c. 3
   d. 4

10. To conduct a peppermint test, add _____ ounce(s) of peppermint oil into each roof opening of the system.
    a. ½
    b. 2
    c. 3
    d. 5

Summary

Cast iron refers to iron products that are formed using a mold. Cast-iron pipe and fittings are available in two types of cast iron: gray iron and ductile iron. Either type can be used to form hub-and-spigot and no-hub cast-iron pipes in lengths of 5 or 10 feet. The pipes are made in two wall thicknesses: service and extra-heavy.

Hub-and-spigot pipes are joined using either a compression joint or a lead and oakum joint to produce a watertight seal. No-hub cast-iron pipes are joined using a coupling consisting of a neoprene gasket, a stainless steel shield, and clamps.

Planning ahead saves time and money when you are installing cast-iron pipe. Accurate measurements reduce the chances that you will waste pipe by making the wrong cut. Several techniques are used to measure pipe correctly before the pipe is cut and joined. Always measure and mark before you cut.

You must support cast-iron pipe to prevent it from bending, sagging, or otherwise losing its alignment. Hangers or supports such as brackets, clamps, and pipe trapezes secure the pipe in either a horizontal or vertical position. Supports can be fastened to wooden, concrete, and masonry frames using anchors. The type of anchor you use depends on the building frame material.

As with all aspects of plumbing, the selection and installation of cast-iron pipe is strictly regulated by building and plumbing codes. Codes vary widely among locations. It is critical to check applicable codes that affect your work. Incorrect selection or installation will cause the job to fail inspection by plumbing code officials.

Notes

Trade Terms Quiz

Fill in the blank with the correct trade term that you learned from your study of this module.

1. A(n) _______________ is a type of bend with small openings that allows you to connect a smaller pipe to the main line.
2. The spoon-like tool used to dip molten lead from the pot and pour it into a joint when making lead and oakum joints is the _______________.
3. _______________ are fittings designed to hold a water seal, which prevents sewer gas from escaping into the building.
4. A vessel called a(n) _______________ holds the lead during the melting process; it is placed on the head of the propane melting furnace.
5. The _______________ used to join no-hub pipes have three components: a gasket, a stainless steel shield, and clamps.
6. You must use _______________ when suspending cast-iron pipe more than 18 inches between nonrigid hangers, or as specified by applicable codes.
7. The fitting that connects the water closet to the closet bend is a(n) _______________.
8. The _______________, also called the _______________, keeps the hot, molten lead in the hub during the pouring operation.
9. Cast-iron pipe that has no enlargement or bell on either end is called _______________.
10. A(n) _______________ allows you to remove the lead and oakum from a joint.
11. When fastening support hangers to masonry or concrete with anchors, use a(n) _______________ to drill into the masonry.
12. A(n) _______________ connects the water closet to the main drainage piping.
13. The most commonly used cast-iron pipe is lightweight, or _______________, cast iron.
14. Horizontal runs of pipe must be installed with the proper _______________ to allow waste to drain away.
15. To join two drains into one pipe at a 45-degree angle, use a(n) _______________.
16. A(n) _______________, also called a(n) _______________, is a fitting that changes the direction of flow from horizontal to vertical and has openings for two branch lines.
17. Cast-iron pipe that has a bell or enlargement at one end of the pipe where the smooth end of the next pipe slides in to form a joint is called _______________.
18. In connecting long runs containing many sections of hub-and-spigot pipe, you must add the _______________ to the length of pipe to get a correct measurement.
19. The _______________ has a length of chain, each link of which contains a cutting wheel.
20. Using a(n) _______________ allows for smoother flow of wastes through the piping system.
21. The _______________ is the vertical main pipe that runs between the building drain and the highest horizontal drain in the system.
22. Molten lead is poured over _______________ to caulk a joint.
23. A(n) _______________ is a fitting that changes the direction of piping.
24. A(n) _______________ provides a fast method of joining hub-and-spigot cast-iron pipe and fittings.
25. The tool used to shape the lead to the inside of the hub in a lead and oakum joint is a(n) _______________.
26. A(n) _______________ is the tool used to pack oakum fiber into a joint by hand.
27. To connect pipe at right-angle intersections, use a(n) _______________.
28. A(n) _______________ is used to drive lead firmly into a lead and oakum joint to ensure that the seal is watertight.

29. The specially shaped tool used to caulk a lead and oakum joint near the ceiling is a(n) _______________.

30. The _______________ uses a 0.22-caliber shell to drive special fasteners into the framework of a structure.

31. A(n) _______________is often used above the heated area of a building to prevent frost buildup from closing the opening in the vent stack.

32. To join three drains into one pipe at a 45-degree angle, use a(n) _______________.

33. A(n) _______________ has a broad, thick blade used to pack oakum into a joint with a hammer.

34. A(n) _______________ is a fitting that changes the direction of piping and has one leg longer than the other.

35. A(n) _______________ is used to clamp the running rope around the pipe next to the hub.

36. The _______________ is used to tighten the coupling clamps when you are joining no-hub cast-iron pipe.

37. _______________ iron, which is made from magnesium and molten gray iron, is less brittle than gray iron.

38. The more complex method of joining hub-and-spigot cast-iron pipe with molten lead is a(n) _______________.

39. Used to melt the lead, a(n) _______________ burns liquefied petroleum (LP) gas.

40. The tool used to shape the lead near the spigot in a lead and oakum joint is a(n) _______________.

Trade Terms

Bend
Caulking iron
Ceiling caulking iron
Closet bend
Closet flange
Compression joint
Coupling
Double sanitary tee
Double wye
Ductile iron
Heel inlet
Hub-and-spigot cast-iron pipe
Increaser
Insertion length
Inside caulking iron
Joint runner
Ladle
Lead and oakum joint
Lead pot
Long bend
Main stack
No-hub cast-iron pipe
Oakum fiber
Outside caulking iron
Packing iron
Pickout iron
Powder-actuated fastening tool
Propane melting furnace
Rotary hammer drill
Running rope
Sanitary cross
Sanitary tee
Service (SV)
Slope
Soil pipe cutter
Spring clamp
Sway brace
Sweep
Torque wrench
Trap
Wye
Yarning iron

Profile in Success

Jason Spengler

Spengler Plumbing Company, Inc.
President
O'Fallon, Illinois

Jason Spengler was born and raised in O'Fallon, Illinois, and attended O'Fallon High School. At age 11, he went to work at his father's shop after school. He swept the floor, put away fittings, and cleaned trucks. He began his plumbing apprenticeship after high school. Jason went to junior college and received an associate's degree in mechanical engineering as well as an associate's degree in business. Jason began the company's residential service component, which has become a major part of the business. Having made his start sweeping floors, Jason has become president of Spengler Plumbing Company, Inc.

How did you become interested in this industry?
I was interested in the field. I knew from around junior high age that this was something I wanted to do. I grew up around the family business. I was proud of our reputation in the town. It was exciting for me as a child to travel with my father through town. It was noticeable to me that he knew a lot of people and that people thought highly of him. I recognized this from an early age and thought it was something I would like to do.

What path did you take to your current position?
I went from an apprentice to president. In the beginning, I was helping in the office and I would help on service calls. I immediately became the resident computer tech in the early to mid-1990s, so I started helping more in the office, with projects in the warehouse, and with making deliveries and service calls. My father became ill, so that's when I quit college and came back to help run the operations of the company full-time. I've been here ever since. I became vice president and then moved to president.

The state requires only a journeyman's license. I did get that license, but I was mainly doing estimating, project management, and daily management. Now I sit on two committees on the Plumbing-Heating-Cooling Contractors–National Association (PHCC–NA) Education Foundation. I became involved in estimating and project management training to develop that aspect of our business. My field training was from the plumber who was with us the longest. My mentor trained me until I took the plumbing license test. I spent time in the field with him for several summers and tried to work with him one-on-one.

Primarily we were a new commercial construction plumbing company. We always had a service component as a convenience to our commercial customers. But it was not a profit vehicle. The service component was there to serve large commercial customers who would call us for their homes. Business owners would call us for problems at their homes.

But then we realized we were getting more and more calls. We started to analyze the financial end and realized that we were not running the business in a profitable way. I started to research other service companies and became involved with PHCC–NA's Quality Service Contractors and with Contractors 2000, which is another industry service group. I became educated on how to provide profitable plumbing service. Today, our residential and commercial service and repair is the most profitable part of our business.

What does it take to be successful in your trade?
Education is the key. The plumbing industry is an uneducated industry. I mean that in an affectionate way. My father was an outstanding plumber, but he had no formal business education. That's what 95 percent of the industry is about. Plumbers who thought they could do better than their bosses went into business for themselves. Many businesses are operating on cash flow, running a business paycheck to paycheck. There are outstanding plumbers with no business education. PHCC–NA has many

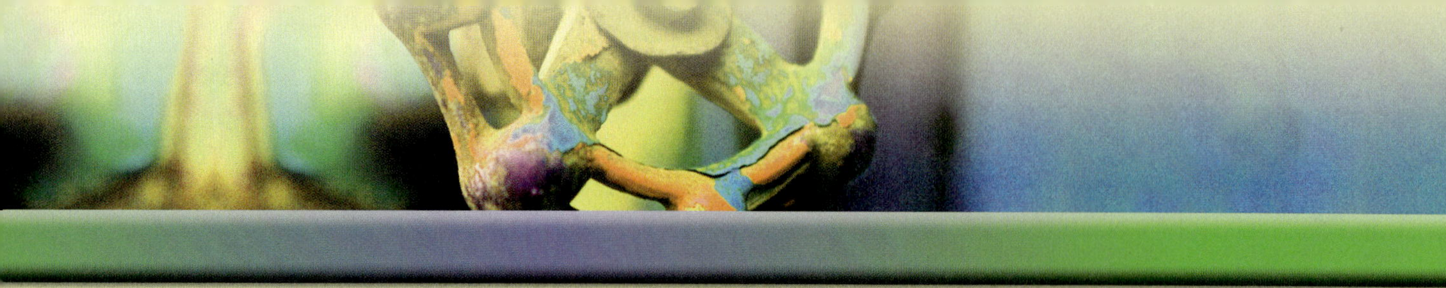

opportunities, but few people take advantage of them. The training I've had changed the way I can calculate and analyze a profit-and-loss statement. I know how to look at a balance sheet. We are businesspeople; we are entrepreneurs and have to operate as such. Education is the key to success.

The next thing is to be ethical. Every time you turn around you see another fly-by-night business that took advantage of a customer. We have to break that stigma. Be professional, run the business by the numbers, and always be ethical.

What are some things you do on the job?
I still do project management and estimating. As president, I am the chief operations officer. I have to make sure that I have a full understanding of where we are financially and how we are operationally. I have to ensure that the service department, for example, is operating within standards and is meeting or exceeding budgets and goals. I have to know where receivables, materials, and purchasing stand, so that different aspects of the business don't get out of hand. We are in the labor business. Every cost and expense has to hinge on a specific level of business. We can't overbuy or oversell the workforce. Many people are uneducated about this, and this is so important.

And I still do a lot of management by walking around. I get out and try to see everyone in the office and make contact with them. I try to do that in the morning, to be visible. It's hard to do with everything else. You have to determine what else is going to give.

What is the most interesting aspect of your profession? What makes your trade stand out from others?
I believe that what is overlooked is how important we are to the nation. The plumber is the ambassador to the public health of the nation. We take it for granted that we have clean water to drink and that the waste is taken away. People in developing countries die from waterborne illnesses every day. The industry undervalues itself. That's what gives me passion about the industry.

There are so many possibilities and different opportunities that cover the broad spectrum of work. Every project is different in some way. You don't get bored in this industry.

What do you like most—and least—about your job?
I most like seeing a project that we did completed. I like helping people. I completed a junior high school in my hometown that was named after a teacher that I had as a student in the school. It was great to be part of that. Those are the best things.

I least like dealing with personnel problems. I'm fighting to raise the bar in the industry. It's a challenge but also a frustration. It's difficult to bring up our industry as a professional industry. Many competitors undervalue their services. This makes the industry look cheaper. I don't like dealing with competitors who aren't professionals and who make the industry look bad.

What would you say to someone entering the trade today?
You need to educate yourself about business. Plumbing is what you do, but a business is what you are running. You have to be a businessman if you are going to run a business. That is so important. I can't convey that enough. So many people just think they can start a business, but they don't educate themselves. Know whether you want to be a plumber or whether you want to run a business. You can't do both. Something has to give, but it had better not be the business that does.

Understand that there is a big difference between running a business and being a plumber. There are always different ways to look at different situations. Every person's perception is going to be different. The owner looks with an owner's set of eyes, the journeyman with a journeyman's set of eyes, and the trainer with a trainer's set of eyes. Just be aware of how other people could be looking at the situation. How you choose to react to any situation is 100 percent in your control. A positive attitude will make anything go more smoothly. Remember that all personalities are unique and different.

What can an apprentice expect to earn in his/her first years on the job in your area? What can he/she expect to earn after 10-plus years in the industry?
Our first-year apprentice makes $13.00 per hour. We have six-month steps up the journeyman's pay scale. It starts with a 10 percent increase every six months for five years. After about 10 years, a foreman can make almost $32.00 per hour plus benefits.

Trade Terms Introduced in This Module

Bend: A fitting that changes the direction of piping.

Caulking iron: A tool used to drive the lead firmly into a lead and oakum joint when working with hub-and-spigot cast-iron pipe and fittings.

Ceiling caulking iron: A specially shaped tool used to drive the lead firmly into a lead and oakum joint when working with hub-and-spigot cast-iron pipe and fittings next to a ceiling.

Closet bend: A fitting that connects the water closet to the main drainage piping.

Closet flange: A fitting used to connect the water closet to the closet bend.

Compression joint: A joint formed using a neoprene gasket inserted into the bell end of a hub-and-spigot pipe. The pressure applied between the two joined pipes forces the gasket to compress and fill in any air gaps in the joint.

Coupling: A fitting used to connect two lengths of no-hub cast-iron pipe.

Double sanitary tee: A tee fitting that changes the direction of flow from horizontal to vertical and that has openings for two branch lines. Also called a sanitary cross.

Double wye: A fitting that changes direction at 45 degrees. It connects horizontal waste pipes and branch drains to the building main. It has openings for two branch lines.

Ductile iron: A type of cast iron in which magnesium is added to the molten gray iron to reduce brittleness.

Heel inlet: A bend that has an inlet to connect a smaller pipe to the main line or a branch line. The inlet is located at the base of the curve or heel of the bend.

Hub-and-spigot cast-iron pipe: Pipe that has a bell or enlargement at one end where the spigot (smooth end) of the next pipe slides in to form a joint. Also called bell-and-spigot pipe.

Increaser: A fitting used to increase the size of a straight-through line of pipe. It is often used for the vent stack before it goes through the roof to reduce the chance of frost clogging the vent opening in very cold climates.

Insertion length: The length of a pipe that fits into the fitting or other joint when assembling a pipe run. The measurement must be figured into the calculation before cutting a piece of pipe for joining.

Inside caulking iron: A tool used to shape the lead near the spigot in a lead and oakum joint.

Joint runner: Fiber rope that is clamped around the pipe using a spring clamp to keep the hot, molten lead in the hub when making a lead and oakum joint. Also called running rope.

Ladle: A spoon-like tool used to dip the molten lead from the lead pot and pour it into the joint.

Lead and oakum joint: A joint for hub-and-spigot cast-iron pipe that involves pouring molten lead over oakum fiber to form a watertight seal between the two pieces of pipe, then caulking the joint after the lead has cooled.

Lead pot: A vessel used to hold molten lead. It is placed on the head of the propane melting furnace during the melting process.

Long bend: A bend fitting with one leg longer than the other.

Main stack: The principal pipe artery to which branches may be connected.

No-hub cast-iron pipe: A pipe with no enlargement or bell on either end.

Oakum fiber: A caulking material of old hemp rope or jute fibers that have been treated with tar or a tar derivative.

Outside caulking iron: A tool used to shape the lead to the inside of the hub or bell of a hub-and-spigot pipe.

Packing iron: A tool with a broad, thick blade used to pack oakum into a joint with a hammer.

Pickout iron: A tool with a diamond-shaped flat point used to remove the lead and oakum from a joint.

Powder-actuated fastening tool: A tool that uses a 0.22-caliber shell to drive nails or fasteners into the framework of a structure. It may be used only by certified workers.

Propane melting furnace: A device used to melt lead. It burns liquefied petroleum (LP) gas.

Rotary hammer drill: A drill with a pounding action that lets you drill into concrete, brick, or tile. It rotates and hammers at the same time and drills much faster than regular drills.

Running rope: Fiber rope that is clamped around the pipe using a spring clamp to keep the hot, molten lead in the hub when making a lead and oakum joint. Also called joint runner.

Sanitary cross: A tee fitting that changes the direction of flow from horizontal to vertical and that has openings for two branch lines. Also called a double sanitary tee.

Sanitary tee: A tee fitting for DWV systems that has a slight curve in the 90-degree inlet side of the fitting; this curve helps to channel the flow of wastewater or sewage from a branch line to the main line. This fitting is always used in the vertical position.

Service (SV): A lightweight kind of cast-iron pipe. Service refers to the pipe's thickness, which can vary.

Slope: Measurement of the fall of a length of pipe from level. The change in level is expressed in degrees of an angle. Also referred to as percent of grade.

Soil pipe cutter: A heavy-duty tool used for cutting cast-iron pipe. Soil pipe cutters can be of the snap or ratchet type.

Spring clamp: A tool used to clamp the running rope around the pipe next to the hub when making lead and oakum joints.

Sway brace: A support that keeps lengths of cast-iron pipe from swaying when they are suspended.

Sweep: A fitting with a radius of the curve greater than 90 degrees. Used for a smooth change of direction.

Torque wrench: A wrench with a gauge or other means to indicate the amount of rotating force applied to a fastener, such as a nut, as it is turned.

Trap: A fitting or device which provides a liquid seal to prevent the emission of sewer gases without materially affecting the flow of sewage or wastewater through it.

Wye: A fitting used to make a 45-degree direction change in horizontal drain and waste pipes when connecting to the building main drain.

Yarning iron: A tool used to pack the oakum fiber into the joint when making lead and oakum joints to connect hub-and-spigot cast-iron pipes.

Additional Resources and References

ADDITIONAL RESOURCES

This module is intended to present thorough resources for task training. The following reference works are suggested for further study. These are optional materials for continued education rather than for task training.

Basic Plumbing with Illustrations, Revised Edition, 1994. Howard C. Massey. Carlsbad, CA: Craftsman Book Company.

Cast Iron Soil Pipe Institute website, www.cispi.org, "Cast Iron Soil Pipe and Fittings Handbook," www.cispi.org/handbook.htm, reviewed August 2003.

Plumber's Handbook, Revised Edition, 1998. Howard C. Massey. Carlsbad, CA: Craftsman Book Company.

Plumbing: Design and Installation, Second Edition, 2002. L. V. Ripka. Homewood, IL: American Technical Publishers.

REFERENCES

Cast Iron Soil Pipe Institute website, www.cispi.org, "Cast Iron Soil Pipe and Fittings Handbook," www.cispi.org/handbook.htm, reviewed August 2003.

Code Check Plumbing: A Field Guide to the Plumbing Codes, 2000. Redwood Kardon, Michael Casey, and Douglas Hansen. Newton, CT: Taunton Press, Inc.

Illinois Plumbing Code. "Section 890.1930, page (m-2)." Printed by Authority of the State of Illinois. P.O. #48250 6/99 25M Illinois Plumbing License Law (225 ILCS 320/). Phone (217) 524–0791.

James Wall Plumbing, Ltd., website, www.jameswall.com (information on cast-iron pipe classification).

National Standard Plumbing Code, 2003. Chapter 1, Definitions (definitions of *main stack* and *trap*). Falls Church, VA: Plumbing-Heating-Cooling Contractors—National Association.

Swafford, Tommy, personal communication (information about smoke and peppermint tests from the *Illinois Plumbing Code*).

2003 International Plumbing Code. "Table 706.3 Fittings for Change in Direction." Falls Church, VA: International Code Council.

NCCER CURRICULA — USER UPDATE

NCCER makes every effort to keep its textbooks up-to-date and free of technical errors. We appreciate your help in this process. If you find an error, a typographical mistake, or an inaccuracy in NCCER's curricula, please fill out this form (or a photocopy), or complete the online form at **www.nccer.org/olf**. Be sure to include the exact module ID number, page number, a detailed description, and your recommended correction. Your input will be brought to the attention of the Authoring Team. Thank you for your assistance.

Instructors – If you have an idea for improving this textbook, or have found that additional materials were necessary to teach this module effectively, please let us know so that we may present your suggestions to the Authoring Team.

NCCER Product Development and Revision

13614 Progress Blvd., Alachua, FL 32615

Email: curriculum@nccer.org
Online: www.nccer.org/olf

❑ Trainee Guide ❑ AIG ❑ Exam ❑ PowerPoints Other ______________________________

Craft / Level: Copyright Date:

Module ID Number / Title:

Section Number(s):

Description:

Recommended Correction:

Your Name:

Address:

Email: Phone:

Carbon Steel Pipe
and Fittings
02109-05

02109-05

Carbon Steel Pipe and Fittings

Topics to be presented in this module include:

Overview

Carbon steel pipe is used for hot and cold water distribution; steam and water heating systems; gas and air piping systems; and drain, waste, and vent (DWV) systems. Plumbers commonly use black steel pipe for gas or air pressure pipe, and galvanized pipe for venting. Steel pipe is manufactured with threads on both ends, according to standards established by the American Standard Pipe and Pipe Thread Dimensions, and is categorized by size and strength.

Plumbers work with threaded and grooved steel pipe. Pipe fittings for steel pipe are made of malleable or cast iron. Common valves for threaded pipe include globe, gate, ball, and stop and waste valves. Fittings and valves are also available for pipe with grooved ends. There are different methods for joining both types of steel pipe. To thread steel pipe, plumbers use hand and power threaders; to join threaded pipe, plumbers use pipe joint compound. Steel pipe is grooved by rolling or by cutting.

Plumbers install and support steel pipe so that the pipe and joints are leakproof and properly aligned. Pipe attachments, connectors, and structural attachments must be installed at sufficient intervals and to the appropriate surfaces. Because the installation of carbon steel pipe is strictly regulated, plumbers always must follow applicable code requirements.

Focus Statement

The goal of the plumber is to protect the health, safety, and comfort of the nation job by job.

Code Note

Codes vary among jurisdictions. Because of the variations in code, consult the applicable code whenever regulations are in question. Referencing an incorrect set of codes can cause as much trouble as failing to reference codes altogether. Obtain, review, and familiarize yourself with your local adopted code.

Objectives

When you have completed this module, you will be able to do the following:

1. Recognize proper applications of carbon steel piping.
2. Identify the material properties, storage, and handling requirements of carbon steel piping.
3. Identify the various techniques used in hanging and supporting carbon steel piping.
4. Properly measure, cut, groove, thread, and join carbon steel piping.

Key Trade Terms

Ball valve
Black steel pipe
Chain vise
Chain wrench
Cross
Die
Double extra strong (double extra heavy)
Drainage fitting (recessed fitting)
Extra strong (extra heavy)
Flange union
Galvanized pipe
Gate valve
Globe valve
Ground joint union
Nipple
Pipe-joint compound (pipe dope)
Pipe tongs
Power-threading machine
Pressure fitting
Sealant tape
Standard weight
Stock
Stop-and-waste valve
Strap wrench
Thread engagement
Yoke vise

Required Trainee Materials

1. Appropriate personal protective equipment
2. Pencils and paper
3. Copy of local adopted code
4. *Carbon Steel Pipe and Fitting Dimensions* book

Prerequisites

Before you begin this module, it is recommended that you successfully complete *Core Curriculum; Plumbing Level One*, Modules 02101-05 through 02105-05.

This course map shows all of the modules in the first level of the *Plumbing* curriculum. The suggested training order begins at the bottom and proceeds up. Skill levels increase as you advance on the course map. The local Training Program Sponsor may adjust the training order.

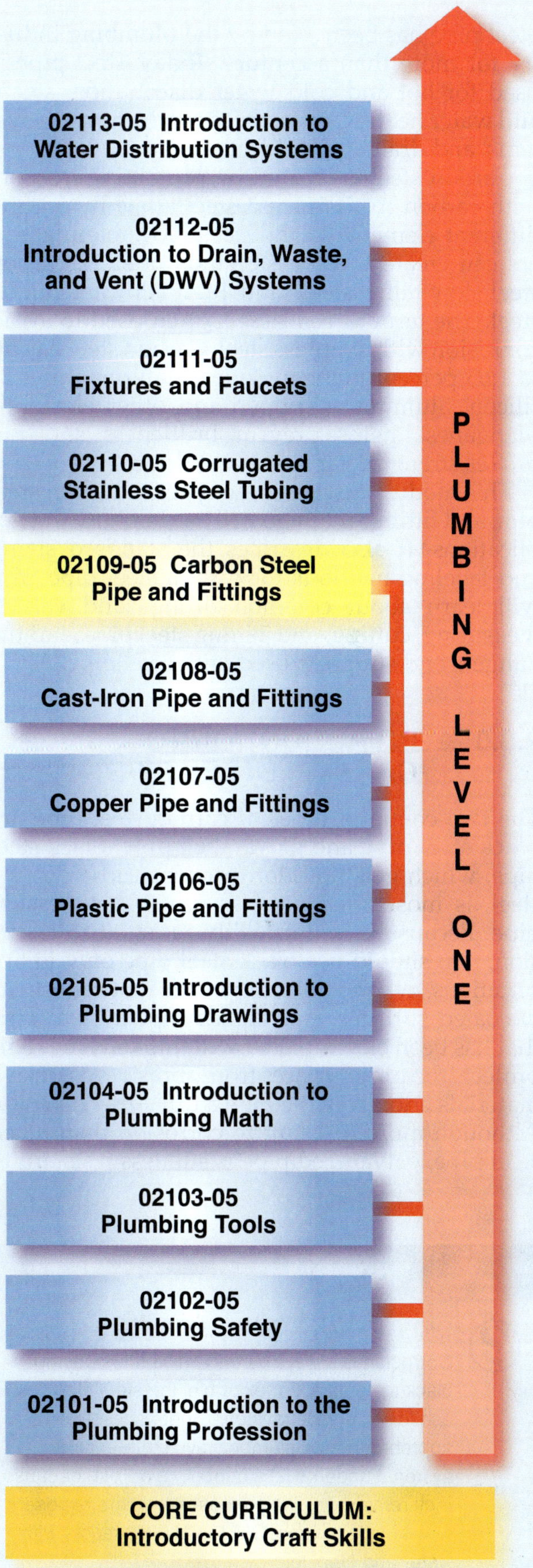

PLUMBING LEVEL ONE

02113-05 Introduction to Water Distribution Systems

02112-05 Introduction to Drain, Waste, and Vent (DWV) Systems

02111-05 Fixtures and Faucets

02110-05 Corrugated Stainless Steel Tubing

02109-05 Carbon Steel Pipe and Fittings

02108-05 Cast-Iron Pipe and Fittings

02107-05 Copper Pipe and Fittings

02106-05 Plastic Pipe and Fittings

02105-05 Introduction to Plumbing Drawings

02104-05 Introduction to Plumbing Math

02103-05 Plumbing Tools

02102-05 Plumbing Safety

02101-05 Introduction to the Plumbing Profession

CORE CURRICULUM: Introductory Craft Skills

109CMAP.EPS

1.0.0 ◆ INTRODUCTION

Steel pipe has been used by the plumbing industry for more than a century. Today, steel pipe is used for hot and cold water distribution, steam and water heating systems, gas and air piping systems, and drainage and vent systems.

Carbon steel is created by combining iron ore with carbon to produce a steel alloy. By adding different elements to the steel alloy, manufacturers can create three basic kinds of steel: carbon steel, low-alloy steel, and high-alloy (stainless) steel. The term *carbon steel* is often used to avoid confusion with stainless steel. Carbon steel has up to 1.65 percent manganese or small quantities of silicon, aluminum, copper, and other elements. Manganese helps prevent brittleness and improves strength, hardness, and flexibility.

This module discusses the types of carbon steel pipe and fittings that you will use in plumbing applications. It also describes the standard sizing and labeling methods for carbon steel pipe. You will learn about common fittings and valves; measuring, cutting, and joining steel pipe; and the hangers and supports used with steel pipe.

2.0.0 ◆ INTRODUCTION TO CARBON STEEL PIPE

The two common types of carbon steel pipe are **black steel pipe** and **galvanized pipe.** Black steel pipe, which gets its color from the carbon in the steel, is most often used as gas or air-pressure pipe. Because it is not highly resistant to corrosion, you should use black steel pipe only in applications where corrosion will not affect its uncoated surfaces. Galvanized pipe is steel pipe that has been electroplated with molten zinc. This protects the pipe surfaces from abrasive, corrosive materials and gives the pipe a dull, grayish color. Although more resistant to corrosion than black steel pipe, galvanized pipe is still less resistant to corrosion than other piping materials. Galvanized pipe is commonly used for venting systems.

Carbon steel pipe can be threaded or grooved, but threaded pipe is more common. Plumbing codes usually specify which type of steel you can use for a particular piping system.

> **DID YOU KNOW?**
>
> Corrosion is always a concern when choosing black steel pipe. For example, the steel pipe you use for a sprinkler system is not necessarily appropriate for a plumbing system. Sprinkler systems undergo less corrosion, or rust, because they remain wet, which means they are exposed to very little oxygen. Oxygen is what reacts with water and steel to create rust.

2.1.0 Domestic and Imported Steel

Steel is manufactured in the U.S. as well as in other countries. You will probably use both domestic (made in the U.S.) and imported (made outside the U.S.) steel pipe. The U.S. steel industry must comply with trade regulations that control the amount of imported steel that companies can purchase and distribute domestically.

As always, the safety and quality of the pipe you use for plumbing systems are critical. U.S. pipe manufacturers must adhere to U.S. manufacturing codes and standards that may not apply elsewhere. For example, the U.S. government requires that pipe adhere to the standards of ASTM International (formerly known as the American Society for Testing and Materials). ASTM-approved pipe is marked with the letters *ASTM.* As a plumber, you are ultimately responsible for adhering to all codes that apply to your installation. Therefore, when purchasing steel, you must know the manufacturer and be able to rely on the quality of its materials and on its compliance with standards.

2.2.0 Threads

Steel pipe is typically manufactured with threads on both ends. It is usually connected with threaded fittings. The length of pipe that gets screwed into the fitting is called the **thread engagement.** Pipe manufacturers adopted the *American Standard Pipe and Pipe Thread Dimensions* in 1886. This decision standardized the thread size on all manufactured pipe products. The standard thread used on pipe is V-shaped, with an angle of 60 degrees (see *Figure 1*). The taper is 1⁄32 inch per inch of length. Tapered pipe threads produce a pressure-tight connection. In addition, when pipes are sealed tightly together, they produce a mechanically rigid piping system.

Note in *Figure 1* that the first seven threads are perfect threads. They are sharp at both the top and bottom. The next two threads are imperfect at the top and perfect at the bottom, and the rest of the threads are imperfect at the top and bottom. Imperfect threads result when threads are not cut completely or when perfect threads are marred and broken. These threads have no sealing power. If perfect threads are marred or broken, they will lose their sealing power as well.

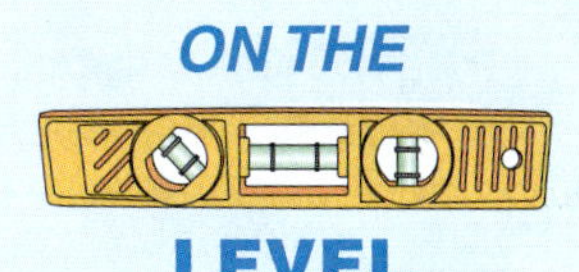

Producing Steel

Carbon steel is produced by one of four processes: Bessemer, open-hearth, electric furnace, or basic oxygen. The Bessemer process, developed in the 1850s, produced steel by blowing air through molten iron at high pressure. The process used oxygen to draw off impurities, but it was not efficient enough to remove all the unwanted elements. This process is no longer used in the United States.

The open-hearth process was the main method of producing steel for many years. An open-hearth furnace is a huge, shallow tank lined with firebrick. A fuel, such as gas, oil, or powdered coal, is mixed with hot air in burners and blasted across the tank. Molten iron, recycled steel, and limestone are melted together in the tank. After the removal of impurities collected by the limestone, the steel is ready.

The electric-furnace process uses tremendous charges of electricity to produce steel. The furnace is loaded with scrap iron. An electrical arc is produced using electrodes. The resulting heat (6,000°F or 3,312°C) melts the scrap iron and produces the steel alloy.

The basic-oxygen process is the method most often used today. Molten iron from a blast furnace (a giant oven lined with firebrick) is poured into the basic-oxygen furnace. Pure oxygen is blown onto the metal through a tube called an oxygen lance. The oxygen combines with the impurities and is then removed from the furnace, leaving the steel.

After the liquid steel has been created in the furnaces, it must be further processed and cast into solid form before manufacturers can use it.

See the number of threads per inch required for each pipe size in *Table 1*. Note that the total number of threads required is different from the number of usable threads required.

2.3.0 Carbon Steel Pipe Sizing

Steel pipe is measured by the nominal size. For pipe that is larger than 6 inches, this size is based on the approximate inside diameter (ID) of the pipe. The nominal size is not necessarily equal to the exact ID. Pipe larger than 12 inches is measured by its outside diameter (OD). Steel pipe is manufactured in the following nominal sizes (in inches): ⅛, ¼, ⅜, ½, ¾, 1, 1¼, 1½, 2, 2½, 3, 4, 5, 6, 8, 10, and 12. See the ID and OD for 1-inch nominal-size steel pipe in *Figure 2*.

Steel pipe is manufactured in lengths of 21 feet, threaded or unthreaded. Pipe with a nominal size of 6 inches or larger is available in random lengths between 18 and 22 feet. In addition to size and length, you need to understand the different weights of steel pipe. Steel pipe for threading is manufactured in three weights or strengths: **standard weight, extra strong,** and **double extra strong.** Some manufacturers refer to extra strong

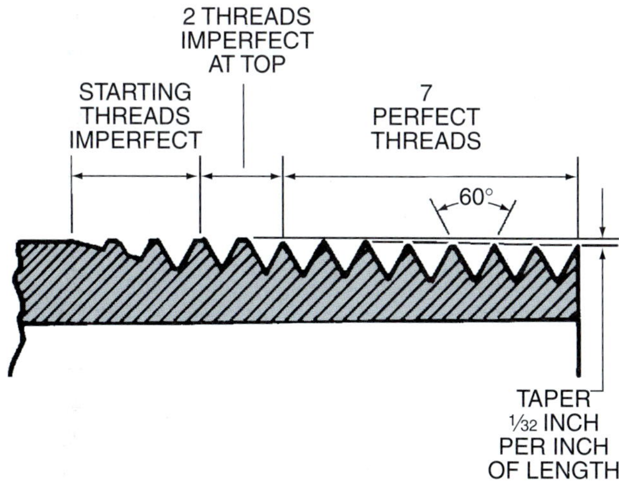

Figure 1 ◆ American standard thread.

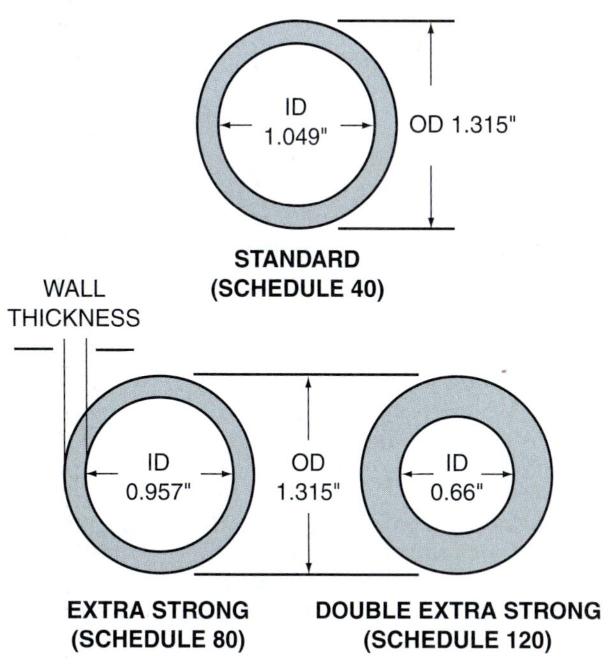

Figure 2 ◆ ID and OD of steel pipe.

Table 1 American Standard Pipe and Pipe Thread Dimensions

Nominal Pipe Size (Inches)	Threads per Inch	Number of Usable Threads*	Hand-Tight Engagement** (Inches)	Thread Makeup Wrench and Hand Engagement** (Inches)	Total Length of External Threads (Inches)
⅛	27	7	3/16	¼	⅜
¼	18	7	¼	⅜	9/16
⅜	18	7	¼	⅜	⅝
½	14	7	5/16	½	¾
¾	14	8	5/16	9/16	13/16
1	11½	8	⅜	11/16	1
1¼	11½	8	7/16	11/16	1
1½	11½	8	7/16	¾	1
2	11½	9	7/16	1¾	1 1/16
2½	8	9	11/16	1⅛	1 9/16
3	8	10	¾	1 3/16	1⅝
3½	8	10	13/16	1¼	1 11/16
4	8	10	13/16	1 5/16	1¾
5	8	11	15/16	1⅜	1 13/16
6	8	12	15/16	1½	1 15/16
8	8	14	1 1/16	1 11/16	2⅛
10	8	15	1 3/16	1 15/16	2⅜
12	8	17	1⅜	2⅛	2 9/16

*Rounded to the nearest thread.

**Rounded to the nearest 1/16 inch.

as extra heavy and double extra strong as double extra heavy. Because all weights for a particular size of pipe have the same OD, the same threading **dies** (devices used to cut the threads) fit all three weights of each size of pipe.

The wall thickness of the pipe and the ID differ with each weight. The thicker the wall, the smaller the ID, and the more pressure the pipe will withstand. Standard weight is adequate in most plumbing situations. However, the two stronger weights must be used when pressures require it. Always check applicable codes for weight requirements. See the dimensions of Schedule 40 (standard weight) and Schedule 80 (extra strong) steel pipe in *Table 2*.

2.4.0 Labeling

For ease of handling, manufacturers ship steel pipe in bundles. The number of lengths per bundle is determined by the size of the pipe. Each bundle has a tag that identifies the number of feet in the bundle. While there is no specific standard for labeling, steel pipe must be labeled with the name or brand of the manufacturer, its length, and *ASTM A 120,* which indicates that the pipe meets ASTM International standards.

2.5.0 Material Storage and Handling

To ensure maximum productivity on the job, it is important to use common sense when storing and handling carbon steel pipe. Before starting work, always plan ahead. Organize pipe and fittings into groups by pipe size and type. Store pipe and fittings near where you will be working so that they are easy to access. When handling pipe and fittings, be careful not to bend or damage them. Pipe that is bent or otherwise damaged not only wastes money, but makes you less productive.

3.0.0 ◆ FITTINGS AND VALVES

The purpose of fittings is to change the direction of flow in water supply and distribution systems and in drain, waste, and vent (DWV) systems. Valves regulate the flow of water in a system. They are often attached to the ends of pipe to provide access to fluid flowing through the system. Fittings are designed so they do not block or slow the flow of materials in the pipe.

Pipe fittings for steel pipe are generally made of malleable iron or cast iron. Annealing ordinary cast iron produces malleable iron. The annealing process, which consists of heating and slowly

Table 2 Dimensions of Schedule 40 and Schedule 80 Steel Pipe

	Dimensions of Schedule 40 (Standard Weight) Steel and Wrought Iron Pipe			Dimensions of Schedule 80 (Extra Strong) Steel and Wrought Iron Pipe		
Nominal Diameter (Inches)	**Actual Inside Diameter (Inches)**	**Actual Outside Diameter (Inches)**	**Weight per Foot (Pounds)**	**Actual Inside Diameter (Inches)**	**Actual Outside Diameter (Inches)**	**Weight per Foot (Pounds)**
⅛	0.269	0.405	0.25	0.215	0.405	0.31
¾	0.364	0.540	0.43	0.302	0.540	0.54
⅜	0.493	0.675	0.57	0.423	0.675	0.74
½	0.622	0.840	0.86	0.546	0.840	1.09
¾	0.824	1.050	1.14	0.742	1.050	1.47
1	1.049	1.315	1.68	0.957	1.315	2.17
1¼	1.380	1.660	2.28	1.278	1.660	3.00
1½	1.610	1.900	2.72	1.500	1.900	3.63
2	2.067	2.375	3.66	1.939	2.375	5.02
2½	2.469	2.875	5.80	2.323	2.875	7.66
3	3.068	3.500	7.58	2.900	3.500	10.25
3½	3.548	4.000	9.11	3.364	4.000	12.50
4	4.026	4.500	10.80	3.820	4.500	14.98
5	5.047	5.583	14.70	4.810	5.563	20.78
6	6.065	6.625	19.00	6.760	6.625	28.57

cooling the iron, makes the iron stronger so that it can be bent or pounded to a reasonable extent without breaking. This is unlike copper, which becomes flexible when heated.

Iron pipe fittings are manufactured in two types: **pressure fittings** and **drainage fittings,** also called recessed drainage fittings (see *Figure 3*). Drainage fittings have a drainage pattern design and recessed threads. They should be used only in soil and waste piping systems.

Pressure pipe fittings cannot be used in drainage systems because their inside surfaces are not smooth enough. Drainage fittings differ from pressure fittings in several ways. The insides of drainage fittings are smooth and are shaped for easy flow. Their shoulders are recessed so that when pipe is screwed in, the pipe fills the recessed area and forms an unbroken, internal contour. Ideally, a pipe screwed into a drainage fitting completely fills the recessed cavity, but this rarely occurs. Drainage fittings are also designed so that horizontal lines entering them will have a slope of ¼ inch per foot. Refer to *Figure 3* for illustrations of the normal and ideal conditions of most drainage joints.

The following sections discuss the different fittings and valves you can use with threaded and grooved pipe. The fittings for threaded and grooved pipe are not interchangeable.

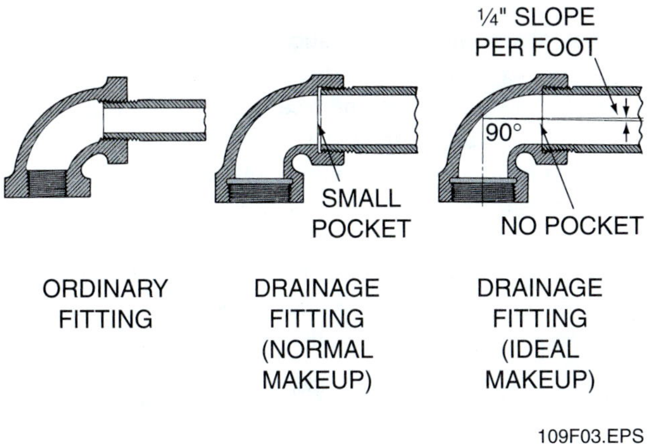

Figure 3 ◆ Pressure and drainage fittings.

3.1.0 Fittings for Threaded Pipe

This section discusses the common pressure fittings used for hot and cold water distribution systems, steam and hot water heating systems, and gas and air piping systems. Slight obstructions on the inside surface of the pipe and fitting joint do not affect the flow of materials in any of these systems.

3.1.1 Tees

Tees are available in a number of sizes and patterns, and they are used to connect branches to a pipe at a 90-degree angle. If all three outlets are the same size, the fitting is called a regular tee (see *Figure 4*). If outlet sizes differ, the fitting is called a reducing tee (see *Figure 5*). Tees are identified first by the dimensions of the straight-through run, then by the side-opening dimensions. For example, a tee with a run outlet of 2 inches, a run outlet of 1 inch, and a branch outlet of ¾ inch is a 2 × 1 × ¾ tee. Always state the large run size first, the small run size next, and the branch size last. Tees are also available with male threads (on the outside of a pipe) on a run or branch outlet.

3.1.2 Elbows

Elbows, or ells, are installed to change the direction of the pipe. The most common elbows, or ells, are the 90-degree ell; the 45-degree ell; the street ell, which has a male thread on one end; and the reducing ell, which has outlets of different sizes (see *Figure 6*). Ells are also available to make 11¼-degree, 22½-degree, and 60-degree bends.

109F04.TIF

Figure 4 ◆ Regular tee.

109F05.TIF

Figure 5 ◆ Reducing tee.

3.1.3 Unions

Unions make it possible to disassemble a threaded piping system. After disconnecting the union, you can unscrew the length of pipe on either end of the union. The two most common steel pipe unions are the ground joint and the flange.

The **ground joint union** connects two pipes by screwing the thread and shoulder pieces of the union onto the pipes. The collar then draws the shoulder and thread parts together. This union creates a gastight or watertight joint (see *Figure 7*).

The **flange union** allows you to screw the flanges to the two pipes you are joining. You then pull the flanges together using nuts and bolts. A gasket between the flanges also makes this connection gastight or watertight (see *Figure 8*).

CAUTION

Never use dielectric unions in hot water systems. The unions cannot handle the temperature changes, and they may leak.

Source: University California–Santa Barbara Design, Construction, and Physical Facilities website. http://facilities.ucsb.edu/Standards/15000Plumbing.pdf, reviewed March 9, 2004.

3.1.4 Couplings and Other Fittings

Couplings are used to connect two lengths of pipe in making straight runs. Couplings cannot be

Dielectric Unions

If you attach unlike metals such as copper and galvanized pipe, it results in a process called electrolysis (producing chemical changes). Transitions fittings, such as dielectric unions, prevent electrolysis. Dielectric (nonconducting) unions isolate the dissimilar materials by using a mixture of brass, plastic, or rubber. Dielectric unions, however, can cause other problems. For example, calcium deposits can build up, which eventually will lead to blockages. Therefore, you should never use dielectric unions unless they are plastic lined.

Source: Peace of Mind Home Inspection Services, LLC, website. www.getpeace.com/plumbing.htm, reviewed March 9, 2004.

Figure 6 ◆ Elbows.

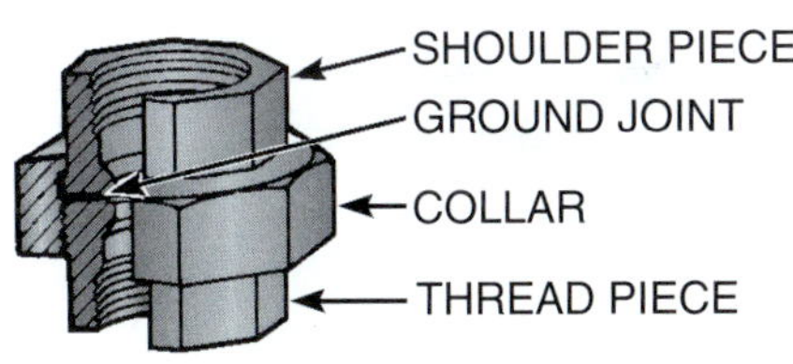

Figure 7 ◆ Ground joint union.

used in place of unions because they cannot be disassembled (see *Figure 9*).

Other fittings for steel pipe include **nipples, crosses,** plugs, caps, and bushings. Nipples are short pieces of pipe, 12 inches long or less, that are threaded on both ends and are used to make extensions from a fitting or to join two fittings (see *Figure 10*). Nipples are manufactured in many

Figure 8 ◆ Flange union.

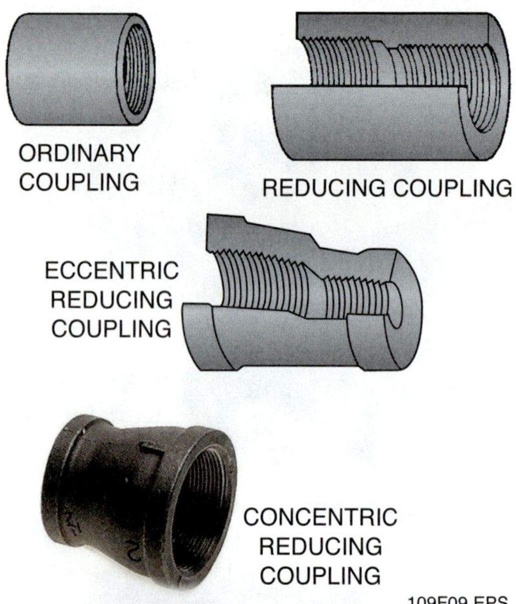

Figure 9 ◆ Couplings.

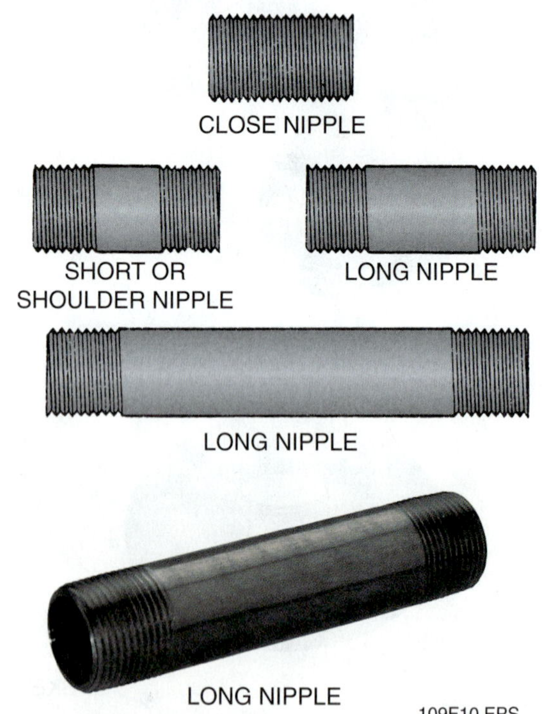

Figure 10 ◆ Nipples.

sizes, beginning with the close or all-thread nipple. Unlike most fittings, which are typically female, nipples are male pieces that connect female components.

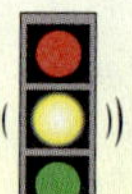

CAUTION

Thread protectors are installed on some steel piping although they are not meant for pressure protection during shipping and handling. Thread protectors may only be used for protection.

CAUTION

Thread protectors are not tapered and may not be used as couplings. Use caution when working with thread protectors because they can easily be mistaken for couplings.

Crosses are fittings that connect four lengths of pipe in a cross-like shape in a distribution system. This fitting allows you to place the horizontal runs and vertical runs centrally (see *Figure 11*). Plugs are available with a variety of heads, such as square, slotted, and hexagon (see *Figure 12*). Caps

Figure 11 ◆ Cross.

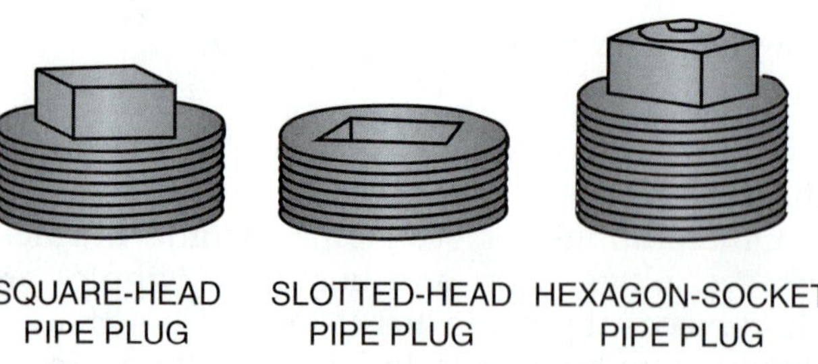

Figure 12 ◆ Plugs.

are used to close openings in other fittings (see *Figure 13*). Bushings are usually used to connect the male end of a pipe to a larger fitting. The ordinary bushing has a hexagon nut at the female end (see *Figure 14*). Some couplings have an extension piece, which is a fitting with a female opening on one end and a male connection on the other.

3.2.0 Valves for Threaded Pipe

All common valves are manufactured for threaded pipe. The most commonly used valves for threaded pipe are the **globe valve, gate valve, ball valve,** and **stop-and-waste valve** (see *Figure 15*). The valves used with carbon steel are not necessarily made of steel.

Globe valves are recommended for general service on steam, water, gas, and oil, where frequent operation and precise control of flow are required. To close the valve, turn the handle clockwise until the valve stem firmly seats the washer or disc between the valve stem and the valve seat. This stops the flow of gas or liquid.

Gate valves are designed to be fully open or fully closed and are best suited as stop valves. They can be used in lines for steam, water, gas, oil, or air. If they are installed as service valves (that is, they are not often used), then they need to be operated periodically to prevent them from freezing shut because of mineral and scale buildups.

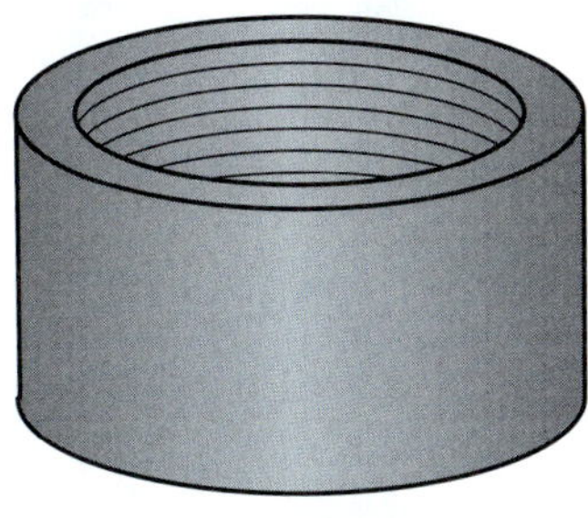

109F13.EPS

Figure 13 ◆ Cap.

109F14.TIF

Figure 14 ◆ Bushing.

DID YOU KNOW?

Gate valves are primarily used as on/off valves. Gate valves used as service valves are used infrequently. Because these valves are operated infrequently, the valve stems can freeze or build up minerals and scale.

Ball valves are used to control the flow of gases and liquids. Ball valves are best suited for terminations on main supply lines and pump lines. They are also used where quick cutoffs for line maintenance are required. The ball part of the valve rotates into the open or closed position.

Stop-and-waste valves are opened or closed by raising or lowering a horizontal disc using a threaded stem. An elastomeric, or rubberized, washer on the end of the stem seals the valve seat, closing off water flow. This valve is most commonly used for draining and freeze protection above ground.

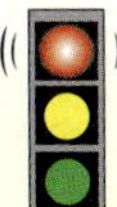

WARNING!

Stop-and-waste valves are prohibited on water supply lines. Do not use stop-and-waste valves underground. There is a risk of cross-contamination from the possibility of a cross-connection.

3.3.0 Fittings and Valves for Grooved Pipe

A wide variety of fittings and valves are available with grooved ends. Special gaskets are also available for various service conditions. It is important to consult the manufacturer's design specifications before using a grooved pipe system.

A grooved pipe joint consists of the following components:

- Specially grooved pipe ends cut by special machines to exact manufacturer's standards
- A synthetic rubber gasket
- Lubricant
- A coupling consisting of two housings

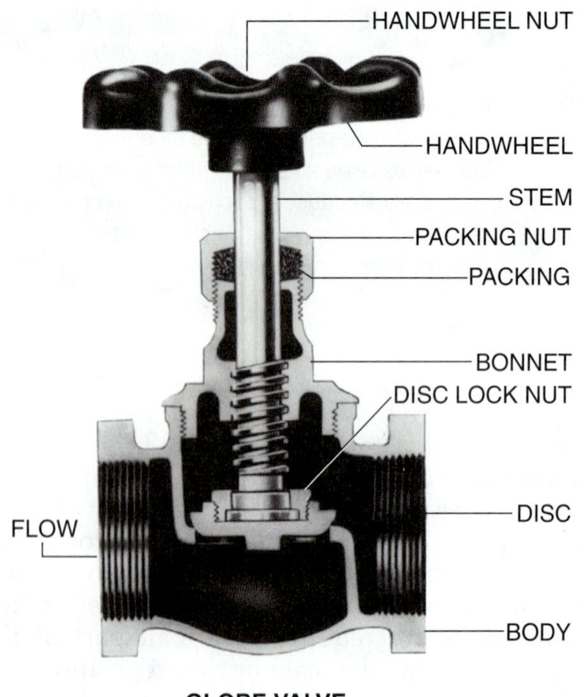

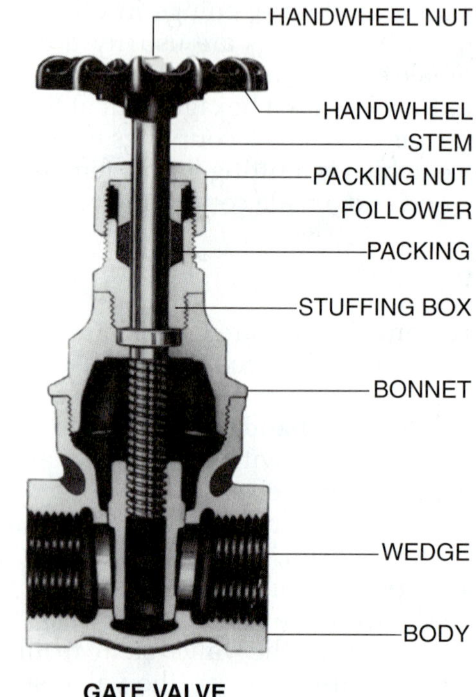

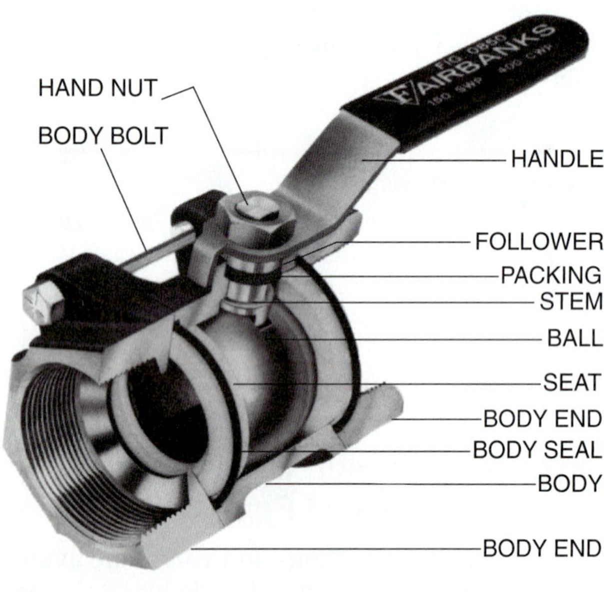

Figure 15 ◆ Valves.

DID YOU KNOW?

The concept of grooved piping dates back to World War I, when armies in the field needed to join pipe quickly and efficiently. Since then, grooved piping has undergone continuous development. Grooved pipe is now a primary means of joining pipe. Steel pipe, copper pipe, cast-iron pipe, and ductile-iron pipe may have grooved joints.

Review Questions

Sections 2.0.0–3.0.0

1. The U.S. government requires that you use carbon steel pipe that meets _____ standards and is marked accordingly.
   a. American Thread
   b. ANSI
   c. ASTM
   d. ADA

2. All weights for a particular size pipe have the same _____.
   a. outside diameter
   b. inside diameter
   c. thread diameter
   d. nominal size

3. Steel pipe that is 1½ inches or larger must be labeled with all of the following *except* _____.
   a. name or brand of the manufacturer
   b. weight
   c. length
   d. *ASTM A 120*

4. _____ have recessed threads and should be used only on soil and waste piping systems.
   a. Drainage fittings
   b. Plugs
   c. Caps
   d. Pressure fittings

5. _____ valves are best suited for main supply lines and pump lines.
   a. Gate
   b. Globe
   c. Ball
   d. Stop-and-waste

4.0.0 ◆ MEASURING, CUTTING, THREADING, AND JOINING STEEL PIPE

You will need to become familiar with many tools and processes to measure, cut, thread, and join materials. Working with threaded steel pipe is very different from working with grooved steel pipe. In addition to requiring different types of fittings, each type of pipe is joined using different methods. You do, however, cut threaded and grooved pipe using the same process. The following sections explain the methods for working with threaded and grooved steel pipe.

4.1.0 Working with Threaded Pipe

When you are working with threaded pipe, you must know how to measure a length of pipe for proper installation. You must take the thread engagement into account or your pipe will be cut too short (see *Table 3*). You must also calculate the size of the fitting.

4.2.0 Measuring

It is extremely important to measure pipe carefully. If you cut a pipe improperly, it cannot make

Table 3 Thread Engagement

Size of Pipe (Inches)	Outside Diameter (Inches)	Number of Threads per Inch	Total Length of Threads (Inches)	Effective Length (Inches) (Approx.)	Thread Engagement (Inches) (Approx.)
¼	0.540	18	⅝	⅜	⅜
⅜	0.675	18	⅝	7⁄16	⅜
½	0.840	14	13⁄16	9⁄16	½
¾	1.050	14	13⁄16	9⁄16	½
1	1.315	11½	1	11⁄16	9⁄16
1¼	1.660	11½	1	11⁄16	⅝
1½	1.900	11½	1	¾	⅝
2	2.375	11½	1 1⁄16	¾	11⁄16
2½	2.875	8	1 9⁄16	1⅛	15⁄16
3	3.500	8	1⅝	1 3⁄16	1
4	4.500	8	1¾	1 5⁄16	1 1⁄16
5	5.560	8	1 13⁄16	1⅜	1 3⁄16
6	6.625	8	1 15⁄16	1½	1¼

a secure joint. You can measure threaded steel pipe using a variety of methods (see *Figure 16*):

- *End-to-end* – Measure the full length of the pipe, including both threaded ends.
- *End-to-center* – Use for pipe that has a fitting screwed on one end only. Pipe length is equal to the measurement minus the end-to-center dimension of the fitting, plus the length of thread engagement.
- *Center-to-center* – Use with a length of pipe that has fittings screwed onto both ends. Pipe length is equal to the measurement minus the sum of the end-to-center dimensions of the fittings, plus twice the length of the thread engagement.
- *End-to-face* – Use for pipe that has a fitting screwed on one end only. Pipe length is equal to the measurement plus the length of the thread engagement.
- *Face-to-face* – Use for the same situation as center-to-center measurement. Pipe length is equal to the measurement plus twice the length of the thread engagement.
- *Face-to-back* – Use with pipe that has fittings screwed onto both ends. Pipe length is equal to the measurement plus the distance from the face to the back of one screwed-on fitting, plus twice the length of the thread engagement.
- *Center-to-face* – Use with pipe that has fittings screwed onto both ends. Pipe length is equal to the measurement from the center of one of the screwed-on fittings to the face of the opposite fitting, plus twice the length of the thread engagement.
- *End-to-back* – Use for pipe that has a fitting screwed on one end only. Pipe length is equal to the measurement plus the length of the screwed-on fitting, plus the length of the thread engagement.

4.3.0 Cutting

The methods you use to cut threaded pipe also apply to grooved pipe. You can cut steel pipe with pipe cutters, which may have from one to four cutting wheels (see *Figure 17*). A single cutting wheel requires enough room for the cutter to be rotated all the way around the pipe. Use a cutter with more wheels when space is limited. The more wheels a cutter has, the fewer rotations it requires to cut the pipe. In addition to the cutting wheels, pipe cutters have an adjusting screw and at least two guiding wheels.

To operate the pipe cutter, revolve it around the pipe and tighten the cutting wheel ¼ revolution with each turn. Avoid overtightening the cutting wheel; this could damage or break the wheel.

Always use sharp cutting wheels. If the tube or pipe looks mashed after it has been cut, replace the cutting wheel(s). Apply lubricating oil periodically to all movable parts on the pipe cutter so it will operate smoothly.

4.4.0 Reaming

After you cut a piece of pipe, use a reamer to remove the burr that remains on the inside of the pipe. If the burr is not reamed off, it will collect deposits and slow the flow of liquid through the pipe. Because reamers are tapered, one reamer can deburr many sizes of pipe. Reamers may be straight or spiral (see *Figure 18*). Reaming is always required for pipe installations.

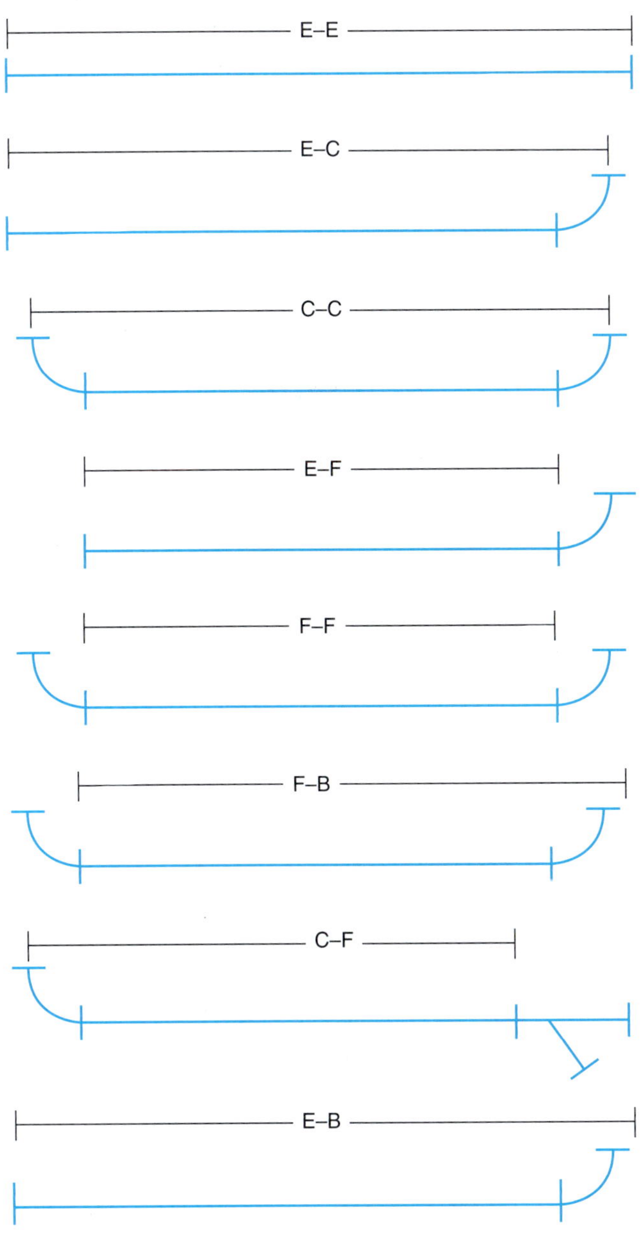

Figure 16 ◆ Measuring methods.

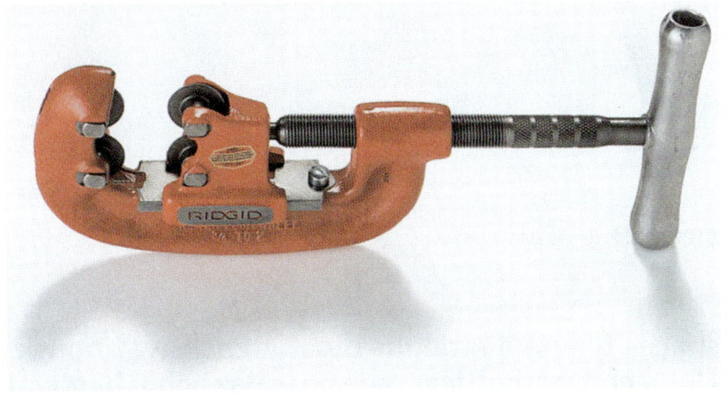

Figure 17 ◆ Pipe cutter.

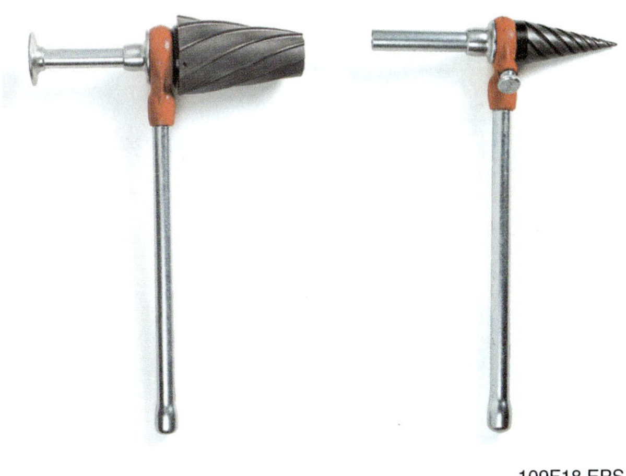

Figure 18 ◆ Reamers.

CAUTION

Be careful not to overream the pipe. Doing so may damage the wall thickness and weaken the pipe, causing the wall to collapse.

4.5.0 Threading

There are two types of pipe threaders: hand and power. Hand threaders consist of the pipe die and the **stock** (*Figure 19*). The die cuts the threads. Apply cutting oil to protect the die segments. The stock holds the die. Although you usually operate a hand threader manually, you can use it with a power drive, which threads automatically (see *Figure 20*).

To thread large quantities of pipe, use a **power-threading machine** (*Figure 21*). This machine rotates, threads, cuts, and reams pipe. To use this machine, mount the pipe through the machine chuck (the piece that centers and grips the pipe), and then tighten the chuck. The oil-pumping trigger allows you to control the application of oil from the cutting oil pan. The oil lubricates the die. Direct the threading process by using the foot switch. Service the foot switch often to ensure that it always works as effectively as possible. Ensure that you use the proper oil for the foot switch.

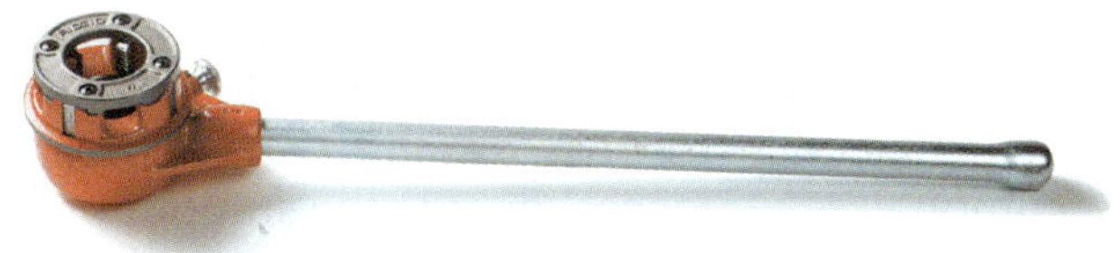

Figure 19 ◆ Hand threader.

109F20.TIF

Figure 20 ◆ Power drive.

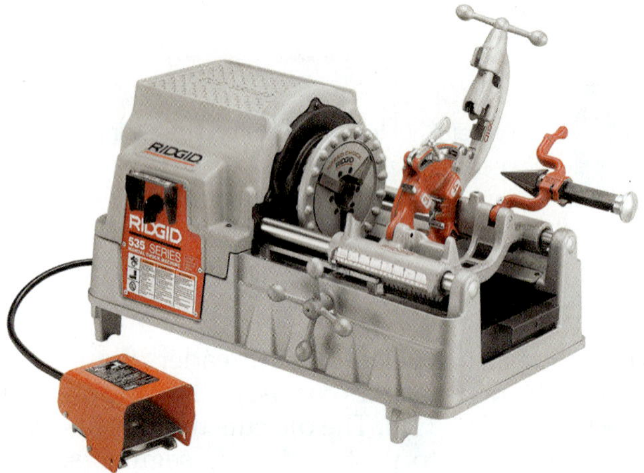

109F21.TIF

Figure 21 ◆ Power-threading machine.

Refer to the manufacturer's specifications for the proper types of oil.

You can use a variety of vises and wrenches to hold pipe while threading it. Vises have jaws that hold the pipe firmly and prevent it from turning. The teeth on vises may leave marks on the pipe, so you should use a vise only on pipe that will not be visible after installation. Use the standard **yoke vise** to hold pipe that is small in diameter (*Figure 22*). Use the **chain vise** to hold much larger pieces of pipe (*Figure 23*). Keep the chain on this vise oiled to prevent it from becoming stiff. Always make sure that you use the correct oil. A stiff chain will make the chain vise operate poorly.

Use pipe wrenches to grip and turn around stock. Pipe wrenches have teeth that are set at an angle. This angle allows the teeth to grip and permits the wrench to turn in only one direction. Common pipe wrenches include the straight pipe wrench (*Figure 24*), the offset pipe wrench (*Figure 25*), and the compound-leverage pipe wrench (*Figure 26*). Use an offset pipe wrench in confined

109F22.TIF

Figure 22 ◆ Yoke vise.

109F23.TIF

Figure 23 ◆ Chain vise.

spaces where a straight pipe wrench will not fit. Use a compound-leverage pipe wrench when you need extra strength to turn pipe assemblies. As the hook jaw turns the pipe one way, an offset-chain-

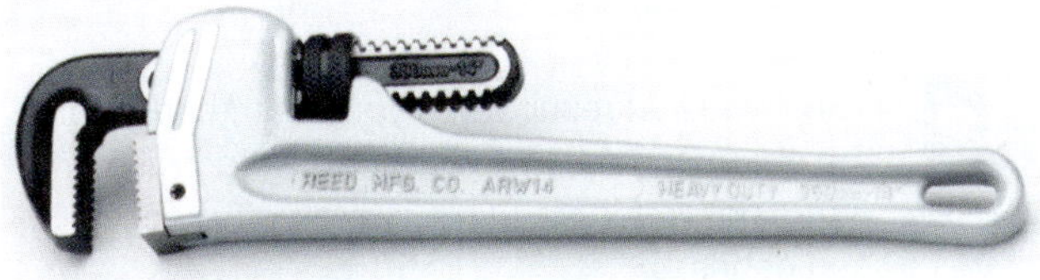

109F24.TIF

Figure 24 ◆ Straight pipe wrench.

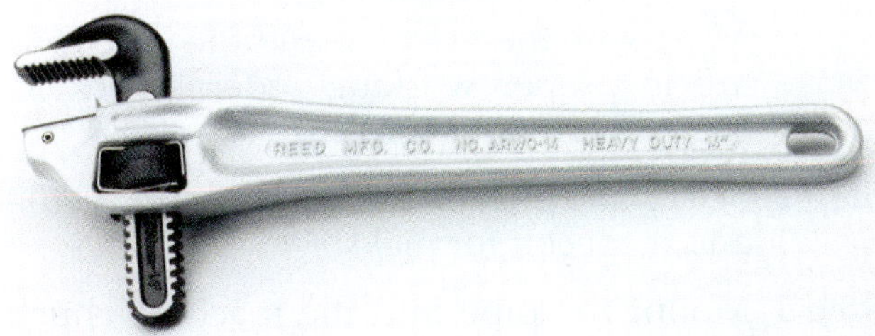

109F25.TIF

Figure 25 ◆ Offset pipe wrench.

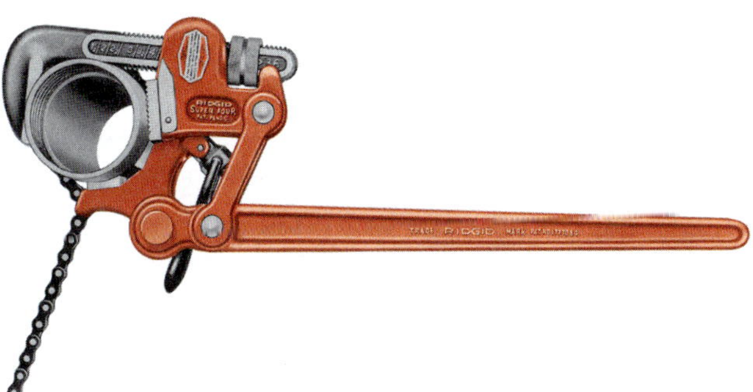

109F26.TIF

Figure 26 ◆ Compound-leverage pipe wrench.

trunnion (pivot) assembly applies pressure in the opposite direction.

To hold large pipes, use **pipe tongs,** also called a **chain wrench** (*Figures 27* and *28*). Pipe tongs are wrench-like tools with a length of chain that holds the pipe. You must oil the chain in pipe tongs and chain wrenches often to prevent the chain from becoming stiff or rusty.

To hold chrome-plated or other types of finished pipe, use a **strap wrench** (*Figure 29*). The strap allows the wrench to grip the pipe without leaving jaw marks or scratches on the pipe. You can apply resin to the strap to increase the wrench's holding power and reduce slippage.

To cut threads using a hand die and stock, follow these steps (see *Figure 30*):

Step 1 Select the correct size of die for the pipe you are threading.

Step 2 Inspect the die to make sure the cutters are free of nicks and wear.

109F27.TIF

Figure 27 ◆ Pipe tongs.

109F28.TIF

Figure 28 ◆ Chain wrench.

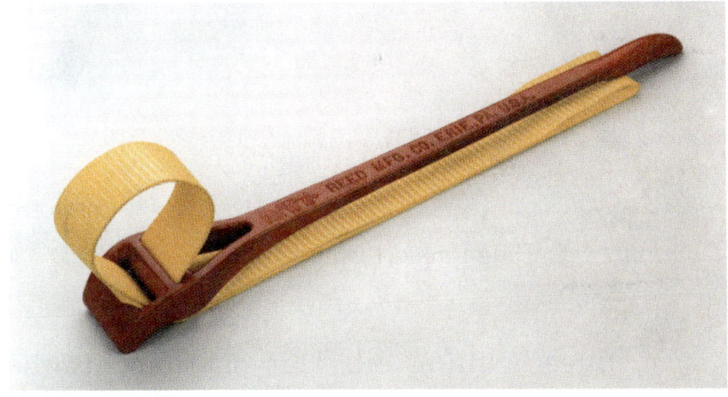

109F29.TIF

Figure 29 ◆ Strap wrench.

Step 3 Lock the pipe securely in a vise.

Step 4 Slide the die over the end of the pipe, guide-end first.

Step 5 Push the die against the pipe with the heel of one hand. Take three or four short, slow, clockwise turns. Be careful to keep the die pressed firmly against the pipe.

Step 6 When you have cut enough thread to keep the die firmly against the pipe, apply some

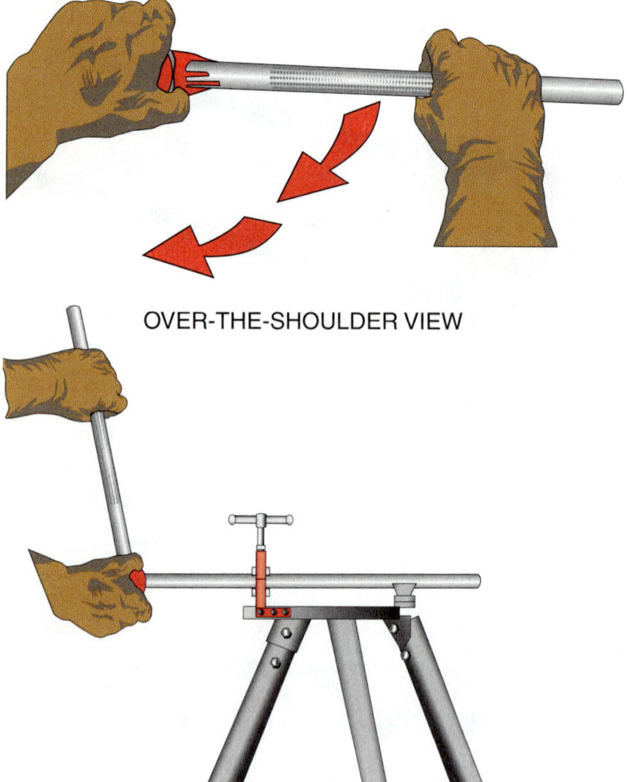

Figure 30 ◆ Cutting threads using a hand die and stock.

thread-cutting oil. This oil prevents the pipe from overheating from friction, and it lubricates the die. Oil the threading die every two or three downward strokes.

Step 7 Back off one quarter-turn after each full turn forward to clear out the metal chips. Continue until the pipe projects one or two threads from the die end of the stock. Having too few threads is as bad as having too many threads.

Step 8 Remove the die by rotating it counterclockwise.

Step 9 Wipe off excess oil and any chips.

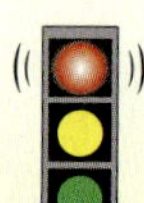

WARNING!

The chips that result when you are threading pipe are sharp and can cut you. Use a wire brush and a rag, not your bare hand, to wipe off excess oil and debris.

To cut threads using a power-threading machine, follow these steps (see *Figure 31*):

Step 1 Make sure the machine and the foot pedal are in proper working order. Check for frayed wires and cords.

Step 2 Select and install the correct size of die, and inspect it for nicks.

Step 3 Mount the pipe into the machine chuck. Make sure to provide additional support for long pipe.

Step 4 Check the pipe and die alignment.

Step 5 Cut threads until two threads appear at the other end of the die. Stop the threading action. Apply the correct cutting oil to the threads and pipe during the threading operation to reduce heat and friction.

Step 6 Back off the die until it is clear of the pipe.

Step 7 Remove the pipe from the machine chuck. Be careful not to mar the threads.

Step 8 Wipe the pipe clean of oil and metal chips.

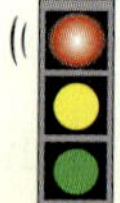

WARNING!

Before using any power machine, become familiar with the machine's maintenance and safety instructions. Foot pedals for safety cutoffs are now standard. A poorly maintained machine is a safety hazard. Dirt and metal chips can get stuck in the cutting dies and produce inadequate threads. Clogged components may also send debris flying—these flying objects can blind you or a co-worker.

Each threading machine is slightly different. Become familiar with the manufacturer's operating procedures before you operate the machine.

4.6.0 Joining Threaded Pipe

To join threaded pipe, you must use **pipe-joint compound,** often called pipe dope, or **sealant tape** to provide lubrication for assembly.

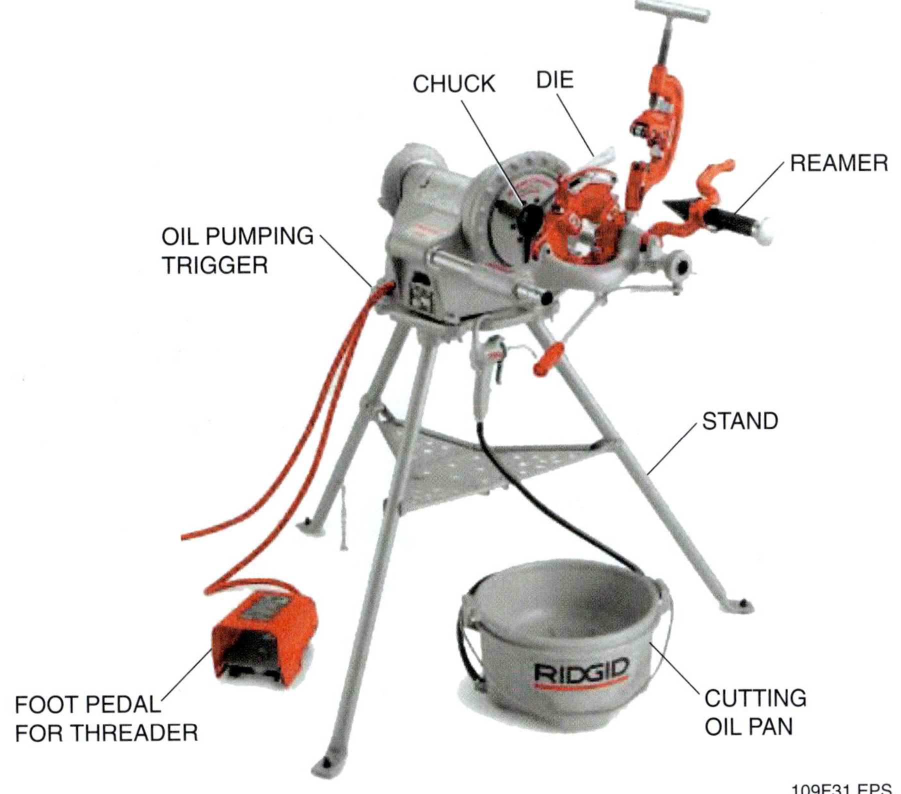

Figure 31 ◆ Cutting threads using a power-threading machine.

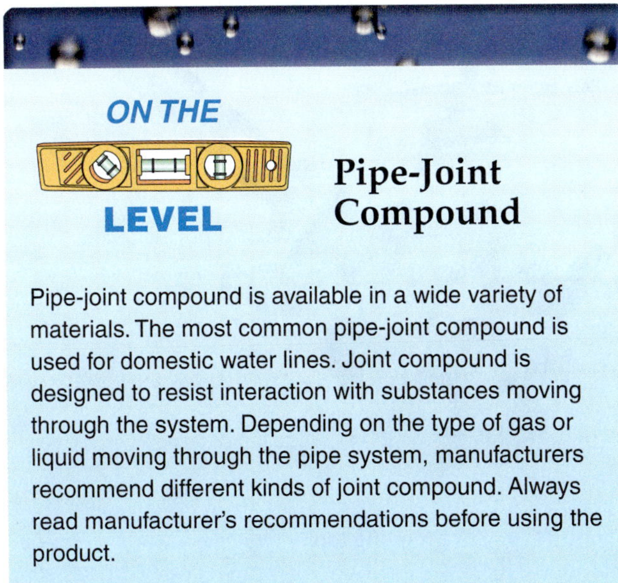

Pipe-Joint Compound

Pipe-joint compound is available in a wide variety of materials. The most common pipe-joint compound is used for domestic water lines. Joint compound is designed to resist interaction with substances moving through the system. Depending on the type of gas or liquid moving through the pipe system, manufacturers recommend different kinds of joint compound. Always read manufacturer's recommendations before using the product.

To join threaded steel pipe, follow these steps (see *Figure 32*):

Step 1 Apply pipe-joint compound or sealant tape only to the male threads of the pipe or fittings before you assemble a pipe connection. Do not apply the compound or tape to the female threads of either a pipe or a fitting. The twisting motion used to join the pipes will ball up the pipe dope or tape if it is applied to the female threads, and the debris will get stuck inside the pipe system.

Step 2 If you are using sealant tape, apply the tape in a clockwise direction, the same direction as the fitting turns. Before using sealant tape, be sure to check all applicable gas codes to make sure the use of tape is permitted.

Step 3 Start the fitting into the threaded pipe by hand, making sure to leave the starter threads free of tape and compound. Turn the fitting clockwise. Finish tightening the fitting with a pipe wrench.

4.7.0 Grooving Pipe

You can groove pipe in two ways: by rolling or cutting. Roll grooving involves cold forming pipe; it does not remove any metal from the pipe. Cut grooving removes metal from the outside

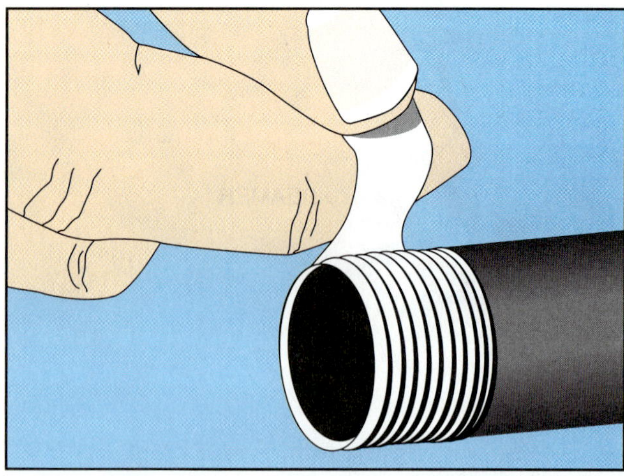

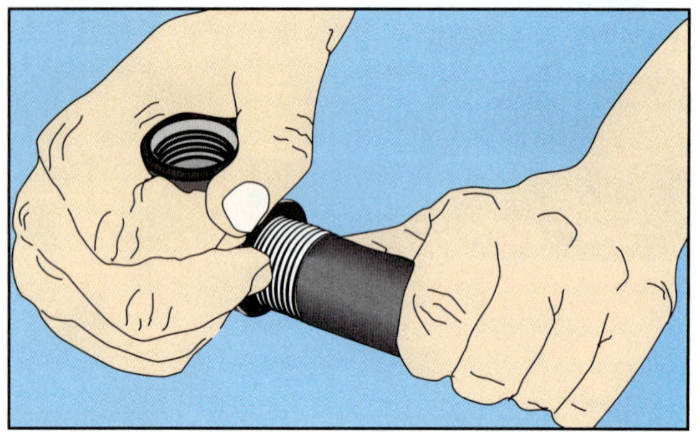

109F32.EPS

Figure 32 ◆ Joining threaded steel pipe.

109F33.TIF

Figure 33 ◆ Roll-grooving tool.

109F34.TIF

Figure 34 ◆ Cut-grooving tool.

diameter of the pipe. You will learn more about grooving pipe elsewhere in this curriculum. See examples of a roll-grooving and cut-grooving tool in *Figures 33* and *34*.

4.8.0 Joining Grooved Pipe

Joining grooved steel pipe is very different from joining threaded steel pipe. To join grooved pipe, follow these steps (see *Figure 35*):

Step 1 Check the pipe ends. To make a leakproof seal, the ends must be free from indentations, projections, or roll marks.

Step 2 Check to make sure that the gasket is suitable for the intended use. Some manufacturers color-code their gaskets. Apply a thin coat of lubricant to the lips and the outside of the gasket.

Step 3 Install the gasket over the pipe end. Be sure the gasket lip does not hang over the pipe end.

Step 4 Align and bring the two pipe ends together. Slide the gasket into position and center it between the grooves on each pipe. Be sure that no part of the gasket extends into the groove on either pipe.

Step 5 Assemble the housing segments loosely, leaving one nut and bolt off to allow the housing to swing over the joint.

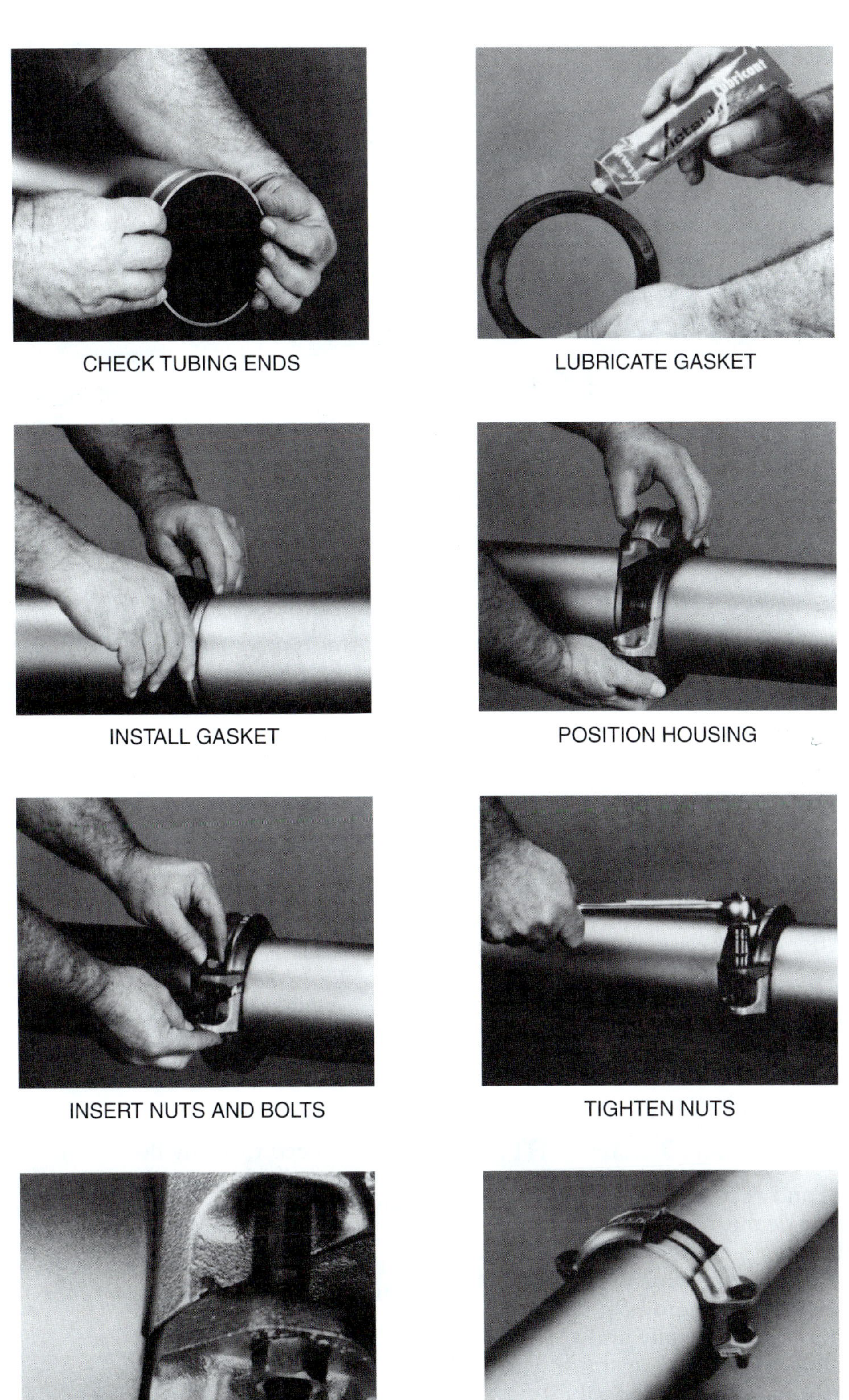

Figure 35 ◆ Joining grooved pipe.

Step 6 Install the housing, swinging it over the gasket and into position in the grooves on both pipes.

Step 7 Insert the remaining bolt and nut. Be sure that the bolt track head engages into the recess in the housing.

Step 8 Tighten the nuts alternately and equally to maintain metal-to-metal contact at the angle bolt pads.

Review Questions

Section 4.0.0

1. When measuring pipe, you must take the _____ into account or your pipe will be cut too short.
   a. diameter
   b. weight of pipe
   c. type of pipe
   d. thread engagement

2. If a tube or pipe looks mashed after it has been cut, you should _____ the cutting wheel or wheels.
   a. sharpen
   b. replace
   c. polish
   d. adjust

3. Use a _____ to clean the burr out of the inside of a pipe.
   a. pipe cutter
   b. threader
   c. reamer
   d. yoke vise

4. A(n) _____ has two parts: a die to cut threads and a stock to hold the die.
   a. hand threader
   b. pipe cutter
   c. offset wrench
   d. chain vise

5. Never put pipe-joint compound or sealant tape on the female threads of a fitting.
   a. True
   b. False

5.0.0 ◆ HANGERS AND SUPPORTS

You must install and support all pipe so the pipe and joints remain leakproof. Improper support can cause joints to sag, and the added stress increases the chances that the pipe will leak, break, or crack. Plumbing codes require pipe to be securely fastened both horizontally and vertically.

If you are installing pipe in a seismically active area (where earthquakes are a possibility) local codes will require seismic restraints. The purpose of these restraints is to ensure that the pipe is securely fastened to the structure in the event of excessive vibration. For example, some codes require hangers and supports to be used at closer intervals than in non-seismically active areas. In addition, they may require that you leave extra spacing for pipes where they meet walls and floors to allow for anticipated movement.

You need to know how to attach pipe to wood, masonry (including tile and concrete), and steel surfaces. You should use one or more hangers to support carbon steel pipe. The basic components of hangers are pipe attachments, connectors, and structural attachments.

5.1.0 Pipe Attachments

Pipe attachments touch or connect directly to the pipe. Rings, clamps, and clevises are among the attachments that can support piping horizontally from the ceiling (*Figure 36*). To support piping on wood-frame construction, you can use pipe hooks, U-hooks, J-hooks, tube straps, tin straps, plumber's tape (also called strap iron or band iron), half clamps, suspension clamps, hold-down clips, and pipe clamps (*Figure 37*). Vertical pipe

Code Requirements

Consult local applicable codes and manufacturer's recommendations for spacing intervals when installing pipe hangers and supports. Codes vary widely from one area to another. Spacing requirements are often found in the specifications section of the building plans. Failing to support pipe adequately will get a rejection from the plumbing inspector. Poor workmanship or misunderstandings will earn a rejection as well.

You must also check the codes to know which types of fasteners are allowed in your area. This information is often in the specifications as well. Double check both sources. If they do not agree, ask your supervisor.

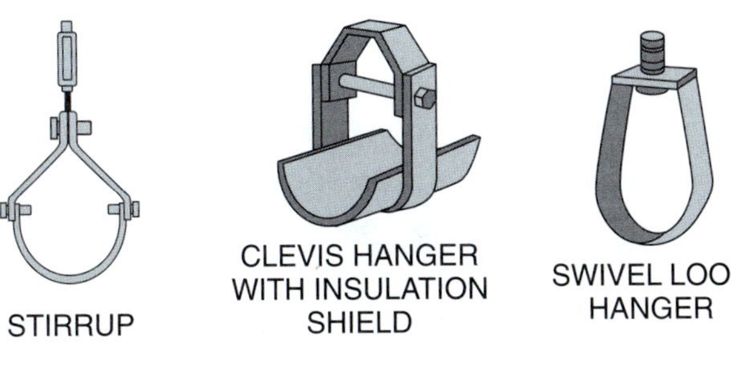

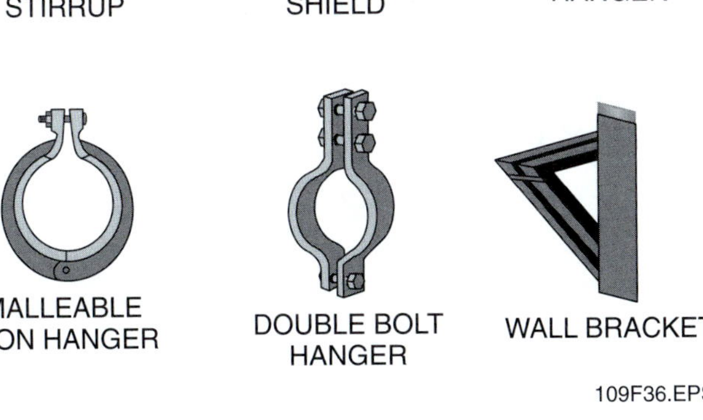

Figure 36 ◆ Hangers used for horizontal support.

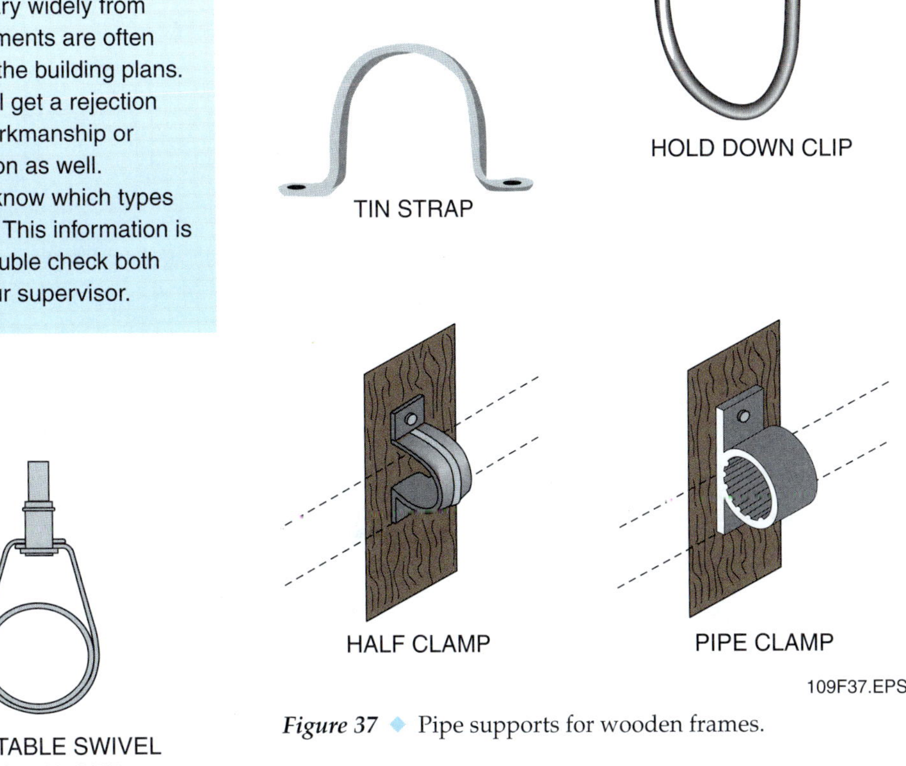

Figure 37 ◆ Pipe supports for wooden frames.

should be supported at each floor level using riser clamps (*Figure 38*).

Other pipe attachments include universal pipe clamps and pipe channels (*Figures 39* and *40*). The clamps are available in standard finishes of mild steel and electro-galvanized steel.

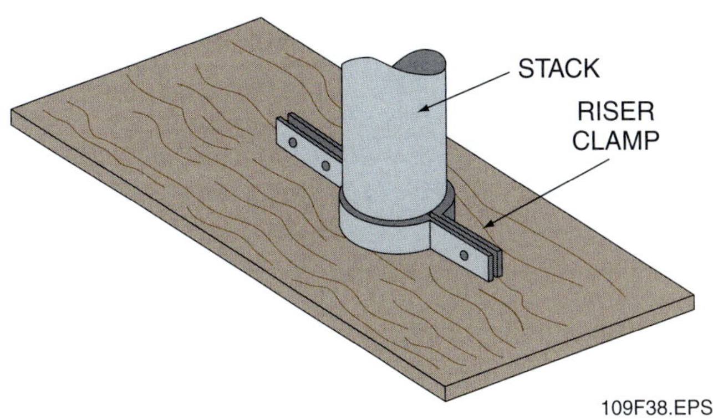

Figure 38 ◆ Vertical pipe supports.

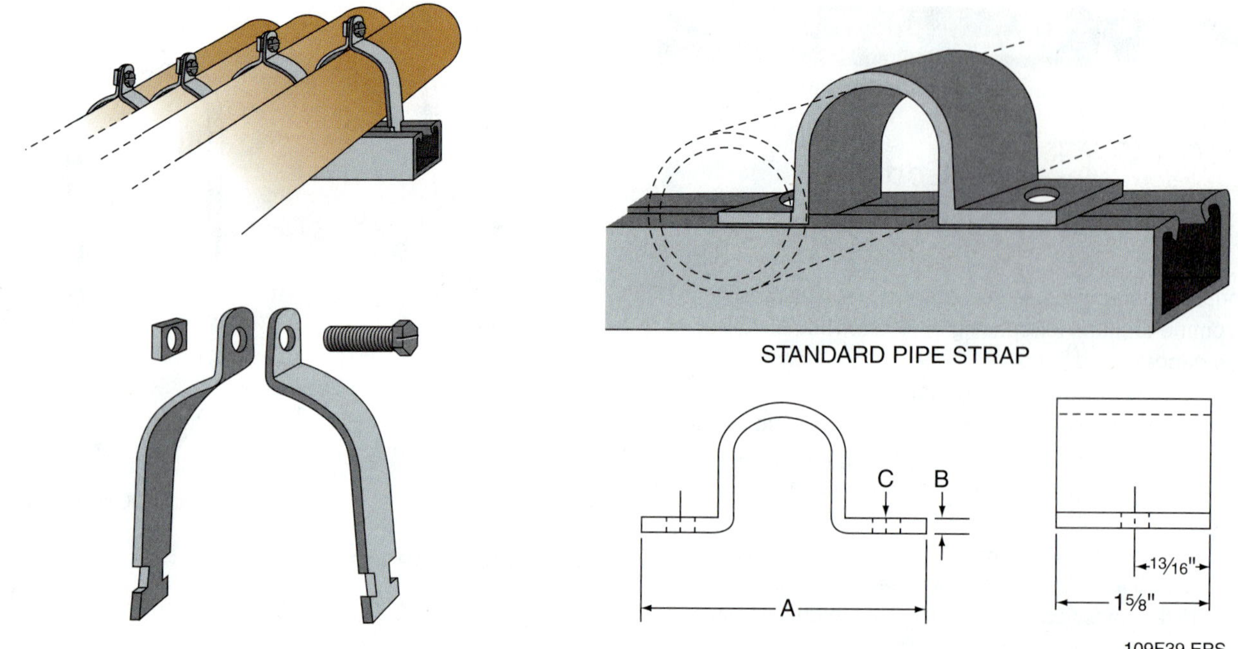

Figure 39 ◆ Universal pipe clamps.

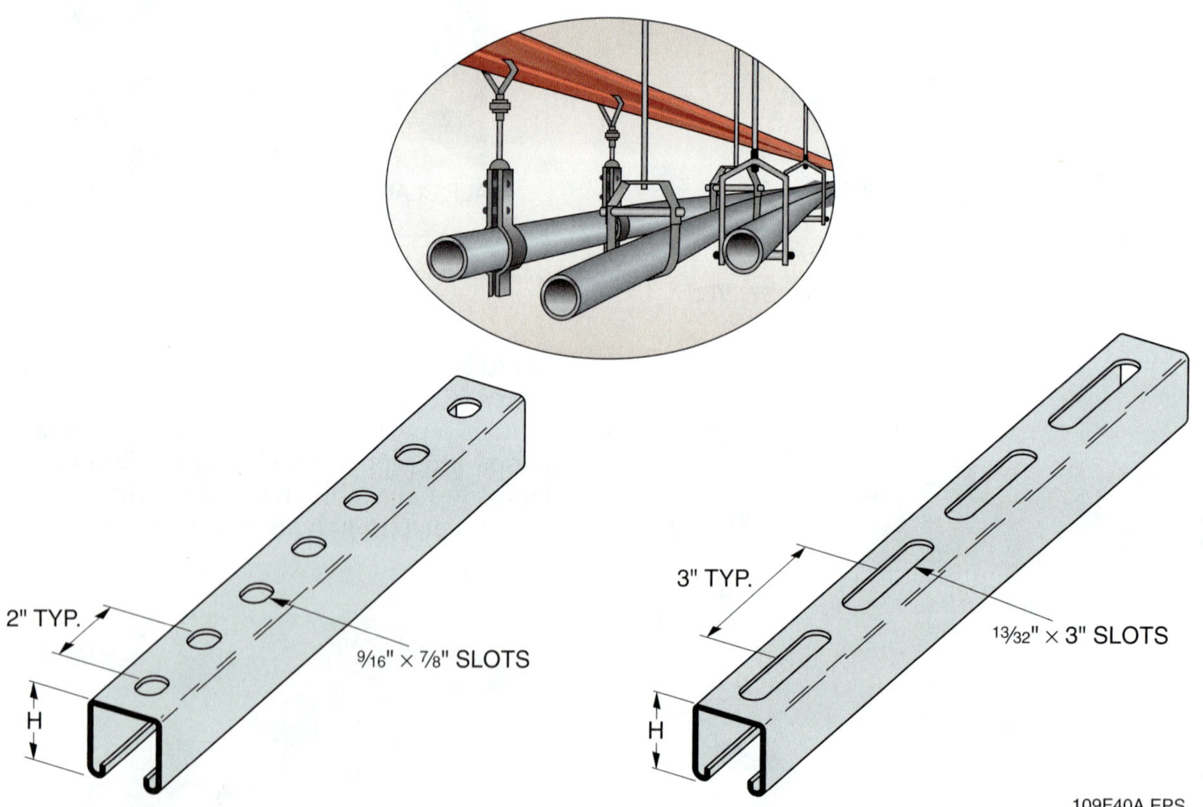

Figure 40 ◆ Pipe channels. (1 of 2)

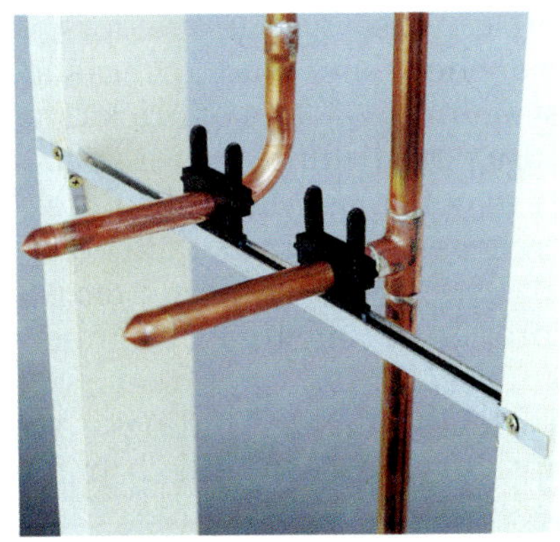

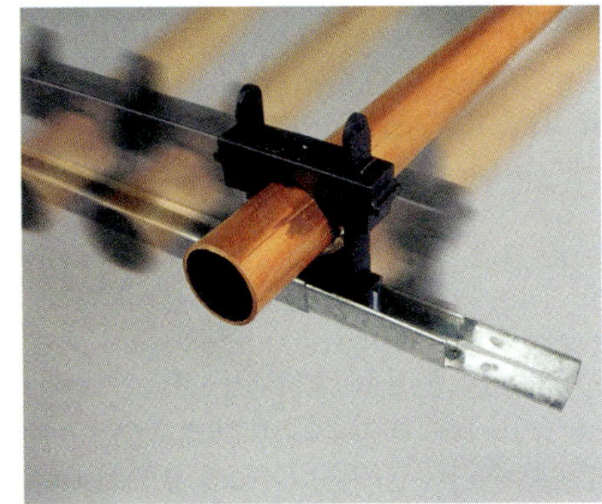

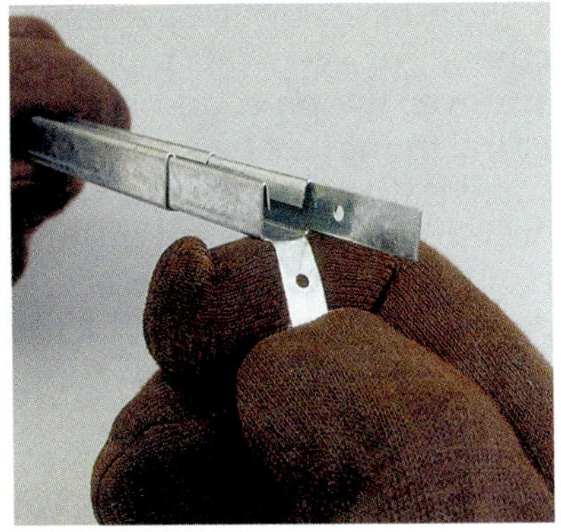

109F40B.EPS

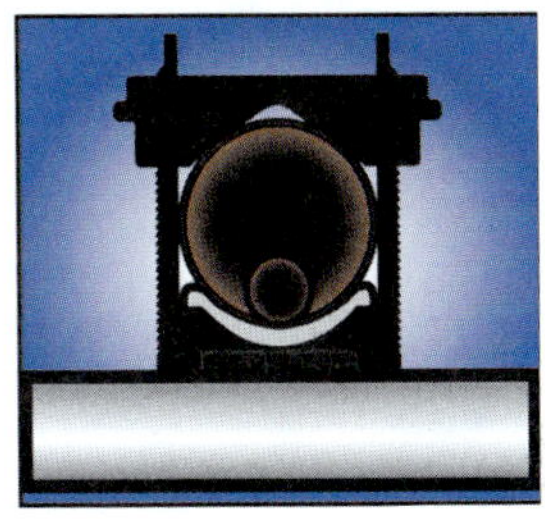

109F40C.EPS

Figure 40 ◆ Pipe channels. (2 of 2)

5.2.0 Connectors

Connectors link the pipe attachment to the structural attachment. The two main groups of connectors are (1) rods and bolts and (2) other rod attachments.

Rod attachments, which are available in several sizes, include eye sockets, extension pieces, rod couplings, hanger adjusters, turnbuckles, clevises, and eye rods (*Figure 41*). Eye sockets are used with hanger rods. You will use extension pieces when you attach hanger rods to beam clamps and other types of structural attachments. Rod couplings and reducing rod couplings support pipe runs where it is possible to connect to an existing stud.

5.3.0 Structural Attachments

Structural attachments are the anchors and anchoring devices that hold the pipe hanger assembly securely to the building structure. They include powder-actuated anchors, concrete inserts, beam clamps, C-clamps, beam attachments, various brackets, ceiling flanges and plates, plate washers, and lug plates.

You can use wedge anchors, sleeve anchors, drop anchors, and stud anchors to fasten the piping system to concrete or brick (*Figure 42*). You can also use non-drilling anchors and self-drilling anchors in concrete. Make absolutely sure that new concrete has had time to cure before you insert anchors. Check with your supervisor to determine the proper curing time; plumbers do not make this judgment.

In steel-framed buildings, you should use clamp-like devices to attach hangers for smaller pipes. These devices include beam clamps, I-beam clamps, and steel C-clamps (*Figure 43*).

In wood-framed buildings, use nails and screws to attach pipe hangers and supports to the structure. For vertical pipe runs, place riser clamps over the pipe, and secure the pipe to the structural frame (see *Figure 44*). If using plumber's tape, nail it directly onto the floor joists.

You must support vertical piping at sufficient intervals to keep the pipe in alignment. How you support vertical piping depends on the size of pipe and whether it needs to be supported at or between the floor lines. Support vertical pipe at each floor line with riser clamps (refer to *Figure 44*). As always, follow applicable code requirements.

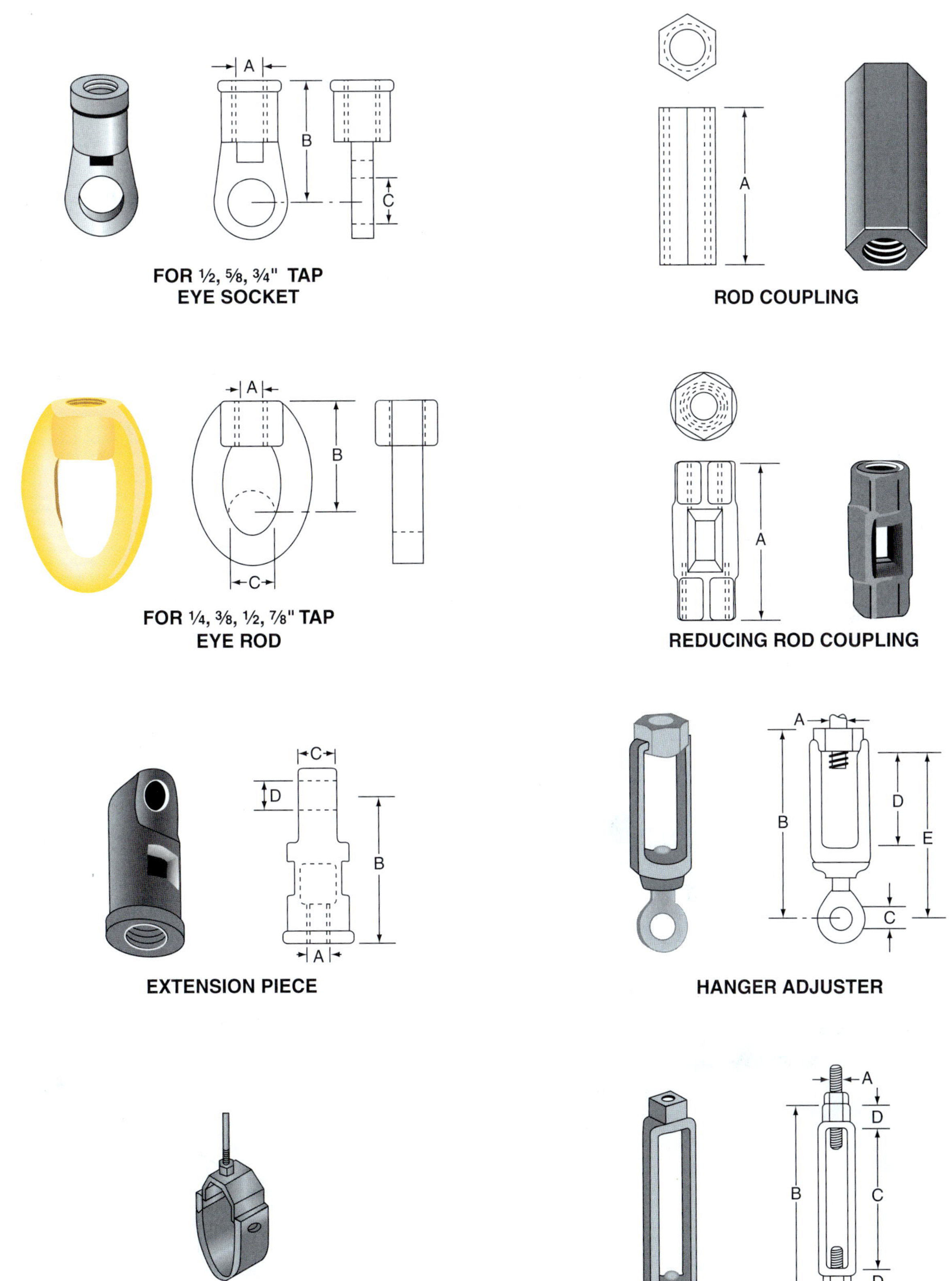

Figure 41 ◆ Connectors.

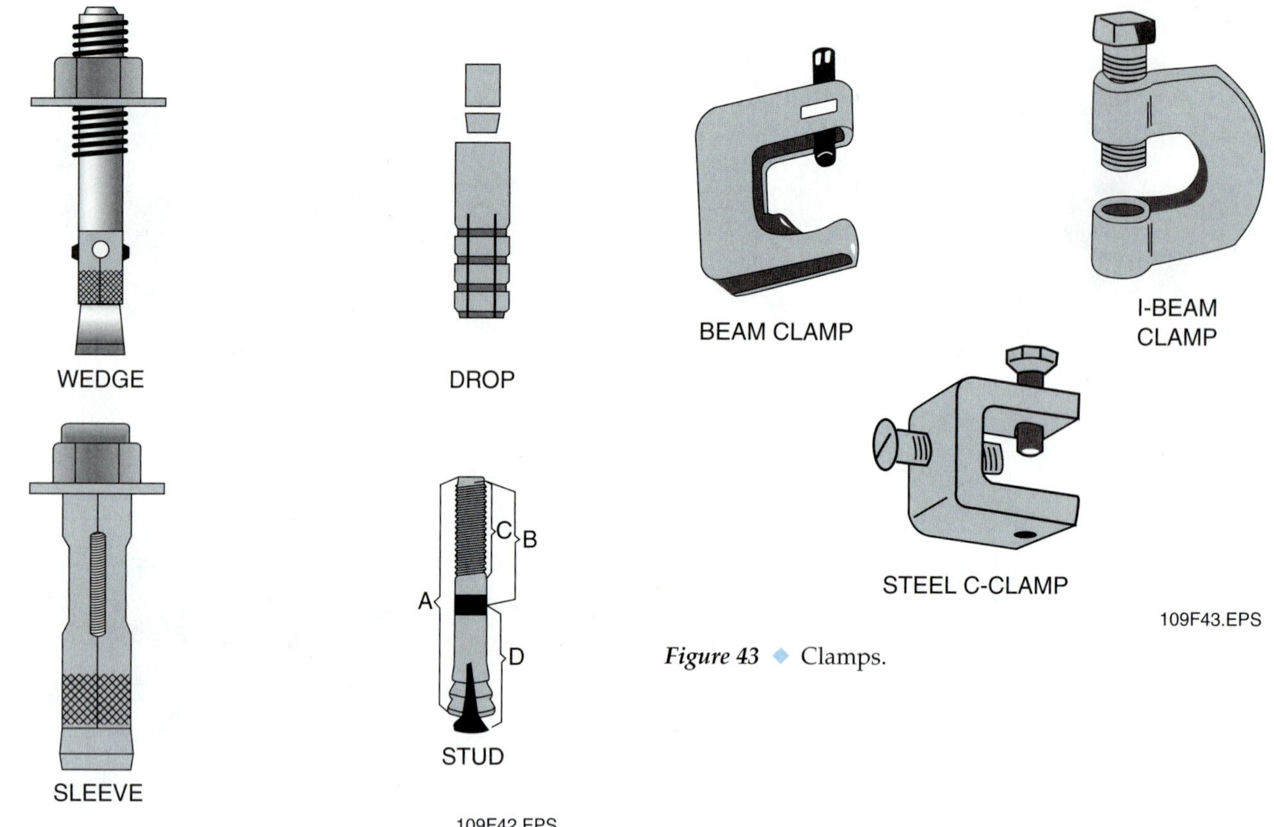

Figure 42 ◆ Anchors for concrete or brick.

Figure 43 ◆ Clamps.

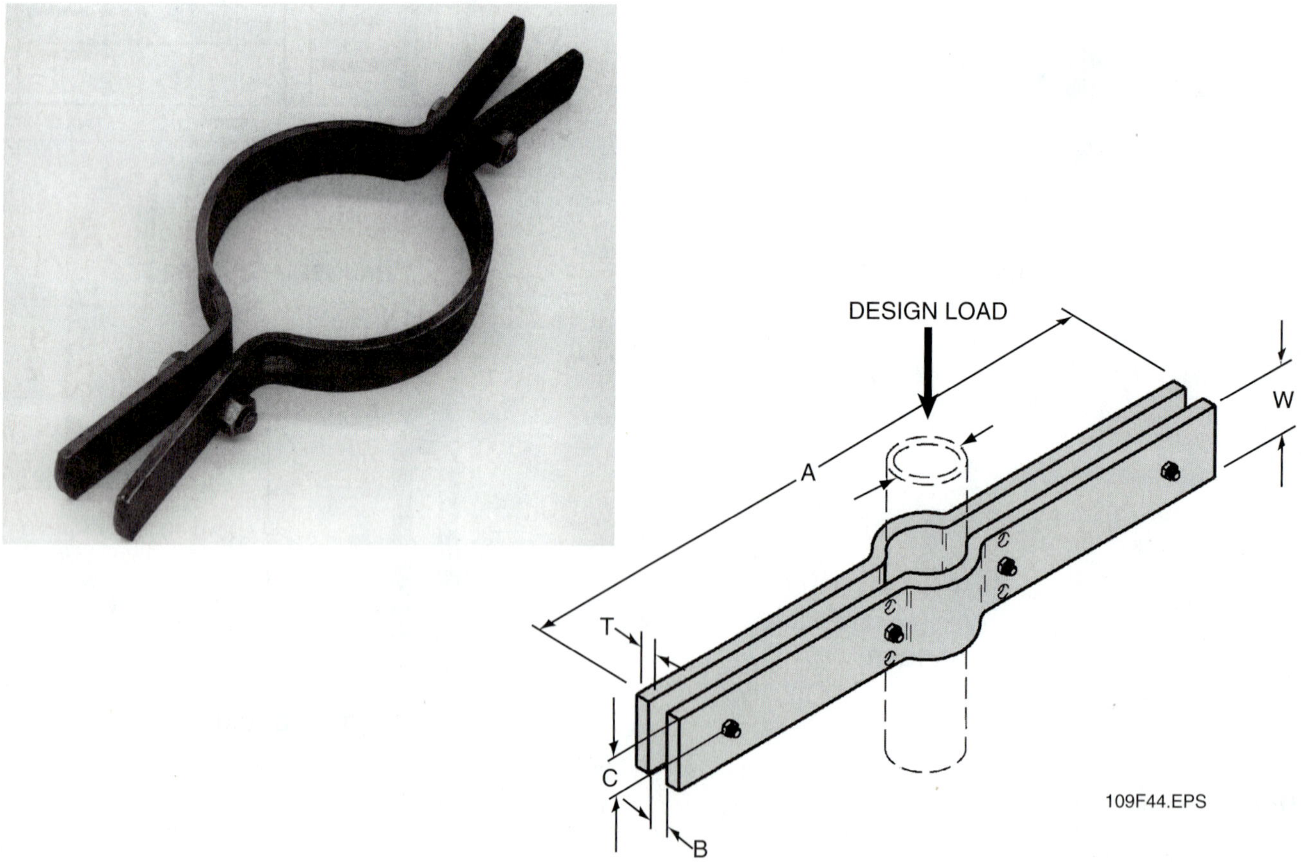

Figure 44 ◆ Riser clamp.

Review Questions

Section 5.0.0

1. Proper _____ are required for pipes not to sag, bend, or move out of alignment.
   a. anchors
   b. attachments
   c. supports
   d. slopes

2. The basic components of hangers are _____.
   a. pipe attachments, connectors, and structural attachments
   b. pipe hooks, U-hooks, and J-hooks
   c. connectors, couplings, and anchors
   d. clamps, pipe attachments, and pipe hooks

3. A pipe attachment is a type of hanger that touches or connects directly to the pipe itself.
   a. True
   b. False

4. _____ are used to fasten the piping system to concrete.
   a. Clamps
   b. Anchors
   c. Risers
   d. Connectors

5. Vertical pipe should be supported at each floor level using _____.
   a. tin straps
   b. riser clamps
   c. turnbuckles
   d. anchors

Summary

Carbon steel pipe is used in many modern plumbing applications. Black steel pipe is used most often for gas and air-pressure applications. Galvanized steel pipe is often used for venting systems, and it is electroplated with zinc to protect the pipe from abrasive or corrosive materials.

Steel pipe is categorized by both its size (inside or outside diameter) and strength. Schedule 40 is a standard-weight pipe, Schedule 80 is an extra-strong weight, and Schedule 120 is a double extra-strong weight.

Steel pipe can be threaded or grooved. Threads used on carbon steel pipe are standardized. This means that thread sizes on all manufactured pipe products are the same. Threaded pipe can be measured in several ways, depending on whether there are fittings on one or two ends. Incorrect measuring will mean your pipe may be cut too long or too short for the installation. A variety of tools are available for working with and cutting carbon steel pipe. Familiarity with tool use and maintenance will ensure that your workmanship is of the best quality.

Iron pipe fittings are made for both pressure and drainage pipe systems. Pressure pipe fittings cannot be used in drainage systems because their inside surfaces are not smooth enough. Drainage fittings are specifically designed for smooth flow.

All pipe requires adequate support during installation to keep it from sagging, bending, and losing proper alignment. Codes and building specifications indicate which hangers, fasteners, and connectors are appropriate for different applications. Seismically active areas may require additional supports.

As with all aspects of plumbing, the selection and installation of carbon steel pipe is strictly regulated by building and plumbing codes. Codes vary widely among locations, so you must always check local applicable codes before installing a piping system.

Notes

Trade Terms Quiz

Fill in the blank with the correct trade term that you learned from your study of this module.

1. When joining threaded pipe, use _______________ or _______________ to provide lubrication for assembly.
2. _______________ are best suited as stop valves.
3. Where frequent operation and close control of flow are required, use a(n)_______________.
4. The purpose of the _______________ is to allow you to screw the flanges to the two pipes you are joining.
5. _______________, or _______________ are wrench-like tools with a length of chain that holds the pipe.
6. A(n)_______________ connects two pipes by screwing the thread and shoulder pieces of the union onto the pipes.
7. You can use the same _______________, or devices used to cut the threads, with all three weights of each size of pipe.
8. _______________ are most commonly used for draining and freeze protection aboveground.
9. Hand threaders consist of two parts: the pipe die and the _______________, the device that holds the die.
10. Pieces of pipe that are used to make extensions from a fitting or to join two fittings are called _______________.
11. A(n) _______________ allows you to place the horizontal runs and vertical runs centrally.
12. When threading pipe that is smaller in diameter, use a(n) _______________ to hold the pipe in place.
13. The length of pipe that gets screwed into the fitting is called the _______________.
14. To avoid leaving jaw marks or scratches on pipe, use a(n) _______________ to grip the pipe.
15. _______________ is most often used as gas or air-pressure pipe.
16. Although more resistant to corrosion than black steel pipe, _______________ is still less resistant to corrosion than other piping materials.
17. To thread large quantities of pipe, use a(n) _______________, which rotates, threads, cuts, and reams pipe.
18. _______________ cannot be used in drainage systems because their inside surfaces are not smooth enough.
19. When threading pipe that is larger in diameter, use a(n) _______________ to hold the pipe in place.
20. _______________ have a drainage pattern design and recessed threads that should be used only on soil and waste piping systems.
21. Where quick cutoffs for line maintenance are required, use a(n) _______________.
22. Steel pipe for threading is manufactured in three weights or strengths: _______________, _______________, and _______________.

Trade Terms

Ball valve
Black steel pipe
Chain vise
Chain wrench
Cross
Die
Double extra strong (double extra heavy)
Drainage fitting (recessed fitting)
Extra strong (extra heavy)
Flange union
Galvanized pipe
Gate valve
Globe valve
Ground joint union
Nipple
Pipe-joint compound (pipe dope)
Pipe tongs
Power-threading machine
Pressure fitting
Sealant tape
Standard weight
Stock
Stop-and-waste valve
Strap wrench
Thread engagement
Yoke vise

Profile in Success

Charles Owenby

JF Ingram Technical College
Plumbing Instructor
Deatsville, Alabama

Charles was born in West Virginia and raised in Montgomery, Alabama. He watched his older brother, a plumber, leave for work early each day and return home early, too. That piqued his interest in plumbing. He got a job working on the utility crew of a company that was putting in subdivisions and has been a plumber ever since. Charles still loves the plumbing profession, as demonstrated by his devotion to teaching. Today, in addition to teaching at local vocational schools in his spare time, he chairs the Educational Planning Committee for the Alabama Association of Plumbing, Gas, and Mechanical Inspectors.

How did you become interested in this industry?
I started working in 1967 with a company building subdivisions. After a bit, I moved on to the new construction part of the company and became a registered apprentice. I went to school one night a week and learned gas, water, and drainage piping installation, and I kept moving up.

What path did you take to your current position?
In 1969, I joined the Seabees, a branch of the Navy that's a mobile construction battalion. I continued plumbing while doing my military service. We worked like a large construction company. Steelworkers, masons, carpenters, site developers, and plumbers all worked on building bases, landing strips, and recreation centers; we also performed camp maintenance. I worked in plumbing, HVAC, water treatment, and sewage treatment. For a time, I was assigned to Puerto Rico, where we provided the outlying islands with fresh water systems.

In 1971, after my discharge, I went back to my previous job. I took my journeyman's exam and was then qualified to run plumbing jobs—working with apprentices. I also joined the Navy Reserves and served 28 years. During Desert Shield and Desert Storm, I was assigned to the Reservist Brigade staff as the training chief for the assigned units. I was in charge of reviewing their annual training plans.

In the meantime, I passed my master plumber exam that qualified me to perform contract work and bid on jobs. Alabama law requires a company to have at least one master plumber when bidding on contracts.

In 1981, I earned my bachelor's degree at Troy State University, which provides educational training and opportunities for military personnel throughout the world. In 1982, I became an instructor at JF Ingram Technical College, and in 1993, I received my master's degree in counseling. As of 2000, I was also certified by NCCER as a master trainer.

Meanwhile, as a reservist, I switched to the environmental unit of the Seabees. We conducted environmental audits, performed hazardous waste planning and control, monitored underground storage tanks, and helped with compliance issues related to government regulations.

What does it take to be successful in your trade?
Plumbing is an art and a science. Too many people think of plumbing as just fixing leaky faucets and repairing toilets. Nothing could be further from the truth. To be a success as a plumber, you have to be flexible enough to adapt to ever-changing situations or different jobs. No two jobs are exactly alike. Some companies may specialize in one aspect of plumbing (water supply piping, DWV piping, maintenance and repair), but you have to know something about each facet of the trade. One sure route to success is to find a mentor—someone you can work with who will help you learn and grow in your career.

You must be prepared to work your way up through the skill levels required to be a great plumber. Apprentices don't make decisions; they have to do what they are told to do. Journeymen make decisions. Master plumbers run jobs and figure prominently in a company's achievements. Your work attitude,

willingness to learn, and your work ethic are all important to your success.

Patience is a virtue. Never try to hurry through a job. Lack of proper planning or not understanding what is needed can cause big problems. As the old saying goes, "If you don't have time to do it right the first time, when are you going to have time to correct it?"

What are some things you do on the job?

I teach courses in plumbing using NCCER's *Contren®* curriculum. We emphasize safety, reading construction drawings, job ethics, and technical skills and techniques. In Alabama, all apprentices must register with the state to be allowed to work in plumbing and to use the tools. After about two years as an apprentice, a student can take the journeyman's exam.

I emphasize safety on many different levels. As an instructor, I am responsible for equipping my students with the knowledge of how to work safely with tools and equipment and how to protect themselves in the work environment. I also stress the effect that their work has on the safety of others. For example, working on medical gas installations can cause serious injury and death. Of course, I also remind students to know and adhere to state and local codes.

I also work to correct the stereotypes of plumbers. This is a great profession with a wonderful potential for making a living. People who are willing to dedicate themselves to learning the skills, knowledge, and art of plumbing will be successful.

What is the most interesting aspect of your profession? What makes your trade stand out from others?

Plumbers protect the health of the nation. We really do, although people don't realize it. It's not just about digging ditches and fixing toilets; that's just a part of it. It's also about preventing backflow, contamination, and other health problems.

There are great opportunities in this industry. High schools don't always advocate the industry as a career because teachers don't understand that there are so many outstanding possibilities. College is just one option. The trades are an excellent option as well!

What do you like most—and least—about your job?

I enjoy sharing the profession with students. I like informing people about the art and science that plumbing requires. I don't think enough people understand the art of plumbing; it is often covered up by wallboard. The scientific element of plumbing is equally challenging and fascinating.

I also love the profession. I could talk for days about what I like most about plumbing. There really isn't anything that I don't like about it!

What would you say to someone entering the trade today?

Learning a trade is a process. After almost 37 years in plumbing, I am still striving and still learning. It's the challenge that makes it all so interesting. Plumbing is a good profession that will allow you to make a good living. The market is growing and there are a lot of jobs available. I tell people entering the trades today that there are so many opportunities.

What can an apprentice expect to earn in his/her first years on the job in your area? What can he/she expect to earn after 10-plus years in the industry?

Apprentice rates are based on a percentage of the shop salary. In the first year, apprentices make about 50 percent of the shop salary and, depending on their level, will receive a percentage increase every six months or so.

Trade Terms Introduced in This Module

Ball valve: A valve used to control the flow of gases and liquids. It is installed in piping systems where quick shutoffs for in-line maintenance may be necessary.

Black steel pipe: Steel pipe with a black color used for gas and air-pressure pipe. It should not be used where corrosion will affect its uncoated surface.

Chain vise: Tool used to cut pipe by holding the pipe in a vise while cutting with a chain that has cutting wheels in each link.

Chain wrench: Another term for pipe tongs.

Cross: Fitting that connects four lengths of pipe in a cross-like shape in a distribution system.

Die: Device used to cut the threads on steel pipe.

Double extra strong (double extra heavy): An industry standard measurement of the weight or strength of steel pipe. Also referred to as Schedule 120.

Drainage fitting (recessed fitting): Fitting designed with smooth inside surfaces and shaped for easy flow to form an unbroken, internal contour for use on drainage systems.

Extra strong (extra heavy): An industry standard measurement of the weight or strength of steel pipe. Also referred to as Schedule 80.

Flange union: Fitting that connects two pipes. A flange is screwed onto the end of each pipe that needs to be joined and then bolted together with nuts and bolts. A gasket in between flanges makes it a leakproof seal.

Galvanized pipe: Pipe electroplated with zinc to provide a protective coating that resists corrosion or oxidation (rust).

Gate valve: A valve in which the flow is controlled by moving a gate or disc that slides in machined grooves at right angles to the flow. This type of valve is designed to be fully open or fully closed.

Globe valve: A valve in which the flow is controlled by moving a circular disc against a metal seat that surrounds the flow opening.

Ground joint union: Fitting that joins two pieces of pipe by screwing the thread and shoulder pieces of the fitting onto the pipe. The collar piece is then tightened to join the sections of pipe into a watertight joint.

Nipple: Short length of pipe with threads at both ends, used to make extensions from a fitting or to join two fittings.

Pipe-joint compound (pipe dope): Sealing material used when joining lengths of pipe.

Pipe tongs: Wrench-like tool with a length of chain, used to tighten large pipe. Also referred to as a *chain wrench.*

Power-threading machine: Machine used to thread large quantities of pipe. The machine rotates, threads, cuts, and reams pipe.

Pressure fitting: Any type of fitting used for steam, gas, or water supply piping. Not designed for use in drainage piping.

Sealant tape: Tape used to wrap the threads of a pipe before joining to ensure a watertight, secure fit.

Standard weight: An industry standard measurement of the weight or strength of steel pipe. Also referred to as Schedule 40.

Stock: Component of a hand threader or power threader that holds the die.

Stop-and-waste valve: Valve that is opened or closed by raising or lowering a horizontal disc using a threaded stem. An elastomeric, or rubberized, washer on the end of the stem seals the valve seat, closing off water flow. This valve is most commonly used for water faucets.

Strap wrench: Wrench used to hold chrome-plated or other types of finished pipe so that there are no jaw marks or scratches left on the pipe.

Thread engagement: The length of pipe that gets screwed into the fitting when joining threaded pipe.

Yoke vise: Tool with jaws that firmly hold a piece of pipe and prevent it from turning.

Additional Resources and References

ADDITIONAL RESOURCES

This module is intended to present thorough resources for task training. The following reference works are suggested for further study. These are optional materials for continued education rather than for task training.

Basic Plumbing with Illustrations, Revised Edition, 1994. Howard C. Massey. Carlsbad, CA: Craftsman Book Company.

Plumber's Handbook, Revised Edition, 1998. Howard C. Massey. Carlsbad, CA: Craftsman Book Company.

Plumbing: Design and Installation, Second Edition, 2002. L. V. Ripka. Homewood, IL: American Technical Publishers.

REFERENCES

Cast Iron Soil Pipe Institute website, www.cispi.org, *"Cast Iron Soil Pipe and Fittings Handbook,"* www.cispi.org/handbook.htm, reviewed August 2003.

National Standard Plumbing Code, 2003. Falls Church, VA: Plumbing-Heating-Cooling Contractors–National Association.

Plumbing and Mechanical website, www.pmmag.com, *"Forecasting the Lifespan of a Sprinkler System,"* Mark Bromann. www.pmmag.com, reviewed August 2003.

Plumbing and Mechanical website, www.pmmag.com, *"Point/Counterpoint: Domestic vs. Imported Cast-Iron Pipe,"* Joe Christiansen and Paula M. Bowe. www.pmmag.com/pm/cda/articleinformation/features/bnp__features__item/0,2379,81934,00.html, reviewed August 2003.

Victaulic website, www.victaulic.com, *"Pipe Preparation Tools and Parts,"* www.victaulic.com/servlet/RetrievePage?site=victaulic&page=tools_text, reviewed November 2003.

Water and Plumbing, Volume 3, 2000. Ifte Choudhury and J. Trost. Upper Saddle River, NJ: Prentice Hall.

NCCER CURRICULA — USER UPDATE

NCCER makes every effort to keep its textbooks up-to-date and free of technical errors. We appreciate your help in this process. If you find an error, a typographical mistake, or an inaccuracy in NCCER's curricula, please fill out this form (or a photocopy), or complete the online form at **www.nccer.org/olf**. Be sure to include the exact module ID number, page number, a detailed description, and your recommended correction. Your input will be brought to the attention of the Authoring Team. Thank you for your assistance.

Instructors – If you have an idea for improving this textbook, or have found that additional materials were necessary to teach this module effectively, please let us know so that we may present your suggestions to the Authoring Team.

NCCER Product Development and Revision

13614 Progress Blvd., Alachua, FL 32615

Email: curriculum@nccer.org
Online: www.nccer.org/olf

❏ Trainee Guide ❏ AIG ❏ Exam ❏ PowerPoints Other ____________________

Craft / Level: Copyright Date:

Module ID Number / Title:

Section Number(s):

Description:

Recommended Correction:

Your Name:

Address:

Email: Phone:

Corrugated Stainless Steel Tubing
02110-05

02110-05

Corrugated Stainless Steel Tubing

Topics to be presented in this module include:

Overview

Corrugated stainless steel tubing (CSST) is commonly used to deliver fuel gases for cooking, heating, and power generation. CSST is a flexible stainless steel tube encased in a polyethylene plastic jacket. Plumbers use CSST as an alternative to rigid gas pipe because it offers several advantages, including mobility, light weight, and flexibility.

CSST is sold as a complete system, including pipe, fittings, and connections. CSST is sized for each installation and labeled with the manufacturer's name and part number and the testing organization that certified its use. Maximum operating pressure is marked at specific intervals. To ensure that gas systems operate at this allowable pressure, plumbers install regulators to decrease or increase the gas flow. To quickly shut off gas flow, plumbers install ball valves and gas cocks.

CSST does not require special tools for cutting and joining. Instead, plumbers connect CSST to appliances or other pipe using special threaded fittings. Straight fittings connect one end of a CSST pipe to an appliance, the gas supply, or a manifold. Plumbers can splice together lengths of CSST by using mechanical coupling fittings. CSST can be supported along walls, ceilings, and flows, but it may never be hung from utility installations. CSST gas piping systems are visually inspected and pressure-tested in accordance with the *National Fuel Gas Code.*

Focus Statement

The goal of the plumber is to protect the health, safety, and comfort of the nation job by job.

Code Note

Codes vary among jurisdictions. Because of the variations in code, consult the applicable code whenever regulations are in question. Referencing an incorrect set of codes can cause as much trouble as failing to reference codes altogether. Obtain, review, and familiarize yourself with your local adopted code.

Objectives

When you have completed this module, you will be able to do the following:

1. Identify the common manufacturers of corrugated stainless steel tubing.
2. Recognize proper and improper applications of corrugated stainless steel tubing.
3. Identify the various techniques used in hanging and supporting corrugated stainless steel tubing.
4. Explain how to properly measure, cut, join, and groove corrugated stainless steel tubing.
5. Identify the material properties, storage, and handling requirements of corrugated stainless steel tubing.

Key Trade Terms

Corrugated stainless steel tubing (CSST)
Coupling fitting
Elevated pressure system
Gas cock
Manifold
Pounds per square inch gauge (psig)
Regulator
Regulator stub-out
Straight fitting
Striker plate
Stub-out
Tee fitting
Water column (w.c.)

Required Trainee Materials

1. Appropriate personal protective equipment
2. Pencils and paper
3. Calculator
4. Copy of local adopted code

Prerequisites

Before you begin this module, it is recommended that you successfully complete *Core Curriculum; Plumbing Level One,* Modules 02101-05 through 02109-05.

This course map shows all of the modules in the first level of the *Plumbing* curriculum. The suggested training order begins at the bottom and proceeds up. Skill levels increase as you advance on the course map. The local Training Program Sponsor may adjust the training order.

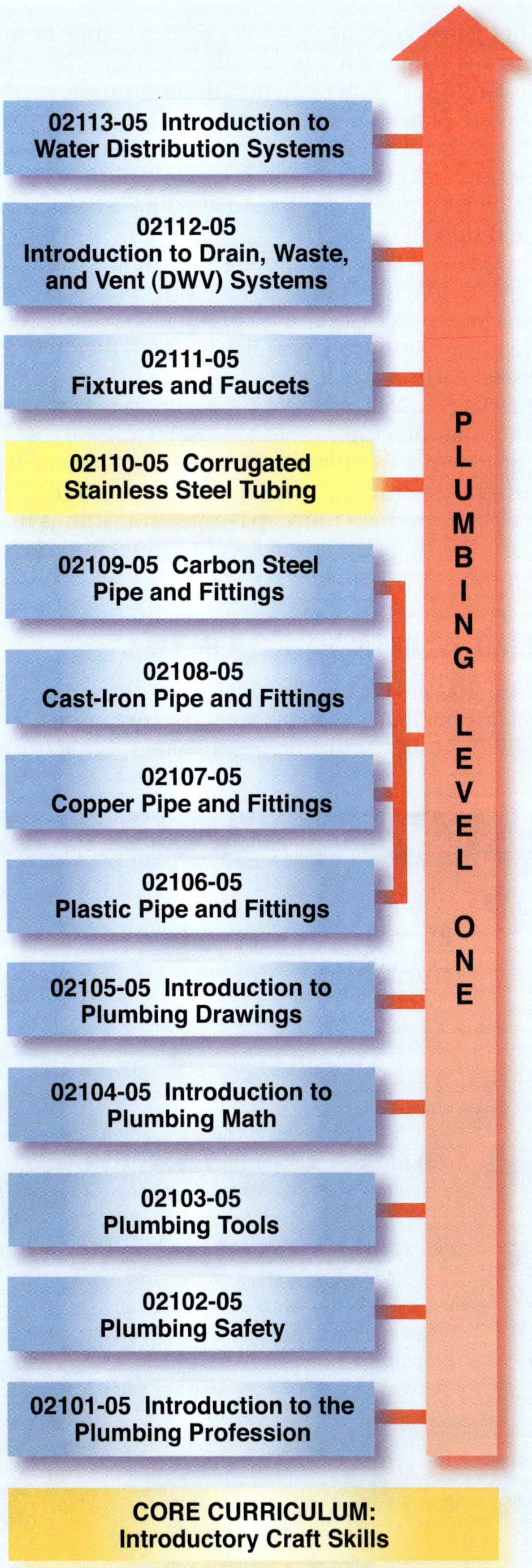

PLUMBING LEVEL ONE

02113-05 Introduction to Water Distribution Systems
02112-05 Introduction to Drain, Waste, and Vent (DWV) Systems
02111-05 Fixtures and Faucets
02110-05 Corrugated Stainless Steel Tubing
02109-05 Carbon Steel Pipe and Fittings
02108-05 Cast-Iron Pipe and Fittings
02107-05 Copper Pipe and Fittings
02106-05 Plastic Pipe and Fittings
02105-05 Introduction to Plumbing Drawings
02104-05 Introduction to Plumbing Math
02103-05 Plumbing Tools
02102-05 Plumbing Safety
02101-05 Introduction to the Plumbing Profession
CORE CURRICULUM: Introductory Craft Skills

110CMAP.EPS

1.0.0 ◆ INTRODUCTION

Fuel gases such as natural gas and liquid petroleum (LP) gas are commonly used for cooking, heating, and power generation throughout the United States. Plumbers are responsible for installing safe and efficient fuel gas systems. One of the most popular methods among plumbers for delivering fuel gas is to use **corrugated stainless steel tubing (CSST).**

CSST is a flexible stainless steel tube encased in a jacket made of polyethylene (PE) plastic. As an alternative to rigid gas pipe, CSST has been steadily gaining in popularity over the past 20 years. As of 2003, all national model codes recognize CSST as an approved gas piping material. In 2002, manufacturers sold almost 50 million feet of CSST, which represents nearly 40 percent of the current gas piping market. In this module, you will be introduced to CSST pipe and fittings and learn the basics of cutting, connecting, and hanging this type of pipe.

2.0.0 ◆ INTRODUCTION TO CSST

CSST has several advantages over other types of gas piping, such as black steel, PE, or copper. It is delivered in precut lengths coiled on drums, which allows the pipe to be loaded and carried more easily (see *Figure 1*). It is lighter than rigid pipe and does not require special tools to cut or install. Because it can be bent, CSST does not require angle fittings, such as elbows, to go around corners (see *Figure 2*). These factors all reduce the labor time and costs required to install CSST to as much as half the costs of other types of pipe. CSST costs more per foot than other pipes, but labor and time cost savings offset at least some of this expense.

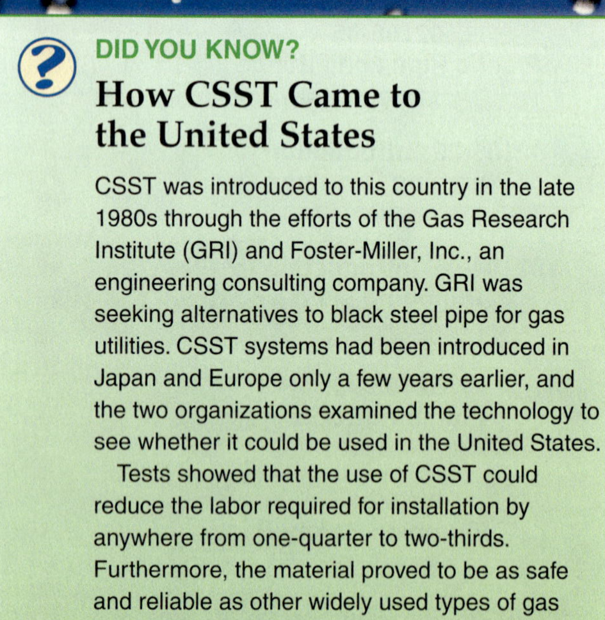

DID YOU KNOW?

How CSST Came to the United States

CSST was introduced to this country in the late 1980s through the efforts of the Gas Research Institute (GRI) and Foster-Miller, Inc., an engineering consulting company. GRI was seeking alternatives to black steel pipe for gas utilities. CSST systems had been introduced in Japan and Europe only a few years earlier, and the two organizations examined the technology to see whether it could be used in the United States.

Tests showed that the use of CSST could reduce the labor required for installation by anywhere from one-quarter to two-thirds. Furthermore, the material proved to be as safe and reliable as other widely used types of gas piping. The 1989 *National Fuel Gas Code* was the first edition of the code to include CSST as an approved gas piping material, and the use of CSST in the United States has continued to skyrocket ever since.

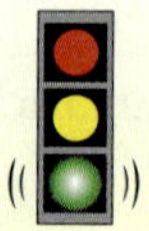

NOTE

Always use the proper fittings when connecting rigid black steel pipe to CSST. Refer to the pipe manufacturer's specifications and your local applicable code for approved fitting materials.

Unlike other types of gas pipe, CSST is sold as a complete system, including pipe, fittings, and connections. The components of CSST systems, however, are not interchangeable among manufacturers. In other words, you cannot use connectors made by one manufacturer on pipe made by another. Yet one advantage is that manufacturers are able to offer warranties on their entire systems. *Table 1* lists the major manufacturers of CSST systems and the trade names of their pipe.

2.1.0 Sizes and Labeling

CSST is manufactured in sizes ranging from 3/8 inch to 2 inches in diameter. The designer deter-

110F01.TIF

Figure 1 ◆ CSST is delivered in coils.

110F02.TIF

Figure 2 ◆ CSST can be installed in tight spaces without angle fittings.

Table 1 CSST Names and Manufacturers

CSST System Trade Name	Manufacturer
Gastite®	Titeflex® Corporation
TracPipe®	OMEGAFLEX®, Inc.
Parker Gas Piping (PGP™)	© Parker Hannifin Corporation
Pro-Flex™	Tru-Flex™ Metal Hose Corporation

mines the proper pipe size for each installation. Regardless of the size, each length of pipe is labeled with the manufacturer's name and part number, as well as the label of the testing organization that certified its use in gas piping installations. The maximum operating pressure and the words *fuel gas piping* are marked at regular intervals. Some manufacturers also provide measurement markings at 1-foot intervals (see *Figure 3*). Refer to your local applicable code for sizing specifications.

110F03.TIF

Figure 3 ◆ Typical markings on a length of CSST.

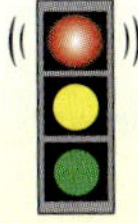

WARNING!

You must be certified by particular CSST manufacturers to install CSST. Also, some states require a gas fitter's certification. If you are interested in doing this type of work, check your local and state regulations first.

Qualified installers are responsible for sizing and installing CSST piping systems. Either the CSST manufacturer or one of the manufacturer's licensed representatives can train installers. Installers must follow the manufacturer's instructions for proper installation of the pipe and fittings. Qualified installers ensure that the pipe

Fuel Gas Codes and Standards

The *National Fuel Gas Code* is approved by the American National Standards Institute (ANSI) and the National Fire Protection Association (NFPA). The code is also referred to as *ANSI Z223.1* and *NFPA 54.* It applies to pipe that is installed between the meter outlet and all appliance inlets in a building. You can find out more about the *National Fuel Gas Code* at the ANSI website, www.ansi.org, and at the NFPA website, www.nfpa.org.

ANSI and the American Gas Association (AGA) publish another important standard related to CSST. *ANSI/AGA LC 1, Interior Fuel Gas Piping Systems Using Corrugated Stainless Steel Tubing,* sets the construction, installation, and performance requirements for CSST systems. You can learn more about this standard at the ANSI website or at the AGA website, www.aga.org.

runs are placed safely, supported properly, and connected securely.

When bending CSST, allow a generous bending radius. A tight bending radius can stress the pipe and fittings, causing leaks. Never kink, twist, or stretch CSST. Refer to the manufacturer's instructions for the allowable bending radius of the pipe you are using.

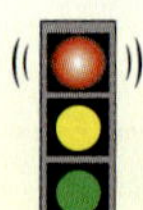

WARNING!

Never connect CSST directly to a meter. Use a length of steel pipe with a minimum of a 1-inch diameter between the meter and the CSST. Attach the steel pipe securely to the exterior wall of the building to avoid accidents and injury. Regulations are based on gas company and code requirements. Consult your applicable local code and the manufacturer's specification for sizing information.

2.2.0 Regulators and Valves

Most residential, commercial, and industrial gas systems operate at a pressure of 7 inches of **water column (w.c.)** measured at the system's meter. Some installations require higher operating pressures. These are called **elevated pressure systems.** These systems operate within a range of ½ to 2 **pounds per square inch gauge (psig)** pressure. The maximum allowable operating pressure for gas systems, according to standards developed by ANSI, is 5 psig. Elevated pressure systems require **regulators** where the piping connects to an appliance or other device (see *Figure 4*). Regulators decrease or increase the gas flow to the device using the gas.

Ball valves and **gas cocks** are installed in CSST systems to shut off gas flow quickly (see *Figure 5*). A handle on the outside of the valve body rotates the valve's ball into the opened or closed position. Ball valves provide positive, quick-flow control on gas piping systems. They are widely used in liquid plumbing systems as well. Install the valve

110F04.TIF

Figure 4 ◆ Regulators are used to lower gas pressure in elevated pressure systems.

Figure 5 ◆ Ball valve and gas cock used in CSST system.

upstream from appliances or connectors. Some codes require the valve to be located within 6 feet of the appliance. Other codes require a distance of only 3 feet and require the valve to be in the same room as the appliance.

2.3.0 Types of CSST Installations

CSST can be run directly from the building meter to each appliance by using connectors. Use tees to connect branches to the main gas line (see *Figure 6*). Multiple appliances, such as unit heaters, can be connected this way (see *Figure 7*). Ensure that all connectors and lines are the same size; otherwise the difference will increase or decrease the gas pressure. Refer to your local code and the manufacturer's installation instructions. You will learn about fittings elsewhere in this module.

CSST can also be connected using **manifolds** (*Figure 8*). Manifolds connect a single gas line to multiple appliances without repeated cuts along the line for each branch. Install the manifold in a central location in the building, and then connect the CSST lines from the appliances to the manifold using approved connectors. Manifolds allow the piping for all appliances to be joined at a single, easily accessible location (see *Figure 9*).

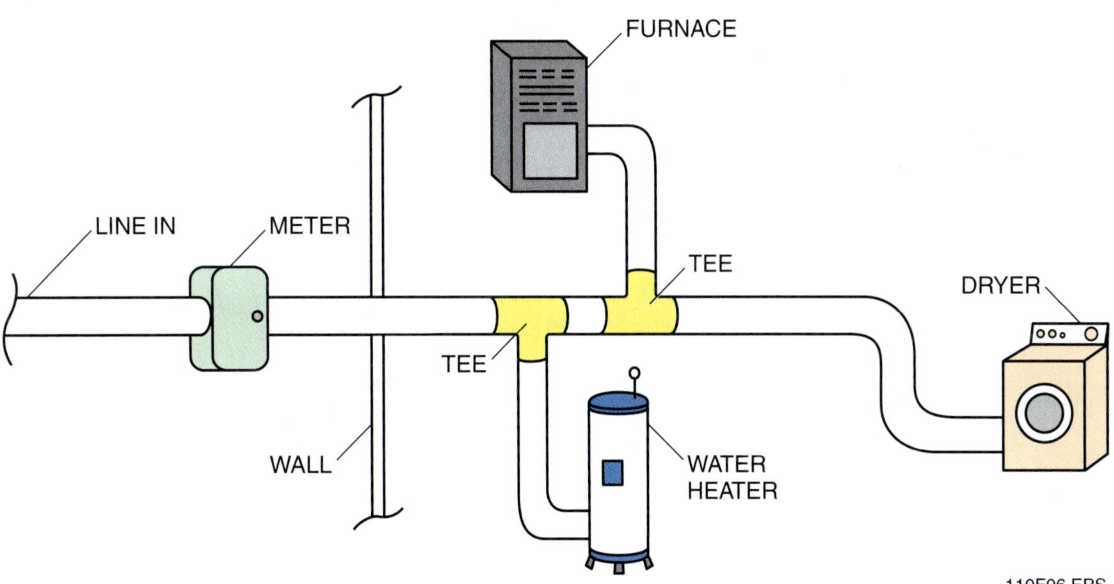

Figure 6 ◆ Simple CSST installation.

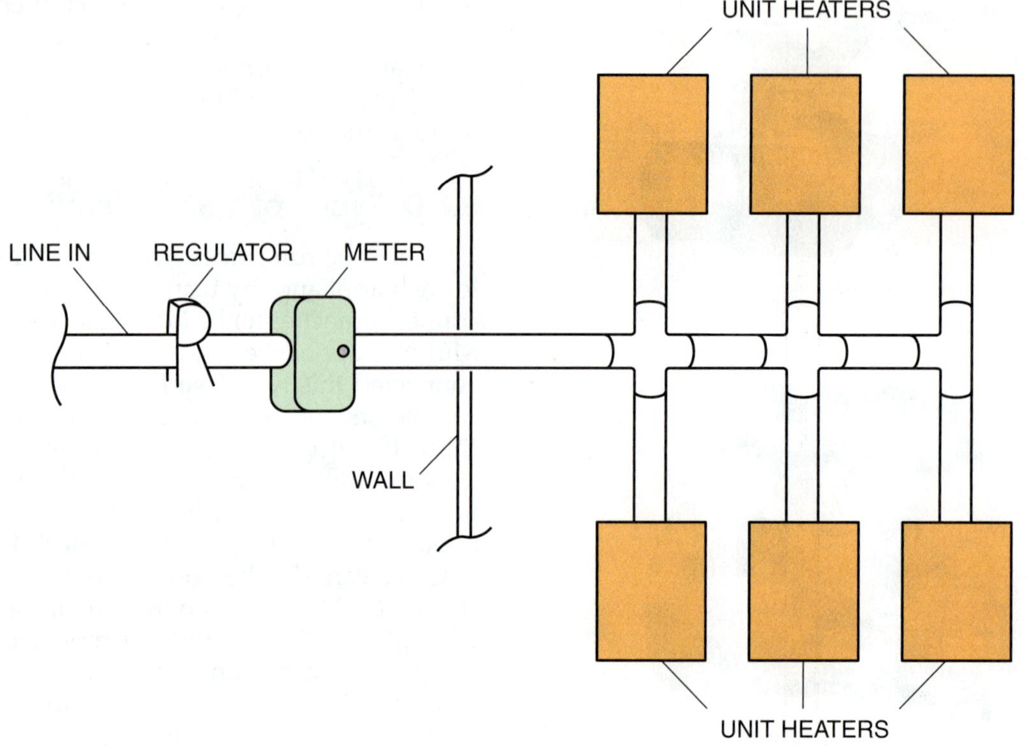

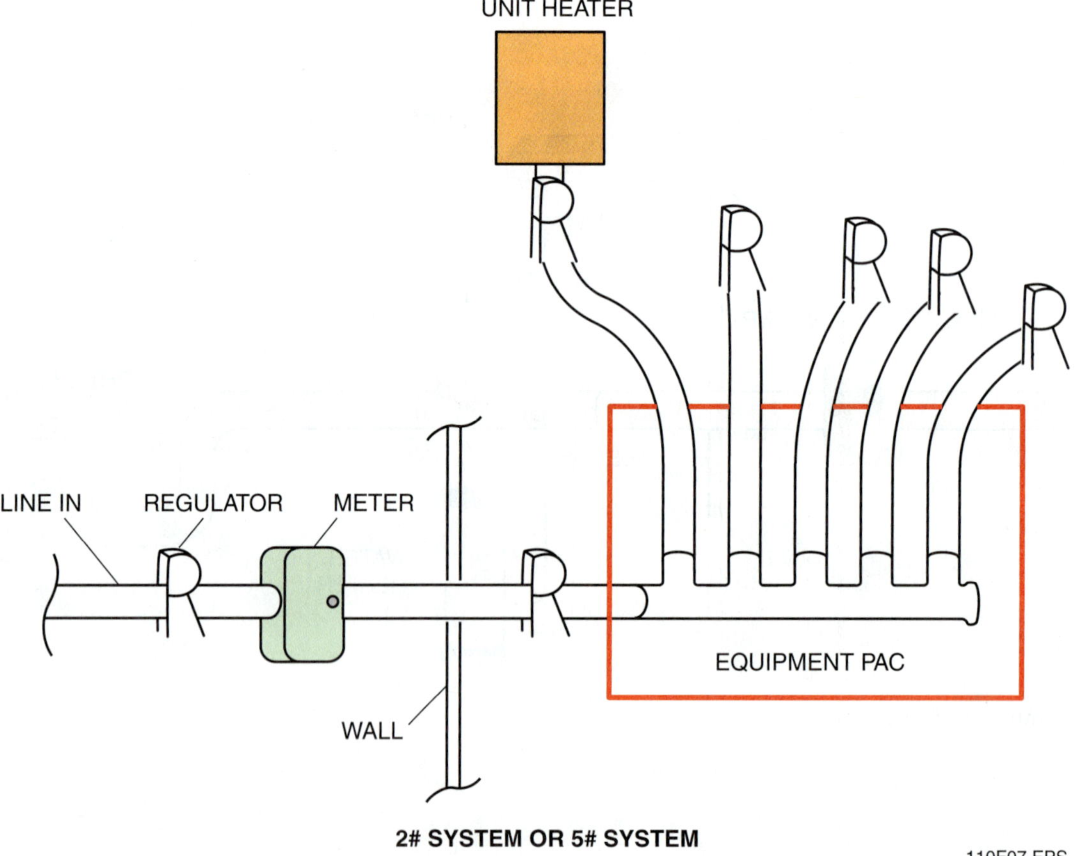

Figure 7 ◆ Typical parallel installation for multiple unit heaters.

110F08.TIF

Figure 8 ◆ Manifolds connect a single gas line to multiple appliances.

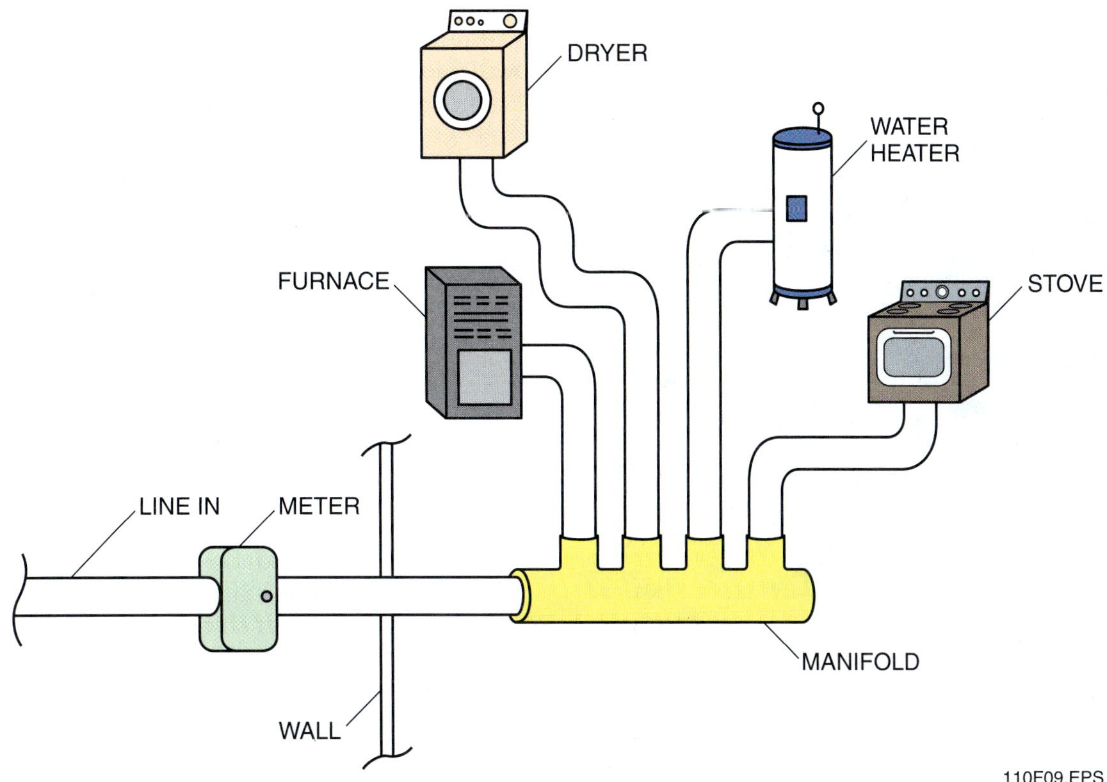

110F09.EPS

Figure 9 ◆ Typical installation with a manifold.

3.0.0 ◆ CUTTING AND CONNECTING CSST

One of the advantages of CSST is that it does not require special tools to cut and join. Many of the tools you will use are already familiar to you. The cutters you use for CSST are similar to pipe cutters you use to cut copper pipe (see *Figure 10*). To use a tube cutter on a length of CSST, place the cutter along the pipe at the place where you want to cut. Tighten the adjuster knob, forcing the cutting wheel against the CSST. Rotate the cutter around the tube, maintaining constant pressure (see *Figure 11*). After the cut has been made, remove any burrs, and ensure that the cut is clean and straight.

Always make sure that pipe is free from dirt or obstructions. Do not use pipe that is damaged,

110F10.TIF

Figure 10 ◆ CSST tube cutters.

110F11.TIF

Figure 11 ◆ Using a tube cutter to cut CSST.

kinked, or defective in any way. Replace any defective fittings on gas piping. Do not attempt to repair damaged fittings.

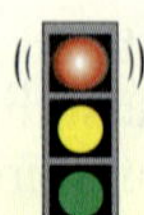

WARNING!

Always follow the manufacturer's instructions when using a pipe-cutting tool. Wear appropriate personal protective equipment. Never wipe the cut edges of a pipe with your hand. The sharp edges and burrs can cut you.

Once you have cut CSST to the proper length, you can connect it to appliances or other pipe using special threaded fittings. Nipples (short pieces of pipe with threads on each end), saddles (floor mounts), and cast-iron fittings are not allowed in CSST systems. Right-left nipples and couplings are permitted only between shutoff valves and appliances, between regulator outlets and manifold inlets, or when installing underground residential pipes. Threaded fittings can be made from the following materials:

- Steel
- Brass
- Bronze
- Black malleable iron

The following sections review the procedures for connecting CSST. Always follow the manufacturer's instructions when cutting and connecting pipe. Use only the tools and materials that are permitted according to your local adopted code. When running CSST and fittings through openings in walls, floors, or ceilings, ensure that the hole is at least ½ inch larger than the pipe diameter.

3.1.0 Connecting CSST to Appliances

Use **straight fittings** (*Figure 12*) to connect CSST pipe to a fixed appliance or other device. Begin by connecting the bushing to the CSST pipe according to the manufacturer's instructions. Then, install the fitting adapter to the device to which the pipe will be connected. Finally, tighten the adapter on the appliance connection using the nut (see *Figure 13*).

Stub-outs for movable appliances such as stoves, ranges, and gas dryers need to be terminated. Use approved flexible connectors to connect the fixed termination point directly to the movable appliance. If shutoff valves and drip legs

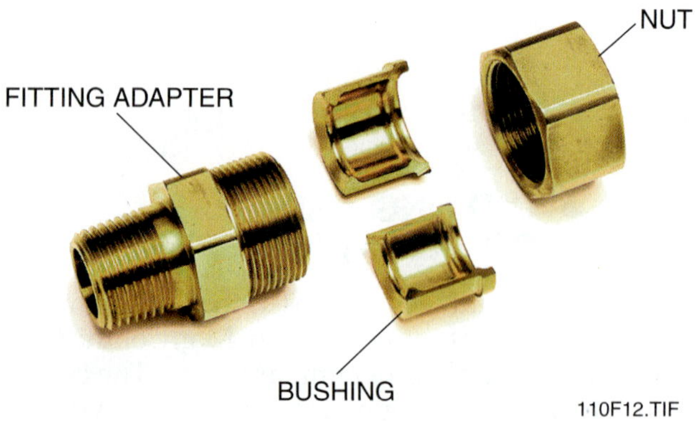

110F12.TIF

Figure 12 ◆ Straight fitting.

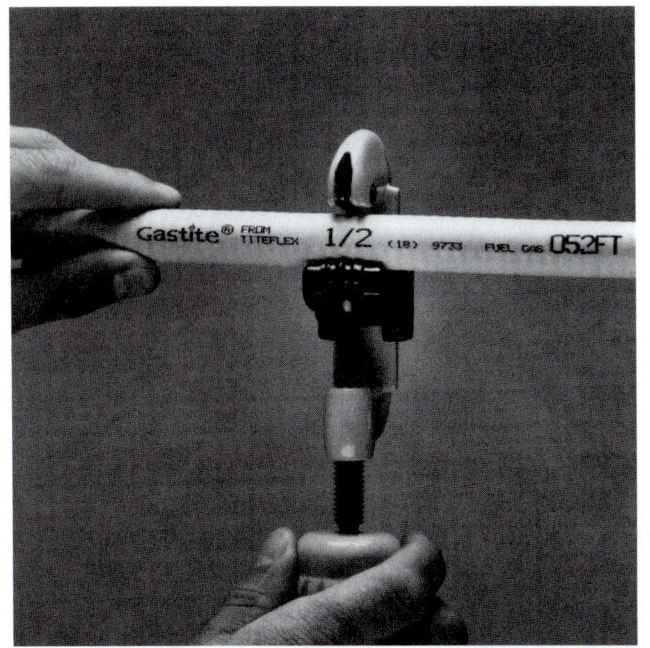

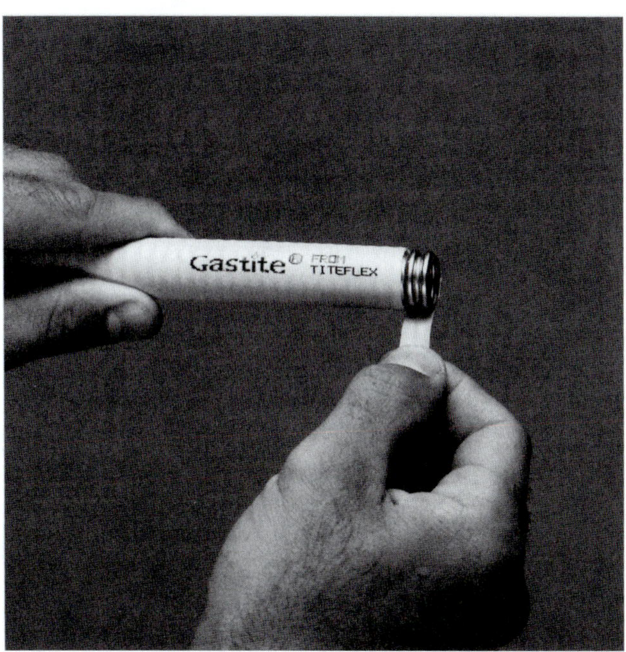

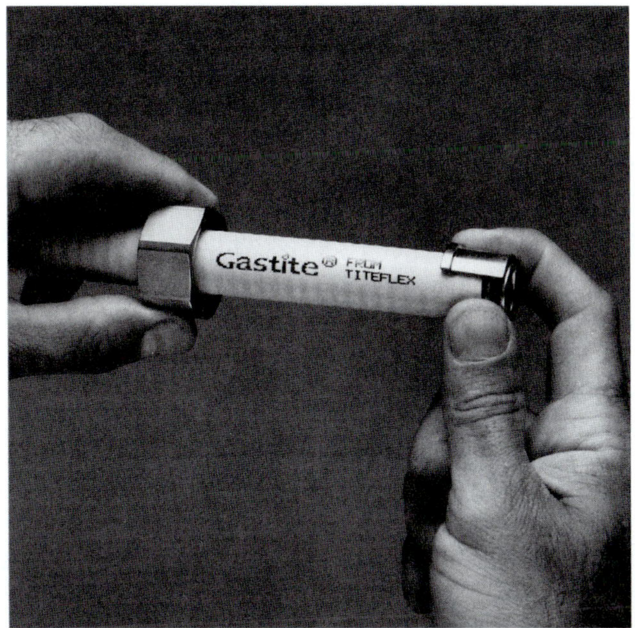

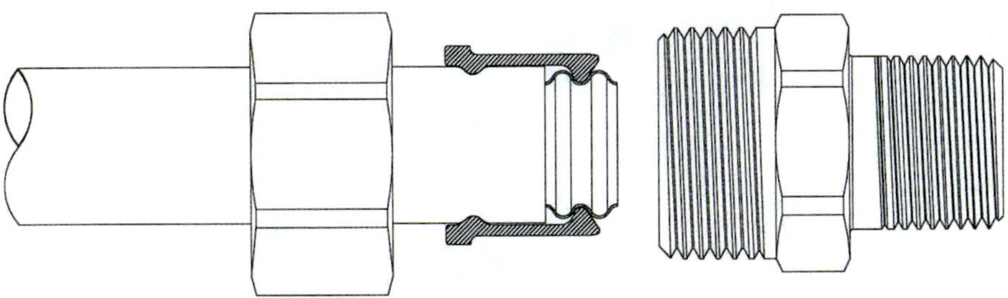

110F13A.EPS

Figure 13 ◆ Connecting CSST to an appliance using a straight fitting. (1 of 2)

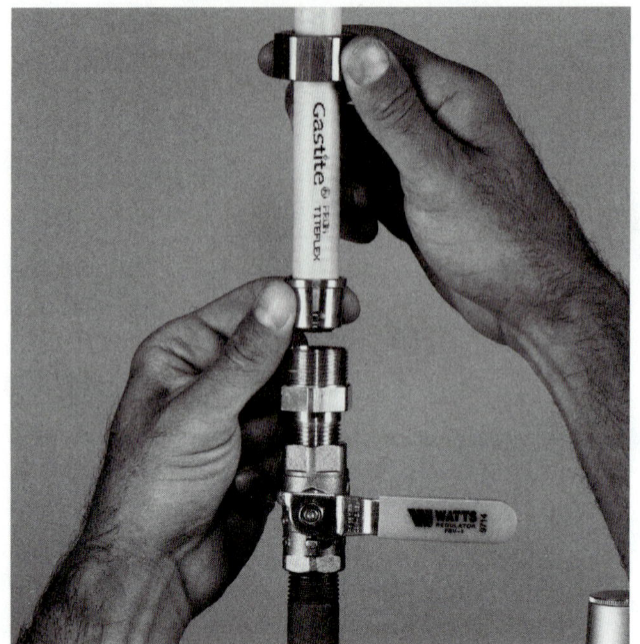

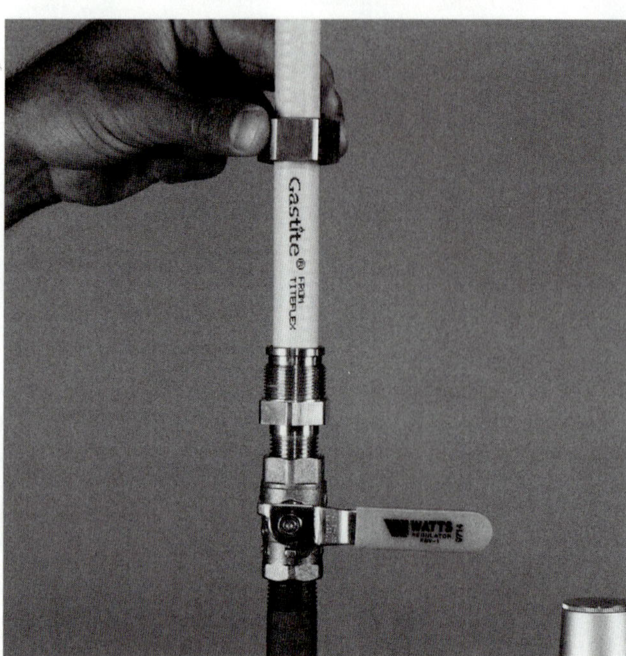

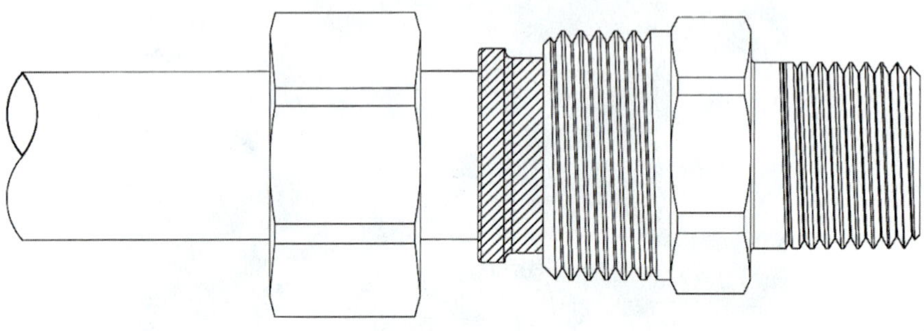

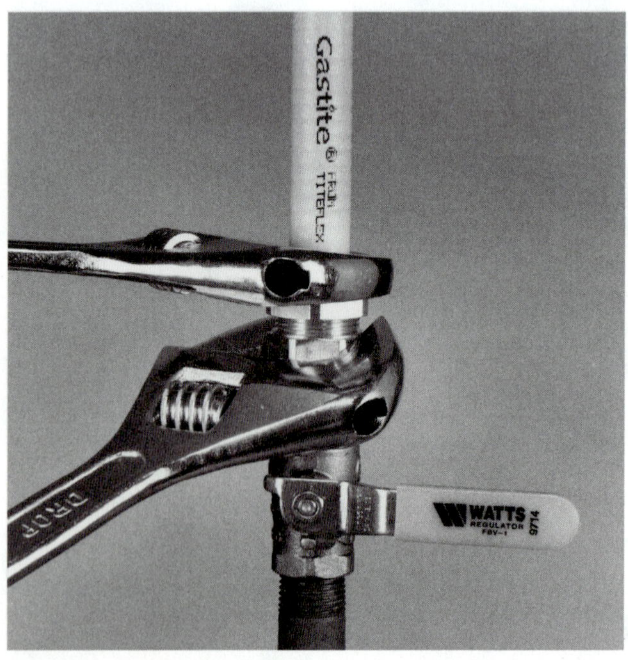

110F13B.EPS

Figure 13 ◆ Connecting CSST to an appliance using a straight fitting. (2 of 2)

are required, install them on the fixed termination point rather than on the flexible connector. Refer to the manufacturer's specifications to ensure that you are using the correct connectors.

Like other types of pipe, CSST can be run behind walls, in ceilings, and under floors. When done, the pipes will be hidden. Install a stub-out at each point where a hidden pipe exits a finished wall or floor (see *Figure 14*). The stub-out serves as a fixed connection point to the hidden pipe. The plate on the stub-out is mounted to the wall or floor, and the threaded end can be connected to an appliance, regulator, or meter as needed. Stub-outs reduce the number of joints in the system by eliminating the need to fabricate rigid-pipe nipple stub-outs. They also give a more finished look to the installation when it is completed.

Manufacturers offer special stub-outs for specific applications. For example, **regulator stub-outs** provide a fixed connection point for regulators in elevated pressure systems (see *Figure 15*). Regulator stub-outs are typically installed in ceilings. Many stub-outs can be prefabricated before installation and assembled on site in minutes. Prefabricated stub-outs can reduce installation time.

110F15.TIF

Figure 15 ◆ Regulator stub-out.

3.2.0 Connecting CSST to CSST

Lengths of CSST can be spliced together using mechanical **coupling fittings** (*Figure 16*). Couplings consist of a fitting adapter, bushings, and a nut. Like straight fittings, coupling fittings are simple to assemble. Begin by connecting the bushings to the ends of the CSST pipes according to the manufacturer's instructions. Then, connect the bushings to the fitting adapter and tighten the adapter using the nut.

110F16.TIF

Figure 16 ◆ CSST coupling fitting.

110F14.TIF

Figure 14 ◆ Appliance stub-outs.

Tee fittings (*Figure 17*) can be used to connect a branch line to a pipe run at a 90-degree angle. The

110F17.TIF

Figure 17 ◆ Tee fitting.

installation process for tee fittings is the same as for straight fittings and coupling fittings. The fitting is designed so that, as the gas changes directions, the pressure and velocity remain the same as when the gas entered the tee.

4.0.0 ◆ HANGING AND RUNNING CSST

CSST can be supported along walls, floors, and ceilings using metal pipe straps, bands, and hangers that are designed to support the size and weight of pipe. You have learned about the wide variety of available hangers and supports elsewhere in *Plumbing Level One.* Never support CSST from other utility installations such as drain, waste, and vent (DWV) pipe; heating, ventilating, and air conditioning (HVAC) duct; or electrical wire conduits. The distance between hangers and supports will vary depending on the jurisdiction. Always follow the applicable standards when installing hangers and supports.

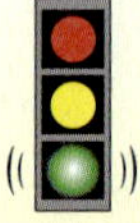

NOTE

Research has demonstrated that CSST has greater durability under seismic loads compared to conventional black steel piping. Because CSST uses fewer fittings, it also has a lower risk of leaking and creates a more secure system.

Some manufacturers offer specially designed pipe supports that can be used to mount CSST pipe runs using common hangers, such as universal pipe supports and struts (*Figure 18*) or pipe straps (*Figure 19*). These supports, designed for installation on walls, floors, ceilings, and roofs, can handle multiple pipe runs. Some supports come with predrilled holes for strut mountings.

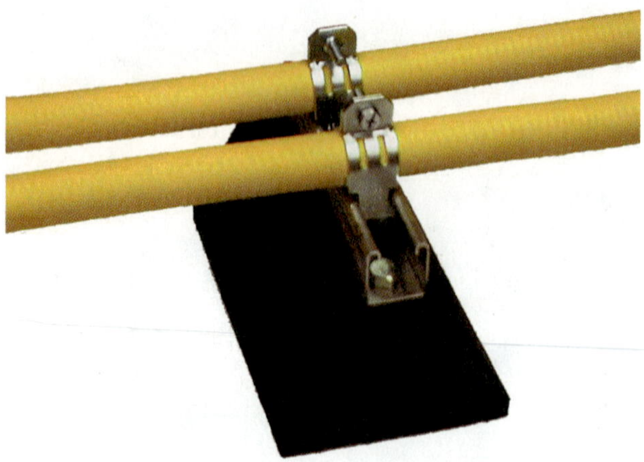

110F18.TIF

Figure 18 ◆ CSST supported with universal pipe clamps on a strut.

110F19.TIF

Figure 19 ◆ CSST supported with pipe straps.

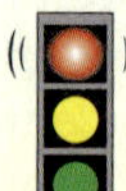

WARNING!

Never pass CSST through an air duct system unless all of the following conditions are met:

- The gas equipment is approved for use in ducts and meets all applicable local codes.
- It is being used to preheat makeup air from the outside.
- It is not installed in residential applications.

Never constrict CSST that is concealed in a hollow wall cavity. When running CSST behind walls, protect it from punctures by hanging the pipe so that it has room to flex if hit by a nail. Another option is to run the pipe through steel conduits installed behind the walls or to install **striker plates** to prevent punctures. CSST can be run outside a building without any additional protection. Buried CSST pipe must be run through conduits that are nonmetallic and watertight. Fittings are not allowed on CSST in conduit.

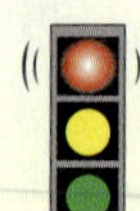

WARNING!

Gas pipe cannot be used as a grounding conductor or as a grounding electrode in most electrical systems. An electrical charge on a gas line could cause a fire or even an explosion. Exceptions to this guideline are low-voltage electrical, control, ignition, and flame-detection device circuits.

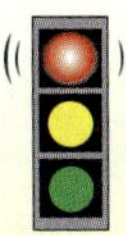

WARNING!

Check all applicable codes and product information supplied by the manufacturer before using striker plates.

5.0.0 ◆ INSPECTING AND TESTING CSST

Gas piping systems are inspected and tested in accordance with local applicable codes. If the local code does not specify inspection and testing standards, use the requirements in the current edition of the *National Fuel Gas Code, ANSI Z223.1/NFPA 54.* CSST installations are visually inspected and subjected to a pressure test before they are certified safe for use. The entire system must be visible for the inspector; underground pipe, as well as pipe located behind walls, floors, and ceilings, may not be covered up or sealed until it has been inspected and approved.

CSST gas piping is tested using a pressure test. All outlets are sealed, and air is pumped into the system at a pressure of at least 10 psig for at least 15 minutes. The pressurized system is then inspected for leaks. This testing method is used for both standard pressure and elevated pressure systems.

Local Governments

Local government agencies are usually responsible for inspecting and testing gas piping installations. Local utility companies require approved inspections and testing before they will allow you to connect a gas system to the utility's main supply.

Review Questions

Sections 1.0.0–4.0.0

1. CSST consists of a flexible stainless steel tube encased in a jacket of _____ plastic.
   a. PEX
   b. PE
   c. PVC
   d. CPVC

2. _____ is a type of piping that is delivered in precut lengths coiled on drums.
   a. Black steel
   b. PE
   c. Copper
   d. CSST

3. CSST is _____ than rigid pipe.
   a. lighter
   b. heavier
   c. longer
   d. shorter

4. The components of CSST systems are _____ among manufacturers.
   a. interchangeable
   b. inconsistent
   c. not interchangeable
   d. consistent

Match the types of CSST with their manufacturers.

5. TracPipe®
6. Parker Gas Piping (PGP™)
7. Gastite®
8. Pro-Flex™

a. Titeflex® Corporation
b. Tru-Flex™ Metal Hose Corporation
c. © Parker Hannifin Corporation
d. Ridgid® Tool Company, Inc.
e. OMEGAFLEX®, Inc.

Review Questions

9. CSST installers can be trained by the manufacturer or by a(n) _____.
   a. representative of the local gas utility
   b. representative of the jurisdiction's code authority
   c. official manufacturer's representative
   d. apprenticeship training program

10. According to ANSI standards, the maximum allowable operating pressure for gas systems is _____ psig.
    a. 2
    b. 3
    c. 4
    d. 5

11. Install ball valves upstream from _____.
    a. appliances or connectors
    b. manifolds
    c. straight fittings
    d. meters

12. Each of the following is an acceptable material for threaded CSST fittings *except* _____.
    a. steel
    b. cast iron
    c. black malleable iron
    d. brass

13. To connect a length of CSST to an appliance, you would use a _____ fitting.
    a. coupling
    b. regulator
    c. straight
    d. tee

14. Regulator stub-outs are typically installed in each of the following *except* _____.
    a. ceilings
    b. floors
    c. walls
    d. ducts

15. CSST should never be supported from _____.
    a. DWV pipe
    b. metal pipe straps
    c. bands
    d. hangers

16. Each of the following is an approved method for protecting CSST from nail punctures in a wall *except* _____.
    a. hanging the pipe so that it can flex
    b. wrapping the pipe in protective padding
    c. installing striker plates
    d. running the pipe through a steel conduit

17. Universal pipe supports and struts or pipe straps can handle _____ pipe runs.
    a. multiple
    b. single
    c. universal
    d. combination

18. _____ installations are visually inspected as well as pressure tested before the system can be considered safe for use.
    a. Copper pipe
    b. CSST
    c. PE
    d. Black steel

Summary

CSST, which consists of flexible steel tubing jacketed in PE plastic, is an increasingly popular method for providing natural gas and LP gas for cooking, heating, and power generation applications. CSST is available in precut lengths coiled on drums. It is light, and it can be installed using ordinary plumbing tools and without angle fittings. It is available in sizes ranging from ½ inch to 2 inches in diameter. Systems can operate at pressures ranging from 7 inches w.c. to 5 psig. CSST must be installed by qualified plumbers who are trained and certified by the manufacturer.

CSST is sold as a complete, warrantied system, including pipe, fittings, supports, regulators, valves, and other accessories. Pipe can be run directly to each appliance or outlet using a combination of manifolds, tee fittings, straight fittings, coupling fittings, and other approved connectors. As long as the pipe is properly supported along walls, under floors, and above ceilings, the piping layout can be determined on site by snaking the pipe along the desired route. Pipe cannot be supported from other utility lines such as DWV pipe or air ducts.

CSST is growing in popularity as more plumbers learn about its advantages in saving time, costs, and labor. CSST is poised to become the fuel gas pipe of choice in the 21st century. As a plumber, you will need to know the advantages of and uses for this type of pipe as you advance in your profession.

Notes

Trade Terms Quiz

Fill in the blank with the correct trade term that you learned from your study of this module.

1. Ball valves and _______________ are installed in CSST systems to quickly shut off gas flow.
2. A(n) _______________ is used to connect CSST to an appliance.
3. Use a(n) _______________ to split one CSST gas line into multiple lines.
4. A(n) _______________ protects hidden CSST from puncture damage.
5. Make a connection through closed walls, floors, and ceilings using a(n) _______________.
6. The maximum pressure for gas systems is 5 _______________.
7. A(n) _______________ can be used to connect two lengths of CSST.
8. _______________ is widely used as gas piping because of its ease of installation.
9. A(n) _______________ operates at a pressure greater than standard pressures.
10. Regulators can be connected to hidden piping using a(n) _______________.
11. The standard pressure for gas systems is 7 inches of _______________.
12. A(n) _______________ is used to control gas pressure in a piping system.
13. Use a(n) _______________ to take off a branch from the main line.

Trade Terms

Corrugated stainless steel tubing (CSST)
Coupling fitting
Elevated pressure system
Gas cock
Manifold
Pounds per square inch gauge (psig)
Regulator
Regulator stub-out
Straight fitting
Striker plate
Stub-out
Tee fitting
Water column (w.c.)

Profile in Success

Richard Kerzetski

Universal Plumbing and Heating Company
President
Las Vegas, Nevada

Richard Kerzetski was born in Perth Amboy, New Jersey, and attended Perth Amboy High School. He began his career as an office boy working for a mechanical engineering contractor and doing a little bit of everything. He was drafted into the U.S. Army, served for two years, and returned to work. Taking advantage of the GI Bill, Richard attended the Newark College of Engineering. Upon graduation, he decided to get away from the snow and ice, and moved near his brother in Las Vegas.

Today, Richard still lives and works in Las Vegas, where he runs his own business. He has passed on his passion for the industry to his two sons, who are also working in the trade. One is an apprentice, and the other is a project manager.

How did you become interested in this industry?
My family told me to get a job; I saw an ad in the paper, and I applied. I enjoyed working for a contractor, because you see the finished building. I had worked for a brief stint in a factory, and I never knew where the product went. I had pride working for a contractor. You could drive by a building and say, "I worked on that building." You can be proud to have children in a school that you worked on. We see a finished product and see that it is working because of the effort we put into it.

What path did you take to your current position?
While working for the mechanical engineering contractor, I went from office boy to estimating. My boss let me work in estimating, purchasing, bookkeeping, and drafting. I did everything that it took to get the job done, from billing to project management. The only thing I did not do was collect money. My boss let me run several jobs at one time, so I had all of the responsibilities of the owner but not the worry about getting paid. Cooperation is key. I was always available for work and was ready to take on any challenge. I did what I was told to do. I try to remember the saying, "If you cannot outsmart someone, you'd better learn how to outwork them." You have to be prepared for everything you do and anticipate needs.

I worked for a man who did project management before the word was in fashion—always preparing and planning how the work would be done. Now it's in fashion to do a lot of planning. We didn't call it that, but we always made sure we had a plan. Our plumbers completed work using the planning they did in the shop. Because of this, we were able to complete jobs efficiently and competitively.

After I received my journeyman's card in Las Vegas, I went to work with Universal Plumbing. In the early 1980s, the owner took a different business route, so I reopened the business on my own. I've always been fortunate; everyone I have worked for has treated me well. I've always followed my parents' work ethic that whoever you work for is paying you, so you'd better do a good job. I have always tried to do a good job.

What are some things you do on the job?
I still do everything from sweeping the floor to signing checks at the end of the day. I'm learning to delegate. We do a lot of planning, fabrication, and getting the job ready in the shop, so that the foremen know what they are doing before they get into the field. I'm learning to make changes to stay competitive. We have a lot of commercial customers who are under contract. But we want repeat customers, so we work hard to please them. I have employees who seem to care even more than I do. We are all focused on making the company work and work right.

We work on about six projects at a time. I've learned to take on only what we can handle. We pace ourselves on what we do. Right now we have about

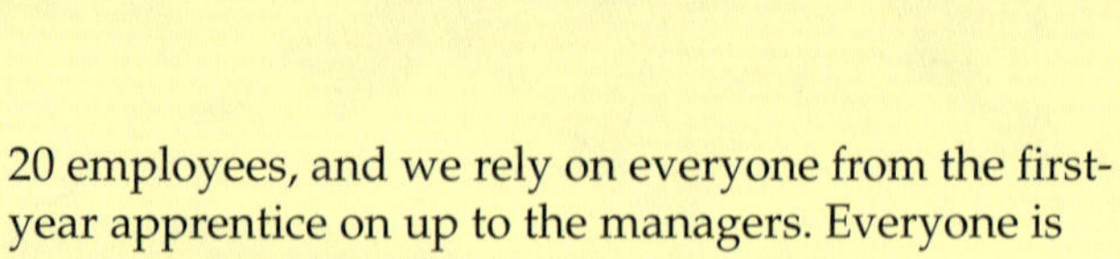

20 employees, and we rely on everyone from the first-year apprentice on up to the managers. Everyone is an important part of our environment.

What does it take to be successful in your trade?
It takes hard work and showing up every day. Be loyal to your employer; don't just do what he or she says, but be on the team. Don't whine, because that's not getting the job done. You're here to work. I learned that if someone said we were going to do something one way, we might end up doing it another. I might be asked for my opinion, but I was also willing to go with the team. You have to be a team player. It's important to be diligent. Always learn and always do what it takes to get the job done well.

What is the most interesting aspect of your profession? What makes your trade stand out from others?
The most interesting aspect is seeing the finished product and feeling proud to put our name on it. We get to see and use the finished product; we work and live in the buildings we build. Our efforts become part of a structure, whether in a house, a school, or an office. We've made a significant part of the building. Although every day is different, there are no real surprises. Every job presents the same challenges. We are uniform in what we do and how we handle it.

What do you like most—and least—about your job?
I most like seeing the finished product, and being part of a team when there are a lot of challenges on a building project. I enjoy coming to work every day, so I can't say that there is something I like least. I guess it would be dealing with employees' personal problems. You should try not to bring problems to work.

What would you say to someone entering the trade today?
When interviewing for an apprenticeship program or when entering the trade, your diplomas and cards are only good for opening the door. They do not keep the job. That piece of paper gets you in the door, but it does not ensure your employment. You are the person who keeps your job by being part of that team, whether sweeping a floor or serving as president. That paper is just the vehicle to get you there.

I try to remember the phrase, "When you are through learning, you are through." Keep on learning and learn to change. You're never done. Work smart, not hard. Some guys are running around and around getting nothing done. Work smart and you'll accomplish more.

What can an apprentice expect to earn in his/her first years on the job in your area? What can he/she expect to earn after 10-plus years in the industry?
First-year apprentices start at $9.30 per hour. We are an open shop, so the sky is the limit for our guys. If you have it in you to do a good job, you can move up to a foreman position with benefits. After 10 years, you can make anywhere between $18 and $33 per hour.

Trade Terms Introduced in This Module

Corrugated stainless steel tubing (CSST): A gas piping material made from flexible steel tubing with a polyethylene (PE) jacket.

Coupling fitting: A fitting that allows two separate lengths of CSST to be connected to each other.

Elevated pressure system: A CSST system that operates above the normal pressure of 7 inches w.c.

Gas cock: A type of valve that provides positive, quick flow control on gas piping systems.

Manifold: A device that allows one CSST pipe to be branched off to multiple outlets.

Pounds per square inch gauge (psig): A unit of measure used to specify the maximum operating pressure for a gas system.

Regulator: A device used in elevated pressure systems to reduce the gas pressure before the pipe outlet.

Regulator stub-out: A stub-out for use with a regulator.

Straight fitting: A fitting used to attach a length of CSST to an appliance or other device.

Striker plate: A manufactured metal plate of a required thickness that is installed behind a wall to protect CSST from penetration by nails.

Stub-out: A device mounted in a finished wall, floor, or ceiling that serves as a fixed connection for an appliance or other device.

Tee fitting: A fitting that allows a branch to be taken off from a main gas line.

Water column (w.c.): The unit of measure used to specify the standard pressure of a gas system.

Additional Resources and References

ADDITIONAL RESOURCES

This module is intended to present thorough resources for task training. The following reference works are suggested for further study. These are optional materials for continued education rather than for task training.

Illustrated Guide to the International Plumbing & Fuel Gas Codes, 2003. Howard C. Massey. Carlsbad, CA: Craftsman Book Company.

National Fuel Gas Code, 2002. Theodore C. Lemoff. Quincy, MA: National Fire Protection Association.

Overview of the International Fuel Gas Code: Based on the 2000 International Fuel Gas Code, 2000. Country Club Hills, IL: Building Officials & Code Administrators International, Inc.

REFERENCES

Cinergy/LG website, www.cinergylg.com. *Gas Installer's Manual*, www.cinergylg.com/Business_Services/service_request/gas_installers_manual/gas_installer_manual.asp, reviewed July 2003.

Dictionary of Architecture and Construction, Third Edition, 2000. Cyril M. Harris, ed. New York: McGraw-Hill.

Foster-Miller, Inc., website, www.foster-miller.com, "Two Foster-Miller Engineers Receive 'Natural Gas Innovator' Award," www.foster-miller.com/innovgasaward.html, reviewed August 2003.

Housing Zone website, www.housingzone.com. "Piping Pick Profitable," July 28, 2003. John O'Reilly. www.housingzone.com/topics/pb/cmaterials/pb02ha020.asp, reviewed August 2003.

Plumbing & Mechanical Magazine website, www.pmmag.com, "More Than a Pipe Dream," June 2000. Steve Smith. www.pmmag.com/CDA/ArticleInformation/features/BNP__Features__Item/0,2379,4489,00.html, reviewed August 2003.

Plumbing & Mechanical Magazine website, www.pmmag.com, "The CSST Battle Is Over," February 2003. Kelly Faloon. www.pmmag.com/pm/CDA/ArticleInformation/features/BNP_Features_Item/0,2379,91474,00.html, reviewed August 2003.

Titeflex Corporation website, www.titeflex.com. Gastite Product Overviews, www.gastite.com/productframe.htm, reviewed August 2003.

NCCER CURRICULA — USER UPDATE

NCCER makes every effort to keep its textbooks up-to-date and free of technical errors. We appreciate your help in this process. If you find an error, a typographical mistake, or an inaccuracy in NCCER's curricula, please fill out this form (or a photocopy), or complete the online form at **www.nccer.org/olf**. Be sure to include the exact module ID number, page number, a detailed description, and your recommended correction. Your input will be brought to the attention of the Authoring Team. Thank you for your assistance.

Instructors – If you have an idea for improving this textbook, or have found that additional materials were necessary to teach this module effectively, please let us know so that we may present your suggestions to the Authoring Team.

NCCER Product Development and Revision

13614 Progress Blvd., Alachua, FL 32615

Email: curriculum@nccer.org

Online: www.nccer.org/olf

❑ Trainee Guide ❑ AIG ❑ Exam ❑ PowerPoints Other ______________________________

Craft / Level: Copyright Date:

Module ID Number / Title:

Section Number(s):

Description:

Recommended Correction:

Your Name:

Address:

Email: Phone:

Fixtures and Faucets

02111-05

02111-05

Fixtures and Faucets

Topics to be presented in this module include:

Overview

Plumbers use a variety of fixtures and faucets for plumbing installations. Fixtures receive water from a water supply line and discharge water or sewage into a connected drainage system. Faucets are fixtures that draw water from pipe. Understanding how fixtures and faucets operate enables plumbers to select the appropriate item for each installation. Plumbing codes determine the types of fixtures and faucets that are allowed in an area. While codes vary, they are all based on the same principles of safety and sanitation.

Fixtures are made from corrosion-resistant, nonabsorbent materials. Plumbers must understand how fixtures are manufactured and be familiar with the properties of the various materials. Fixtures, such as sinks, bathtubs, and toilets, perform different functions and are available in numerous styles. To properly install plumbing fixtures, plumbers must know how they operate and how they connect with water supplies and waste outlets. Applicable plumbing codes specify size, installation requirements, and materials allowed in each area.

Faucets are classified by their operation—compression or noncompression faucets—and their application. Faucets, too, are available in many sizes, shapes, and styles. Bathroom and kitchen faucets are decoratively designed, while utility faucets are basic and functional. Plumbers must be able to choose the right type of faucet for each installation.

Focus Statement

The goal of the plumber is to protect the health, safety, and comfort of the nation job by job.

Code Note

Codes vary among jurisdictions. Because of the variations in code, consult the applicable code whenever regulations are in question. Referencing an incorrect set of codes can cause as much trouble as failing to reference codes altogether. Obtain, review, and familiarize yourself with your local adopted code.

Objectives

When you have completed this module, you will be able to do the following:

1. Identify the basic types of materials used in the manufacture of plumbing fixtures.
2. Discuss common types of sinks, lavatories, and faucets.
3. Identify and discuss common types of bathtubs, bath-shower modules, shower stalls, and shower baths.
4. Discuss common types of toilets, urinals, and bidets.
5. Identify and describe common types of drinking fountains and water coolers.
6. Discuss common types of garbage disposals and domestic dishwashers.

Key Trade Terms

Bibb
China
Diverter
Faucet
Fiberglass
Fixture
Flange
Flood-level rim
Flushometer
Flushometer valve
Flush valve
Laundry tray
Lavatory
Left-hand bathtub
Plumbing fixture
Porcelain enamel
Right-hand bathtub
Seat washer
Terrazzo
Valve seat
Valve stem
Vent pipe
Vent system
Vitrified porcelain
Waste pipe

Required Trainee Materials

1. Appropriate personal protective equipment
2. Pencils and paper
3. Copy of local adopted code

Prerequisites

Before you begin this module, it is recommended that you successfully complete *Core Curriculum; Plumbing Level One,* Modules 02101-05 through 02110-05.

This course map shows all of the modules in the first level of the *Plumbing* curriculum. The suggested training order begins at the bottom and proceeds up. Skill levels increase as you advance on the course map. The local Training Program Sponsor may adjust the training order.

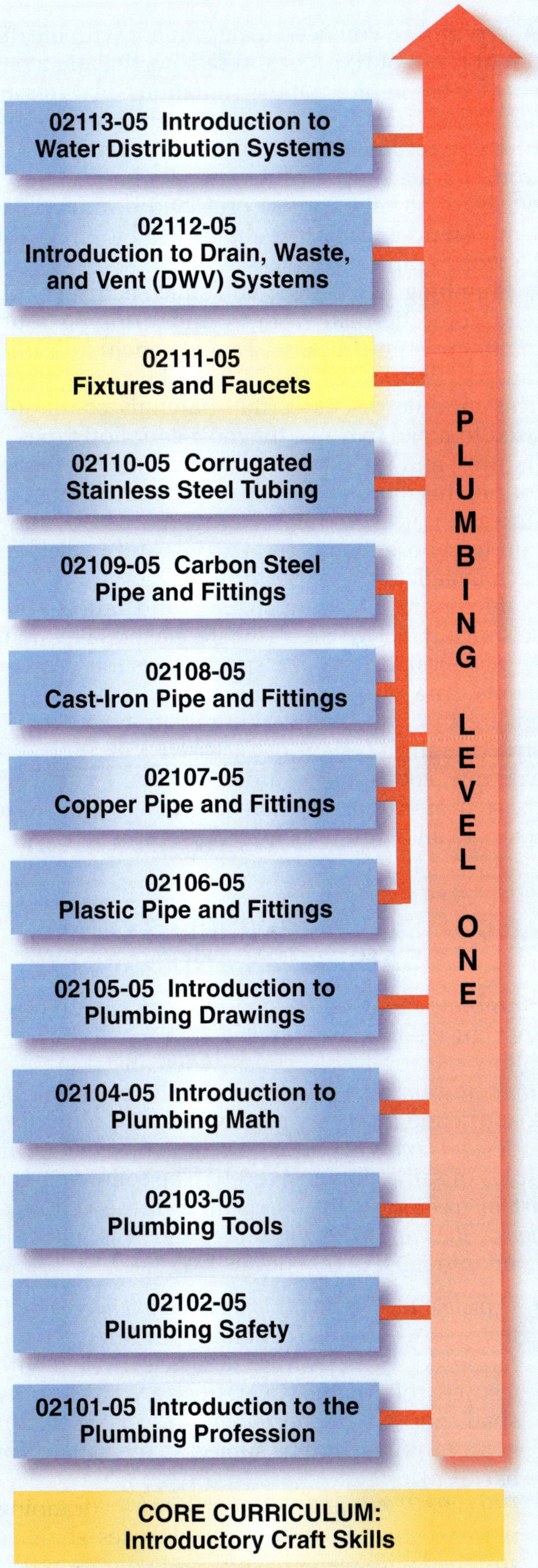

02113-05 Introduction to Water Distribution Systems

02112-05 Introduction to Drain, Waste, and Vent (DWV) Systems

02111-05 Fixtures and Faucets

02110-05 Corrugated Stainless Steel Tubing

02109-05 Carbon Steel Pipe and Fittings

02108-05 Cast-Iron Pipe and Fittings

02107-05 Copper Pipe and Fittings

02106-05 Plastic Pipe and Fittings

02105-05 Introduction to Plumbing Drawings

02104-05 Introduction to Plumbing Math

02103-05 Plumbing Tools

02102-05 Plumbing Safety

02101-05 Introduction to the Plumbing Profession

CORE CURRICULUM: Introductory Craft Skills

PLUMBING LEVEL ONE

111CMAP.EPS

1.0.0 ◆ INTRODUCTION

As a plumber, you need to be familiar with the different types of **fixtures** and **faucets** that are available for use in plumbing installations. Fixtures are also often referred to as **plumbing fixtures.** A fixture is a receptacle or device that receives water from a water supply line and discharges water, waste, or sewage into a connected drainage system. A faucet is a fixture that is used to draw water from a pipe.

Plumbing fixtures include sinks, bathtubs, toilets, domestic dishwashers, and drinking fountains, to name a few. Faucets include utility faucets, bathroom faucets, and kitchen faucets. This module discusses the materials commonly used to make fixtures, the most common types of fixtures, and the types of faucets available. In addition, this module explains how each type of fixture and faucet operates. Knowing this will allow you to choose the proper fixtures and faucets for each installation you perform.

If you are installing fixtures and faucets in a seismically active area (where earthquakes may occur) local codes will specify seismic requirements. The purpose of these requirements is to ensure that the pipe is securely fastened to the structure in the event of excessive vibration. For example, codes may require that you leave extra space around pipes where they meet walls and floors to allow for anticipated movement.

2.0.0 ◆ BASIC PRINCIPLES OF SANITATION AND SAFETY

Plumbing codes determine the types of fixtures and faucets that are allowed in an area. Always follow all applicable codes when installing fixtures and faucets. While the details of plumbing construction may vary among different plumbing codes, all codes are based on the same basic principles of sanitation and safety. The following is a list of basic principles as stated in the *2003 National Standard Plumbing Code,* and it is incorporated into most other codes:

- *Principle 1: All occupied premises shall have potable water.* – All premises intended for human habitation, occupancy, or use shall be provided with a supply of potable water. Such a water supply shall not be connected with unsafe water sources, nor shall it be subject to the hazards of backflow.
- *Principle 2: Adequate water required.* – Plumbing fixtures, devices, and appurtenances shall be supplied with water in sufficient volume and at pressures adequate to enable them to function properly and without undue noise under normal conditions of use.
- *Principle 3: Hot water required.* – Hot water shall be supplied to all plumbing fixtures that normally need or require hot water for their proper use and function.
- *Principle 4: Water conservation.* – Plumbing shall be designed and adjusted to use the minimum quantity of water consistent with proper performance and cleaning.
- *Principle 5: Safety devices.* – Devices for heating and storing water shall be so designed and installed as to guard against dangers from explosion and overheating.
- *Principle 6: Use public sewer where available.* – Every building with installed plumbing fixtures and intended for human habitation, occupancy, or use, and located on premises where a public sewer is on or passes said premises within a reasonable distance, shall be connected to the sewer.
- *Principle 7: Required plumbing fixtures.* – Each family dwelling unit shall have at least one water closet, one lavatory, one kitchen-type sink, and one bathtub or shower to meet the basic requirements of sanitation and personal hygiene. All other structures for human habitation shall be equipped with sufficient sanitary facilities. Plumbing fixtures shall be made of durable, smooth, non-absorbent and corrosion-resistant material and shall be free from concealed fouling surfaces.
- *Principle 8: Drainage system.* – The drainage system shall be designed, constructed, and maintained to guard against fouling, deposit of solids, and clogging, and with adequate cleanouts arranged so that the pipes may be readily cleaned.
- *Principle 9: Durable materials and good workmanship.*–The piping of the plumbing system shall be of durable material, free from defective workmanship and so designed and constructed as to give satisfactory service for its reasonable expected life.
- *Principle 10: Fixture traps.* – Each fixture directly connected to the drainage system shall be equipped with a liquid seal trap.
- *Principle 11: Trap seals shall be protected.* – The drainage system shall be designed to provide an adequate circulation of air in all pipes with no danger of siphonage, aspiration, or forcing of trap seals under condition of ordinary use.
- *Principle 12: Exhaust foul air to outside.* – Each vent terminal shall extend to the outer air and

be so installed as to minimize the possibilities of clogging and the return of foul air to the building.

- *Principle 13: Test the plumbing system.* – The plumbing system shall be subjected to such tests as will effectively disclose all leaks and defects in the work or the material.
- *Principle 14: Exclude certain substances from the plumbing system.* – No substance that will clog or accentuate clogging of pipes, produce explosive mixtures, destroy the pipes or their joints, or interfere unduly with the sewage-disposal process shall be allowed to enter the building drainage system.
- *Principle 15: Prevent contamination.* – Proper protection shall be provided to prevent contamination of food, water, sterile goods, and similar materials by backflow of sewage. When necessary, the fixture, device, or appliance shall be connected indirectly with the building drainage system.
- *Principle 16: Light and ventilation.* – No water closet or similar fixture shall be located in a room or compartment that is not properly lighted and ventilated.
- *Principle 17: Individual sewage disposal systems.* – If water closets or other plumbing fixtures are installed in buildings where there is no sewer within a reasonable distance, suitable provision shall be made for disposing of the sewage by some accepted method of sewage treatment and disposal.
- *Principle 18: Prevent sewer flooding.* – Where a plumbing drainage system is subject to backflow of sewage from the public sewer or private disposal system, suitable provision shall be made to prevent its overflow in the building.
- *Principle 19: Proper maintenance.* – Plumbing systems shall be maintained in a safe and serviceable condition from the standpoint of both mechanics and health.
- *Principle 20: Fixtures shall be accessible.* – All plumbing fixtures shall be so installed with regard to spacing as to be accessible for their intended use and for cleaning.
- *Principle 21: Structural safety.* – Plumbing shall be installed with due regard to preservation of the strength of structural members and prevention of damage to walls and other surfaces through fixture usage.
- *Principle 22: Protect ground and surface water.* – Sewage and other waste shall not be discharged into surface or sub-surface water unless it has first been subjected to some acceptable form of treatment.

3.0.0 ◆ MATERIALS USED TO MAKE FIXTURES

Modern plumbing fixtures are made from a variety of durable, corrosion-resistant, nonabsorbent materials that have smooth, easy-to-clean surfaces. To make plumbing fixtures, manufacturers use **china** (also called **vitrified porcelain**), steel coated with **porcelain enamel,** cast iron coated with porcelain enamel, stainless steel, acrylic plastic, and **fiberglass** reinforced with plastic. Manufacturers may produce fixtures from a single material, such as kitchen sinks made of stainless steel, or a combination of materials, such as shower stalls made of fiberglass-reinforced plastic.

3.1.0 Porcelain

Molded vitrified porcelain, also called china or vitreous clay, is used in more expensive fixtures. Porcelain is made from a mixture of fine clay (called kaolin), quartz, feldspar, and silica. The materials are fused, or vitrified, into a glass-like state. The complete process of fusion, or vitrification, requires a temperature of 2,600°F (1,426°C). The process produces a sanitary surface that is strong, nonporous, and easy to clean when used alone or as a coating for steel or cast iron.

Water Conservation

The most practical way to prevent water shortages is to conserve water. This means using less water. Developing new water sources is very expensive. A variety of devices have been developed to conserve both water and energy in the face of increasing demand worldwide. In the United States, 1.6-gallon water conservation water closets are replacing the old water closets that typically used 3 to 5 gallons to flush the bowl. Old water closets are no longer being manufactured.

Title 18 CFR 803, the *Conservation of Power and Water Resources Act*, which was revised in April 1999, defines the steps that state and local governments and manufacturers must take to conserve water.

3.2.0 Porcelain Enamel

Porcelain is also used in its liquid, or enamel, form in combination with other materials, such as steel or cast iron. Manufacturers use a bonding process that fuses the two materials together. This process creates a more durable and functional material for making fixtures. The resulting material is called porcelain enamel, also referred to as glass lining or vitreous enamel.

3.3.0 Cast Iron

Some plumbing fixtures, such as bathtubs and slop sinks, are made from a type of cast iron called gray iron, a metal that is ideally suited for plumbing fixtures because of its ability to be molded into a variety of shapes. Cast iron is an iron alloy made from molten iron that is formed using a mold. Cast iron can be coated with an enamel powder and then heated at extremely high temperatures. The enamel powder contains pigments that allow manufacturers to produce fixtures of different colors.

3.4.0 Sheet Steel

Manufacturers also produce fixtures from sheet steel. Using this low-cost process, manufacturers form the fixtures by stamping the sheet steel into the desired shapes. The surfaces of these plumbing fixtures are then fused with a porcelain enamel finish, which is glossy, decorative, protective, and sanitary. While these fixtures are usually less expensive than those made of cast iron, they are also less durable, because they chip easily.

3.5.0 Stainless Steel

Stainless steel is another metal commonly used to make plumbing fixtures, particularly kitchen sinks. Stainless steel, a corrosion-resistant steel alloy, is an excellent material for plumbing fixtures. Although it is difficult to form, stainless steel has several advantages over enameled surfaces. It does not require an additional coating to produce a sanitary surface, and it has a durable finish; it will not chip.

3.6.0 Plastics

Plastics are often used for plumbing fixtures because of their versatility, light weight, and relatively low cost, although the finish is not as durable as enameled cast iron. The plastics most commonly used today are acrylic plastic and plastic-reinforced fiberglass. Plastic-reinforced fiberglass is used to make one-piece showers and bathtubs, which include both the fixture and the adjoining walls. The strength of the product comes from the glass fibers. Fiberglass tubs and shower stalls are lightweight and easy to install. A plastic gel coating installed during manufacturing gives these fixtures a shiny, enamel-like surface. Manufacturers can now use marbled, colored acrylic plastic sheets to make decorative fixtures that appear to be made of marble. The fixtures can be just as attractive as marble, but they do not absorb water as marble does.

ON THE LEVEL

Today's Materials

Many of the most popular bathtubs, shower stalls, and hot tubs that manufacturers produce today appear to be made of natural materials but are actually made from acrylic, a type of plastic. One example of such a material is DuPont Corian®, an acrylic-based, premium, solid surface material that resists damage, is easy to maintain, and can be shaped into countertops, work surfaces, sinks and bowls, shower and tub surrounds, and wainscotting and windowsills.

4.0.0 ◆ BASIC TYPES OF FIXTURES

This section describes the basic types of fixtures you will encounter. All fixtures are available in a variety of styles. Fixtures made for commercial installations are designed to withstand heavier use than fixtures for residential installations. For example, a sink in a restaurant needs to be more durable than a sink in a home. Some commercial fixtures are even designed to be vandal-proof.

NOTE

Always consult submittal data to obtain the correct fixture. Shipping containers for fixtures are usually not labeled with the name of the item (for example, "corner sink"). More commonly, they only feature the product's catalog number assigned by the manufacturer.

DID YOU KNOW?

Plumbing codes set requirements for the minimum number of plumbing fixtures in different types of buildings. For example, the *National Standard Plumbing Code*, the Plumbing-Heating-Cooling Contractors–National Association's model code, and most other codes state that each family dwelling must have at least one water closet, one lavatory, one kitchen-type sink, and one bathtub or shower.

111F01.TIF

Figure 1 ◆ Double-compartment sink.

For a public building to comply with the *Americans with Disabilities Act of 1990* (ADA), it must adhere to ADA requirements regarding the location and spacing of plumbing fixtures. The ADA established very explicit requirements to improve the accessibility of facilities to people with disabilities. The ADA requirements are based on space allowances and reach ranges for people with various disabilities. The technical requirements in the *ADA Standards for Accessible Design* apply to all public accommodations, such as restaurants, movie theaters, stores, and medical facilities. The requirements affect everything from sinks and toilets to water coolers and grab bars. For example, ADA-compliant sinks must be mounted with the counter or rim no higher than 34 inches above the finish floor. As a plumber, you must have the appropriate information in order to rough in the dimensions to meet the ADA requirements.

The American National Standards Institute (ANSI) also sets forth similar guidelines for accessible design. The ANSI guidelines are outlined in ANSI document *A117.1-1980.*

4.1.0 Sinks and Lavatories

A sink is a shallow, flat-bottomed basin used for food preparation and dishwashing. Kitchen sinks come in single, double, and triple compartments and are often installed in a countertop (see *Figure 1*). They are manufactured from a variety of materials and are available in different shapes, sizes, colors, and styles.

People often confuse the terms *sink* and ***lavatory.*** A lavatory is a basin designed for the primary purpose of washing hands and face. Lavatories are available in a variety of colors, sizes, and shapes. Installing lavatories requires a small-diameter **vent pipe** and a **waste pipe.** Vent pipes are also often referred to as **vent systems.** The vent pipe allows a flow of air to escape from the drain, waste, and vent (DWV) system and provides a circulation of air within a DWV system. Vent pipes protect trap seals from siphonage and back pressure. The waste pipe, which connects the trap to the vent pipe, conveys only wastewater to the building drain.

Sinks and lavatories are classified by their method of installation. The four major categories of sinks and lavatories are wall hung, self rimming, built-in rim, and undercounter (see *Figure 2*). Additionally, two examples of the roughing-in measurements that you will typically find in fixture catalogs are shown in *Figure 2*. You need roughing-in measurements to ensure that you install the correct fixture at the correct location and height.

A wall-hung lavatory, also called a lavatory carrier, is fastened to the wall by a support bracket. The manufacturer supplies the bracket, which is usually made from stamped steel or cast iron. You typically install this type of lavatory with the **flood-level rim** 31 inches above the floor. The flood-level rim is the top edge of the receptor or fixture. If water overflows the rim, it will spill over onto the countertop and floor.

Several different styles of wall-hung lavatories are available. They include corner, raised back,

WALL HUNG

SELF RIMMING

BUILT-IN RIM

UNDERCOUNTER

111F02A.EPS

Figure 2 ◆ Four major categories of lavatories and sinks. (1 of 2)

ledge or shelf back, and wheelchair accessible (see *Figure 3*).

To comply with the ADA, wheelchair-accessible lavatories must be installed with the flood-level rim or counter surface no higher than 34 inches above the floor. This makes the lavatory accessible to people in wheelchairs and permits the wheelchair to slide easily under the lavatory. The drain fittings should be offset or moved back toward the wall rather than dropping directly down from the fixture. This placement of the fittings allows room for the user's legs to fit under the lavatory. Wheelchair-accessible lavatories are fitted with special faucets that have long blade handles that extend toward the user and do not require grasping or twisting to turn on the water (see *Figure 4*). This feature is particularly necessary for users who have restricted mobility or who suffer from arthritis.

The self-rimming lavatory or sink is placed directly on the countertop. This fixture is manufactured with the mounting rim as part of the overall structure. It requires no external mounting frame to secure it to the countertop. Kitchen and bar sinks are often self-rimming designs. The built-in rim lavatory has a metal frame that is set in a stainless steel rim. This rim supports the bowl in the countertop.

The undercounter lavatory is attached to the underside of a countertop. It has no rim. It requires brackets placed underneath the counter to support it. The vanity-top lavatory is an undercounter lavatory that has a one-piece washbasin and countertop set on top of a vanity cabinet (see *Figure 5*). Vanity tops may be manufactured from marbleized acrylic plastic and are available in a variety of colors and sizes.

Pedestal lavatories are decorative lavatories that sit on pedestals rather than on countertops or vanities (see *Figure 6*). They may be mounted to the floor or to the wall. They must be firmly mounted according to the manufacturer's specifications.

4.2.0 Bathtubs

Bathtubs are plumbing fixtures that serve as containers for water. They are shaped to fit the human

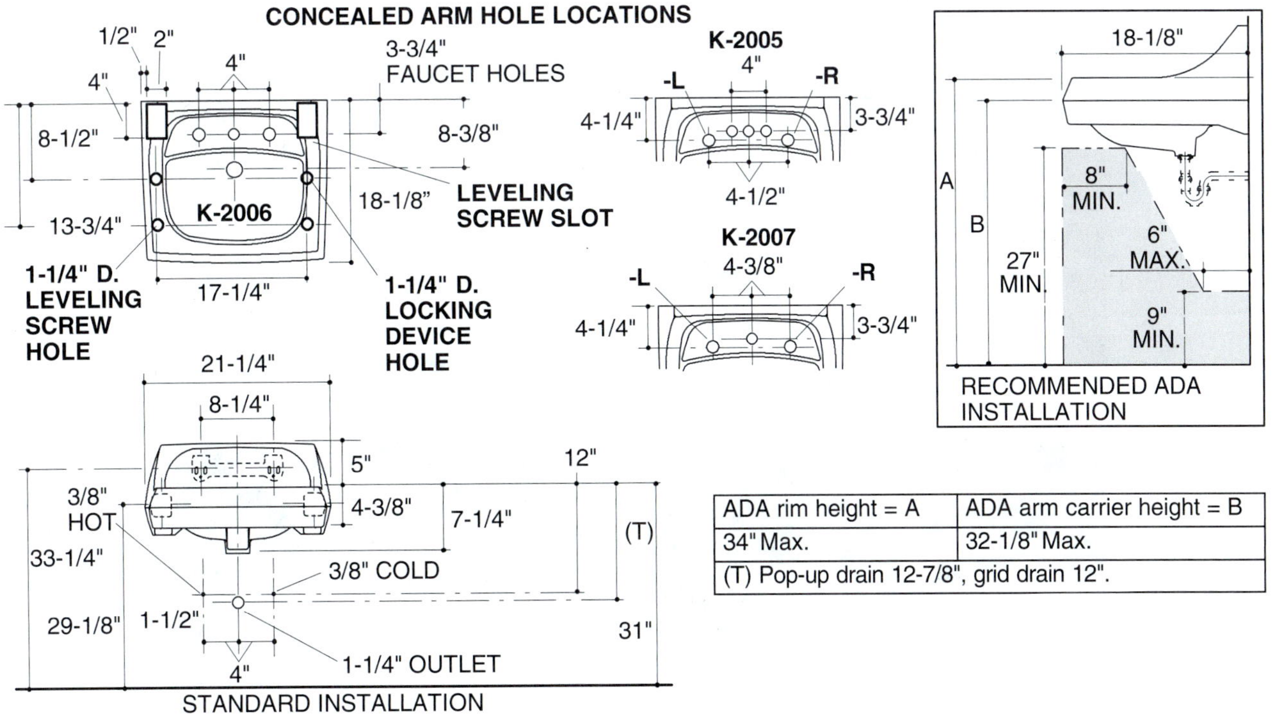

ADA rim height = A	ADA arm carrier height = B
34" Max.	32-1/8" Max.
(T) Pop-up drain 12-7/8", grid drain 12".	

WALL-HUNG ROUGH-IN

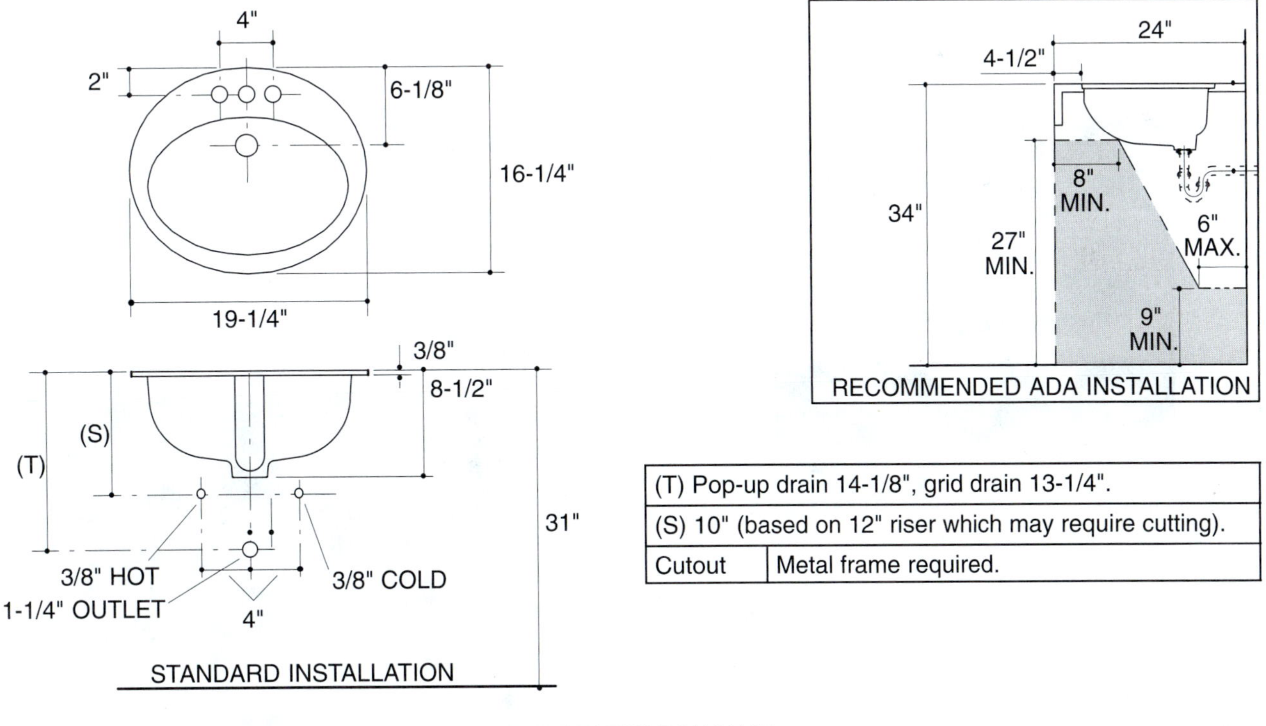

(T) Pop-up drain 14-1/8", grid drain 13-1/4".	
(S) 10" (based on 12" riser which may require cutting).	
Cutout	Metal frame required.

BUILT-IN RIM ROUGH-IN

111F02B.EPS

Figure 2 ◆ Four major categories of lavatories and sinks. (2 of 2)

CORNER

RAISED BACK

WHEELCHAIR ACCESSIBLE

LEDGE OR SHELF BACK

111F03.EPS

Figure 3 ◆ Wall-hung lavatories.

111F04.TIF

Figure 4 ◆ Faucet for wheelchair-accessible lavatory.

body and are used for bathing. Bathtubs are described as **right-hand bathtubs** or **left-hand bathtubs,** depending on where the drain is. A right-hand tub has the drain on the right end of the tub as you face the length of its finished side (see *Figure 7*). A left-hand tub has the drain on the left end of the tub.

To provide proper sanitation and to reduce maintenance, the exterior surface of the tub must be durable enough to withstand frequent cleanings.

111F05.TIF

Figure 5 ◆ Vanity-top lavatory.

111F06.TIF

Figure 6 ◆ Pedestal lavatory.

111F07.TIF

Figure 7 ◆ Right-hand bathtub.

Bathtubs are available in several sizes and shapes, but the standard bathtub is usually 5 feet long. Some tubs have a special nonslip or slip-resistant safety feature to help reduce the danger of falls. Specialty bathtubs are available with jets to provide high-pressure outlets for water therapy massage (see *Figure 8*). These tubs are commonly called whirlpools.

Bathtubs are available in freestanding, built-in, or bath-shower module styles (see *Figure 9*).

111F08A.EPS

Figure 8a ◆ Whirlpool bathtub.

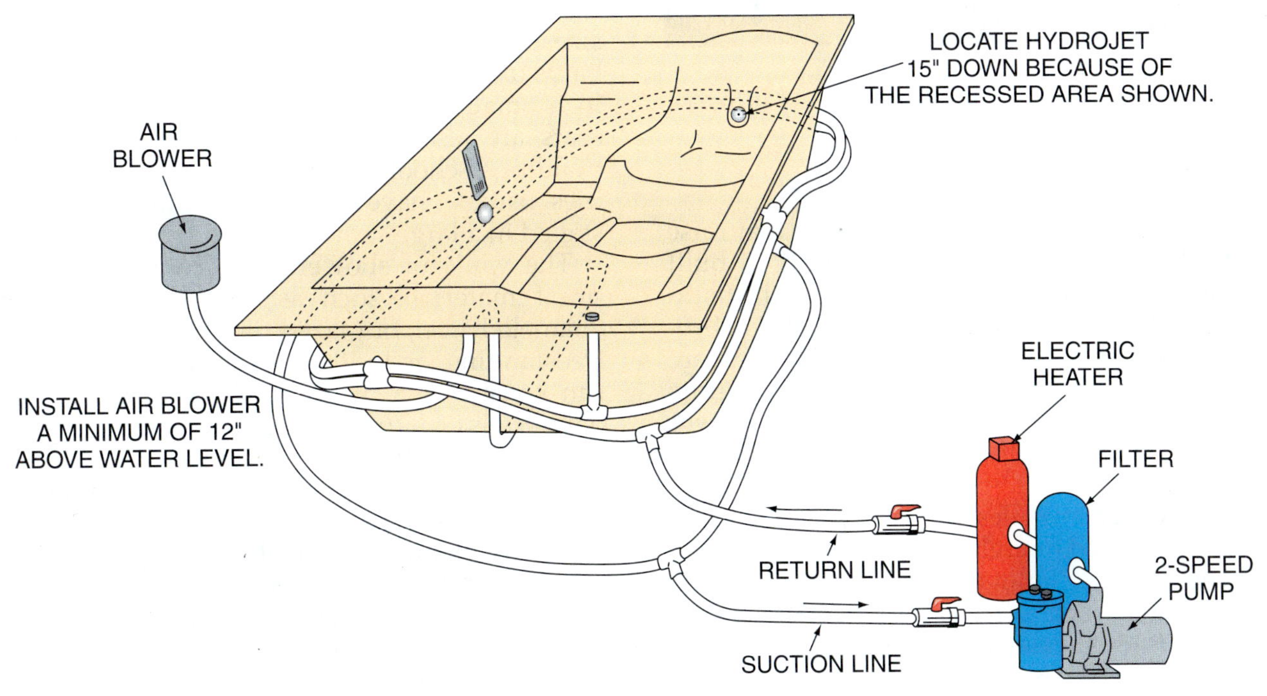

111F08B.EPS

Figure 8b ◆ Whirlpool bathtub.

Figure 9a ◆ Bathtubs.

Figure 9b ◆ Bathtubs.

The freestanding bathtub rests on feet that raise it off the floor. The built-in bathtub is built permanently into the walls and floor of the bathroom. Bath-shower modules are usually manufactured as one piece, although some models are available with the walls and tub as separate units. Because of their large size, bath-shower modules are generally installed only in new, larger homes.

Bath-shower modules are manufactured from plastic-reinforced fiberglass. The walls are usually high enough to eliminate the need for conventional wall material such as ceramic tile. Because these modules have no joints, cleaning the tub after installation is easy.

Figure 9c ◆ Bathtubs.

4.3.0 Shower Stalls

Shower stalls, also called shower baths, are stalls or baths with faucets that spray water from above onto the user's body. They are available in many sizes, shapes, and colors. Residential shower baths usually consist of one showerhead installed in a small, enclosed space. Shower stalls with seats are also available. Showers in industrial buildings, schools, gymnasiums, and similar facilities usually have a series of showerheads installed in a large shower room.

The walls of shower stalls consist of waterproof materials such as gel-coated fiberglass, enameled steel, or glazed tile. They can be either freestanding or built-in stalls. Built-in showers can consist of a premade shower enclosure or have walls constructed of glazed ceramic tile (see *Figure 10*). In either case, plumbing codes require a safety pan under the installation. You should always insulate the shower stall. You can use non-expanding foam, a relatively new material, when installing a fiberglass shower. Simply inject the foam between the wall and the floor and the outside of the shower enclosure. Check applicable plumbing code requirements regarding size, in-

111F10.TIF

Figure 10 ◆ Premade built-in shower.

stallation of shower stalls and floors, and materials allowed in your area.

Shower enclosures are available as one-piece modules or with the walls and base as separate units. You must assemble shower enclosures that consist of separate units on the job site. The dimensions of the shower base are usually 36 inches by 36 inches or 36 inches by 48 inches, with a slope of ¼ inch per foot toward a center drain. Bases for shower stalls are usually made of fiberglass, cast stone, enameled steel, or **terrazzo.** Terrazzo consists of small pieces of marble or other hard stone embedded in a mortar base.

4.4.0 Water Closets

Water closets, commonly referred to as toilets or W.C.s, are water-flushed plumbing fixtures designed to carry away solid, organic waste. They are usually manufactured from vitreous china and are available in both round-front and elongated shapes. Elongated water closets are used in public facilities. Both round-front and elongated shapes are used in residences.

The ADA sets forth very specific requirements for accessible water closets, including the height of the fixture and the size and location of grab bars, flush controls, and toilet paper dispensers. If you are installing a water closet for individuals with disabilities, you must adhere to these requirements. An accessible water closet must be between 17 and 19 inches high, measured to the top of the toilet seat. The grab bar behind the water closet must be a minimum of 36 inches long. Controls for **flush valves** must be mounted on the wide side of toilet areas, no more than 44 inches above the floor. Flush valves are used to flush water closets and similar fixtures.

Water closets are available in two styles: floor mounted and wall hung (see *Figure 11*). Floor-mounted water closets are secured directly to the floor and connected to the drainage system pipe by a closet **flange,** the rim at the end of a pipe that provides a connection point to another pipe. Wall-hung water closets are suspended from the wall by supporting chair carriers.

New toilets installed today must comply with the *Conservation of Power and Water Resources Act of 1999.* Newer models reduce water use from 3 to 5 gallons per flush to 1.6 gallons.

4.4.1 Basic Operating Principles

The water closet is designed to carry away waste. When you flush a toilet, the water moves through passages in the bowl under pressure or gravity. The force of the water cleans the bowl. The water then moves the solid waste through the trap and into the drain. A siphoning action removes the solid waste from the bowl. This siphoning action occurs when the water moves through the downward leg of the trap, creating a drop in the atmospheric pressure at the trap outlet. This partial vacuum, along with the head of water from the bowl, sweeps out the wastes.

The siphoning action continues to draw water out of the bowl even after it slows to a trickle. As the water flow through the trap lessens, air is allowed to rush into the trap. This causes a break in the vacuum and equalizes the pressure on both sides of the trap. The flush valve delivers enough additional water to reseal the trap.

The flush valve controls the flow of water from the water closet flush tank into the bowl (see *Figure 12*). When you depress the tank lever, the valve lifts above the tank outlet and floats there. This allows the water in the tank to flow rapidly into the bowl. When the water level in the tank drops to the point where the flush valve no longer floats, the valve reseats on the tank drain and the tank refills.

FLOOR MOUNTED

WHEELCHAIR ACCESSIBLE

WALL HUNG

111F11.EPS

Figure 11 ◆ Water closets.

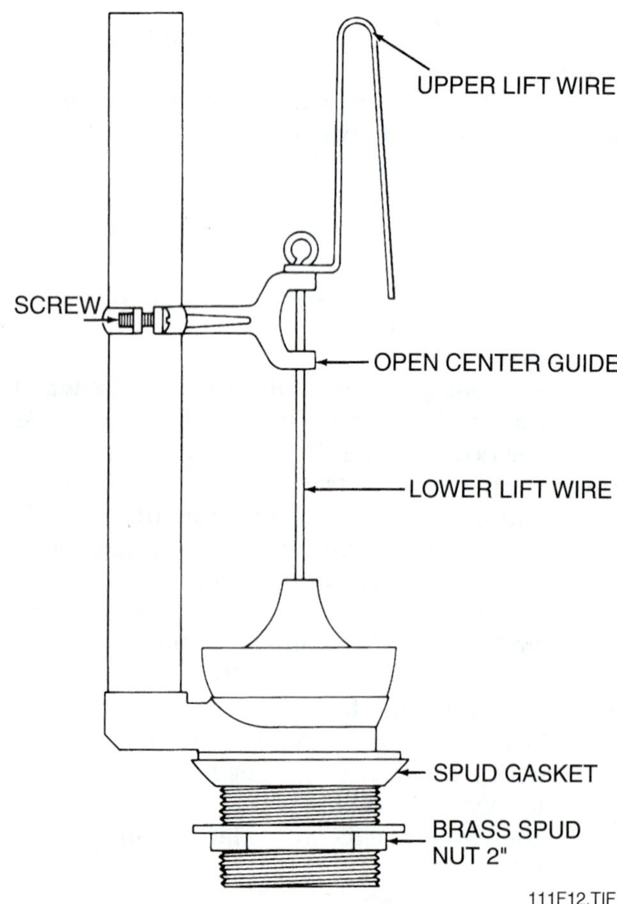

Figure 12 ◆ Flush valve.

4.4.2 Older-Style Water Closets

Generally, you are not permitted to sell or install water closets that do not conform to the *Conservation of Power and Water Resources Act*. However, you may still find older styles of water closets in commercial and residential buildings, so you need to be familiar with them. You may even come across such old styles as the wall-mounted chain-pull tank and the corner tank. Older-style toilets that you are most likely to encounter are washdown water closets and siphon water closets.

The washdown water closet is inexpensive and simple in construction (see *Figure 13*). It is the least efficient and the noisiest of all water closets. The waste passageway is located at the front of the bowl, which causes the front portion of the closet to protrude over the passageway. It is more sus-

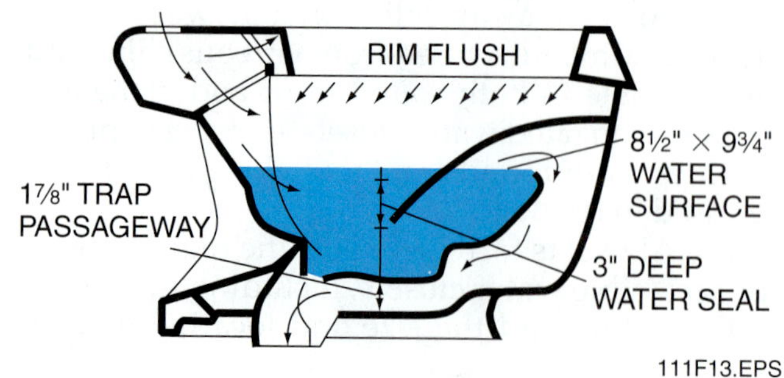

Figure 13 ◆ Washdown water closet.

ceptible to staining and contamination because the flat surface at the front of the bowl area is above the water level.

Siphon water closets use a jet of water to increase the speed of the siphon action and have a downstream leg that is longer than the upstream leg. This design creates a partial vacuum, which helps pull the waste from the bowl. Siphon water closets can have one of four main designs: the reverse trap, the siphon jet, the siphon action, and the blowout.

Although the reverse-trap water closet looks and flushes like the siphon-action water closet, its water surface area, passageway, and water seal are smaller than those of the other siphon water closet models (see *Figure 14*). The waste passageway is 2 inches in diameter, the water surface area is 10¼ inches by 10 inches, and the water seal is 2½ inches deep. These features allow the reverse-trap water closet to operate on less water than the other models.

The siphon-jet water closet has a larger passageway and greater water surface area that reduce the tendency for waste articles to become stuck and clog the passageway (see *Figure 15*). It has a waste passageway that measures 2⅝ inches in diameter, a water area of 12 inches by 10½ inches, and a 3-inch-deep water seal.

This type of water closet does not require a head of water in the bowl to flush out solids. Instead, the stream of water that the closet delivers spins to the outlet of the trap.

The siphon-action water closet is usually manufactured as a one-piece closet that combines the closet and flush tank in one unit, which makes it more compact than other types (see *Figure 16*). It has a water surface area that is 12 inches by 11½ inches, and it flushes quietly. It has a trap passageway diameter of 2 inches and a 3-inch-deep water seal.

The blowout water closet, which is usually a wall-hung bowl, is found only in commercial plumbing installations (see *Figure 17*). It is flushed by the action of a large jet of water directed to the inlet of the trap passageway. The force of the jet of water draws the contents of the bowl into the inlet trap passageway. The jet then forces or blows the waste into the outlet passageway to flush the bowl. The blowout water closet's main advantage is its ability to flush large amounts of toilet paper without becoming clogged. The blowout water closet has a waste passageway that is 2¾ inches in diameter, a water surface area of 11¼ inches by 14⅛ inches, and a 3¼-inch-deep water seal.

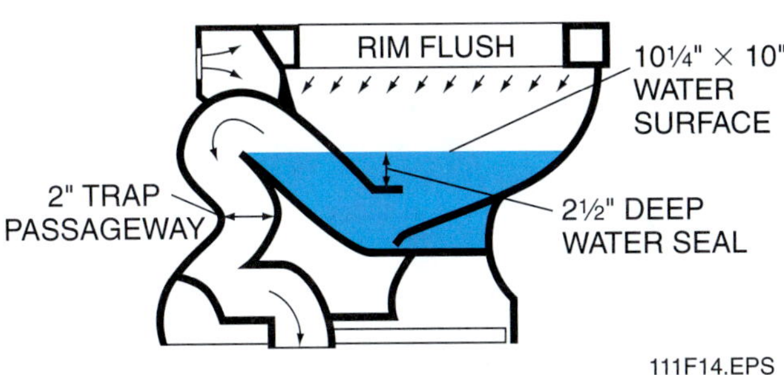

Figure 14 ◆ Reverse-trap water closet.

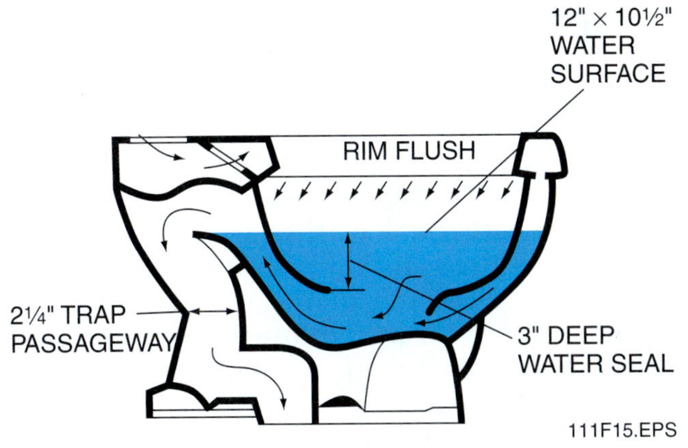

Figure 15 ◆ Siphon-jet water closet.

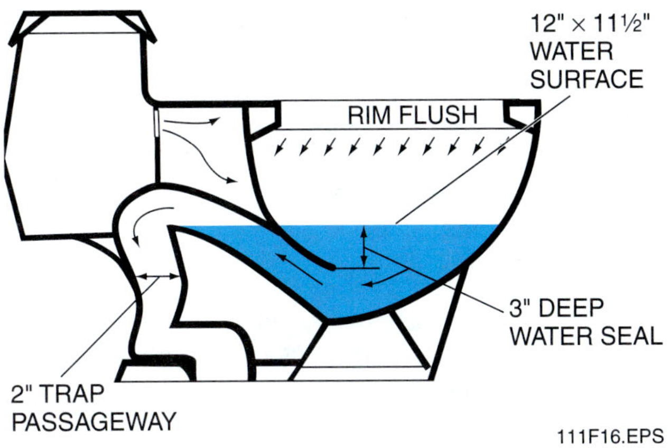

Figure 16 ◆ Siphon-action water closet.

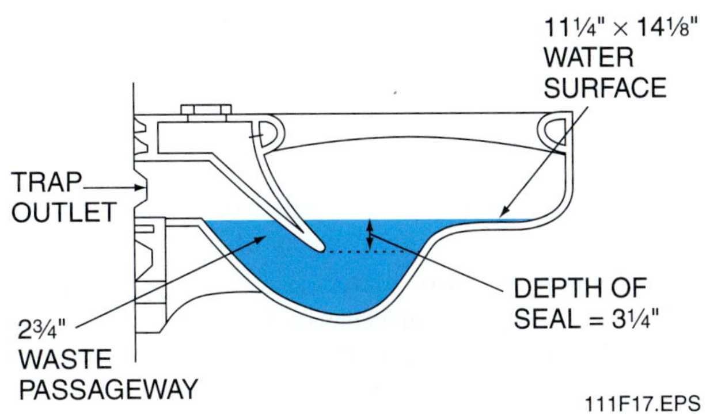

Figure 17 ◆ Blowout water closet.

Review Questions

Sections 1.0.0–4.4.0

1. A(n) __________ is a device that receives water from a water supply line and discharges it into a connected drainage system.
   a. fixture
   b. vent
   c. outlet
   d. inlet

2. A(n) __________ is a fixture that is used to draw water from a pipe.
   a. outlet
   b. faucet
   c. inlet
   d. drain

3. Molded __________, also called china or vitreous clay, is used in more expensive fixtures.
   a. vitrified porcelain
   b. porcelain enamel
   c. cast iron
   d. sheet steel

4. __________ fixtures are less expensive than cast iron, but they are also less durable, because they chip easily.
   a. Porcelain enamel
   b. Sheet steel
   c. Stainless steel
   d. Plastic

5. __________ is often used for plumbing fixtures because of its versatility, light weight, and relatively low cost.
   a. Porcelain
   b. Cast iron
   c. Stainless steel
   d. Plastic

6. The *ADA Standards for Accessible Design* apply to __________.
   a. medical facilities, private restrooms, and restaurants
   b. public restrooms, private property, and medical facilities
   c. restaurants, public restrooms, and private property
   d. medical facilities, public restrooms, and restaurants

7. Corner lavatories, raised-back lavatories, and ledge lavatories are examples of __________ lavatories.
   a. wall-hung
   b. wheelchair-accessible
   c. self-rimming
   d. undercounter

8. Left-hand bathtubs are designed specifically so that the faucet and drain are located where they are easiest for a left-handed person to use.
   a. True
   b. False

9. Plumbing code requires that built-in showers have __________ under the installation.
   a. safety pans
   b. glazed tile
   c. mounting frames
   d. enameled steel

10. A floor-mounted water closet is secured directly to the floor and connected to the drainage system pipe by a __________.
    a. trap
    b. closet flange
    c. flush valve
    d. chair carrier

4.5.0 Urinals

Urinals are water-flushed plumbing fixtures designed to receive liquid waste directly. They are manufactured from vitreous china or stainless steel. Urinals are commonly installed in public restrooms for men. To comply with the ADA, urinals must have an elongated rim at a maximum of 17 inches above the floor. The two main types are washout and siphon-jet urinals.

4.5.1 Washout Urinals

The washout urinal washes waste out of the trap rather than flushing it out. The washing water enters the top of the urinal, flows out through openings in the top rim of the urinal into the fixture, spreads across the back of the urinal, and flows by gravity out through the urinal trap at the base and into the sanitary drainage system.

Washout urinals are characterized by the restricted opening over the trap inlet at the base of the urinal (see *Figure 18*). The purpose of these small openings over the trap is to prevent the trap from becoming plugged by debris in the urinal. These restricted openings may be of two designs: either a beehive or small openings cast in the china.

Wall-hung washout urinals have built-in, 2-inch P-traps. The restrictive trap openings are cast into the china at the bottom of the urinal. Wall-hung washout urinals require a 2-inch waste pipe and a small-diameter vent pipe.

Some wall-hung washout urinals have a bottom outlet. They have a beehive strainer over a drain opening that is attached to a separate, exposed 1½-inch trap. The bottom-outlet, wall-hung washout urinal requires a small-diameter vent pipe. Some codes prohibit its use because it has an invisible trap seal. Always check applicable codes before installing a urinal.

4.5.2 Siphon-Jet Urinals

Siphon-jet urinals use a flushing action similar to the action used in siphon-jet water closets. This type of urinal has a large opening over the trap inlet located in the bottom of the urinal. To comply with the *Conservation of Power and Water Resources Act*, siphon-jet urinals are designed to flush 1 gallon or less of water per use. The most common type of siphon-jet urinal is wall hung (see *Figure 19*). Wall-hung, siphon-jet urinals usually have a 2-inch, built-in trap. They require a small-diameter vent pipe.

4.5.3 Flushing Devices

A urinal must have an adequate flushing device. The flushing device must be able to remove the urine completely from the fixture after it is used, to prevent the spread of disease, fouling of the urinal, and offensive urine odor. Because the turnover in users is so rapid, and because the bulk of a water tank is not practical in these installations, urinals are usually fitted with **flushometers** (*Figure 20*). Flushometers are also often referred to as **flushometer valves.** A flushometer is a flushing device that is connected directly to the water supply, discharges a predetermined amount of water, and requires no storage tank. This device closes by direct water pressure or other mechanical means. Alternatively, some urinals are equipped with a

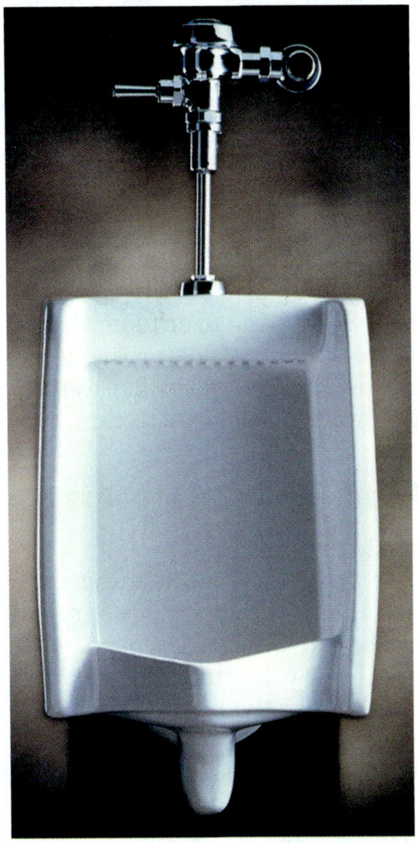

111F18.TIF

Figure 18 ◆ Washout urinal.

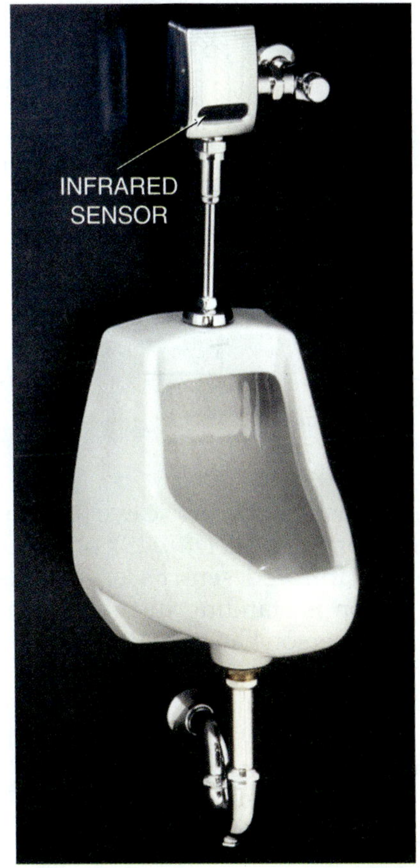

111F19.EPS

Figure 19 ◆ Wall-hung siphon-jet urinal with electric eye.

111F20A.TIF

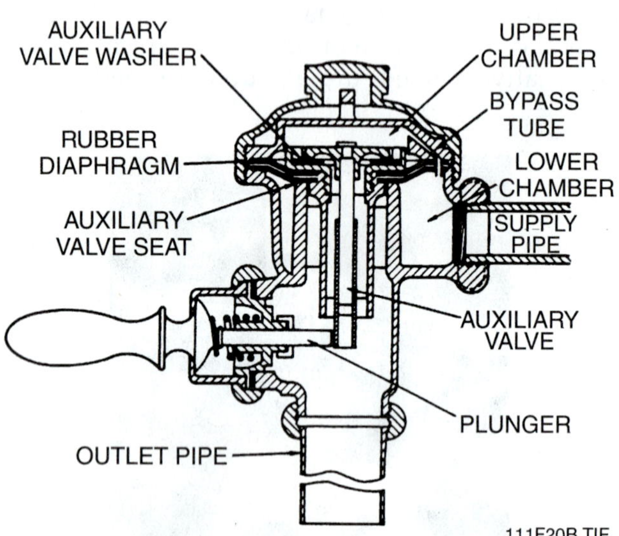

111F20B.TIF

Figure 20 ◆ Flushometer.

hand valve, flush tank, electric eye (which uses an infrared sensor; refer to *Figure 19*), or an automatic flush device. Infrared sensors on urinals detect when a person is standing in front of the urinal and when the person has stepped away, which automatically triggers the flushing action. Infrared urinals are most common in commercial settings, such as airports and movie theaters, where the number of users is very high.

4.6.0 Bidets

Bidets (pronounced bih-*days*) are companion fixtures to water closets (*Figure 21*). They are used for personal hygiene in bathing the external genitals and posterior parts of the body. Because of their low height and overall convenience, they are particularly useful for older persons or persons recovering from an illness who have difficulty using a shower or tub.

111F21.TIF

Figure 21 ◆ Bidet.

The bidet has long been a popular plumbing fixture in European and South American bathrooms, and it is becoming increasingly popular in modern bathrooms in the United States. Bidets are fitted with cold and hot water faucets so that the user can regulate water temperature and rate of flow. The user sits astride the bidet, facing the faucets. The bidet is also fitted with a pop-up stopper that allows the user to let the water accumulate in the bowl. For rinsing the bowl, the bidet has a rim-flushing action similar to that used in a water closet and has a 1½-inch trap.

4.7.0 Food-Waste Disposers

A food-waste disposer (*Figure 22*) is an electric device used to grind food waste into a pulp and discharge it into the drainage system. Food-waste disposers should always be used with running water supplied by the kitchen sink faucet. They are mounted under the kitchen sink (or under one compartment of a two-compartment sink) in place of the basket strainer assembly that would otherwise lead to the drainage fixture. Disposers may be connected so that the waste is discharged into a separate P-trap or to a continuous waste fitting. The fitting allows the wastewater to flow from the

111F22.TIF

Figure 22 ◆ Food-waste disposer.

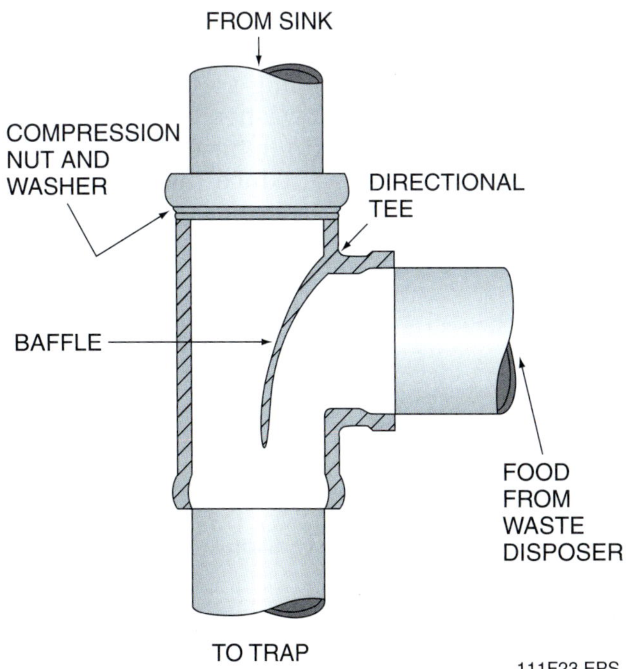

Figure 23 ◆ Directional tee and baffle on a food-waste disposer.

dishwasher through the food-waste disposer. Most plumbing codes require this process for dishwashers. Always check applicable codes in your area for specific requirements.

When installing a food-waste disposer in a double sink, you must use a directional tee with an internal baffle that joins the disposal waste and the waste from the other sink compartment (see *Figure 23*). The tee and baffle prevent the disposal waste from backing up into the other sink compartment.

4.8.0 Domestic Dishwashers

A domestic dishwasher (*Figure 24*) is an electric appliance used in the home for washing dishes.

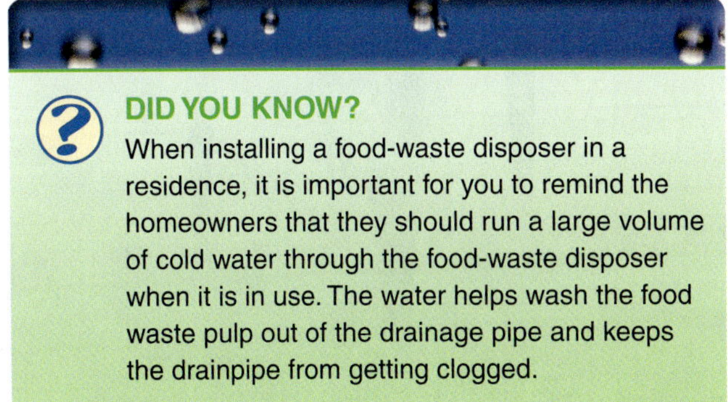

DID YOU KNOW?

When installing a food-waste disposer in a residence, it is important for you to remind the homeowners that they should run a large volume of cold water through the food-waste disposer when it is in use. The water helps wash the food waste pulp out of the drainage pipe and keeps the drainpipe from getting clogged.

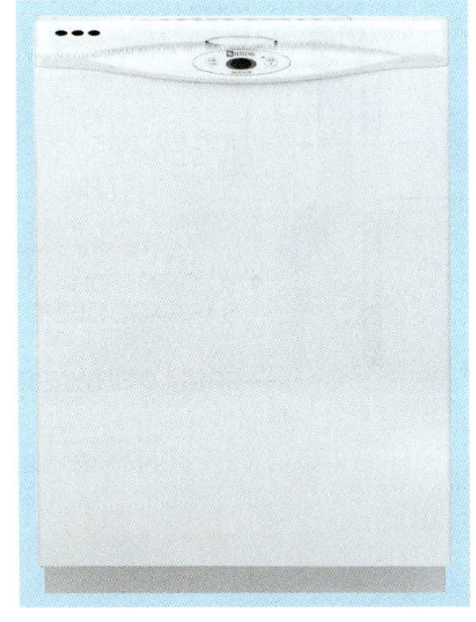

111F24.TIF

Figure 24 ◆ Domestic dishwasher.

The most common sizes of dishwashers are 18-inch and 24-inch units. Most dishwashers installed in homes use a pump-out waste disposal system. This system is installed with a rubber hose or copper tubing to connect the dishwasher waste unit to a dishwasher tailpiece under the kitchen sink basket strainer. If the sink is equipped with a food-waste disposer unit, you can connect the dishwasher into the dishwasher drain connection on the food-waste disposer.

In either case, some codes require a vacuum breaker or air gap assembly (*Figure 25*) in the dishwasher discharge piping so that waste will not back up into the dishwasher. Other codes require that you loop the drain hose above the fixture's flood-level rim to reduce the chance of a backup in the dishwasher if the waste line clogs (see *Figure 26*).

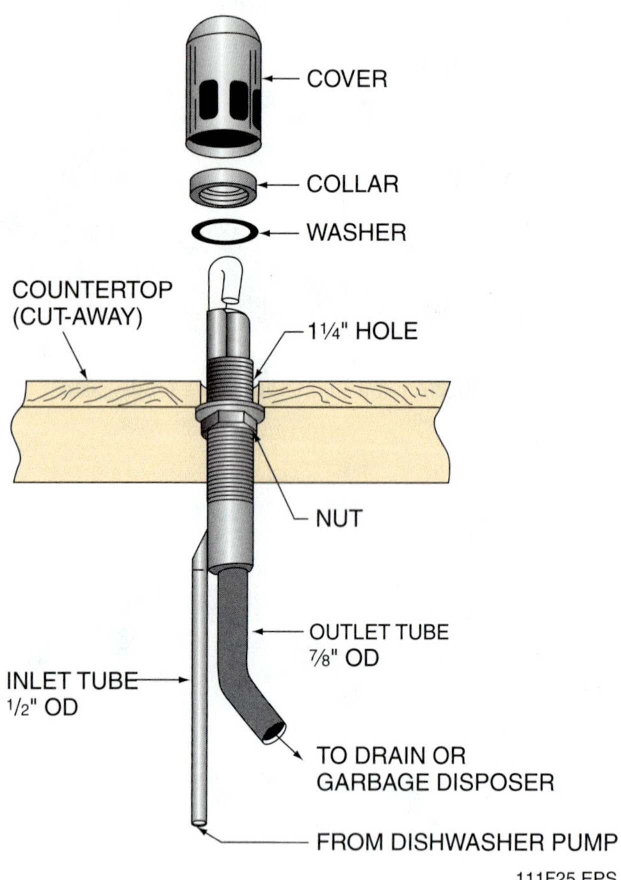

Figure 25 ◆ Air gap assembly for a domestic dishwasher.

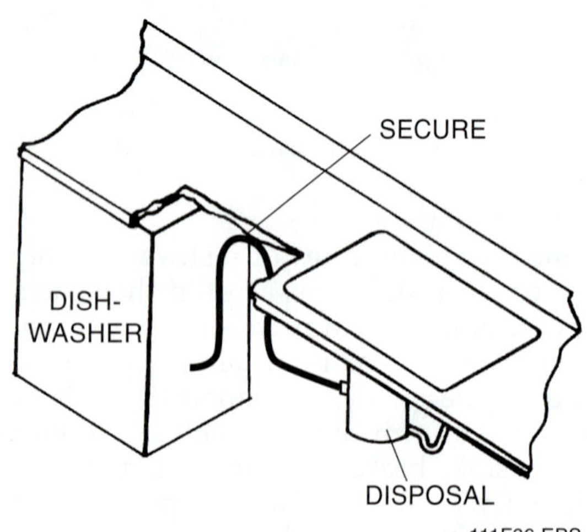

Figure 26 ◆ Looping the dishwasher drain hose.

There is a third option for venting a dishwasher. You may connect a discharge hose from a kitchen sink or domestic dishwasher through a single 1½-inch trap. Use a discharge line that is at least ¾ inch in diameter. Loop the discharge line and fasten it underneath the counter. You can also connect the discharge line to a deck-mounted dishwasher air gap fitting. To do so, use a wye fitting to connect the drain hose between the sink waste outlet and the trap. Always follow all applicable codes for your installation.

4.9.0 Laundry Trays

Laundry trays (*Figure 27*) are tub fixtures fitted with cold water, hot water, and drain connections. Much like a sink but with one or two extra-deep basins, a laundry tray is used for washing clothes and other household items and for receiving waste from automatic clothes washers. They can also be used to store water from automatic clothes washers that are equipped with a water reuse cycle. Laundry trays are usually installed in the laundry room or utility area of homes.

Laundry trays are available in either single-compartment or double-compartment tubs. They are available as floor models set on steel legs or as wall-hung models mounted on a bracket that attaches to the wall. Manufacturers supply both models with the proper mounting hardware.

4.10.0 Service Sinks and Mop Basins

Service sinks and mop basins are plumbing fixtures that are installed in commercial buildings

Figure 27 ◆ Floor-type laundry tray.

More Than Meets the Eye

Plumbers are also called upon to install other fixtures and appliances. Typically, a plumber is required to install the washer box for hooking up the washing machine, as shown in this figure. Plumbers rough in the hot and cold water connections and a drain. A laundry tray provides an alternative method for draining a washing machine.

It is also fairly common for plumbers to install refrigerator icemaker connections. Plumbers may also help install private swimming pool water supply and disposal systems.

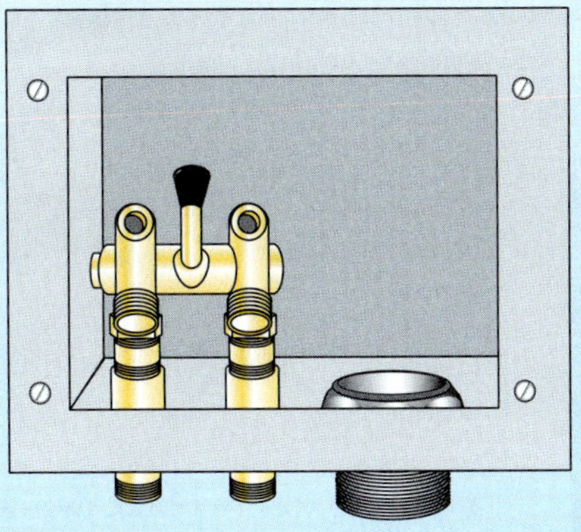

111SA01.EPS

for building maintenance personnel to use (see *Figures 28* and *29*). They are installed in janitors' closets and building maintenance areas.

A service sink, also called a slop sink, has a deep basin that is large enough to hold a scrub pail. These sinks are used to fill and empty scrub pails, rinse mops, and dispose of cleaning water. The sinks are usually made of enameled cast iron, and they come supplied with a standard P-trap.

Mop basins, also called mop receptors, are service sinks that are set on the floor of a janitors' closet. Their low position can make them more convenient to use than service sinks because users need not lift the mop pail or other items in and out of the fixture. Mop basins can be made of enameled cast iron, terrazzo, fiberglass, or ceramic tile.

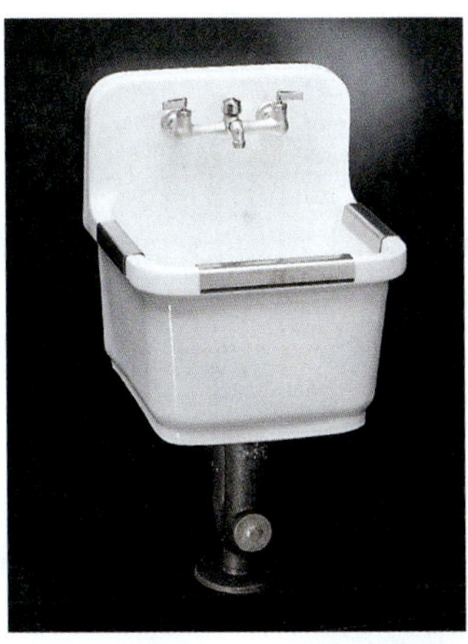

111F28.TIF

Figure 28 ◆ Service sink.

4.11.0 Floor Drains and Floor Sinks

A floor drain (*Figure 30*) is a fixture installed in the floor that drains water and liquid waste into a plumbing system. A floor sink (*Figure 31*) is a type of floor drain that is coated with a porcelain or other easy-to-clean finish. Often fitted with a deep seal trap, a floor sink reduces the amount of contaminant that can collect in the grate and body of the drain. Plumbing codes generally require floor drains in certain facilities that require a high level of sanitation. For example, codes may require floor drains to be installed in commercial kitchens, common laundry rooms in commercial buildings, and bathrooms in commercial buildings containing two or more water closets or urinals.

4.12.0 Drinking Fountains and Water Coolers

Drinking fountains (*Figure 32*) are plumbing fixtures that deliver a stream or jet of water through

111F29.TIF

Figure 29 ◆ Mop basin.

111F30.TIF

Figure 30 ◆ Floor drain.

111F31.TIF

Figure 31 ◆ Floor sink.

111F32.TIF

Figure 32 ◆ ADA-compliant drinking fountain.

a nozzle or bubbler without cooling it. They provide a convenient source of sanitary drinking water. A drinking fountain that includes a water-cooling unit is called a water cooler (*Figure 33*). These units require an electrical hookup to power the water-cooling system. The drinking water is cooled to the desired temperature, usually around 50°F, before it is delivered to the nozzle. Both drinking fountains and water coolers are available in a variety of styles, but the wall-hung fixture is the most common. Both types of fixtures can be manufactured from vitreous china, enameled cast iron, or stainless steel.

To be ADA-compliant, drinking fountains and water coolers must adhere to many requirements. For example, they must have spouts no higher than 36 inches from the ground to the spout out-

111F33.TIF

Figure 33 ◆ ADA-compliant water cooler.

let. In addition, the spouts must be located at the front of the fixture, and there must be clear knee space between the bottom of the fixture's apron and the floor or ground that is at least 27 inches high, 20 inches wide, and 17 to 19 inches deep.

Plumbing codes specify the minimum number of drinking water facilities required for various types of buildings. For example, according to the *National Standard Plumbing Code,* theaters, public auditoriums, restaurants, and office buildings must have one drinking water facility per 1,000 people. Schools must have one per 100 people. As always, check the applicable codes for your particular installation.

Review Questions

Sections 4.5.0–4.12.0

1. A __________ urinal is characterized by the restricted opening over the trap inlet at the base of the urinal.
a. washout
b. wall-hung
c. siphon-jet
d. water-flushed

2. A siphon-jet urinal is designed to flush with __________ gallon(s) per use.
a. 0.5
b. 1
c. 1.6
d. 2

3. __________ are companion fixtures to water closets and are used for personal hygiene.
a. Urinals
b. Bidets
c. Lavatories
d. Bath-shower modules

4. To prevent disposal waste from backing up into the other compartment in a double sink, you must install a __________.
a. vacuum breaker
b. directional tee with an internal baffle
c. strainer with a directional tee
d. waste valve

5. A(n) __________ prevents waste from backing up into the dishwasher.
a. internal baffle
b. air gap assembly
c. directional tee
d. discharge pipe

6. You may connect a discharge hose from a kitchen sink or domestic dishwasher through a single __________ trap.
a. 1-inch
b. 1½-inch
c. 2-inch
d. 2½-inch

7. __________ can be used to store water from washers that are equipped with a water reuse cycle.
a. Slop sinks
b. Service sinks
c. Mop basins
d. Laundry trays

8. A __________ typically has a deep basin that is large enough for a scrub pail.
a. utility sink
b. mop basin
c. laundry tray
d. service sink

9. A __________ is coated with a porcelain or other easy-to-clean finish and is installed in the floor to drain water and liquid waste into a plumbing system.
a. mop basin
b. floor sink
c. laundry tray
d. utility sink

10. __________ are plumbing fixtures that deliver a stream or jet of water through a nozzle or bubbler without cooling it.
a. Drinking fountains
b. Water coolers
c. Sinks
d. Lavatories

Local Plumbing Codes

Local plumbing codes govern construction and installation of drinking fountains and water coolers. Typical plumbing codes require the following sanitary features:

- The bowl of the fountain must be constructed of a nonabsorbent material.
- A mouth protector must be provided over the mouthpiece.
- The drinking water must be delivered at an angle so that no water can fall back onto the nozzle.
- The nozzle must be located above the flood-level rim of the fountain.
- The water supply to the fountain must be entirely separated from the waste.

5.0.0 ◆ FAUCETS

Because faucets are used in all residential, commercial, and industrial buildings, it is important to understand what kinds of faucets are available, how they work, and how and where they are used. Familiarity with the different sizes, shapes, and styles of faucets is essential so that you can choose the right kind for each installation.

Faucets fall into two basic classifications: compression and noncompression. Both types can be further divided into bathroom, kitchen, and utility faucets.

5.1.0 Compression and Noncompression Faucets

Compression faucets control the flow of water by compressing a **seat washer** between the **valve stem** and the **valve seat** (see *Figure 34*). The valve seat is the actual opening in the faucet through which the water flows. The valve stem is the part of the faucet that rises and lowers when you turn the handle to open and close the faucet outlet. The valve stem has an enlarged threaded portion that threads into the faucet body so that the stem moves up or down as the handle is turned. The seat washer is a disc that sits at the end of the valve stem.

Noncompression faucets regulate the flow of water by obstructing the water. When the faucet is open, the openings in two discs are matched, allowing the water to flow through. When the faucet is closed, the discs rotate and cut off the flow of water.

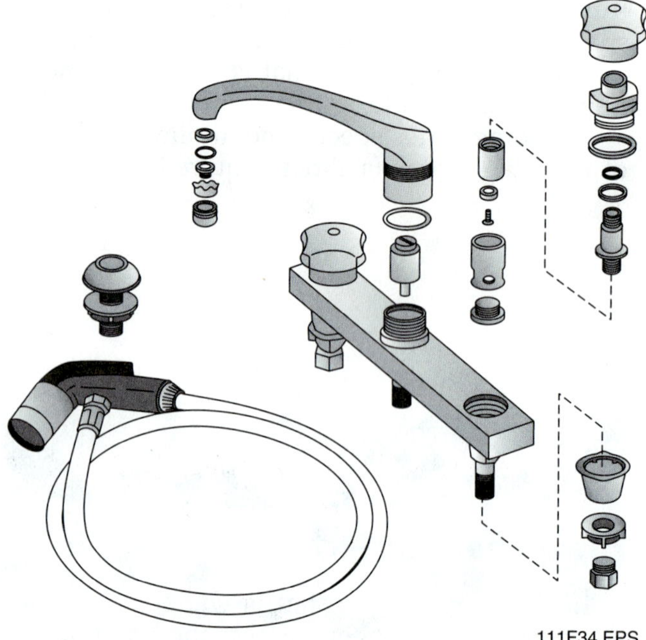

Figure 34 ◆ Exploded drawing of a compression faucet with sprayer hose.

5.1.1 Compression Faucets

Compression faucets for kitchen sinks or bathroom lavatories usually combine two compression valves and a mixer that combines the hot and cold water and delivers it through a mixing spout common to both valves (see *Figures 35* and *36*). A plunger located between supply valves controls the drain valve (see *Figure 37*). Some older dwellings may have sinks and lavatories that use individual compression valves and faucets for the hot and cold water. These faucets are installed in pairs (one for hot water and one for cold water).

Figure 35 ◆ Compression faucet for the kitchen with sprayer hose.

111F36.TIF

Figure 36 ◆ Compression faucet for the bathroom.

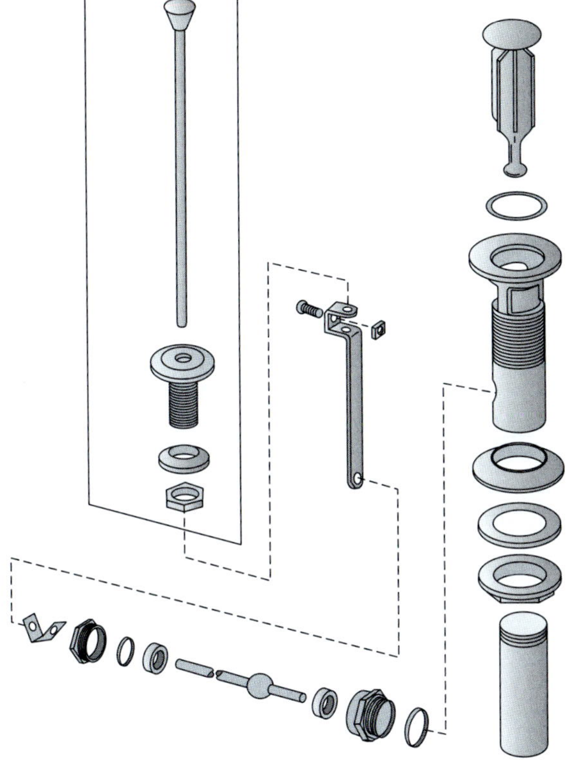
111F37.EPS

Figure 37 ◆ Exploded drawing of a drain valve.

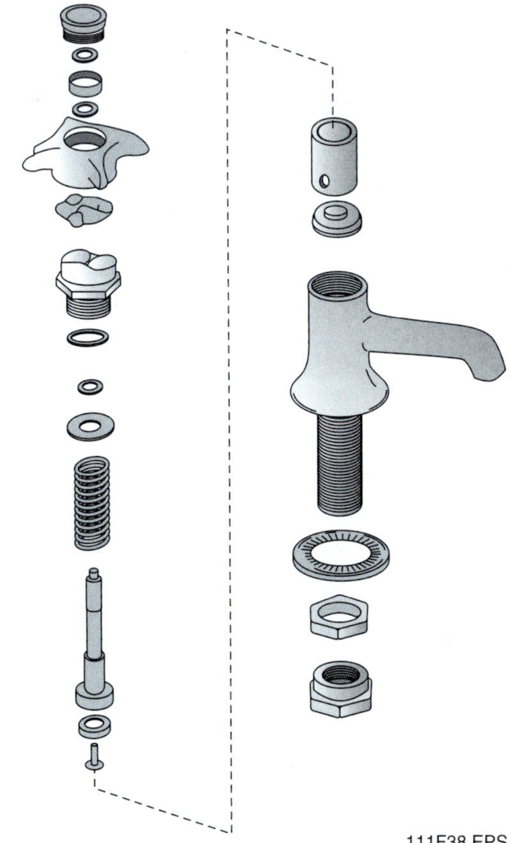
111F38.EPS

Figure 38 ◆ Exploded drawing of a self-closing compression faucet.

In some cases, it may be more practical to install a self-closing compression faucet to conserve water (see *Figure 38*). The self-closing faucet automatically closes the valve as soon as the handle is released. A self-closing faucet may present some problems over time, though. It has a tendency to receive too little water flow, which can cause lime and other minerals to build up in the system. As a result, the faucet begins to malfunction.

5.1.2 Noncompression Faucets

Noncompression faucets do not rely on the compression of a washer to control the flow of water. Water flow is controlled by rotating balls, rotating cylinders, or rotating discs. Because these faucets do not use the repeated compression of a washer, they usually require less maintenance than compression faucets.

The most common styles of noncompression faucets are washerless, single control, and push button. Washerless faucets control water flow by matching openings in two separate discs located in the valve. While one disc remains stationary, the other disc rotates with the band control or lever (see *Figure 39*). Although these faucets are called by the term *washerless,* you may use an O-ring to prevent water leakage from around the valve assembly.

Single-control faucets (see *Figure 40*) control both the hot and cold water with one hand control. These faucets have become popular because of their convenience. The most popular designs include the rotating ball faucet and the rotating cylinder faucet. The first type uses a rotating ball in place of a compression washer. The ball is made of metal or plastic and contains several different sizes of holes or openings. As the user rotates the single-lever control, these openings align with the hot and cold water ports. The position of the lever controls water mixture and flow. Moving the lever

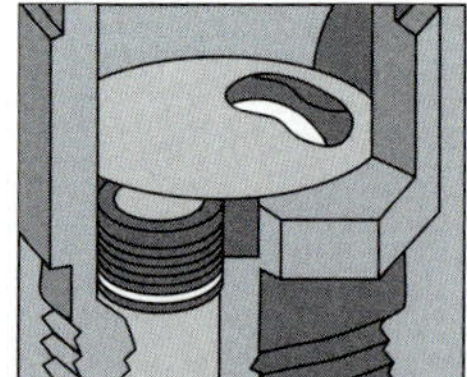

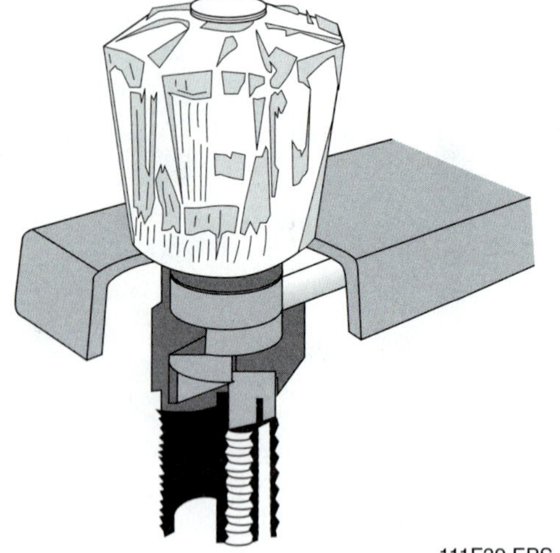

Figure 39 ◆ How a washerless faucet controls water flow.

Figure 40 ◆ Single-control faucet.

to the right causes cold water to flow. Moving it to the left causes hot water to flow. Pushing the lever back increases the rate of flow.

The rotating cylinder faucet, also called a cartridge faucet, controls the water temperature and the rate of water flow by using a balancing piston. The piston adjusts its position when pressure of either hot or cold water changes (see *Figure 41*). As you make changes in the water temperature and pressure by either turning or pulling the single handle, the balancing piston moves to the right (cold) or to the left (hot). This ensures that the flow of water remains constant.

Another type of single-control faucet uses a ceramic disc to control the flow of water. The ceramic disc is sealed inside the cartridge. Because

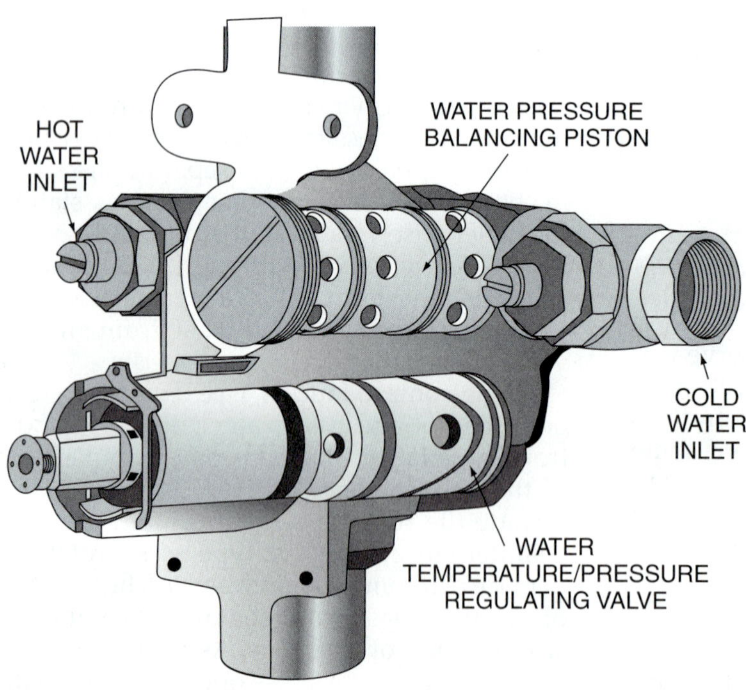

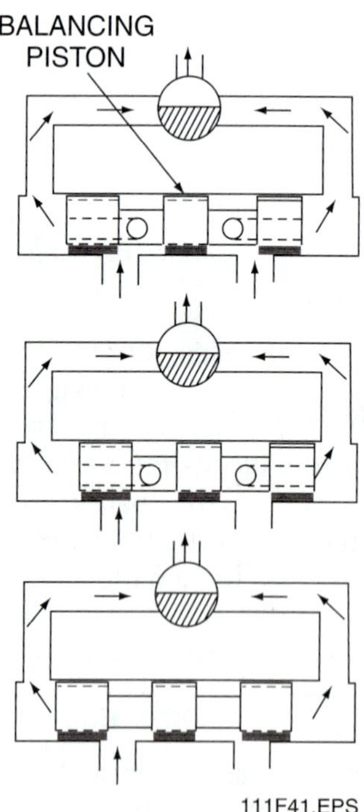

Figure 41 ◆ Rotating cylinder faucet.

ceramic is very durable and unlikely to erode, it provides long-wearing, trouble-free service.

5.2.0 Kitchen and Bathroom Faucets

Faucets and accessories for kitchen sinks, showers, and bathroom lavatories and fixtures are made of many different materials and come in a variety of decorator shapes and sizes. Materials used for making faucets include brass, white metallic alloys, and different types of plastics, such as nylons and acrylics.

The distance between the hot and cold water inlets on faucets varies depending on where the openings are in the sink or lavatory. This distance, usually measured in even numbers, ranges from 4 to 12 inches. The most common center for a kitchen faucet is 8 inches (see *Figure 42*). The most common centers for bathroom faucets are 4 or 8 inches.

5.2.1 Combination Shower and Bath Fittings

A combination shower and bath fitting (*Figure 43*) holds both a faucet and showerhead. A **diverter,** located in the spigot or spout, sends the water to the showerhead. When you lift the diverter, it blocks the flow of water to the spout and forces the water up into the showerhead. Dual hot and cold valves control water pressure and temperature. Some units use a single-control pressure-mixing valve to adjust the water's flow rate and temperature (see *Figure 44*).

Another type of combination shower uses a flexible water line. A flexible water line allows you to raise, lower, or turn the showerhead in any direction. The head can either be handheld or placed on a wall-mounted bracket (see *Figure 45*).

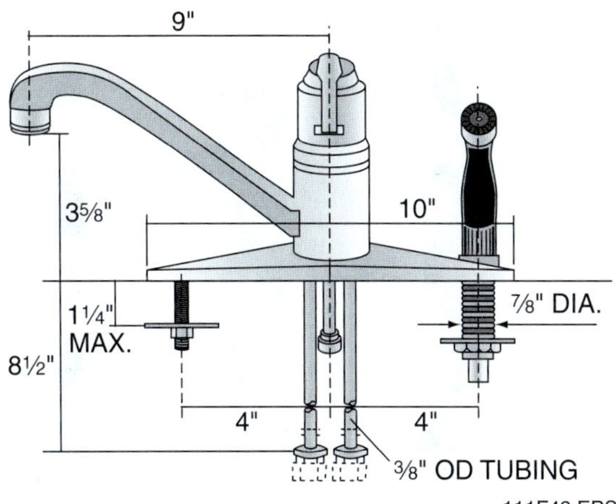

111F42.EPS

Figure 42 ◆ Kitchen faucet with an 8-inch center.

111F43.TIF

Figure 43 ◆ Combination shower and bath fitting.

111F44.TIF

Figure 44 ◆ Single-control pressure-mixing valve.

5.3.0 Utility Faucets

Utility faucets are found in boiler rooms and laundry rooms and on a building's outside walls. They are usually manufactured from extra-grade brass, white metallic alloy, or thermoplastic materials. Most utility faucets are not plated because appearance is generally not important for these.

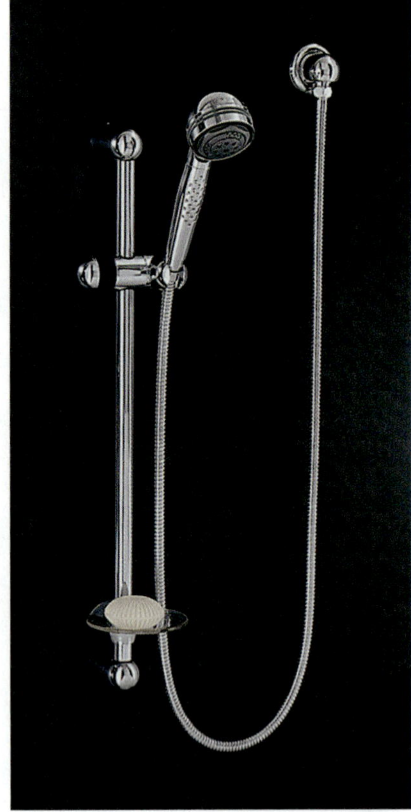

111F45.TIF

Figure 45 ◆ Combination fitting with a flexible water line.

Common utility faucets include laundry faucets used for laundry tubs, utility sinks, service sinks, and mop sinks, as well as sediment and boiler drains (see *Figure 46*).

The water outlet end for a utility faucet, referred to as a **bibb,** can be one of two types: plain (threadless) or hose (threaded) (see *Figure 47*). The bibb is angled to direct the water downward. Most common utility compression faucets have female iron pipe threads on the end that attach to the water pipe during installation. The plain bibb faucet has a flange designed to fit flush against siding or wall covering. It has hidden female pipe fittings on the water inlet end.

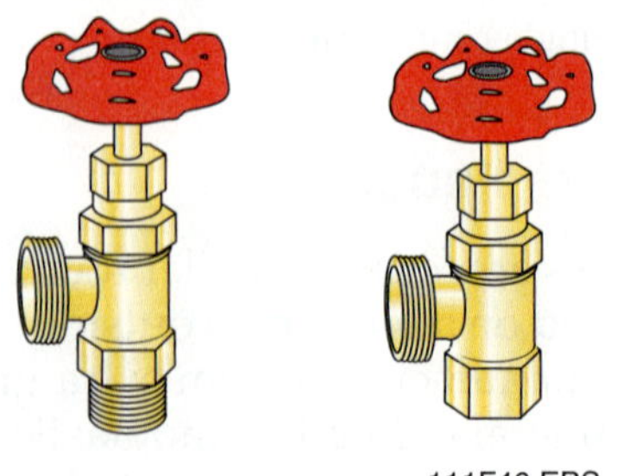

111F46.EPS

Figure 46 ◆ Boiler drains.

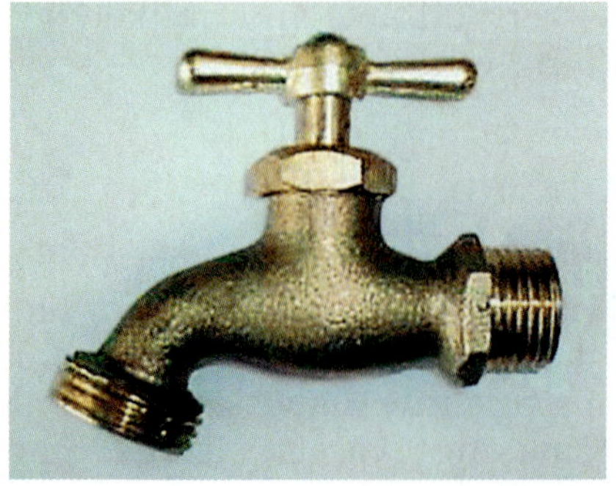

111F47.TIF

Figure 47 ◆ Threaded hose bibb.

Another term for a hose bibb is a sill cock. In commercial applications, sill cocks are often specified as frost-resistant wall hydrants (*Figure 48*). The weatherproof yard hydrant, or freeze-proof compression faucet (*Figure 49*), is used to provide water access outside a building. These faucets attach to the water pipe with male threads. The faucets are run through the outside wall of the structure, leaving only the spigot, or faucet head, exposed to the weather. When you turn off the water, it drains from the spigot through the extension tube to the valve seat, which is inside the

111F48.TIF

Figure 48 ◆ Wall hydrant.

111F49.EPS

Figure 49 ◆ Yard hydrants.

building. It is important to install the yard or freeze-proof faucet properly so that it does not freeze from exposure to the weather. The valve seat should be well within the protection of the structure and installed according to the manufacturer's specifications. On new installations, vacuum breakers are required on hose bibbs for backflow prevention.

Review Questions

Section 5.0.0

1. __________ faucets control the flow of water by pressing a seat washer between the valve stem and the valve seat.
   a. Valve
   b. Rotating ball
   c. Cylinder
   d. Compression

2. __________ faucets have a tendency to receive too little water flow, which can cause lime and other minerals to build up in the system.
   a. Plain bibb
   b. Compression
   c. Self-closing
   d. Weatherproof

3. A rotating ball faucet uses a rotating ball in place of a __________.
   a. valve
   b. compression washer
   c. disc
   d. piston

4. Washerless faucets control water flow by matching openings in two separate discs located in the piston.
   a. True
   b. False

5. The rotating cylinder faucet, also called a __________ faucet, controls the water temperature and the rate of water flow by using a balancing piston.
   a. single-control
   b. rotating ball
   c. cartridge
   d. ceramic disc

6. The distance between the hot and cold water inlets on faucets usually ranges from __________ inches.
   a. 1 to 9
   b. 2 to 10
   c. 3 to 11
   d. 4 to 12

7. A __________, located in the spigot or spout, sends the water to the showerhead.
   a. diverter
   b. divider
   c. fixture riser
   d. water line

8. Utility faucets are usually manufactured from __________, white metallic alloy, or thermoplastic materials.
   a. gray iron
   b. sheet steel
   c. copper
   d. extra-grade brass

9. There are two types of water outlet ends for a utility faucet: a plain bibb and a(n) __________ bibb.
   a. hose
   b. flanged
   c. threadless
   d. angled

10. Weatherproof yard hydrants run through the outside wall of the structure, leaving only the __________ exposed to the weather.
    a. sill cock
    b. spigot
    c. hose bibb
    d. valve seat

Summary

Fixtures in plumbing installations are devices that receive water from a water supply line. Fixtures are used in every kind of building—residential, commercial, and industrial. In this module, you have learned how these fixtures are manufactured, how they operate, the advantages and disadvantages of each, and how they connect with water supplies and waste outlets.

You also have learned the basics about faucets. It is important to be able to identify various types of faucets, how they operate, and which faucets are used for different plumbing applications.

Notes

Trade Terms Quiz

Fill in the blank with the correct trade term that you learned from your study of this module.

1. _____________ is a durable material used for fixtures that is created by combining porcelain in its liquid state with other materials, such as steel or cast iron.
2. The _____________, which connects the trap to the vent pipe, conveys only wastewater to the building drain.
3. When reinforced with plastic, _____________ can be used to create one-piece showers and bathtubs that are lightweight and easy to install.
4. The actual opening in a faucet through which the water flows is the _____________.
5. A(n) _____________ is a basin designed for the primary purpose of washing hands and face.
6. One of the materials used for the base of shower stalls is _____________, small pieces of marble or other hard stone embedded in a mortar base.
7. A(n) _____________ has the drain on the right end of the tub as you face the length of its finished side.
8. Although more expensive than other materials, _____________, also called _____________, has many benefits when used in plumbing fixtures, including its glass-like surface, which is strong and easy to clean.
9. The device that controls the flow of water from the water closet flush tank into the bowl is called the _____________.
10. Because the turnover in users of urinals is so rapid, urinals are usually fitted with _____________, also called _____________, which are flushing devices that are connected directly to the water supply and discharge a predetermined amount of water.
11. The _____________ is the threaded portion of the faucet that rises and lowers when you turn the handle to open and close the faucet outlet.
12. Common _____________, also called _____________, are receptacles or devices that receive water from a water supply line. They include toilets, bathtubs, and water coolers.
13. Much like a sink but with one or two extra-deep basins, a(n) _____________ is used for washing clothes and other household items and for receiving waste from automatic clothes washers.
14. The device that redirects the flow of water in a combination faucet, such as a combination bath and shower fitting, is called a(n) _____________.
15. One method of installing a water closet is by using a floor mount, which involves securing the water closet directly to the floor and connecting it to the drainage system pipe by a closet _____________.
16. The water outlet end for a utility faucet, called a(n) _____________, is angled downward and is available in a threadless or hose style.
17. A fixture that is used to draw water from a pipe is called a(n) _____________.
18. To install a lavatory, you must have a(n) _____________ or _____________ to allow air to escape from, and provide the circulation of air within the DWV system.
19. A(n) _____________ has the drain on the left end of the tub as you face the length of its finished side.

20. When installing lavatories, you should ensure that the lavatory's _______________ is 31 inches above the floor, which allows excess water to spill over onto the countertop and floor.

21. The _______________ is a disc that sits at the end of the valve stem.

Trade Terms

- Bibb
- China
- Diverter
- Faucet
- Fiberglass
- Fixture
- Flange
- Flood-level rim
- Flushometer
- Flushometer valve
- Flush valve
- Laundry tray
- Lavatory
- Left-hand bathtub
- Plumbing fixture
- Porcelain enamel
- Right-hand bathtub
- Seat washer
- Terrazzo
- Valve seat
- Valve stem
- Vent pipe
- Vent system
- Vitrified porcelain
- Waste pipe

Profile in Success

Thomas J. Swafford

Ivey Mechanical Company, LLC
Training Manager, Corporate Office
Kosciusko, Mississippi

Thomas Swafford was born and raised in Jackson, Mississippi, and after high school, he served as a corpsman in the U.S. Army medical service. When he returned from the army, he entered Mississippi State University on a scholarship awarded by the Mississippi Valley Gas Company. He was transferred by the gas company to Kosciusko, Mississippi, and was unable to complete his education at that time. Some 30 years later, he completed his education at Belhaven College in Jackson, where he received his bachelor's degree in business, with a 3.87 grade point average.

How did you become interested in this industry?
My first job in construction, which was also my first job period, was with Mississippi Valley Gas Company, working on underground gas utility construction. I was promoted to a job that handled transmission lines in seven counties. I developed plans for the distribution lines in each community, along with handling sales and service for this division. After 11 years with this company, I joined Ivey Plumbing and Electrical Company and served in many corporate capacities. Thirty-three years later, I am training manager in the corporate office.

What path did you take to your current position?
Mr. Kermit Ivey was the founder of Ivey Plumbing and Electrical, and when his son J. Marlin Ivey expanded the company, I was part of that transition. As Ivey Mechanical and Electrical, Inc., grew and we began to bid jobs out of state, we needed licenses to bid this work. I was asked to take the tests for these licenses, and I ordered the books to study and prepare for these tests. I had already passed the Mississippi master's test as well as the tests for plumbing; heating, ventilating, and air conditioning (HVAC); electrical; and pipe fitting. Passing the Mississippi tests qualified me to take other state license exams. As we continued to grow, I continued to add additional licenses, and today I have my master's and contractor's licenses in 23 states, and I have just taken exams in West Virginia, Indiana, and Illinois.

What does it take to be successful in your trade?
If one is willing to learn, one can learn a trade and eventually earn a good income. You have to be willing to work and willing to stay in high school and graduate. You must have a high school diploma or general education development (GED) certificate to open doors for your future education and training. Though you may start as a laborer on a job, if you are disciplined and have a desire to be trained, you have a great opportunity to learn many construction skills. Success is up to the individual.

Ivey Mechanical offers a four-week boot camp; we take trainees through the *Core Curriculum* of the *Contren® Learning Series* and through essential aspects of working in the field. The classroom is a working shop, and trainees conduct performance testing while they are in camp. When trainees arrive on the job site, working in their trade, a performance test is conducted with each of the modules in the curriculum being taught. Trainees receive a $2,000 bonus if they finish the written program in two years.

What are some things you do on the job?
I train and certify instructors, and I handle advertising to hire new apprentices. Our company is an equal opportunity employer and follows the rules for affirmative action. I am also required by many states to complete continuing education each year to renew my license.

Here at Ivey Mechanical Company, LLC, I have a large group of certified trainers instructed by me as a master trainer with NCCER, which enables them to instruct apprentices. We have apprenticeship programs in Florida, South Carolina, North Carolina, Georgia, Mississippi, Alabama, and Texas. I oversee the procedure for hiring at these locations and the training for these apprentice programs.

What is the most interesting aspect of your profession? What makes your trade stand out from others?
The most rewarding aspect in my job is seeing someone finish his or her apprenticeship education and seeing the smile that comes with the satisfaction of a completed goal. A teacher once told me that she knew when I understood and solved a problem, and I smiled to hear her say that. Now that I teach others, it is wonderful to see an apprentice understand and comprehend his or her work, and to see the satisfaction on that person's face.

I would choose the plumbing trade because plumbers are the first ones on the job and the last ones to leave the job site. They are able to read specifications and plans and instructions, and they have a good general knowledge of construction. Our plumbing curriculum includes sheet metal and pipe fitting, which enables plumbers to cross over to these other trades easily. This is important to us as a merit shop contractor (providing us uninterrupted time on the job) because we have employees (plumbers who are trained in three trades) with the ability to do it all, so to speak.

What do you like most—and least—about your job?
It is great to see our company offer a training program that can help people who have not had the opportunity to get a college education but can, through hard work, make a very comfortable living.

It saddens me the most to see someone with great potential, but who lacks the desire to work or to learn.

The most exciting project that I was associated with was the F-117A Stealth Fighter Bomber complex in New Mexico. I was the senior project manager on that very successful design-build project. I was also the senior project manager on the B-2 Bomber complex at Whitman Air Force Base. Later, it was rewarding to see the results of the two airplane groups in action in Desert Storm and Iraq. As senior project manager on the Shriner's Children's Hospital in Houston, Texas, I saw construction that eventually would be able to meet myriad health needs for children.

What would you say to someone entering the trade today?
I would encourage them to get into an apprentice program offered by their company or take classes in their desired trade at a local community college. Stay committed and take every opportunity to learn. Construction companies are always looking for qualified tradesmen.

What can an apprentice expect to earn in his/her first years on the job in your area? What can he/she expect to earn after 10-plus years in the industry?
Apprentices make $8.50 per hour and work up to $17 per hour for a beginning journeyman in plumbing and $16 per hour for a beginning journeyman in sheet metal.

As a journeyman, you can pass the local tests and be eligible for a foreman position. That salary depends on your ability to manage people, read construction drawings, and follow instructions. The person who is able to read specifications and understand the overall project is the person given the leadership position. There is quite a range of pay for the foreman position; one can make from $22 to $30 per hour. The next step would be the superintendent of the project and then, as his/her skills develop, ownership in one's own company.

Trade Terms Introduced in This Module

Bibb: A faucet with an outlet at a downward angle, generally used for hose connections on the outside of a building.

China: A mixture of fine clay, quartz, feldspar, and silica that is heated to 2,600°F (1,426°C) to create a nonporous, glass-like finish. Also referred to as vitrified porcelain.

Diverter: A device that redirects the flow of water, as in a combination shower and bath fitting. When the diverter is engaged, water flows through the showerhead instead of through the spigot.

Faucet: A fixture that is used to draw water from a pipe.

Fiberglass: Spun filaments of glass woven into yarn or roving (twisted) strands and textile materials such as cloth or mats. Fiberglass cloth saturated with a plastic (vinyl resin or epoxy) can be pressed into molds to make products such as bathtubs and shower enclosures.

Fixture: A receptacle or device which is either permanently or temporarily connected to the water drainage system of the premises, and demands a supply of water therefrom, or discharges used water, liquid-borne waste materials, or sewage either directly or indirectly to the drainage system of the premises, or which requires both a water supply connection and a discharge to the drainage system of the premises. Plumbing appliances as a special class of fixture are further defined. Also referred to as plumbing fixture.

Flange: A rim on one end of a length of pipe that provides a connection point to another length of pipe or piece of equipment, such as a water closet. Bolts or studs with nuts are used to hold two flanges together, with some type of gasket between them. The flange is soldered or solvent-welded to the drainpipe stub to connect the fixture to the drain, waste, and vent system.

Flood-level rim: The edge of the receptor or fixture which water overflows.

Flushometer: A device which discharges a predetermined quantity of water to fixtures for flushing purposes and is closed by direct water pressure or other mechanical means. Also referred to as flushometer valve.

Flushometer valve: Another term for flushometer.

Flush valve: A device located at the bottom of a tank for flushing water closets and similar fixtures.

Laundry tray: A fixed tub, usually installed in the laundry room or utility area of homes. It is used for washing clothes and for receiving wastewater from automatic clothes washers.

Lavatory: A basin designed for installation in bathrooms and other locations, primarily for washing the hands and face.

Left-hand bathtub: A bathtub that has the drain on the left end of the tub as you face its length.

Plumbing fixture: Another term for fixture.

Porcelain enamel: A liquid used to coat materials such as steel and cast iron to create a vitrified porcelain, or china, finish.

Right-hand bathtub: A bathtub that has the drain on the right end of the tub as you face its length.

Seat washer: A disc that fits at the end of the valve stem and is used to regulate the flow of water through a faucet.

Terrazzo: A type of floor surface (commonly used for showers and other plumbing fixtures) made by embedding small pieces of marble or other hard stone in a mortar base. When hardened, the surface is ground and polished smooth, although it may be left slightly rough to create a nonslip surface.

Valve seat: An opening in a faucet through which the water flows. When the handle is turned to open the faucet, the valve stem raises the seat washer up and clears the opening for water to flow.

Valve stem: The part of a faucet that rises and lowers when the handle is turned; it opens or closes the faucet outlet.

Vent pipe: A pipe, or pipes, installed to provide a flow of air to or from a drainage system, or to provide a circulation of air within such a system to protect trap seals from siphonage and back pressure. Also referred to as vent system.

Vent system: Another term for vent pipe.

Vitrified porcelain: A mixture of fine clay, quartz, feldspar, and silica that is heated to 2,600°F (1,426°C) to create a nonporous, glass-like finish. Also referred to as china.

Waste pipe: A pipe which conveys only waste.

Additional Resources and References

ADDITIONAL RESOURCES

This module is intended to present thorough resources for task training. The following reference works are suggested for further study. These are optional materials for continued education rather than for task training.

ANSI 117A.1-1980. New York: American National Standards Institute.

Plumbing Installation and Design, second edition, 1987. L.V. Ripka. Homewood, IL: American Technical Publishers.

REFERENCES

ADA Standards for Accessible Design, 28 CFR Part 36, 1994. Washington, DC: Department of Justice.

Architectural Handbook, 1979. Alfred M. Kemper. New York: John Wiley & Sons.

Dictionary of Architecture and Construction, 1993. Cyril M. Harris, ed. New York: McGraw-Hill.

Kohler website, www.kohler.com. "Kohler Tech Terms," www.serviceparts.kohler.com/glossA.html. Reviewed September 2003.

National Standard Plumbing Code, 2003. Falls Church, VA: Plumbing-Heating-Cooling Contractors–National Association.

National Standard Plumbing Code, 2003. Chapter 1, Definitions (definitions of *fixture, flood-level rim, flushometer, flush valve, vent pipe,* and *waste pipe*). Falls Church, VA: Plumbing-Heating-Cooling Contractors–National Association.

NCCER CURRICULA — USER UPDATE

NCCER makes every effort to keep its textbooks up-to-date and free of technical errors. We appreciate your help in this process. If you find an error, a typographical mistake, or an inaccuracy in NCCER's curricula, please fill out this form (or a photocopy), or complete the online form at **www.nccer.org/olf**. Be sure to include the exact module ID number, page number, a detailed description, and your recommended correction. Your input will be brought to the attention of the Authoring Team. Thank you for your assistance.

Instructors – If you have an idea for improving this textbook, or have found that additional materials were necessary to teach this module effectively, please let us know so that we may present your suggestions to the Authoring Team.

NCCER Product Development and Revision

13614 Progress Blvd., Alachua, FL 32615

Email: curriculum@nccer.org
Online: www.nccer.org/olf

❑ Trainee Guide ❑ AIG ❑ Exam ❑ PowerPoints Other ______________________________

Craft / Level: Copyright Date:

Module ID Number / Title:

Section Number(s):

Description:

Recommended Correction:

Your Name:

Address:

Email: Phone:

Introduction to Drain, Waste, and Vent (DWV) Systems
02112-05
P – 6B
TRANSFER TO CLARIFIER
PUMPS

02112-05

Introduction to Drain, Waste, and Vent (DWV) Systems

Topics to be presented in this module include:

Overview

To design, install, and maintain drain, waste, and vent (DWV) systems, plumbers must be familiar with the factors that affect them. Sanitary drainage systems include the pipes inside the building, the drain pipe buried outside the building, and the public sewer. Knowing how drains, fittings, vents, and pipe move waste out of a building enables plumbers to prevent system malfunctions.

Fixture drains connect fixtures to the building's DWV piping system. Traps and vents protect people from pathogens, odors, and sewer gases. While codes govern specific requirements for these drains and traps, plumbers must understand the installation requirements, including dimensions, and they must be able to troubleshoot trap failure. Plumbers install vents to provide a free flow of air and to maintain equalized pressure throughout the drainage system. To properly size drains and vents, plumbers must thoroughly understand the mechanics of fluid flow in the pipes.

Drainage systems are categorized as storm water drains and building drains. Because these systems rely on gravity to move solid and liquid waste, plumbers install them on a slope, or grade. The local municipality installs, maintains, and controls the public or municipal sewer systems. These systems collect waste, treat it at a sewage plant, and then discharge it back into the ecosystem. Properly designed, installed, and maintained DWV systems are essential to public safety.

Focus Statement

The goal of the plumber is to protect the health, safety, and comfort of the nation job by job.

Code Note

Codes vary among jurisdictions. Because of the variations in code, consult the applicable code whenever regulations are in question. Referencing an incorrect set of codes can cause as much trouble as failing to reference codes altogether. Obtain, review, and familiarize yourself with your local adopted code.

Objectives

When you have completed this module, you will be able to do the following:

1. Explain how waste moves from a fixture through the drain system to the environment.
2. Identify the major components of a drainage system and describe their functions.
3. Identify the different types of traps and their components, explain the importance of traps, and identify the ways that traps can lose their seals.
4. Identify the various types of drain, waste, and vent (DWV) fittings and describe their applications.
5. Identify significant code and health issues, violations, and consequences related to DWV systems.

Key Trade Terms

Adapter
Back pressure
Backpressure backflow
Branch interval
Building sewer
Capillary attraction
Cleanout
Crown weir/trap weir
Double ¼ bend
Double trapping
Drainage fitting
DWV system
Elevation
Evaporation
Fall
Fixture drain
Grade
Hydraulic gradient
Interceptor
Inverted wye
Long sweep ¼ bend
P-trap
Pipe scale
Run
S-trap
Sanitary combination
Sanitary fitting
Sanitary increaser
Sanitary upright wye
Sanitary wye
Short sweep ¼ bend
Side inlet
Siphonage
Slope (percent of grade)
Sludge
Stack
Test tee
Velocity
Vent branch (branch vent)
Vent ell
Vent tee
Weir (trap or crown)

Required Trainee Materials

1. Appropriate personal protective equipment
2. Pencils and paper
3. Copy of local adopted code

Prerequisites

Before you begin this module, it is recommended that you successfully complete *Core Curriculum; Plumbing Level One,* Modules 02101-05 through 02111-05.

This course map shows all of the modules in the first level of the *Plumbing* curriculum. The suggested training order begins at the bottom and proceeds up. Skill levels increase as you advance on the course map. The local Training Program Sponsor may adjust the training order.

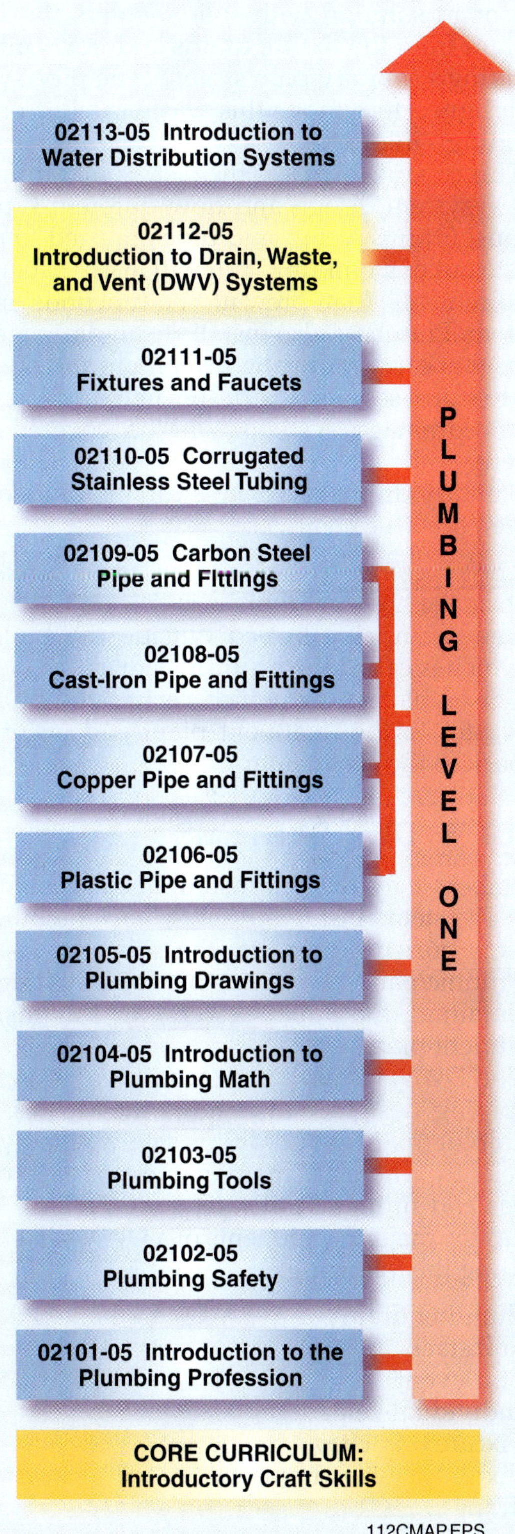

1.0.0 ◆ INTRODUCTION

As a plumber, you need to understand how drainage systems work. This module describes the flow of waste products from a building, to the treatment facilities, and back into the ecosystem (streams, rivers, and lakes). This cycle begins at the fixture drains that connect to the drain, waste, and vent (DWV) piping system. Waste, either liquid or in solution with solids, enters the **DWV system** from the fixture drains and flows into the building's sanitary pipe system. The pipe system is designed to remove this waste safely from the building's interior.

This module explains the factors that influence DWV system design and how different types of drains, fittings, vents, and pipe are used to move waste out of a building. You will learn installation requirements that prevent malfunctions in the system. Plumbers also install storm drainage systems that carry rainwater from roofs and open areas to storm sewers. The design and installation of storm drainage systems will be covered elsewhere.

Sanitary drainage systems can be divided into three parts (see *Figure 1*):

- The pipes inside the building, usually referred to as the DWV system
- The drain pipe buried outside the building, which is called the **building sewer**
- The public sewer, which carries the building wastes to the treatment plant and eventually back to the ecosystem

2.0.0 ◆ DWV SYSTEMS

Plumbers may design, install, and maintain the DWV systems inside buildings and the building sewers buried outside on the property. Usually, the municipality is responsible for installing and maintaining the public sewers, lift stations, and treatment plants.

The DWV system inside a building is a circuit of piping designed to remove the wastes from plumbing fixtures and drains safely, reliably, and efficiently. There are many names for each of the pipes and fittings in this network. *Figure 2* illustrates the major components of a DWV system, including the following:

- Building drain
- Soil stack
- Stack vent
- Individual vents
- Fixture branches
- Fixture drain or trap arm
- Traps
- Bends or elbows
- Tees
- Wyes
- Couplings, reducers, and adapters

3.0.0 ◆ FIXTURE DRAINS

Fixture drains connect fixtures to the building's DWV piping system. Many fixtures have drains that strain the wastewater before it enters the drainage piping. Examples of fixture drains include a basket strainer for a kitchen sink, PO (pop-up) plugs for lavatories, and other strainers for bidets and showers (see *Figure 3*).

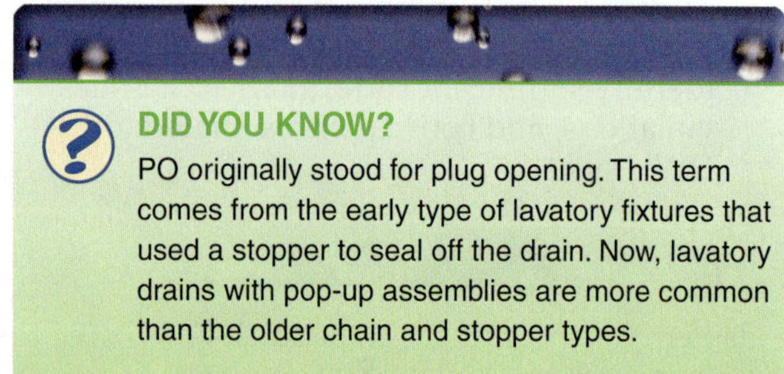

DID YOU KNOW?

PO originally stood for plug opening. This term comes from the early type of lavatory fixtures that used a stopper to seal off the drain. Now, lavatory drains with pop-up assemblies are more common than the older chain and stopper types.

4.0.0 ◆ TRAPS

Traps and vents protect the safety of homes and other buildings. The water seal in a trap protects people from airborne pathogens (germs), foul odors, and potentially explosive sewer gases.

Traps are important components of the DWV piping system. As you have already learned, a trap is a fitting or device that provides a liquid seal of 2 to 4 inches. This seal prevents sewer gases from leaking back into the building but should not affect the flow of sewage or wastewater through the drain. As will be explained later in this module, trap seals are protected by vents.

Modern codes require every plumbing fixture to have a trap that protects the fixture and users from the sanitary drain system. Some fixtures, such as water closets and many urinals, have integral or built-in traps.

A fixture trap is a vital part of any DWV system. To function properly, a fixture trap must flush completely, be self-cleaning, have a smooth interior waterway, and be accessible for cleanout. The depth of the seal and the amount of water normally held in the trap are important factors in trap design. The fixture trap, which creates a water seal, requires a vent system to protect it from **siphonage, back pressure,** wind, and aspiration. Back pressure is often referred to as **backpressure backflow.**

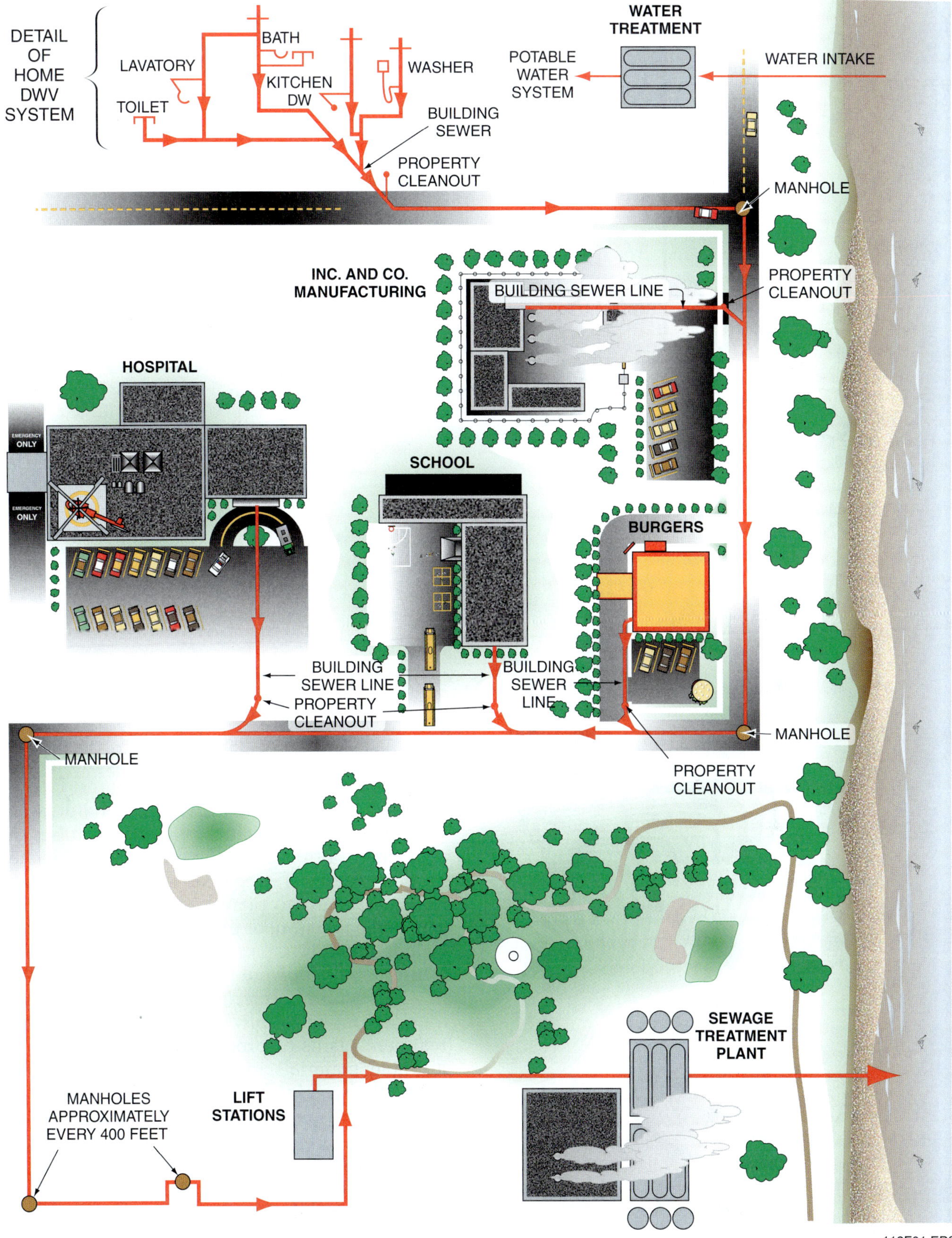

Figure 1 ◆ Overview of a typical community sewer system.

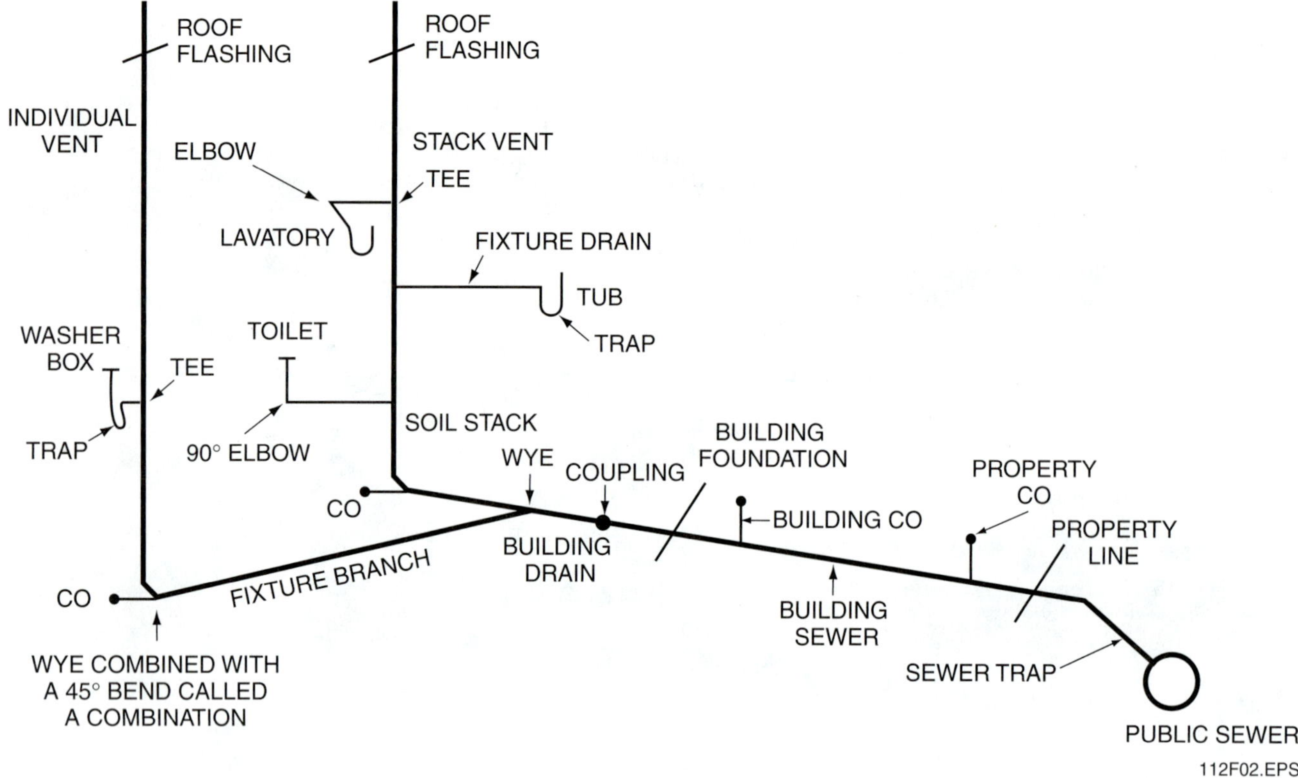

Figure 2 ◆ Major components of a DWV system.

Figure 3 ◆ Fixture drains.

4.1.0 Types of Traps

The P-trap is the most commonly used trap. **P-traps** can be one-piece or two-piece with a union nut. P-traps can also have a **cleanout.** Several styles of P-traps are shown in *Figure 4*. P-traps designed to be attached directly to fixtures are usually 1¼ to 1½ inches for sinks and lavatories, 1½ to 2 inches for tubs and showers, and 2 to 4 inches for floor drains. P-traps can be made of brass, brass with chrome plating, plastic, copper, cast iron, malleable iron, or glass. Floor drains may be designed with P-traps.

Interceptors prevent hazardous or undesirable materials from entering building drainage systems, public or private sewers, and sewage treatment plants or processes. Hazardous or undesirable materials include hair, lint, fats, oils, grease, flammable liquids, sand, solids, acid or alkaline waste, and chemicals. Interceptors are available for specific applications. Hair interceptors, for example, may be installed in beauty salons, barbershops, hospitals, or pet grooming shops. Grease interceptors (*Figure 5*) may be installed in restaurants, commercial kitchens, or auto repair shops. Although they differ in design, all interceptors operate on similar principles. Wastewater flows through a chamber where harmful materials are separated before the waste-

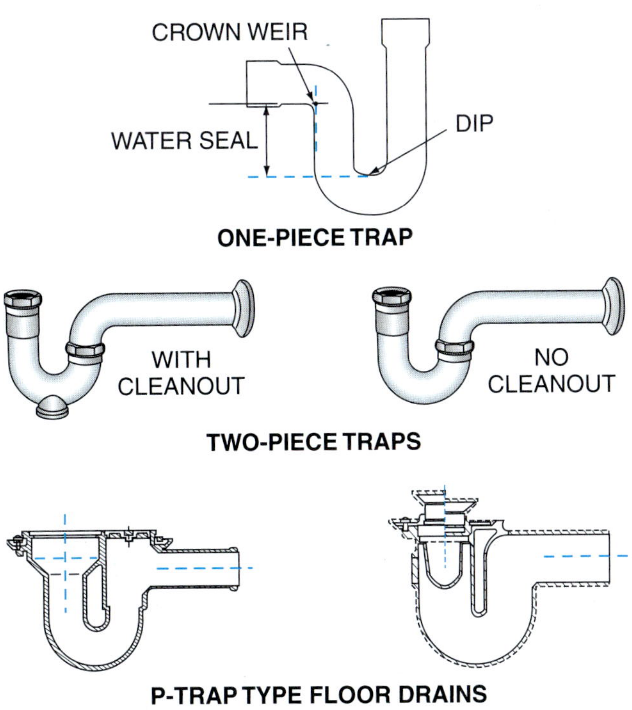

Figure 4 ◆ P-traps.

water flows out again. For example, in interceptors designed to capture solid wastes such as grease, wastewater flows into a chamber through screens. Because the solids are heavier than the wastewater, gravity causes them to fall to the bottom of the chamber, where they are retained until the chamber is cleaned out.

Current plumbing codes prohibit the use of **S-traps** (*Figure 6*). P-traps are more efficient and effective. The long downstream leg of the S-trap tends to promote siphonage. Plumbers could make the trap more effective by increasing the depth of the seal, but this led to other problems. With greater depth, there was a greater chance that solids would stay in the trap. Fungus growth was also a problem in traps that were too deep. Two types of S-traps were used: the full S-trap and the ¾ S-trap. Many older homes still have S-traps. Plumbing supply stores stock these traps for service but not for installation in new construction.

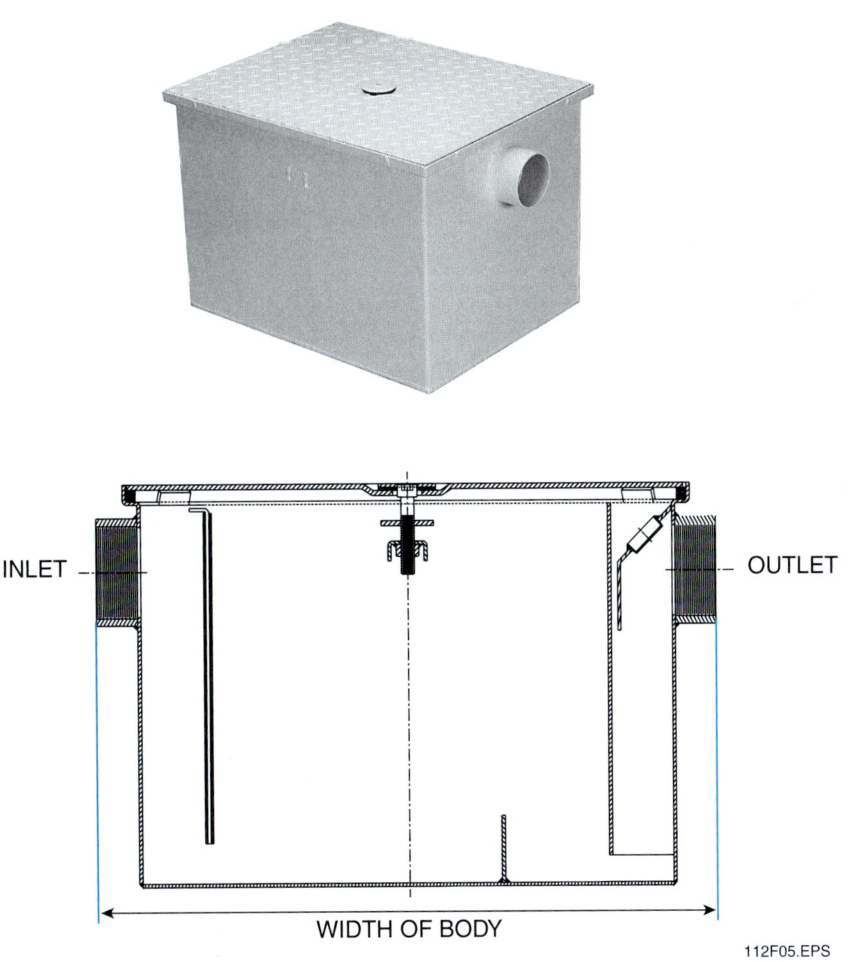

Figure 5 ◆ Grease interceptor.

Figure 6 ◆ S-trap.

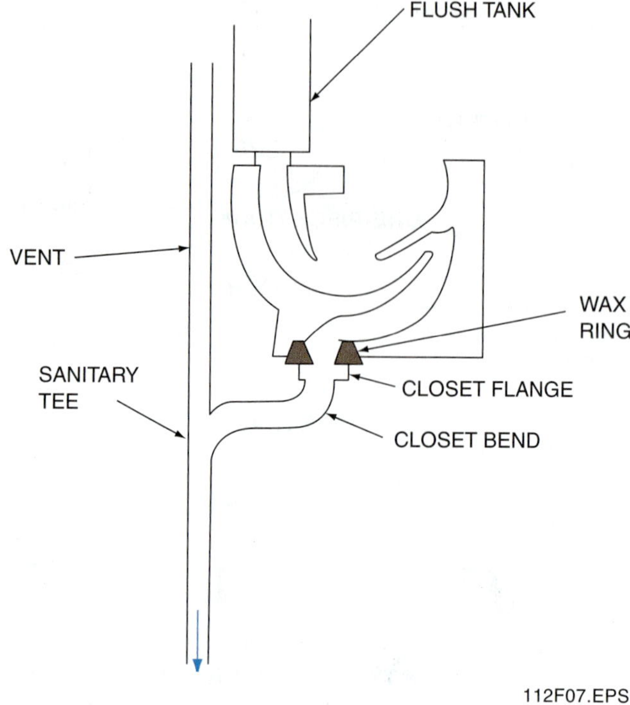

Figure 7 ◆ Integral or built-in trap.

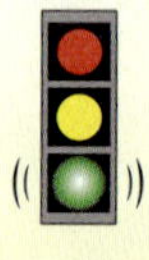

NOTE

Nonsiphon traps are available with deeper trap seals. These are used where the plumbing system is subjected to abnormal changes in pressure or to a lot of evaporation.

Water closets (toilets) have integral traps as part of their design. Water closets (see *Figure 7*) should never be attached to another trap. This is called **double trapping.** Double trapping creates a pressure that will stop the flow of drainage.

4.2.0 Parts of Traps

The following are the basic parts of a trap (see *Figure 8*):

- *Inlet* – where water enters from the fixture
- *Top dip* – the inside curve of the pipe under the inlet
- *Bottom dip* – the bottom of the lowest curve of the pipe beneath the inlet
- ***Crown weir*** *(sometimes called the* ***trap weir****)* – the highest point in the seal of the trap. Crown weirs and trap weirs are often referred to simply as **weirs.**
- *Fixture drain (also called the trap arm)* – the point where wastewater leaves the trap and goes into the drainage piping

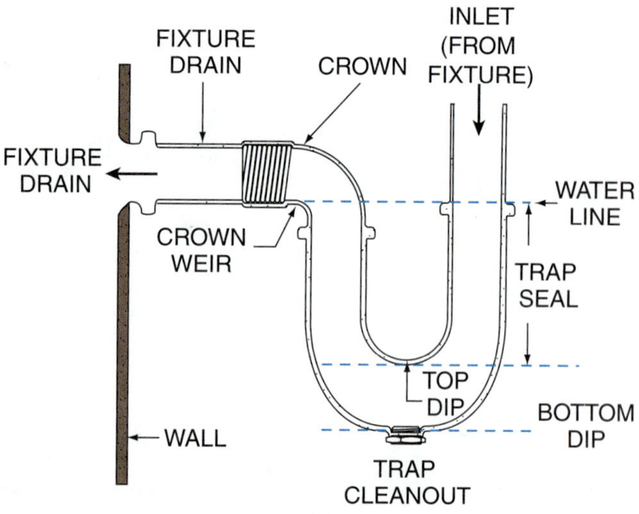

Figure 8 ◆ Parts of a trap.

4.3.0 Trap Installation Requirements

Although applicable code governs specific installation requirements, such as dimensions and location of traps, there are some typical trap installation requirements (see *Figure 9*).

Generally, the vertical distance from the fixture outlet to the crown weir may not exceed 24 inches. The second critical dimension is the horizontal distance from the crown weir to the trap vent. This distance varies depending on the diameter of the trap (see *Table 1*). The third important dimension

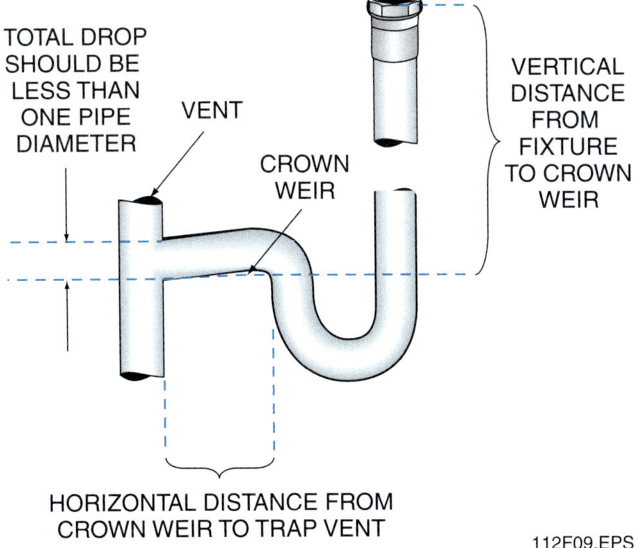

Figure 9 ◆ Critical dimensions of a trap vent.

is the total drop (or **fall**) in the horizontal pipe from the crown weir to the vent. It may not exceed one pipe diameter. Usually, the fall is ¼ inch per foot. If the horizontal leg is installed with greater drop, the trap is likely to siphon.

Many P-traps are manufactured in two pieces. The J-bend piece joins with the fixture tailpiece and is secured with a gasket and compression nut called a slip-joint washer and a slip-joint nut. The outlet end, called the trap arm, joins the horizontal drain, which connects to the vent.

Generally, traps should be installed for each plumbing fixture that does not have a built-in trap. Some codes permit one trap to serve more than one fixture. For example, three lavatories that are 30 inches or less apart may be connected to one trap. Sinks containing two or three compartments also may be connected to one trap.

DID YOU KNOW?

In the earliest home plumbing fixtures (in the 1850s), the main safeguards against odors and sewer gases were handmade traps that the plumber installed in the drains of individual fixtures. These traps often lost their water seals because of siphonage and back pressure (described in this section) and became ineffective. Efforts to prevent seal loss failed, because the principle of venting fixture drains (covered later in this module) was not known at the time.

In the early 1900s, the problems with fixture traps led health officials to require a secondary safeguard: the installation of building traps on each sanitary or combined building sewer. Without this additional safeguard, rats were able to travel freely from one building to another. Building traps became the second line of defense against rats in the sewer systems.

This requirement was a big advance at the time. However, since the development of modern collection, drainage, and venting systems, most model codes don't require building traps. In fact, many codes actually prohibit building traps. The only exceptions are in areas where sewer gases are extremely corrosive or where the sewer gases contain high explosive gas content, creating a risk of explosions in the public sewer system that might, for instance, blow off manhole covers and cause considerable damage.

Table 1 Horizontal Distance of Fixture Trap from Vent at Slope of ¼ Inch per Foot

		Distance from Trap		
		Standard Plumbing Code **(1994)** ***International Plumbing Code®*** **(1997)**	***Uniform Plumbing Code*** **(1997)**	***National Standard Plumbing Code***
Size of Fixture Drain (1996)	**Size of Trap**			
1¼"	1¼"	3'6"	2'6"	3'6"
1½"	1½"	5'0"	3'6"	5'0"
2"	2"	6'0"	5'0"	8'0"
3"	3"	10'0"	6'0"	10'0"
4" and larger*	4"	12'0"	10'0"	12'0"

*Size of fixture drain specification is 4" only (not 4" and larger) for *Standard Plumbing Code* (1994) and *International Plumbing Code* (1997).

112T01.EPS

4.4.0 Why a Trap Loses Its Seal

A thorough understanding of how traps function will help you understand how important it is to install traps correctly. Also, you can use this knowledge to determine what causes a trap to malfunction.

When a trap functions properly, waste from the fixture flows into and through the trap (see *Figure 10*). The trap is refilled with the last of the wastewater to leave the fixture. This water provides the necessary liquid seal. For the trap to function this way, the pressure on both sides of the trap must remain nearly equal. Water tends to flow in a level line, called the **hydraulic gradient.** The crown weir must always be installed lower than the top of where the fixture drain enters the vent line.

A trap may lose its seal in a number of ways—through siphonage, aspiration, momentum, oscillation (wind effect), back pressure, evaporation, capillary attraction, or cracks. Properly designing and installing the DWV system can prevent siphonage and back pressure.

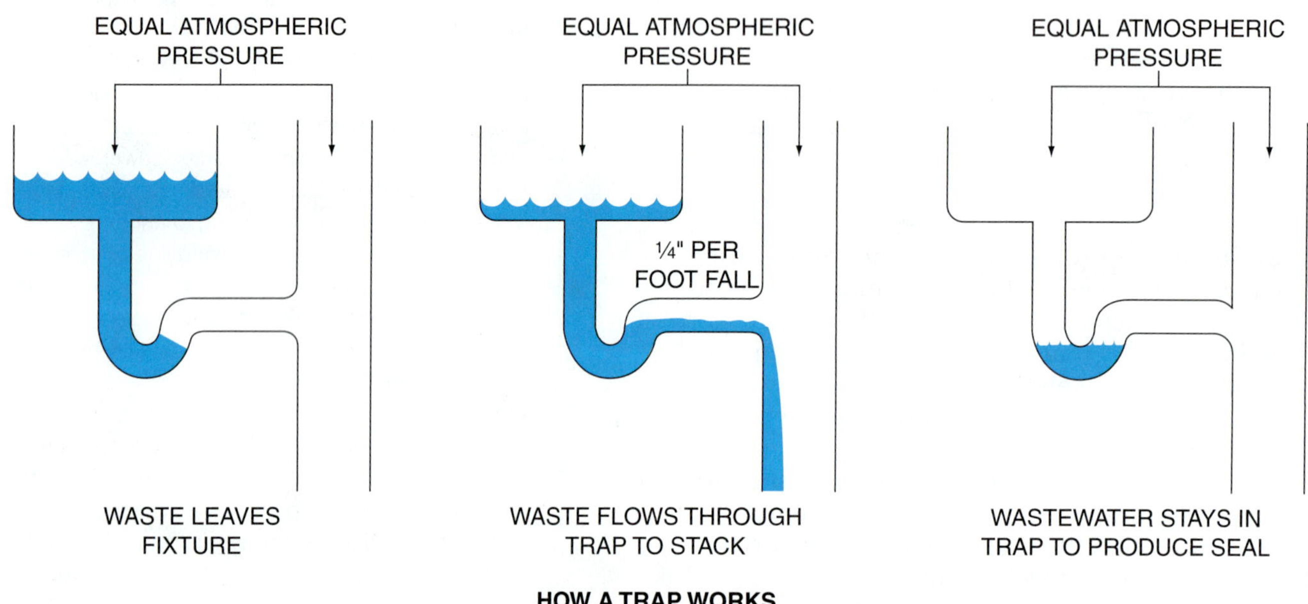

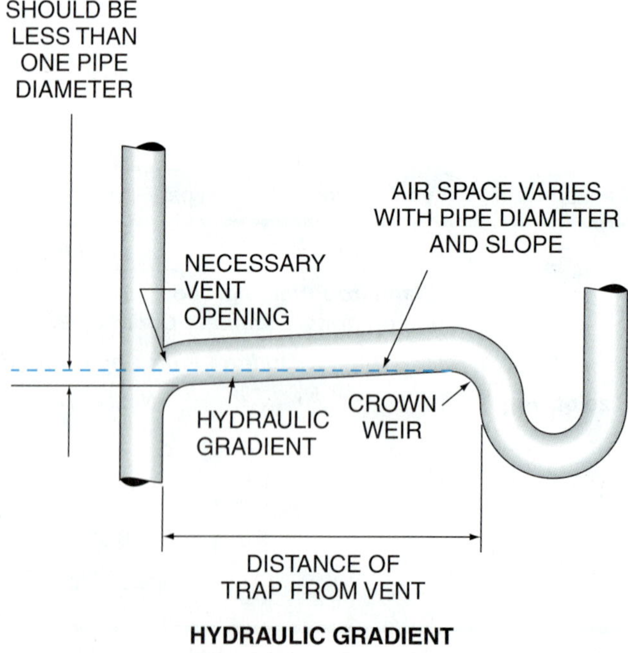

Figure 10 ◆ How a trap works.

4.4.1 Siphonage

If the trap is not properly vented, it is likely to siphon. Siphonage occurs when there is negative pressure inside the DWV piping. This pressure difference pushes the water that is normally held in the trap into the DWV piping system. Generally, siphonage occurs when the DWV piping is improperly vented or the vent is blocked (see *Figure 11*). As the waste leaves the trap, an area of reduced pressure is created in the drainage piping. Because of the difference in pressure, the water is forced from the trap. This destroys the trap seal. Siphonage happens when there is too much fall and the crown weir is higher than the top of where the fixture drain enters the vent.

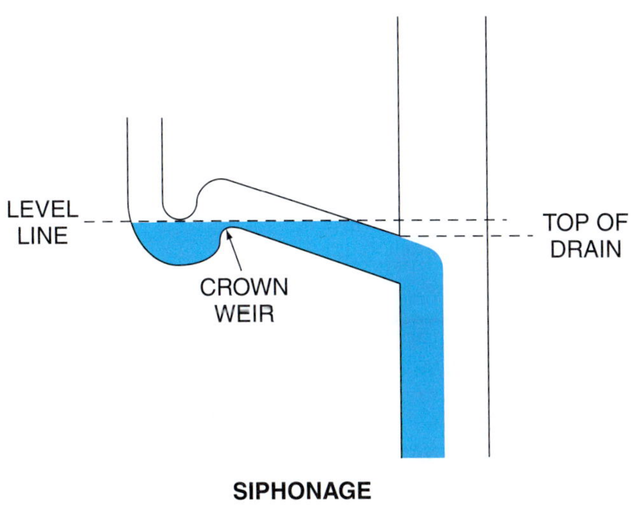

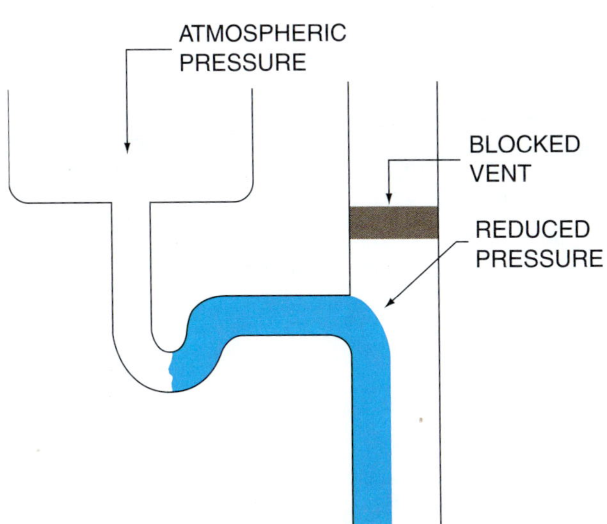

Figure 11 ◆ Siphonage.

4.4.2 Aspiration

The term aspiration means the drawing in, out, or up of something, usually a fluid. In piping, aspiration takes place when a large volume of water flows near the trap, creating negative pressure. This negative pressure draws the water from the trap and causes the seal to fail. Because of aspiration, trap seals can absorb gases and odors. When the trap seal is saturated with the gas or odor, it will emit the same, often unpleasant, odor into the building.

Fixtures with S-traps are the most vulnerable to aspiration. Although modern plumbing codes prohibit S-traps, you may still encounter them in older buildings. Where an S-trap is installed, you can detect a failed seal by the smell of sewer gases or by a gurgling sound in the pipe. Drainage systems must provide adequate circulation of air in the piping to prevent siphonage and aspiration and to protect trap seals.

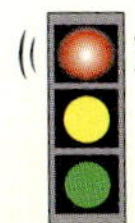

WARNING!

Many toxic gases do not emit an odor. Always test for gas before working in an area where toxic gases may be present. Use a smoke, water, or peppermint test to locate gas that may be entering the system through a leak in the piping. Use a toxic gas detector to monitor potentially dangerous levels of gas before working in an exposed area.

4.4.3 Momentum

The momentum, or speed, of water rushing through a pipe can force the standing water out of a trap and empty it, thus breaking the seal. Water can gain enough speed to empty a trap when the vertical distance between the fixture outlet and the trap is too long. In most cases, that distance should be limited to around 12 inches, although longer vertical distances may be required for certain types of standpipes. Refer to applicable codes to determine the correct distance.

4.4.4 Oscillation

Oscillation, or wind effect, is one of the least likely ways a trap can lose its seal. Where there are strong upward or downward air currents, the pressure or suction of the moving air may cause the water in the trap to rise or fall. If it rises enough to spill over into the waste pipe, less water remains in the trap and the seal is weakened. Lower than normal back pressures could break the seal.

4.4.5 Back Pressure

Back pressure (*Figure 12*) can cause a trap seal to break. Back pressure is pressure inside the DWV piping that is greater than atmospheric pressure. If enough wastewater from a fixture enters the **stack** so that a slug of water forms a moving plug, the air in the stack below the plug is compressed. This excess pressure tries to escape through the trap in fixture B illustrated in *Figure 12*. To prevent normal back pressure from destroying the trap seal, the stack must be properly sized, and the trap must be properly sized and protected by a vent. Also, the seal must be deep enough; generally a trap seal of 2 to 4 inches is required.

4.4.6 Evaporation

A trap may lose its seal as a result of **evaporation.** This is most likely to happen in traps that are seldom used. The water evaporates, causing the seal to break. If the DWV piping is properly designed, evaporation will become a problem only during long periods of nonuse. When sewer gas enters a structure, unused floor drains are often the cause. If you anticipate long periods of nonuse, you can install extra-deep traps. Many codes require trap primers where evaporation of the trap seal is likely. Trap primers are connected to a regularly used water line. Water flows through a small tube into the trap so that it can keep its seal.

4.4.7 Capillary Attraction

Capillary attraction (*Figure 13*) may cause a trap seal to break if a porous material, such as string or

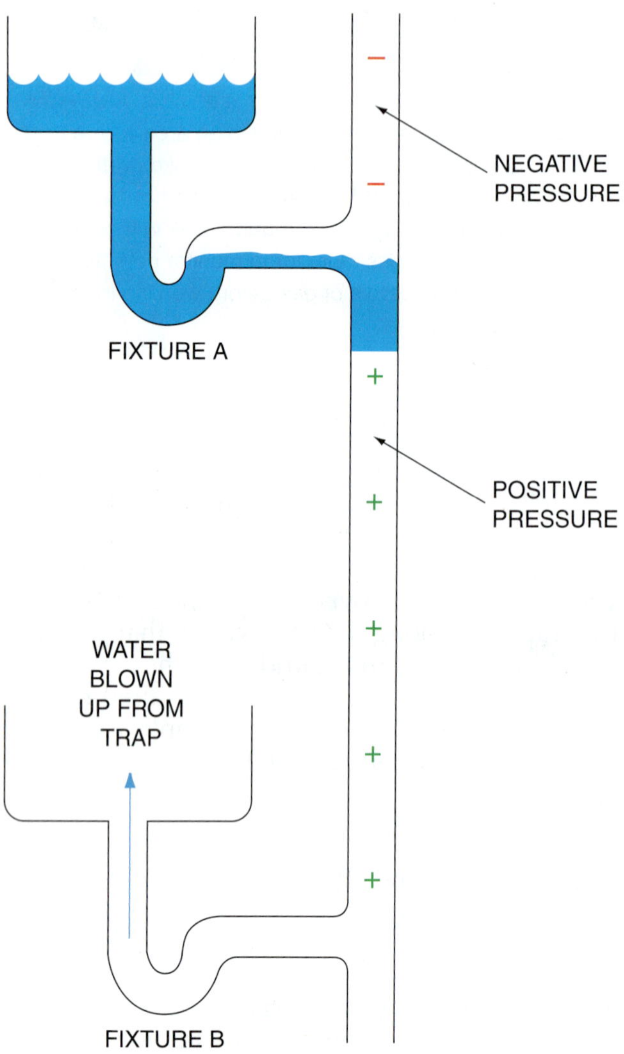

Figure 12 ◆ Back pressure.

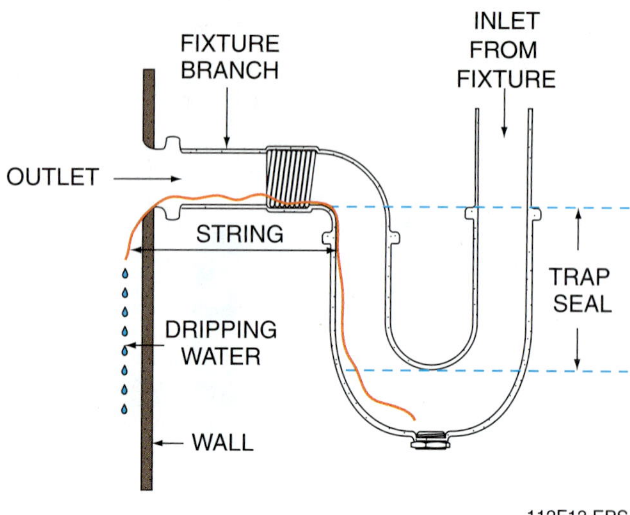

Figure 13 ◆ Capillary attraction.

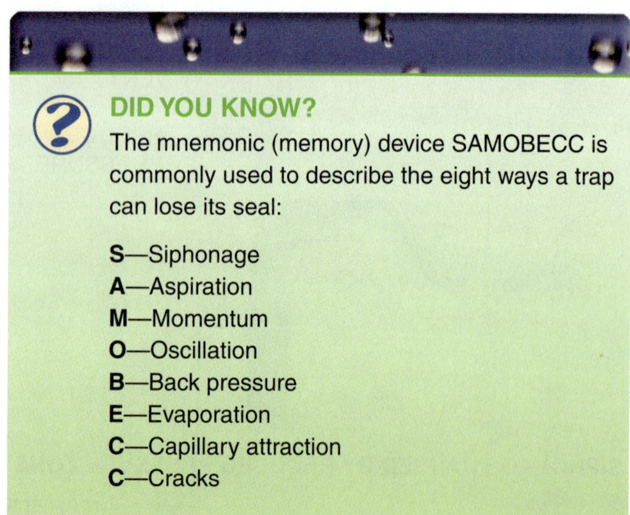

DID YOU KNOW?

The mnemonic (memory) device SAMOBECC is commonly used to describe the eight ways a trap can lose its seal:

S—Siphonage
A—Aspiration
M—Momentum
O—Oscillation
B—Back pressure
E—Evaporation
C—Capillary attraction
C—Cracks

paper, is caught in the trap. The porous material acts as a wick and draws the water out of the trap by capillary action. Cleaning the trap will solve this problem.

4.4.8 Cracks

A more common cause of waste and sewer gas leaking into a building is a crack in the trap. Cracks can be caused by worn washers, or by a broken nut, solder joint, or glue joint.

Review Questions

Sections 1.0.0–4.0.0

1. Sanitary drainage systems can be divided into three parts: the pipes inside the building, the drain pipe buried outside the building, and the _____.
   a. building sewer
   b. public sewer
   c. treatment plant
   d. lift station

2. Usually the _____ is responsible for installing and maintaining the public sewers, lift stations, and treatment plants.
   a. municipality
   b. federal government
   c. contractor
   d. land owner

3. The _____ inside a building is a circuit of piping designed to remove the wastes from plumbing fixtures and drains safely, reliably, and efficiently.
   a. drainage system
   b. sewage-disposal system
   c. DWV system
   d. vent system

4. Many fixtures have _____ that strain the wastewater before it enters the drainage piping.
   a. drains
   b. seals
   c. traps
   d. vents

5. The _____ seal in a trap protects people from airborne pathogens (germs), foul odors, and potentially explosive sewer gases.
   a. gas
   b. air
   c. pressure
   d. water

6. A fixture trap must _____, be self-cleaning, have a smooth interior waterway, and be accessible for cleanout.
   a. flush completely
   b. contain an antibacterial filter
   c. be constructed of at least four separate pieces
   d. have a nonrusting hinge

7. To avoid double trapping, _____ should never be attached to another trap.
   a. sinks
   b. tubs
   c. water closets
   d. shower assemblies

8. The highest point in the seal of the trap is the _____.
   a. crown weir
   b. inlet
   c. bottom dip
   d. fixture

9. Siphonage occurs when there _____.
   a. is negative pressure outside the DWV piping
   b. is negative pressure inside the DWV piping
   c. are not enough bends in the DWV piping
   d. are two or more vents in a vertical piping installation

10. A properly sized stack will prevent _____ from breaking the trap seal.
    a. wind effect
    b. back pressure
    c. backflow
    d. capillary attraction

5.0.0 ◆ VENTS

Every trap requires a vent of some type. Vent pipes are critical for plumbing fixtures to function correctly as part of the sanitary drainage system. Venting prevents back pressure or siphonage from breaking the water trap seals that serve the fixtures. All the vent pipes of a building create the vent system and are connected to the drain pipes. The system may include one or more pipes. Vents are installed to provide a free flow of air and to maintain equalized pressure throughout the drainage system. There are many types of vents, one of which is shown in *Figure 14*. Different types of vents will be discussed later in your training.

5.1.0 Distance from Trap to Vent

If a vent is placed too far from the trap, the vent opening will fall below the crown weir because of the fall of the waste pipe. This situation could cause the trap to self-siphon and lose its seal. To prevent this, most codes publish tables based on pipe fall and pipe diameter. These tables give the maximum distances allowed between the pipe and the vent. Check applicable code for details.

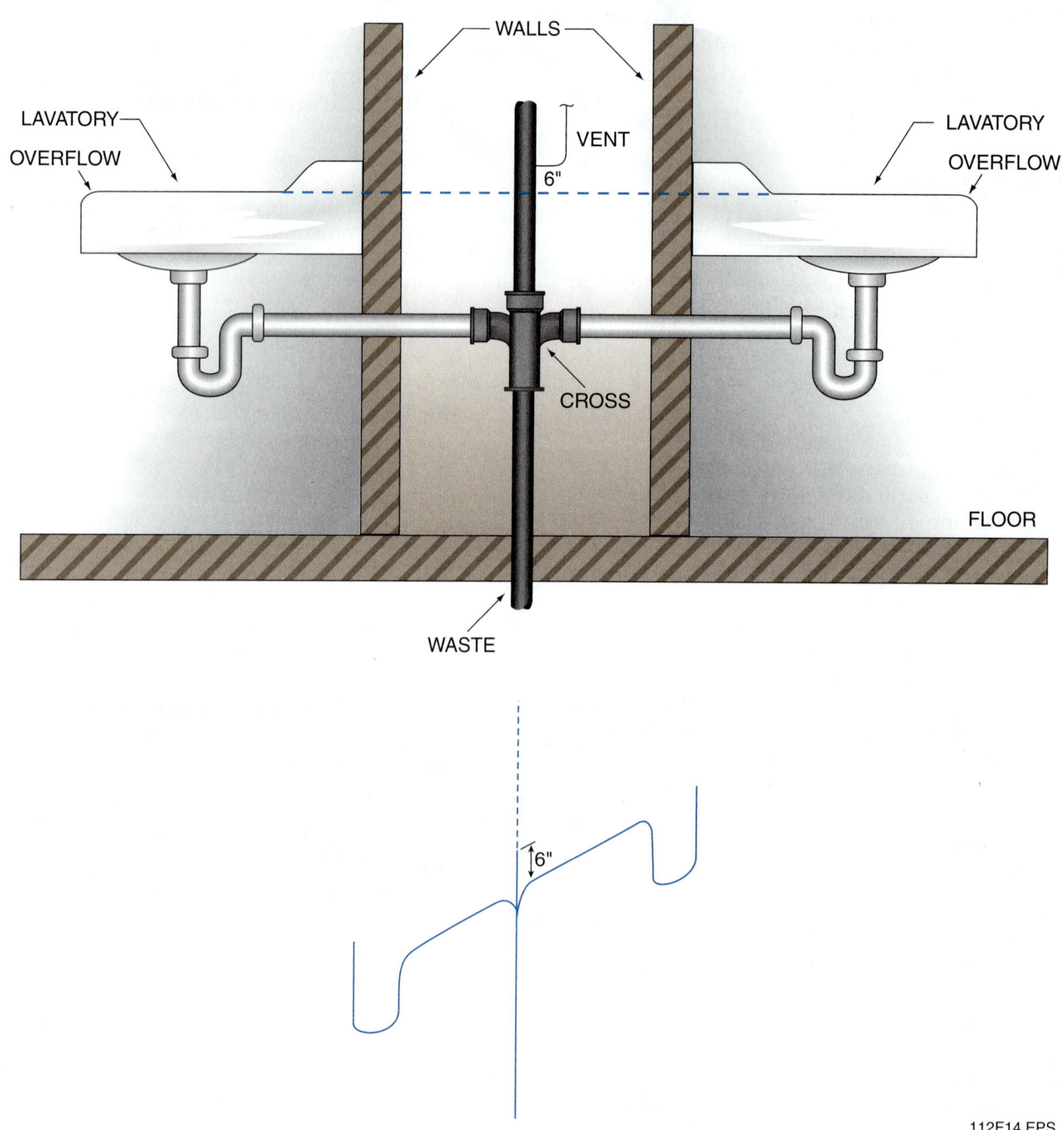

Figure 14 ◆ Vent.

6.0.0 ◆ SIZING DRAINS AND VENTS

Drainage systems fall into two major categories: storm water drains and building drains. Storm water drains collect storm water from roofs and pavement. The water is either held for on-site disposal or is disposed of at a certain rate into a storm sewer system. The sizing of storm drainage systems is based on expected rainfall. Building drainage systems must be appropriately sized based on the expected water use in the building. Plumbers must understand the mechanics of fluid flow in pipes to properly size drains and vents.

Vents are used in plumbing systems to balance pressure in the piping network. This balance of pressure is necessary to prevent the fixture traps from losing their seals. For the vents to function properly, plumbers must size the vents correctly according to how many fixtures are connected, the number of drainage fixture units (how much water discharges into the drain per minute), and the length of the vent pipe. If inadequately sized vents are installed, the plumbing system will not work properly. You will learn how to size drainage and venting systems as you advance through the levels of the plumbing curriculum.

Review Questions

Sections 5.0.0–6.0.0

1. _____ balance the pressure within the drainage system to maintain the water seals in fixture traps.

a. Fixture drains
b. Vents
c. P-traps
d. Cleanouts

2. Vents use _____ to maintain equalized pressure throughout the drainage system.

a. air
b. sewage gas
c. liquid wastes
d. water

3. Tables based on fall and pipe diameter give the maximum _____ allowed between the crown weir and the vent.

a. sizes
b. distances
c. fittings
d. drains

4. If the total drop between the crown weir and the fixture vent exceeds one pipe diameter, the trap is likely to _____.

a. evaporate
b. crack
c. siphon
d. pressurize

5. Plumbers must size the vents correctly according to how many _____ are connected, the number of drainage fixture units, and the length of the vent pipe.

a. fittings
b. fixtures
c. drains
d. cleanouts

7.0.0 ◆ FITTINGS AND THEIR APPLICATIONS

Fittings are devices used to connect pipe. Those used in DWV systems are called **drainage fittings.** This section describes the various types of drainage fittings and their uses within the DWV piping system.

7.1.0 DWV Fitting Materials

DWV fittings are made from many different materials, including copper, brass, lead, steel, cast iron, clay, glass, and various types of plastic. Cast iron and plastic are the most commonly used materials (see *Figure 15*). Some codes may prohibit the use of certain materials for certain applications, so be sure to check applicable code requirements before installing a DWV piping system. Not all fittings are available in all materials.

CAST-IRON PIPE AND FITTINGS

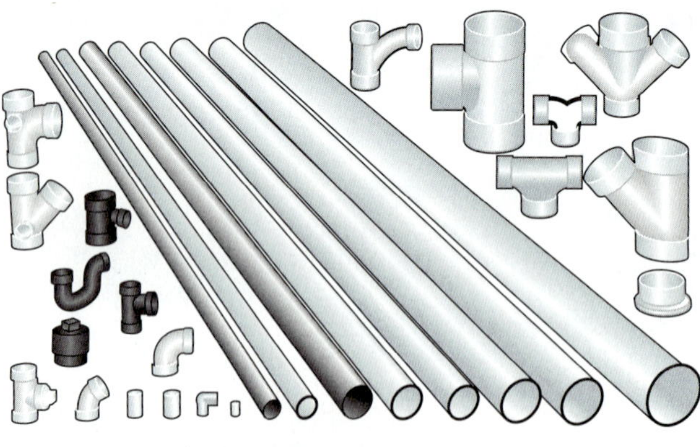

PLASTIC PIPE AND FITTINGS

112F15.EPS

Figure 15 ◆ DWV fittings.

Although the fitting materials may vary, fittings of the same design have the same names. For example, a plastic sanitary tee and a cast-iron sanitary tee are basically identical, even though they are made from different materials.

A number of fittings are available for copper pipe for DWV purposes. Those fittings include 90-degree ells, 45-degree ells, 22½-degree ells, male adapters, tees, cleanout tees, reducing tees, and reducers.

7.2.0 General DWV Fitting Requirements

Most codes state that drainage fittings may not slow or block the flow of materials in the pipe. Because of this, sanitary drainage fittings are made with a sweeping design (*Figure 16*) to allow for the smooth flow of material in the system.

Another code requirement is that the direction of hub-type fittings should not go against the flow of the system—that is, the wastes should flow from the bell to the spigot end of the pipe.

7.3.0 DWV Fittings

Sanitary fittings are used to connect DWV branches to the main DWV system. The branch inlets of these fittings may be reducing (going from a larger pipe to a smaller pipe). If so, they can be joined to the system without reducers.

7.3.1 Bends

The term bend is often used in reference to cast-iron fittings. With other types of fittings, the term elbow is more common. Bends are used to change the direction of a **run** of pipe. A run is one or more lengths of pipe in a straight line. Bends are available in different sizes. *Figure 17* shows 1⁄16, ⅛, ⅙, ⅕, and ¼ bends. The ⅕ bend is available only in cast-iron pipe.

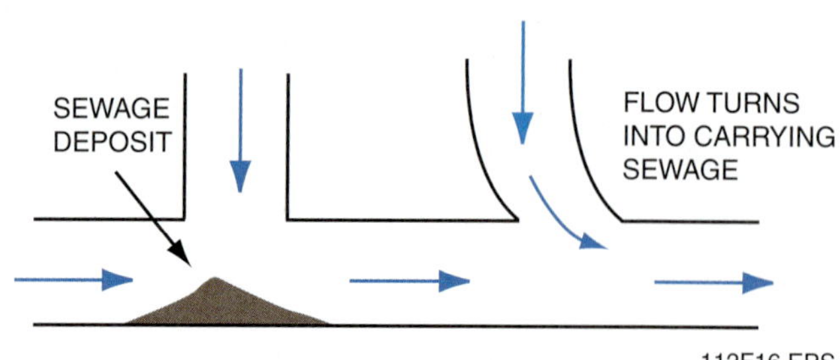

112F16.EPS

Figure 16 ◆ Sweeping design.

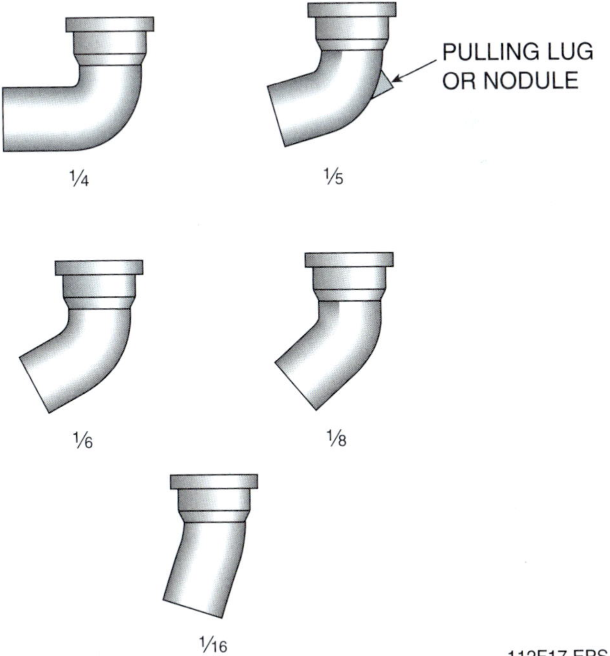

Figure 17 ◆ Bends.

The plumbing design tells you which bends are used and where they are placed in the system. Bends are expressed as fractions of a complete circle. A circle contains 360 degrees. You can easily determine the number of degrees a given bend turns by multiplying the bend fraction by 360 degrees twice. For example, to determine the bend angle of a ¼ bend, multiply the bend type (¼) by 360 (¼ × 360 = 0.025). Next, multiply that answer by 360 to determine the number of degrees in a ¼ bend (0.025 × 360 = 90 degrees). Bends are available with heel inlets and **side inlets** to allow smaller lines to be connected to the bend (see *Figure 18*). Be sure to check applicable code requirements. It is important to note that a high- or low-pattern side inlet bend cannot be used as a vent if the inlet is horizontal.

Bends with side inlets are available with single and double side inlets. To determine whether the inlets are right or left inlets, place the spigot of the bend down and look through the hub (or bell) end (this is the same direction water would be flowing down the drain) (see *Figure 19*). Left inlets will be on the left side, right inlets on the right side.

Three patterns of ¼ bends are available (see *Figure 20*). The basic fitting is simply called a ¼ bend.

Short sweep ¼ bends and **long sweep ¼ bends** may be used at the base of DWV stacks. They are

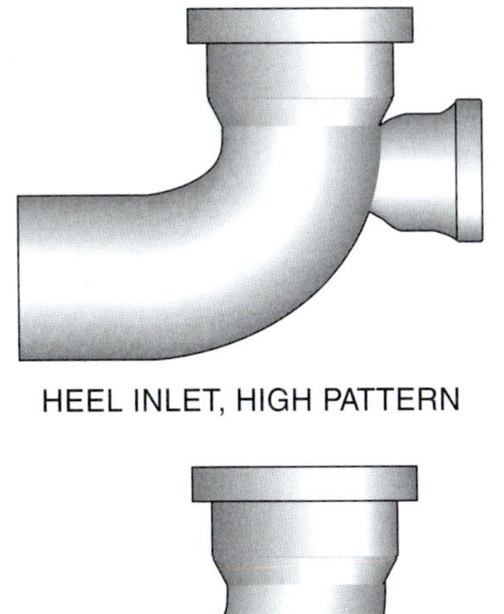

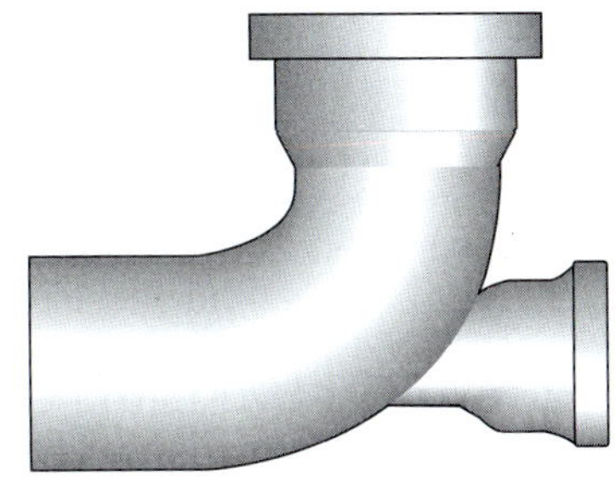

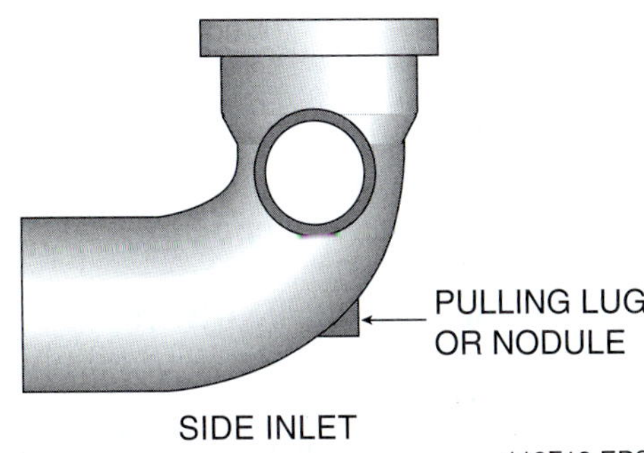

Figure 18 ◆ Bends with heel and side inlets.

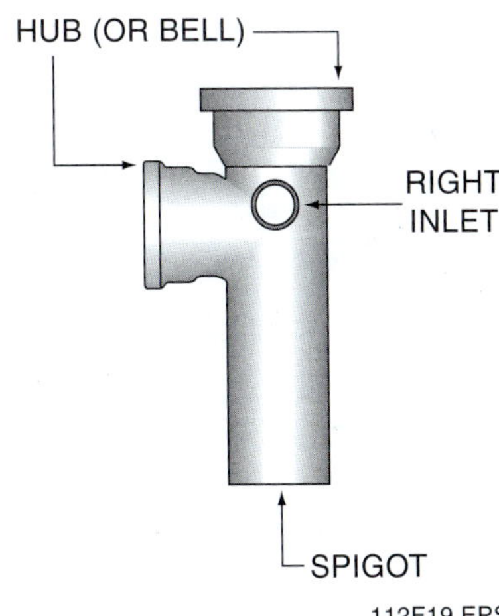

Figure 19 ◆ Determining left or right side inlet.

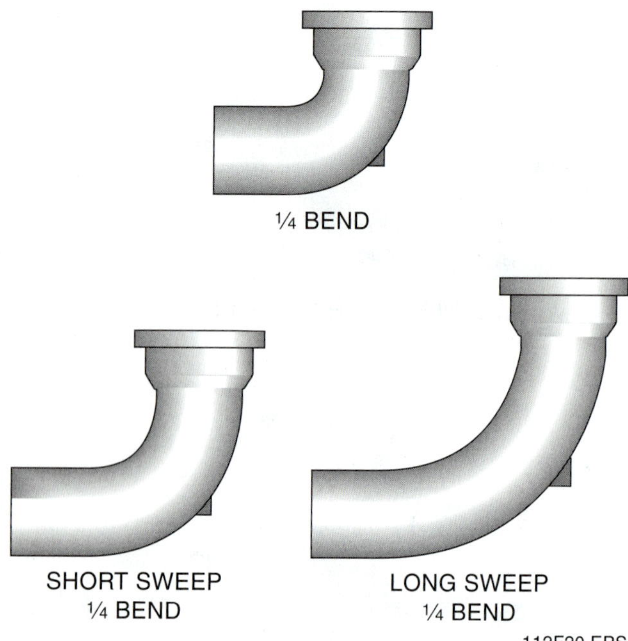

Figure 20 ◆ Available ¼ bends.

required by various codes because the longer radius of their turn greatly decreases flow resistance and back pressure. Always check the applicable codes to determine which bend must be used in your area.

Double ¼ bends (also called twin ells), shown in *Figure 21*, are used to collect and combine the flow from two opposite runs into a single run of pipe. Note the direction of flow in *Figure 21*. Be sure to check the applicable code requirements.

Codes allow the use of **vent ells** (*Figure 22*) only in vent lines. If they were placed in drainage or waste lines, their sharp turn radius would severely restrict the flow of materials. They are available only in plastic fittings.

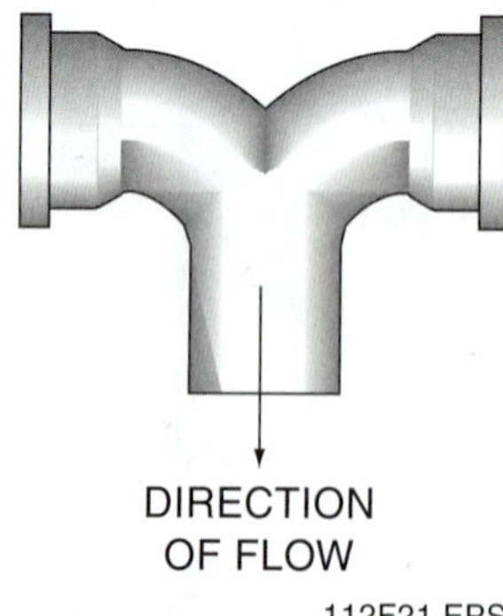

Figure 21 ◆ Double ¼ bend.

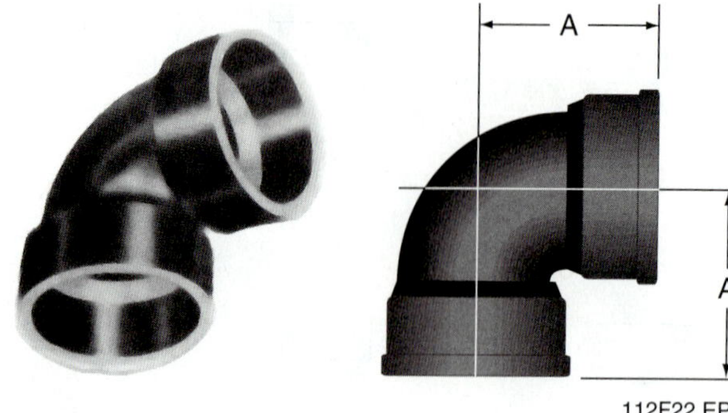

Figure 22 ◆ Vent ells.

7.3.2 Adapters

DWV fittings are most often used with pipe made from the same material. However, **adapters** (*Figure 23*) can be used to join pipe of different materials. For instance, adapters can be used to join copper tubing to galvanized iron pipe or plastic to cast-iron pipe.

7.3.3 Cleanouts

Fittings are also available for cleanouts. Cleanout fittings have internal threads on the branch fitting to accept a threaded cleanout plug. A cleanout adapter (*Figure 24*) may be installed in one end or branch of a fitting to provide a cleanout access. In both cases, a cleanout plug is provided to permit access to the DWV piping system to remove blockage and to prevent leakage.

Sanitary fittings are available in single and double patterns. The double pattern is used to connect two branch lines entering the system from opposite directions. This allows for central placement of horizontal runs and vertical runs. Sanitary branch fittings consist of tees, wyes, and combinations of the two.

7.3.4 Tees

Sanitary tees, shown in *Figure 25*, are used for branches that run from horizontal to vertical. Model codes restrict their use to sanitary drainage systems where the flow of material is from the horizontal to the vertical.

Sanitary tees are available with side inlets. The side inlet allows smaller drains to be connected from the right or left.

Most codes restrict the use of **vent tees** (*Figure 26*) and they may not be used to vent lines or as cleanout fittings. They are prohibited for use in drainage systems because their design restricts the flow of material. This design restriction may allow wastes and **pipe scale** to collect within the fitting.

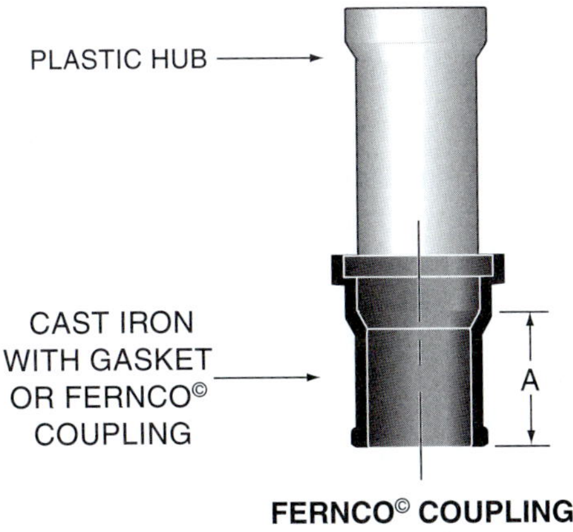

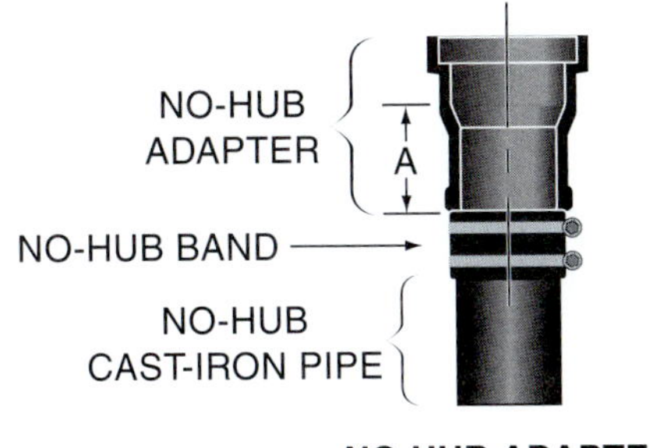

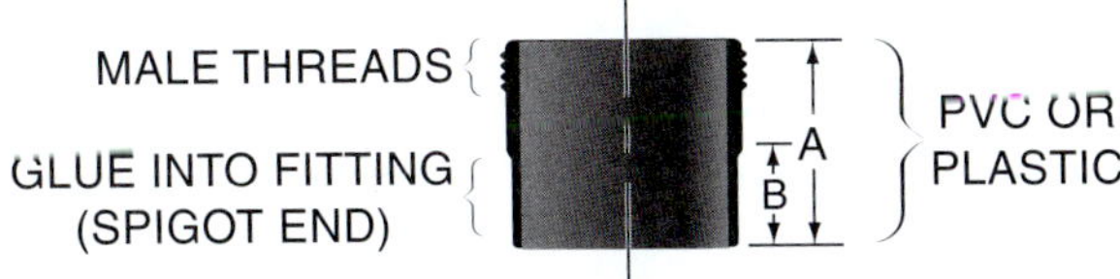

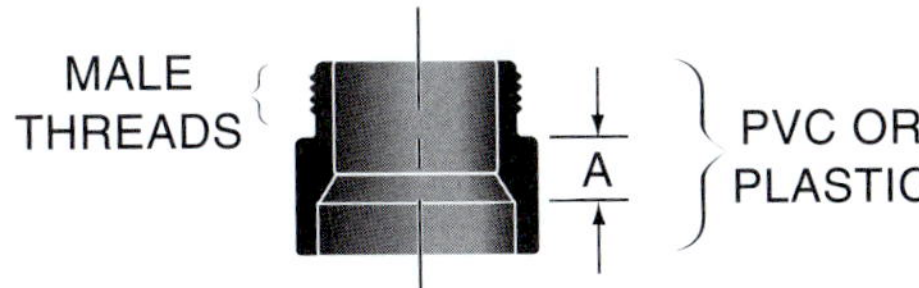

Figure 23 ◆ Adapters.

Figure 24 ◆ Cleanout adapters.

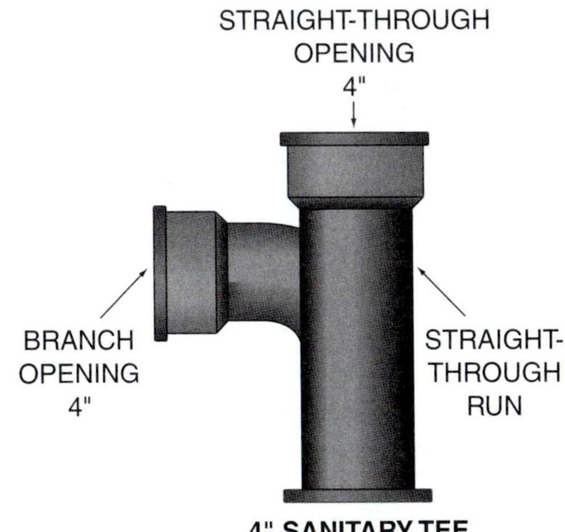

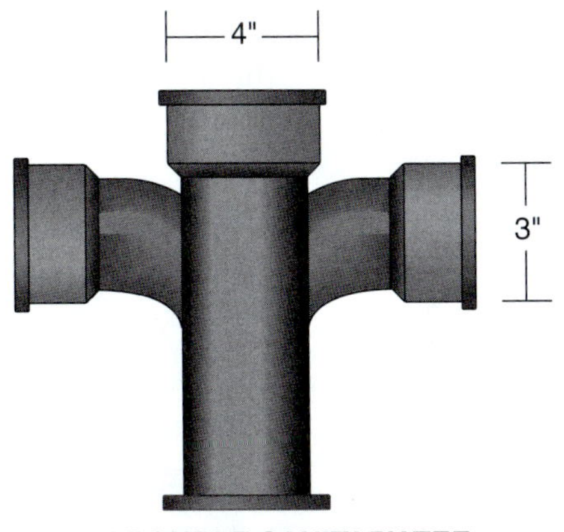

Figure 25 ◆ Sanitary tee and cross.

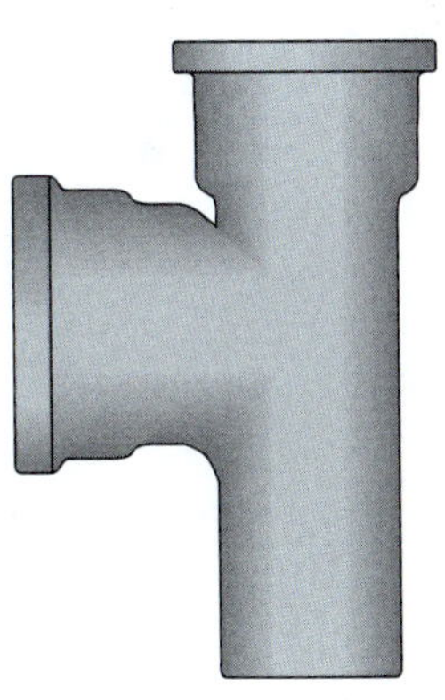

Figure 26 ◆ Vent tee.

Test tees (*Figure 27*) are installed in the DWV plumbing system as required by various codes. You will learn about placement of these and other fittings for cleanout purposes later in your training. These tees serve as test locations from which the system is pressurized to test for leaks.

7.3.5 Wyes

Sanitary wyes (*Figure 28*) are used to provide a smooth-flowing DWV system, in keeping with code requirements. Codes require vertical branch lines that intersect with horizontal branch lines to connect with long turn fittings, such as wyes, with 45-degree angles or sweeps. If the correct fittings are not used, soil and wastes may collect on the pipe wall opposite the branch.

Sanitary upright wyes (*Figure 29*) are used to connect the vent stack to the lower end of the soil and waste stacks.

Vent branches (*Figure 30*) are used to join the upper end of the vent to the top of the soil and waste stacks.

Inverted wyes (*Figure 31*) may be used in place of vent branches. A **sanitary combination** is a fitting that combines a wye and a ⅛ bend. Also called a tee-wye, this fitting is available with single or double inlets (see *Figure 32*). Sanitary combinations are used to connect horizontal branch lines that intersect at 90 degrees with other horizontal branch lines. They also are used to connect a vertical stack with a horizontal drain. They are used

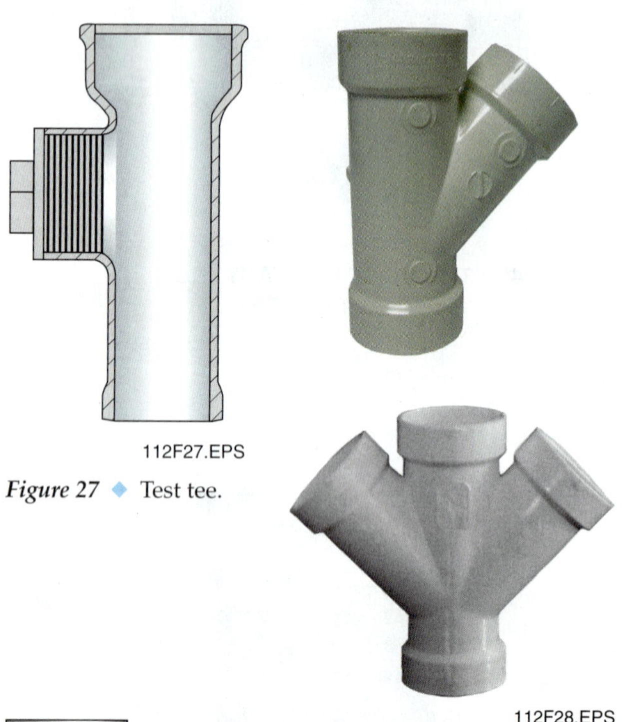

112F27.EPS

Figure 27 ◆ Test tee.

112F28.EPS

Figure 28 ◆ Sanitary wyes.

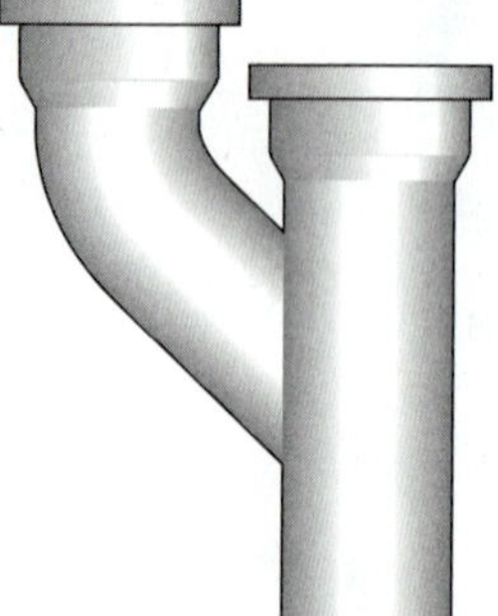

112F29.EPS

Figure 29 ◆ Sanitary upright wye.

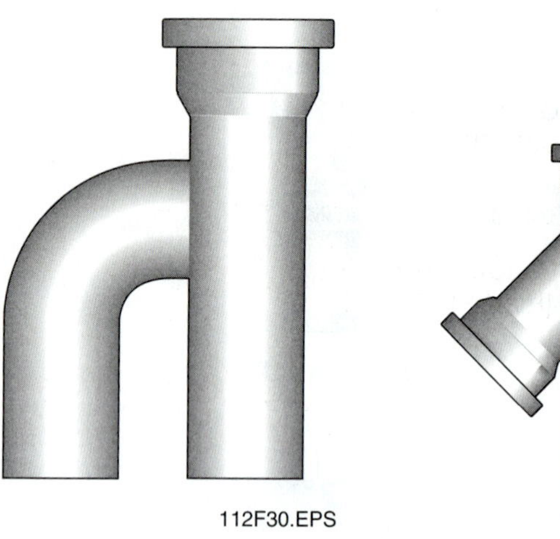

112F30.EPS

Figure 30 ◆ Vent branch.

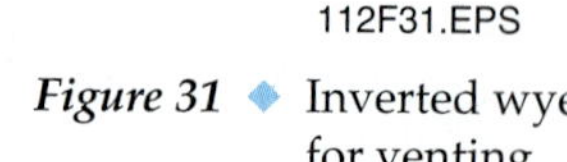

112F31.EPS

Figure 31 ◆ Inverted wye for venting.

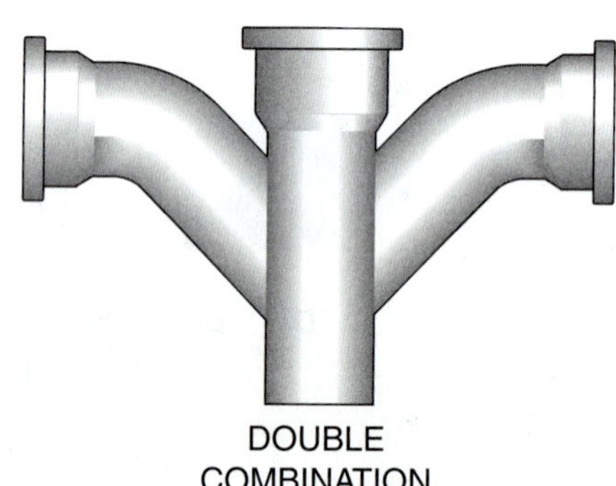

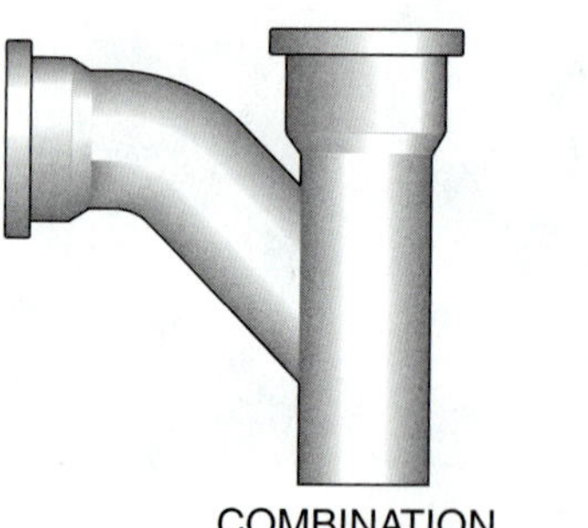

112F32.EPS

Figure 32 ◆ Sanitary combinations.

in place of sanitary tees whenever space permits because they offer less resistance to the flow of materials than sanitary tees do. Combination fittings also reduce the number of fittings needed. *Figure 33* shows that a sanitary wye and a ⅛ bend would be needed to do the same job as a combination fitting. Reducing the number of fittings also reduces the number of joints to be made and the amount of time needed to make them.

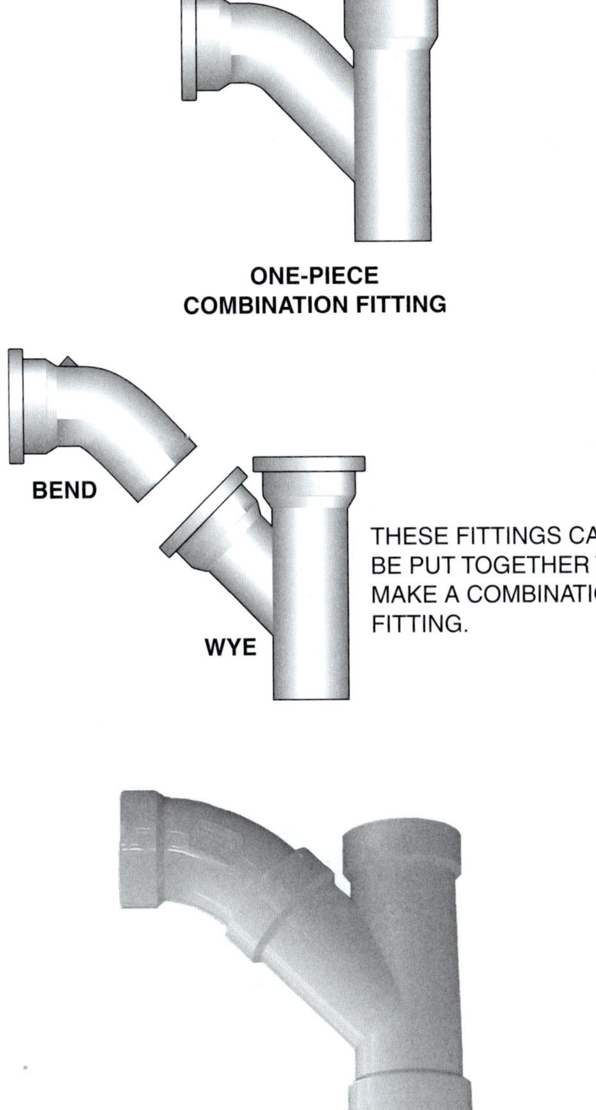

Figure 33 ◆ Combination fittings.

7.3.6 Miscellaneous Fittings

Sanitary increasers (*Figure 34*) are used to enlarge the diameter of vent stacks and are usually placed at least 1 foot below the stack's intersection with the roof, although this distance may vary by code. Sanitary increasers are necessary in cold climates to keep condensing water vapor from freezing and gradually closing the vent opening. The loss of the vent could cause the loss of the trap seals, allowing sewer gas to enter the building.

As you have already learned, offsets (*Figure 35*) are used to change the path of the pipe to avoid obstruction. They can offset the run of the pipe from 2 to 12 inches.

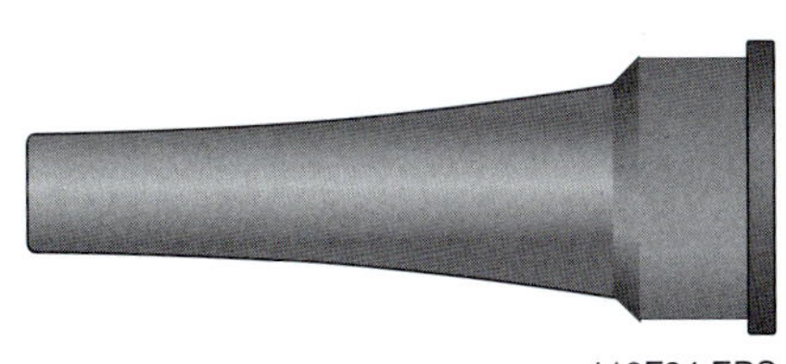

Figure 34 ◆ Sanitary increaser.

Figure 35 ◆ Offset.

Review Questions

Section 7.0.0

1. __________ and plastic are the most commonly used materials for DWV fittings.
   a. Brass
   b. Cast iron
   c. Clay
   d. Glass

2. Fittings of the same design will have the same name unless they are made of different materials.
   a. True
   b. False

3. Sanitary drainage fittings are made with a __________ design to allow for the smooth flow of material in the system.
   a. circular
   b. straight
   c. sweeping
   d. downward

4. The direction of hub-type fittings should flow from the __________ of the pipe.
   a. bell to the spigot end
   b. spigot end to the bell
   c. elbow to the spigot end
   d. bell to the elbow

5. If the branch inlets of sanitary fittings are __________, they can be joined to the DWV system without reducers.
   a. reducing
   b. expanding
   c. increasing
   d. decreasing

6. A ¼ bend turns __________ of a complete circle.
   a. 45 degrees
   b. 90 degrees
   c. 30 degrees
   d. 22½ degrees

7. Use a(n) __________ to join pipes of different materials in the DWV piping system.
   a. gasket
   b. increaser
   c. adapter
   d. tee

8. A __________ fitting has internal threads on the branch fitting to accept a threaded plug.
   a. cleanout
   b. sanitary
   c. wye
   d. reducer

9. For branches that run from horizontal to vertical, use a(n) __________.
   a. sanitary tee
   b. vent tee
   c. sanitary increaser
   d. adapter

10. Sanitary __________ are used to enlarge the diameter of vent stacks.
    a. combinations
    b. expanders
    c. increasers
    d. wyes

8.0.0 ◆ GRADE

Drainage and waste systems (see *Figure 36*) rely on gravity to move solid and liquid wastes, so these piping systems must be installed at a slope toward the point of disposal. In the plumbing industry, this slope is called **grade.** Grade is also often referred to as **slope** or percent of grade. Drainage and waste piping systems are designed with the grade engineered into the system. Grade determines the velocity of the liquid waste flowing through the piping. If the grade is too shallow, the liquid waste moves too slowly and will not scour the pipe and remove solid wastes. If the grade is too steep, the liquid waste will flow too fast and leave solids behind in the piping. The architects and engineers who design the piping system normally determine the grade. However, in some cases, such as in residential plumbing, the plumber selects the grade according to applicable code.

8.1.0 The Importance of Grade

In a system with the proper grade, the liquid wastes will flow at the right **velocity,** or speed, to scour the insides of the pipe, and the solids will be carried away. Proper grade is essential. If too much grade is used, the liquid wastes may flow too fast, leaving the solids behind. If too little grade is used, the liquid wastes will not flow fast enough to scour the pipe and remove the solid wastes. If the grade of a pipe does not remain constant, the velocity of the liquid wastes will change at the point where the grade changes. In any of these cases, the pipe will soon become blocked with solid wastes.

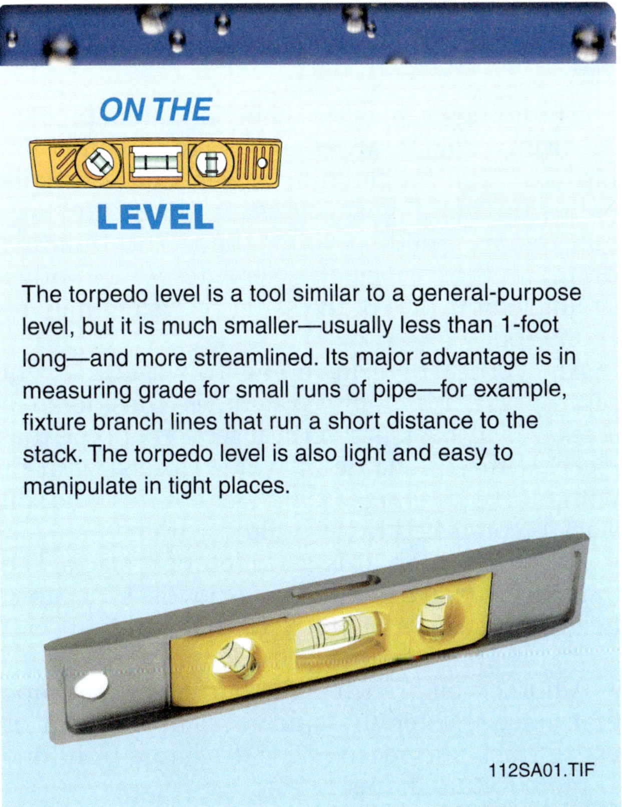

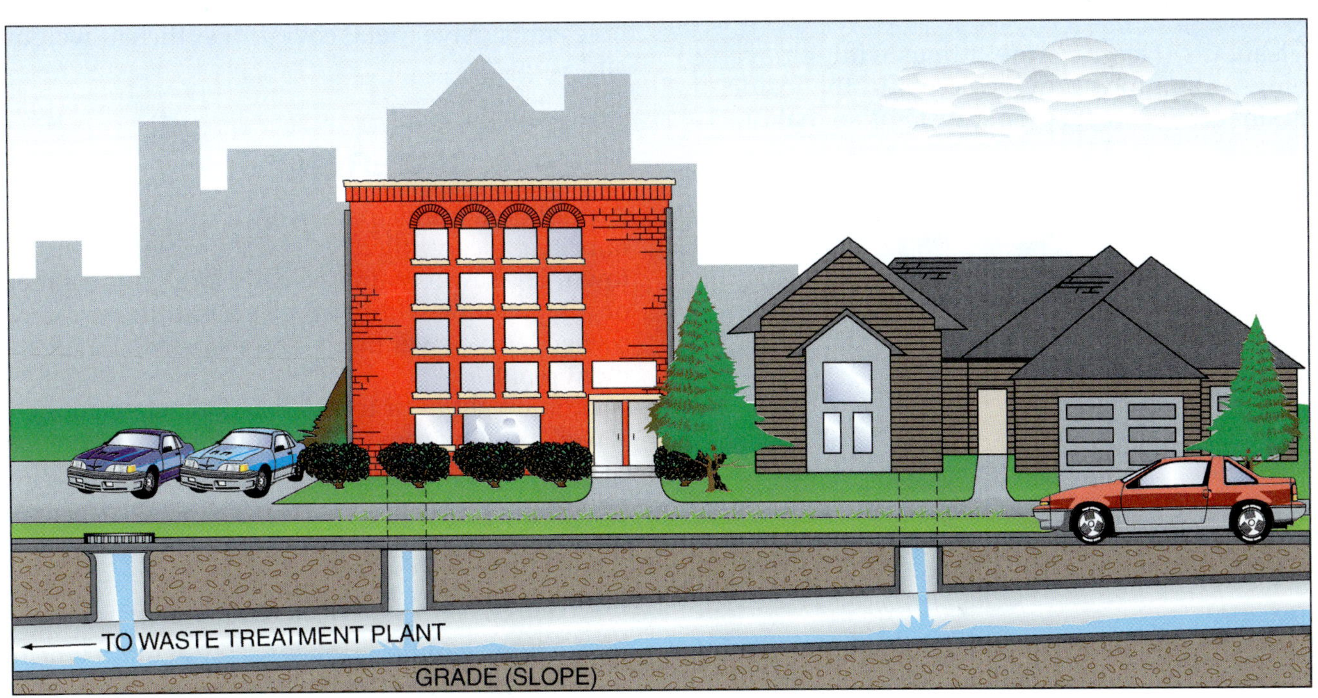

Figure 36 ◆ Grade.

Plumbers must determine the grade before they begin their work. This information may be in the local plumbing codes, in the specifications (specs) for the structure, or in the construction drawings. If it is not in any of these places, contact the local plumbing inspector for a decision.

9.0.0 ◆ BUILDING DRAIN

The building drain is the main horizontal pipe inside a building. It carries all sewage and other liquid wastes to the building sewer. Codes usually define the building sewer as standing 2 to 3 feet outside the building foundation. The building drain is the principal artery to which other drainage branches of the sanitary system may be connected.

Any vertical pipe, including the waste and vent piping of a plumbing system, is considered a stack. A soil stack is a vertical section of pipe that receives the discharge of water closets, with or without the discharge from other fixtures. The soil stack is connected to the building drain.

A **branch interval** is a section of a stack. The branch interval usually corresponds to a story height (the height of one floor in a building), but it can never be less than 8 feet long.

A horizontal branch is the part of a drain pipe that extends laterally (sideways) from a soil or waste stack and receives the discharge from one or more fixture drains.

9.1.0 Cleanouts

Cleanouts (*Figure 37*) are fittings with removable plugs. The plugs provide access to the inside of drainage and waste piping systems so that blockages can be removed. As mentioned earlier, cleanout adapters may be used to convert other fittings so they can be used as cleanouts.

PVC DWV TWISTLOK™ PLUG

112F37.TIF

Figure 37 ◆ Cleanout adapter with recessed hubs.

10.0.0 ◆ BUILDING SEWER

The building sewer or house sewer is the drainage piping that runs from the building's foundation to the sewer main or septic tank (private waste disposal system). It normally starts approximately 2 to 3 feet outside the building.

Building sewers are commonly made using ABS (acrylonitrile-butadiene-styrene) or PVC (polyvinyl chloride) plastic, cast-iron, or vitrified clay (a hard, nonporous clay) pipe. When sewers are laid, the ground must be tamped to keep the pipes from settling and losing their grade. Sometimes the pipes are installed over a bed of gravel that supports them. In some parts of the country, plumbers lay the entire sewer from the foundation wall to the sewer main. In other areas, the municipal sewer crew may be in charge of the installation from the property line to the sewer main. All operations associated with building sewers are regulated by code. Always check the applicable codes when you are working on a sewer system.

Manholes must be provided for underground piping that is 8 inches or larger in diameter. They should be located at intervals not more than 400 feet apart and at every major change in direction, grade, **elevation,** or pipe size. To meet applicable codes for traffic and loading conditions, the manholes must have metal covers of sufficient weight and strength.

11.0.0 ◆ SEWER MAIN

A public or municipal sewer is installed, maintained, and controlled by the local municipality or town. The sewer main is usually located in a street or alleyway or within an easement on privately owned land. Sewer mains carry waste to the treatment plant.

Municipalities or towns usually install a 6-inch sewer laterally from the sewer main to the edge of each building lot. This lateral pipe connects the building sewer to the public sewage system.

12.0.0 ◆ WASTE TREATMENT

Many municipalities have sewer systems in which wastes are collected and treated at a sewage plant, then discharged back into the ecosystem. Other municipalities require households to treat their waste individually in private waste disposal systems.

12.1.0 Municipal Waste Treatment Systems

Municipal waste treatment plants (see *Figure 38*) are highly sophisticated facilities, as public health and safety depend on their operation.

These systems are designed to handle thousands of gallons of sewage each day. The treatment facilities receive sewage into huge holding tanks, where heavier substances settle to the bottom and lighter substances float to the top. The heavier layer is called **sludge.** Both the sludge and the wastewater are then treated.

12.2.0 Private Waste Disposal Systems

Private waste disposal systems are designed to meet the needs of individual households and the requirements of applicable codes and health departments. As in municipal waste treatment systems, but on a much smaller scale, the waste flows into a holding tank, where the sludge settles out and is digested by bacteria. Liquid waste flows through a distribution box into a leachfield, where it seeps into the earth in a natural purification cycle.

Plumbers must know about different types of private disposal systems and the advantages and disadvantages of each. Plumbers also must be able to install private waste disposal systems correctly and according to code requirements. You will learn more about these systems as you advance through the plumbing curriculum. The following is a description of one basic component of a private waste disposal system—the conventional septic tank system.

A conventional septic system consists of a septic tank, a distribution box, a leachfield, and piping between those parts (see *Figure 39*). A septic tank system provides partial treatment of raw wastewater. It protects the soil absorption system from becoming clogged by solids that are suspended in the raw wastewater. Applicable codes strictly regulate the use of these systems.

112F38.EPS

Figure 38 ◆ Municipal sewage treatment plant.

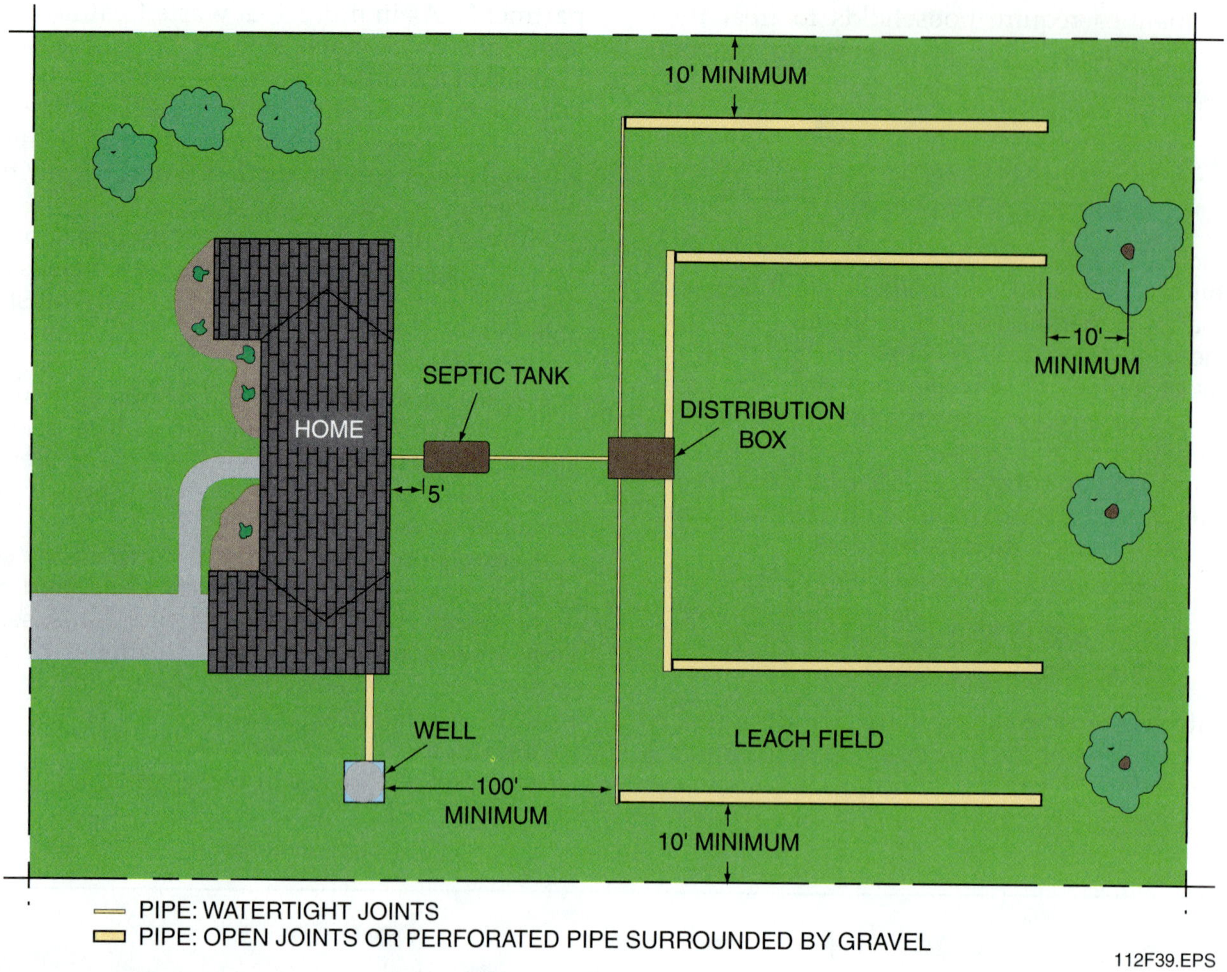

10' MINIMUM
SEPTIC TANK
HOME
5'
DISTRIBUTION BOX
10' MINIMUM
WELL
100' MINIMUM
LEACH FIELD
10' MINIMUM
PIPE: WATERTIGHT JOINTS
PIPE: OPEN JOINTS OR PERFORATED PIPE SURROUNDED BY GRAVEL
112F39.EPS

Figure 39 ◆ Typical layout for a septic tank system.

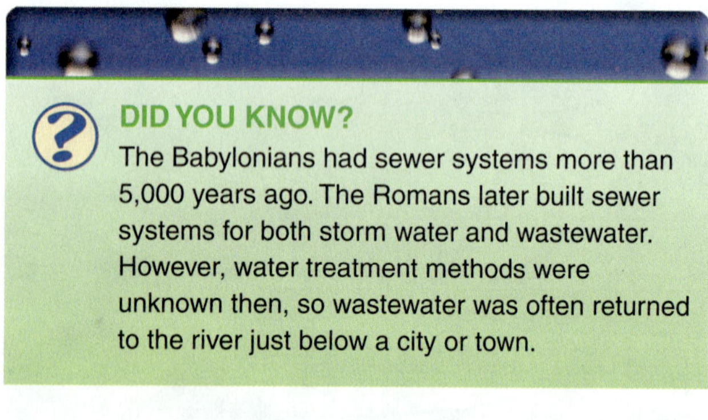

DID YOU KNOW?

The Babylonians had sewer systems more than 5,000 years ago. The Romans later built sewer systems for both storm water and wastewater. However, water treatment methods were unknown then, so wastewater was often returned to the river just below a city or town.

13.0.0 ◆ CODE AND HEALTH ISSUES

Properly designed and installed DWV systems are essential to public safety. Without DWV systems, the public would be at great risk of waste-borne illness and disease. The plumbing profession has improved public health, safety, and comfort over the past 150 years. Many serious health risks have been dramatically reduced as a direct result of good plumbing, especially properly designed and installed DWV systems, and the enforcement of plumbing codes.

13.1.0 DWV Plumbing Codes

Plumbing codes protect the safety, health, and welfare of the public. Although code requirements vary, all codes are based on principles of sanitation and safety.

There is no single national plumbing code whose requirements are adopted by all states and localities, but model codes are developed and revised on a regular basis. States and other jurisdictions can use these model codes as a basis for developing their own plumbing codes. DWV piping requirements are also defined by counties and municipalities and enforced by their inspectors.

You must always check the codes in the area where you are working. Model codes, for example, vary widely on how far vents have to be from traps and on the size of fixture drains. Consult applicable code before you start a job. Some variations in model codes are shown in *Table 1*.

Almost all codes require you to install cleanouts to provide access to all parts of the drainage system so that obstructions can be removed. Cleanouts range from removable plugs in horizontal drainage piping to manhole covers in building sewers. Codes vary widely on the specifics, so you must check applicable code requirements.

Under most codes, you may not use a cleanout plug opening to install new fixtures, unless there is another cleanout of equal accessibility and capacity and you have written approval from the plumbing inspector. On pipes that are 4 inches or less in diameter, many codes specify that cleanouts must be the same size as the pipe they serve. For larger pipes, the size of the cleanout must be at least 4 inches in diameter.

The various plumbing codes require that cleanout fittings be installed at specified locations within the DWV piping system and that the fittings be accessible. Most codes have requirements for cleanout locations in horizontal runs of pipe. In horizontal drain lines 4 inches in diameter or less, cleanouts generally must be installed no more than 50 feet apart. For larger lines, cleanouts cannot be more than 100 feet apart. Local codes may differ regarding these distances. Always refer to the applicable code.

Some codes also require cleanouts at or near the foot of each waste or soil stack and near the junction of the building drain and the building sewer.

13.2.0 Health Issues

The health issues related to the improper design, installation, and maintenance of DWV systems are significant. In the past, diseases such as cholera, typhoid fever, typhus, and dysentery have been traced to failures in DWV systems. These diseases can spread rapidly when bacteria from the sewer system enter buildings through damaged or improperly installed vents and traps.

Another health issue related to DWV systems is the accumulation of toxic sewer gases. Explosions, fires, and suffocation can occur when sewer gases are not properly released through good ventilation. Control of sewage through DWV systems can eliminate many public sanitation and health problems.

DID YOU KNOW?

Good plumbing saves lives. Thus, good plumbing systems are extremely important to the health of our nation. In 2003, the World Health Organization commissioned an investigation to find out if plumbing systems contributed to the outbreak of severe acute respiratory syndrome (SARS) in Hong Kong. The investigation stated that poor installation, operation, and improper plumbing applications were likely contributors. Virus-rich droplets re-entered apartments through sewage and drainage systems where there were strong upward airflows, inadequate traps, and nonfunctioning water seals. Investigators also noted that plumbing, when properly designed, installed, and maintained, is an important tool in stopping transmission of disease through a building's drainage system.

Source: 2003 World Health Organization Press Release, www.who.int/mediacentre/releases/2003/pr70/en/print.html.

Review Questions

Sections 8.0.0–13.0.0

1. Drainage and waste systems depend on __________ to move solid and liquid wastes through the pipes.
   a. seepage
   b. gravity
   c. specifications
   d. adapters

2. The main horizontal drain pipe located inside a building is called a __________.
   a. DWV stack
   b. building drain
   c. soil stack
   d. branch interval

3. Fittings that work as removable plugs used to provide access to the inside of DWV piping are called __________.
   a. ports
   b. cleanouts
   c. inlets
   d. outlets

4. The __________ is the drainage piping that goes from the building's foundation to the sewer main or septic tank (private waste disposal system).
   a. horizontal branch
   b. building sewer
   c. building drain
   d. vent stack

5. In a septic tank, __________ settles out of the raw wastewater.
   a. seepage
   b. bacteria
   c. sewage
   d. sludge

6. Plumbing codes do not vary among jurisdictions.
   a. True
   b. False

Summary

The DWV system of a building is part of a plumbing system designed to protect the health and safety of the people who use its facilities. The system carries wastewater out of the building, treats it, and returns it to the ecosystem.

Drain and waste piping removes wastewater from a building. The type and size of the piping system selected depends on critical factors such as the amount of fluids expected to flow through the piping, the types of fluids that will be carried, and the grade. Drainage and waste systems rely on gravity to move solid and liquid wastes, so the plumber must install piping at the correct grade toward the building sewer where the wastes leave the building and enter the public or private waste disposal system. Grade determines the velocity of the liquid waste flowing through the piping. If the grade is too shallow, the liquid waste moves too slowly and will not scour the pipe and remove solid wastes. If the grade is too steep, the liquid waste will flow too fast and leave solids behind in the piping.

The vent system is an important part of the overall DWV system. Vent piping provides for the free flow of air in the drainage system. This air equalizes the atmospheric pressure inside the pipes and prevents back pressure or siphoning from destroying the water trap seals in the fixtures. The water trap seals keep sewer gas and odors out of the building. Traps provide a way for the wastewater or sewage to flow through the fixture and into the piping system while protecting the occupants of the building from bacteria and potentially explosive gases.

The DWV piping system relies on different kinds of fittings to join the lengths of pipe. The fittings may be made of copper, brass, lead, steel, cast iron, clay, glass, or plastic. The type of installation determines the best piping and fitting material to use. The fittings are designed so they do not block or slow the flow of materials in the pipe. The sweeping design of DWV fittings allows for the smooth flow of material within the system. Different fitting shapes serve specific purposes, depending on whether a pipe runs horizontally or vertically. Correct selection and installation of fittings are vital for the system to function properly.

Notes

Trade Terms Quiz

Fill in the blank with the correct trade term that you learned from your study of this module.

1. DWV branches are connected to the main DWV system using _______________.
2. _______________ describes the attachment of one trap to another trap in a water closet.
3. The substance that settles on the bottom of holding tanks is _______________.
4. The underground drain pipe that carries waste from the building to the public sewer is called a(n)_______________.
5. _______________ allow pipes made from different materials to be connected.
6. The flow from two opposite runs into a single run of pipe are collected and combined in _______________.
7. _______________ are bend fittings with a short radius that are used at the base of a DWV stack.
8. In cold climates, _______________ can prevent vent openings from closing as a result of frozen condensation in pipes.
9. _______________ may not be used in drainage systems because their design can cause waste to collect in the fitting.
10. The sharp turning radius of _______________ severely restricts the flow of materials, making their use allowable only in vent lines.
11. _______________ are designed to keep undesirable or hazardous materials from entering a building drainage system, a public or private sewer, or sewage treatment plant or process.
12. Vertical branch lines that intersect with horizontal branch lines must be joined at an angle with _______________.
13. _______________ serve as locations to conduct leakage tests.
14. _______________ collects on fittings, especially on iron or steel, as a result of metal corrosion.
15. Horizontal branch lines that intersect with other horizontal branch lines at a 90-degree angle are connected using _______________.
16. _______________ is one of four factors that must be considered when determining the intervals for manholes.
17. The _______________, also called _______________, of the piping system works with gravity to move solid and liquid waste through DWV systems.
18. The discharge overflow of the trap outlet is the _______________, which is also simply known as a(n) _______________.
19. _______________ strain the wastewater before it enters the drainage piping.
20. A(n) _______________ is one or more lengths of pipe that continue in a straight line.
21. The pressure inside the DWV piping that is greater than atmospheric pressure is called _______________, or _______________.
22. A(n) _______________ is the access point to all parts of the drainage system for the removal of blockages.
23. _______________ allow smaller lines to be connected to a bend.
24. A(n) _______________ is the section of stack that connects branch pipes to the main DWV stack.
25. An imbalance in pressure between the inside and outside DWV piping can cause _______________.
26. Code generally requires the use of _______________ at the base of stacks.
27. _______________ is a general term for most vertical line including offsets of soil, waste, vent, or inside conductor piping.
28. A one-piece or two-piece trap with a union nut is called a(n) _______________.
29. A(n) _______________ consists of a circuit of piping inside a building.

30. Improvements in plumbing have replaced the _______________ with the P-trap.

31. Porous material caught in a trap can cause a seal to be broken by _______________.

32. The infrequent use of a trap can cause _______________ of the seal.

33. _______________ are devices used to connect pipe in DWV systems.

34. The upper ends of vents are joined to the top of soil and waste stacks by _______________.

35. Vent stacks are attached to the lower end of the soil and waste stacks by _______________.

36. Connecters that may be used in place of vent branches are called _______________.

37. The level line in which water tends to flow through the trap is the _______________.

38. _______________ is the speed at which waste flows through pipes.

39. The _______________ in a horizontal pipe from crown weir to vent must not exceed one pipe diameter.

Trade Terms

Adapter
Back pressure
Backpressure backflow
Branch interval
Building sewer
Capillary attraction
Cleanout
Crown weir/trap weir
Double ¼ bend
Double trapping
Drainage fitting
DWV system
Elevation
Evaporation
Fall
Fixture drain
Grade
Hydraulic gradient
Interceptor
Inverted wye
Long sweep ¼ bend
P-trap
Pipe scale
Run
S-trap
Sanitary combination
Sanitary fitting
Sanitary increaser
Sanitary upright wye
Sanitary wye
Short sweep ¼ bend
Side inlet
Siphonage
Slope (percent of grade)
Sludge
Stack
Test tee
Velocity
Vent branch (branch vent)
Vent ell
Vent tee
Weir (trap or crown)

Profile in Success

Jonathan Byrd

Jonathan Byrd was born in Fort Collins, Colorado, and grew up in the nearby Boulder area. After moving to Georgia in 1992, his dissatisfaction with the retail industry led him to a job in the plumbing trade, where he worked for the same plumber for several years.

Jonathan then earned his journeyman classification and joined Kosciusko, Mississippi-based Ivey Mechanical Company. Taking advantage of the company's training programs, he enrolled in a prep class that led to his Unrestricted Master's License, a Georgia license that allows a plumber to perform any type of plumbing work within the state.

Throughout his career, Jonathan has worked on dozens of projects in the Southeast, from sports arenas to broadcasting studios. Most notably, he participated in the restoration of the Georgia State Capitol Building. Built in the 1890s, the capitol building was in desperate need of new fixtures and plumbing. Jonathan assisted in the removal of the existing plumbing and the installation of the new plumbing and fixtures, often through terra cotta walls that were several feet thick.

While at Ivey Mechanical, Jonathan was offered a position teaching the apprenticeship program for the Construction Education Foundation of Georgia (CEFGA). It was there that he discovered his love of teaching. He taught the apprenticeship program at CEFGA part time for two years before moving to a full-time position. Eventually, he returned to a full-time position at Ivey, where he now teaches the prep class for the journeyman's license and arranges training opportunities for Ivey employees. As well as teaching continuing-education classes for the Plumbing and Mechanical Association of Georgia, Jonathan continues to teach CEFGA's apprenticeship program part time.

According to Jonathan, the plumbing trade offers more freedom today than it did years ago. He says, "Today you can specialize. Back in the day, a plumber had to do a little of everything in order to survive. There's a lot of things a person can do."

Beyond the training opportunities that companies offer to their employees, craftspeople can go to associations to continue their training. "Associations are the ones who are going to assure that craft training is done right," Jonathan says. "Associations have a vested interest in training. Some are just for continuing education rather than apprenticeship, but almost every association is going to offer some training."

Jonathan advises, "Take the time to learn the trade. That might sound trite, but some people out there believe that they can be a helper for three months and then start calling themselves a plumber. I was a helper for years. I might have cut that time in half if I had some formal training."

He believes that textbook learning is also important. "The things you learn in a textbook you might not need for several years, but when you finally come across different materials on a job site, you'll know what to do." Jonathan has been in the trade long enough to see too many untrained people make some serious mistakes. "They'll try to take shortcuts, but somebody down the road will end up having to fix them. That just makes the property owner upset. Plumbing has to be done right the first time. Eighty-five percent of the work a plumber does is behind a wall. So, if a mistake has been made, someone has to go back, cut the wall, fix the mistake, and patch up the wall. It's time-consuming."

When asked what he likes most about his job, Jonathan replied, "Although I do the same thing every day, I'm always doing it somewhere new. So it's the same, and it's not the same. Once you know a trade, there's going to be enough to keep it interesting."

Trade Terms Introduced in This Module

Adapter: A fitting that joins pipes of different sizes or materials, such as copper and galvanized pipe or cast-iron and plastic pipe.

Back pressure: A condition which may occur in the potable water distribution system, whereby a higher pressure than the supply pressure is created causing a reversal of flow into the potable water piping. Also referred to as backpressure backflow.

Backpressure backflow: Another term for back pressure.

Branch interval: A distance along a soil or waste stack corresponding, in general, to a story height, but in no case less than 8 feet within which the horizontal branches from one floor or story of a building are connected to the stack.

Building sewer: That part of the drainage system which extends from the end of the building drain and conveys its discharge to a public sewer, private sewer, individual sewage-disposal system, or other point of disposal.

Capillary attraction: The tendency of water to be drawn by porous material. If a piece of paper or string is stuck in a pipe, the water is drawn out of the trap and the seal is broken.

Cleanout: An access point to all parts of the drainage system for the removal of blockages.

Crown weir/trap weir: Discharge overflow of the trap outlet. Also referred to as weir, trap, or crown.

Double ¼ bend: A fitting used to collect and combine the flow from two opposite runs into a single run of pipe.

Double trapping: A situation in which one trap is attached to another, creating negative pressure that will stop the flow of drainage.

Drainage fitting: Any of a variety of fittings used in the DWV system to remove waste from a building.

DWV system: An acronym for "drain-waste-vent" referring to the combined sanitary drainage and venting systems. This term is technically equivalent to "soil-waste-vent" (SWV).

Elevation: The height above an established reference point, such as a grade reference point on a construction drawing.

Evaporation: Loss of water, especially in a drainage system, into the atmosphere.

Fall: The amount of slope given to horizontal runs of pipe.

Fixture drain: The drain from the trap of a fixture to the junction of that drain with any other drain pipe.

Grade: The slope of a horizontal run of pipe. Also referred to as slope (percent of grade).

Hydraulic gradient: The level line in which water tends to flow in a pipe.

Interceptor: A device designed and installed so as to separate and retain deleterious, hazardous, or undesirable matter from normal waste while permitting normal sewage or liquid wastes to discharge into the drainage system by gravity.

Inverted wye: A fitting used to join the upper end of the vent to the top of the soil and waste stacks. It looks like an upside-down letter Y.

Long sweep ¼ bend: A bend used at the base of DWV stacks because the longer radius of its design greatly decreases flow resistance and back pressure. Use is regulated by code.

P-trap: A P-shaped trap that provides a water seal in a waste or soil pipe, used mostly at sinks and lavatories.

Pipe scale: Flaky material resulting from corrosion of metals, especially iron or steel. Also, a heavy oxide coating on copper or copper alloys resulting from exposure to high temperatures and an oxidizer.

Run: One or more lengths of pipe that continue in a straight line.

S-trap: A trap with a long downstream leg, which tends to promote siphonage. S-traps are no longer installed but are still found in older buildings.

Sanitary combination: A fitting that combines a wye and ⅛ bend. It is used to connect horizontal branch lines that intersect other horizontal branch lines. It offers less resistance to the flow of material than a sanitary tee. Also called a tee-wye.

Sanitary fitting: A fitting used to connect DWV branches to the main DWV system and to serve as a cleanout.

Sanitary increaser: A fitting used to enlarge the diameter of the vent stack. It is usually placed at least 1 foot below the intersection of the stack and the roof. Use is regulated by code.

Sanitary upright wye: A fitting used to connect the vent stack to the lower end of the soil and waste stacks.

Sanitary wye: A drainage fitting, shaped like the letter Y, that joins the main run of pipe at an angle.

Short sweep ¼ bend: A bend fitting with a short radius used at the base of a DWV stack. Use is regulated by code.

Side inlet: An opening in an ell or tee fitting at right angles to the line of the run, used to connect smaller lines to the main line.

Siphonage: Loss of water in a trap seal caused by unequal pressure inside and outside DWV piping.

Slope (percent of grade): Another term for grade.

Sludge: Semi-liquid matter that settles out in a holding tank during the waste treatment process.

Stack: A general term for any vertical line including offsets of soil, waste, vent, or inside conductor piping. This does not include vertical fixture and vent branches that do not extend through the roof or that pass through not more than two stories before being reconnected to the vent stack or stack vent.

Test tee: A tee installed as a test location for pressurizing the system to test for leaks.

Velocity: Speed of motion, such as the speed of sewage or wastewater through the drainage piping.

Vent branch (branch vent): A vent connecting one or more individual vents with a vent stack or stack vent.

Vent ell: A plastic fitting with a sharp turn radius, used only in vent piping systems. Use is regulated by code.

Vent tee: A fitting used in venting systems or as a cleanout. It may not be used in the drainage system because it restricts the flow of material. Use is regulated by code.

Weir (trap or crown): Also referred to as crown weir or trap weir.

Additional Resources and References

ADDITIONAL RESOURCES

This module is intended to present thorough resources for task training. The following reference works are suggested for further study. These are optional materials for continued education rather than for task training.

Basic Plumbing with Illustrations, Revised Edition, 1994. Howard C. Massey. Carlsbad, CA: Craftsman Book Company.

Plumbing Systems: Analysis, Design and Construction, 1996. Tim Wentz. Upper Saddle River, NJ: Prentice Hall.

REFERENCE

2003 National Standard Plumbing Code. Chapter 1, Definitions (definitions of back pressure, branch interval, building sewer, crown weir or trap, DWV system, fixture drain, interceptor, stack, and vent branch). Falls Church, VA: Plumbing-Heating-Cooling Contractors–National Association.

NCCER CURRICULA — USER UPDATE

NCCER makes every effort to keep its textbooks up-to-date and free of technical errors. We appreciate your help in this process. If you find an error, a typographical mistake, or an inaccuracy in NCCER's curricula, please fill out this form (or a photocopy), or complete the online form at **www.nccer.org/olf**. Be sure to include the exact module ID number, page number, a detailed description, and your recommended correction. Your input will be brought to the attention of the Authoring Team. Thank you for your assistance.

Instructors – If you have an idea for improving this textbook, or have found that additional materials were necessary to teach this module effectively, please let us know so that we may present your suggestions to the Authoring Team.

NCCER Product Development and Revision

13614 Progress Blvd., Alachua, FL 32615

Email: curriculum@nccer.org
Online: www.nccer.org/olf

❑ Trainee Guide ❑ AIG ❑ Exam ❑ PowerPoints Other ____________________________

Craft / Level: Copyright Date:

Module ID Number / Title:

Section Number(s):

Description:

Recommended Correction:

Your Name:

Address:

Email: Phone:

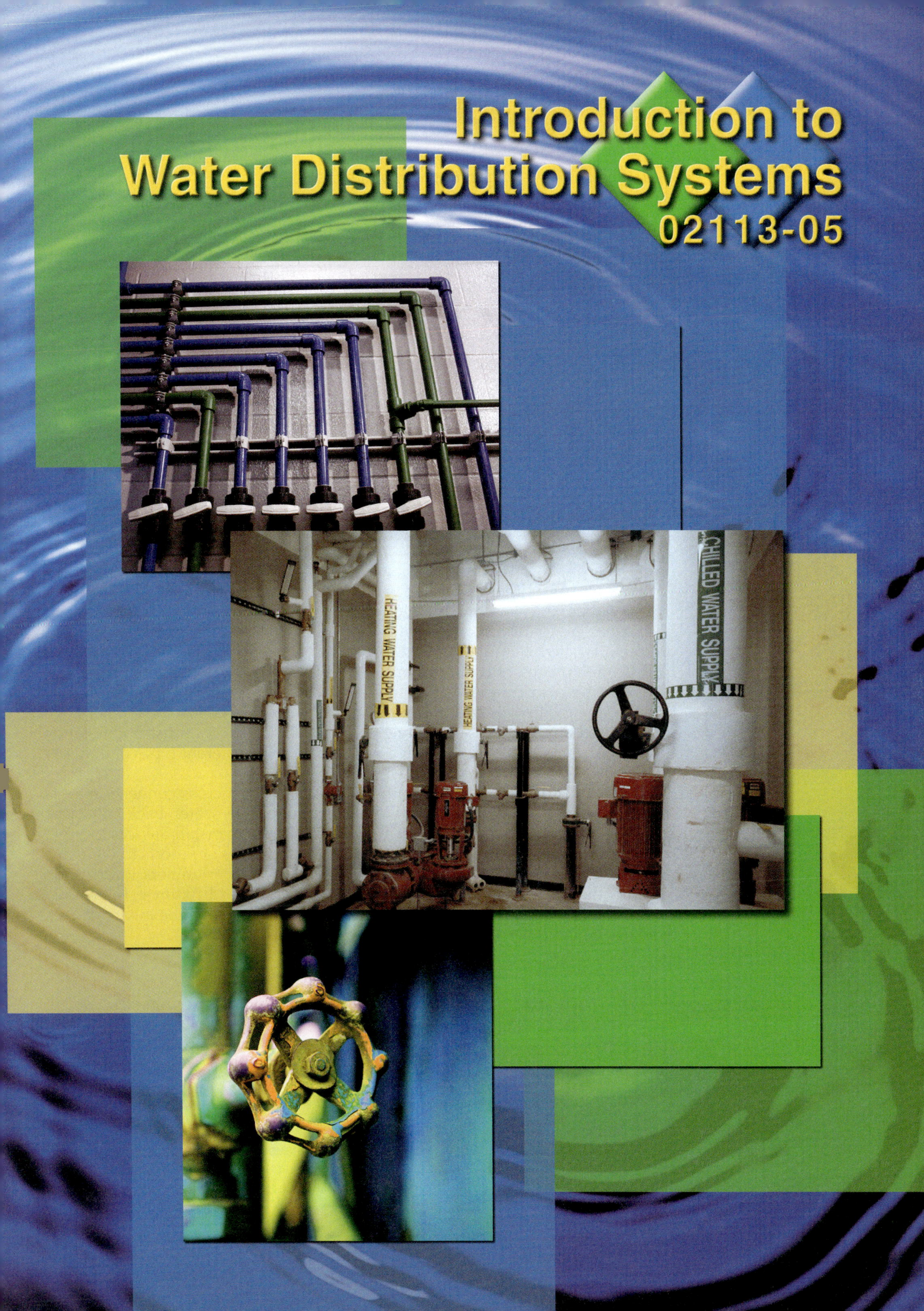
Introduction to
Water Distribution Systems
02113-05
HEATING WATER SUPPLY
HEATING WATER SUPPLY
CHILLED WATER SUPPLY

02113-05

Introduction to Water Distribution Systems

Topics to be presented in this module include:

Overview

Water supply and distribution play an important part in plumbing systems. Water supply is either public or private; components include pipes and fittings, valves, and water heating and treatment equipment. Water that comes from open reservoirs, lakes, streams, and some wells usually must be treated before it is safe for human use. In municipal systems, city water undergoes a purification process. Private water treatment often includes water softeners.

The water distribution system moves water from its source to the building or structure where it is needed. The way the water moves and the type of materials used depend on the building or structure. Plumbers must understand how these water distribution systems work and how different types of materials are used in these systems.

Plumbers protect water distribution systems against cross-connection and backflow, the unwanted flow of used or nonpotable water back into the potable water distribution system. To do this, plumbers install devices to protect against backpressure and backsiphonage, and valves or faucets to regulate the flow of water. After the supply piping is installed, plumbers locate and connect the water heater, hose bibbs, water softener, and fixtures. To ensure that the water supply is properly connected, plumbers must be aware of the construction and working principles of each fixture and faucet.

Focus Statement

The goal of the plumber is to protect the health, safety, and comfort of the nation job by job.

Code Note

Codes vary among jurisdictions. Because of the variations in code, consult the applicable code whenever regulations are in question. Referencing an incorrect set of codes can cause as much trouble as failing to reference codes altogether. Obtain, review, and familiarize yourself with your local adopted code.

Objectives

When you have completed this module, you will be able to do the following:

1. Describe the process by which water is distributed in municipal, residential, and private water systems.
2. Identify the major components of a water distribution system, and describe the function of each component.
3. Explain the relationships between components of a water distribution system.

Key Trade Terms

Angle valve
Backing board
Branch
Buffalo box
Check valve
Coagulation
Corporation stop
Curb box
Curb stop
Feeder line
Fixture risers
Full flow
Galvanic corrosion
Hammer arrester
Hose bibb
pH
Precipitate
Pressure regulator valve
Pressure relief valve
Reservoir
Service line
Straight-through flow
Supply stop valve
Temperature and pressure (T and P) relief valve
Thermostatic/pressure balancing valve, combination
Throttled flow
Trimming out
Turbidity
Vacuum breaker
Water meter
Water supply fixture unit (WSFU)
Water table
Well casing

Required Trainee Materials

1. Appropriate personal protective equipment
2. Pencils and paper
3. Copy of local adopted code
4. Calculator

Prerequisites

Before you begin this module, it is recommended that you successfully complete *Core Curriculum; Plumbing Level One,* Modules 02101-05 through 02112-05.

This course map shows all of the modules in the first level of the *Plumbing* curriculum. The suggested training order begins at the bottom and proceeds up. Skill levels increase as you advance on the course map. The local Training Program Sponsor may adjust the training order.

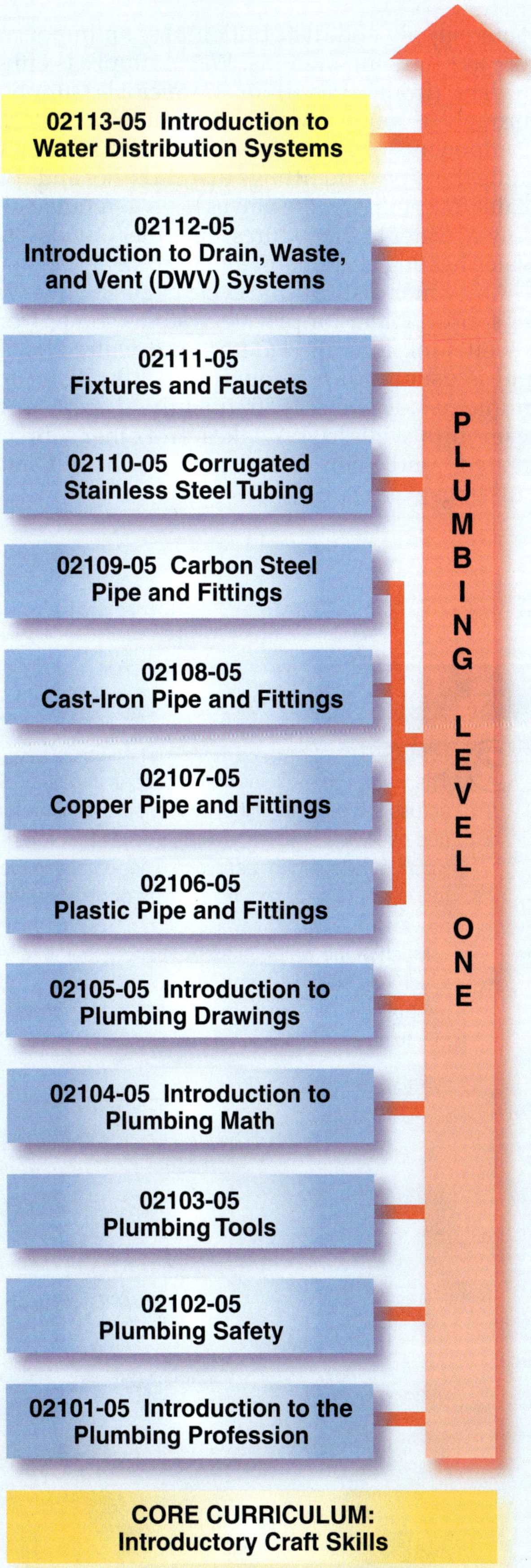

PLUMBING LEVEL ONE

02113-05 Introduction to Water Distribution Systems

02112-05 Introduction to Drain, Waste, and Vent (DWV) Systems

02111-05 Fixtures and Faucets

02110-05 Corrugated Stainless Steel Tubing

02109-05 Carbon Steel Pipe and Fittings

02108-05 Cast-Iron Pipe and Fittings

02107-05 Copper Pipe and Fittings

02106-05 Plastic Pipe and Fittings

02105-05 Introduction to Plumbing Drawings

02104-05 Introduction to Plumbing Math

02103-05 Plumbing Tools

02102-05 Plumbing Safety

02101-05 Introduction to the Plumbing Profession

CORE CURRICULUM: Introductory Craft Skills

113CMAP.EPS

1.0.0 ◆ INTRODUCTION

Water supply and distribution play an important role in plumbing systems. Water supply is either private (from a well) or municipal (supplied through a public water distribution system). Components of the water distribution system include the pipes and fittings that carry hot and cold water in a building, the valves used to regulate the flow of water to the fixtures and other outlets, and water heating and treatment equipment.

Any water distribution cycle begins with a water source. Water for private systems comes from a well sunk into an underground water supply and is usually pure enough to drink. Water for municipal systems is frequently drawn from **reservoirs,** wells, rivers, lakes, and other sources, treated, and then distributed to homes and buildings.

In municipal systems, city water undergoes a purification process before it reaches the faucet. Water is pumped to a treatment plant where harmful impurities are removed through processing. Chlorine, alum, and activated charcoal are added to the water during this cycle. The water, with its added chemicals, flows into a mixing basin where paddles mix the chemicals into the water. From here, the water moves to a settling basin where impurities are separated out. Moving from the basin, the water is filtered through sand and gravel to screen out most of the remaining impurities. Chlorine may be added again to make sure that the water is free of harmful bacteria. Some cities also add fluoride to help prevent tooth decay in the general population. The filtered water is held in a reservoir until it is pumped into the main supply pipes that lead to building **service lines.** The water is then delivered to sinks, bathtubs, showers, dishwashers, icemakers, hoses, and any other water outlet (see *Figure 1*).

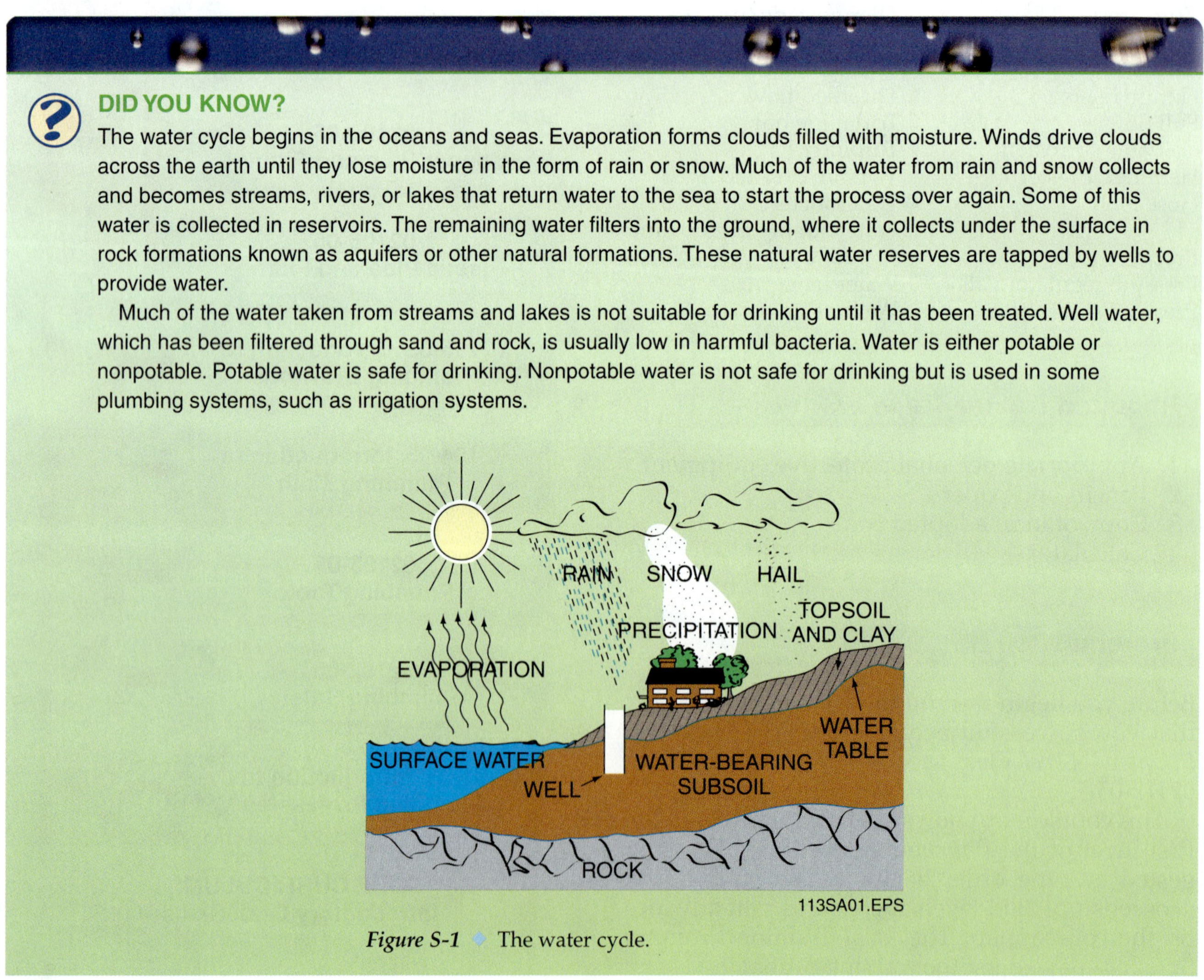

Figure S-1 ◆ The water cycle.

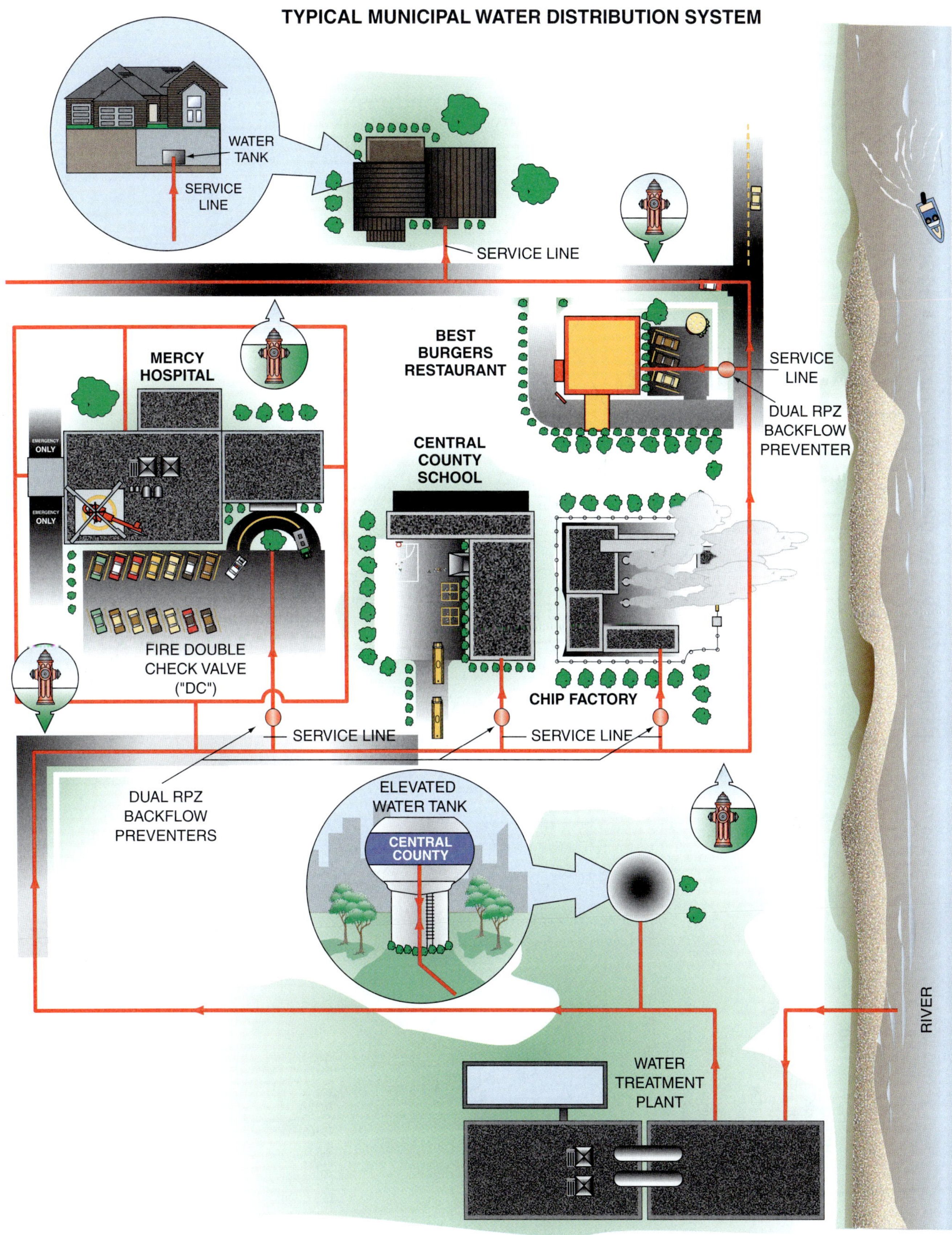

Figure 1 ◆ Water distribution. (1 of 2)

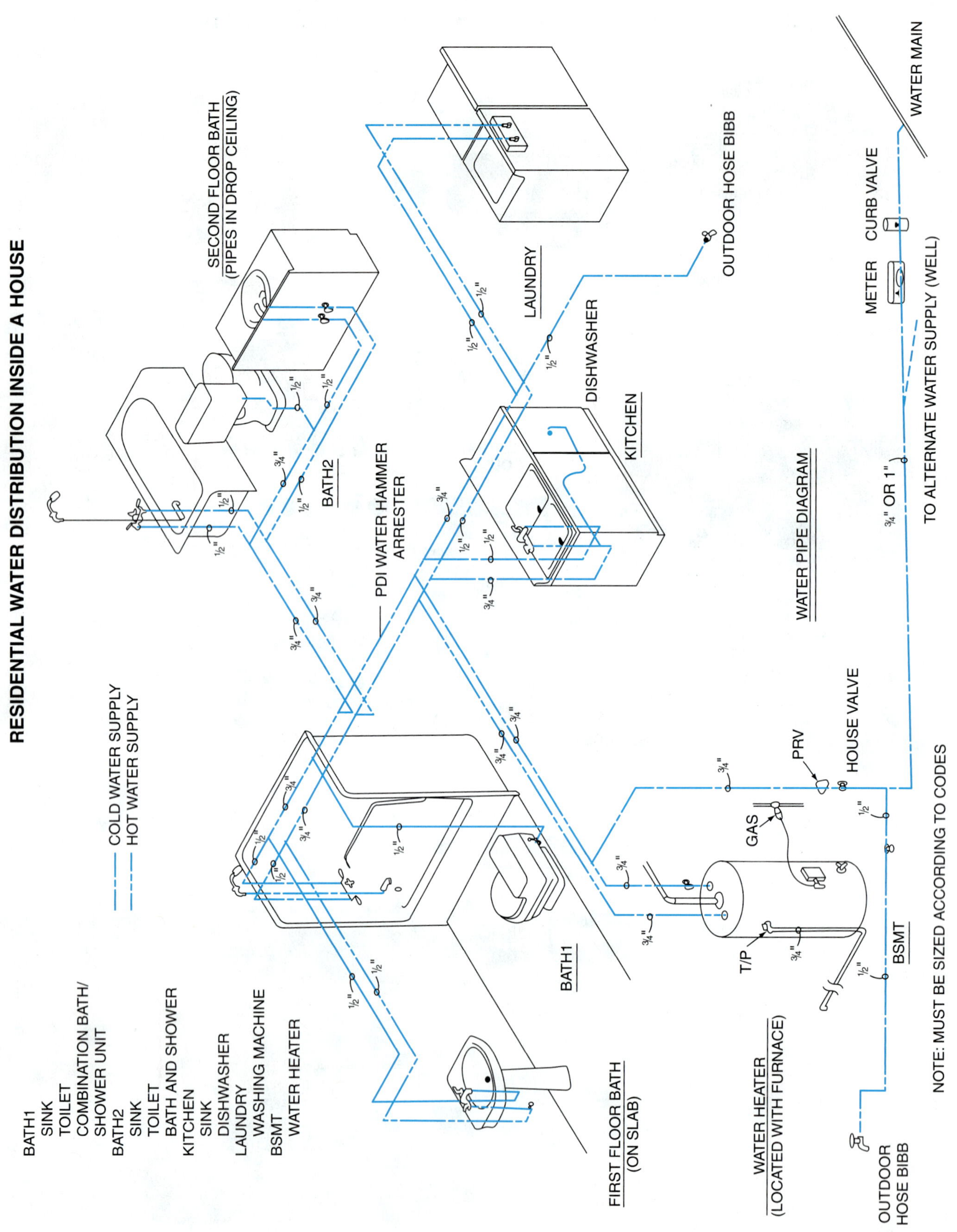

Figure 1 ◆ Water distribution. (2 of 2)

2.0.0 ◆ SOURCES OF WATER

Nearly 18 percent of the U.S. population gets its water from private sources. Most of this water comes from wells. Wells are sunk (dug, driven, drilled, or bored) into the earth to extract the water (see *Figure 2*). Wells that are dug or driven are considered shallow wells. This type of well is generally used where the **water table** is within 20 to 50 feet of the earth's surface. Dug wells are easily contaminated by surface water. Driven wells are made by forcing a well point into the earth. These wells are practical only where the soil is fairly free of rocks.

Wells that are drilled are considered deep wells. They may extend hundreds of feet down through the earth. They are made using a drilling rig that penetrates subsurface materials. Bored wells are made using an earth auger (drill). Once the water table is reached and an adequate amount of water volume is found, a **well casing** is inserted to protect the well from contamination. Casings are made of various materials and in different diameters (1½ to 2 inches) and depths, depending on how far the water is beneath the surface.

After the well is established, water is pumped to the surface to a storage tank inside or near the building, where it can be treated and used by the owners. Generally, the deeper the well, the more gallons per minute can be pumped. The pump pressurizes the water in the storage tank so that it provides enough water for each of the fixtures to work properly.

Reservoirs are another source of water. Most reservoirs are made by building dams across rivers or streams. Municipalities can then collect and store water for future use. Reservoirs are particularly important in areas that receive little rainfall for part of the year. A pump is used to move the water from the reservoir to the water treatment plant.

ON THE LEVEL

Locating and Drilling a Well

The first consideration before drilling or driving a well is state and local regulations. Wells may be approved for household, domestic, or commercial use. A local well driller is the best source of answers to the critical questions of what type of well is needed, what permits are required, and the history of water wells in the area. A variety of different aquifers (water channels) and water tables exist in various formations below the earth's surface. A history of wells drilled in a particular area provides important information for the well driller to locate the best spot to drill.

Other considerations are the distance from property lines and the position of septic tanks and leachfields in relation to the well site. Most codes require at least 100 feet between the well and the leachfield. Because the plumber will make connections to both the well and the wastewater disposal, the plumber must work closely with the owner, the well driller, and code officials.

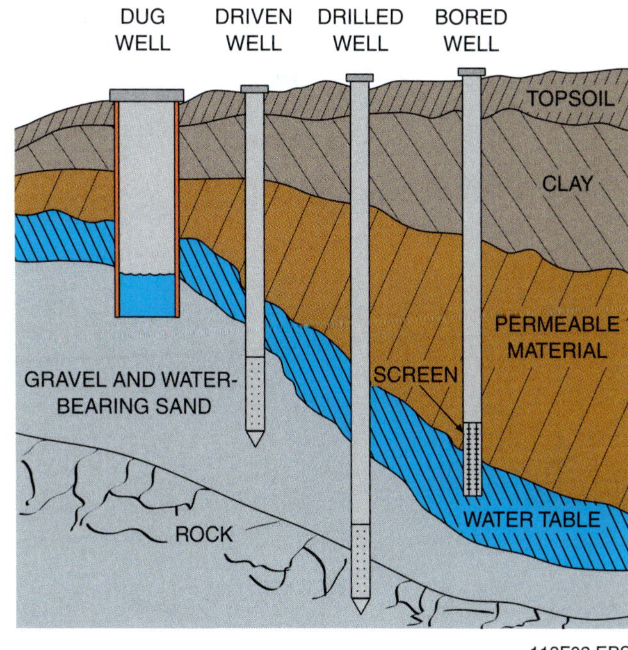

Figure 2 ◆ Types of wells.

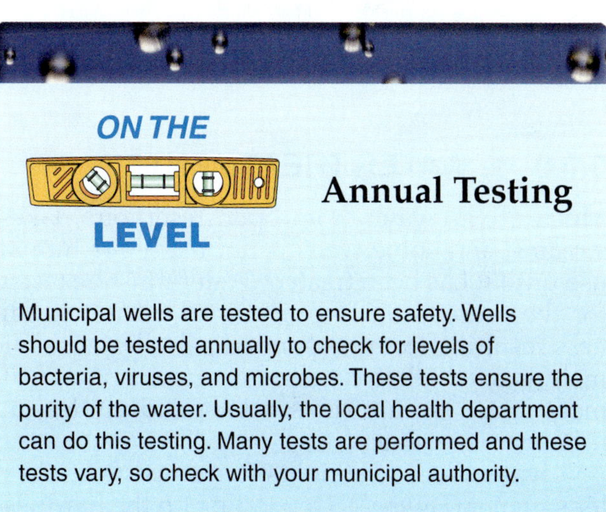

ON THE LEVEL

Annual Testing

Municipal wells are tested to ensure safety. Wells should be tested annually to check for levels of bacteria, viruses, and microbes. These tests ensure the purity of the water. Usually, the local health department can do this testing. Many tests are performed and these tests vary, so check with your municipal authority.

DID YOU KNOW?

Water Resources

Ninety-nine percent of Earth is ocean water, but fresh water is scarce and water resources are rapidly depleting. According to the United Nations Environment Program (UNEP), Earth has a total water volume of 1.4 billion km^3. Despite this large volume of water, fresh water only accounts for 2.5 percent of the total. Of the fresh water, 69 percent is ice and snow cover, 30 percent is stored as ground water, and 0.3 percent is in fresh water lakes and rivers. This means that the total amount of usable fresh water is less than 1 percent of all fresh water resources and is but 0.1 percent of the total water on Earth.

Source: United Nations Environment Program (UNEP) website at www.unep.org. Igor A. Skiklomanov, State Hydrological Institute (SHI, St. Petersburg) and United Nations Educational, Scientific, and Cultural Organization (UNESCO Paris) 1999, at www.unep.org/vitalwater/01.htm, reviewed March 23, 2004.

Selected Examples of Aquifer Depletion

Country	Region	Description of Depletion
China	North China Plain	Water table is falling by 2–3 meters per year under much of the Plain. As pumping costs rise, farmers are abandoning irrigation.
United States	Southern Great Plains	Irrigation is heavily dependent on water from Ogallala aquifer, largely a fossil aquifer. Irrigated areas in Texas, Oklahoma, and Kansas are shrinking as aquifer is depleted.
Pakistan	Punjab	Water table is falling under the Punjab and in the provinces of Baluchistan and Northwest Frontier.
India	Punjab, Haryana, Rajasthan, Andhra Pradesh, Maharashtra, Tamil Nadu, and other states	Water tables are falling by 1–3 meters per year in some parts. In some states, extraction is double the recharge. In the Punjab, India's breadbasket, water table is falling by nearly 1 meter per year.
Iran	Chenaran Plain, northeastern Iran	Water table was falling by 2.8 meters per year but, in 2001, drought and drilling of new wells to supply nearby city of Mashad dropped it by 8 meters.
Yemen	Entire country	Water table is falling by 2 meters per year throughout country and 6 meters a year in Sana'a basin. Nation's capital, Sana'a, could run out of water by end of this decade.
Mexico	State of Guanajuato	In this agricultural state, the water table is falling by 1.8–3.3 meters per year.

Source: China from Michael Ma, "Northern Cities Sinking as Water Table Falls," *South China Morning Post,* 11 August 2001; United States from Sandra Postel, *Pillar of Sand* (New York: W.W. Norton & Company, 1999), and from Bonnie Terrell and Phillip N. Johnson, "Economic Impact of the Depletion of the Ogallala Aquifer: A Case Study of the Southern High Plains of Texas," presented at the American Agricultural Economics Association annual meeting in Nashville, TN, 8–11 August 1999; Pakistan, India, and Mexico from Tushaar Shah et al., *The Global Groundwater Situation: Overview of Opportunities and Challenges* (Colombo, Sri Lanka: International Water Management Institute, 2000); Postel, op. cit. this note; Iran from Chenaran Agricultural Center, Ministry of Agriculture, according to Hamid Taravati, publisher, Iran, e-mail to author, 25 June 2002; Christopher Ward, "Yemen's Water Crisis," based on a lecture to the British Yemeni Society in September 2000, July 2001.

3.0.0 ◆ WATER TREATMENT

Much of the water from open reservoirs, lakes, streams, and some wells is not ready for human use until it has been treated. Water must be tested for the presence of chemicals, **turbidity** (cloudiness resulting from suspended particles), organic materials, or other types of contaminants. Treatment removes impurities, odors, and unpleasant taste from the water.

Private water treatment often consists of a water softener (*Figure 3*). Depending on the hardness and quality of the water, some homes may have a sediment filter or carbon filter.

Municipal water treatment is more complex because of the high volume of water the general public uses. Millions of gallons of water must be treated in municipal water treatment plants each day to ensure an acceptable level of safety for the public. Many municipal water treatment plants treat water using the following sequence (see *Figure 4*):

Step 1 Water is pumped from a river or lake.

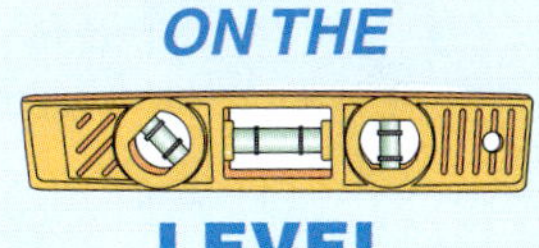

Hotel Water Conservation

Many hotels now place water-saving notices in their guestrooms. These notices commonly request guests to leave their towels on the floor only on the days that they want them washed. Hotels use these notices to encourage guests to reuse towels and bed linens. Guests are also being reminded to conserve water by turning off faucets when brushing their teeth and when shaving.

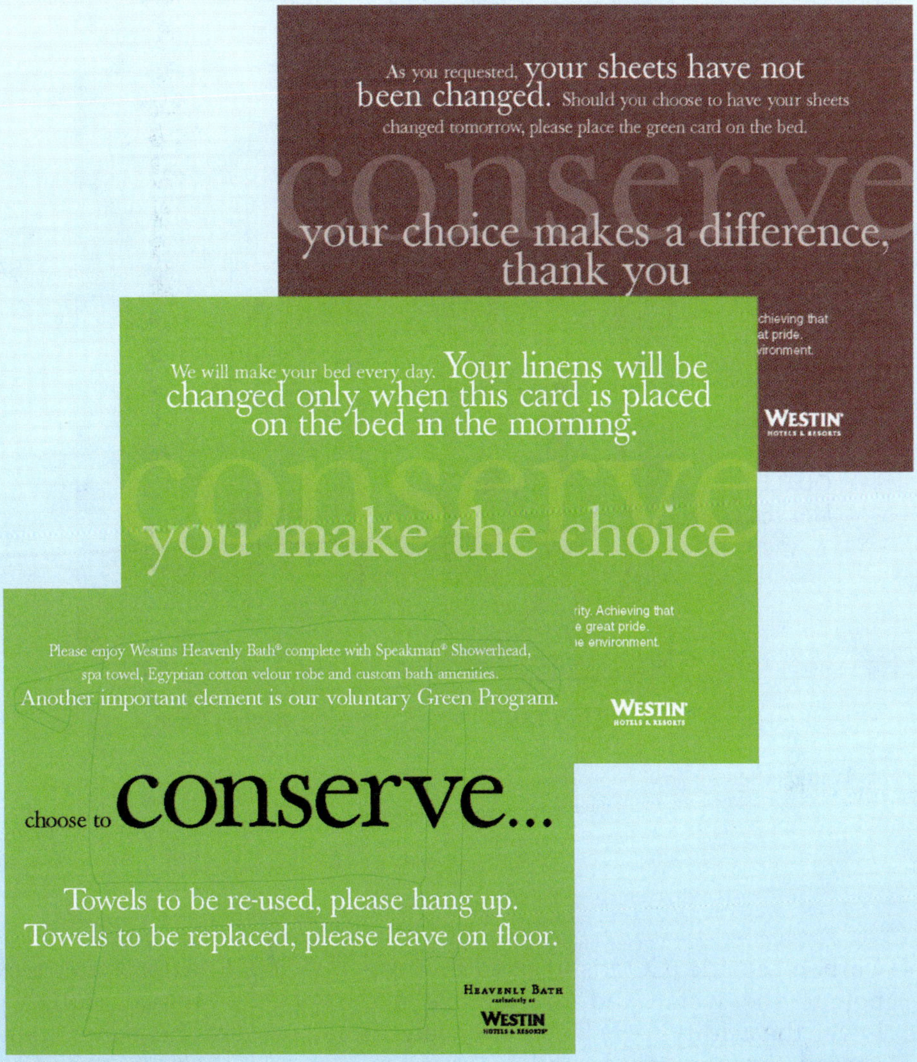

113SA02.EPS

Step 2 The water is sent to the aerators, where dissolved carbon dioxide is removed. The aerators also remove iron and manganese by oxidizing the minerals and filtering them out of the water.

Step 3 The aerated water is then pumped to the clarifier, where lime and soda ash are mixed in to cause **coagulation,** or thickening, which removes **precipitates** of calcium and magnesium. Precipitates are deposits containing these elements. This process softens the water.

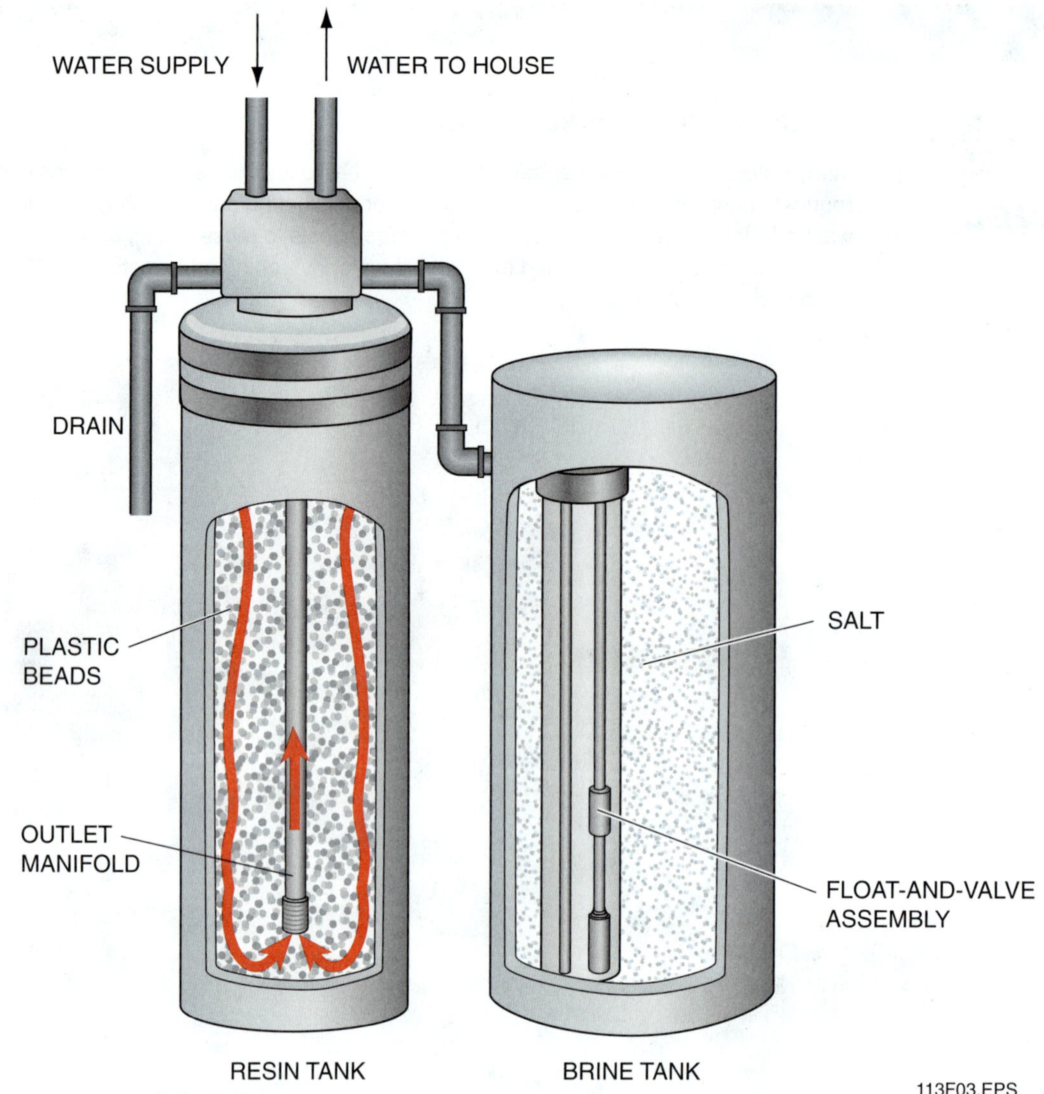

Figure 3 ◆ Water softener.

Step 4 Carbon dioxide (CO_2) is injected to recarbonate the water and to stabilize and lower the acidity (**pH**) level. The water is passed through rapid sand filters to remove any remaining particles.

Step 5 After the filtering process, the water passes to the tanks, where a chlorine solution (sodium hydrochloride) disinfects it. At this stage, fluoride is usually added to reduce the chance of tooth decay. This treated water is then stored in reservoirs until needed.

Step 6 After water leaves the treatment plant, it flows through pipes called water mains. The water mains usually run under the streets and serve many buildings. Permits are required to make connections to the water mains. In some areas, only municipal workers are authorized to make these connections. In other areas, licensed contractors or plumbers install the connections. The connections bring water from the main to the individual buildings through the building water service line.

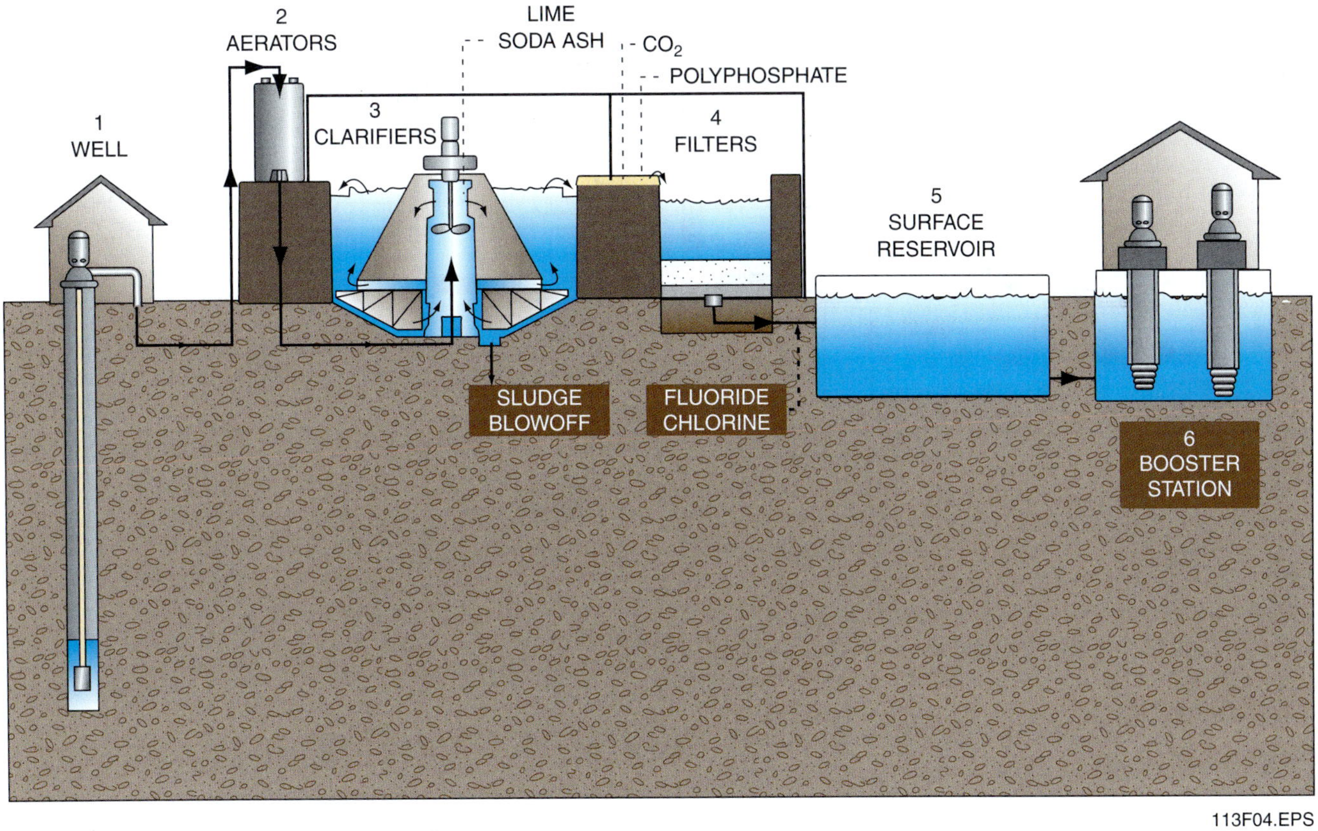

1 WELL
2 AERATORS
3 CLARIFIERS
LIME
SODA ASH
CO_2
POLYPHOSPHATE
4 FILTERS
5 SURFACE RESERVOIR
6 BOOSTER STATION
SLUDGE BLOWOFF
FLUORIDE CHLORINE
113F04.EPS

Figure 4 ◆ Municipal water treatment plant.

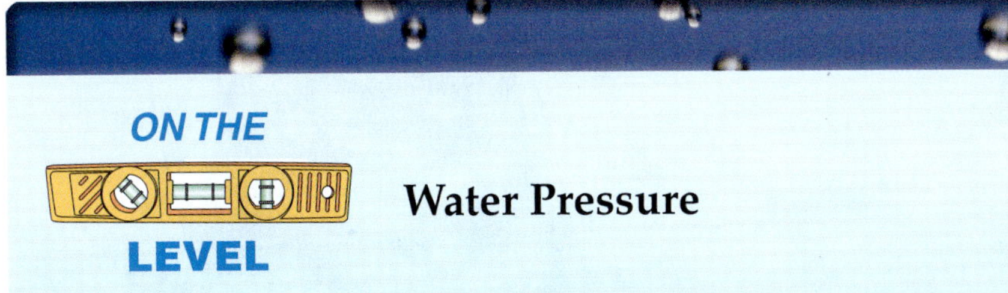

Water Pressure

The water pressure in some municipal systems may be either too low or too high for the plumbing system in a building. The plumber has ways to compensate. If the water pressure is too low, a booster system can be installed. One kind of booster has variable capacities to meet increased or decreased demand for water (see *Figure 5*). For example, it might have two pumps—one that functions at 33 percent of system capacity, and one that functions at 66 percent. When demand is low, the smaller pump operates. As demand increases, it shuts down and the larger pump starts up. At peak demand periods, both pumps operate.

If the water pressure in the system is too high, the plumber can use valves to lower it and to eliminate violent pressure changes (see *Figure 6*).

Valves are located according to the needs of the system. Some codes may require a pressure-reducing valve where pressure buildup at any time of the day exceeds 80 pounds per square inch (psi). A certain appliance (such as a dishwasher) may require a pressure-reducing valve because the operating pressure of the appliance is very different from the system's line pressure. Wherever they are located, the valves must be accessible for maintenance and must be protected from abuse or tampering.

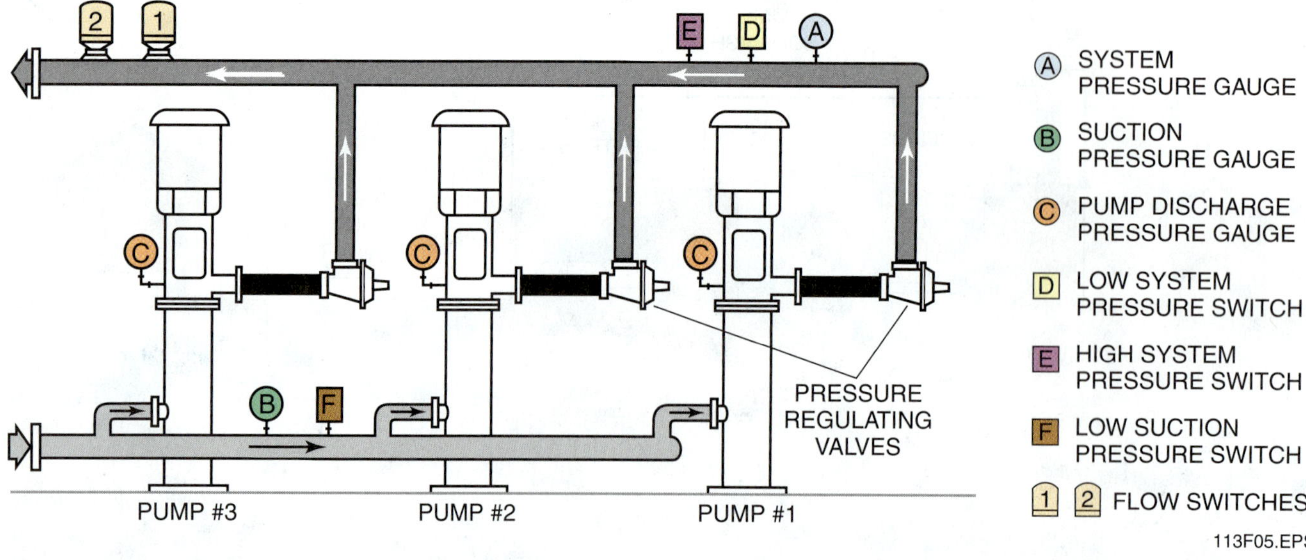

Figure 5 ◆ Variable-capacity system.

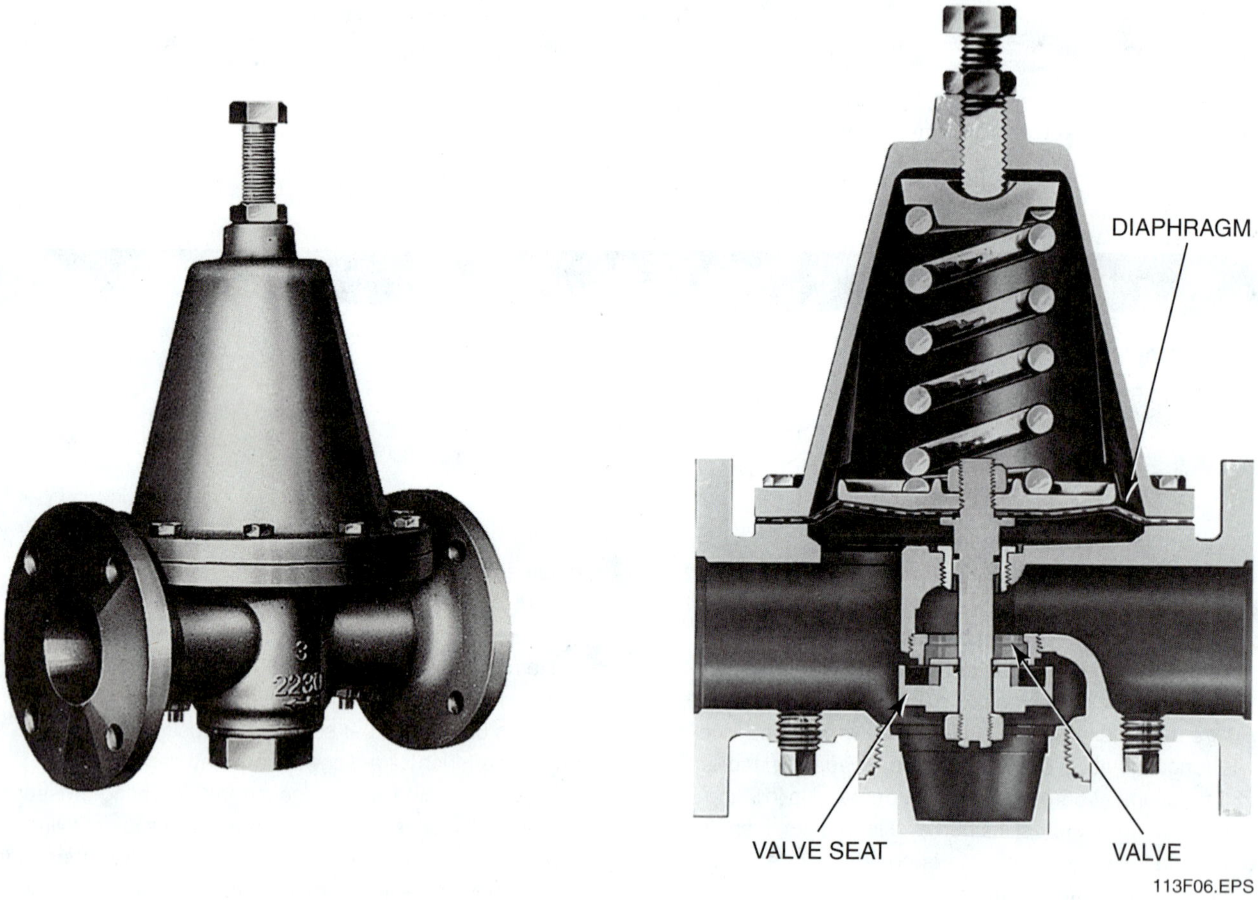

DIAPHRAGM

VALVE SEAT

VALVE

113F06.EPS

Figure 6 ◆ Water pressure regulator valve.

Review Questions

Section 3.0.0

1. Untreated water that comes from natural lakes and streams is _____ for drinking _____ it has been treated.
   a. safe; before
   b. unsafe; until
   c. never safe; before or after
   d. unsafe; even after

2. _____ are usually the only types of water treatment equipment used in a private water treatment system.
   a. Carbon filters
   b. Sediment filters
   c. Sediment and carbon filters
   d. Organic filters

3. One reason that municipal water treatment is more complex than private treatment is that _____.
   a. millions of gallons of water per day are processed
   b. it requires more water softener
   c. permits are required to make the connections
   d. fluoride must be added to reduce tooth decay

4. After municipal water is aerated, it is pumped into a clarifier to be _____.
   a. recarbonated
   b. softened
   c. oxidized
   d. filtered

5. The pipes that carry municipal water from the treatment plant to buildings are called _____.
   a. sewers
   b. branch feeders
   c. reservoirs
   d. water mains

4.0.0 ◆ SUPPLY AND DISTRIBUTION

The water distribution system moves water from its source to the building or structure where it is needed. The way the water moves and the type of materials used can vary depending on the building or structure. It is important to understand how the water distribution system works for both private wells and municipal water systems. You should also know how different types of plumbing materials are used in these systems.

4.1.0 Materials

Water service lines are made of one of three materials: copper, plastic, or steel. Mains are made from cast iron. The cost of pipe and the length of time it is expected to last often determine the choice of materials. Other considerations include corrosion resistance, chance of freezing weather, working pressure, local plumbing codes, and ease of installation. For example, galvanized steel pipe generally has a life expectancy of 30 years. Copper can last more than twice that long. Plastic pipe lasts longest but can absorb chemicals, such as insecticides, from the ground. In many cases, the life expectancy of galvanized and copper pipe depends on soil and water conditions. Galvanized pipe is electroplated with zinc to provide a protective coating that resists corrosion or oxidation (rust). Acidic soils, however, tend to act on galvanized pipe. Soil with carbon dioxide tends to limit the life of copper pipe. Hydrogen sulfide also has harmful effects on copper pipe. The corrosion caused by this gas breaks down the walls of metal pipe and causes the metal to thin or pit.

Another consideration in selecting the water supply pipe is the possibility of **galvanic corrosion,** which can develop when pipes of different materials, such as copper and galvanized steel, are joined. It is caused by a weak electrical current that flows between the different metals. Plumbers must take steps to reduce the likelihood of galvanic corrosion. You will learn more about corrosion prevention as you advance in your training.

4.2.0 Service Line from a Private Water Supply

In a private water supply, the service line runs from the well to the house, and the pump can be located at the well or at the house (see *Figure 7*).

The installation of a private water supply system usually requires a well, although it may mean getting water from a spring or a cistern. A private water source is required when access to municipal water supplies is not available, such as in rural areas.

The plumber may be responsible for setting (installing) the pump and completing the water service to the structure, or a certified well driller may do it. In some cases, a pump installer is required to set the pump, provide electrical service, and complete the water service to the pressure tank. These responsibilities overlap a great deal. Each state and plumbing code has specific requirements for water service from a well. In most cases, a water sample must be drawn and tested by appropriate authorities after the well is purged and before the final tie-in.

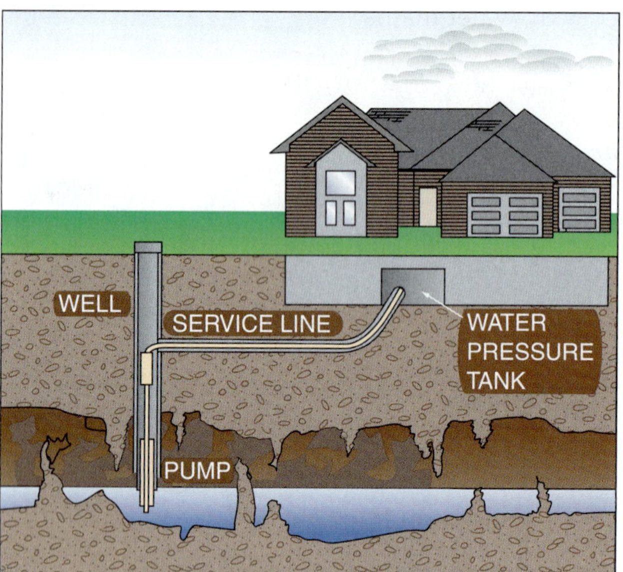

PUMP LOCATED AT THE WELL

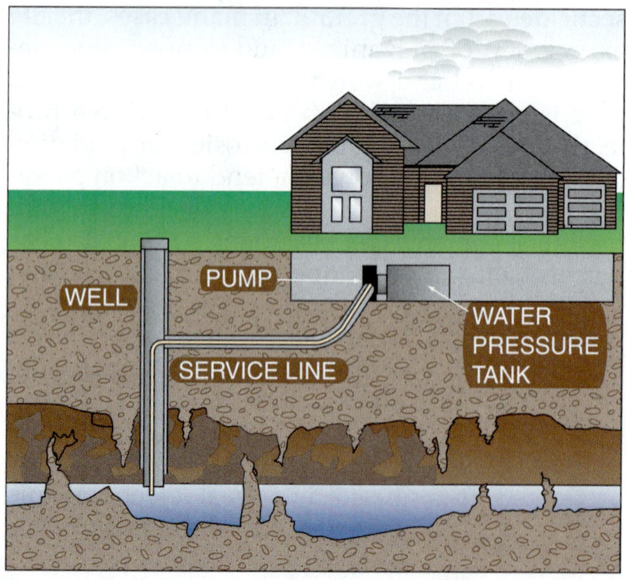

PUMP LOCATED AT THE HOUSE

113F07.EPS

Figure 7 ◆ Well point.

4.3.0 Service Line from a Public Water Main

Water distribution and supply systems in buildings connected to the public water main differ from those used in private water supplies. With private water supplies, pumps are the primary distribution mechanism. For buildings, the building service line is tapped into the water main using a **corporation stop** (*Figure 8*). This is a valve that is threaded into the main without interrupting service to other locations. A short piece of pipe leads from the corporation stop to the **curb stop** (*Figure 9*). The curb stop is a control valve installed in the building water supply line between the corporation stop and the building. The **curb box** is a round casing placed in the ground over the curb stop. The top opening is

113F08.TIF

Figure 8 ◆ Corporation stop.

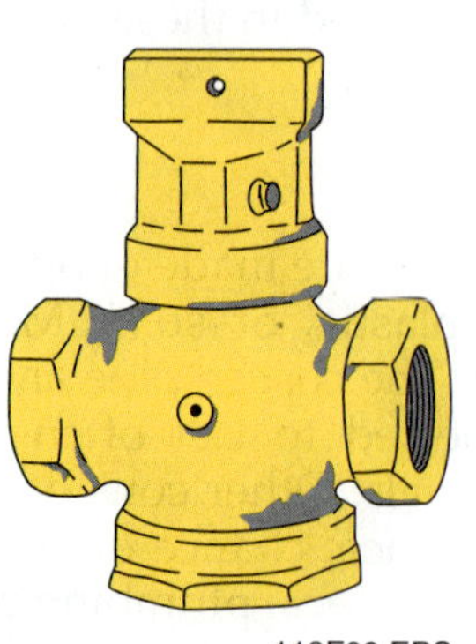

113F09.EPS

Figure 9 ◆ Curb stop.

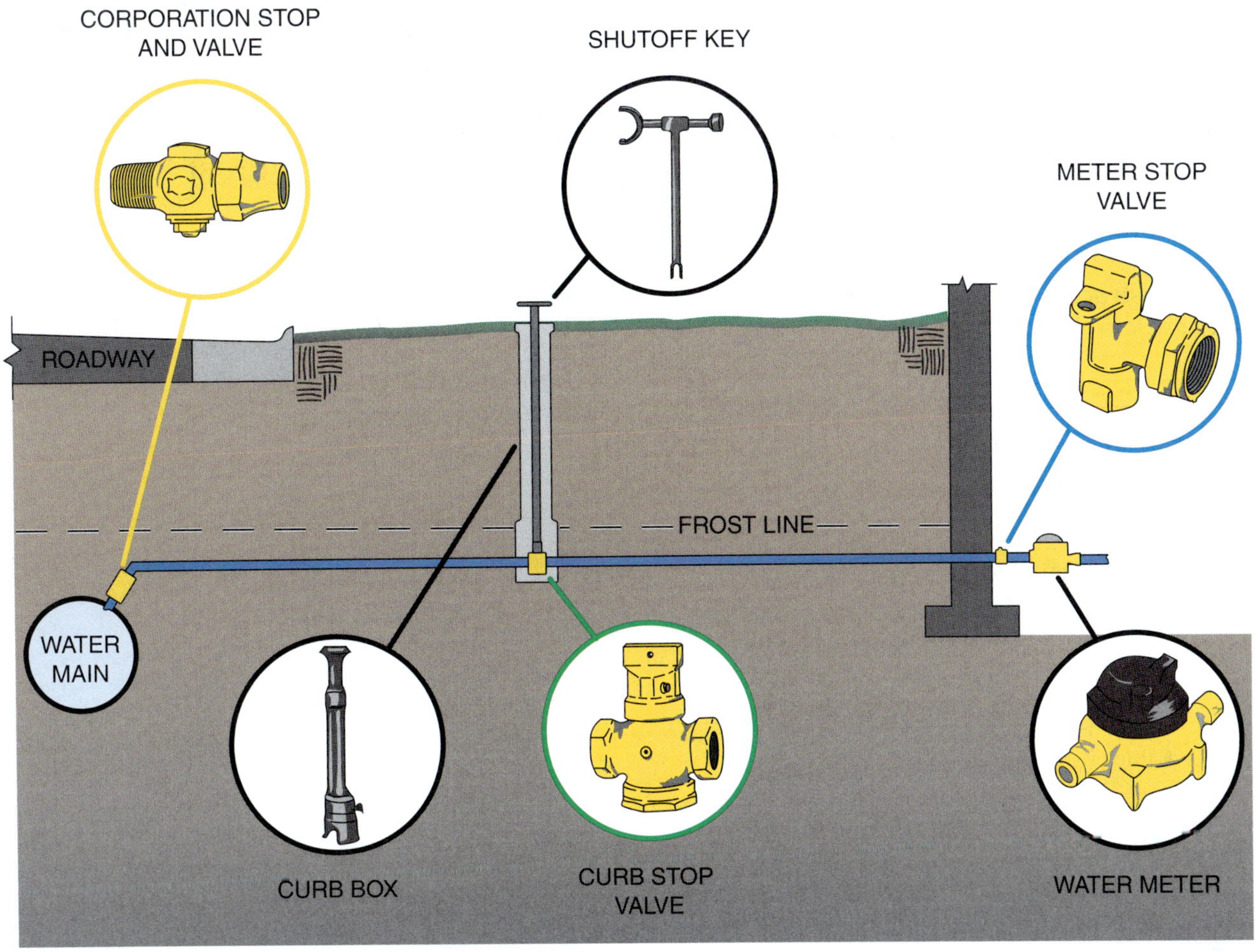

Figure 10 ◆ Municipal water supply connection.

at ground level. A special key can be inserted into the curb box to turn off the curb stop in emergencies or for service (see *Figure 10*). The curb box, sometimes called a **buffalo box,** is marked on city plans and drawings. On the property, it can be located by finding the metal marker (*Figure 11*). Because of differences in local procedures and applicable codes, a plumber should contact local municipal, township, or village authorities about connecting to the water main.

The plumber runs the water supply piping from the curb stop to the building's interior and installs a water line between the curb stop and the **water meter** connection (*Figure 12*). The water

113F11.EPS

Figure 11 ◆ Metal marker for the curb box.

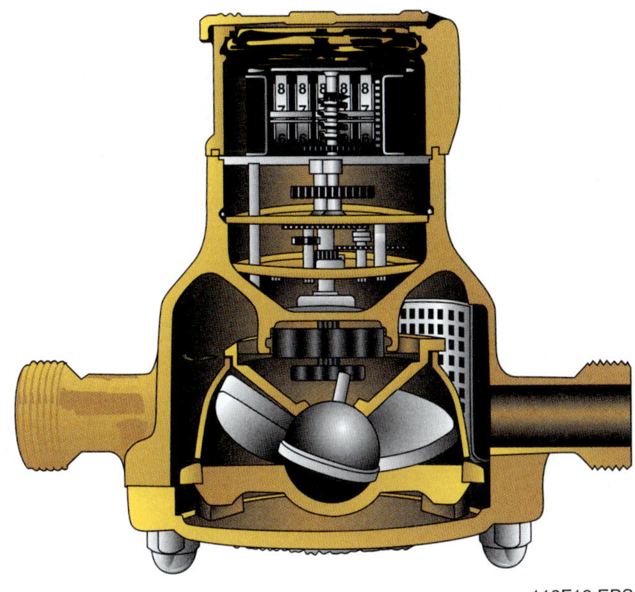

113F12.EPS

Figure 12 ◆ Water meter.

meter measures water use (in gallons or cubic feet). Usually, water meters belong to the municipality, and a city or county employee must install them. A meter stop valve (*Figure 13*) is installed in front of the water meter to allow cutoff of the water service to the entire building. The building main shutoff valve is installed inside the building after the meter is installed (see *Figure 14*). This valve allows the service to be turned off and on during construction. After this point, the building supply system delivers water to the fixtures.

Frost Protection

Frost protection is an important consideration in many parts of the country. Frost depths, called frost lines, are different for each area. The plumber must install the water service line below the frost line to make sure that the piping is not affected by freezing and thawing temperatures in the earth. If the water supply line enters the building above grade, it must be insulated to prevent freezing. Check your local code for the frost depth in your area.

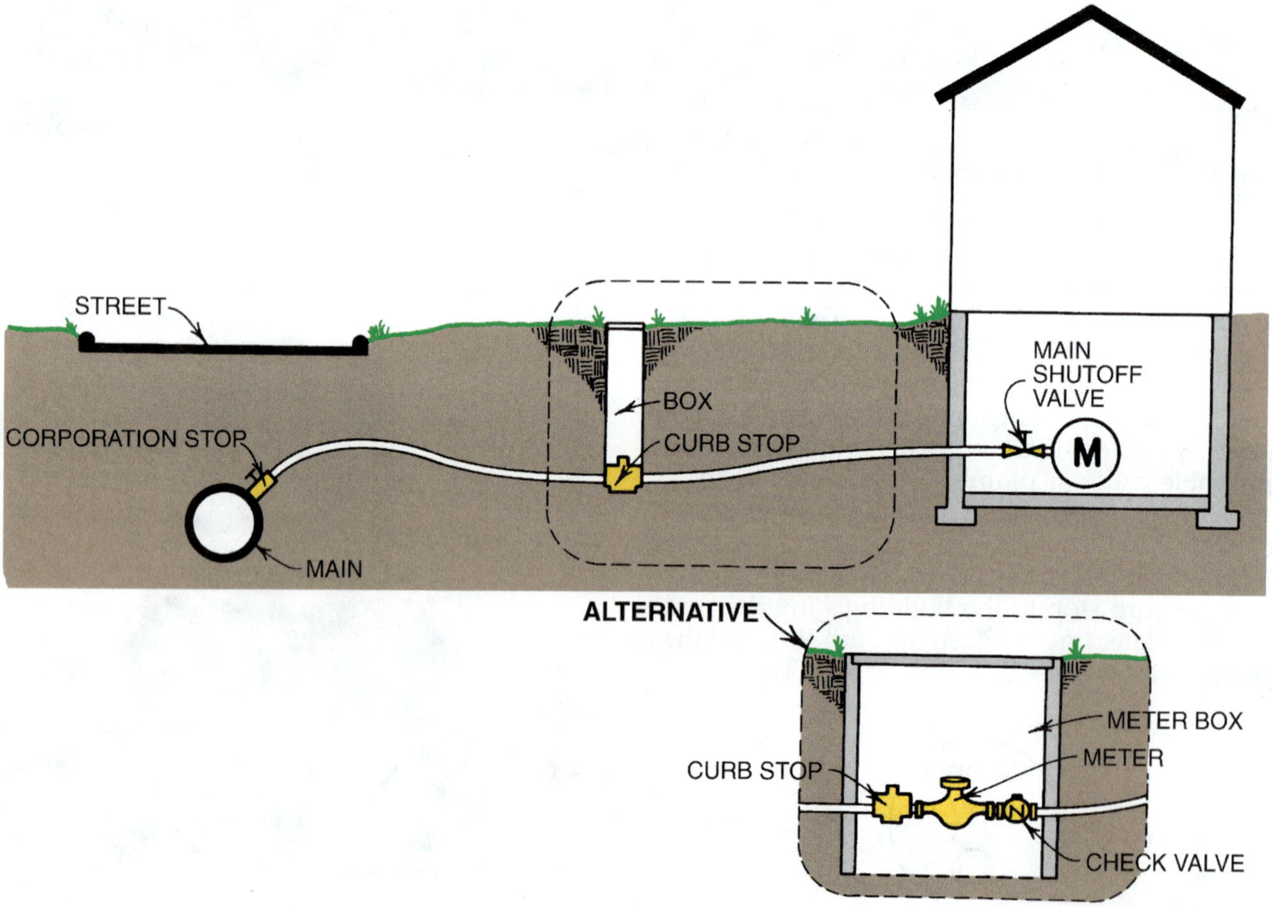

113F13.EPS

Figure 13 ◆ Meter stop valve.

STREET
CORPORATION STOP
MAIN
BOX
CURB STOP
MAIN SHUTOFF VALVE
M
ALTERNATIVE
CURB STOP
METER BOX
METER
CHECK VALVE

113F14.EPS

Figure 14 ◆ Main shutoff valve.

Review Questions

Section 4.0.0

1. Water service lines are made of copper, plastic, or _____.
   a. steel
   b. cast iron
   c. brass
   d. clay

2. _____ pipe lasts the longest of the pipes used for water service lines, but it can absorb chemicals from the ground.
   a. Copper
   b. Plastic
   c. Brass
   d. Steel

3. Soil with carbon dioxide tends to limit the life of _____ pipe.
   a. clay
   b. plastic
   c. copper
   d. steel

4. _____ corrosion is caused by a weak electrical current that flows between the different metals.
   a. Electric
   b. Magnetic
   c. Acidic
   d. Galvanic

5. In a private water supply, the service line runs from the _____ to the house.
   a. sewer
   b. municipal water supply
   c. well
   d. water pressure tank

6. The building service line is tapped into the main using a _____.
   a. curb box
   b. corporation stop
   c. water meter
   d. pressure valve

7. The _____ is a round casing placed in the ground over the curb stop.
   a. curb box
   b. water main
   c. water meter
   d. meter box

8. The _____ measures water use in gallons or cubic feet.
   a. check meter
   b. pressure valve
   c. water meter
   d. corporation stop

9. Usually, the water meter is installed by _____.
   a. the person who installs the curb box
   b. a plumbing contractor
   c. a meter reader
   d. a municipal employee

10. Curb stops are the only way the building main water supply can be turned off and on.
    a. True
    b. False

5.0.0 ◆ CROSS-CONNECTION

Backflow is any unwanted flow of used or nonpotable water back into the potable water distribution system. This reverse flow occurs as a result of cross-connection, which is a direct link between a contaminated liquid and a potable water supply. Backflow can occur as a result of an improper or altered plumbing hookup or a garden hose being left in a pool of contaminated water.

These conditions do not in themselves cause a significant hazard, but they create the potential for serious hazards to the public, especially in terms of waterborne diseases. The danger occurs when a break in the water main or other vacuum potential exists somewhere in a water line. If this happens, the water could be drawn backward through the water supply line. Backflow prevention is required in many plumbing installations to keep contaminated water or other liquids from flowing back into the potable water system.

Water supply lines should be protected from back siphonage by air gaps and by the overflow level. Air gaps are measured vertically from the lowest end of the potable water outlet to the flood-level rim of the fixture into which it discharges. The minimum required air gap is two times the size of the potable water outlet. Consult your local code for minimum air gap requirements.

Basic backflow prevention devices are designed to safeguard against dangerous cross-connections. In a bathroom sink, for example, the faucet line is set above the sink's flood-level rim (the point at which water begins to overflow the top of the sink). In addition, the top drain returns water to the bottom drain before the water can rise to the faucet. The **vacuum breaker** (*Figure 15*) in a **hose bibb** connection acts as a backflow preventer by interrupting the vacuum that could cause a backflow.

There must be a physical separation between wells and municipal water. Wells must not be connected to municipal water supplies. Codes specify the degree of danger and the appropriate device that should be used for various applications (see *Table 1*).

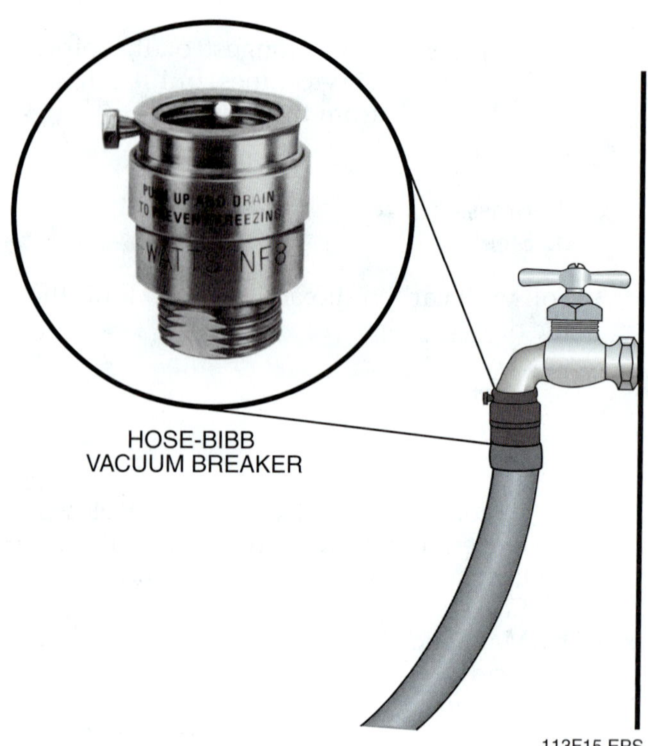

Figure 15 ◆ Vacuum breaker.

Table 1 Application for Backflow Preventers

Device	Degree of Hazard[a]	Application[b]	Applicable Standards
2003 International Plumbing Code® *Table 608.1, Application of Backflow Preventers*			
Air gap	High or low hazard	Backsiphonage or backpressure	*ASME A112.1.2*
Air gap fittings for use with plumbing fixtures, appliances and appurtenances	High or low hazard	Backsiphonage or backpressure	*ASME A112.1.3*
Antisiphon-type fill valves for gravity water closet flush tanks	High hazard	Backsiphonage only	*ASSE 1002, CSA-B125*
Barometric loop	High or low hazard	Backsiphonage only	See *Section 608.13.4*
Reduced pressure principle backflow preventer and reduced pressure principle fire protection backflow preventer	High or low hazard	Backpressure or backsiphonage Sizes ⅜"–16"	*ASSE 1013, AWWA C511, CAN/CSA B64.4*
Reduced pressure detector fire protection backflow prevention assemblies	High or low hazard	Backsiphonage or backpressure (Fire sprinkler systems)	*ASSE 1047*
Double check backflow prevention assembly and double check fire protection backflow prevention assembly	Low hazard	Backpressure or backsiphonage Sizes ⅜"–16"	*ASSE 1015, AWWA C510*
Double check detector fire protection backflow prevention assemblies	Low hazard	Backpressure or backsiphonage (Fire sprinkler systems) Sizes 2"–16"	*ASSE 1048*
Dual-check-valve-type backflow preventer	Low hazard	Backpressure or backsiphonage Sizes ¼"–1"	*ASSE 1024*
Backflow preventer with intermediate atmospheric vents	Low hazard	Backpressure or backsiphonage Sizes ¼"– ¾"	*ASSE 1012, CAN/CSA-B64.3*
Backflow preventer for carbonated beverage machines	Low hazard	Backpressure or backsiphonage Sizes ¼"– ⅜"	*ASSE 1022*
Pipe-applied atmospheric-type vacuum breaker	High or low hazard	Backsiphonage only Sizes ¼"–4"	*ASSE 1001, CAN/CSA-B64.1.1*
Pressure vacuum breaker assembly	High or low hazard	Backsiphonage only Sizes ½"–2"	*ASSE 1020*
Hose-connection vacuum breaker	High or low hazard	Low head backpressure or backsiphonage Sizes ½", ¾", 1"	*ASSE 1011, CAN/CSA-B64.2*
Vacuum breaker wall hydrants, frost-resistant, automatic draining type	High or low hazard	Low head backpressure or backsiphonage Sizes ¾", 1"	*ASSE 1019, CAN/CSA-B64.2.2*
Laboratory faucet backflow preventer	High or low hazard	Low head backpressure and backsiphonage	*ASSE 1035, CSA B64.7*
Hose connection backflow preventer	High or low hazard	Low head backpressure, rated working pressure backpressure or backsiphonage Sizes ½"–1"	*ASSE 1052*
Spillproof vacuum breaker	High or low hazard	Backsiphonage only Sizes ¼"–2"	*ASSE 1056*

For SI: 1 inch = 25.4 mm

a. Low hazard–See Pollution (Section 202). High hazard–See Contamination (Section 202).

b. See Backpressure (Section 202). See Backpressure, low head (Section 202). See Backsiphonage (Section 202).

113T01.EPS

5.1.0 Backflow Preventers

Some backflow preventers protect against both back pressure and back siphonage, while others can handle only one type of backflow. Backpressure occurs in the water distribution system when a pressure higher than the supply pressure causes a reverse flow into the potable water piping. Backsiphonage occurs when contaminated or polluted water flows from a plumbing fixture back into the potable water piping. Backsiphonage is caused by a negative pressure in the plumbing fixture. The five most commonly used mechanical backflow preventers are as follows (*Figure 16*):

- Atmospheric vacuum breakers
- Pressure-type vacuum breakers
- Dual-check valve backflow preventers
- Double-check valve assemblies
- Reduced-pressure zone principal backflow preventers

Each type of backflow preventer has specific applications and limitations, which you will learn more about as you continue your training. These factors must be considered in any type of system design using backflow preventers. Before you select and install any backflow prevention or siphonage prevention device, consult local codes.

Backflow prevention devices must be well-maintained to work effectively. Most devices are field tested before installation and then tested annually. When devices are not working properly, they must be repaired and tested according to the manufacturer's instructions. Testing must be performed by certified personnel.

113F16A.TIF

113F16B.TIF

Figure 16 ◆ Backflow prevention devices.

CAUTION

Some codes require installation of backflow preventers in every new structure. Always be sure to check all applicable codes before any installation.

5.2.0 Valves

A valve or faucet regulates the flow of water in a water distribution system. It may be used to turn water service on and off, act as a throttling device that controls the rate of water flow, regulate the pressure, or prevent a reversal of flow through a line.

It is important for plumbers to understand the terms related to valves. The following are commonly used valve terms:

- ***Straight-through flow*** – This flow is not restricted as it passes through a valve. The element that closes the valve is pulled back entirely clear of the passage.
- ***Full flow*** – This describes the flow capacities of a valve.
- ***Throttled flow*** – This flow rate can be controlled through the valve. Not all types of valves are suitable for throttled flow.

Trimming out is the task of placing attachments on fixtures to finish a job. These attachments, also known as the trim, are the parts of a valve that receive the most wear. Stems, discs, seat rings, disc holders or guides, wedges, and bushings are all considered trim.

Backflow Case Studies

Always install proper backflow prevention devices where they are required by your local code. Be sure to install them according to the manufacturer's instructions. Don't let your community become a horror story like the ones discussed in the case studies in this module.

Backflow Case Study 1

During a routine maintenance check of Colorado Middle School's hot water heating boiler, maintenance workers left open a valve between the potable water line and the boiler. This allowed boiler water containing antifreeze to backflow into the school's potable water system. There was no backflow preventer on the feed line to the boiler.

Nine children were sent to the hospital with flu-like symptoms. The hospital treated the students for ethylene glycol (the ingredient in the antifreeze) poisoning. The school was closed while workers flushed the potable water piping and repaired the leak. Workers were instructed to install a backflow preventer in the potable water line to the hot water heating boiler.

Source: University of Florida TREEO Center website. "Case Histories of Selected Backflow Incidents," www.treeo.ufl.edu/backflow/casehist.asp#34, reviewed March 22, 2004.

Backflow Case Study 2

A municipal office inspected a cross-connection at a block of high-rise apartments. Officials found that a check valve failed to protect three water heaters. The pressure in a high-pressure cold water pump was fluctuating. When the pressure dropped, antifreeze contaminated the potable water system.

Source: Backflow Prevention website. "Cross Connection Control," www.rbamech.ab.ca/backflowpreventors/, reviewed March 22, 2004.

Backflow Case Study 3

A Florida home was mistakenly hooked up to treated wastewater lines instead of drinking water lines. As a result, the resident experienced severe stomach cramping and diarrhea. According to a city report, workers accidentally switched the lines when they failed to dig deep enough to expose the identification tape distinguishing the two water lines.

The city was asked to sanitize the home plumbing over a three-month period and monitor the resident's health for 10 years. As part of an agreement with the state health department, the city was also required to spend nearly $75,000 to improve training, installation procedures, and public education efforts.

Source: Associated Press. www.bradenton.com/mld/bradenton/news/8092345.htm "Man sues city over wastewater pumped through his water line," reviewed March 22, 2004.

5.3.0 Types of Valves

Plumbers must be familiar with a wide variety of valves. The following are common types of valves:

- Gate valves
- Globe valves
- **Angle valves**
- Ball valves
- **Check valves**
- **Pressure regulator valves**
- **Supply stop valves**
- **Temperature and pressure (T and P) relief valves** (also referred to as **thermostatic/pressure balancing valves**)

These valves are used in different parts of the water distribution system (see *Figure 17*).

5.3.1 Gate Valves

A gate valve (*Figure 18*) is a valve in which the flow is controlled by moving a gate or disc that slides in machined grooves at right angles to the flow. The gate is moved by the action of the threaded stem on the control handle.

There are two basic designs of gate valves: rising-stem gate valves (also known as inside screw stem valves) or nonrising-stem valves (also known as outside screw stem valves). A rising-stem gate valve has the stem attached to the gate. The gate and stem rise and lower together as the valve is operated. The stem on a nonrising-stem gate valve is threaded on the lower end into the gate. When the handwheel on the stem is rotated, the gate moves up or down the stem.

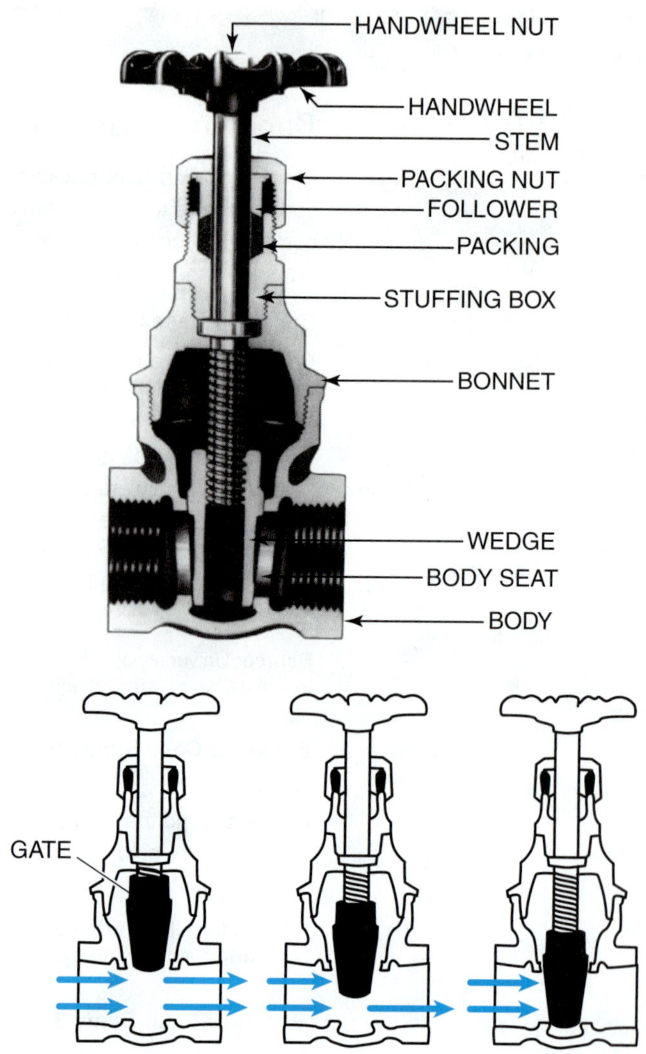

Figure 18 ◆ Gate valve.

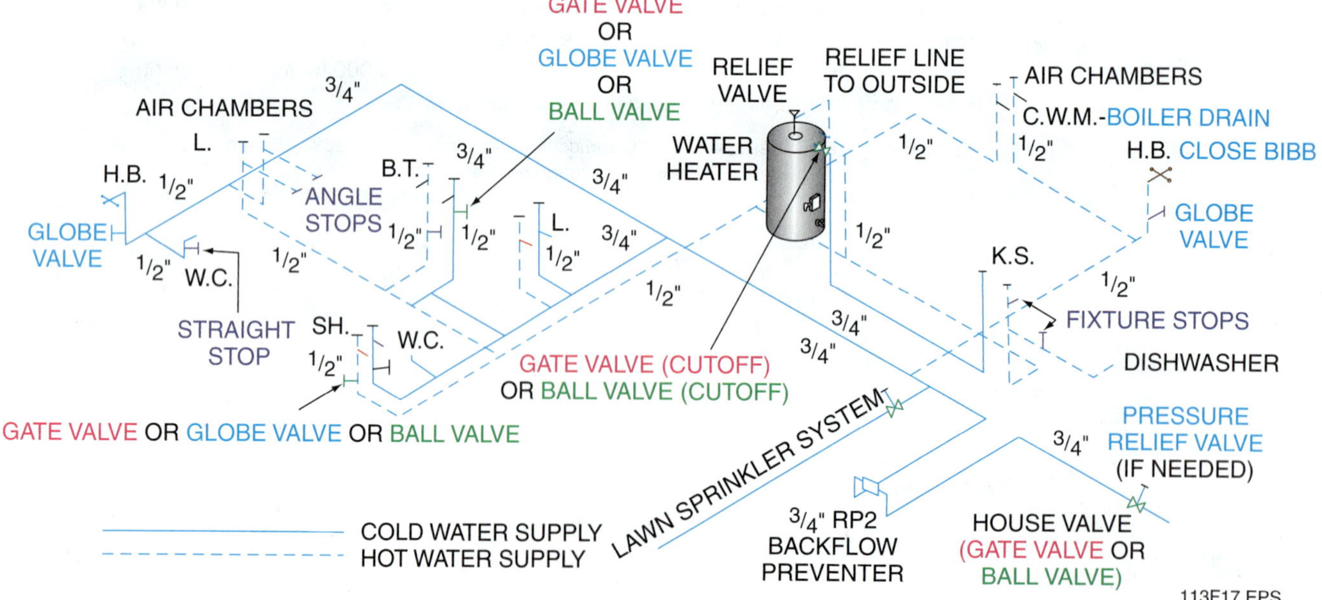

Figure 17 ◆ Valves in the distribution system.

Gate valves are best suited for main supply lines and pump lines. They provide an unobstructed passageway when the gate is fully opened. They can be used on lines containing steam, water, gas, oil, or air.

5.3.2 Globe Valves

In a globe valve (*Figure 19*), the flow is controlled by moving a circular disc against a metal seat that surrounds the flow opening. The disc is forced onto the seat or withdrawn from it by screw action as the handle is turned.

Globe valves are recommended for general service on lines containing steam, water, gas, or oil, where frequent operation and close control of flow is required. Inside the globe-shaped body of the valve is a partition. This partition closes off the inlet side of the valve from the outlet side, except for a circular opening called the valve seat. The upper side of the valve seat is ground smooth so that a proper and complete seal can be made when the valve is in the closed position. Turning the handle clockwise until the valve stem firmly seats the washer or disc between the valve stem and the valve seat closes the valve. This stops the flow of gas or liquid.

5.3.3 Angle Valves

The angle valve (*Figure 20*) is similar to the globe valve, but it can serve as both a valve and a 90-degree elbow. Because flow changes direction only twice through an angle valve, the angle valve is less resistant to flow than the globe valve, in which flow must change direction three times.

There are three types of angle valves: conventional, plug-type, or composition discs. Conventional angle valves have a disc that is in close contact with the body seat. These valves are suitable for less severe services, but may not be used for close throttling. Plug-type angle valves have a long taper and a wide seat bearing. These valves are ideal for severe throttling and are effective in resisting erosion. Composition discs have a tight seating against a raised crown. These valves are well suited for moderate pressure service.

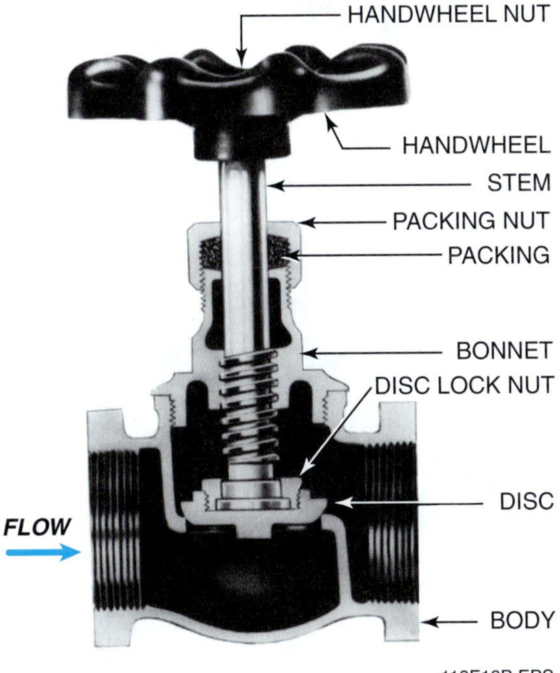

Figure 19 ◆ Globe valve.

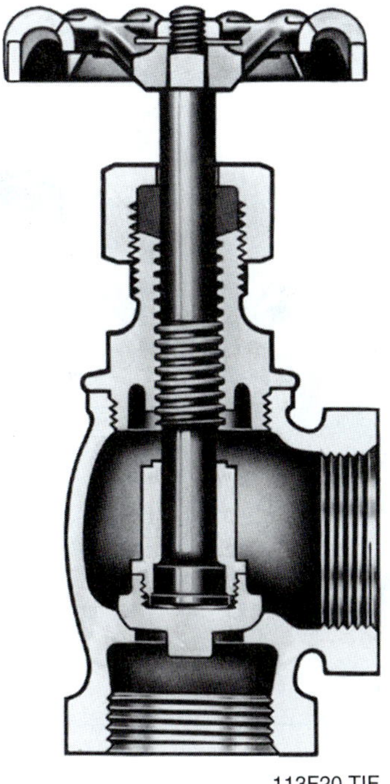

Figure 20 ◆ Angle valve.

5.3.4 Ball Valves

The ball valve (*Figure 21*) is used to control the flow of gases and liquids. Ball valves are installed in piping systems where quick shutoffs for in-line maintenance may be necessary, or in lines used for mixing various liquids and gases. The ball part of the valve is rotated into the open or closed position by a handle on the outside of the valve body. These valves allow quick action in controlling the flow in piping systems.

5.3.5 Check Valves

A check valve is used to prevent reversal of flow in a piping system. Pressure in the line keeps the valve open. The valve is automatically closed by the reversal of flow or by the weight of the disc mechanism. Three types of check valves are available: the ball-check valve, the swing-check valve, and the lift-check valve.

The ball-check valve (*Figure 22*) allows one-way flow in water supply or drainage lines and can be used with extremely low backpressure.

The swing-check valve (*Figure 23*) has a low flow resistance that makes it well suited for lines containing liquids or gases with low to moderate pressure. The swing-check valve is available in up to three different types, depending on the manufacturer: single disc, conventional; dual disc, split swing discs; and single disc, angle seating.

The lift-check valve (*Figure 24*) can be used for gas, water, steam, or air. Lift-check valves are recommended for lines that have frequent fluctuations in flow. These valves are available in either horizontal or vertical styles. The horizontal type has an integral construction similar to the globe valve, while the vertical type has a straight-through flow.

5.3.6 Pressure Regulator Valves

The pressure regulator valve (*Figure 25*) is used to reduce water pressure in a building. The valve is

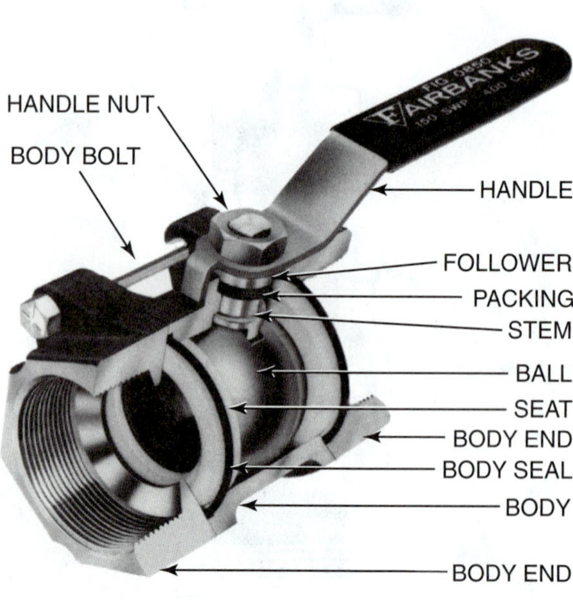

Figure 21 ◆ Ball valve.

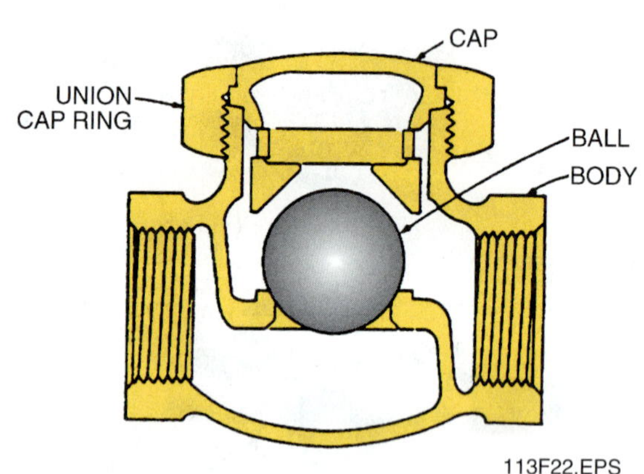

Figure 22 ◆ Ball-check valve.

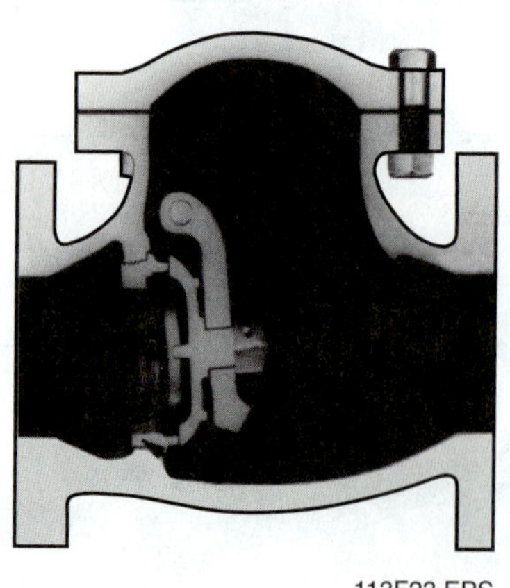

Figure 23 ◆ Swing-check valve.

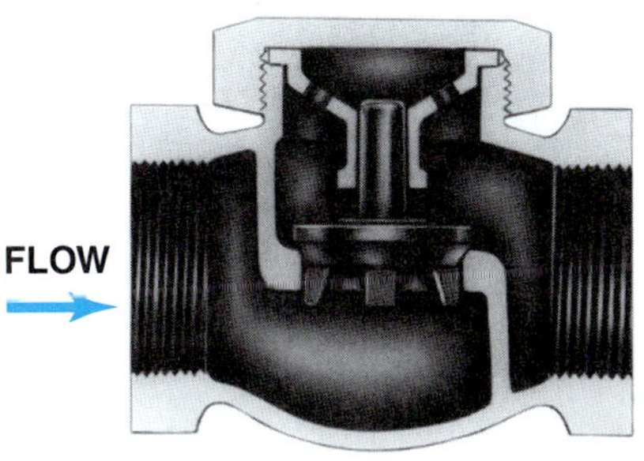

113F24.EPS

Figure 24 ◆ Lift-check valve.

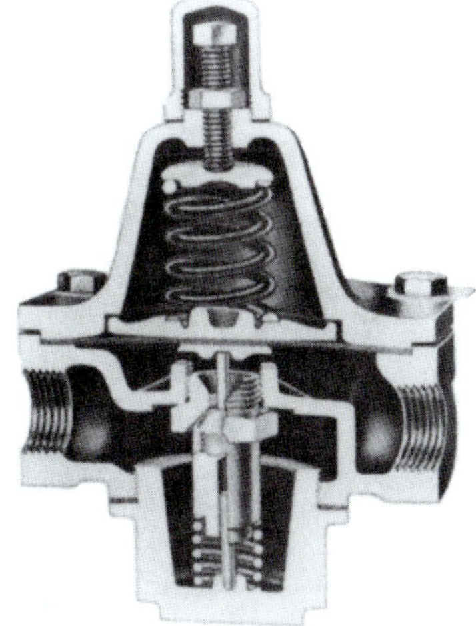

113F25.EPS

Figure 25 ◆ Pressure regulator valve.

activated by changes in pressure within the system. As the pressure changes, a spring located in the dome of the valve acts on the diaphragm to move the valve up or down. The valve is in the open position when the diaphragm is pressed down away from the valve seat, and it remains open until the pressure in the building reaches a set level. The valve then closes and remains closed until the pressure in the building begins to drop.

5.3.7 Supply Stop Valves

Supply stop valves, or supply valves, are commonly used to disconnect the hot or cold water supply to water closets and sinks. These valves make it easy to control the water connection at an individual fixture for repair work. They are available in either right angle or straight design (see *Figure 26*).

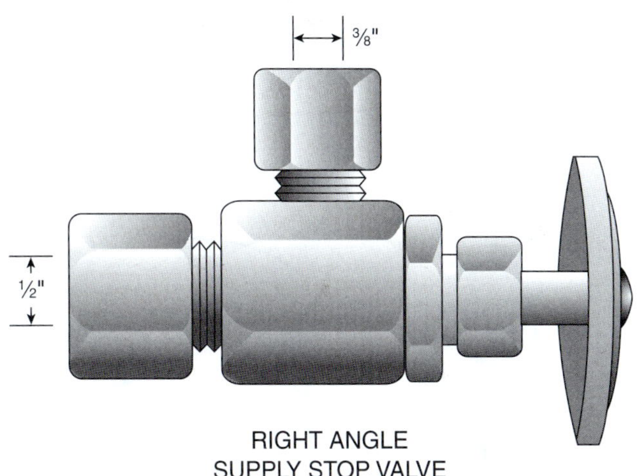

113F26.EPS

Figure 26 ◆ Supply stop valve.

5.3.8 Temperature and Pressure (T and P) Relief Valves

Temperature and pressure (T and P) relief valves (*Figure 27*) are normally used for liquid service, although safety valves also may be used. Ordinarily, **pressure relief valves** do not have a chamber or a regulator ring for varying or adjusting blowdown, so they operate with a relatively lazy motion. As pressure increases, they slowly open; as pressure decreases, they slowly close.

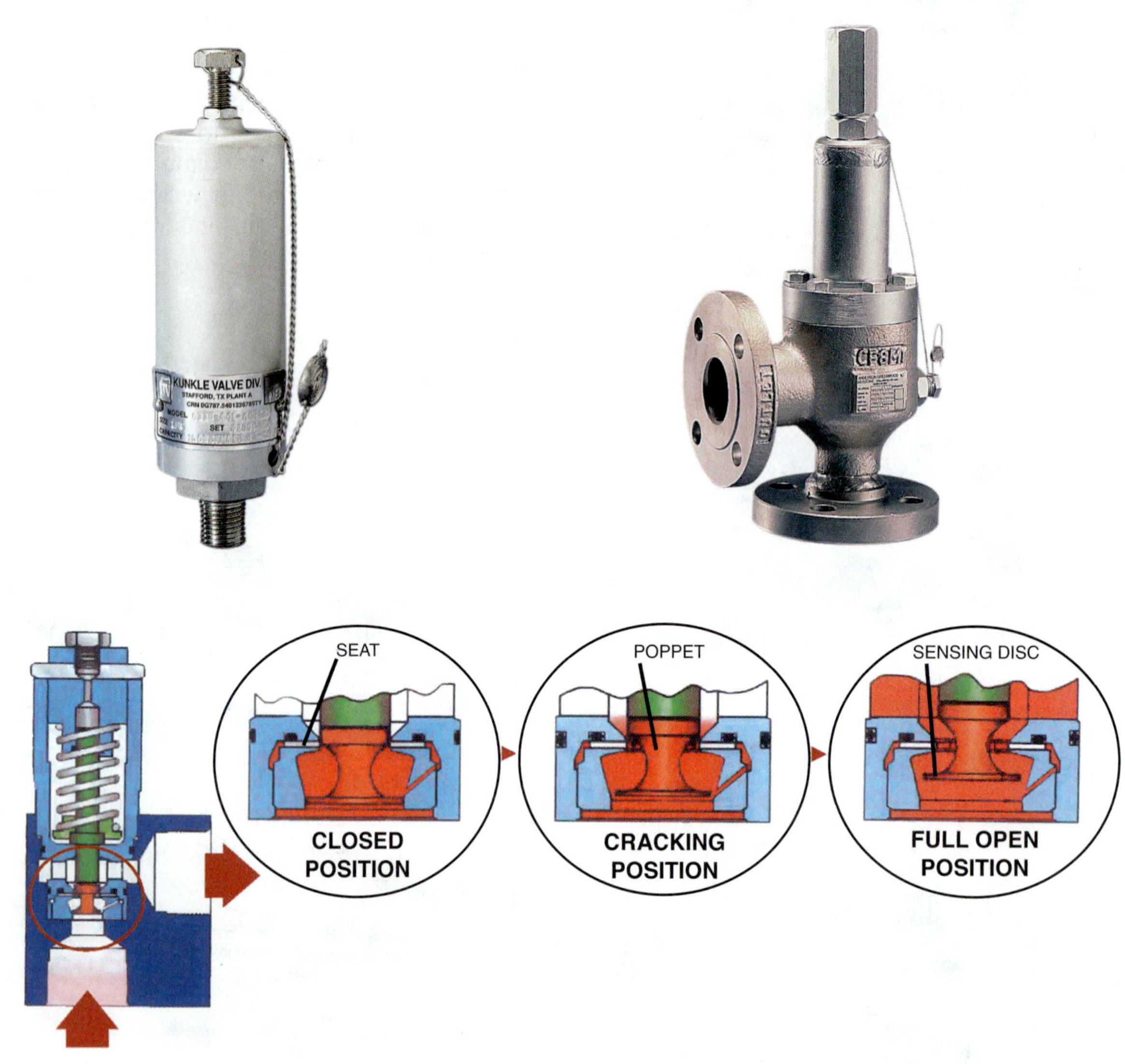

Figure 27 ◆ Pressure relief valve.

Review Questions

Section 5.0.0

1. The reverse flow of contaminated water or other liquids into the potable water system is caused by _____.
   a. faulty throttling devices
   b. connection pressure
   c. cross-connection
   d. backflow connection

2. Leaving a garden hose in contaminated water is one cause of water supply contamination when _____ occurs.
   a. backflow
   b. backpressure
   c. vacuum
   d. cross-pressure

3. If a break in the water main or other potential for a vacuum exists in a water line, _____ could be drawn backward into the water supply system.
   a. contaminated water
   b. dust particles
   c. pressure
   d. excess air

4. An atmospheric vacuum breaker is not considered a backflow preventer.
   a. True
   b. False

5. _____ are best suited for main supply lines and pump lines because they provide an unobstructed passageway for the water.
   a. Globe valves
   b. Angle valves
   c. Check valves
   d. Gate valves

6. For quick shutoffs for in-line maintenance or for lines mixing various liquids and gases, use a(n) _____ valve.
   a. ball
   b. check
   c. gate
   d. angle

7. A _____ valve is used to prevent the reverse flow of liquid in a piping system.
   a. ball
   b. check
   c. gate
   d. globe

8. For lines that have frequent changes in direction, a _____ valve can be used in either a vertical or horizontal style.
   a. lift-check
   b. swing-check
   c. gate-check
   d. ball-check

9. A _____ valve is commonly used as a cutoff valve for the hot or cold water supply to water closets and sinks.
   a. gate
   b. globe
   c. supply stop
   d. swing-check

10. As pressure increases, a _____ valve opens slowly; as pressure decreases, it closes slowly.
    a. gate-check
    b. temperature and pressure
    c. supply stop
    d. swing-check

6.0.0 ◆ BUILDING DISTRIBUTION

Once the water supply piping is installed to bring the water from the main into the building, the next step is to install the water heater, hose bibbs, water softener (if needed), and fixtures. These components complete the water distribution system (see *Figure 28*).

6.1.0 Locating the Water Heater

Place the water heater in the most efficient location, which is usually as close as possible to the greatest number of hot water outlets. This minimizes the length of hot water piping that runs between the heater and the fixtures. You should also consider the location of the gas supply if installing a gas water heater, or the power supply if installing an electric water heater.

6.2.0 Locating the Water Softener

If the plumbing system includes a water softener, you must first determine the expected use of various fixtures. Not all fixtures or outlets require access to softened water. For example, the hot water supply should always be softened, but water flowing to the hose bibbs may not need to be softened. This decision affects the placement of pipe runs.

6.3.0 Locating Hose Bibbs

An exterior water outlet, or hose bibb (*Figure 29*), is frequently required in residential structures. The piping runs serving hose bibbs may be long, and they should bypass the water softener.

6.4.0 Locating Fixtures

After locating the water heater, hose bibbs, and water softener, locate the water supply piping that serves individual fixtures and appliances in the

Figure 29 ◆ Hose bibb.

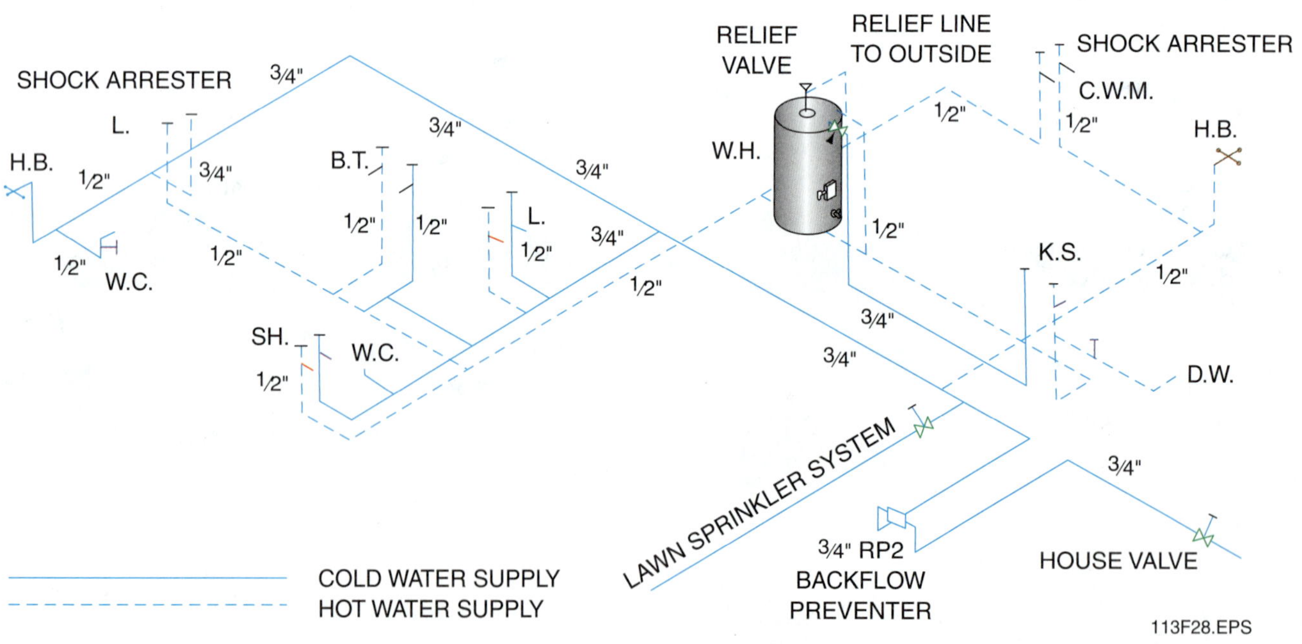

Figure 28 ◆ Sample water distribution piping diagram.

building. Then locate the **fixture risers** and fixture stub-outs. When you are installing the plumbing for a structure, you must first install the DWV system. Because of the large size of the pipe and limited flexibility in installation, and because it cannot be moved once it is installed, proper layout (done first) of the DWV can eliminate most problems. Water supply piping is located in relation to the already installed DWV piping, not to the walls or corners of the building. Water supply piping is also located in relation to drains and fixture controls for the fixtures. When you are installing water supply piping, recheck the location of the drains to make sure that you are positioning the piping correctly.

You construct a vertical section of pipe (also known as a fixture riser) that goes inside the wall as the connection from the fixture to the supply pipe beneath the flooring. Assemble fixture risers and stub-outs, and mount them on a **backing board** inside the wall behind the fixture (see *Figure 30*). When the water supply lines have been located at the fixtures and the stub-outs assembled, the assembly is placed through the access hole in the floor and connected to the **feeder lines** or **branches.** Feeder lines are the main water supply line, to which branches are connected. These lines are also known as service lines. Branches are any part of the piping system other than the riser, main, or stack. Always use approved submittal data to determine measurements when installing fixture risers and stub-outs.

In large water lines or where fixture controls have quick-closing valves, **hammer arresters** or shock arresters (*Figure 31*) must be installed to absorb the energy when the flow of water suddenly stops. They are usually placed near the fixture as an extension of the water pipe riser. If hammer arresters are not installed, are installed incorrectly, or fail, the pipe may hammer (vibrate and rattle) when a quick-closing fixture control is used. Hammering could eventually lead to pipe, joint, or valve failure in the water distribution system.

CAUTION

Keep the water supply piping as clean as possible before installing and using it. Clean, dry storage of pipes helps prevent contamination. Cap all ends of pipe at the end of each workday.

6.5.0 Main Supply Lines

You must correctly size the supply pipes to the fixtures and appliances and use sizing tables supplied by the state or local plumbing codes (see *Tables 2* and *3*). These tables estimate the anticipated demand for water as measured by **water supply fixture units (WSFUs).** Sizing pipes must take the following into account:

- Type of flush devices used on different fixtures
- Water pressure in pounds per square inch (psi) at the source
- Length of pipe in the building
- Types and number of different fixtures installed
- Total number of fixtures in use at any one time

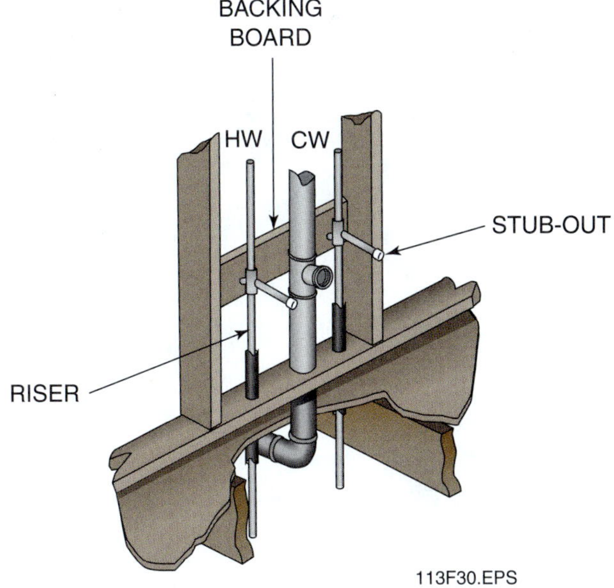

Figure 30 ◆ Installation of fixture riser and stub-outs.

Figure 31 ◆ Shock arrester.

Table 2 Water Supply Fixture Units and Minimum Fixture Branch Pipe Sizes

2003 National Standard Plumbing Code
Table 10.14.2A Water Supply Fixture Units and Minimum Fixture Branch Pipe Sizes

	Minimum Branch Pipe Size	Individual Dwelling Units	Serving Three or More Dwelling Units	Other Than Dwelling Units	Heavy-Use Assembly
Bathroom Groups Having 1.6 GPF Gravity-Tank Water Closets					
Half-bath or Powder Room		3.5	2.5		
1 Bathroom Group		5.0	3.5		
1½ Bathrooms		6.0			
2 Bathrooms		7.0			
2½ Bathrooms		8.0			
3 Bathrooms		9.0			
Each Additional ½ Bath		0.5			
Each Additional Bathroom Group		1.0			
Bathroom Groups Having 1.6 GPF Pressure-Tank Water Closets					
Half-bath or Powder Room		3.5	2.5		
1 Bathroom Group		5.0	3.5		
1½ Bathrooms		6.0			
2 Bathrooms		7.0			
2½ Bathrooms		8.0			
3 Bathrooms		9.0			
Each Additional ½ Bath		0.5			
Each Additional Bathroom Group		1.0			
Bathroom Groups Having 3.5 GPF Pressure-Tank Water Closets					
Half-bath or Powder Room		4.0	3.0		
1 Bathroom Group		6.0	5.0		
1½ Bathrooms		8.0			
2 Bathrooms		10.0			
2½ Bathrooms		11.0			
3 Bathrooms		12.0			
Each Additional ½ Bath		0.5			
Each Additional Bathroom Group		1.0			
Bath, Kitchen, and Laundry Groups					
Bath Group (1.6 GPF flushometer valve)		6.0	4.0		
Bath Group (3.5 GPF flushometer valve)		8.0	6.0		
Kitchen Group (sink and dishwasher)		2.0	1.5		
Laundry Group (sink and clothes washer)		5.0	3.0		
Individual Fixtures					
Bathtub or Combination Bath/Shower	½"	4.0	3.5		
Bidet	½"	1.0	0.5		
Clothes Washer, Domestic	½"	4.0	2.5	4.0	
Dishwasher, Domestic	½"	1.5	1.0	1.5	
Drinking Fountain or Watercooler	⅜"			0.5	0.75
Hose Bibb	½"	2.5	2.5	2.5	

Table 2 Continued

Hose Bibb, Each Additional	½"	1.0	1.0	1.0	
Kitchen Sink, Domestic	½"	1.5	1.0	1.5	
Laundry Sink	½"	2.0	1.0	2.0	
Lavatory	⅜"	1.0	0.5	1.0	1.0
Service Sink or Mop Basin	½"			3.0	
Shower	½"	2.0	2.0	2.0	
Shower, Continuous Use	½"			5.0	
Urinal, 1.0 GPF	¾"			4.0	5.0
Urinal, Greater Than 1.0 GPF	¾"			5.0	6.0
Water Closet, 1.6 GPF Gravity Tank	½"	2.5	2.5	2.5	4.0
Water Closet, 1.6 GPF Flushometer Tank	½"	2.5	2.5	2.5	3.5
Water Closet, 1.6 GPF Flushometer Valve	1"	5.0	5.0	5.0	8.0
Water Closet, 3.5 GPF Gravity Tank	½"	3.0	3.0	5.5	7.0
Water Closet, 3.5 GPF Flushometer Valve	1"	7.0	7.0	8.0	10.0
Whirlpool Bath or Combination Bath/Shower	½"	4.0	4.0		

Table 3 Table for Concerting Demand in Water Supply Fixture Units and Gallons Per Minute

2003 National Standard Plumbing Code
Table for Concerting Demand in Water Supply Fixture Units and Gallons Per Minute[1]

WSFU	GPM Flush Tanks[2]	GPM Flush Valves[3]	WSFU	GPM Flush Tanks[2]	GPM Flush Valves[3]
3	3		120	49	74
4	4		140	53	78
5	4.5	22	160	57	83
6	5	23	180	61	87
7	6	24	200	65	91
8	7	25	225	70	95
9	7.5	26	250	75	100
10	8	27	300	85	110
11	8.5	28	400	105	125
12	9	29	500	125	140
13	10	29.5	750	170	175
14	10.5	30	1,000	210	210
15	11	31	1,250	240	240
16	12	32	1,500	270	270
17	12.5	33	1,750	300	300
18	13	33.5	2,000	325	325
19	13.5	34	2,500	380	380
20	14	35	3,000	435	435
25	17	38	4,000	525	525
30	20	41	5,000	600	600
40	25	47	6,000	650	650
50	29	51	7,000	700	700
60	33	55	8,000	730	730
80	39	62	9,000	760	760
100	44	68	10,000	790	790

Notes:

1. This table converts water supply demand in water supply fixture units (WSFUs) to required flow in gallons per minute (GPM) for the purpose of pipe sizing.

2. This column applies to portions of piping systems where the water closets are the flush tank type (gravity or pressure) or there are no water closets, and to hot water piping.

3. This column applies to portions of piping systems where the water closets are the flush valve type.

For residential installations, the main feeder line beyond the water heater must be sized to supply the required flow and pressure. Smaller pipe sizes can save both energy and money. Larger pipe sizes carry more water, but water left in the piping system when the fixture is turned off cools down. When the user turns on the faucet again, more cool water must clear the pipe before hot water reaches the fixture. This wastes water and energy. Always consult applicable code to determine the water- and energy-saving measures possible for each job.

At the end of the hot water line, a tee and a pump can be installed to return water into the cold side or into the drain valve opening of the heater. A check valve needs to be installed on the incoming cold water so the return water will flow in only one direction. This is called a hot water recirculating system and is needed when the fixtures are a long distance from the hot water source. The complete loop of a recirculating system should be installed.

When you are installing cold water lines, you must consider possible branches to hose bibbs and plan efficiently to save pipe and time. Cold water that is used at the hose bibbs for lawn sprinklers and other outdoor applications should not be softened.

7.0.0 ◆ FIXTURES AND FAUCETS

Various types of fixtures are used in all residential, commercial, and industrial buildings. For this reason, you need to know more than just installation techniques. You should be aware of the construction and working principles of each type of fixture to make the proper water supply connections. The same is true for different types of faucets. Fixtures and faucets are covered in depth in *Fixtures and Faucets.* This section contains information about regulating the temperature of water coming out of a faucet.

Hot and cold water can be mixed to reduce the risk of scalding at faucets, such as shower valves or sweating at water closet tanks. The best way to do this is to use a tempering valve (*Figure 32*). Tempering valves are available for both residential and commercial/industrial applications. These valves mix the water coming from the hot water heater to a predetermined temperature. The temperature is set using an adjustable thermostat incorporated into the valve.

Figure 32 ◆ Tempering valves.

Review Questions

Sections 6.0.0–7.0.0

1. Water heaters, hose bibbs, water softeners, and fixtures make up a _____.
   a. water distribution system
   b. DWV system
   c. vent system
   d. sewage-disposal system

2. _____ should be placed as close as possible to the greatest number of hot water outlets.
   a. Fixtures
   b. Hose bibbs
   c. Water heaters
   d. Water softeners

3. If a plumbing system includes a(n) _____, the plumber should first determine the expected use of various fixtures.
   a. exterior water outlet
   b. hose bibb
   c. water heater
   d. water softener

4. The piping runs serving _____ may be long, and they will probably bypass the water softener.
   a. fixtures
   b. hose bibbs
   c. water heaters
   d. gas supplies

5. Water supply piping is located in relation to the already installed _____ piping.
   a. cold water
   b. PVC
   c. DWV
   d. hot water

6. The plumber constructs a vertical section of pipe, known as a _____, to connect the fixtures to the supply pipe.
   a. backboard
   b. stub-out
   c. feeder line
   d. fixture riser

7. Sizing tables estimate the anticipated demand for water as measured by water _____.
   a. supply fixture units
   b. pressure fixture counts
   c. supply fixture risers
   d. pressure fixture valves

8. To make sure the water flows in only one direction, install _____ in a hot water recirculating system.
   a. a water softener
   b. a check valve
   c. branches
   d. hose bibbs

9. Cold water that is used at the hose bibbs for lawn sprinklers and other outdoor applications should not be _____.
   a. softened
   b. heated
   c. treated
   d. filtered

10. A check valve is used to mix hot and cold water to prevent scalding.
    a. True
    b. False

Summary

Water comes from a private source such as a well or a municipal source, a public water distribution system. While well water is usually drinkable, water supplied by municipalities must be treated to remove harmful impurities. Both private and public water service lines move water from its source to homes and other buildings. These lines, or pipes, may be made of copper, plastic, or steel. Factors such as weather, resistance to corrosion, working pressure, applicable codes, ease of installation, cost, and life expectancy determine the choice of pipe materials. Pumps move water from wells into the water distribution system. Plumbers choose the correct pump based on the depth of the well, the rate at which the water is pumped, and the total distance the water must travel. The flow of water in a water distribution system is regulated by valves or faucets, which are used to turn water on and off, to control the rate of water flow, to regulate the pressure, or to prevent a reversal of flow through a line. After a water supply to a home or other building is set up, plumbers install the water heater, hose bibbs, water softener (if needed), and fixtures. Plumbers must locate and connect these items efficiently to avoid softening water that does not need to be softened and to save on materials and time.

Notes

Trade Terms Quiz

Fill in the blank with the correct trade term that you learned from your study of this module.

1. A(n) _______________ is a device that measures the amount of water used.
2. The use of a(n) _______________ , which acts as a backflow preventer, prevents a vacuum in a water supply from causing backflow.
3. _______________ are any part of the piping system other than the riser, main, or stack.
4. A(n) _______________ is a threaded outlet that connects to a hose on the outside of a building.
5. _______________ are inserted into drilled or bored wells to protect the well from contamination.
6. Curb boxes are sometimes called _______________.
7. _______________, or _______________, are the main water supply lines, to which branches are connected.
8. _______________ is the task of placing attachments on fixtures to finish a job.
9. A special key can be inserted into a(n) _______________ to stop the flow of water during servicing or in case of an emergency.
10. _______________ thickens liquids into soft or solid masses.
11. A(n) _______________ connects the building water service line to the water main.
12. Fixture risers and stub-outs are mounted on a(n) _______________.
13. Deterioration to metal pipe caused by an electric current or chemical reaction that occurs between different types of metal pipe is _______________.
14. _______________ are a measure used to properly size pipes.
15. The acidity of water is identified by its _______________ level.
16. When water is allowed to flow freely through a valve, the flow is typically referred to as _______________.
17. _______________ are pipes on the interior portion of the wall that connect the fixture to the supply pipe.
18. The distance from the dry soil at the surface to the saturated soil underground is the _______________.
19. A(n) _______________ is commonly used to disconnect the hot or cold water supply to water closets and sinks.
20. A(n) _______________ shuts off the water supply based on temperature and pressure levels.
21. A(n) _______________ is used to reduce water pressure in a building.
22. A(n) _______________ is similar to the globe valve but can serve as both a valve and a 90-degree elbow.
23. A(n) _______________ allows liquid to flow in only one direction.
24. A(n) _______________ controls the rate of flow through a valve.
25. _______________ are natural or artificial ponds or lakes used for water storage.
26. _______________ are particles that are removed from water during the water softening process.
27. The flow capacity of different types of valves can be described as _______________.
28. _______________ is the sediment or foreign particles that are stirred up or suspended in water, making the water look cloudy.
29. The loud noise that results from the sudden stop of water flow can be eliminated by using _______________.
30. The valve that controls the distribution of water from the building water service to the public water main is the _______________.

Trade Terms

Angle valve
Backing board
Branch
Buffalo box
Check valve
Coagulation
Corporation stop
Curb box
Curb stop
Feeder line
Fixture risers
Full flow
Galvanic corrosion
Hammer arrester
Hose bibb
pH
Precipitate
Pressure regulator valve
Pressure relief valve
Reservoir
Service line
Straight-through flow
Supply stop valve
Temperature and pressure (T and P) relief valve
Thermostatic/pressure balancing valve, combination
Throttled flow
Trimming out
Turbidity
Vacuum breaker
Water meter
Water supply fixture unit (WSFU)
Water table
Well casing

Profile in Success

Dale L. Powell

Northeast Regional Manager, NAPTFI Field Team
Copper Development Association, Inc.
New York, New York

Dale Powell was born in Harrisburg, Pennsylvania, and attended Central Dauphin High School in Harrisburg. He studied architectural drafting, architecture, and building construction technology (BCT) at the Harrisburg Area Community College, the University of Kentucky, and Penn State - Middletown. He received an associate's degree from Harrisburg Area Community College and continued studies for a bachelor's degree, but decided it was not for him. The work he did for a plumbing contractor called him back to the trade. He moved up from plumbing apprentice to estimating, project management, and minor system and component design. Today, Dale is a regional manager for Copper Development Association, Inc. (CDA), and Dale says that, as far as the trade goes, "It's been very good to me."

How did you become interested in this industry?
When I was still in high school, I worked for a plumbing contractor in the evenings, cleaning up the shop. Then I was provided an opportunity to work in the field as a plumber's helper. I worked for this contractor all through high school, college, and my summer vacations. When I left college, I realized that I was making more as a plumber's helper than I would make otherwise, and I was afforded the opportunity to continue learning by entering an apprenticeship training program through the United Association of Plumbers and Pipefitters of the United States and Canada (UA). I felt that this would be a great way to continue my education in a field that had already captured my interest. Upon completing my apprenticeship, I was lucky enough to continue to grow into positions of more and more responsibility in the trade. Each position allowed me to use more of the education that I received in college, in the apprenticeship program, and on the job. I found that the knowledge and skills that I had gained in each of these areas were irreplaceable, especially as I worked as an estimator and project manager for a major mechanical contracting firm. Here critical thinking skills, time management, and overall building construction knowledge gained from my college studies, combined with the in-depth knowledge of piping system design and installation, were necessary to design, bid, and manage major design/build construction projects.

What path did you take to your current position?
My first construction job was working as a plumber's helper with Francis R. Mummert, Inc., through high school and college, to help pay for my tuition. Following college, I joined the Local Union 520 Plumbers and Pipe Fitters Apprenticeship Training Program, where I spent time working as an apprentice with York Air Cooling & Heating, and finally with G.R. Sponaugle & Sons, Inc. After four years of apprenticeship, I received my journeyman's card, and following an additional four years of on-the-job training, I successfully tested to become a master plumber.

Upon receiving my journeyman's card, I continued working with G.R. Sponaugle & Sons, where, with the exception of working out of the union hall on a variety of jobs for one year in the early 1980s, I remained for almost 16 years. G.R. Sponaugle & Sons is a full-line electrical, mechanical, sheet metal, service, and telecommunications firm. I spent time in a number of functions there and eventually became an estimator and project coordinator. In that capacity, I was also responsible for the majority of the CADD (computer-assisted design and drafting) drafting work for the pipefitting, pipe welding, and prefabricated piping systems shop, and was instrumental in the development of the CADD-assisted coordinated pipe drawings and the implementation of a piping prefabrication shop.

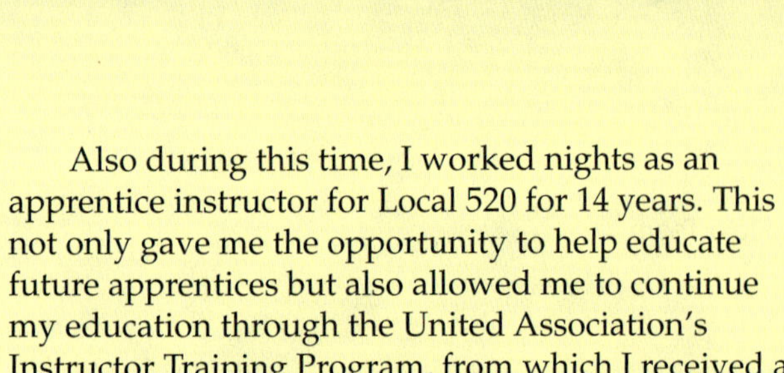

Also during this time, I worked nights as an apprentice instructor for Local 520 for 14 years. This not only gave me the opportunity to help educate future apprentices but also allowed me to continue my education through the United Association's Instructor Training Program, from which I received a certificate from Purdue University as a UA Certified Instructor of Journeymen and Apprentices.

This is where I first became interested in working for CDA, a not-for-profit trade association for the copper, brass, and bronze industry, as a regional manager. One of the things that drew me to my current position was CDA's emphasis on educating and training the designers, specifiers, and installers of copper piping systems. I began with CDA as northeast regional manager in 1995 and have been responsible for all aspects of our copper piping systems efforts in the northeastern United States since that time.

What are some things you do on the job?

My primary function is to grow and support the copper piping systems market in the Northeast, and in most cases this revolves around providing support to the end users of our industry's products, including engineers, contractors, building owners, inspectors, and home builders. The challenge is that this needs to be approached in a different way, and with a different audience each and every day. Some days this may involve providing technical information with respect to codes and standards for copper and copper alloy tube, pipe, and fittings to an engineer designing a major health care facility, while the next day it may be helping the contractor decipher these requirements so that the piping system can be installed correctly. And the day following that may be providing troubleshooting assistance to help a building owner determine why a copper system may not be functioning to his or her satisfaction. Many days include meeting with, or fielding phone calls from, engineers, architects, and designers located within my region, composed of the 13 states from Maine to Virginia, West Virginia, and the District of Columbia, to provide information regarding installations and design criteria for piping systems. Some days and many evenings are spent conducting installation training workshops and seminars with contractors and professional organizations regarding soldering, brazing, and other joining procedures for copper and copper alloy piping systems.

What does it take to be successful in your trade?

The first thing you have to have is a positive attitude. Be open to, and accepting of, new technology and be willing to change with the times. You have to really have a desire to get ahead and to learn something new every day. You have to have a personal desire to better yourself within the trade. If you feel comfortable getting into your truck and going to work every day, then you've found your niche and you should concentrate more on getting better at that niche, not moving out of it. But defining that niche is the key. For me, that niche is the plumbing and piping industry, not whether or not you are turning wrenches or jockeying a keyboard in that industry. I think that you should look at moving from mechanics to a managerial position, or vice versa, based on what would give you the most job satisfaction, make you the most useful to your company, and be the best use of your skills, and not necessarily on the pay scale. I strongly feel that those educated in the trade should not be afraid to take the skills that have been successful in the field and apply them to office and managerial positions. Those skills will continue to serve the worker well in a new role and may provide both the worker and the company even greater advantages.

I took as much schooling as my employers were willing to give me. I was given the best educational opportunities, and, in many ways, they were far and above those that I received in college. I would not have realized the professional opportunities that I have been afforded without the excellent trade-based training I received from UA. For the past 12 years, I've been an instructor at the week-long UA Instructor Training Program, teaching soldering and brazing techniques and how-to-teach methods to the instructors that are tasked each year with educating the new apprentices who will carry our trade into the future. This year (2003) was the program's 50th anniversary.

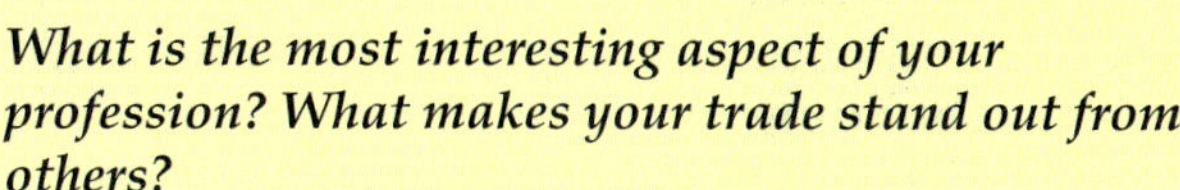

What is the most interesting aspect of your profession? What makes your trade stand out from others?
Our trade's motto is that the plumber protects the health and safety of the nation. The United States and Canada provide the cleanest water, and people expect that. When people walk into a restaurant, school, or any public place for that matter, they assume they will get clean, potable water every time. People take it for granted that plumbers' day-to-day work provides for safe, sanitary facilities and potable water for the public. You just don't see that same emphasis in many other places in the world.

What do you like most—and least—about your job?
I most like the people that I meet at training and the contacts and friendships that I am able to make and maintain. I am able to use my knowledge and experience to positively influence others on the use and installation of copper and copper alloy piping systems, and, in many ways, help our trade move forward.

I least like the amount of travel that is required to adequately provide the necessary services to as many people as I can reach in the 14 states and District of Columbia that comprise my region.

What would you say to someone entering the trade today?
Keep your nose to the grindstone and learn as much as you can. If you do, then you'll probably be working every day. Very few of the people who give 100 percent every day, or 8 for 8 (that is, 8 hours of work for 8 hours of pay), are ever out of work for long.

What can an apprentice expect to earn in his/her first years on the job in your area? What can he/she expect to earn after 10-plus years in the industry?
I can only speak from my past experience, but I believe the rate with Local 520 for a journeyman is $24 to $25 per hour in the paycheck plus paid benefits. An apprentice is paid fairly close to $11 or $12 an hour to start. The average plumber can probably earn between $45,000 and $55,000 per year, although I know of a few in supervisory positions that are earning in the $60,000 to $70,000 range. After 10 years in the industry, that is tough to predict. With a good education, a good work ethic, the willingness to learn and change each day, and sometimes a little luck, your earnings will likely increase. Beyond that, your potential depends on the direction you choose within the industry and the vigor with which you pursue that route.

Trade Terms Introduced in This Module

Angle valve: A valve that is similar to the globe valve but can serve as both a valve and a 90-degree elbow.

Backing board: The unfinished side of the inside wall, used to mount fixture risers and stub-outs behind the fixture.

Branch: Any part of a piping system other than a riser, main, or stack.

Buffalo box: Another term for curb box.

Check valve: A valve that allows liquid to flow in only one direction. Pressure within the line keeps the valve open. Closure of the valve is automatically activated by the reversal of flow or by the weight of the disc mechanism.

Coagulation: Thickening of a liquid into a soft, semi-solid, or solid mass.

Corporation stop: A valve installed in the building water service line to connect to the water main.

Curb box: A cylindrical casing placed in the ground over the curb stop, into which a special key can be inserted to turn off the curb stop. Also referred to as a buffalo box.

Curb stop: A control valve installed in building water supply lines between the corporation stop and the building.

Feeder line: The main water supply line, to which branches are connected. Also referred to as a service line.

Fixture risers: Vertical sections of pipe located inside the wall to connect the fixture to the supply pipe beneath the flooring.

Full flow: Describes the flow capacities of various valves based on the size of the valve's interior.

Galvanic corrosion: Corrosion caused by a weak electrical current that occurs when two different metals are joined.

Hammer arrester: A device installed in a piping system to absorb hydraulic shock waves and eliminate water hammer, a loud thumping that results when the flow of water is suddenly stopped.

Hose bibb: A faucet with a threaded outlet used to connect a hose. Usually located on the outside of a building.

pH: Potential of hydrogen; a measure of the acidity or alkalinity of a solution. A pH number of 7 is considered neutral; higher numbers are more alkaline, lower numbers are more acidic.

Precipitate: A solid that is chemically separated from a solution.

Pressure regulator valve: A valve used to reduce water pressure in a building. The valve is activated by changes in pressure within the system.

Pressure relief valve: A valve normally used for liquid service. As pressure increases it slowly opens, and as pressure decreases it slowly closes.

Reservoir: A body of water collected and stored in a natural or artificial (man-made) lake.

Service line: Another term for feeder line.

Straight-through flow: A flow that is not restricted as it passes through a valve. The element that closes the valve is retracted entirely clear of the passage.

Supply stop valve: A valve that is commonly used to disconnect the hot or cold water supply to water closets and sinks. They are available in either right angle or straight design.

Temperature and pressure (T and P) relief valve: A mixing valve which senses outlet temperature and incoming hot and cold water pressure and compensates for fluctuations in hot and cold water temperatures and/or pressures to stabilize its outlet temperatures. Also referred to as thermostatic/pressure balancing valve, combination.

Thermostatic/pressure balancing valve, combination: Another term for temperature and pressure (T and P) relief valve.

Throttled flow: The controlled rate of flow through a valve, which does not impede flow but causes some friction loss.

Trimming out: The task of placing attachments on fixtures to complete a job.

Turbidity: Presence of particles (sand, mud, silt) suspended in water that give the water a cloudy appearance.

Vacuum breaker: A type of backflow preventer that inhibits backflow caused by low pressure in a water supply system.

Water meter: A device to measure water flow, in gallons or cubic feet, into an individual building.

Water supply fixture unit (WSFU): A design factor to determine the load that different plumbing fixtures produce on the supply side of a plumbing system.

Water table: The distance below the ground's surface at which the soil becomes saturated.

Well casing: Outer tube or pipe sunk into the ground after drilling or driving a well.

Additional Resources and References

ADDITIONAL RESOURCES

This module is intended to present thorough resources for task training. The following reference works are suggested for further study. These are optional materials for continued education rather than for task training.

Advanced Home Plumbing, 1997. Black & Decker Home Improvement Library. Minnetonka, MN: Cowles Creative Publishing, Inc.

Basic Plumbing with Illustrations, 1980. Howard C. Massey. Carlsbad, CA: Craftsman Book Company.

2003 International Plumbing Code. Falls Church, VA: International Code Council.

2003 National Standard Plumbing Code. Falls Church, VA: Plumbing-Heating-Cooling Contractors–National Association.

REFERENCE

2003 National Standard Plumbing Code. Chapter 1, Definitions (definitions of *branch, temperature and pressure (T and P) relief valve,* and *vacuum breaker*). Also Tables 10.14.2A and B. Falls Church, VA: Plumbing-Heating-Cooling Contractors–National Association.

NCCER CURRICULA — USER UPDATE

NCCER makes every effort to keep its textbooks up-to-date and free of technical errors. We appreciate your help in this process. If you find an error, a typographical mistake, or an inaccuracy in NCCER's curricula, please fill out this form (or a photocopy), or complete the online form at **www.nccer.org/olf**. Be sure to include the exact module ID number, page number, a detailed description, and your recommended correction. Your input will be brought to the attention of the Authoring Team. Thank you for your assistance.

Instructors – If you have an idea for improving this textbook, or have found that additional materials were necessary to teach this module effectively, please let us know so that we may present your suggestions to the Authoring Team.

NCCER Product Development and Revision
13614 Progress Blvd., Alachua, FL 32615

Email: curriculum@nccer.org
Online: www.nccer.org/olf

❏ Trainee Guide ❏ AIG ❏ Exam ❏ PowerPoints Other ______________________________

Craft / Level: Copyright Date:

Module ID Number / Title:

Section Number(s):

Description:

Recommended Correction:

Your Name:

Address:

Email: Phone:

Glossary of Key Trade Terms

Aboveground rough-in: The second phase of a plumbing project. During this phase, holes are cut in walls, ceilings, and floors. Then, supply and waste pipes are attached or hung so they can be connected to fixtures.

ABS (acrylonitrile-butadiene-styrene): Plastic pipe and fittings used extensively in drain, waste, and vent (DWV) systems.

ACR: Air conditioning and refrigeration system.

ACR tubing: Annealed copper tubing that is manufactured specifically for use in air conditioning and refrigeration work.

Adapter: A fitting that joins pipes of different sizes or materials, such as copper and galvanized pipe or cast-iron and plastic pipe.

Amperage: A measure of electrical current.

Angle valve: A valve that is similar to the globe valve but can serve as both a valve and a 90-degree elbow.

Annealing: Heating a material and slowly cooling it to relieve internal stress. This process reduces brittleness and increases toughness.

Apparatus: One tool, which combines a variety of functions to perform a particular job.

Approved submittal data: The fixtures and fittings in catalog drawings (or submittal data) that have been approved by the architect or engineer.

Appurtenances: Accessories or apparatus.

Aqueduct: A man-made channel used to carry water.

Architect's scale: A measuring device that uses smaller units, such as ½ inch or ¼ inch, to represent 1 foot. Architect's scales are also issued in metric units. The units are used so that all building measurements are in proportion to their actual measurements but able to fit in the drawings.

Asbestos: A fibrous, fire-resistant substance used in pipe insulation, shingles, wallboard, floor coverings, and certain types of insulation. Now banned by government regulation as a health hazard.

Atmospheric hazard: A potential danger in the air or a condition of poor air quality.

Back: Part of the fitting that is opposite to the side with an opening or face.

Back pressure: A condition which may occur in the potable water distribution system, whereby a higher pressure than the supply pressure is created causing a reversal of flow into the potable water piping. Also referred to as backpressure backflow.

Backflow preventer: A potable device that prevents nonpotable water from entering a freshwater supply system.

Backflow: The flow of contaminated water into the freshwater system resulting from a cross-connection between potable and nonpotable water systems.

Backing board: The unfinished side of the inside wall, used to mount fixture risers and stubouts behind the fixture.

Backpressure backflow: Another term for back pressure.

Ball valve: A valve used to control the flow of gases and liquids. It is installed in piping systems where quick shutoffs for in-line maintenance may be necessary.

Bar stock: Uncut bar material on hand for use on a project.

Bell-and-spigot pipe: Pipe that has a bell, or enlargement, also called a hub, at one end of the pipe and a spigot, or smooth end, at the other end. The bell and spigot of two different pipes slide together to form a joint. Also called hub-and-spigot pipe.

Benching: A method of protecting workers from cave-ins by excavating the sides of an excavation to form one or a series of horizontal levels or steps, usually with vertical or near-vertical surfaces between levels.

Bend: A fitting that changes the direction of piping.

Bibb: A faucet with an outlet at a downward angle, generally used for hose connections on the outside of a building.

Black steel pipe: Steel pipe with a black color used for gas and air-pressure pipe. It should not be used where corrosion will affect its uncoated surface.

Bladed tools: Tools that use sharp edges to accomplish their tasks. Bladed tools include saws, knives, scissors, tin snips, and wire cutters.

Branch: Any part of a piping system other than a riser, main, or stack.

Branch interval: A distance along a soil or waste stack corresponding, in general, to a story height, but in no case less than 8 feet within which the horizontal branches from one floor or story of a building are connected to the stack.

Buffalo box: Another term for curb box.

Building sewer: That part of the drainage system which extends from the end of the building drain and conveys its discharge to a public sewer, private sewer, individual sewage-disposal system, or other point of disposal.

Bullhead tee: A tee fitting used on a branch that is longer than the main line, or that has one outlet larger than the run openings.

Burr: Uneven or jagged edge left on metal by certain cutting tools.

Capillary action: The process during soldering in which the molten solder flows into the narrow gap between the pipe and the joint, regardless of whether the solder is flowing up, down, or horizontally.

Capillary attraction: The tendency of water to be drawn by porous material. If a piece of paper or string is stuck in a pipe, the water is drawn out of the trap and the seal is broken.

Catalog drawing: A drawing of a plumbing fixture that is found in manufacturers' catalogs. Also referred to as submittal data.

Caulking iron: A tool used to drive the lead firmly into a lead and oakum joint when working with hub-and-spigot cast-iron pipe and fittings.

Ceiling caulking iron: A specially shaped tool used to drive the lead firmly into a lead and oakum joint when working with hub-and-spigot cast-iron pipe and fittings next to a ceiling.

Cellular core wall: Plastic pipe wall that is low-density, lightweight plastic containing entrained (trapped) air.

Center: A point exactly halfway between two other points or surfaces.

Center point: Point created where the centerline of the pipe and the centerline of the fitting meet within the fitting. The center point is used to determine the correct length of a pipe.

Centerline: On a drawing, a line that shows the center of an object.

Chain vise: Tool used to cut pipe by holding the pipe in a vise while cutting with a chain that has cutting wheels in each link.

Chain wrench: Another term for pipe tongs.

Chamfer: To bevel the edge of construction material to a 45-degree angle.

Check valve: A valve that allows liquid to flow in only one direction. Pressure within the line keeps the valve open. Closure of the valve is automatically activated by the reversal of flow or by the weight of the disc mechanism.

Chemically inert: Does not react with other chemicals.

China: A mixture of fine clay, quartz, feldspar, and silica that is heated to 2,600°F (1,426°C) to create a nonporous, glass-like finish. Also referred to as vitrified porcelain.

Chlorine: A heavy, greenish-yellow gas used as a disinfectant in water treatment. Chlorine should be handled only when wearing appropriate personal protective equipment.

Cleanout: An access point to all parts of the drainage system for the removal of blockages.

Clevis: An iron, or link in a chain, bent into the form of a horseshoe, stirrup, or letter U, with holes in the ends to receive a bolt or pin.

Closet bend: A fitting that connects the water closet to the main drainage piping.

Closet flange: A fitting used to connect the water closet to the closet bend.

Coagulation: Thickening of a liquid into a soft, semi-solid, or solid mass.

Code: A requirement published by state and local governments to establish minimum standards for various types of construction. A code carries the force of law.

Combustible: Air or materials that can explode and cause a fire.

Competent person: An individual who is capable of identifying existing and predictable hazards or working conditions that are hazardous, unsanitary, or dangerous to employees, and who has authorization to take prompt, corrective measures to eliminate or control these hazards and conditions.

Compression collar: A piece of hardware that uses compression force to connect sections of polyethylene piping.

Compression joint: A joint formed using a neoprene gasket inserted into the bell end of a hub-and-spigot pipe. The pressure applied between the two joined pipes forces the gasket to compress and fill in any air gaps in the joint.

Compression joint: A method of connection in which tightening a threaded nut squeezes a compression ring to seal the joint.

Computer-aided drafting (CAD): A sophisticated design program used on computers. It allows designers to create drawings in two or three dimensions.

Confined space: A space that, by design and/or configuration, has limited openings for entry and exit, has unfavorable natural ventilation, may contain or produce hazardous substances, and is not intended for continuous employee occupancy.

Construction drawing: A drawing that shows the design, location, and dimensions of a building and its various components.

Coordination drawing: A dimensioned drawing with elevations and sections that indicate the proposed routing of system components. The dimensioned coordination drawing includes the actual dimensioned equipment as well as the access space this equipment will need.

Corporation stop: A valve installed in the building water service line to connect to the water main.

Corrugated stainless steel tubing (CSST): A gas piping material made from flexible steel tubing with a polyethylene (PE) jacket.

Coupling: A fitting used to connect two lengths of no-hub cast-iron pipe.

Coupling fitting: A fitting that allows two separate lengths of CSST to be connected to each other.

CPVC (chlorinated polyvinyl chloride): Plastic pipe and fittings used extensively in hot and cold water distribution systems.

Cross: Fitting that connects four lengths of pipe in a cross-like shape in a distribution system.

Cross-connection: An arrangement between a potable water system and a nonpotable water system in which an accidental pressure differential between the two systems causes backflow of contaminated water into the freshwater system.

Crown weir/trap weir: Discharge overflow of the trap outlet. Also referred to as weir, trap, or crown.

Curb box: A cylindrical casing placed in the ground over the curb stop, into which a special key can be inserted to turn off the curb stop. Also referred to as a buffalo box.

Curb stop: A control valve installed in building water supply lines between the corporation stop and the building.

Cutaway drawing: A section drawing that shows the construction elements of a particular part of a building or fixture.

Decibels (dB): A measure of sound intensity or loudness. The higher the decibel level, the louder and more potentially damaging the sound is.

Details: Sections of construction drawings that are enlarged to make them clearer.

Diameter: The distance across the center of a circle.

Die: Device used to cut the threads on steel pipe.

Dimension line: A line on a drawing with a measurement indicating actual length.

Disinfection: The process of destroying harmful organisms in potable water.

Diverter: A device that redirects the flow of water, as in a combination shower and bath fitting. When the diverter is engaged, water flows through the showerhead instead of through the spigot.

Double ¼ bend: A fitting used to collect and combine the flow from two opposite runs into a single run of pipe.

Double extra strong (double extra heavy): An industry standard measurement of the weight or strength of steel pipe. Also referred to as Schedule 120.

Double sanitary tee: A tee fitting that changes the direction of flow from horizontal to vertical and that has openings for two branch lines. Also called a sanitary cross.

Double trapping: A situation in which one trap is attached to another, creating negative pressure that will stop the flow of drainage.

Double wye: A fitting that changes direction at 45 degrees. It connects horizontal waste pipes and branch drains to the building main. It has openings for two branch lines.

Drain, waste, and vent (DWV): A piping system that combines sanitary drainage with venting.

Drainage fitting: Any of a variety of fittings used in the DWV system to remove waste from a building.

Drainage fitting (recessed fitting): Fitting designed with smooth inside surfaces and shaped for easy flow to form an unbroken, internal contour for use on drainage systems.

Drawn copper: Tubing produced by pulling the tube through dies to reduce its diameter. The drawing process hardens the copper and makes it very rigid.

Drop forged: A characteristic of a product made when heated metal is pounded or shaped between dies with a drop hammer or press.

Ductile iron: A type of cast iron in which magnesium is added to the molten gray iron to reduce brittleness.

DWV system: An acronym for "drain-waste-vent" referring to the combined sanitary drainage and venting systems. This term is technically equivalent to "soil-waste-vent" (SWV).

Easement: A designated right-of-way, such as the access guaranteed to utility companies for repair of utilities that are located on, or cross over, private land, or for vehicles to cross private property.

Elastomeric: Rubberized. Made of an elastic substance such as a polyvinyl elastomer.

Electrical drawing: A drawing that shows the location of outlets, switches, and electrical fixtures. An electrical drawing may be superimposed on the floor plan.

Electrical ground: A conductive connection that provides a path for electrical current to pass from an electrical component into the earth.

Electrically powered tools: Tools that use electrical current to operate.

Elevated pressure system: A CSST system that operates above the normal pressure of 7 inches w.c.

Elevation: The height above an established reference point, such as a grade reference point on a construction drawing.

Elevation drawing: A drawing of a structure showing a side, front, or back view.

Energy source: Any source of electrical, mechanical, hydraulic, pneumatic, chemical, thermal, or other energy.

Energy-isolating device: Any mechanical device that physically prevents the transmission or release of energy. Can include manually operated electrical circuit breakers, disconnect switches, line valves, and blocks.

Ethics: A set of principles and values that guide an individual's conduct.

Evaporation: Loss of water, especially in a drainage system, into the atmosphere.

Exploded drawing: A drawing that shows how to assemble a complex product. It also shows the relationship of the individual parts to the object as a whole. Also referred to as an assembly drawing.

Extension line: A line used on a drawing to locate a dimension away from the actual points of the dimension. This method is used when a drawing would be too crowded or cluttered if the dimension were shown within the two points.

Extra strong (extra heavy): An industry standard measurement of the weight or strength of steel pipe. Also referred to as Schedule 80.

Face: The open end of a fitting where a pipe is joined to the fitting, such as the opening of the inlet or either end.

Fall: The amount of slope given to horizontal runs of pipe.

Faucet: A fixture that is used to draw water from a pipe.

Feeder line: The main water supply line, to which branches are connected. Also referred to as a service line.

Ferrous: Containing iron.

Ferrule: A brass compression ring used for joining.

Fiberglass: Spun filaments of glass woven into yarn or roving (twisted) strands and textile materials such as cloth or mats. Fiberglass cloth saturated with a plastic (vinyl resin or epoxy) can be pressed into molds to make products such as bathtubs and shower enclosures.

Filtration: The process of cleansing water to remove particles and chemicals.

Finish: The third phase of a plumbing project. During the finish phase, plumbers install fixtures, appliances, water purification systems, water heaters, and controls.

Fitting allowance: The distance from the end of the pipe that goes into a fitting to the center of the fitting. Also called fitting takeoff.

Fixture: A receptacle or device which is either permanently or temporarily connected to the water drainage system of the premises, and demands a supply of water therefrom, or discharges used water, liquid-borne waste materials, or sewage either directly or indirectly to the drainage system of the premises, or which requires both a water supply connection and a discharge to the drainage system of the premises. Plumbing appliances as a special class of fixture are further defined. Also referred to as plumbing fixture.

Fixture: A device that receives water from a water supply line. Common fixtures include sinks, shower stalls, and toilets.

Fixture drain: The drain from the trap of a fixture to the junction of that drain with any other drain pipe.

Fixture drawing: A drawing that shows the components of a fixture in detail.

Fixture risers: Vertical sections of pipe located inside the wall to connect the fixture to the supply pipe beneath the flooring.

Flange: A rim on one end of a length of pipe that provides a connection point to another length of pipe or piece of equipment, such as a water closet. Bolts or studs with nuts are used to hold two flanges together, with some type of gasket between them. The flange is soldered or solvent-welded to the drainpipe stub to connect the fixture to the drain, waste, and vent system.

Flange union: Fitting that connects two pipes. A flange is screwed onto the end of each pipe that needs to be joined and then bolted together with nuts and bolts. A gasket in between flanges makes it a leakproof seal.

Flare joint: A fitting in which one end of each tube to be joined is flared outward using a special tool. The flared tube ends mate with the threaded flare fitting and are secured to the fitting with flare nuts.

Flood-level rim: The edge of the receptor or fixture which water overflows.

Floor plan: A construction drawing of a building looking down at the floor from above (bird's-eye view). The plan shows at least the outline of the wall locations and lengths to scale. Normally, this drawing is oriented such that the top of the page represents north.

Flush valve: A device located at the bottom of a tank for flushing water closets and similar fixtures.

Flushometer: A device which discharges a predetermined quantity of water to fixtures for flushing purposes and is closed by direct water pressure or other mechanical means. Also referred to as flushometer valve.

Flushometer valve: Another term for flushometer.

Flux: A water-soluble substance that facilitates the fusion (joining) of metals and helps prevent surface oxidation (rusting, tarnishing) during welding, brazing, and soldering. Also called soldering paste.

Formability: The ease with which a material can bend.

Foundation plan: A construction drawing showing the placement and dimensions of a building foundation.

Full flow: Describes the flow capacities of various valves based on the size of the valve's interior.

Fusion fitting: A fitting with a butt that has the same outside diameter and inside diameter as the pipe. It is usually joined to a pipe by heat.

Galvanic corrosion: Corrosion caused by a weak electrical current that occurs when two different metals are joined.

Galvanized pipe: Pipe electroplated with zinc to provide a protective coating that resists corrosion or oxidation (rust).

Gas cock: A type of valve that provides positive, quick flow control on gas piping systems.

Gassy operations: Working conditions in which one or more of the following conditions exist: higher than minimum levels of methane or explosive gases are present; a gas ignition has previously occurred there; or the area is connected to an underground area designated a gassy operation.

Gate valve: A valve in which the flow is controlled by moving a gate or disc that slides in machined grooves at right angles to the flow. This type of valve is designed to be fully open or fully closed.

Globe valve: A valve in which the flow is controlled by moving a circular disc against a metal seat that surrounds the flow opening.

Grade: The slope of a horizontal run of pipe. Also referred to as slope (percent of grade).

Ground joint union: Fitting that joins two pieces of pipe by screwing the thread and shoulder pieces of the fitting onto the pipe. The collar piece is then tightened to join the sections of pipe into a watertight joint.

Guards: Devices that protect tool operators from dangerous moving parts, such as blades, gears, and pulleys.

Guy wires: Ropes, chains, cables, or rods attached to something as a brace or guide.

Hammer arrester: A device installed in a piping system to absorb hydraulic shock waves and eliminate water hammer, a loud thumping that results when the flow of water is suddenly stopped.

Hazard Communication (HazCom) Standard: A federal OSHA regulation requiring employers to educate and inform workers about chemical hazards on the job site (*29 CFR 1910.1200*).

Head: The height of a water column, measured in feet. One foot of head is equal to 0.433 pounds per square inch (psi).

Heel inlet: A bend that has an inlet to connect a smaller pipe to the main line or a branch line. The inlet is located at the base of the curve or heel of the bend.

Hose bibb: A faucet with a threaded outlet used to connect a hose. Usually located on the outside of a building.

Hub-and-spigot cast-iron pipe: Pipe that has a bell or enlargement at one end where the spigot (smooth end) of the next pipe slides in to form a joint. Also called bell-and-spigot pipe.

HVAC (heating, ventilating, and air conditioning) drawing: A construction drawing that shows the placement of the furnace and air conditioning equipment and the location of ducts and registers or pipes and radiators.

Hydraulic gradient: The level line in which water tends to flow in a pipe.

Hydronic: A system that heats and cools by circulating water or steam through a closed piping system.

Hydrostatically pressure-test: To fill a pipe with water and bleed all air out from the highest and farthest points in the run.

Hypothermia: A life-threatening condition caused by exposure to very cold temperatures.

Impact tools: Tools that must strike or be struck to accomplish their task. They include hammers, chisels, and taps.

Increaser: A fitting used to increase the size of a straight-through line of pipe. It is often used for the vent stack before it goes through the roof to reduce the chance of frost clogging the vent opening in very cold climates.

Insertion length: The length of a pipe that fits into the fitting or other joint when assembling a pipe run. The measurement must be figured into the calculation before cutting a piece of pipe for joining.

Inside caulking iron: A tool used to shape the lead near the spigot in a lead and oakum joint.

Inside diameter (ID): The distance between the inner walls of a pipe; the standard measure of piping used in heating and plumbing.

Insulation: A substance that retards the flow of heat.

Interceptor: A device designed and installed so as to separate and retain deleterious, hazardous, or undesirable matter from normal waste while permitting normal sewage or liquid wastes to discharge into the drainage system by gravity.

Interference fit: Fit that tightens as the pipe is pushed into the socket.

Inverted wye: A fitting used to join the upper end of the vent to the top of the soil and waste stacks. It looks like an upside-down letter Y.

Isometric drawing: A pictorial drawing that creates the illusion of a three-dimensional object. All horizontal lines are projected at a 30-degree angle.

Joint runner: Fiber rope that is clamped around the pipe using a spring clamp to keep the hot, molten lead in the hub when making a lead and oakum joint. Also called running rope.

Joist: A piece of lumber used horizontally as a support for a ceiling or a floor.

Journey plumber: A plumber who has successfully completed an apprenticeship training program.

Keel crayon: A waxy crayon used to mark a cutting line on the surface of a tube. Also called soapstone.

Kerf: The cut or groove made by a saw blade, determined by the way the teeth are set on the blade.

Ladle: A spoon-like tool used to dip the molten lead from the lead pot and pour it into the joint.

Laundry tray: A fixed tub, usually installed in the laundry room or utility area of homes. It is used for washing clothes and for receiving wastewater from automatic clothes washers.

Lavatory: A basin designed for installation in bathrooms and other locations, primarily for washing the hands and face.

Lead and oakum joint: A joint for hub-and-spigot cast-iron pipe that involves pouring molten lead over oakum fiber to form a watertight seal between the two pieces of pipe, then caulking the joint after the lead has cooled.

Lead pot: A vessel used to hold molten lead. It is placed on the head of the propane melting furnace during the melting process.

Left-hand bathtub: A bathtub that has the drain on the left end of the tub as you face its length.

Level: Straight on a horizontal plane.

Liquid-fuel tools: Tools that use a liquid fuel, such as gasoline or liquid propane, to operate.

Lockout: The placement of a lockout device on an energy-isolating device, in accordance with an established procedure, ensuring that the energy-isolating device and the equipment being controlled cannot be operated until the lockout device is removed.

Lockout device: Any device that uses positive means such as a lock to hold an energy-isolating device in a safe position, thereby preventing the energizing of machinery or equipment.

Lockout/tagout procedure: A process for identifying hazardous equipment, locking it so that no workers can use it until it is certified for safe use, and placing a tag on the equipment that describes the problem and warns against use.

Long bend: A bend fitting with one leg longer than the other.

Long sweep ¼ bend: A bend used at the base of DWV stacks because the longer radius of its design greatly decreases flow resistance and back pressure. Use is regulated by code.

Main stack: The principal pipe artery to which branches may be connected.

Manifold: A device that allows one CSST pipe to be branched off to multiple outlets.

Material safety data sheet (MSDS): A document that must accompany any hazardous material. The MSDS identifies the material and gives the exposure limits, the physical and chemical characteristics, the kind of hazard it presents, precautions for safe handling and use, and specific control measures.

Miter: A surface forming the beveled end or edge of a piece where a joint is made by cutting two pieces at an angle and then fitting these pieces together.

Model code: A construction ordinance that is written by a national construction organization according to suggested national plumbing standards. Model codes do not have the force of law.

NFPA warning diamond: A four-color diamond label placed on containers or doors to alert people to specific safety hazards in a product, room, or building.

Nipple: Short length of pipe with threads at both ends, used to make extensions from a fitting or to join two fittings.

No-hub cast-iron pipe: A pipe with no enlargement or bell on either end.

Nominal size: Approximate measurement in inches of the inside diameter (ID) of most pipe from 6 inches up to 12 inches. Pipe larger than 12 inches is measured by its outside diameter (OD). The designation used to specify the size of pipe is not necessarily equal to the exact inside diameter. The nominal size of ACR tubing is based on the outside diameter (OD).

Nonferrous: Not containing iron and therefore not magnetic.

Nonpermit-required confined space: A confined workspace free of any atmospheric, physical, electrical, and mechanical hazards that can cause injury or death.

Oakum fiber: A caulking material of old hemp rope or jute fibers that have been treated with tar or a tar derivative.

Oblique drawing: A pictorial drawing that shows the shape of an object. It shows the front of the object with the body of the object at a slight angle.

Occupational Safety and Health Administration (OSHA): The division of the U.S. Department of Labor mandated to ensure a safe and healthy environment in the workplace.

Offset: A combination of elbows or bends which brings one section of the pipe out of line but into a line parallel with the other section.

On-the-job training (OJT): Field experience used in conjunction with classroom lessons in an apprenticeship program. ATELS requires 144 hours of classroom instruction per year and 2,000 hours of OJT per year.

Orthographic drawing: A construction drawing that shows straight-on views of the different sides of an object. Orthographic drawings are used for elevation drawings.

Outside caulking iron: A tool used to shape the lead to the inside of the hub or bell of a hub-and-spigot pipe.

Outside diameter (OD): The distance between the outer walls of a pipe.

Oxygen-deficient atmosphere: An atmosphere in which there is not enough oxygen to support life. Usually considered less than 19.5 percent oxygen by volume.

Oxygen-enriched atmosphere: An atmosphere in which there is too much oxygen. Usually considered more than 23.5 percent oxygen by volume.

Packing iron: A tool with a broad, thick blade used to pack oakum into a joint with a hammer.

PB (polybutylene): Plastic piping that was formerly used for plumbing pipe; it is no longer used but is still found in some residences.

PE (polyethylene): Flexible plastic pipe, tubing, and fittings, usually used for water distribution, that do not deteriorate when exposed to sunlight.

Permit-required confined space: A confined space that has actual or possible hazards. These hazards can be atmospheric, physical, electrical, or mechanical.

PEX (cross-linked polyethylene): Tubing and fittings made with heat and high pressure that resist high temperatures, pressure, and chemicals.

pH: Potential of hydrogen; a measure of the acidity or alkalinity of a solution. A pH number of 7 is considered neutral; higher numbers are more alkaline, lower numbers are more acidic.

Pickout iron: A tool with a diamond-shaped flat point used to remove the lead and oakum from a joint.

Pictorial drawing: A drawing that shows a three-dimensional view of an object.

Pipe riser clamp: A vertical extension of pipe hanger that provides support for pipe and tubing.

Pipe scale: Flaky material resulting from corrosion of metals, especially iron or steel. Also, a heavy oxide coating on copper or copper alloys resulting from exposure to high temperatures and an oxidizer.

Pipe tongs: Wrench-like tool with a length of chain, used to tighten large pipe. Also referred to as a *chain wrench.*

Pipe-joint compound (pipe dope): Sealing material used when joining lengths of pipe.

Plot plan: A drawing of a structure that includes the dimensions of the building site, location of the structure in relation to the property boundaries, elevation of key points, existing and finish contour lines, utility services, and compass directions. Also referred to as a site plan.

Plumb: Straight on a vertical plane.

Plumbarius: The Roman term for someone who works with lead. The root of the modern word *plumber.*

Plumber: One who installs or repairs plumbing systems and fixtures.

Plumbing: According to the *National Standard Plumbing Code,* plumbing is "the practice, materials, and fixtures within or adjacent to any building structure or conveyance, used in the installation, maintenance, extension, alteration, and removal of all piping, plumbing fixtures, plumbing appliances, and plumbing appurtenances connected with the following: (a) sanitary drainage systems and related vent systems; (b) storm water drainage facilities and venting systems; (c) public or private potable water supply systems; (d) the initial connection to a potable water supply upstream of any required backflow prevention devices and the final connection that discharges indirectly into a public or private disposal system; (e) medical gas and medical vacuum systems; (f) indirect waste piping, including refrigeration and air conditioning drainage; and (g) liquid waste or sewage, and water supply, of any premises to their connection with the approved water supply system or to an acceptable disposal facility."

Plumbing drawing: A construction drawing that shows the location of fixtures and pipe runs, and gives the size and type of pipe to be installed.

Plumbing fixture: Another term for fixture.

Plumbum: Latin word for lead.

Polyvinyl chloride (PVC): A thermoplastic material frequently used in tubing for cold water systems and the first type of plastic approved for use in plumbing.

Porcelain enamel: A liquid used to coat materials such as steel and cast iron to create a vitrified porcelain, or china, finish.

Potable: Water that is safe for cooking and drinking.

Pounds per square inch gauge (psig): A unit of measure used to specify the maximum operating pressure for a gas system.

Powder-actuated fastening tool: A tool that uses a 0.22-caliber shell to drive nails or fasteners into the framework of a structure. It may be used only by certified workers.

Power tools: Tools that require a power source, such as electricity, hydraulics, or pneumatics, to operate.

Power-threading machine: Machine used to thread large quantities of pipe. The machine rotates, threads, cuts, and reams pipe.

Precipitate: A solid that is chemically separated from a solution.

Pressure drop: A decrease in pressure from one point to another caused by friction losses in a fluid system, such as a water system.

Pressure fitting: Any type of fitting used for steam, gas, or water supply piping. Not designed for use in drainage piping.

Pressure rating: The maximum pressure at which a component or system may be operated continuously.

Pressure regulator valve: A valve used to reduce water pressure in a building. The valve is activated by changes in pressure within the system.

Pressure relief valve: A valve normally used for liquid service. As pressure increases it slowly opens, and as pressure decreases it slowly closes.

Propane melting furnace: A device used to melt lead. It burns liquefied petroleum (LP) gas.

Protective system: A method of protecting employees from cave-ins, from material that could fall or roll from an excavation face or into an excavation, or from the collapse of adjacent structures. Protective systems include support systems, sloping and benching systems, and shielding systems.

psi (pounds per square inch): A measurement of pressure.

P-trap: A P-shaped trap that provides a water seal in a waste or soil pipe, used mostly at sinks and lavatories.

PVC (polyvinyl chloride): Plastic pipe and fittings used for cold water distribution and for industrial water and chemicals, as well as for drain, waste, and vent (DWV) systems.

Reaming: A process that removes burrs from pipes after they have been cut.

Regulator: A device used in elevated pressure systems to reduce the gas pressure before the pipe outlet.

Regulator stub-out: A stub-out for use with a regulator.

Remainder: The leftover amount in a division problem. For example, in the problem $34 \div 8$, 8 goes into 34 four times ($8 \times 4 = 32$) and 2 is left over or, in other words, it is the remainder.

Reservoir: A body of water collected and stored in a natural or artificial (man-made) lake.

Right-hand bathtub: A bathtub that has the drain on the right end of the tub as you face its length.

Ring-tight gasket fitting: Fitting with a rubber O-ring or gasket in the socket.

Riser diagram: A drawing that shows vertical and horizontal piping along with sizes and a riser number that refers back to the full set of plumbing drawings.

Roll grooved: A type of copper piping that is compatible with grooved copper fittings.

Rotary hammer drill: A drill with a pounding action that lets you drill into concrete, brick, or tile. It rotates and hammers at the same time and drills much faster than regular drills.

Run: One or more lengths of pipe that continue in a straight line.

Running rope: Fiber rope that is clamped around the pipe using a spring clamp to keep the hot, molten lead in the hub when making a lead and oakum joint. Also called joint runner.

Sanitary combination: A fitting that combines a wye and ⅛ bend. It is used to connect horizontal branch lines that intersect other horizontal branch lines. It offers less resistance to the flow of material than a sanitary tee. Also called a tee-wye.

Sanitary cross: A tee fitting that changes the direction of flow from horizontal to vertical and that has openings for two branch lines. Also called a double sanitary tee.

Sanitary fitting: A fitting used to connect DWV branches to the main DWV system and to serve as a cleanout.

Sanitary increaser: A fitting used to enlarge the diameter of the vent stack. It is usually placed at least 1 foot below the intersection of the stack and the roof. Use is regulated by code.

Sanitary tee: A tee fitting for DWV systems that has a slight curve in the 90-degree inlet side of the fitting; this curve helps to channel the flow of wastewater or sewage from a branch line to the main line. This fitting is always used in the vertical position.

Sanitary upright wye: A fitting used to connect the vent stack to the lower end of the soil and waste stacks.

Sanitary wye: A drainage fitting, shaped like the letter Y, that joins the main run of pipe at an angle.

Scale: The relationship of the dimensions on a drawing to the actual dimensions of the structure. For example, in a ¼ scale, 0.25 inches represent 1 foot. Scale is often provided in both English and metric units.

Schedule: A measurement that describes pipe wall thickness.

Schematic drawing: A single-line drawing of a plumbing system, electrical wiring routing, or circuit.

Schematic drawings: Simple, single-line representations (drawings) of pipe and fittings.

Scribe: A sharply pointed and hardened steel tool used for marking a surface to be cut by etching a line or a point into the surface.

Sealant tape: Tape used to wrap the threads of a pipe before joining to ensure a watertight, secure fit.

Seat washer: A disc that fits at the end of the valve stem and is used to regulate the flow of water through a faucet.

Sepia: A print or construction drawing with dark reddish-brown lines on a light background.

Service (SV): A lightweight kind of cast-iron pipe. Service refers to the pipe's thickness, which can vary.

Service line: Another term for feeder line.

Setback: The distance a code requires between a building and a property line, such as the street.

Shield: A structure that is able to withstand the forces imposed on it by a cave-in and thereby protect employees within the excavation. Shields can be permanent structures or portable and moved along as work progresses. Shields can be either premanufactured or job-built in accordance with *29 CFR 1926.652 (c)(3)* or *(c)(4)*.

Shoring: A structure such as a metal hydraulic, mechanical, or timber system that supports the sides of an excavation and is designed to prevent cave-ins.

Short sweep ¼ bend: A bend fitting with a short radius used at the base of a DWV stack. Use is regulated by code.

Side inlet: An opening in an ell or tee fitting at right angles to the line of the run, used to connect smaller lines to the main line.

Side yards: The spaces along the sides of a structure that provide access to rear yards, reduce the possibility of fire jumping from one building to the next, and promote ventilation around the structure.

Single-line drawing: A plumbing drawing that uses a single line to represent the centerline of a pipe. Single-line drawings can be used to represent pipe of any diameter.

Siphonage: Loss of water in a trap seal caused by unequal pressure inside and outside DWV piping.

Site plan: Another term for plot plan.

Size dimension ratio (SDR): A measurement of pipe size that relates pipe wall thickness to pipe diameter.

Sizing tool: A tool consisting of a plug and a sizing ring that is used to reshape a deformed pipe back to roundness.

Slope: Measurement of the fall of a length of pipe from level. The change in level is expressed in degrees of an angle. Also referred to as percent of grade.

Slope (percent of grade): Another term for grade.

Sludge: Semi-liquid matter that settles out in a holding tank during the waste treatment process.

Soapstone: Another term for keel crayon.

Softening: The process of removing magnesium and sodium salts that cause scale on the inside of pipes and fittings.

Soil pipe cutter: A heavy-duty tool used for cutting cast-iron pipe. Soil pipe cutters can be of the snap or ratchet type.

Solder: An alloy (tin plus antimony, copper, and silver) with a low melting point used to join metals or seal joints.

Soldering: A method of joining metals or sealing joints using solder and heat.

Solid wall: Plastic pipe wall that does not contain trapped air.

Solvent weld: A joint created by joining two pipes using solvent cement that softens the material's surface.

Specifications: Written requirements included with the drawings or blueprints of a construction project. They provide more details or descriptions of the technical standards that must be met during construction. Specifications usually override drawings, but are overridden by the contract. Also referred to as specs.

Specs: Another term for specifications.

Spirit level: A level in which the adjustment to the horizon is shown by the position of a bubble in liquid contained in a nearly horizontal glass tube or a circular box with a glass cover.

Spring clamp: A tool used to clamp the running rope around the pipe next to the hub when making lead and oakum joints.

Stack: A general term for any vertical line including offsets of soil, waste, vent, or inside conductor piping. This does not include vertical fixture and vent branches that do not extend through the roof or that pass through not more than two stories before being reconnected to the vent stack or stack vent.

Stack-out: Another term for aboveground rough-in.

Standard weight: An industry standard measurement of the weight or strength of steel pipe. Also referred to as Schedule 40.

Stock: Component of a hand threader or power threader that holds the die.

Stop-and-waste valve: Valve that is opened or closed by raising or lowering a horizontal disc using a threaded stem. An elastomeric, or rubberized, washer on the end of the stem seals the valve seat, closing off water flow. This valve is most commonly used for water faucets.

Straight fitting: A fitting used to attach a length of CSST to an appliance or other device.

Straightedge: A length of wood or metal that does not bow or twist along its length.

Straight-through flow: A flow that is not restricted as it passes through a valve. The element that closes the valve is retracted entirely clear of the passage.

Strap wrench: Wrench used to hold chrome-plated or other types of finished pipe so that there are no jaw marks or scratches left on the pipe.

S-trap: A trap with a long downstream leg, which tends to promote siphonage. S-traps are no longer installed but are still found in older buildings.

Striker plate: A manufactured metal plate of a required thickness that is installed behind a wall to protect CSST from penetration by nails.

Stub-out: A device mounted in a finished wall, floor, or ceiling that serves as a fixed connection for an appliance or other device.

Submittal data: Another term for catalog drawings.

Subsidence: A depression in the earth that is caused by unbalanced stresses in the soil surrounding an excavation.

Supply stop valve: A valve that is commonly used to disconnect the hot or cold water supply to water closets and sinks. They are available in either right angle or straight design.

Sway brace: A support that keeps lengths of cast-iron pipe from swaying when they are suspended.

Sweat joint: A pipe joint made by applying solder to the joint and heating it until it flows into the joint.

Sweep: A fitting with a radius of the curve greater than 90 degrees. Used for a smooth change of direction.

Symbol: A mark or drawing used to indicate a specific object, material, class, or entity. A legend shows the symbols used on a drawing and their meanings.

Tagout device: Any prominent warning device, such as a tag and a means of attachment, that can be fastened securely to an energy-isolating device in accordance with an established procedure. The tag indicates that the machine or equipment to which it is attached is not to be operated until the tagout device is removed in accordance with the energy-control procedure.

Takeoff: The process by which detailed lists are compiled, based on drawings and specifications, of all the material and equipment necessary to construct a project. Such a list is also called a material takeoff.

Tee fitting: A fitting that allows a branch to be taken off from a main gas line.

Temper: The strength and resilience of a metal.

Temperature and pressure (T and P) relief valve: A mixing valve which senses outlet temperature and incoming hot and cold water pressure and compensates for fluctuations in hot and cold water temperatures and/or pressures to stabilize its outlet temperatures. Also referred to as thermostatic/pressure balancing valve, combination.

Terrazzo: A type of floor surface (commonly used for showers and other plumbing fixtures) made by embedding small pieces of marble or other hard stone in a mortar base. When hardened, the surface is ground and polished smooth, although it may be left slightly rough to create a nonslip surface.

Test tee: A tee installed as a test location for pressurizing the system to test for leaks.

Thermoplastic: A plastic material used in plumbing and sanitary systems that is soft and pliable when heated and hard and rigid when cooled.

Thermoplastic pipe: Pipe that can be repeatedly softened by heating and hardened by cooling. When softened, thermoplastic pipe can be molded into desired shapes.

Thermoset: A plastic material used in plumbing and sanitary systems that becomes substantially infusible and insoluble when treated by heat or chemicals.

Thermosetting pipe: Pipe that changes chemically when heated, so that once hardened by heat or chemicals, it is hardened permanently.

Thermostatic/pressure balancing valve, combination: Another term for temperature and pressure (T and P) relief valve.

Thread engagement: The length of pipe that gets screwed into the fitting when joining threaded pipe.

Thread makeup: The distance that a pipe screws into a fitting. Also called thread engagement or thread-in.

Throat: The part of the fitting where you thread in another pipe or fitting.

Throttled flow: The controlled rate of flow through a valve, which does not impede flow but causes some friction loss.

Tolerance: Allowable variation in a given measurement or quantity.

Top-out: Another term for aboveground rough-in.

Torque: Twisting or turning force applied in a rotating motion. Measurements are given in either inch-pounds or foot-pounds.

Torque wrench: A wrench with a gauge or other means to indicate the amount of rotating force applied to a fastener, such as a nut, as it is turned.

Transition fitting: A special fitting used to connect plastic pipe to pipe of a dissimilar material, as specified by applicable code.

Trap: A fitting or device which provides a liquid seal to prevent the emission of sewer gases without materially affecting the flow of sewage or wastewater through it.

Trim finish: Another term for finish.

Trimming out: The task of placing attachments on fixtures to complete a job.

Trim-out: Another term for finish.

Tube stock: Uncut tubing material on hand for use on a project.

Turbidity: Presence of particles (sand, mud, silt) suspended in water that give the water a cloudy appearance.

Underground rough-in: The phase of a plumbing project during which the plumber locates all supply and waste connections from the building systems to public utilities, and establishes where these systems will enter or leave the building.

Vacuum breaker: A type of backflow preventer that inhibits backflow caused by low pressure in a water supply system.

Valve seat: An opening in a faucet through which the water flows. When the handle is turned to open the faucet, the valve stem raises the seat washer up and clears the opening for water to flow.

Valve stem: The part of a faucet that rises and lowers when the handle is turned; it opens or closes the faucet outlet.

Velocity: Speed of motion, such as the speed of sewage or wastewater through the drainage piping.

Vent branch (branch vent): A vent connecting one or more individual vents with a vent stack or stack vent.

Vent ell: A plastic fitting with a sharp turn radius, used only in vent piping systems. Use is regulated by code.

Vent pipe: A pipe, or pipes, installed to provide a flow of air to or from a drainage system, or to provide a circulation of air within such a system to protect trap seals from siphonage and back pressure. Also referred to as vent system.

Vent system: Another term for vent pipe.

Vent tee: A fitting used in venting systems or as a cleanout. It may not be used in the drainage system because it restricts the flow of material. Use is regulated by code.

Vitrified porcelain: A mixture of fine clay, quartz, feldspar, and silica that is heated to 2,600°F (1,426°C) to create a nonporous, glass-like finish. Also referred to as china.

Voltage drop: The tendency of electricity traveling through the length of an extension cord to lose voltage.

Waste pipe: A pipe which conveys only waste.

Water column (w.c.): The unit of measure used to specify the standard pressure of a gas system.

Water hammer: An extreme change in water pressure within a pipe that can cause a loud, banging sound and even damage the system.

Water meter: A device to measure water flow, in gallons or cubic feet, into an individual building.

Water supply fixture unit (WSFU): A design factor to determine the load that different plumbing fixtures produce on the supply side of a plumbing system.

Water table: The distance below the ground's surface at which the soil becomes saturated.

Weir (trap or crown): Also referred to as crown weir or trap weir.

Well casing: Outer tube or pipe sunk into the ground after drilling or driving a well.

Wye: A fitting used to make a 45-degree direction change in horizontal drain and waste pipes when connecting to the building main drain.

Yarning iron: A tool used to pack the oakum fiber into the joint when making lead and oakum joints to connect hub-and-spigot cast-iron pipes.

Yoke vise: Tool with jaws that firmly hold a piece of pipe and prevent it from turning.

Figure Credits

Module 02101-05

Ivey Mechanical Company, Module divider
LeDuc & Dexter Plumbing, TOC art

Module 02102-05

Accuform Signs, 102F19, 102F26 (two "danger" tags)
Adobe® Image Library, Module divider
Bosch Power Tools and Accessories, 102F32 (hammer drills)
Thomas P. Burke, 102F29
DeWalt Power Tools, 102F32 (portable band saw, reciprocating saw)
DBI/SALA and Protecta, 102F10
Draeger Safety, Inc., 102F43 (PAC III detection meter)
ESAB Welding and Cutting Products, 102F35
Hornell, Inc., 102F14
IDESCO Corporation, 102F26 ("start-up instructions," "do not start," "out of order," and "authorized personnel" tags)
Richard Keretzski, 102F44
Brigid McKenna, 102F20, 102F27
Anna Meade, 102F07
MCR Safety, 102F06
Courtesy of North Safety Products USA, TOC art, 102F04, 102F05, 102F11, 102F41
Occupational Safety and Health Administration, Table 3
RAE Systems, Inc., 102F43 (RAE detection meter)
Courtesy of Ridge Tool Company, 102F30 (hacksaw, tube cutter, PVC saw), 102F31 (chisels)
Chuck Rogers, 102F08, 102F09
Scott Health and Safety, 102F12, 102F13, 102F15
The Stanley Works, 102F30 (utility knife), 102F31 (punches, wedge)
Trench Shoring Services, Inc., 102F36, 102F37, 102F39
Virginia KMP Corporation, 102F18
Veronica Westfall, 102F33

Module 02103-05

BernzOmatic, 103F49
Bosch Power Tools and Accessories, 103F37
Calculated Industries, Inc., 103F01 (calculator), 103F75
Courtesy of Cooper Hand Tools, 103F01 (keyhole saw), 103F02, 103F03 (plumber's rule), 103F58, 103F63
DeWalt Power Tools, 103F22, 103F23, 103F35, 103F36, 103F39
Digital Vision LTD, TOC art
ESAB Welding and Cutting Products, 103F01 (torch kit)
Klein Tools, Inc., 103F57
Laser Reference, Inc., 103F14
LENOX, 103F19
Milwaukee Electric Tool Corporation, 103F01 (magnetic drivers), 103F24, 103F40
Courtesy of North Safety Products USA, 103F01 (safety glasses)
Reed Manufacturing Company, 103F01 (pliers, pipe wrenches, Crescent® wrenches), 103F50 (heavy-duty pipe wrench, 45-degree offset wrench, 90-degree pipe wrench, straight pipe wrench), 103F54, 103F55, 103F59, 103F62, 103F71
Courtesy of Ridge Tool Company, 103F01 (hacksaw, tube cutters, screwdrivers, PVC saw), 103F10, 103F25, 103F29, 103F32, 103F33, 103F34, 103F41, 103F44 (die-tool set), 103F45, 103F47, 103F50 (heavy-duty end wrench), 103F53, 103F56, 103F60, 103F61, 103F69 (screwdrivers), 103F72, 103F73, 103F74
Sokkia Corporation, 103F13 (transit level, leveling rod)
The Stanley Works, 103F01 (torpedo level, claw hammer, ball-peen hammer, chalk box, plumb bob, utility knife), 103F04, 103F05, 103F06, 103F07, 103F08, 103F09, 103F11, 103F12, 103F15, 103F17, 103F18, 103F27, 103F64, 103F66, 103F67, 103F68, 103F70
Veronica Westfall, Module divider
Courtesy of Wheeler Rex Pipe Tools, 103F43
David White, Inc., 103F13 (builder's level)

Module 02104-05

Calculated Industries, Inc., 104F02
Ivey Mechanical Company, Module divider
LeDuc & Dexter Plumbing, TOC art

Module 02105-05

Adobe® Image Library, Module divider
American Standard Companies, Inc., 105F31
Courtesy of Cooper Hand Tools, 105SA01
Corbis, TOC art
D&E Services, Inc., 105F02

Moen, Incorporated, 105F29
Ritterbush-Ellig-Hulsing P.C., 105F01, Appendix (plot plan, floor plan, interior elevations, piping plan)
Sloan Valve Company, 105F30
Tommy Swafford, Appendix (riser diagram)

Module 02106-05

Adjustable Clamp Company, 102F22 (C-clamp)
Charlotte Pipe and Foundry Company, 106F02, 106F03, 106F04, 106F09 (sanitary tee, wye), 106F16
Consumer Plumbing Recovery Center, Inc., 106F07
Genova Products, Inc., 106F09 (closet bend)
Hudson Extrusions, Inc., 106F05
International Code Council, Inc., Table 4, Table S-1. *International Plumbing Code 2003, Falls Church, VA: International Code Council, Inc. Reproduced with permission.*
Ivey Mechanical Company, Module divider, TOC art
LASCO Fittings, Inc., 106F08 (union, elbow 90, elbow 45, tee, coupling, cap, plug), 106F09 (coupling)
Plastic Pipe and Fittings Association, Table 1
Reed Manufacturing Company, 106F11
Courtesy of Ridge Tool Company, 106F10
Sioux Chief Manufacturing Co., Inc., 106F21, 106F22 (suspension, I-beam, beam clamps), 106F23
Tempil, Inc., An Illinois Tool Works, Inc., Company, 106F20 (temperature indicator stick)
Uponor Wirsbo AB, 106F08 (manifold)
Vanguard Piping Systems, Inc., 106F18 (crimpsert)
Zurn Plumbing Products Group, 106F06

Module 02107-05

Adobe® Image Library, Module divider
BernzOmatic, 107F13 (soldering torch)
LeDuc & Dexter Plumbing, TOC art
Mueller/B&K Industries, Inc., 107F03 (stop and waste valve)
NIBCO, 107F21 (pipe strap)
Oatey Co., 107F13 (flux brush and solder paste, wire solder)
Courtesy of Ridge Tool Company, 107F08, 107F09, 107F12, 107F13 (fitting brush), 107F16
Sioux Chief Manufacturing Co., Inc., 107F21 (pipe hooks, J-hooks, locking tube strap, plumber's tape)
Becky Swinehart, 107F02 (top)
T-Drill Industries, Inc., 107F05, 107F06
Victaulic Company of America, 107F02 (bottom)
Watts Regulator Company, 107F03 (gate valve, globe valve, check valve)
Weil-McLain, 107F04
Wheeler Rex Pipe Tools, 107F28

Module 02108-05

Adobe® Image Library, Module divider
Charlotte Pipe and Foundry Company, 108F02, 108F36 (heavy-duty coupling)
Fernco®, Inc., 108F36 (Fernco® coupling)
International Code Council, Inc., Table 1. *International Plumbing Code 2003, Falls Church, VA: International Code Council, Inc. Reproduced with permission.*
Ivey Mechanical Company, TOC art
Milwaukee Electric Tool Corporation, 108F46
Oatey Co., 108F09
Radiator Specialty Company / Douglas Tools, 108F37
Courtesy of Ridge Tool Company, 108F19, 108F29 (cold chisels)
The AntimonyMan.com, 108F25
The Cast Iron Soil Pipe Institute, 108F03 (logo), 108SA02
Tyler Pipe Company, 108F03 (pipe), 108F36 (regular coupling, six-banded coupling)
Veronica Westfall, 108F47
Zurn Plumbing Products Group, 108SA01 (right)

Module 02109-05

Cooper B-Line, Inc., 109F40A (1 of 3), 109F44B
Ivey Mechanical Company, Module divider, TOC art
JMF Company, 109F14
Mueller / B&K Industries, Inc., 109F04, 109F05, 109F06, 109F08, 109F09, 109F10, 109F11
Reed Manufacturing Company, 109F24, 109F25, 109F27, 109F29
Courtesy of Ridge Tool Company, 109F17, 109F18, 109F19, 109F20, 109F21, 109F22, 109F23, 109F26, 109F28, 109F31
Sioux Chief Manufacturing Co., Inc., 109F40B, C (2 of 3, 3 of 3), 109F43
Victaulic Company of America, 109F33, 109F34

Module 02110-05

Gastite Division of Titeflex Corporation, Module divider, TOC art, 110F01, 110F02, 110F03 (ball valve), 110F08, 110F10, 110F11, 110F12, 110F13, 110F14, 110F15, 110F16, 110F17, 110F18, 110F19
Watts Regulator Company, 110F05 (gas cock)

Module 02111-05

Adobe® Image Library, Module divider
American Standard Companies, Inc., 111F01, 111F02 (self rimming), 111F03 (corner, raised back, wheelchair accessible), 111F05, 111F06,

111F07, 111F08, 111F11 (floor mounted, wheelchair accessible), 111F18, 111F21
Courtesy of Delta Faucet Company, 111F34, 111F43 (Michael Graves Collection)
E.L. Mustee & Sons, Inc., 111F27
Halsey Taylor, 111F32, 111F33
In-Sink-Erator, 111F22
Josam Company, 111F30
Kings Supply Company, 111F47
Courtesy of Kohler Co., 111F02 (wall hung, built-in rim, undercounter), 111F03 (ledge or shelf back), 111F04, 111F09, 111F10, 111F11 (wall hung), 111F19, 111F28, 111F29, 111F35, 111F36, 111F40, 111F44, 111F45
LeDuc & Dexter Plumbing, TOC art
Maytag Corporation, 111F24
MIFAB Manufacturing Inc., 111F31
Murdock, Inc., 111F49
Sloan Valve Company, 111F20
Woodford Manufacturing Company, 111F48

Module 02112-05

Charlotte Pipe and Foundry Company, 112F24, 112F28 (top)
Fernco®, Inc., 112F23
Genova Products, Inc., 112F03 (top), 112F28 (bottom), 112F37
Ivey Mechanical Company, Module divider, TOC art
The Stanley Works, 112SA01
Watts Regulator Company, 112F03 (bottom)
Zurn Plumbing Products Group, 112F05

Module 02113-05

Cary Mandeville, 113F01 (2 of 2)
David J. Dachtera/DJE Systems, 113F16
Holby Valve Company, 113F32 (right)
International Code Council, Inc., Table 1. *International Plumbing Code 2003, Falls Church, VA: International Code Council, Inc. Reproduced with permission.*
Ivey Mechanical Company, Module divider
LeDuc & Dexter Plumbing, TOC art
Marotta Controls Inc., 113F27 (bottom)
Mueller/B&K Industries, Inc., 113F08, 113F29
Plumbing-Heating-Cooling Contractors–National Association, Table 2, Table 3
Sioux Chief Manufacturing Co., Inc., 113F31
Starwood Hotels and Resorts Worldwide, Inc., 113SA02
Tyco Valves & Controls, 113F27 (top)
United Nations Environment Programme, Table S-1
Watts Regulator Company, 113F06, 113F19 (top), 113F21, 113F26 (bottom), 113F32 (right)

Other Credits

Corbis; Index Stock Imagery, Inc.

The designation "*f*" refers to figures.

C

D

E

G

H

I

J

K

L

M

N

O

P

R

S

T

U

V